Microbiological Activity for Soil and Plant Health Management

Ravindra Soni • Deep Chandra Suyal •
Prachi Bhargava • Reeta Goel
Editors

Microbiological Activity for Soil and Plant Health Management

Springer

Editors
Ravindra Soni
Department of Agricultural Microbiology, College of Agriculture
Indira Gandhi Krishi Vishwa Vidyalaya
Raipur, Chhattisgarh, India

Deep Chandra Suyal
Department of Microbiology
Akal College of Basic Sciences, Eternal University
Sirmaur, Himachal Pradesh, India

Prachi Bhargava
Department of Bioscience & Technology
Shri Ramswaroop Memorial University
Barabanki, Uttar Pradesh, India

Reeta Goel
Institute of Applied Sciences & Humanities
GLA University
Mathura, Uttar Pradesh, India

ISBN 978-981-16-2921-1 ISBN 978-981-16-2922-8 (eBook)
https://doi.org/10.1007/978-981-16-2922-8

This Springer imprint is published by the registered company Springer Nature Singapore Pte Ltd.
The registered company address is: 152 Beach Road, #21-01/04 Gateway East, Singapore 189721, Singapore

Preface

Food is one out of the three basic necessities for any living form on this biosphere. Therefore, human race has strived hard to fulfil its demands mostly at the cost of nature. With the growing population on this planet, productivity enhancement with limited arable land resource has become the major challenge for the agriculture communities. Be it the Green revolution or blue revolution, current agricultural practices have resulted a huge amount of toxic effluents directly or indirectly into the soil, air, and water. Therefore, the need of the hour is to increase the arable land with sustainable agriculture practices and judiciously involve microorganisms as the major stakeholders. These microorganisms being indigenous can beautifully interact with their micro environment surroundings either synergistically or antagonistically making plant–microbe synergism as ecological sustainable. Being at the receiving end of agricultural products, man has tried to curtail the losses occurred due to biotic and abiotic stresses. It is, therefore, important to explore the dynamic microbe–plant–soil interactions going on at every fractions of second. Keeping this perspective in mind, this book is a brainchild to recapitulate the labyrinthine mechanisms involved in microbe abetted sustainable management of soil environment. It consists of chapters focusing on challenges and opportunities of microbes in sustainable agriculture, the various factors governing the soil ecosystem affecting the plant mineral nutrition, usage of microbes to deal with biotic and abiotic stress, etc. Innovations and recent trends in current agriculture have been highlighted with explicit reference to new strategies for commercialization of microbial technologies and futuristic approaches for indigenous microbial resource conservation and management.

While accomplishing higher goals, it is always WE and never ME. The editors nimbly acknowledge the overwhelming support and encouragement received from all the well-wishers. The editors express their heartfelt gratitude to all the authors who have contributed in shaping this book. It is their cooperation, understanding, patience, and timely response that have made this dream come true. Due to the predefined page limitation, all the acknowledgement cannot be added here. Any suggestions to improvise the book is welcomed.

Raipur, Chhattisgarh, India Ravindra Soni
Sirmaur, Himachal Pradesh, India Deep Chandra Suyal
Barabanki, Uttar Pradesh, India Prachi Bhargava
Mathura, Uttar Pradesh, India Reeta Goel

Contents

About the Editors

Ravindra Soni is an Assistant Professor at the Department of Agricultural Microbiology, I.G.K.V., and Raipur, India. He has more than 12 years of teaching & research experience in the field of agricultural microbiology, microbial ecology, molecular biology & biotechnology. He is a recipient of the DST- Young Scientist award from SERB, Department of Science & technology, India. He was also selected for UGC-DS Kothari Postdoc Fellowship and Insa summer Fellowship. He has published edited books, lab manual, several research articles and chapters in reputed international journals and books.

Deep Chandra Suyal is an Assistant Professor at the Department of Microbiology, Akal College of Basic Sciences, Eternal University, Baru Sahib, Himachal Pradesh, India. He has been actively engaged in research for 11 years, and has experience in the fields of agricultural microbiology, molecular biology & biotechnology and metagenomics. He is currently investigating the genomics & proteomics of cold adapted microorganisms. He has published numerous articles and books with reputed international journals and publishers. He has three patents in his credit.

Prachi Bhargava has over 16 years of teaching and research experience in microbial molecular biology and is currently working as an Associate Professor at Shri Ramswaroop Memorial University, Lucknow. She has worked as Principal Investigator on various major Government of India sponsored projects, and was recently involved in a project sponsored by SERB, New Delhi. As an active researcher, Dr Bhargava has published numerous research and review articles in respected journals, along with several book chapters. She has also co-edited two books with renowned publishers.

Reeta Goel is a Professor and former Head of the Department of Microbiology, G. B. Pant University of Agriculture & Technology, Uttrakhand, India. She has also served as visiting scientist at various institutes in the USA and has more than 35 years of teaching and research experience in the area of plant-microbe interactions, microbial ecology, bio-degradation and metagenomics, especially with reference to higher altitudes. She has worked extensively in the field of genomics and proteomics of cold adapted microorganisms. She has published over 100 research articles, book chapters and edited books. She has six patents to her credit, including one U.S. Patent, and received numerous awards for her contributions to science & technology. Presently, serving as distinguished Professor at GLA UNIVERSITY, MATHURA, UP (India) after her retirement.

Agriculturally Important Microbes: Challenges and Opportunities

1

Pooja Joshi, Aditi Saini, Sabyasachi Banerjee, Ratnaboli Bose, M. S. Bhandari, Amit Pandey, and Shailesh Pandey

Abstract

Burgeoning world population has pressurized the agricultural sector immensely. Fertilizer and pesticide usage injudiciously in conventional farming has adversely affected environment and human health. Globally coordinated sustainable agriculture is the way forward. Sustainable agriculture aims to maintain productivity, feed billions, and yet conserve the environment. Productivity devoid of environmental degradation, maintenance of agro-ecosystem health, and agro-biodiversity are essential for proper management of cultivated ecosystems. Soil dwelling microbial communities are key to resolving these concerns. A prominent rhizospheric microbial assemblage termed plant growth-promoting rhizobacteria contributes significantly to plant growth promotion and development. PGPR bioinoculants release and modify endogenic phytohormone levels, other biologically active molecules, solubilize or fix minerals facilitating nutrient uptake, helping crops overcome abiotic stresses. Further they control plant pathogenesis/disease by deploying antibiosis, rhizospheric competence, enzyme secretion, and induction of systemic resistance in host plants. Soil health and fertility are maintained owing to biofortification of nutrients. This translates to greater food security, social well-being, and human health. Bioinoculant viability and effectiveness are determined by the carrier material. Persistent research developing nontoxic, user-friendly bioformulations that can remain functionally viable during storage is pertinent to the continued large-scale application of PGPR. This chapter focuses upon the potential of four agriculturally important

P. Joshi · A. Saini · S. Banerjee · R. Bose · A. Pandey · S. Pandey (✉)
Forest Pathology Discipline, Forest Protection Division, Forest Research Institute, Dehradun, India
e-mail: pandeysh@icfre.org

M. S. Bhandari
Genetics and Tree Improvement Division, Forest Research Institute, Dehradun, India

R. Soni et al. (eds.), *Microbiological Activity for Soil and Plant Health Management*,
https://doi.org/10.1007/978-981-16-2922-8_1

microbes—*Azotobacter*, *Serratia*, *Bacillus*, and *Pseudomonas*. Furthermore, the challenges to crop production by the usage of PGPR, problems in shelf life of bioformulations for commercial agriculture, and directions for future of PGPR research are highlighted.

Keywords

Agricultural sustainability · Plant growth-promoting rhizobacteria · Biocontrol · Bio-fertilizers · Environmental-remediation · Soil health · Stress management

1.1 Introduction

Agriculture has formed the economic basis of society for as long as man has existed. Owing to the ongoing COVID-19 pandemic, the current count of undernourished people would shoot up by 132 million in 2020 (FAO 2020a). Current agricultural policies must harmonize with UN-Sustainable Development Goals facilitating less hazardous and nature friendly food production. Sustainable agriculture is upheld by United Nations (UN) Sustainable Development Goals like SDG 1 (no poverty), SDG 2 (zero hunger), SDG 3 (good health and well-being), and SDG 12 (responsible consumption and production) (Priyadarshini and Abhilash 2020; Ferdous et al. 2020). Global population is 7.8 billion and growing, inhabiting about 6.38 billion hectares, and 1.31 billion are primarily dependent on agriculture (Gouda et al. 2018). Recent statistics suggest that nearly 690 million or 8.9% of the world population go to bed with an empty stomach. In 2019 alone, 750 million people were stricken with dire food insecurity. Current projections dictate that more than 840 million people would be affected by hunger by 2030, should current trends persist (FAO 2020b). During his Nobel Peace Prize 2020 acceptance speech, Executive Director David Beaseley UN World Food Programme (WFP) said "Because of so many wars, climate change, the widespread use of hunger as a political and military weapon, and a global health pandemic that makes all of that exponentially worse—270 million people are marching towards starvation". Global achievement of Zero Hunger by 2030 seems a distant dream under these circumstances.

Food security needs to be addressed with concerted international efforts, with defined long-term challenges and possibilities. Food production must increase by 70–110% to cope with a 40% increase in world population and to raise average food consumption to 3130 kcal per person per day by 2050 (Bruinsma 2009; Royal Society of London 2009; Tilman et al. 2011). Meanwhile, land under crop production has risen marginally from 1961 to 2018 since total agricultural area expanded by 6.43% from 4.51 billion ha to 4.8 billion ha. Simultaneously the world population has grown from 3 to 7.8 billion (160% hike) (FAOSTAT 2009, 2020; Bologna and Aquino 2020). A growing world population would require an additional 2.7–4.9 million hectare of cropland per year on average (Lambin and Meyfroidt 2011). The demand and supply of food is inversely proportional since agricultural production is arithmetic in progression while population grows in geometric progression (Malthus

2013). While his pessimistic views have been criticized and falsified by various researchers, if left unchecked, world population would likely outrun its subsistence (Royal Society of London 2009).

Presently some 11% (1.5 billion ha) of the global land area (13.4 billion ha) comes under cultivation (arable land and land under permanent crops). Global area under cultivation is about 36%, of the total land estimated to be suitable for crop production (Bruinsma 2003). India has 60% of the land area under cultivation for the production of cereals, vegetables, pulses upon which half the population of India depends (Gouda et al. 2018). Degraded land area is about 25% of total world area with steady degradation on the rise (Abhilash et al. 2016). Global arable land lost per annum to soil erosion is between 2 and 5 million hectares (Bringezu et al. 2010). About 50 tonnes per hectare soil is lost in a year which may be 100 times faster than the rate of soil formation (Banwart 2011). No exact consensus exists upon the actual extent or spatial distribution of degraded land, with estimates ranging from less than 1 billion hectares to greater than 6 billion hectares (Gibbs and Salmon 2015). Various land uses, such as cropland, (for agriculture) pastures, natural forests, planted forests, urban built-up area, unused, productive land, industrial forestry, protected areas, and land lost to land degradation, compete for existence (Lambin and Meyfroidt 2011). However, land is a finite resource and agricultural use of land will have to contend with other land uses for habitation, infrastructure, and industry. Shifts in the existing patterns of land use would potentially affect the resilience of ecological and socioeconomic systems (Anderson 2010).

This means, agricultural land cannot expand beyond its current limits. Public funded research must support the sustainable intensification of food farming practices (Royal Society of London 2009). The pressure on land area to meet the demand for food production is immense, the world shall reach nearly 10 billion by 2050. To cope up with the demands of a growing human population and yet maintain soil health and fertility, sustainable agriculture based on scientific farming methods is imperative. Sustainable agriculture depends on sustainable management of soils and intensive farming. The identity, location, and manner of cropping are guided by various factors such as soil nutrients (nitrogen, phosphorus, and potassium), water, climate (temperature, radiant energy for photosynthesis), pests (vertebrates and invertebrates), diseases (fungal, bacterial, or viral), weeds, and beneficial organisms which are disease suppressive (Royal Society of London 2009).

Agricultural productivity depends upon abiotic factors, viz. nutrient cycling, energy fluxes, and carbon fluxes between soil organic matter, lithosphere, hydrosphere, and atmosphere (Lehmann and Kleber 2015). As climate change creates harsher scenarios, abiotic and biotic stresses would continue to assail crop productivity (Cavicchioli et al. 2019). Wheat and barley yield declined by 1.9% and 4.8%, respectively, between 1980 and 2015, owing to changes in climate conditions, a Central Asia based research group reported (Schierhorn et al. 2020). As per the International Panel on Climate Change (IPCC 2019), Agriculture, Forestry and Other Land Use (AFOLU) contributed to 22% anthropogenic greenhouse gas (GHG) between 2007 and 2016. In 2017, agriculture alone accounted for 42% of total CH_4 (methane) and 75% of global total N_2O (nitrous oxide) emissions

(FAOSTAT 2020). Nitrous oxide is the major emission by-product of inorganic nitrogen fertilizers (IPCC 2007). Conventional agriculture utilizing agrochemicals in the form of fertilizers and pesticides has substantially damaged soil owing to soil acidification and exchangeable base reduction, alienating these chemicals from standing crops causing loss to productivity. The continuous soil application of chemical inputs as fertilizers or pesticides to agricultural land degrades soil texture, productivity, native microflora and surrounding environment. Pesticide residues on fruits, vegetables, crops that enter the food chain have mutagenic, carcinogenic, cytotoxic, genotoxic among other health problems in humans (Singh et al. 2020; Kumari et al. 2019). Plants cannot locomote which heightens their vulnerability to a variety of biotic (pathogen attack) as well as abiotic stresses (salinity, drought, flooding, heavy metal toxicity, extreme temperatures). Desirable qualities in crop cultivars include disease resistance, salt tolerance, drought tolerance, heavy metal stress tolerance, and higher nutritional value. Beneficial microbes are a viable route to achieve these purposes (Gouda et al. 2018; Kenneth et al. 2019) and an essential component of organic agriculture (Zarb et al. 2005). Organic agriculture is a method of agriculture where no synthetic fertilizers and pesticides are used. The International Federation of Organic Agriculture Movements (IFOAM) states that organic farming is a production system which maintains soil health, ecosystems, and humans.

Agricultural biotechnology has responded by way of genetic engineering. For instance, genetically engineered *Escherichia coli* capable of diazotrophic (ability to fix nitrogen) growth was created by introgression of 18 essential nif genes from *Klebsiella oxytoca* (a free-living nitrogen fixer) into its genome. Genetically modified crop cultivars are being developed that are capable of forming artificial symbioses or associations with diazotrophs (nitrogen fixers) by introgression of *nif* genes into heterologous hosts to reduce the demand for chemical nitrogen fertilizers, a major contributor to global climate change (Burén et al. 2018; Sutton et al. 2011). Excellent reviews on genetically engineered non-legume crops, capable of nitrogen fixation in the absence of rhizobacteria ere done by Mus et al. (2016) and Pankievicz et al. (2019). For instance, transgenic tobacco plants were developed by integration of the active iron (Fe) subunit of nitrogenase gene sets (the primary bacterial enzyme for biological nitrogen fixation) into the tobacco chloroplast genome (Ivleva et al. 2016). Extensive transgenic crop acceptance is fraught with obstacles regarding possible toxicity, allergy risks to humans, transgene introgression to wild relatives, deleterious effects to non-target organisms, and development of weed/pathogen/pest resistance (Clark and Maselko 2020; Snow 2002). To bolster environmental sustainability and tackle food insecurity, agrobiology must support organic agriculture, an important component of which are beneficial agriculturally important microorganisms (AIM). Beneficial microbes could potentially mitigate associated difficulties in agricultural practices. Nature provides for a plethora of viable ubiquitous microbes inhabiting the rhizosphere, phyllosphere, endosphere of plants which are potentially agriculturally important microorganisms (AIM). Cyanobacteria or blue-green algae (BGA) are prokaryotic, photosynthetic, diazotrophs posing great potential as biofertilizer, especially in paddy cultivation (Múnera-Porras et al. 2020), owing to their unique heterocysts which are nitrogen-fixing cells (Verma and Patel 2019). Important filamentous cyanobacteria possessing the ability to fix atmospheric nitrogen include *Nostoc*, *Anabaena*, and *Cylindrospermum* (Baweja et al. 2019). Two-thirds of the annual turnover of biospheric nitrogen estimated to be between 100 and 200 million metric tonnes is of biological origin where BGA play an important role (Pabbi 2015). Carboxysomes in cyanobacteria which are specialized cellular microcompartments capable of concentrating CO_2 confer greater photosynthetic efficiency to them. This is because the efficiency of RuBisCO (ribulose bisphosphate carboxylase oxygenase) enzyme, the vital photosynthetic enzyme, is enhanced manifold at higher intracellular CO_2 concentration. RuBisCO is the most abundant enzyme on Earth, yet lethargic since it also reacts with oxygen, thus wasting energy in this side reaction (Ferraroni 2019). German and Australian research groups have collaborated in order to transfer functioning carboxysomes to higher plants for more efficient photosynthesis and hence greater biomass production (Wang et al. 2019). Cyanobacteria are important candidates in agricultural practices for reduction of global warming by reducing atmospheric CO_2 (Chittora et al. 2020). Apart from their use as biofertilizers, they possess plant growth promotion activities such as nutrient solubilization (phosphorus), phytohormone production (auxins, ethylene, cytokinin, and gibberellin) as well as siderophore production that hampers growth of phytopathogenic fungi (Chittora et al. 2020).

However, light is a limiting factor in extensive cyanobacterial use as agriculturally important bioinoculants. The rhizosphere comprises the soil region directly influenced by roots of higher plants. This soil zone harbours diverse fungal, bacterial communities recruited by host plants. Compared to bulk soil, microbial diversity is higher in this region. Research since the 1980s has focussed upon screening of cultural microbiome of root associated microbes, screening their potential as plant growth promoters or biocontrol, and finally application in field crops.

Prominently, the soil associated microbes are most easily isolated and abundantly available, making them abundantly acceptable in green agriculture. Versatile AIM include (rhizospheric/endophytic/epiphytic) non-pathogenic, saprotrophic, plant growth-promoting fungi (PGPF) (Hossain and Sultana 2020). About 20 species of notable PGPF are widely reported to belong to phylum Ascomycota, Basidiomycota, and Zygomycota (Hossain et al. 2017; Hossain and Sultana 2020). Plant growth-promoting fungal interactions positively influence root and shoot organs of plants. This includes increase in seed germination, seedling vigour, biomass production, root hair development, photosynthetic efficiency, flowering, and yield. Certain strains improve plant biochemical composition. PGPF can also control numerous foliar and root pathogens by antibiosis or by triggering induced systemic resistance (ISR) in the host plants (Navarro et al. 2019; Naziya et al. 2020). For instance, in *Triticum aestivum* rhizospheric PGPF stimulated induced systemic resistance (ISR) against wilt disease by *Rhizoctonia solani* R43, resulting in a restriction of symptoms and disease development. The PGPF strains activated the pathogenesis-related gene (PR-1, 2), plant defensive chitinase (Chit-1), and β-1, 3-glucanase

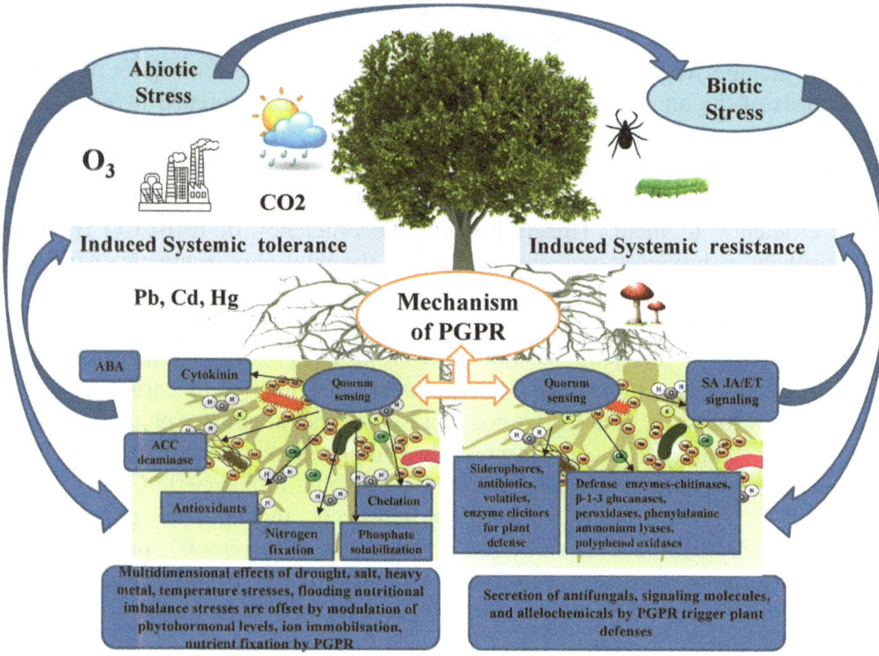

Fig. 1.1 Multifarious mechanisms for PGPR actions in plant rhizosphere

(Glu-2) genes and increased the plant-specific defensive proteins against wilt pathogen *Rhizoctonia solani* (El-Maraghy et al. 2020). Their ability to enhance nutrient uptake, release phytohormones, reprogram plant gene expression, by differential activation of plant signalling pathways contributes to the above physiological benefits they confer upon plants they are associated with (Hossain et al. 2017).

According to Hiltner, the portion of soil influenced by plant roots was termed as the rhizosphere. The rhizomicrobiome comprises diverse microbial groups that derive nutrition obtained through the root exudates and plant cell debris (Backer et al. 2018; Kenneth et al. 2019). In turn, rhizospheric microbial groups support plant functions by assisting in nutrient uptake and providing protection against disease (Berendsen et al. 2012). Root associated microbes, such as PGPR (plant growth-promoting rhizobacteria) show antagonistic and synergistic interactions improving plant growth (Kannojia et al. 2019).

The PGPR, constituting 2–5% of the rhizosphere competent bacteria, when present in a large number enhances plant growth by performing direct and indirect actions (Fig. 1.1). Direct actions include important biochemical and physiological functions such as nutrient abstraction (K and P solubilization) and assimilation (Fe and Zn), enhancement of soil texture, phytohormone secretion (auxins such as indole acetic acid), release of secondary metabolites (siderophores, ammonia), and biological nitrogen fixation (BNF). Indirect actions include antagonistic tendencies towards plant pathogens like antibiotic production, HCN production, various extracellular signalling chemical for induction of systemic resistance (ISR) eliciting defence mechanism in distal plant cells (Braud et al. 2009; Hayat et al. 2010; Nagargade et al. 2018; Backer et al. 2018; Antoun 2013). Since Kloepper et al. (1980), first reported the role of fluorescent pseudomonads a prominent group of PGPR, possessing antagonism against, phytopathogen *Erwinia carotovora*, owing to their siderophore activity (reducing available iron through chelation) and hence curbing the growth of *E. carotovora*, research has focussed on the biocontrol aspects of PGPR as well (Yasmin et al. 2017). Their role as bio-stimulants is paramount in modulating plant stress responses. Drought/water stress occurs due to multitude of causes like scant rainfall, salinity, extremes of temperature, and light intensity (Salehi-Lisar and Bakhshayeshan-Agdam 2016). Drought stress leads to detrimental effects like ethylene build-up in plants, which in turn reduces root extension and development, thus impeding plant growth and yield. ACC is the immediate biochemical precursor to ethylene in the methionine pathway of vascular plants. Some PGPR can produce ACC (1-aminocyclopropane-1-carboxylate) deaminase. This enzyme catalyses ACC to α-ketoglutarate and ammonia instead of ethylene (Danish and Zafar-ul-Hye 2019), thus circumventing the ill effects of drought stress. Studies have revealed that co-application of *Bacillus amyloliquefaciens,* an ACC deaminase producing PGPR and timber-waste biochar perks up pigment formation (chlorophyll a, chlorophyll b), growth and yield of wheat under drought stress (Danish and Zafar-ul-Hye 2019). Further PGPR support the growth and development of hyperaccumulator plants of heavy metals like Cu, Cd, Pb, Cr, Hg, and Ni, to clean up toxicity ridden landscapes. PGPR aid in bioremediation of contaminated soil by accumulation, adsorption, immobilization, or transformation of contaminants, thereby enhancing plant growth and development (Verma et al. 2017). Multiple omics–metagenomics, metatranscriptomics and metaproteomics, phenomics, metabolomics strategies to decipher microbial community functions in microbe mediated abiotic stress (salinity, temperature, drought, high light, flood, heavy metal) management in crops—have emerged as promising tools. Functional (gene-enzyme function, biochemical pathways) and structural (16rRNA sequencing) characterization of unculturable populations supporting crop growth, from crop rhizospheres in perturbed soil are a possibility owing to data rich high throughput approaches like NGS (next generation sequencing). Model plant *Arabidopsis* and crops like rice, wheat, maize soybean, chickpea have been subjected to rhizospheric metagenome sequencing (Bulgarelli et al. 2012; Bhattacharyya et al. 2016; Srivastava et al. 2020; Enebe and Babalola 2020; Mendes et al. 2014; Zhou et al. 2020) revealing gene catalogues of root systems, influence of root exudates in recruitment of microbial communities, manure (organic and inorganic) treatment effects on rhizospheric composition. Stringent methodological validation of metagenomic analysis is a prerequisite before confirmation of microbial presence and activity in any microbiome (Goel et al. 2018).

Wheat rhizosphere metagenomic snapshot using 16SrRNA sequencing revealed phyla composition to be Proteobacteria (68%), Firmicutes (13%) followed by Bacteroidetes (3%), Actinobacteria (3%), Chloroflexi (2%), and Cyanobacteria (1%) (Srivastava et al. 2020), where Proteobacteria are majorly unculturable

(Arjun and Harikrishnan 2011). The great plate count anomaly is the observation where $<1\%$microbial load from a particular microbiome say rhizosphere can be cultured in vitro, precluding unculturable microbes (Staley and Konopka 1985). This is why metagenomic studies to decipher taxonomic diversity is so important for bioprospection of important AIM.

Culturable microbial resources which are conveniently tapped strengthen crop tolerance to abiotic stresses (drought, heat, and salinity, heavy metal contamination) as well as biotic stress (disease fungal or bacterial) as climate change conditions become more and more pronounced. Among the mentioned major groups of AIM, we shall focus attention upon PGPR, as our choice in this article. Specifically, four internationally recognized, readily isolated, culturable documented AIM which are PGPRs as well, reputed to support and promote crop growth and health shall come into our scrutiny—*Azotobacter*, *Serratia*, *Bacillus,* and *Pseudomonas.* Scientists are focussing on these AIM as a way to reap the immense benefits they bring on account of sustainable agriculture. Opportunities afforded by their roles in promotion of crop growth and health will come under our purview. However, the use of PGPR's area also seriously limited due to variability and inconsistency of result observed under laboratory, greenhouse, and field trails (Gouda et al. 2018). The longevity and persistence of PGPRs in soil need to be addressed to make soils healthy, disease suppressive, and sustainably productive again. These challenges posed to the field application of these AIM are discussed as well.

1.2 *Azotobacter*

Beijerinck in 1901 discovered the genus *Azotobacter,* and was the first person who isolated and cultured *Azotobacter chroococcum* and *A. agilis*. *Azotobacter* is free-living gram negative, heterotrophic, and nitrogen-fixing diazotroph and found in agricultural soils playing different beneficial roles (Tejera et al. 2005). *Azotobacter* has gainful consequences upon crop growth and yield through biosynthesis of biologically active substances, encouragement of rhizospheric microbes, producing phytopathogenic inhibitors (Jnawali et al. 2015), and alteration in nutrient uptake and ultimately boosting up the biological nitrogen fixation (Somers et al. 2004). Several bacterial strains belonging to *Azotobacter* have been successfully proved as nitrogen fixing in agriculture applications (Islam et al. 2013). In addition, production of varieties of vitamins, amino acids, plant growth hormones, antifungal substances, hydrogen cyanide, and siderophores and growth-promoting substances (indole acetic acid, gibberellic acid, arginine, etc.) has direct influence on shoot and root length as well as seed germination of several agricultural crops (Gauri et al. 2012; Gurikar et al. 2016). *Azotobacter chroococcum* present in soil rhizosphere can enhance the plant growth when roots inoculated on tubers. Many researchers have experimentally proved that PGPR (Table 1.1) inoculation elevates soil nitrogen levels, plant dry weight, development, enhances root architecture, grain yield, protein, and mineral nutrient content (Vikhe 2014).

Table 1.1 Different species of agriculturally important *Azotobacter and* their applications

Azotobacter species and their metabolites	Crops/ plants	Response	References
Azotobacter chroococcum	Maize	The grain yield increased in three maize hybrids	Hajnal-Jafari et al. (2012)
Indole-3-acetic acid	Maize	Maximum length and weight of shoot	Zahir et al. (2000)
Indole-3-acetic acid	Sesbania	Root elongation	Ahmad et al. (2005)
Azotobacter sps.	Tomato	Greater lycopene antioxidant activity, enhanced potassium content in shoot and fruit	Ordookhani et al. (2010)
Azotobacter chroococcum (strains AC1 and AC10)	*Gossypium hirsutum* (Cotton)	Both strains are capable of fixing nitrogen, solubilizing phosphorus, synthesizing indole compounds and producing hydrolytic enzymes	Romero-Perdomo et al. (2017)
Azotobacter chroococcum + arbuscular mycorrhizal fungi (AMF)	*Jatropha curcas*	Gain in the shoot height, shoot diameter, fruit yield/plant, and seed yield (g)/plant	Kumar et al. (2016)
Azotobacter	Garden thyme	Maximum effect on growth and total phenolic content of garden thyme.	Naseri and Sharafzadeh (2013)

1.2.1 Action Mechanism of Plant Growth-Promoting Rhizobacteria

The plant growth regulated by PGPR is direct and indirect. Direct mechanisms (nitrogen fixation, phytohormones production such as indole acetic acid, gibberellic acid, siderophore production and lowering of ethylene concentration and solubilization of phosphorus) are exhibited in the absence of plant pathogens or other soil microorganisms. Moreover, indirect mechanisms (antibiotic synthesis of antifungal metabolites, synthesis of antifungal cell wall lysis enzymes, competition for the sites on the roots and induced systemic resistance) involve the ability of PGPR to reduce the detrimental effects of plant pathogens on crop yield (Nelson 2004; Lenin and Jayanthi 2012).

1.2.2 *Azotobacter* as Biofertilizers and Biocontrol Agents

Antagonistic microorganisms are used as biocontrol against phytopathogens that are alternative to chemical compounds for crop protection (El-Katatany et al. 2003). *Azotobacter* species control plant diseases through the mechanisms that involve competition for niches and nutrients antibiosis, predation, and induction of plant defence responses. *Azotobacter* sp. had drawn worldwide attention because of production of secondary metabolites such as siderophore, antibiotics, enzymes, and phytohormones and involving in nitrogen fixation. The action of siderophores produced from *Azotobacter* sp. is called Azotobactin. Three different types of coordinating moieties containing azotobactin, namely a hydroxamate, an R-hydroxy acid, and a catechol, making it a delegate of several classes of siderophores simultaneously (Muthuselvan and Balagurunathan 2013). The recognized intrusion of pyoverdine-mediated uptake of Fe III gives the indirect evidence that azotobactin shows antibacterial effect against root colonizing bacterial and fungal pathogens (Schalk 2008). *Azotobacter salinestris* studied antifungal efficiency against *Fusarium* species (Chennappa et al. 2018). *Azotobacter nigricans* applied as biocontrol on *Fusarium* infection in maize, sorghum, and wheat *Fusarium sporotrichioides*, *Fusarium graminearum*, *Fusarium poae,* and *Fusarium equiseti* reduced *Fusarium* infection incidence up to 50% in all the three treated cereals (Nagaraja et al. 2016). Hindersah et al. (2018) also *reported the activity of Azotobacter as biofertilizer and biocontrol against damping off on long bean.*

Cell count is the essential parameter in any microbial biofertilizer formulation, while the ability of AIM to fix nitrogen or solubilize phosphorus, potassium, zinc, etc. are important characters being considered in formulating the quality standards for biofertilizers (Pandey and Chandra 2016). Talc, liquid, and secondary metabolite-based formulations of soil microorganisms available in global market range (Keswani et al. 2016). El_Komy et al. (2020) used a mixture of rhizobacterial strains (*Azotobacter chroococcum* ZCR, *Azospirillum brasilense* SBR, and *Klebsiella pneumoniae* KPR) to control the root-rot disease complex of sunflower caused by *Macrophomina phaseolina*, *Rhizoctonia solani*, and *Fusarium solani*. Interestingly, the bioinoculants showed successful rhizoplane colonization and persistence up to 60 days. Importantly, in field conditions, significant reduction in disease severity was recorded with seed treatment and soil drenching with a bioformulation containing a bacterial consortium (HALEX bioformulation). Other beneficial effects included improved plant growth, yield, and oil content.

1.3 *Serratia spp.*

Serratia was named thus in honour of an Italian physicist Serafino Serrati by Italian pharmacist Bartolomeo Bizio. *Serratia* (Enterobacteriaceae) is ubiquitous in nature, can thrive in soil, water, plants, and vertebrates including humans. Interestingly, *Serratia* has dual characteristics of appearing as a boon as well as a bane. Owing to its long association of beneficial contributions in agriculture, *Serratia* is majorly considered as a boon, such as plant growth promotion, antagonism, induced systemic resistance (ISR), pesticide, heavy metal tolerance, phytoremediation, etc. (Banerjee et al. 2020; El-Esawi et al. 2018; Singh and Jha 2016; Natasha et al. 2012).

1.3.1 Action Mechanism as Plant Growth-Promoting Rhizobacteria

Members of the genus *Serratia* exist as a potent plant growth promoter associated with several plant species, such as black pepper (Dastager et al. 2011); wheat (Singh and Jha 2016); Nothofagus alpine (Martínez et al. 2018); *Dalbergia sissoo* (Banerjee et al. 2020), etc. Indole acetic acid (IAA) influences plant cell division and differentiation; stimulates seed and tuber germination; speeds up xylem and root development; controls vegetative growth; initiates lateral and adventitious root formation; affects photosynthesis, biosynthesis of various metabolites, and resistance to stressful conditions. Rhizobacteria interfere with the above physiological processes of plants by altering the plant auxin pool. Low concentrations of IAA stimulate primary root elongation, while high IAA levels stimulate the lateral root formation, decrease primary root length, and increase root hair formation (Spaepen et al. 2007; Vacheron et al. 2013). Further bacterial IAA increases root surface area and length providing the plant more access to soil nutrients. Also, rhizobacterial IAA loosens plant cell walls and as a result, facilitates an increasing amount of root exudation that provides additional nutrients to support the growth of rhizosphere bacteria. IAA is a plant phytohormone that stimulates the overproduction of root hairs (Zavattieri et al. 2016). *S. marcescens* CDP-13 isolated from rhizospheric sample of *Capparis decidua* produced 0.34 ± 0.02μg/ml IAA, upon inoculation as wheat develops highly organized root system (Singh and Jha 2016). Genome sequence of *Serratia* sp. S119 isolated from peanut root nodules contains dhaS and ipdC genes. Translational products of these genes are indole-3-acetaldehyde dehydrogenase and indolepyruvate decarboxylase, respectively, are responsible for tryptophan dependent IAA synthesis (Ludueña et al. 2018).

Generally, phosphate (P) is abundant in mineral phosphate form, not suitable for plant uptake. Plants can absorb only mono- and dibasic phosphate which is the soluble form of phosphate (Jha et al. 2012; Jha and Saraf 2015). Phosphate solubilization is done by secretion of low molecular weight organic acid by microbes as a by-product sugar metabolism (Goswami et al. 2014). *S. marcescens* UENF-22GI genome revealed several genes involved in the production of gluconic acid from glucose and has a conserved pstABCS operon which codes for a phosphate-specific transport system. Gluconic acid production is the major mechanism of phosphate solubilization exerted by PGPR. The co-existence of pyrroloquinoline quinone (PQQ) coenzyme with GDH and PST genes is responsible for solubilization of P through soil acidification (Matteoli et al. 2018). Phosphate solubilization activity also been shown by *Serratia* sp. S119, contributes growth promotion of peanut and maize plants (Taurian et al. 2010). Plant growth promotion of *Serratia* sp. S119 has been depended upon phosphate mineralization phenotype (Anzuay et al. 2017). During the stationary phase *Serratia* sp. S119 shows highest phosphate solubilization activity owing to gluconic acid production (Anzuay et al. 2013).

1.3.2 *Serratia* as Biocontrol Agents

Siderophore-producing PGPR support plant health at various levels by improving iron nutrition, thus hindering the growth of pathogens by limiting the iron available for the pathogen, generally fungi, which are unable to absorb the iron–siderophore complex (Shen et al. 2013). Siderophore acts as a cofactor for the nitrogenase enzyme, which actively participates in atmospheric nitrogen fixation (Rahi et al. 2009). *S. marcescens* SM6 and SR41-8000 demonstrated in - vitro siderophore production activity 12 h post incubation, maximum at 30 h of growth, and remained constant (Khilyas et al. 2016). *S. marcescens* AL2-16 showed maximum siderophore production after 48 hours of incubation in the presence of 1 mM of ferric chloride. Apex siderophore production was associated with oxalic acid (86.9%) followed by citric acid (83.4%) (Devi et al. 2016). Chitinase lyses chitin, an insoluble linear polymer of β-1, 4-N-acetyl-glucoseamine, the major fungal cell wall component, thus affecting the structural integrity of the target pathogen. Chitinolytic and antifungal activities of a potent biocontrol strain of *S. marcescens* B2 possessing multiple chitinase genes (ChiA, ChiB, Chic) were seen against soil borne pathogens *Rhizoctonia solani* and *Fusarium oxysporum* (Someya et al. 2000). Fungal pathogen mycelia co-inoculated with this strain showed varied abnormalities ranging from partial swelling in the hyphae at the tip, hyphal curling, or bursting of the hyphal tip. Examples of protection from phytopathogenic infection as a result of the activity of cell wall-degrading enzymes include control of *S. rolfsii* and *F. oxysporum on* beans (Felse and Panda 2000). Chitinase producing bacteria attain a rich source of nutrients by catabolism of chitin into soluble oligosaccharides which enter into periplasm via a chitoporin channel, where they further break into mono- and di-saccharides (Paspaliari et al. 2017; Hayes et al. 2017). Babashpour et al. 2012 developed and purified recombinant bacterial chitinase from *S. marcescens* B4A. The enzyme activity was stable below 50 C for 20 min. The K_m and V_{max} values of chitinase were 3.30 mg ml^{-1} and 0.92 mM min^{-1}, respectively. This enzyme showed a wide range of pH activity (5.0–8.0) and temperature range at 30–60 °C. Recombinant chitinase demonstrated highly fungal antagonism by hyphal growth inhibition of *Bipolaris* sp., *Alternaria raphani*, *Alternaria brassicicola*. The wide range of pH and temperature stability of chitinase has great industrial and commercial importance. It can be useful in transgenic resistance application in crop plants. *S. marcescens* strain JPP1 is isolated from peanut hulls, China has been found to possess chitinolytic activities. For quantitative antagonism against plant pathogenic fungal mycelial growth and aflatoxin (a major fungal toxin involved in plant pathogenesis) production estimation, visual agar plate assay and tip culture method were employed where *S. marcescens* JPP1 exhibited a prompt inhibitory effect on mycelia growth (antifungal ratio >95%) and aflatoxin production (antiaflatoxigenic ratio >98%). Real time (RT)-PCR analysis showed that crude chitinase repressed the transcription of the aflatoxin (a major fungal toxin involved in plant pathogenesis) gene cluster primarily aflR, aflC (pksL1), and aflO (dmtA) (Wang et al. 2013). Besides that, seed bio-priming with *S. marcescens* JPP1 antagonized fungal growth and aflatoxin production on peanut seeds. Interestingly the effect was superior to chemical fungicide, carbendazim (Wang et al. 2013). *S. marcescens* NASC 1 isolated from the soil samples showed maximum chitinase activity at 0.5% chitin concentration, pH 9, and temperature of 37 °C. The enzyme could be used in bioremediation of chitinous waste (Abdhul et al. 2018). Induced systemic resistance (ISR) is salicylic acid signalling dependent plant defence mechanism, where plants build up resistance against bacterial, fungal pathogens through various structural (cell wall structure) and physiological modification changes such as lytic enzymes and synthesis of phytoalexins (Heil and Bostock 2002). N-acyl-homoserine lactones (AHL), the quorum sensing signalling molecule in *S. marcescens* MG1 was important in mediating ISR in tomato plants. This induced antifungal defence against the leaf pathogen *A. alternata* (Schuhegger et al. 2006). AHL produced by *S. plymuthica* strains isolated from rhizospheric soil also plays a role in ISR (Pang et al. 2008). *S. marcescens* strain CDP-13 generated induced systemic resistance in cucumber against wilt causing pathogen *F. oxysporum* (Kloepper et al. 1980).

The typical blood red colour of *Serratia* is associated with prodigiosin (2-methyl-3-pentyl-6- methoxy prodiginine) pigment, secreted as water-insoluble secondary metabolite, bound to bacterial cell envelope found in species like *S. marcescens, S. plymuthica, S. nematodiphila,* and *S. rubidaea*. However, the pigment is soluble in organic solvents (Elkenawy et al. 2017; Jafarzade et al. 2013; Darshan and Manonmani 2015, Grimont and Grimont 2004). The ideal condition for prodigiosin production in peptone glycerol broth is 30 °C at neutral pH and 87% glycerol concentration, during late log phase of growth (Slater et al. 2003; Pore et al. 2016). The antimicrobial activity exerted by this pigment is to be more against gram-positive than gram-negative bacteria. Prodigiosin shows better antimicrobial activity at the acidic pH than the basic pH. Media augmentation with maltose and peptone yields a maximal amount of prodigiosin (Gulani et al. 2012). pigA-O, a cluster of operonic genes regulates the production of prodigiosin (Ruiz et al. 2010). Purified prodigiosin extracts of *S. marcescens* B2 and *S. marcescens* B10 VKM are active against *Staphylococcus*, *Pseudomonas*, *Bacillus*, *Candida* (Pore et al. 2016). The pigment helps to invade *Fusarium oxysporum* hyphae, causes increasing permeability in fungal cell membrane, resulting in mycelial death (Hazarika et al. 2020). *S. marcescens* strain ETR17 is another example of prodiginines releasing bacteria, isolated from tea rhizosphere, which exhibits antagonistic activity against tea root-rot disease-causing pathogen *R. solani*. In some cases, prodigiosin has been proved as a better pathogen suppresser in comparison to chitinase. Prodigiosin produced by *S. marcescens* B10 VKM is a more potent suppressor of the fungus *Didymella applanata* (Duzhak et al. 2012).

1.3.3 *Serratia* in Abiotic Stress Tolerance

Heavy metal contamination, a major agricultural hazard, has grown manifold recently. An approach has been initiated by employing beneficial soil microbes to abolish the detrimental effects of metal stress in plants. Exposure of heavy metal like cadmium in a small dose can alter many physiological parameters such as gas exchange attributes, nutrients uptake, antioxidant capacity, the contents of chlorophyll, total phenolics, flavonoids, soluble sugars. Inoculation of Cd-stressed soybean plants with *S. marcescens* BM1 strain enhanced all these above discussed physiological parameters. *S. marcescens* BM1 inoculation not only reduced cadmium uptake by 21% but also significantly induced the antioxidant enzyme activity and stress-related genes expression and reduced H_2O_2 level by 15.27% in Cd-stressed plants. In addition to that Cd-stressed *S. marcescens* BM1 treated plants increased shoot length, root, and shoot fresh weight. A similar example can be noted *in S. nematodiphila* LRE07 strain. This strain enhanced photosynthetic pigments biosynthesis of *Solanum nigrum* L. plants under Cd stress conditions (Wan et al. 2012). *Serratia* sp. RSC-14 inoculation conferred Cd stress tolerance by improving chlorophyll biosynthesis (Khan et al. 2017).

Salinity is another common abiotic stressor that most agricultural crops contend with. It severely affects plant growth and productivity. As salt stress increases, oxidative damage to lipid results in increased malondialdehyde (MDA) content increases, which results in membrane permeability, exosmosis of electrolytes, and ends with cell death (Mittler 2002). Oxidative damage can be mitigated by eliminating reactive oxygen species (ROS) with antioxidant enzymes is the salinity tolerance mechanism obtained by plants (Singh and Jha 2016). Superoxide dismutase (SOD), catalase (CAT), and peroxidase (POX) act as scavengers for superoxide radicals during salinity stress. POX activity also contributes to plant growth and development as this enzyme is involved in lignin formation during plant growth. *S. marcescens* CDP-13 induces the accumulation of proline, which nullifies the salt stress-induced reduction in the activities of antioxidant enzymes, stabilizes proteins, membranes, and sub-cellular structures, and protects cellular functions by scavenging ROS (Ozden et al. 2009; Singh and Jha 2016). Other properties that protect plants from stress condition are auxin concentration which is severely affected by the stressors. *S. marcescens* CDP-13 significantly increased the auxin content in wheat plants by 25 to 29%. Moreover, *S. liquefaciens* KM4 significantly induced leaf gas exchange attributes, photosynthesis process, and osmolytes biosynthesis of maize plants grown under saline stress conditions (El-Esawi et al. 2020). Some instances where *Serratia* spp. counter abiotic and biotic stresses upon crop plants are higlighted in Table 1.2.

1.4 *Bacillus* spp

Each year, agricultural research reports *Bacillus* sp. novel supporters of plant growth. Gram positive in nature, these ePGPR are used singly or as consortia. For instance, *Bacillus sp* was co-inoculated *Azospirillum brasilense and Frankia* which significantly enhanced tolerance of durum wheat (*Triticum durum*) to water deficit (Benmati et al. 2020). Halotolerant bacteria closely related to *Bacillus* sp. and *B. licheniformis* showed prominent PGPR traits such as IAA (22.41,18.28,13.63μg/ml) production, nitrogen fixation, and siderophore production that resulted in improved growth of wheat plants under saline conditions. *Bacillus subtilis* was a prominent IAA producer, showed ACC deaminase activity, *Bacillus* sp. *Bacillus subtilis* at 10^6, 10^9,10^{11} CFU/ml concentration showed up to 64% disease suppression against sheath blight in rice incited by *Rhizoctonia solani* (Zhou et al. 2020). Soybean cyst nematode population *Heterodera glycines* was reduced and suppressed, respectively, by *B. velezensis* in greenhouse, microplot, and field trials while for *B. mojavensis* in greenhouse trial. *B. subtilis* ssp. *subtilis* and

Table 1.2 Different species of agriculturally important *Serratia* and their applications

Name of Microbe	Source/host plant	Application	References
S. marcescens S1	Shisham	Exhibited PGPR activity (chitinase, siderophore production, phosphate solubilisation activity) and antagonism against *Fusarium solani*	Banerjee et al. (2020)
S. liquefaciens KM4	Maize rhizosphere	Salinity stress tolerance	El-Esawi et al. (2018)
S. marcescens CDP-13	Capparis decidua	Induced systemic resistance and enhanced salinity tolerance of wheat	Singh and Jha (2016)
Serratia sp. S119	Peanut root	PGPR (IAA activity)	Ludueña et al. (2018)
S. marcescens UENF-22GI	vermicompost	PGPR (PSA)	Matteoli et al. (2018)
S. marcescens SM6	Michael Benedik (Texas A&M University)	PGPR (siderophore production)	Khilyas et al. (2016)
S. marcescens SR41-8000	Michael Benedik (Texas A&M University)	PGPR (siderophore production)	Khilyas et al. (2016)
*S. marcescens*AL2-16	Leaves and stems of *A. aspera*	PGPR (siderophore production)	Devi et al. (2016)
*S. marcescens*B2	Phylloplane of tomato	PGPR (chitinolytic activity)	Someya et al. (2000)
S. marcescens B4A	soils, water	PGPR (chitinolytic activity)	Babashpour et al. (2012)
S. marcescens strain JPP1	peanut hulls	PGPR (chitinolytic activity)	Wang et al. (2013)
S. marcescens NASC 1	soil samples	PGPR (chitinolytic activity)	Abdhul et al. (2018)
S. plymuthica	Rhizospheric soil	Induced systemic resistance (ISR)	Pang et al. (2009)
S. marcescens MG1	Tomato	Induced systemic resistance (ISR)	Schuhegger et al. (2006)
*S. marcescens*B10 VKM	NCIM (National Collection of Industrial Microorganisms), India	Prodigiosin associated antagonism	Pore et al. (2016)
*S. marcescens*strain ETR17	Tea rhizosphere	Prodigiosin associated antagonism	Dhar Purkayastha et al. (2018)

(continued)

Table 1.2 (continued)

Name of Microbe	Source/host plant	Application	References
S. marcescens BM1	Maize rhizosphere	Cadmium stress tolerance	El-Esawi et al. (2018)
S. nematodiphila LRE07	*Solanum nigrum*	Cadmium stress tolerance	Wan et al. (2012)

B. velezensis enhanced soybean plant height and biomass at early growth phase (Xiang et al. 2017). *B. subtilis* showed 54.7% antagonism in vitro against peanut pathogens *Sclerotium rolfsii* and 47.1% against *R. solani. B. subtilis* reduced damping off in field when mixed infection of both pathogens to peanut was given with 90.95% apparent healthy pods per plot and formulated *B. subtilis* enhanced pod dry weight (27.90 gm/pot) (Ahmad et al. 2019). Copper resistant *Bacillus* spp. PGPR enhanced Cu phytoextraction by ryegrass and fescue growing on Cu contaminated sites through antioxidant release and Cu solubilization. Two strains of *Bacillus* sp. (EhS5, EhS7) were most competent; EhS5% application increased root biomass of ryegrass up by 21.95% and fescue up by 51.06% in Cu contaminated soil at 200 mg/kg; mixed application enhanced by 54.63% root biomass of ryegrass (Ke et al. 2020). Inoculation of Sarju rice (*Oryza sativa*) with *B. amyloliquefaciens* under suboptimal nutrient conditions enhanced plant pigments by 72% (chlorophyll a), 1.92% (chlorophyll b), and 73% (carotenoids). Nutrient content analysis of rice seedlings inoculated with *B. amyloliquefaciens* showed macronutrient increase between 12% and 95% (in shoot), 5% and 70% (in root); micronutrient content went up by 95% and 195% (in shoot), and 70% and 132% (in root). Gene expression analysis by qRT PCR confirmed that this PGPR strain metabolically reprograms this rice variety under nutrient stress. For instance, gene expression of tricarboxylic acid pathway genes (isocitrate dehydrogenase) LOC_Os05g49760, (succinate dehydrogenase) LOC_Os08g02640, and (malate dehydrogenase) LOC_Os04g46560 was altered to aid survival by sugar accumulation in nutrient deprived conditions (Bisht and Chauhan 2020). Plant pathogen *Alternaria alternata* causes significant losses in China to the tobacco industry. *Bacillus siamensis* isolated from rhizosphere showed 81.96% inhibition to *A. alternata.* Mycelial and spore development were hindered by antifungal protein and VOC (volatile organic compound) secretion by *B. siamensis.* Further enhanced production of defence enzymes like peroxidase (POD) and polyphenol oxidase (PPO) in tobacco revealed that this PGPR could suppress disease severity and occurrence (Xie et al. 2020). Maize crop, subjected to salt stress and phosphorus deficiency was root inoculated with strains of both *Arthrobacter* sp. and *Bacillus* sp. Nutrient supply to thus root amended maize was increased, as well growth promotion through phytohormone production by the bacterial strains (Vanissa et al. 2020). Recent developments suggest that using modern visualization techniques like FISH (fluorescence in situ hybridization) and confocal laser scanning microscopy (CLSM) for real time capture and imaging of banana root colonization dynamics by *B. amyloliquefaciens* and *Pseudomonas palleroniana* showed them to be synergistic, hence an effective replacement to

Table 1.3 Different species of agriculturally important *Bacillus* and their applications

Name of microbe	Source/host plant	Application	References
Bacillus xiamenensis	*Saccharum officinarum* L.	Antagonism against sugarcane pathogens revealed the following rates of inhibtion *Colletotrichum falcatum* (53.81), *Fusarium oxysporum* (68.24), *Fusarium moniliforme* (69.70), *Rhizoctonia solani* (71.62), *Macrophomina phaseolina* (67.50), and *Pythium splendens* (77.58).	Xia et al. (2020)
Bacillus sp	Tomato	In vitro PGPR traits like nutrient solubilization(P,Zn), auxin(IAA) production, siderophore, hydrogen cyanide (HCN), and 1-aminocyclopropane-1-carboxylate (ACC) deaminase production; antagonism to other microbes seen and formed biofilm.	Kalam et al. (2020)
Bacillus velezensis	*Juncus effusus*	Biocontrol against *Botrytis cinerea* infection in tomato and strawberry plants; reduced infection by 50% and severity by 60%. Foliar and root application of this PGPR was done which increased biomass of plants, reduced oxidative damage and induce callose deposition in plant cells	Toral et al. (2020)
Bacillus proteolyticus, Bacillus velezensis	*Scripus grossus*	Tolerate lead toxicity (300 mg/L).	Kamaruzzaman et al. (2020)
Bacillus licheniformis	*Solanum tuberosum*	Polyphenol biosynthetic activity was mediated by gene activation and systemic resistance in potato plants against Alfalfa mosaic virus (AMV), caused 86.79% reduction in viral load; pyrrolo(1,2-a) pyrazine-1,4-dione was the elicitor molecules for induction of systemic acquired resistance in potato plants	Abdelkhalek et al. (2020)

chemical fertilizer in banana cultivation (Gamez et al. 2020). This suggests that *Bacillus* is an important bioresource, abundantly available in the rhizosphere and the benefit to agriculture is immense with important instances of biotic and abiotic stress management when applied to crop plants shown in Table 1.3.

1.5 *Pseudomonas*

Kloepper and Schroth (1978) were the first to announce PGPR as plant growth promoters that promoted radish crop growth. The worldwide interest in this group of rhizobacteria was sparked off by the studies initiated at the University of California, Berkeley, USA during 1970s. Fluorescent pseudomonas, gram negative, aerobic, motile, yielding water miscible yellow-green pigment (Palleroni 2015) is the largest and highly promising group of PGPR as they grow fast, are diverse, and can colonize a variety of substrates. Since *Pseudomonas* is ubiquitous, and agricultural soil hosts them in large numbers, this PGPR is a preferred species across a variety of crop systems. *Pseudomonas* is a non-spore forming microorganism, thus making stable bioformulations for commercial applications is challenging. *P. fluorescens* and *P. synxantha* both show *PGPR* activity. These bioformulations in talc were found to be stable for 180 days for *P. synxantha* and 90 days for *P. fluorescens* (Novinscak and Filion 2020). Co-application of 25% vermicompost along with *Pseudomonas* sp. to chickpea seed showed lower induction of defence enzymes peroxidase (POD) and phenylalanine ammonia lyase (PAL) as compared to pathogen infected (*S. rolfsii*) chickpea plants. This showed lower need for defence enzymes, hence it aided in conservation of cellular resources in the PGPR amended plants for tackling collar rot pathogen *S. rolfsii* (Sahni and Prasad 2020). *Pseudomonas* is a widely researched PGPR with far reaching effects ranging from bioremediation, stress tolerance, and biological control in agriculture, with some examples being highlighted in Table 1.4.

1.6 Challenges to the Use of Agriculturally Important Microbes

1.6.1 Screening of Microbes and Poor Shelf Life of Bioformulation

The present agronomy faces many challenges, for example, loss of soil fertility, fluctuating environmental factors, and greater pathogen and pest attacks. Sustainability and environmental safety of agricultural production rests upon eco-friendly approaches like biofertilizers, biopesticides, and crop residue return (Gopalakrishnan et al. 2015). Selection of efficient PGPR strains from among thousands of rhizospheric bacteria is an arduous process. For potentially aiding crop yield and disease management, sampling strategy is paramount, since diverse substrates would ensure greater viability in harsh conditions. Screening of new isolates on the basis of morphology, physiology, and biochemistry (IMVic tests, HCN, siderophore production based on Bergey's Manual of Bacteriology) is the foremost step. Further molecular characterization by way of DNA sequencing to confirm identity is now convenient and even more affordable. Biofertilizers and biocontrol developers face many problems to grow crops under varied climatic and environmental conditions mainly temperature, rainfall, soil type, crop variety which vary from farm to farm or even within one field, and such differences cause

Table 1.4 Different species of agriculturally important *Pseudomonas* and their applications

Name of microbe	Source or host plants	Application	References
Pseudomonas atacamensis	rhizosphere of desert bloom plant	Genes associated with plant growth-promoting rhizobacteria were observed, and is a potential candidate to be used for recovery of contaminated soils	Poblete-Morales et al. (2020)
Pseudomonas aeruginosa	Rice rhizosphere	Antagonism was seen against *Xanthomonas oryzae* (Xoo) rice pathogen . Siderophores (1-hydroxy-phenazine, pyocyanin, and pyochellin), rhamnolipids, 4-hydroxy-2-alkylquinolines (HAQs) as well as novel 2,3,4-trihydroxy-2-alkylquinolines and 1,2,3,4-tetrahydroxy-2-alkylquinolines were found in crude extract of *Pseudomonas* responsible for antibacterial activity of against Xoo pathogen. This strain pf *Pseudomonas* acts as a biological control agent against Xoo possessing PGPR properties to enhance yield of Super Basmati rice	Yasmin et al. (2017)
Pseudomonas aeruginosa	Rhizosphere	*Ganoderma boninense* disease of palm was managed using this strain. Crude extract of *P. aeruginosa* revealed 3-demethylubiquinone-9. P. aeruginosa and 3-demethylubiquinone-9 as antimicrobials against *Ganoderma boninense*	Lim et al. (2019)
Pseudomonas sp.	Rhizosphere of potato	Biocontrol against *Phytophthora infestans* causing late blight of potato	De Vrieze et al. (2018)
Pseudomonas fluoroscens	*Solanum tuberosum*	Suppressed the fungal pathogen *Fusarium sambucinium* causing dry rot of potato (35% reduction)	Al-Mughrabi (2010)

inconsistency in the potentiality of PGPR-based biofertilizers (Kamilova et al. 2015). Although researchers are developing efficient strains of bioinoculants with improved shelf life, the use of plant growth-promoting rhizobacteria in sustainable agriculture for augmenting agricultural productivity is still in its infancy (Glick 2014; Goswami et al. 2016). Carriers define the physical form of the bioinoculant. Coal, clay, charcoal, volcanic ash, pumice, diatomaceous earth, organic material such as wheat bran, sawdust and non-reactive substances like vermiculite, perlite, kaolin, bentonite, silicates are all possible carrier matrix materials (Smith 1976). Alginate sourced from kelps (*M. pyrifera, Laminaria digitata, L. hyperborean, and Eklonia cava*) encapsulation for degradation-based release of microbes is another method for ensuring shelf life (Schoebitz et al. 2013). However, these carriers based bioformulations do not have viability of more than 3 months. Further development of liquid based bioformulations has extended shelf life for upto 18 months (Gopi et al. 2019).

Biopesticides and biofertilizers production is a time worthy process involving careful selection of suitable strain for formulation, its stable mass production of selected strain, devising suitable carrier formulation, shelf life analysis for selected formulation, product efficacy, and viability in field condition and the changeable seasonal nature requires the capable storage for biopesticides (Keswani et al. 2016).

1.6.2 Lack of Field Reproducibly of PGPR Performance

Greenhouse or field experiments have furnished effective results of biocontrol against phytopathogenic diseases, for instance, late blight disease of potato (Caulier et al. 2018). Crop protection, however, is not reproducible for this disease my means of BCA (Axel et al. 2012). Biocontrol agents applied as foliar sprays, unlike synthetic molecules must compete with native microflora and colonize it, survive despite UV, temperature changes, and humidity gradients. Only upon establishment, they can produce bioactive molecules, eventually triggering a cascade of signals finally triggering the plant's immune defence, hindering the pathogen's development. Pathogens such as the oomycete *Phytophthora infestans* during infection possess various developmental stages such as oospores and zoosporangia or mycelia. Biocontrol measures using PGPR must circumvent the problem of successfully attacking different developmental stages of pathogens by the utilization of mixtures of strains rather than single agents. Functional polyvalence (targeting different stages of the pathogen life cycle) and redundancy (maximizing the chances of successful host plant colonization in various environmental conditions) are important qualities that a PGPR must embody. This is laden with contradictions since using five commercially available biocontrol agents (two based on *Bacillus*, one on *Streptomyces*, and two on *Trichoderma* strains) against *Phytophthora ramorum* in a detached leaf assay, Elliott et al. (2009) observed lower efficacy of the mixture compared to some of its individual components, suggesting antagonistic effects between the different strains composing the mixture.

Serious efforts are being made globally to instil faith among the farmers towards the utilization of alternative methods to chemical inputs. That PGPR biostimulants cannot be used solely for agroecological management is well agreed upon and must be integrated into fertilization and biocontrol regimes, to complement chemical inputs, reducing their dosage and frequency of application. Despite wide acclaim of agriculturally important microorganisms like PGPR farmers are sceptical about making the switch to PGPR applications, since field performance is highly variable (Beckers and Conrath 2007). Importantly, farmers, regulatory authorities, investors require greater access to knowledge about how to use these tools in agricultural practices as well as their potential benefits to promote sustainable agriculture.

1.6.3 Skewed Perception

While the Green Revolution (1960–2000) led to the commercialization and commodification of subsistence agriculture (Harwood 2019), wide consensus points out that its effects were soil fertility exhaustion, owing to the preponderance of mineral fertilizers and pesticides (Pimentel 1996; Shiva 1991). Yet scientific opinion is divided as to which mode of farming is the way forward the conventional manner, as recommended by the Green Revolution, or organic agriculture that has parallelly existed, albeit inconspicuously taken the limelight in recent years. A pioneering review comparing organic farming with conventional farming in the twenty-first century by a USA based research group underlines the constant tussle since the 1970s between sceptics and proponents of organic ecological/ecological agriculture (Reganold and Wachter 2016). Critics debate that organic farming is less efficient when compared to conventional agriculture, yielding lesser food produce for the same area of land with no climate benefits (Leifeld et al. 2013; Kirchmann 2019). Further, some workers have quantified nutrient influx of N, P, K to be 23%, 73%, and 53% from conventional farming to organic farming stations, thus acknowledging the need to evaluate the nutrient input of conventional farming while appraising the ecological organic cultivation performance (Nowak et al. 2013).

1.6.4 Challenges in Product Commercialization

The bioinoculant development is dependent on state-of-the-art lab facility, field trials, production, product commercialization, storage and industrial manufacture, registration, and regulatory matters. This entire chain of events concerns complex toxicological proofing which is very costly and lengthy, also requiring huge scientific and legal expertise.

1.6.5 Challenges in Products Registration and Patent Filing

The registration of biocontrol agents is not an easy process, as they are limited by environmental protection agencies. The registration of biocontrol agents is extensive, difficult, and convoluted. The conversion of lab screened isolates to final commercially accepted and patented BCAs/bioinoculants is full of bottlenecks. Very few patents become registered for agricultural application.

1.7 Future Aspects

Conversely, champions of organic farming believe that in-depth research could help narrow the gulf between conventional and organic yields. Further they argue that the major sustainability goals intrinsic to sustainable agriculture—SDG 1 (no poverty), SDG 2 (zero hunger), SDG 3 (good health and well-being), and SDG 12 (responsible consumption and production) find greatest consonance with transformative practices like organic farming (Eyhorn et al. 2019). Presently 1.4% of global agricultural area is under organic cultivation, making it an under exploited resource sufficient to feed an expected 10 billion people, by 2050 (IFOAM). Global sales in organic food and beverage were valued at 100 billion US $ in 2018, as per Ecovia Intelligence (Sahota 2019). Leaders in organic farming are Oceania (8.5%), Europe (2.9%; European Union 7.2%). Encouraging developments suggest that the largest increase organic farmland increased by 11.7 million hectares or 20% in 2017. Australia recorded the highest increase of about 8.5 million hectares. India has a net increase of 0.3 million hectares, of organic farmland (Willer and Lernoud 2019). Therefore organic agriculture or green farming reliant upon AIM for disease management and plant growth promotion can safely be heralded as the sustainable way forward. Nanotechnology involves nanoparticles possessing dimensions in the order of 100 nm or less (Hamad et al. 2020). Nanomaterials find application in plant protection, nutrition, and management of farm practices due to small size, high surface to volume ratio, and unique optical properties (Souza et al. 2019). The use of microorganisms with nanoparticles can be effective as nanopesticide. *Bacillus thuringiensis* finds use as an eco-friendly biopesticide, against larval and pupal stages of *Spodoptera littoralis*. Enhancement of pesticide rate of efficacy is done by synthesizing a nanocomposite of sodium titanate with *Bacillus thuringiensis* (Zaki et al. 2017). The viability and efficacy of two phosphate solubilsing bacterial species—*Pseudomonas putida* and *Ps. kilonensis* in solubilizing phosphate containing subtratum viz. tricalcium phosphate and hydroxyapatite with nano-carriers such as nanoclay, natural char micro-particles (NCMPs), nanoclay + alginate, NCMPs + alginate, and natural char nano-particles (NCNPs)+alginate were investigated. Results for efficacy were promising for natural char micro-particles (NCMPs) + alginate and nanoclay + alginate carriers at temperatures 4 °C and 28 °C (Safari et al. 2020). Polymeric nanoparticles for coating biofertilizers are also a recent advancement in nanoagroparticle synthesis. AIM like fungal mycorrhizae, *Rhizobium*, *Azotobacter*, *Azospirillum*, and blue-green algae are eminent biofertilizers converting organic matter into essential elements for plant nutrition. Nanoemulsions help distribute microorganisms through liquid formulations, although sedimentation problems can decrease their efficiency. To overcome this problem, hydrophobic silica nanoparticles can be used in improving the cell viability (Duhan et al. 2017).

Natural variations make it difficult to predict how PGPR fare when applied to field conditions. In addition, many factors are responsible for these techniques such as high cost, laborious work, and viability chances of microbes in field. In order to improve the efficiency and shelf life of biocontrol agents, various formulations based on solid and liquid carriers should be developed. Future research in optimizing growth condition and increased shelf life of PGPR bioformulations not phytotoxic to crop plants, tolerating adverse environmental conditions, ensuring higher yield and yet being cost effective PGPR inputs for the farmer deserves sincere investigation. Hence, PGPR must be propagated artificially to optimize their viability and biological activity under field applications. Further, PGPR must be reinoculated every year/season because they will not live forever in the soil.

Research must help comprehend environmental parameters affecting the efficiency of these products, especially for field crops. The screening methodology, preparation of bioformulation, its influence upon the environment deserve careful, rigorous, and sincere attention. Commercialization of biostimulants and biocontrol agents is a major challenge. Research is helpless when not in tandem with policy, as well as enhanced societal awareness. Pertinent to agricultural important microbe usage, biocontrol is well integrated in actual agroecological systems while biofertilizer usage is still scant (Wezel et al. 2014). The ultimate goal is to become pesticide or chemical fertilizer free. Agrochemicals leave undesirable environmentally burdensome residues. Currently, various agrochemicals (i.e., herbicides, fungicides, insecticides, nematicides, molluscicides, rodenticides, chemical fertilizers) are being used non-judiciously (Meena et al. 2016).

Developing nations use nanopesticides as magic bullets which make the problem worse. Intentional application of nanoparticle-based pesticides and fertilizers is an issue of great concern, as it may pose serious hazards (Pandey et al. 2018). Prolonged and indiscriminate use of chemical inputs in agri-enterprises adversely affects the soil agricultural sustainability, soil microbiota, and food safety. This in turn has harmful effects upon nutritional security, human, and animal health. Global climate change, increasing environmental concerns, and population increase, necessitate the utilisation of AIM in raising crops. This opens up important a multitude of potential opportunities for achieving sustainable food production thus enabling the world to strive towards global food security and banishment of hunger.

References

Abdelkhalek A, Al-Askar AA, Behiry SI (2020) Bacillus licheniformis strain POT1 mediated polyphenol biosynthetic pathways genes activation and systemic resistance in potato plants against Alfalfa mosaic virus. Sci Rep 10(1):1–6

Abdhul K, Sakthinarayanan K, Murugan M (2018) Isolation and identification of *Serratia marcescens* NASC 1 and optimization of its Chitinase production. J Pharm Sci Res 10:1195–1197

Abhilash PC, Tripathi V, Edrisi SA, Dubey RK, Bakshi M, Dubey PK, Singh HB, Ebbs SD (2016) Sustainability of crop production from polluted lands. Energy Ecol Environ 1(1):54–65

Ahmad AG, Attia AZ, Mohamed MS, Elsayed HE (2019) Fermentation, formulation and evaluation of PGPR Bacillus subtilis isolate as a bioagent for reducing occurrence of peanut soil-borne diseases. J Integr Agric 18(9):2080–2092. https://doi.org/10.1016/S2095-3119(19)62578-5

Ahmad F, Ahmad I, Khan MS (2005) Indole acetic acid production by the indigenous isolates of Azotobacter and fluorescent pseudomonas in the presence and absence of tryptophan. Turk J Biol 29(1):29–34

Al-Mughrabi KI (2010) Biological control of fusarium dry rot and other potato tuber diseases using Pseudomonas fluorescens and Enterobacter cloacae. Biol Control 53(3):280–284. https://doi.org/10.1016/j.biocontrol.2010.01.010

Anderson K (2010) Globalization's effects on world agricultural trade, 1960–2050. Philos Trans R Soc Lond Ser B Biol Sci 365:3007–3021

Antoun H (2013) Plant-growth-promoting rhizobacteria. In: Brenner's encyclopedia of genetics, vol 5, 2nd edn. Academic Press, San Diego, CA. https://doi.org/10.1016/B978-0-12-374984-0.01169-4

Anzuay MS, Frola O, Angelini JG, Ludueña LM, Fabra A, Taurian T (2013) Genetic diversity of phosphate-solubilizing peanut (*Arachis hypogaea* L.) associated bacteria and mechanisms involved in this ability. Symbiosis 60:143–154

Anzuay MS, Ruiz Ciancio MG, Ludueña LM, Angelini JG, Barros G, Pastor N, Taurian T (2017) Growth promotion of peanut (*Arachis hypogaea* L.) and maize (*Zea mays* L.) plants by single and mixed cultures of efficient phosphate solubilizing bacteria that are tolerant to abiotic stress and pesticides. Microbiol Res 199:98–109

Arjun JK, Harikrishnan K (2011) Metagenomic analysis of bacterial diversity in the rice rhizosphere soil microbiome research article. Biotechnol Bioinformatics Bioeng 1(3):361–367

Axel C, Zannini E, Coffey A, Guo J, Waters DM, Arendt EK (2012) Ecofriendly control of potato late blight causative agent and the potential role of lactic acid bacteria: a review. Appl Microbiol Biotechnol 96:37–48. https://doi.org/10.1007/s00253-012-4282-y

Babashpour S, Aminzadeh S, Farrokhi N, Karkhane A, Haghbeen K (2012) Characterization of a chitinase (Chit62) from *Serratia marcescens* B4A and its efficacy as a bioshield against plant fungal pathogens. Biochem Genet 50:722–735

Backer R, Rokem JS, Ilangumaran G, Lamont J, Praslickova D, Ricci E et al (2018) Plant growth-promoting rhizobacteria: context, mechanisms of action, and roadmap to commercialization of biostimulants for sustainable agriculture. Front Plant Sci 9:1473. https://doi.org/10.3389/fpls.2018.01473

Banerjee S, Singh S, Pandey S, Bhandari MS, Pandey A, Giri K (2020) Biocontrol potential of *Pseudomonas azotoformans*, *Serratia marcescens* and *Trichoderma virens* against Fusarium wilt of *Dalbergia sissoo*. For Path 50:e12581. https://doi.org/10.1111/efp.12581

Banwart S (2011) Save our soils. Nature 474:151–152

Baweja P, Kumar S, Kumar G (2019) Organic fertilizer from algae: a novel approach towards sustainable agriculture. In: Biofertilizers for sustainable agriculture and environment. Springer, Cham, pp 353–370. https://doi.org/10.1007/978-3-030-18933-4_16

Beckers GJM, Conrath U (2007) Priming for stress resistance: from the lab to the field. Curr Opin Plant Biol 10(4):425–431

Benmati M, Mouellef A, Belbekri N, Djekoun A (2020) Effects of durum wheat (Triticum durum Desf.) Inoculation with PGPR (Azospirillum brasilense, Bacillus sp and Frankia CcI3) and its tolerance to water deficit. Indian J Agric Res 54(4):437–444. https://doi.org/10.18805/IJARe.A-516

Berendsen RL, Pieterse CMJ, Bakker PAHM (2012) The rhizosphere microbiome and plant health. Trends Plant Sci 17:478–486. https://doi.org/10.1016/j.tplants.2012.04.001

Bhattacharyya P, Roy KS, Das M, Ray S, Balachandar D, Karthikeyan S, Nayak AK, Mohapatra T (2016) Elucidation of rice rhizosphere metagenome in relation to methane and nitrogen metabolism under elevated carbon dioxide and temperature using whole genome metagenomic approach. Sci Total Environ 542:886–898. https://doi.org/10.1016/j.scitotenv.2015.10.154

Bisht N, Chauhan PS (2020) Comparing the growth-promoting potential of Paenibacillus lentimorbus and Bacillus amyloliquefaciens in Oryza sativa L. var. Sarju-52 under suboptimal nutrient conditions. Plant Physiol Biochem 146:187–97. https://doi.org/10.1016/j.plaphy.2019.11.023

Bologna M, Aquino G (2020) Deforestation and world population sustainability: a quantitative analysis. Sci Rep 10:7631. https://doi.org/10.1038/s41598-020-63657-6

Braud A, Jézéquel K, Bazot S, Lebeau T (2009) Enhanced phytoextraction of an agricultural Cr-and Pb-contaminated soil by bioaugmentation with siderophore-producing bacteria. Chemosphere 74(2):280–286

Bringezu S, O'Brien M, Pengue W, Swilling M, Kauppi L (2010) Assessing global land use and soil management for sustainable resource policies. Scoping paper for the International Panel for Sustainable Resource Management. UNEP

Bruinsma J (ed) (2003) World agriculture: towards 2015/2030: an FAO perspective. Earthscan, London

Bruinsma J (2009) The Resource Outlook to 2050: by How Much Do Land, Water and Crop Yields Need to Increase by 2050? Paper for the EM

Bulgarelli D, Rott M, Schlaeppi K, van Themaat EVL, Ahmadinejad N, Assenza F, Rauf P, Huettel B, Reinhardt R, Schmelzer E, Peplies J (2012) Revealing structure and assembly cues for Arabidopsis root-inhabiting bacterial microbiota. Nature 488(7409):91–95. https://doi.org/10.1038/nature11336

Burén S, López-Torrejón G, Rubio LM (2018) Extreme bioengineering to meet the nitrogen challenge. Proc Natl Acad Sci 115(36):8849–8851. https://doi.org/10.1073/pnas.1812247115

Caulier S, Gillis A, Colau G, Licciardi F, Liépin M, Desoignies N et al (2018) Versatile antagonistic activities of soil-borne Bacillus spp. and Pseudomonas spp. against Phytophthora infestans and other potato pathogens. Front. Microbiol 9:143. https://doi.org/10.3389/fmicb.2018.00143

Cavicchioli R, Ripple WJ, Timmis KN, Azam F, Bakken LR, Baylis M, Behrenfeld MJ, Boetius A, Boyd PW, Classen AT, Crowther TW (2019) Scientists' warning to humanity: microorganisms and climate change. Nat Rev Microbiol 17(9):569–586. https://doi.org/10.1038/s41579-019-0222-5

Chennappa G, Sreenivasa MY, Nagaraja H (2018) Azotobacter salinestris: a novel pesticide-degrading and prominent biocontrol PGPR bacteria. In: Microorganisms for green revolution. Springer, Singapore, pp 23–43. https://doi.org/10.1007/978-981-10-7146-1_2

Chittora D, Meena M, Barupal T, Swapnil P, Sharma K (2020) Cyanobacteria as a source of biofertilizers for sustainable agriculture. Biochem Biophys Rep 22:100737. https://doi.org/10.1016/j.bbrep.2020.100737

Clark M, Maselko M (2020) Transgene biocontainment strategies for molecular farming. Front Plant Sci 11:1–11

Danish S, Zafar-ul-Hye M (2019) Co-application of ACC-deaminase producing PGPR and timber-waste biochar improves pigments formation, growth and yield of wheat under drought stress. Sci Rep 9(1):1–13. https://doi.org/10.1038/s41598-019-42374-9

Darshan N, Manonmani HK (2015) Prodigiosin and its potential applications. J Food Sci Technol 52:5393–5407

Dastager SG, Deepa CK, Pandey A (2011) Potential plant growth-promoting activity of *Serratia nematodiphila* NII-0928 on black pepper (Piper nigrum L.). World J Microbiol Biotechnol 27: 259–265

De Vrieze M, Germanier F, Vuille N, Weisskopf L (2018) Combining different potato-associated Pseudomonas strains for improved biocontrol of Phytophthora infestans. Front Microbiol 9:2573. https://doi.org/10.3389/fmicb.2018.02573

Devi KA, Pandey P, Sharma GD (2016) Plant growth-promoting endophyte *Serratia marcescens* AL2-16 enhances the growth of *Achyranthes aspera* L., a medicinal plant. Hayati J Biosci 23: 173–180

Dhar Purkayastha G, Mangar P, Saha A, Saha D (2018) Evaluation of the biocontrol efficacy of a *Serratia marcescens* strain indigenous to tea rhizosphere for the management of root rot disease in tea. PLoS One 13(2):e0191761

Duhan JS, Kumar R, Kumar N, Kaur P, Nehra K, Duhan S, Nanotechnology (2017) The new perspective in precision agriculture. Biotechnol Rep 15:11–23. https://doi.org/10.1016/j.btre.2017.03.002

Duzhak AB, Panfilova ZI, Duzhak TG, Vasyunima EA, Shternshis MV (2012) Role of prodigiosin and chitinases in antagonistic activity of the bacterium *Serratia marcescens* against the fungus Didymella applanata. Biochemistry 77:910–916. https://doi.org/10.1134/S0006297912080123

El-Esawi MA, Alaraidh IA, Alsahli AA, Alzahrani SM, Ali HM, Alayafi AA, Ahmad M (2018) Serratia liquefaciens KM4 improves salt stress tolerance in maize by regulating redox potential,

ion homeostasis, leaf gas exchange and stress-related gene expression. Int J Mol Sci 19 (11):3310. https://doi.org/10.3390/ijms19113310
El-Esawi MA, Elkelish A, Soliman M, Elansary HO, Zaid A, Wani SH (2020) *Serratiamarcescens* BM1 enhances cadmium stress tolerance and phytoremediation potential of soybean through modulation of osmolytes, leaf gas exchange, antioxidant machinery, and stress-responsive genes expression. Antioxidants 9:43. https://doi.org/10.3390/antiox9010043
El-Katatany M, Hetta A, Sliaban G, ElKomy H (2003) Improvement of cell wall degrading enzymes production by alginate encapsulated Trichoderma spp. Food Technol Biotechnol 41 (3):219–225
Elkenawy NM, Yassin AS, Elhifnawy HN, Amin MA (2017) Optimization of prodigiosin production by *Serratia marcescens* using crude glycerol and enhancing production using gamma radiation. Biotechnol Rep 14:47–53
El_Komy MH, Hassouna MG, Abou-Taleb EM, Al-Sarar AS, Abobakr Y (2020) A mixture of Azotobacte r, Azospirillum, and Klebsiella strains improves root-rot disease complex management and promotes growth in sunflowers in calcareous soil. Eur J Plant Pathol 156(3):713–726. https://doi.org/10.1007/s10658-019-01921-w
Elliott M, Shamoun SF, Sumampong G, James D, Masri S, Varga A (2009) Evaluation of several commercial biocontrol products on European and North American populations of Phytophthora ramorum. Biocontrol Sci Technol 19(10):1007–1021. https://doi.org/10.1080/09583150903243870
El-Maraghy SS, Tohamy TA, Hussein KA (2020) Role of plant-growth promoting fungi (PGPF) in defensive genes expression of Triticum aestivum against wilt disease. Rhizosphere 15:100223. https://doi.org/10.1016/j.rhisph.2020.100223
Enebe MC, Babalola OO (2020) Effects of inorganic and organic treatments on the microbial community of maize rhizosphere by a shotgun metagenomics approach. Ann Microbiol 70:1–8. https://doi.org/10.1186/s13213-020-01591-8
Eyhorn F, Muller A, Reganold JP, Frison E, Herren HR, Luttikholt L, Mueller A, Sanders J, Scialabba NE, Seufert V, Smith P (2019) Sustainability in global agriculture driven by organic farming. Nat Sustain 2:253–255. https://doi.org/10.1038/s41893-019-0266-6
FAO (2020a) FAOSTAT Emissions shares, http://www.fao.org/faostat/en/#data/EM
FAO (2020b) World Food and Agriculture - Statistical Yearbook 2020. Rome. https://doi.org/10.4060/cb1329en
FAOSTAT F (2009) Agriculture Organization of the United Nations. Statistical Database
FAOSTAT FAO (2009) Statistical databases. Food and Agriculture Organization of the United Nations. Available online at http://www.fao.org/3/i0680e/i0680e.pdf
FAOSTAT FAO (2020) Crops. Food and Agriculture Organization of the United Nations. Available online at. http://www.fao.org/faostat/cn/#home
Felse PA, Panda T (2000) Production of microbial chitinases–a revisit. Bioprocess Eng 23:127–134. https://doi.org/10.1007/PL00009117
Ferdous Z, Zulfiqar F, Datta A, Hasan AK, Sarker A (2020) Potential and challenges of organic agriculture in Bangladesh: a review. J Crop Improv. https://doi.org/10.1080/15427528.2020.1824951
Ferraroni M (2019) γ-Carbonic anhydrases. In: Supuran CT Nocentini A (eds) Carbonic anhydrases. Biochemistry and pharmacology of an evergreen pharmaceutical target. Academic Press, pp 79–105. https://doi.org/10.1016/B978-0-12-816476-1.00004-6
Gamez RM, Ramirez S, Montes M, Cardinale M (2020) Complementary dynamics of Banana root colonization by the plant growth-promoting rhizobacteria Bacillus amyloliquefaciens Bs006 and Pseudomonas palleroniana Ps006 at spatial and temporal scales. Microb Ecol 80 (3):656–668
Gauri SS, Mandal SM, Pati BR (2012) Impact of Azotobacter exopolysaccharides on sustainable agriculture. Appl Microbiol Biotechnol 95:331–338. https://doi.org/10.1007/s00253-012-4159-0
Gibbs HK, Salmon JM (2015) Mapping the world's degraded lands. Appl Geogr 57:12–21. https://doi.org/10.1016/j.apgeog.2014.11.024

Glick BR (2014) Bacteria with ACC deaminase can promote plant growth and help to feed the world. Microbiol Res 169:30–39. https://doi.org/10.1016/j.micres.2013.09.009

Goel R, Kumar V, Suyal DC, Soni R (2018) Toward the unculturable microbes for sustainable agricultural production. In: Role of rhizospheric microbes in soil. Springer, Singapore, pp 107–123. https://doi.org/10.1007/978-981-10-8402-7_4

Gopalakrishnan S, Sathya A, Vijayabharathi R, Varshney RK, Gowda CL, Krishnamurthy L (2015) Plant growth promoting rhizobia: challenges and opportunities. 3 Biotech 5(4):355–377. https://doi.org/10.1007/s13205-014-0241-x

Gopi GK, Meenakumari KS, Nysanth NS, Subha P (2019) An optimized standard liquid carrier formulation for extended shelf-life of plant growth promoting bacteria. Rhizosphere 11:100160. https://doi.org/10.1016/j.rhisph.2019.100160

Goswami D, Dhandhukia P, Patel P, Thakker JN (2014) Screening of PGPR from saline desert of Kutch: growth promotion in *Arachis hypogea* by *Bacillus licheniformis* A2. Microbiol Res 169: 66–75. https://doi.org/10.1016/j.micres.2013.07.004

Goswami D, Thakker JN, Dhandhukia PC (2016) Portraying mechanics of plant growth promoting rhizobacteria (PGPR): a review. Cogent Food Agric 2(1):1–19. https://doi.org/10.1080/23311932.2015.1127500

Gouda S, Kerry RG, Das G, Paramithiotis S, Shin HS, Patra JK (2018) Revitalization of plant growth promoting rhizobacteria for sustainable development in agriculture. Microbiol Res 206: 131–140. https://doi.org/10.1016/j.micres.2017.08.016

Grimont P, Grimont F (2004) Genus SerratiaBizio 1823, 288AL. In: Krieg NR, Holt JG (eds) Bergey's manual of systematic bacteriology. Williams and Wilkins, Baltimore

Gulani C, Bhattacharya S, Das A (2012) Assessment of process parameters influencing the enhanced production of prodigiosin from *Serratiamarcescens* and evaluation of its antimicrobial, antioxidant and dyeing potentials. Malays J Microbiol 8:116–122

Gurikar C, Naik MK, Sreenivasa MY (2016) Azotobacter: PGPR activities with special reference to effect of pesticides and biodegradation. In: Singh D, Singh H, Prabha R (eds) Microbial inoculants in sustainable agricultural productivity. Springer, New Delhi, pp 229–244

Hajnal-Jafari T, Latković D, Đurić S, Mrkovačkiand N, Najdenovska O (2012) The use of Azotobacter in organic maize production. Res J Agric Sci 44(2):28–32

Hamad HT, Al-Sharify ZT, Al-Najjar SZ, Gadooa ZA (2020) A review on nanotechnology and its applications on fluid flow in agriculture and water recourses. In: IOP conference series: materials science and engineering, vol 870(1). IOP Publishing, p 012038. https://doi.org/10.1088/1757-899X/870/1/012038

Harwood J (2019) Was the green revolution intended to maximise food production? Int J Agric Sustain 17(4):312–325. https://doi.org/10.1080/14735903.2019.1637236

Hayat R, Ali S, Amara U, Khalid R, Ahmed I (2010) Soil beneficial bacteria and their role in plant growth promotion: a review. Ann Microbiol 60:579–598. https://doi.org/10.1007/s13213-010-0117-1

Hayes CA, Dalia TN, Dalia AB (2017) Systematic genetic dissection of chitin degradation and uptake in *Vibrio cholerae*. Environ Microbiol 19:4154–4163

Hazarika DJ, Gautom T, Parveen A, Goswami G, Barooah M, Modi MK, Boro RC (2020) Mechanism of interaction of an endofungal bacterium *Serratiamarcescens* D1 with its host and non-host fungi. PLoS One 15:e0224051. https://doi.org/10.1371/journal.pone.0224051

Heil M, Bostock RM (2002) Induced systemic resistance (ISR) against pathogens in the context of induced plant defenses. Ann Bot 89:503–512

Hindersah R, Kalay M, Talahaturuson A, Lakburlawal Y (2018) Nitrogen fixing bacteria azotobacter as biofertilizer and biocontrol in long bean. Agric 30(1):25–32. https://doi.org/10.24246/agric.2018.v30.i1, https://doi.org/10.4060/ca9692en

Hossain MD, Sultana F (2020) Application and mechanisms of plant growth promoting fungi (PGPF) for Phytostimulation. In: Das SC (ed) Organic agriculture. Intech Open. https://doi.org/10.5772/intechopen.92338

Hossain MM, Sultana F, Islam S (2017) Plant growth-promoting fungi (PGPF): phytostimulation and induced systemic resistance. In: Plant-microbe interactions in agro-ecological perspectives. Springer Singapore, Singapore, pp 135–191. https://doi.org/10.5772/intechopen.92338

IPCC (2007) Climate change 2007: synthesis report. Contribution of Working Groups I, II and III to the Fourth Assessment Report of the Intergovernmental Panel on Climate Change

IPCC (2019) Intergovernmental Panel on Climate Change, Special Report on Land and Climate Change, Summary for Policymakers. https://www.ipcc.ch/srccl/download/

Islam MR, Sultana T, Joe MM, Yim W, Cho JC, Sa T (2013) Nitrogen-fixing bacteria with multiple plant growth-promoting activities enhance growth of tomato and red pepper. J Basic Microbiol 53(12):1004–1015. https://doi.org/10.1002/jobm.201200141

Ivleva NB, Groat J, Staub JM, Stephens M (2016) Expression of active subunit of nitrogenase via integration into plant organelle genome. PLoS One 11(8):e0160951. https://doi.org/10.1371/journal.pone.0160951

Jafarzade M, Yahya NA, Shayesteh F, Usup G, Ahmad A (2013) Influence of culture conditions and medium composition on the production of antibacterial compounds by marine *Serratia* sp. WPRA3. J Microbiol 51:373–379

Jha CK, Saraf M (2015) Plant growth promoting rhizobacteria (PGPR): a review. J Agric Res Dev 5(2):108–119

Jha CK, Patel B, Saraf M (2012) Stimulation of the growth of *Jatropha curcas* by the plant growth promoting bacterium *Enterobacter cancerogenus* MSA2. World J Microbiol Biotechnol 28: 891–899. https://doi.org/10.1007/s11274-011-0886-0

Jnawali AD, Ojha RB, Marahatta S (2015) Role of Azotobacter in soil fertility and sustainability–A review. Adv Plants Agric Res 2(6):1–5. https://doi.org/10.15406/apar.2015.02.00069

Kalam S, Basu A, Podile AR (2020) Functional and molecular characterization of plant growth promoting Bacillus isolates from tomato rhizosphere. Heliyon 6(8):e04734. https://doi.org/10.1016/j.heliyon.2020.e04734

Kamaruzzaman MA, Abdullah SR, Hasan HA, Hassan M, Othman AR, Idris M (2020) Characterisation of Pb-resistant plant growth-promoting rhizobacteria (PGPR) from Scirpus grossus. Biocat Agric Biotechnol 23:101456

Kamilova F, Okon Y, de Weert S, Hora K (2015) Commercialization of microbes: manufacturing, inoculation, best practice for objective field testing, and registration. In: Lugtenberg B (ed) Principles of plant-microbe interactions. Springer, Cham, pp 319–327

Kannojia P, Choudhary KK, Srivastava AK, Singh AK (2019) PGPR bioelicitors: induced systemic resistance (ISR) and proteomic perspective on biocontrol. In: PGPR amelioration in sustainable agriculture. Woodhead Publishing, Cambride, MA, pp 67–84. https://doi.org/10.1016/B978-0-12-815879-1.00004-5

Ke T, Zhang J, Tao Y, Zhang C, Zhang Y, Xu Y, Chen L (2020) Individual and combined application of cu-tolerant Bacillus spp. enhance the Cu phytoextraction efficiency of perennial ryegrass. Chemosphere 263:1–12. https://doi.org/10.1016/j.chemosphere.2020.127952

Kenneth OC, Nwadibe EC, Kalu AU, Unah UV (2019) Plant growth promoting rhizobacteria (PGPR): a novel agent for sustainable food production. Am J Agric Biol Sci 14:35–54. https://doi.org/10.3844/ajabssp.2019.35.54

Keswani C, Bisen K, Singh V, Sarma BK, Singh HB (2016) Formulation technology of biocontrol agents: present status and future prospects. In: Arora N, Mehnaz S, Balestrini R (eds) Bioformulations: for sustainable agriculture. Springer, New Delhi, pp 35–52

Khan AR, Park G-S, Asaf S, Hong S-J, Jung BK, Shin J-H (2017) Complete genome analysis of *Serratia marcescens* RSC-14: a plant growth-promoting bacterium that alleviates cadmium stress in host plants. PLoS One 12:e0171534. https://doi.org/10.1371/journal.pone.0171534

Khilyas IV, Shirshikova TV, Matrosova LE, Sorokina AV, Sharipova MR, Bogomolnaya LM (2016) Production of siderophores by *Serratiamarcescens* and the role of MacAB efflux pump in siderophore secretion. Bio NanoSci 6:480–482. https://doi.org/10.1007/s12668-016-0264-3

Kirchmann H (2019) Why organic farming is not the way forward. Outlook Agric 48(1):22–27

Kloepper JW, Schroth MN (1978) Plant growth promoting rhizobacteria on radishes. In: Proceedings of the 4th international conference on plant pathogenic bacteria, Angers, France 1979, pp 879–882

Kloepper JW, Leong J, Teintze M, Schroth MN (1980) Enhanced plant growth by siderophores produced by plant growth-promoting rhizobacteria. Nature 286:885–886. https://doi.org/10.1038/286885a0

Kumar A, Sharma S, Mishra S (2016) Evaluating effect of arbuscular mycorrhizal fungal consortia and Azotobacter chroococcum in improving biomass yield of Jatropha curcas. Plant Biosyst Int J Dealing All Aspects Plant Biol 150(5):1056–1064. https://doi.org/10.1080/11263504.2014.1001001

Kumari B, Mallick MA, Solanki MK, Solanki AC, Hora A, Guo W (2019) Plant growth promoting rhizobacteria (PGPR): modern prospects for sustainable agriculture. In Ansari RA, Mahmood I (eds) Plant health under biotic stress. Springer, pp 109–127

Lambin EF, Meyfroidt P (2011) Global land use change, economic globalization, and the looming land scarcity. Proc Natl Acad Sci 108(9):3465–3472. https://doi.org/10.1073/pnas.1100480108

Lehmann J, Kleber M (2015) The contentious nature of soil organic matter. Nature 528 (7580):60–68. https://doi.org/10.1038/nature16069

Leifeld J, Angers DA, Chenu C, Fuhrer J, Kätterer T, Powlson DS (2013) Organic farming gives no climate change benefit through soil carbon sequestration. Proc Natl Acad Sci 110(11):984. https://doi.org/10.1073/pnas.1220724110

Lenin G, Jayanthi M (2012) Indole acetic acid, gibberellic acid and siderophore production by PGPR isolates from rhizospheric soils of Catharanthus roseus. Int J Pharmacy Biol Arch 3 (4):933–938

Lim P, Gansau JA, Chong KP (2019) Biocontrol of basal stem rot pathogen Ganoderma boninense by Pseudomonas aeruginosa. Bangladesh J Bot 48(2):209–215. https://doi.org/10.3389/fmicb.2017.01895

Ludueña LM, Anzuay MS, Angelini JG, McIntosh M, Becker A, Rupp O, Goesmann A, Blom J, Fabra A, Taurian T (2018) Strain *Serratia* sp. S119: a potential biofertilizer for peanut and maize and a model bacterium to study phosphate solubilization mechanisms. Appl Soil Ecol 126:107–112. https://doi.org/10.1016/j.apsoil.2017.12.024

Malthus T (2013) An essay on the principle of population (1798). Yale University Press, New Haven, CN, pp 15–30

Martínez OA, Encina C, Tomckowiack C, Droppelmann F, Jara R, Maldonado C, Muñoz O, García-Fraile P, Rivas R (2018) *Serratia* strains isolated from the rhizosphere of raulí (*Nothofagus alpina*) in volcanic soils harbour PGPR mechanisms and promote raulí plantlet growth. J Soil Sci Plant Nutr 18:804–819

Matteoli FP, Passarelli-Araujo H, Reis RJ, da Rocha LO, de Souza EM, Aravind L, Olivares FL, Venancio TM (2018) Genome sequencing and assessment of plant growth-promoting properties of a *Serratia marcescens* strain isolated from vermicompost. BMC Genomics 19:750

Meena H, Meena RS, Rajput BS, Kumar S (2016) Response of bio-regulators to morphology and yield of clusterbean [Cyamopsis tetragonoloba (L.) Taub.] under different sowing environments. J Appl Nat Sci 8:715–718

Mendes LW, Kuramae EE, Navarrete AA, Van Veen JA, Tsai SM (2014) Taxonomical and functional microbial community selection in soybean rhizosphere. ISME J 8(8):1577–1587. https://doi.org/10.1038/ismej.2014.17

Mittler R (2002) Oxidative stress, antioxidants and stress tolerance. Trends Plant Sci 7:405–410
Múnera-Porras LM, García-Londoño S, Ríos-Osorio LA (2020) Action mechanisms of plant growth promoting cyanobacteria in crops in situ: a systematic review of literature Hindawi. Int J Agronomy 2020:1–9. https://doi.org/10.1155/2020/2690410
Mus F, Crook MB, Garcia K, Garcia Costas A, Geddes BA, Kouri ED, Paramasivan P, Ryu MH et al (2016) Symbiotic nitrogen fixation and the challenges to its extension to nonlegumes. Appl Environ Microbiol 82:3698–3710
Muthuselvan I, Balagurunathan R (2013) Siderophore production from Azotobacter sp. and its application as biocontrol agent. Int J Curr Res Rev 5(11):30–42
Nagaraja H, Chennappa G, Rakesh S, Naik MK, Amaresh YS, Sreenivasa MY (2016) Antifungal activity of Azotobacter nigricans against trichothecene-producing fusarium species associated with cereals. Food Sci Biotechnol 25:1197–1204. https://doi.org/10.1007/s10068-016-0190-8
Nagargade M, Tyagi V, Singh MK (2018) Plant growth-promoting rhizobacteria: a biological approach toward the production of sustainable agriculture. In: Role of rhizospheric microbes in soil. Springer, Singapore, pp 205–223. https://doi.org/10.1007/978-981-10-8402-7_8
Naseri S, Sharafzadeh S (2013) Impact of Azotobacter on growth and total phenolic content of garden thyme. Adv Environ Biol 7:113–115
Natasha N, Ramteke PW, Verma OP, Avantika P (2012) Antibiotic susceptibility and heavy metal tolerance pattern of Serratia marcescens isolated from soil and water. J Bioremed Biodegr 3 (7):158
Navarro MO, Barazetti A, Niekawa ET, Dealis ML, Matos JMS, Liuti G et al (2019) Microbial biological control of diseases and pests by PGPR and PGPF. In: Microbial interventions in agriculture and environment. Springer, Singapore, pp 75–122
Naziya B, Murali M, Amruthesh KN (2020) Plant growth-promoting fungi (PGPF) instigate plant growth and induce disease resistance in Capsicum annuum L. upon infection with Colletotrichum capsici (Syd.) Butler & Bisby. Biomolecules 10(1):41. https://doi.org/10.1016/j.rhisph.2020.100223
Nelson LM (2004) Plant growth promoting rhizobacteria (PGPR): prospects for new inoculants. Crop Manage 3(1):1–7. https://doi.org/10.1094/CM-2004-0301-05-RV
Novinscak A, Filion M (2020) Long term comparison of talc-and peat-based Phytobeneficial Pseudomonas fluorescens and Pseudomonas synxantha bioformulations for promoting plant growth. Front Sustain Food Syst 4:262. https://doi.org/10.3389/fsufs.2020.602911
Nowak B, Nesme T, David C, Pellerin S (2013) To what extent does organic farming rely on nutrient inflows from conventional farming? Environ Res Lett:1–8. https://doi.org/10.1088/1748-9326/8/4/044045
Ordookhani K, Khavazi K, Moezzi A, Rejali F (2010) Influence of PGPR and AMF on antioxidant activity, lycopene and potassium contents in tomato. Afr J Agric Res 5(10):1108–1116. https://doi.org/10.5897/AJAR.9000428
Ozden M, Demirel U, Kahraman A (2009) Effects of proline on antioxidant system in leaves of grapevine (*Vitis vinifera* L.) exposed to oxidative stress by H_2O_2. Sci Hortic 119:163–168
Pabbi S (2015) Blue Green Algae: a potential biofertilizer for rice. In: Sahoo D, Seckbach J (eds) The Algae World. Cellular origin, life in extreme habitats and astrobiology, vol 26. Springer, Dordrecht. https://doi.org/10.1007/978-94-017-7321-8_17
Palleroni NJ (2015) Pseudomonas. Bergey's Manual of Systematics of Archaea and Bacteria 2015. https://doi.org/10.1002/9781118960608.gbm01210
Pandey S, Giri K, Kumar R, Mishra G, Rishi RR (2018) Nanopesticides: opportunities in crop protection and associated environmental risks. Proc Natl Acad Sci India Section B: Biol Sci 88 (4):1287–1308
Pandey V, Chandra K (2016) Agriculturally important microorganisms as biofertilizers: commercialization and regulatory requirements in Asia. In: Singh H, Sarma B, Keswani C (eds) Agriculturally important microorganisms. Springer, Singapore, pp 133–145

Pang Y, Liu X, Ma Y, Chernin L, Berg G, Gao K (2008) Induction of systemic resistance, root colonisation and biocontrol activities of the rhizospheric strain of *Serratia plymuthica* are dependent on N-acyl homoserine lactones. Eur J Plant Pathol 124:261–268

Pang YD, Liu XG, Ma YX, Chernin L, Berg G, Gao KX (2009) Induction of systemic resistance, root ssscolonisation and biocontrol activities of the rhizospheric strain of Serratia plymuthica are dependent on N-acyl homoserine lactones. Eur J Plant Pathol 124:261–268. https://doi.org/10.1007/S10658-008-9411-1

Pankievicz VC, Irving TB, Maia LG, Ané JM (2019) Are we there yet? The long walk towards the development of efficient symbiotic associations between nitrogen-fixing bacteria and non-leguminous crops. BMC Biol 17(1):99. https://doi.org/10.1186/s12915-019-0710-0

Paspaliari DK, Kastbjerg VG, Ingmer H, Popowska M, Larsen MH (2017) Chitinase expression in listeria monocytogenes is influenced by lmo0327, which encodes an internalin-like protein. Appl Environ Microbiol 83:e01283–e01217

Pimentel D (1996) Green revolution agriculture and chemical hazards. Sci Total Environ 1996 (188):S86–S98

Poblete-Morales M, Carvajal D, Almasia R, Michea S, Cantillana C, Levican A, Silva-Moreno E (2020) Pseudomonas atacamensis sp. nov., isolated from the rhizosphere of desert bloom plant in the region of Atacama, Chile. Antonie van Leeuwenhoek. https://doi.org/10.1007/s10482-020-01427-0

Pore TS, Khanolkar AB, Nadaf NH (2016) Production, purification, identification of prodigiosin from *Serratia* sp. and its antimicrobial activity. Research journal of life sciences, bioinformatics, pharmaceutical and chemical science 1:1–2

Priyadarshini P, Abhilash PC (2020) Policy recommendations for enabling transition towards sustainable agriculture in India. Land Use Policy 96:104718. https://doi.org/10.1016/j.landusepol.2020.104718

Rahi P, Vyas P, Sharma S, Gulati A (2009) Plant growth promoting potential of the fungus *Discosia* sp. FIHB571 from tea rhizosphere tested on chickpea, maize and pea. Ind J Microbiol 49:128–133

Reganold JP, Wachter JM (2016) Organic agriculture in the twenty-first century. Nature Plants 2 (2):1–8. https://doi.org/10.1038/NPLANTS.2015.221

Romero-Perdomo F, Abril J, Camelo M, Moreno-Galván A, Pastrana I, Rojas-Tapias D, Bonilla R (2017) Azotobacter chroococcum as a potentially useful bacterial biofertilizer for cotton (Gossypium hirsutum): effect in reducing N fertilization. Revista Argentina de microbiologia 49(4):377–383. https://doi.org/10.1016/j.ram.2017.04.006

Royal Society of London (2009) Reaping the benefits: science and the sustainable intensification of global agriculture. Royal Society, London. ISBN 978-0-85403-784-1

Ruiz B, Chávez A, Forero A, García-Huante Y, Romero A et al (2010) Production of microbial secondary metabolites: regulation by the carbon source. Crit Rev Microbiol 36(2):146–167

Safari M, Motamedi E, Dolatabad HK, Sanavy SA (2020) Nano-carriers effects on the viability and efficiency of pseudomonas strains as phosphate solubilizing bacteria. Heliyon 6(10):e05076. https://doi.org/10.1016/j.heliyon.2020.e05076

Sahni S, Prasad BD (2020) Management of collar rot disease using vermicompost and a PGPR strain pseudomonas sp. and their effect on defense-related enzymes in chickpea. Indian Phytopathol 73:301–311. https://doi.org/10.1007/s42360-020-00203-4

Sahota A (2019) The Global Market for Organic Food & Drink: Trends Future Outlook. 2019:59–64. https://www.ecoviant.com/global-organic-food-market-trends

Salehi-Lisar SY, Bakhshayeshan-Agdam H (2016) Drought stress in plants: causes, consequences, and tolerance. In: Drought stress tolerance in plants, vol 1. Springer, Cham, pp 1–16. https://doi.org/10.1007/978-3-319-28899-4_1

Schalk IJ (2008) Metal trafficking via siderophores in gram-negative bacteria: specificities and characteristics of the pyoverdine pathway. J Inorg Biochem 102(5–6):1159–1169. https://doi.org/10.1016/j.jinorgbio.2007.11.017

Schierhorn F, Hofmann M, Adrian I, Bobojonov I, Müller D (2020) Spatially varying impacts of climate change on wheat and barley yields in Kazakhstan. J Arid Environ 178:104164. https://doi.org/10.1016/j.jaridenv.2020.104164

Schoebitz M, López MD, Roldán A (2013) Bioencapsulation of microbial inoculants for better soil–plant fertilization. A review. Agron Sustain Dev 33(4):751–765

Schuhegger R, Ihring A, Gantner S, Bahnweg G, Knappe C, Vogg G, Hutzler P, Schmid M, Van Breusegem F, Eberl LE, Hartmann A (2006) Induction of systemic resistance in tomato by N-acyl-L-homoserine lactone-producing rhizosphere bacteria. Plant Cell Environ 29:909–918

Shen X, Hu H, Peng H, Wang W, Zhang X (2013) Comparative genomic analysis of four representative plant growth–promoting rhizobacteria in *Pseudomonas*. BMC Genomics 14: 271. https://doi.org/10.1186/1471-2164-14-271

Shiva V (1991) The violence of the green revolution: third world agriculture, ecology and politics. University Press of Kentucky

Singh D, Singh SK, Modi A, Singh PK, Zhimo VY, Kumar A (2020) Impacts of agrochemicals on soil microbiology and food quality. In: Prasad MNV (ed) Agrochemicals detection, treatment and remediation. pp 101–116. https://doi.org/10.1016/B978-0-08-103017-2.00004-0

Singh RP, Jha PN (2016) The multifarious PGPR *Serratia marcescens* CDP-13 augments induced systemic resistance and enhanced salinity tolerance of wheat (*Triticum aestivum* L.). PLoS ONE 11:e0155026. https://doi.org/10.1371/journal.pone.0155026

Slater H, Crow M, Everson L, Salmond GP (2003) Phosphate availability regulates biosynthesis of two antibiotics, prodigiosin and carbapenem, in *Serratia* via both quorum-sensing-dependent and-independent pathways. Mol Microbiol 47:303–320

Smith SE (1976) Neuromuscular blocking drugs in man. In: Zaimis E (ed) Neuromuscular junction. Handbook of experimental pharmacology, vol 42. Springer, Heidelberg, pp 593–660

Snow AA (2002) Transgenic crops—why gene flow matters. Nat Biotechnol 20(542)

Somers E, Vanderleyden J, Srinivasan M (2004) Rhizosphere bacterial signalling: a love parade beneath our feet. Crit Rev Microbiol 30(4):205–240. https://doi.org/10.1080/10408410490468786

Someya N, Kataoka N, Komagata T, Hirayae K, Hibi T, Akutsu K (2000) Biological control of cyclamen soilborne diseases by *Serratia marcescens* strain B2. Plant Dis 84:334–340. https://doi.org/10.1094/PDIS.2000.84.3.334

Souza LRR, da Rocha Neto AC, da Silva CR, Franchi LP, de Souza TAJ (2019) Green synthesis approaches of nanoagroparticles. In: Nanobiotechnology in bioformulations. Springer, Cham, pp 353–380. https://doi.org/10.1007/978-3-030-17061-5_15

Spaepen S, Vanderleyden J, Remans R (2007) Indole– 3–acetic acid in microbial and microorganism – plant signaling. FEMS Microbiol Rev 31:425–448. https://doi.org/10.1111/j.1574-6976.2007.00072.x

Srivastava R, Srivastava AK, Ramteke PW, Gupta VK, Srivastava, AK (2020) Metagenome dataset of wheat rhizosphere from Ghazipur region of Eastern Uttar Pradesh. Data in Brief. Elsevier Inc 28:1–8 https://doi.org/10.1016/j.dib.2019.105094

Staley JT, Konopka A (1985) Measurement of in situ activities of nonphotosynthetic microorganisms in aquatic and terrestrial habitats. Annu Rev Microbiol 39(1):321–346

Sutton MA, Howard CM, Erisman JW, Bealey WJ, Billen G, Bleeker A, Bouwman AF, Grennfelt P, van Grinsven H, Grizzetti B (2011) The challenge to integrate nitrogen science and policies: the European Nitrogen Assessment approach. Cambridge University Press, Cambridge. https://doi.org/10.1017/CBO9780511976988.008

Taurian T, Anzuay MS, Angelini JG, Tonelli ML, Ludueña L, Pena D, Ibáñez F, Fabra A (2010) Phosphate-solubilizing peanut associated bacteria: screening for plant growth-promoting activities. Plant Soil 329:421–431

Tejera N, Lluch C, Martinez-Toledo MV, Gonzalez-Lopez J (2005) Isolation and characterization of Azotobacter and Azospirillum strains from the sugarcane rhizosphere. Plant Soil 270 (1):223–232. https://doi.org/10.1007/s11104-004-1522-7

Tilman D, Balzer C, Hill J, Befort BL (2011) Global food demand and the sustainable intensification of agriculture. Proc Natl Acad Sci 108(50):20260–20264. https://doi.org/10.1073/pnas.1116437108

Toral L, Rodríguez M, Béjar V, Sampedro I (2020) Crop protection against Botrytis cinerea by rhizosphere biological control agent bacillus velezensis XT1. Microorganisms 8(7):2–17. https://doi.org/10.3390/microorganisms8070992

Vacheron J, Desbrosses G, Bouffaud ML, Touraine B, Moënne-Loccoz Y, Muller D, Legendre L, Wisniewski-Dyé F, Prigent-Combaret C (2013) Plant growth–promoting rhizobacteria and root system functioning. Front Plant Sci 4. https://doi.org/10.3389/fpls.2013.00356

Vanissa TT, Berger B, Patz S, Becker M, Turečková V, Novák O, Tarkowská D, Henri F, Ruppel S (2020) The response of maize to inoculation with Arthrobacter sp. and Bacillus sp. in phosphorus-deficient, salinity-affected soil. Microorganisms 8:1–18. https://doi.org/10.3390/microorganisms8071005

Verma C, Das AJ, Kumar R (2017) PGPR assisted phytoremediation of cadmium: An advancement towards clean environment. Curr Sci 113(4):715

Verma D, Patel RP (2019) Cyanobacteria and Azolla in rice cultivation/improving biological N_2-fixation system in rice. In: Kaushik BD, Kumar D, Shamim MD (eds) Biofertilizers and Biopesticides in Sustainable Agriculture. Apple Academic Press, pp 221–236

Vikhe PS (2014) Azotobacter species as a natural plant hormone synthesizer. Res J Recent Sci 3: 59–63

Wan Y, Luo S, Chen J, Xiao X, Chen L, Zeng G, Liu C, He Y (2012) Effect of endophyte-infection on growth parameters and cd-induced phytotoxicity of cd-hyperaccumulator *Solanum nigrum* L. Chemosphere 89:743–750

Wang K, Yan PS, Cao LX, Ding QL, Shao C, Zhao TF (2013) Potential of chitinolytic *Serratia marcescens* strain JPP1 for biological control of *Aspergillus parasiticus* and aflatoxin. Biomed Res Int 7:2013. https://doi.org/10.1155/2013/397142

Wang C, Sun B, Zhang X, Huang X, Zhang M, Guo H, Chen X, Huang F, Chen T, Mi H, Yu F (2019) Structural mechanism of the active bicarbonate transporter from cyanobacteria. Nature Plants 5(11):1184–1193. https://doi.org/10.1038/s41477-019-0538-1

Wezel A, Casagrande M, Celette F, Vian JF, Ferrer A, Peigné J (2014) Agroecological practices for sustainable agriculture. A review. Agron Sustain Dev 34(1):1–20

Willer H, Lernoud J (2019) The world of organic agriculture. Statistics and emerging trends 2019 Research Institute of Organic Agriculture FiBL and IFOAM Organics International, pp 1–336. https://doi.org/10.4324/9781849775991

Xia Y, Farooq MA, Javed MT, Kamran MA, Mukhtar T, Ali J, Tabassum T, Ur Rehman S, Munis MF, Sultan T, Chaudhary HJ (2020) Multi-stress tolerant PGPR Bacillus xiamenensis PM14 activating sugarcane (Saccharum officinarum L.) red rot disease resistance. Plant Physiol Biochem 151:640–649. https://doi.org/10.1016/j.plaphy.2020.04.016

Xiang N, Lawrence KS, Kloepper JW, Donald PA, McInroy JA (2017) Biological control of Heterodera glycines by spore-forming plant growth-promoting rhizobacteria (PGPR) on soybean. PloS One 12(7)1–19. https://doi.org/10.1371/journal.pone.0181201

Xie Z, Li M, Wang D, Wang F, Shen H, Sun G, Feng C, Wang X, Chen D, Sun X (2020) Biocontrol efficacy of Bacillus siamensis LZ88 against brown spot disease of tobacco caused by Alternaria alternata. Biol Control 154:104508. https://doi.org/10.1016/j.biocontrol.2020.104508

Yasmin S, Hafeez FY, Mirza MS, Rasul M, Arshad HM, Zubair M, Iqbal M (2017) Biocontrol of bacterial leaf blight of rice and profiling of secondary metabolites produced by rhizospheric Pseudomonas aeruginosa BRp3. Front Microbiol 26; 8:1895. https://doi.org/10.3389/fmicb.2017.01895

Zahir ZA, Abbas SA, Khalid M, Arshad M (2000) Substrate dependent microbially derived plant hormones for improving growth of maize seedlings. Pak J Biol Sci 3(2):289–291

Zaki AM, Zaki AH, Farghali AA, Abdel-Rahim EF (2017) Sodium titanate-bacillus as a new nanopesticide for cotton leaf-worm. J Pure Appl Microbiol 11(2):725–732. https://doi.org/10.1016/B978-0-12-815829-6.00010-3

Zarb J, Ghorbani R, Koocheki A, Leifert C (2005) The importance of microorganisms in organic agriculture. Outlooks Pest Manage 16(2):52–55

Zavattieri MA, Ragonezi C, Klimaszewska K (2016) Adventitious rooting of conifers: influence of biological factors. Trees 30(4):1021–1032. https://doi.org/10.1007/s00468-016-1412-7

Zhou Y, Coventry DR, Gupta VV, Fuentes D, Merchant A, Kaiser BN, Li J, Wei Y, Liu H, Wang Y, Gan S (2020) The preceding root system drives the composition and function of the rhizosphere microbiome. Genome Biol 21:1–19. https://doi.org/10.1186/s13059-020-01999-0

Agriculturally Important Microorganism: Understanding the Functionality and Mechanisms for Sustainable Farming

2

Arun Kumar Rai, Kiran Sunar, and Hemant Sharma

Abstract

Soil is identified to be a complex microhabitat for two distinguishing properties. Firstly, the microbial inhabitants in the soil are enormously diverse and secondly, the soil remains a structured, heterogeneous, and discontinuous system, generally poor in essential nutrients and energy sources. Soil microflora plays the most substantial part in the rhizosphere of the higher plants, where the plant growth-promoting traits of beneficial microorganisms influence the soil and plant health. Beneficial microorganisms used to improve agricultural products are broadly termed as Agriculturally Important Microorganisms (AIMs). AIMs represent a wide range of microorganisms which include Plant Growth-Promoting Rhizobacteria (PGPR), Biocontrol Agents (BCA), Plant Growth-Promoting Fungi (PGPF), Actinomycetes, Mycorrhiza, and Endophytes. AIMs can influence plant growth and the health of the soil through direct and indirect mechanisms along with molecular signaling. Plant signaling molecules play important roles in efficient root colonization, modulation of root system architecture, cell to cell communication, gene regulation, plant immunity development process, and finally, influence plant health. Integration of Agriculturally Important Microorganisms in agriculture is a promising sustainable solution to improve production, however, commercialization of bioformulations will require addressing a number of issues like a selection of broad-spectrum microbial strains, retention of quality and efficacy under field conditions, and product registration. A rational approach to comprehend the key mechanisms associated

A. K. Rai (✉) · H. Sharma
Botany Department, Sikkim University, Gangtok, Sikkim, India
e-mail: akrai@cus.ac.in

K. Sunar
Balurghat Mahila Mahavidyalaya, Balurghat, West Bengal, India

R. Soni et al. (eds.), *Microbiological Activity for Soil and Plant Health Management*,
https://doi.org/10.1007/978-981-16-2922-8_2

with AIMs-Plant interaction and development of model-based inoculum would facilitate productive field application and sustainable agriculture production under the changing climatic conditions.

Keywords

Agriculturally important microorganisms · Endophytes · Mycorrhiza · Plant immunity

2.1 The Concept of Plant Microbiome and the Rhizobiome

Diverse groups of microorganisms inhabiting plants are known for their prominent roles in maintaining plant health and proper growth. Researchers over the previous decade have confirmed that a diverse group of microbial communities is associated with parts of the plant (Hardoim et al. 2015). Plant microbiota (consisting of all the microbes) comprises the microbial constituents of plant holobiont or the plant microbiome which accounts for all the microbial genomes. The rhizosphere, phyllosphere, and endosphere regions of the plants are inhabited by these microbial florae which play an important role in plant growth promotion and health (Lemanceau et al. 2017). Properly understanding the essential roles typically performed by these plant-associated microorganisms and the key features affecting the active community can accurately deliver more valuable understandings into the plant as a meta-organism and the possible benefits conferred earnestly by the microorganism to the plant partners (Hardoim et al. 2015; Hacquard 2016). The natural environment for plant microbiome is typically built around the plant genotype, species of the plants, and edaphic as wells as environmental factors that form an integral part between the plant and associated microbes (Compant et al. 2019). The microbial communities typically inhabiting the rhizosphere are termed as Rhizobiome (Sasse et al. 2018; Olanrewaju et al. 2019). The texture of soil and other environmental factors are responsible for affecting microbial communities in the soil (Bach et al. 2018). Various studies suggest that exudates from the roots influence the diversity and richness of the rhizobiome. Even though there are reports indicating limited functions of the root exudates in influencing the rhizobiome when compared to the remaining rhizodeposits that consist of volatile compounds, root cells that are sloughed-off, mucilages, and lysates (Dennis et al. 2010; Lettice 2019).

Diverse forms of metabolites are secreted by the different parts of roots (Tückmantel et al. 2017) in which tip of roots are known to secrete a profuse number of exudates thereby encouraging the association of a diverse group of microorganisms (Massalha et al. 2017).

2.2 Agriculturally Important Microorganisms (AIMs)

Microorganisms residing in the soil were classified into beneficial and harmful groups based on their influence on the quality of the soil and crop plants. Later, it was understood that microorganisms that aid in the fixation of the atmospheric Nitrogen, decomposition of wastes and pesticides, secrete bioactive metabolites, improve cycling of the nutrients, and assist in plant growth were considered as beneficial microorganisms. A diverse group of microorganisms residing in the soil play a significant role in influencing the cycling of essential molecules such as carbon and nitrogen at a regional as well as global scale. The structure of these microbial communities is regulated by the availability of nutrients in the soil, such as plant debris and organic molecules, that acts as a major growth-limiting factor (Fig. 2.1) (Zak et al. 2011).

Sustainable agricultural practices involving green strategies with the use of very low toxic components as alternatives for managing nutrients and controlling phytopathogens should be of prime focus. Beneficial microorganisms with an array of functionalities could substitute the chemical inputs in agriculture (Ahmad et al. 2018). A diverse group of microorganisms, isolated from different niches, have exhibited immense potential in agriculture as bio-fertilizing agents or biopesticides. The use of different species of fungi and bacteria in consortia has proven to improve

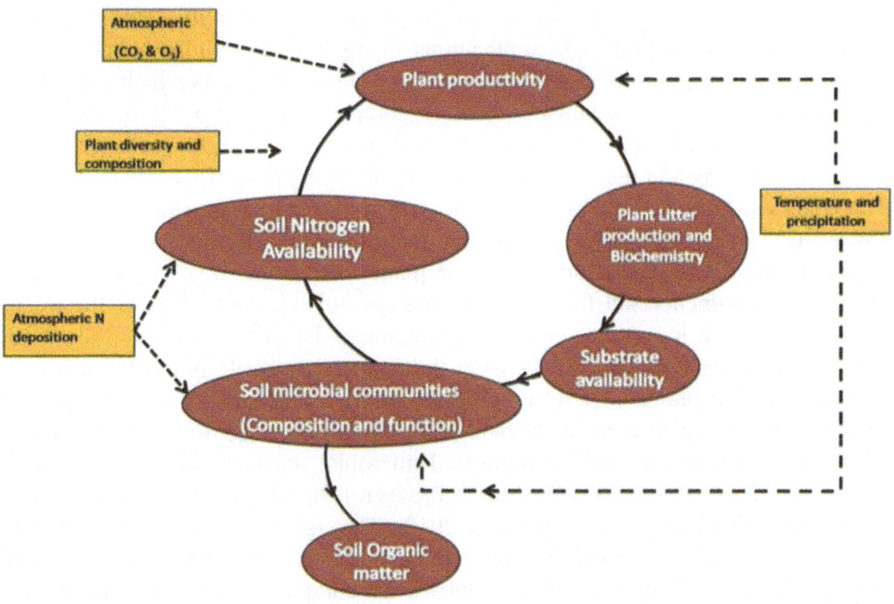

Fig. 2.1 Plants and microorganisms are linked with each other via the limiting factors required for either of their growths. The availability of the organic molecules and plant detritus influences the microbial community. Likewise, the growth of the plant is limited by the availability of nitrogen facilitated through the decaying of dead plant tissues by the microorganisms (Zak et al. 2011)

the growth and produce of horticulture crops (Bagyaraj and Ashwin 2017). The long-term sustainability is possible with the application of biofertilizers in terms of cost as well as its impact on the environment (Kumar et al. 2018). As per a report published in May 2020, the increasing trend of organic farming system, reduced cost for the production of microbial, the increasing cost involved in the production of chemical pesticides, and growth in the implementation of Integrated Pest Management programs, have resulted in the projection of growth of market related to microbial used in agriculture at a Compound Annual Growth Rate (CAGR) of 14.1% up to 2025. Among all the microorganisms, bacterial formulation occupies the highest market share and is expected to dominate the market during the said period. The fruits and vegetable sectors will be the highest consumer of agriculture microbial (Markets and Markets Research Private Ltd. 2020). The organic food market is expected to grow at a CAGR of 16.15% from 2017 to 2022 (Mukherjee et al. 2018).

2.3 Diversity and Functionality of AIMs

2.3.1 Diversity and Interrelationship

Microorganisms in the soil play an important role in the promotion of plant growth, protection from phytopathogens, tolerance to biotic and abiotic stress, etc. Microbial natural products are an essential source of various functionalities with different properties that could find their applications in organic agriculture practices. The rhizosphere is a region of soil in the vicinity of roots of the plants which consists of diverse groups of microorganisms forming complex associations with the plants (Mendes et al. 2013). This region of the soil has a vital role in promoting plant growth, mobilizing and recycling nutrients, and inducing host resistance to the plant pathogens and abiotic factors. A diverse group of microorganisms belonging to prokaryotes and eukaryotes constitutes microflora of the rhizosphere region (Hyakumachi 1994). The exudes from the plant roots is the most important factor for the establishments and flourishing of the type of microbial flora in the rhizosphere region that is influenced by the plant for its benefits (Mendes et al. 2013; Gahan and Schmalenberger 2014). Over the years, reports of composite microbial groups inhabiting plants have been published by several workers and when used for agriculture they are termed as Agriculturally Important Microbial groups. These AIMs form complex as well as organized microbial networks (King et al. 2012; Huang et al. 2020) where individual taxon has its role in maintaining plant health and functioning of the ecosystem (Zhou et al. 2010). Through the study conducted by Shi et al. (2016), it is understood that intricate networks in the rhizosphere have more potential for niche-sharing and interactions when compared with bulk soil. Species of microorganism forms a complex network by regularly interrelating with every component in the plant microbiome (Khan et al. 2019).

Analysis of enormous data obtained from high-throughput DNA sequencing tools has provided possible relations among the microorganisms and their co-occurrence

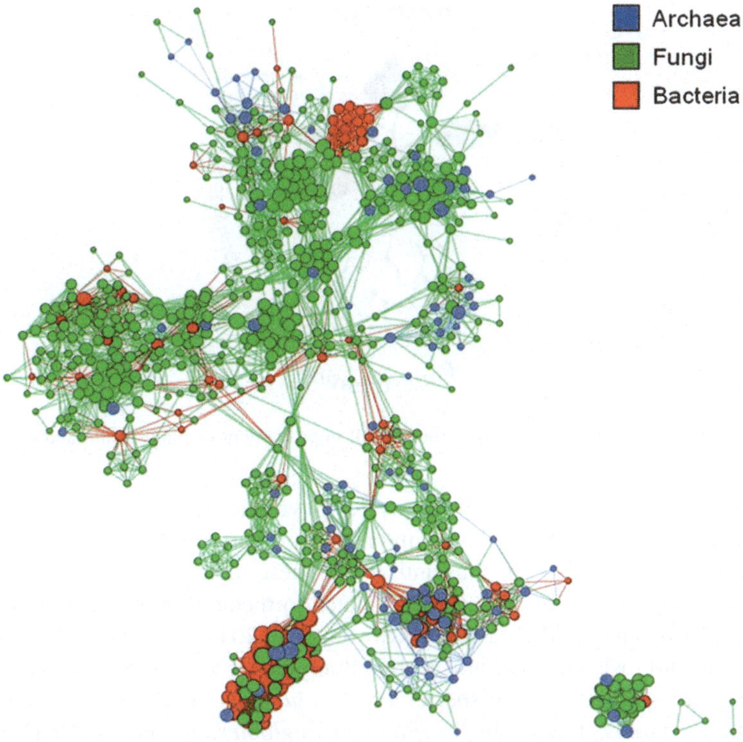

Fig. 2.2 Analysis to show diverse microbial networks within the rhizosphere. Recognition of possible associations between species is aided by co-occurrence networks. The microbial operational taxonomic unit corresponds with the nodes which edged towards the microbial associations. The figure is readapted from Monard et al. (2016)

patterns. The analysis provided in Fig. 2.2, adopted from Monard et al. (2016) shows that diverse microbial groups in a rhizosphere are complex and exist in independent niches but at the same time, they are interdependent with each other. This type of diversity and interrelationship hugely influences their functionality and influence on their host and the coordinated network helps the overall health and growth of the host plant (Fig. 2.3 and Table 2.1).

2.3.2 Diversity and Functionality

2.3.2.1 Biocontrol Agents (BCA)

Biocontrol agents use more than one process in controlling the invasion by phytopathogens. Some of them include competition for available nutrition and space for colonization, production of antimicrobial metabolites, stimulation and induction of resistance in the host plant, etc. (Pandin et al. 2017).

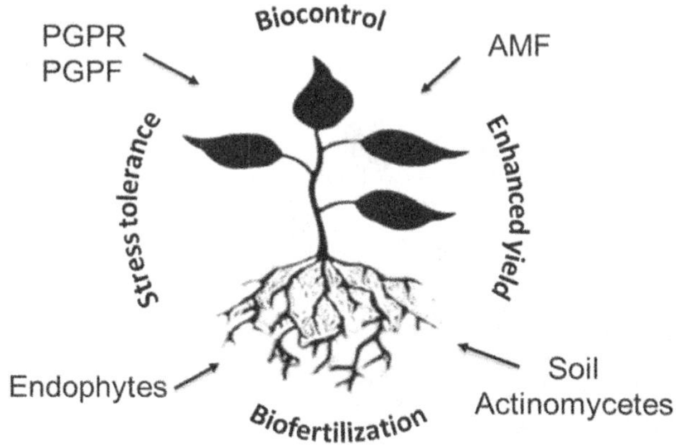

Fig. 2.3 Coordinated effect of AIMs in the overall improvement of plant health and growth

Biocontrol agents influence the growth of phytopathogens by producing certain compounds or metabolites with antimicrobial properties. This process of direct inhibition of pathogen is known as antibiosis. These metabolites inhibit the growth of pathogens by interfering in pathways related to metabolite production or protein synthesis (Keswani and Mishra 2014; Keswani et al. 2017). Some of the compounds involved in antibiosis include Iturin A, Bacillomycin D, Mycosubtilin, Zwittermicin, and Kanosamine produced by species of *Bacillus* genera, Gliotoxin produced by *Trichoderma virens,* Herbicolin by *Pantoea agglomerans*, and many more such compounds (Ram et al. 2018).

Direct control of certain fungal pathogens involves the growth of mycelium of biocontrol fungus towards the pathogens which secretes cell lytic enzymes resulting in the destruction of the cell wall of the invading fungus. Some of the species belonging to the genera of *Trichoderma, Acremonium, Ampelomyces, Gliocladium, Coniothyrium, Sporidesmium, Pythium, Aspergillus,* and *Acrodontium* are involved in mycoparasitism (Altomarc et al. 1999; Kiss 2003; Ram et al. 2018).

Certain microorganisms secrete volatile organic compounds, chelating agents, for instance, siderophores, that rapidly depletes the availability of vital nutrients such as Iron in its vicinity thus limiting the growth of pathogens. This method helps in the control of phytopathogens indirectly through the mode of competition. Microorganisms belonging to the genera of *Pseudomonas* sp., *Streptomyces* sp., *Pennisetum* sp., etc., have been found to secrete siderophores (Mishra et al. 2016; Ram et al. 2018).

Certain species of microorganisms such as *Trichoderma* sp. have been found to induce host resistance against some phytopathogens by influencing the cascade of biochemical reaction in the host plant (Bisen et al. 2016; Rajesh et al. 2016).

Studies on biofilm formation by microorganism and their role as a potential biocontrol agent against phytopathogens is recently gathering interest. Plants are known to secrete certain factors or exude to initiate the process of biofilm formation of microorganisms for their advantage. Surfactin and plant exudates produced by

Table 2.1 Characteristic features of different types of agriculturally important microorganisms

Agriculturally important microorganisms	Types of microorganisms	Site of colonization	Significance
Plant growth promoting rhizobacteria (PGPR)	Bacteria	Colonization limited to roots of host plants	Promotes acquisition of nutrient resources, control hormone levels in plant, reduces influence of phytopathogens, acts as a bio-stimulant.
Plant growth promoting fungi (PGPF)	Fungi	Colonization limited to rhizosphere region of host plants	Secretes plant growth promoting factors, reduces influence of phytopathogens on host plants, induces resistance of plants to range of biotic and abiotic factors.
Mycorrhiza	Fungi	Symbiotic colonization limited to roots of host plants	Secretes plant growth promoting factors, reduces influence of phytopathogens and abiotic factors on the host plants, enhances nutrition uptake, induces secondary metabolites and volatile organic compounds in the host plants.
Endophytes	Bacteria, fungi	Systemic and have been obtained from all the plant parts	Produces host secondary metabolites, clinically important metabolites, plant growth promoting factors, antagonistic against phytopathogens and induces host resistance to biotic and biotic factors.
Soil actinomycetes	Bacteria	Soil and rhizosphere	Produces clinically important metabolites including commercially available antibiotics, plant growth promoting factors, antagonistic against phytopathogens.

plants help in the formation of biofilm (Fan et al. 2011; Zeriouh et al. 2014). In one study by Xu et al. (2014), cucumber plants were found to induce a strain of *Bacillus amyloliquefaciens* to produce Bacillomycin D which eventually triggers the formation of biofilm in the rhizosphere region of the cucumber plant. Biofilm formation by *Paenibacillus* sp. has been found to protect seeds of wheat by *Fusarium graminearum* infection (Díaz et al. 2016).

A study on *B. amyloliquefaciens* as a biocontrol agent has shown that the production of metabolites at the rhizosphere has greater significance in the induction of systemic resistance in plants, a primary mechanism for suppression of phytopathogens, rather than direct production of numerous antimicrobial metabolites in the region (Lugtenberg 2015).

2.3.2.2 Plant Growth-Promoting Rhizobacteria (PGPR)

The microorganisms associated with the rhizosphere region of a plant help in decomposition and mineralization of essential nutrients along with fixation of Nitrogen, protection of plants from phytopathogens, enhance the resistance of the host plants to biotic as well as abiotic stress and acts as a bio-stimulant by facilitating the uptake of nutrients from the soil. PGPR in the soil promotes plant growth through direct or indirect mechanisms, i.e. they are able to produce plant hormones, viz. Auxins, Cytokines, Gibberellins, and ACC deaminase responsible for promoting growth in plants. These microorganisms are equally capable of increasing the availability of Nitrogen in the rhizosphere region by fixing atmospheric Nitrogen. Solubilization of inorganic phosphates with the secretion of extracellular enzymes increase the continuous availability of phosphorus to the plants under stressful conditions. Siderophores producing microorganisms reduce the availability of iron to other microbial pathogens hence limiting their influence on the plants. Some of the microorganisms are also able to produce antimicrobial metabolites while others produce cellulolytic enzymes or Hydrogen Cyanide that have an antagonistic effect against the phytopathogens (Armada et al. 2016; Hashem et al. 2019; Emmanuel and Babalola 2020). Some of the PGPR species of *Pseudomonas, Streptomyces, Paenibacillus,* and *Azotobacter* in association with Plant Growth-Promoting Fungi *Acaulospora, Glomus,* and *Scutellospora* species have shown impressive phytoremediation property in soils contaminated with Fe^{3+}. *Pennisetum glaucum* inoculated with the combination of the above species displayed a positive synergistic effect in the absorption of iron due to the secretion of siderophores (Mishra et al. 2016). Positive synergistic effect of PGPR and Arbuscular Mycorrhizae was observed on *Fragaria vesca* and *Trifolium repens* when inoculated with *B. amyloliquefaciens* and *Rhizophagus irregularis.* The plants showed increased photosynthetic efficiency and shoot weight in presence of the consortia (Xie et al. 2018). *Brassica napus* inoculated with the strains of *Microbacterium oxydans, Burkholderia cepacia,* and *Pseudomonas thivervalensis* was effective in reducing the heavy metal stress on the plant specimens along with substantial increment in the growth of plants. One of the strains of PGPR was also able to increase the uptake of Cu up to 113.38% (Ren et al. 2019). *Pseudomonas* sp., *Bacillus sp.*, and *Stenotrophomonas* sp. have demonstrated antifungal activity against fungal phytopathogens along with the production of factors promoting plant growth (Islam et al. 2016; Chenniappan et al. 2019; Chandra et al. 2020). *Bacillus* sp. produces antimicrobial lipopeptides such as surfactin, fengycin, and iturin of which iturin was found to possess antifungal property. Two isolates of *B. velezensis* have demonstrated antifungal activity against *Ralstonia solanacearum* and *Fusarium oxysporum* (Cao et al. 2018). Microorganisms with simultaneous phosphate solubilizing and biocontrol potential are the best bioinoculants with possible use as plant growth promoters and biocontrol agents (Sunar et al. 2017). PGPR reduces plant disease due to its ability to restrict the multiplication of the pathogens in the soil through properties such as direct antagonism, production of siderophores, or production of HCN. Siderophores are known to restrict the activity of phytopathogens by reducing the availability of iron by sequestering the ferric iron. The antimicrobial activity of secondary metabolites produced by some of the strains of Bacillus and Pseudomonas is well known (Compant et al. 2005; Chakraborty et al. 2014; Ossowicki et al. 2017; Sahni et al. 2020). An increase in the activities of the enzymes like PR proteins-chitinase (CHT), β-1, 3-glucanase (GLU), and defense-related enzymes peroxidase (POX) and phenylalanine ammonia lyase (PAL) have been reported in a number of experiments where these enzymes were directly associated with induction of resistance against the phytopathogens Chakraborty et al. (2004, 2009, 2014), Sunar et al. (2014, 2017, 2020).

2.3.2.3 Plant Growth-Promoting Fungi (PGPF)

Soil fungi inhabiting the rhizosphere region of the plant and responsible for the promotion of plant growth through the release of different functionalities are classified under plant growth-promoting fungi. Fungal microorganisms belonging to the genera of *Trichoderma*, *Phoma*, *Rhizoctonia*, *Penicillium*, *Fusarium*, *Aspergillus*, etc. are known for their ability to produce such factors (Hossain et al. 2017). PGPF has shown enhanced yield in crops and control of phytopathogens Plant Growth-Promoting Fungi: Diversity and Classification. In one such study conducted by Zhang et al. (2016) on wheat seedlings inoculated with *Trichoderma longibrachiatum* T6, the plant showed resistance to nematodes and induced tolerance to stress under high NaCl concentration. Plant's growth and yield are reduced drastically with the infection of phytopathogens. White rot disease in onion caused by *Sclerotium cepivorum* is responsible for the huge loss of onion production worldwide. The pathogen damages the bulb of onion and lasts in soil for 15 years which hinders the cultivation of onion. Further, the growth of pathogen is enhanced by the exudates from the onion. Some of the PGPF including *Penicillium* sp. *and Trichoderma* sp. were examined for its effects on the reduction of disease severity. These PGPF showed holistic plant growth including a significant reduction in the germination of sclerotia, the inactive stage of the fungal pathogen, that lasts in the soil for long-duration (Elsharkawy and El-Khateeb 2019). PGPF is also known for soil remediation contaminated with heavy metals. *Mucor* sp. when inoculated with *Pelargonium hortorum* was able to promote Phyto-availability of Pb and displayed enhanced phytoextraction along with the increase of plant biomass through the production of factors responsible for plant growth (Manzoor et al. 2019).

2.3.2.4 Arbuscular Mycorrhizal Fungi (AMF)

Mycorrhiza is a large number of fungi majorly belonging to Glomeromycota and some from Zygomycota, Ascomycota, and Basidiomycota that are known to infect the roots of the majority of higher plant species, i.e. gymnosperms, angiosperms, and pteridophytes thus forming a robust mutual association or parasitic one (Harley 1989; Johnson et al. 1997; Kirk et al. 2001). Mycorrhiza is broadly classified into two major groups ectomycorrhiza, an external association of fungi with plant roots, and endomycorrhiza, an internal association of fungal species by invading plant cells. The mycorrhizal association has proven to be beneficial for the plants especially for those growing in stressful conditions and soils with less nutritive content which is most likely dependent on mycorrhizal symbiosis (Akbar et al. 2017).

Arbuscular mycorrhizae (AM) play a significant role in plant growth promotion by enhancing nutrition uptake from the soil, provide support to the plants under drought conditions, and offer relief to the plants from phytopathogens. AMs facilitate the uptake of phosphorus in the host plants (Smith et al. 2011). Elliott et al. (2020) ascertained that the commercially available arbuscular mycorrhizae (AM) enhances root colonization of the wheat plants and promotes the uptake of phosphorus by 30% across all the cultivars by influencing the microbial flora within the rhizosphere region of the host plants. In yet another study, the application of AF in wheat plants positively influenced their harvest index under drought conditions while limiting the influence of aphids (Pons et al. 2020). An increase in the leaf weight, root weight, chlorophyll content, enzymatic activity of the Mexican lime seedlings, inoculated with a combination of AMF *Glomus mosseae* and *Glomus intraradices* along with vermicompost, was observed under water stressed conditions in comparison to the non-inoculated seedlings (Nejad et al. 2020). Production of secondary metabolites, uptake of water, Nitrogen, and Phosphorus increased after the inoculation of *Portulaca oleracea* L roots with *Rhizophagus irregularis* under water deficit conditions (Hosseinzadeh et al. 2020). A strain of AM, *Funneliformis mosseae,* induced Volatile Organic Compounds (VOCs) production in one of the cultivars of *Vitis vinifera.* VOCs are known to play a significant part in mechanisms involved in plant defenses and interaction between plants (Velásquez et al. 2020). The severity of charcoal root rot, in soybean plants, infected by a phytopathogen, *Macrophomina phaseolina,* increased with the application of Nitrogen fertilizers. Even though N fertilizers reduced the colonization potential of AM, *Rhizophagus intraradices*, the severity of the disease caused by the phytopathogen reduced significantly in the plants inoculated with the AF (Spagnoletti and Cornero 2020). Phosphorus loss due to leaching is very common during paddy plantation. To investigate the effect of AM to prevent the loss of phosphorus via leaching, S. Zhang et al. (2020a) studied rice inoculated with the AF. They observed that there was a cumulative phosphorus reduction of 10% in the mycorrhizae system when compared with the control system.

2.3.2.5 Endophytes

Over the last couple of decades, there has been considerable interest in the exploration of microorganisms from tissues of plants that have the abilities to secrete host secondary metabolites and various other functionalities with immense significance to agro-pharmaceutical industries (Sudheep et al. 2017; Le Cocq et al. 2017; Rajamanikyam et al. 2017). These microorganisms residing within the plant tissues without causing any harmful effect to the host plant as described by Bacon and White (Bacon and White 2000) or the diverse group of plant symbionts that live within plant tissues without any symptoms for most of their lives are known as endophytes (Wang et al. 2008). They are systemic in nature and have been isolated from various plant tissues such as leaves, stem, roots, etc. (Arnold et al. 2000; Potshangbam et al. 2017). Bacteria and fungi are the only microorganisms that have been isolated as endophytes so far, though the absence of other microorganisms such as mycoplasmas and archaea bacteria cannot be ruled out with certainty

(Tayung and Barik 2011). Several plant growth-promoting factors and plant hormones from fungal and bacterial endophytes have been reported by different research groups (Forchetti et al. 2007; Yanni and Dazzo 2010; Ahmad et al. 2010; Shahabivand et al. 2012; Zhang et al. 2018). Endophytes with possible bio fertilization properties such as siderophore production, phosphate solubilization, and Nitrogen fixation have been isolated from the same plant species (Nyambura Ngamau 2012). Some of the endophytes have been found to increase resistance in plants against a variety of phytopathogens by secreting pathogenesis-related proteins. Numerous fungal species isolated from the leaves of trees growing in the Western Ghats, Tamil Nadu demonstrated chitinolytic properties as these fungi were able to secrete chitinase and chitosanase in laboratory conditions. Chitin and chitosan have been known to trigger host responses against plant pathogens hence chitinolytic enzymes could play a major role in increasing plant defense systems against phytopathogens resulting in increased resistance. Endophytes such as *Xylariaceae* sp., *Aureobasidium pullulans*, *Colletotrichum* sp., *Lasiodiplodia theobromae*, *Phomopsis* sp., and *Fusarium* sp., Botrytis sp., *Trichoderma* sp., *Alternaria* sp., *Nodulisporium gregarium*, Nigrospora *oryzae*, *Drechslera* sp., *Pithomyces* sp. *Sordaria* sp., and *Pestalotiopsis* sp. obtained from leaves of different tree species of Western Ghats showed properties such as pathogenesis-related proteins, phytoalexins, and proteinase inhibitors in plants which are known to act against phytophagous nematodes and plant pathogenic fungi (Govinda Rajulu et al. 2011). Similarly, several fungal isolates belonging to Ascomycota and a few Zygomycota obtained as endophytes from *Panax notoginseng* exhibited antifungal properties against root rot pathogens (Zheng et al. 2017) while others have insecticidal properties as evident from the extracts of *Aspergillus* sp. and *Emericella* sp. obtained from *Rhizophora mucronate* (Abraham et al. 2015). Some of the endophytes amplified the tolerance of plants in heavy metal contaminated soils (Shahabivand et al. 2012; Waqas et al. 2014; Yamaji et al. 2016). Metabolites extracted from two strains of *Streptomyces* sp. isolated from *Artemisia herba-alba,* a white wormwood shrub was tested for their efficacy against *Botrytis cinerea* which is responsible for chocolate spot disease in faba bean, *Vicia faba* cv. Giza 3, under in vivo conditions. The environmentally safe metabolites were able to lessen the severity of the symptoms related to the disease significantly and promoted plant growth (El-Shatoury et al. 2020). Some of the properties attributed to the crops by the endophytes include tolerance to stresses such as drought, salt, and temperature stresses, decrease water consumption by the plants, stability in diverse soil types and different climatic conditions, etc. The formulations are easy to apply in the field and are compatible with commonly used chemicals in agriculture and the endophytes do not compete with soil microorganisms. The product has viability of more than 2 years when stored at 4 °C (Adaptive Symbiotic Technologies 2020).

2.3.2.6 Actinomycetes

Actinomycetes are gram-positive filamentous bacteria that belong to the phylum Actinobacteria. They are ubiquitous in nature and most of them are saprophytic. Although the composition of the cell wall of actinomycetes is similar to that of bacteria, they are classified into different groups due to the presence of peculiar morphological structure, presence of mycelia, just like that of fungi, and cultural characterizes (Bhatti et al. 2017). Some of the important genera under this phylum include *Streptomyces*, *Micromonospora*, and Nocardia (Singh et al. 2019a). Actinomycetes are known for their ability to produce an array of antimicrobial agents (Dimri et al. 2019), enzymes (Mukhtar et al. 2017), and many functionalities with a huge significance in pharmaceutical industries (Nalini and Prakash 2017). *Streptomyces* species are known to produce about 60% of antibiotics used in the agriculture system (Couillerot et al. 2014). Apart from producing metabolites of paramount importance to human health, actinomycetes are well known for producing factors responsible for the promotion of plant growth (Singh et al. 2018), antifungal agents against phytopathogens, phosphate solubilization potential, siderophores production, Nitrogen fixation, decomposition of compounds with high molecular weight, production of plant growth hormones, etc. (Edelvio et al. 2018; Meghvansi and Varma 2020; Dede et al. 2020). Some of the attributes and functions are presented in Table 2.2.

2.4 Mechanisms Involved in Plant and Soil Health Improvement

Microorganisms help in the promotion of plant growth through direct or indirect mechanisms. The direct mechanisms involved in helping the plants with the supply or facilitating nutrients such as N, P, requisite minerals, i.e. iron, phosphorus, etc., or growth improvement through the production or stimulation of different types of phytohormones whereas biocontrol aspect of the microbial inputs and tolerance to abiotic factors are considered under indirect mechanisms (Arora et al. 2013).

2.4.1 Direct Mechanism

Microbial organisms residing in the rhizosphere facilitate the growth of the plant through various mechanisms (Vessey 2003). Apart from aiding plants through the supply of nutrients, rooting patterns are also altered due to the microbial activity in the region (Ahemad and Kibret 2014). Some of the endophytes promote plant growth through the production and secretion of plant growth-promoting factors, production of siderophores, and solubilization of inorganic phosphates as described above. Endophytes transform the physiology in the host plant which directly enhances its growth. The biofertilization properties of the endophytes help certain plants to resist and survive severe environmental conditions. Some of the strains of PGPR help in the phytoremediation of soils contaminated with Fe3+ through the production of siderophores (Mishra et al. 2016). Plants growing in stressful conditions benefit from the symbiotic relationship with the mycorrhizal fungi. AMs promote plant growth by facilitating the uptake of phosphorus (Hosseinzadeh et al. 2020). Microorganisms can be used for the amendment of the soil thereby

Table 2.2 Some of the functionalities of soil Actinomycetes and their prospective properties

Sl. No.	Functionalities	Actinomycetes	Properties	Reference
1	Antifungalmycin N2	*Streptomyces* sp. Strain N2	Antifungal action against *Rhizoctonia solani.*	Zhang et al. (2020b)
3	IAA and ACC deaminase activity	*Arthrobacter arilaitensis* and *Streptomyces pseudovenezuelae*	Plant growth promotion under drought stress.	Chukwuneme et al. (2020)
4	IAA production, Phosphate solubilization, Siderophore production, Ammonia production. Antibacterial activity	*Streptomyces* sps.	Plant growth promotion and antibacterial activity against bacterial phytopathogens, i.e. *Pseudomonas gingeri, Pseudomonas syringae, and Xanthomonas campestris* pv. vesicatoria.	Dede et al. (2020)
5	Antifungal	*Streptomyces griseorubiginosus* FX81	Antagonistic against *Fusarium oxysporum* f. sp. *cubense* that causes wilt in banana.	Lai et al. (2020)
6	Antimycins	*Streptomyces* sp. AN120537	Insecticidal activities *Aedes albopictus* and *Plutella xylostella.*	Kim et al. (2020)
7	Proteases and cellulases	*Streptomyces tsukiyonensis*	Antagonistic against *Colletotrichum dematium* causing leaf disease in *Sarcandra glabra.*	Song et al. (2020)
8	Antifungal and plant growth promotion	*Streptomyces pactum* Act12 and *Streptomyces rochei* D74	Antagonistic against *Sclerotium rolfsii* responsible for causing southern blight and *Fusarium oxysporum* responsible for root rot in (*Aconitum carmichaelii*). The consortia also increased beneficial microbes and reduced harmful microbes in rhizosphere region.	Li et al. (2020b)
9	Volatile organic compounds (VOCs)	*Streptomyces lavendulae* SPS-33	Potential post-harvest fumigant for the control of *Ceratocystis fimbriata* a causative	Li et al. (2020a)

(continued)

Table 2.2 (continued)

Sl. No.	Functionalities	Actinomycetes	Properties	Reference
			organism for black spot disease in sweet potato.	
10	Actinopyrone A, Anguinomycin A and Leptomycin A	*Streptomyces palmae* CMU-AB204	Antagonistic activity against *Ganoderma boninense* causing basal stem rot (BSR) disease in *Elaeis guineensis* (oil palm tree).	Sujarit et al. (2020)
11	Auxin and ACC deaminase	*Streptomyces* sp. CLV45	Promotion of growth in soybean plants.	Horstmann et al. (2020)
12	Solubilization of phosphates, secretion of IAA and production of ACC deaminase	*Streptomyces lydicus* M01	Promotion of beneficial microflora and suppression of foliar disease in cucumbers caused by *Alternaria alternata*.	Wang et al. (2020)
13	Auxins, Polyaminase and ACC deaminase	Consortia of *Streptomyces chartreusis*, *S. tritolerans*, and *S. rochei*	Promotion of growth of *Salicornia bigelovii* in saline soils.	Mathew et al. (2020)
14	Ammonia solubilization of minerals and Chitinase production	*Nocardiopsis dassonvillei* strain YM12	Promote growth of pearl millet (*Pennisetum glaucum*) and inhibition of *Fusarium oxysporum*	Patel and Thakker (2019)
15	Cellulase and protease	*Streptomyces vinaceus* RCS260 and *Kitasatospora aburavienis* RCS252	Antimicrobial activity against phytopathogen and extracellular enzymes production	Singh et al. (2019b)

increasing nutrient content of the soil, reducing pathogens in the soil, protecting soil from droughts, and promoting plant growth (Bharti et al. 2017).

2.4.1.1 Biological Fixation of the Atmospheric Nitrogen

Biological fixation of Nitrogen is carried by microorganisms having a symbiotic or non-symbiotic association with the host plants. *Rhizobia* with leguminous plants and *Frankia* with non-leguminous plants are classified as having a symbiotic relationship while free-living microorganisms such as species of cyanobacteria or endophytes help in Nitrogen fixation without having a symbiotic association with the host plants (Yimer 2019).

2.4.1.2 Solubilization of Phosphates by Microorganisms

Phosphorus is one of the most important nutrients whose deficiency limits the proper growth of the plant (Ezawa et al. 2002). Microorganisms residing in the rhizosphere with phosphate solubilizing potential were reported as early as 1903 (Khan et al. 2009). Among Phosphate solubilizing microorganism (PSM), Phosphate solubilizing bacteria (PSB) belonging to the species of *Bacillus*, *Rhizobium,* and *Pseudomonas* strains; fungal strains majorly belonging to the species of *Penicillium* and *Aspergillus* along with AMF play an important role (Whitelaw 2000; Igual et al. 2001; Fankem et al. 2006; Sunar et al. 2015). The solubility of calcium bound phosphorus in the soil increases with a decrease in soil pH where the PSMs residing in the rhizosphere have a major role to play. These PSMs secrete certain organic acids that release bound phosphorus (Fankem et al. 2006; Joseph and Jisha 2009). Organic acids such as carboxylic acid released by PSMs, viz. *Pseudomonas* sp. are also able to release phosphorus bound to aluminum and iron (Henri et al. 2008).

2.4.1.3 Production of Siderophores by Microorganisms

Siderophore is chelating agents that are composed of protein molecules with a low molecular weight that plays an important role in solubilizing iron from organic compounds or minerals. These types of molecules are produced by microorganisms as well as gramineous plants for obtaining iron from the environment (Hider and Kong 2010). Plants growing in soils contaminated with metal haven have been reported to be deficient in iron molecules. Microorganisms producing siderophores can help the plant to sustain by regulating the availability of siderophores through the rhizosphere (Jing et al. 2007).

2.4.1.4 Production of Phytohormones

Microorganisms have been reported to produce phytohormones. About 80% of microorganisms obtained from the rhizosphere soil of crop plants have been reported to produce auxins such as IAA (Samuel et al. 2017). IAA is considered an important hormone to play a major role during rhizobacteria and plant interactions (Ahemad and Kibret 2014). Ethylene, apart from being one of the most important growth hormones in plants, is also a stress hormone. Certain rhizobacteria are able to produce 1-aminocyclopropane-1-carboxylate deaminase that helps to relieve environmental stress in plants such as temperatures, high salt or heavy metal concentration, waterlogging or predation by insect pests and promote holistic growth of the plant (Saleem et al. 2007).

2.4.2 Indirect Mechanism

Microorganisms indirectly promote plant health through developing a competitive environment for obtaining nutrients, secreting antimicrobial compounds, and inducing systemic resistance in the host plants (Ahemad and Kibret 2014). Some of the microorganisms secrete antimicrobial agents that interfere with the pathways for protein synthesis while others produce lytic enzymes that degrade the cell wall of fungal pathogens (Ram et al. 2018). Volatile organic compounds, chelating agents, and siderophores production also help in controlling pathogens (Mishra et al. 2016; Ram et al. 2018). Antimicrobial properties of actinomycetes from the soil as well as endophytes play an important role in plant protection (Lai et al. 2020). Certain microorganisms, like *B. amyloliquefaciens,* induce systemic resistance in the host plant by producing metabolites in the rhizosphere region (Lugtenberg 2015). Induction of such resistance is brought by changes in the biochemical composition thus preparing the defenses in the host to subsequent infections by various phytopathogens (Thomashow 1996). Beneficial or parasitic microorganisms are responsible for triggering host defenses at a local or systemic level. Plant hormones such as Jasmonic acid, Ethylene, and Salicylic acid are major players in altering the network signals that trigger host defenses in the plant (Han and Kahmann 2019). The collective effect of both the mechanisms plays a major role to enhance the growth of the plant, maintaining plant health including management of disease through the control of phytopathogens. Hence, it is tough to differentiate precisely the exact effect of either of the mechanisms (Bhattacharyya and Jha 2012). Under natural conditions, metabolites released by these microorganisms may play a similar or different role depending on the environmental factors (Arora et al. 2013).

2.4.2.1 Induction of Resistance in Host Plants by AIMs

Various factors such as light, gravity, stress, water molecules, availability of nutrients, chemical metabolites produced from soil, and microorganisms associated with the host plants stimulate different responses in the plant (Vallad and Goodman 2004). Such a stimulus is responsible for the induction of defense responses in the plant as well as improving its resistance during subsequent infections at the local or systemic level (Vos et al. 2015). Several factors and pathways related to the induction of resistance by BCA or non-pathogenic microbes are being characterized (Audenaert et al. 2002; Vallad and Goodman 2004). Pathogenesis-related (PR) proteins are produced after the infection of the plants by pathogens, which is mediated by Salicylic Acid (SA) (Vallad and Goodman 2004; Leonetti et al. 2017), and composed of different types of enzymes which restricts invading pathogens through different mechanisms (Vallad and Goodman 2004). Among beneficial microorganisms, strains of *Bacillus*, *Pseudomonas*, *Trichoderma, Rhizophagus,* etc. including non-pathogenic strains of *Fusarium* sp. are known to activate Induced Systemic Resistance (ISR) (Romera et al. 2019) which has been reported to be mediated by JA or ethylene (Audenaert et al. 2002; Yoo and Sang 2017). Control in the induction of strong resistance could be achieved through synergistic interaction with the endogenous signals (Moyne et al. 2000; Mayo et al. 2015).

2.5 Molecular Signaling in the Rhizosphere and Beyond: The Cross Talk

2.5.1 Microbe Triggered Immunity

The ecosystem in the rhizosphere is greatly altered by the metabolites secreted by the plant roots as well as microorganisms that provide favorable environment for the multiplication and sustenance of different types of microorganisms. These factors are generally responsible for the promotion of growth in plants or combating phytopathogens (Venturi and Keel 2016). Microbe-associated molecular patterns (MAMPs) results in the initiation of weak but effective immune responses throughout the plant tissues. Priming of improved defenses in plants is induced by advantageous microorganisms through signaling pathways similar to jasmonate and ethylene dependent ones (Van Wees et al. 2008). Plants and beneficial microorganisms have been observed to have optimum co-ordination as MAMPs associated immunity does not act against the interacting beneficial AIMs. Plant hormones control the signaling networks responsible for the induction of defense responses (Glazebrook 2005) which are tuned through cross-communication between different pathways according to the type of pest invasion (Koornneef and Pieterse 2008). Transcriptional regulator NPR1 controls both the pathogen induced Systemic Acquired Resistance (SAR) and ISR triggered by beneficial microorganisms (Pieterse and Van Loon 2004). Certain fungal strains have been reported to induce resistance through Jasmonic Acid (JA), ET, and/or NPR-1 dependent pathways or SA signaling pathways or requirements for both ISR and SAR components (Conn et al. 2008).

In Plant–pathogen interaction, a majority of pathogens/microbes are blocked by the host plant through a strategy of non-host resistance which consists of physical barriers, viz. waxy cuticles, rigid cell walls, and antimicrobial compounds. Virulent phytopathogens that are able to cross the primary defense structures must face the plants' innate immune system that might terminate its progression further. Plants are not known to possess somatic adaptive immune system with mobile defender cells which become requisite for them to depend upon the individual cells to exhibit innate immunity, with the signals released from the infected cells which help the plant cells to memorize earlier infections (Reimer-Michalski and Conrath 2016). Two individual branches of molecular defense strategies have been described under zig-zag co-evolutionary model (Jones and Dangl 2006). Firstly, the microorganisms or pathogen associated molecules (MAMP/PAMP) modulate the recognition receptors (PRRs), which ultimately induces MAMP-triggered immunity (MTI), PAMP-triggered immunity (PTI), and DAMP-triggered immunity jointly denoted as pattern-triggered immunity PTI (Saijo et al. 2018). On the other side of the model, host molecules that recognize microbial effectors and the virulence factors suppress MTI along a cascade of resistance proteins (R) and ultimately effector-triggered immunity (ETI) is initiated. As a result of these immune responses, a cascade of signaling events is triggered that leads to protection against the invading pathogen (Nishad et al. 2020) (Fig. 2.4).

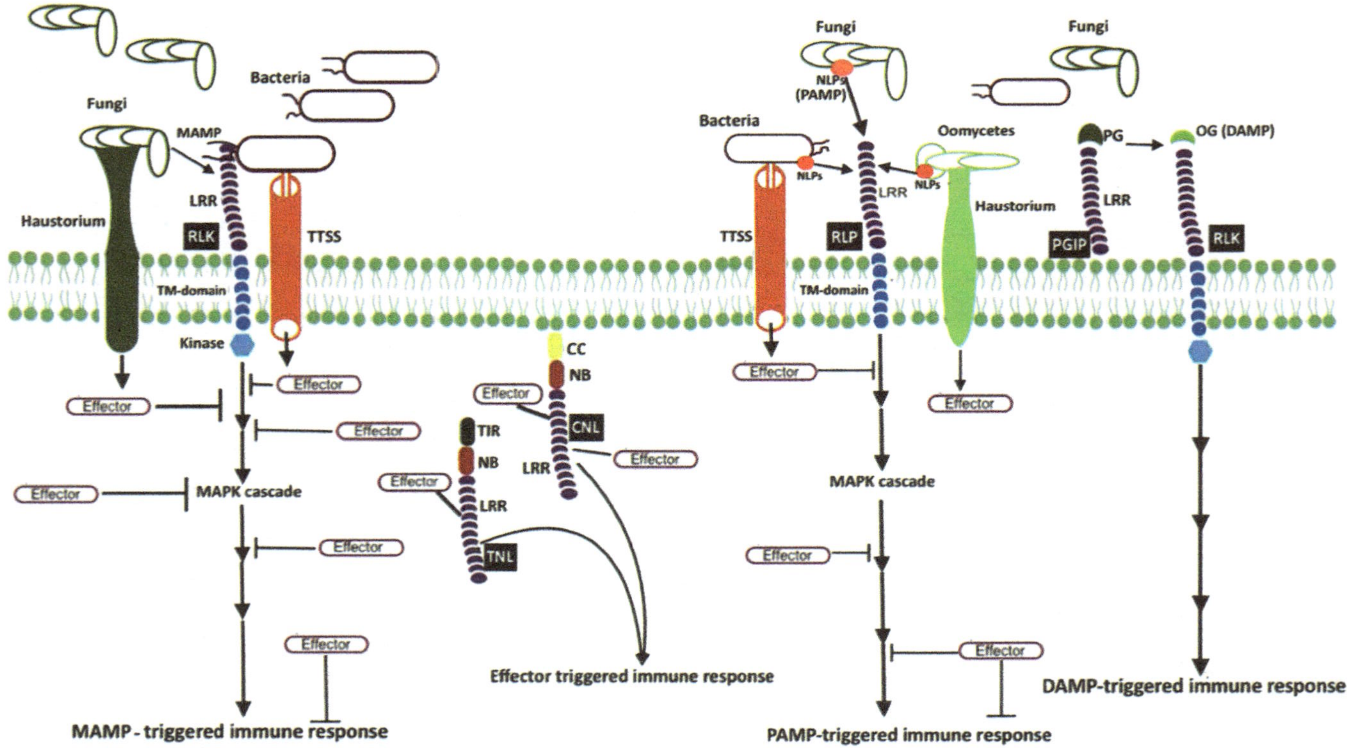

Fig. 2.4 Schematic representation showing MAMP recognition by the Receptor-Like Kinases (RLK; a PRR) that activates cascades of mitogen-activated protein kinase (MAPK), eventually leading to MAMP-triggered immunity (MTI). Another PRR, RLP23 perceives MAMP or pattern-associated molecular patterns (PAMP), and induces pattern-triggered immunity (PTI). Necrosis and ethylene–inducing peptide 1 (NEP1)–like proteins (NLPs) are examples of PAMP, which are recognized by RLP23. Effectors are released into plant cells to counter the defense responses of the plant. Type III secretion system (T3SS) is

Fig. 2.4 (continued) used by bacteria to transport effector proteins inside the cells of the plant while fungal species use haustoria for the same purpose. Effector-triggered immune responses (ETI) are induced as and when plant resistance protein coiled–coil (CC) NLR (CNL) and Toll-interleukin-1 receptor (TIR) NLR (TNL) recognizes the activity of the effector in the cell. Oligogalacturonides (OG) are formed when Extracellular LRR protein—polygalacturonase-inhibiting protein (PGIP) interacts with the microbial polygalacturonase (PG) which results in the slowing down of the process of pectin degradation. Damage-associated molecular pattern (DAMP) triggered immune responses (DPI) is activated as soon as DAMP is perceived by PRR (Nishad et al. 2020)

2.5.2 Microbial Signaling

Certain microorganisms have been identified to produce compounds that act as signals to synchronize the expression of the genes in accordance with the population of the cells. This process is known as Quorum Sensing (QS) (Fuqua et al. 2001). QS signals have been reported from many strains of microorganisms isolated from the rhizosphere indicating the possibility of a cell-to-cell signaling. Gram-positive microorganisms produce short modified peptides, whereas some gram-negative microorganisms associated with the plants including the strains belonging to the genus of *Pseudomonas, Serratia, Erwinia*, etc., are been known to secrete Acylated Homoserine Lactone (AHL). Signals involving pyrones and dialkylresorcinols have also been detected in gram-negative microorganisms that are related to AHL family and are identified by the LuxR protein. QS is mainly involved in the expression of genes, production and regulation of antibiotics, the formation of biofilms, factors involved with virulence, induce plant resistance, and regulate the growth of the plant and its development (Ferluga et al. 2008; Venturi and Fuqua 2013). Some of the gram-negative microorganisms including species belonging to the genus of *Burkholderia* and *Stenotrophomonas maltophilia* produce a Diffusible Signal Factor (DSF) known for eliciting innate immune system in plants (Ryan et al. 2015). Antibiotics too have been proposed to act as molecules for QS at a non-inhibitory concentration within inter and intraspecies level (Andersson and Hughes 2014). Some of the Volatile Organic Compounds (VOCs) produced by microorganisms have been reported to interact at inter and intraspecies level as well as act in plant-microbe interaction at the rhizosphere (Bitas et al. 2013).

2.6 Current and Future Challenges

Agricultural inputs of microbial origin have shown a growing trend with the increase in the interest towards organically derived agriculture produce. The advantages of microbial fertilizers and biopesticides towards human health as well the environment outweigh the returns of chemical inputs. However, some limitations of microbial inputs need to be addressed before replacing the conventional chemicals used in agriculture. These regularly used agrochemicals have broad-spectrum activity with very little interference in its activity from varying environmental conditions. Whereas, microbial inputs are target specific with a narrow spectrum of activity. The environmental factors, both biotic and abiotic, interfere in the efficacy and functionality of such products. Challenges in maintaining consistency of similar products and their registration at a national and regional level is also a major concern (Timmusk et al. 2017). The microbial inputs should be able to generate consistent results in farms, should be easy to handle, and should be cost-effective with a good shelf-life (Murphy et al. 2018). Redressal to these possible drawbacks could undoubtedly help in the considerable success of microbial inputs in the agriculture sector.

2.7 Conclusion

Agriculture produce is most important for the sustenance of human beings on this planet. Chemical inputs have increased agriculture harvest to many folds which have aided many countries to overcome food deficit and provided for the people. However, the continuous and growing use of chemical inputs has inevitably brought its share of grave complications. Apart from affecting human health, the extensive damages caused by hazardous chemicals to the environment including soil, flora, and fauna remain a major concern and require an immediate solution to mitigate the problem. Microorganisms have demonstrated huge potential in promoting plant growth, helping plants in resisting environmental stresses, combating phytopathogens, amending contaminated soils, and improving agricultural yield without affecting the environment adversely. Diverse types of microorganisms through direct or indirect approaches have the enhanced ability to facilitate growth in plants and progressively improve soil fertility. These specific functionalities of the beneficial microorganisms could be tapped for its use as bio-fertilizing agents or as biopesticides to progressively replace or decrease the extensive usage of chemical compounds in the agriculture system. Some of the limitations such as consistency of product and its effect at different environmental conditions, the viability of microorganisms during storage and registration of microbial products require further studies. Screening of beneficial microorganisms, for its novel or better attributes, from various ecological niches and unique habitats should remain a continuous process to fulfill the overgrowing demand for natural products.

References

Abraham S, Basukriadi A, Pawiroharsono S, Sjamsuridzal W (2015) Insecticidal activity of ethyl acetate extracts from culture filtrates of mangrove fungal endophytes. Mycobiology 43:137–149. https://doi.org/10.5941/MYCO.2015.43.2.137

Adaptive Symbiotic Technologies (2020) BioEnsure® Microbial Inoculant. http://www.adaptivesymbiotictechnologies.com. Accessed 30 July 2020

Ahemad M, Kibret M (2014) Mechanisms and applications of plant growth promoting rhizobacteria: current perspective. J King Saud Univ Sci 26:1–20. https://doi.org/10.1016/j.jksus.2013.05.001

Ahmad M, Pataczek L, Hilger TH, Zahir ZA, Hussain A, Rasche F, Schafleitner R, Solberg SØ (2018) Perspectives of microbial inoculation for sustainable development and environmental management. Front Microbiol 9:1–26. https://doi.org/10.3389/fmicb.2018.02992

Ahmad N, Hamayun M, Khan SA, Khan AL, Lee I-J, Shin D-H (2010) Gibberellin-producing endophytic fungi isolated from Monochoria vaginalis. J Microbiol Biotechnol 20:1744–1749. https://doi.org/10.4014/jmb.1005.05018

Akbar A, Sinegani S, Yeganeh ME (2017) The occurrence of arbuscular mycorrhizal fungi in soil and root of medicinal plants in Bu-Ali Sina garden in Hamadan, Iran. Biol J Microorg 5:43–59

Altomare C, Norvell WA, Björkman T, Harman GE (1999) Solubilization of phosphates and micronutrients by the plant-growth-promoting and biocontrol fungus Trichoderma harzianum Rifai 1295-22. Appl Environ Microbiol 65:2926–2933. https://doi.org/10.1128/AEM.65.7.2926-2933.1999

Andersson DI, Hughes D (2014) Microbiological effects of sublethal levels of antibiotics. Nat Rev Microbiol 12:465–478. https://doi.org/10.1038/nrmicro3270

Armada E, Probanza A, Roldán A, Azcón R (2016) Native plant growth promoting bacteria Bacillus thuringiensis and mixed or individual mycorrhizal species improved drought tolerance and oxidative metabolism in Lavandula dentata plants. J Plant Physiol 192:1–12. https://doi.org/10.1016/j.jplph.2015.11.007

Arnold A, Maynard Z, Gilbert G, Coley P, Kursar T (2000) Are tropical fungal endophytes hyperdiverse? Ecol Lett 3:267–274. https://doi.org/10.1046/j.1461-0248.2000.00159.x

Arora NK, Tewari S, Singh R (2013) Multifaceted plant-associated microbes and their mechanisms diminish the concept of direct and indirect PGPRs. In: Plant microbe symbiosis: fundamentals and advances. Springer, New Delhi, pp 411–449

Audenaert K, Pattery T, Cornelis P (2002) Induction of systemic resistance to-Botrytis cinerea in tomato by Pseudomonas aeruginosa 7NSK2: role of salicylic acid, pyochelin and pyocyanin. Mol Plant-Microbe Interact 15:1147–1156. https://doi.org/10.1094/MPMI.2002.15.11.1147

Bach EM, Williams RJ, Hargreaves SK, Yang F, Hofmockel KS (2018) Greatest soil microbial diversity found in micro-habitats. Soil Biol Biochem 118:217–226. https://doi.org/10.1016/j.soilbio.2017.12.018

Bacon CW, White JF (2000) Physiological adaptations in the evolution of endophytism in the Clavicipitaceae. In: Microbial endophytes, 1st edn. M. Dekker, New York, pp 251–276

Bagyaraj DJ, Ashwin R (2017) Soil biodiversity: role in sustainable horticulture. Biodivers Hortic Crop 5:1–18

Bharti VS, Dotaniya ML, Shukla SP, Yadav VK (2017) Managing soil fertility through microbes: prospects, challenges and future strategies. In: Singh JS, Seneviratne G (eds) Agro-environmental sustainability. Springer, Cham, pp 81–111

Bhattacharyya PN, Jha DK (2012) Plant growth-promoting rhizobacteria (PGPR): emergence in agriculture. World J Microbiol Biotechnol 28:1327–1350. https://doi.org/10.1007/s11274-011-0979-9

Bhatti AA, Haq S, Bhat RA (2017) Actinomycetes benefaction role in soil and plant health. Microb Pathog 111:458–467

Bisen K, Keswani C, Patel JS, Sarma BK, Singh HB (2016) Trichoderma spp.: efficient inducers of systemic resistance in plants. In: Choudhary DK, Varma A (eds) Microbial-mediated induced systemic resistance in plants, 2016th edn. Springer, Singapore, pp 185–195

Bitas V, Kim HS, Bennett JW, Kang S (2013) Sniffing on microbes: diverse roles of microbial volatile organic compounds in plant health. Mol Plant-Microbe Interact 26:835–843. https://doi.org/10.1094/MPMI-10-12-0249-CR

Cao Y, Pi H, Chandrangsu P, Li Y, Wang Y, Zhou H, Xiong H, Helmann JD, Cai Y (2018) Antagonism of two plant-growth promoting bacillus velezensis isolates against Ralstonia solanacearum and fusarium oxysporum. Sci Rep 8:1–14. https://doi.org/10.1038/s41598-018-22782-z

Chakraborty U, Chakraborty BN, Tongden C, Roychowdhury P (2004) Assessment of Tea rhizosphere microorganisms as plant growth promoters. J Tea Res 25:38–47

Chakraborty U, Chakraborty BN, Basnet M, Chakraborty AP (2009) Evaluation of Ochrobactrum anthropi TRS-2 and its talc based formulation for enhancement of growth of tea plants and management of brown root rot disease. J Appl Microbiol 107: 625–634. https://doi.org/10.1111/j.1365-2672.2009.04242.x

Chakraborty BN, Chakraborty U, Dey PL Sunar K (2014) Evaluation of PGPR traits of native bacterial strains from Darjeeling and analysis of their diversity. J Botanical Soc Bengal 68:159–176

Chandra H, Kumari P, Bisht R, Prasad R, Yadav S (2020) Plant growth promoting Pseudomonas aeruginosa from Valeriana wallichii displays antagonistic potential against three phytopathogenic fungi. Mol Biol Rep 47:6015–6026. https://doi.org/10.1007/s11033-020-05676-0

Chenniappan C, Narayanasamy M, Daniel GM, Ramaraj GB, Ponnusamy P, Sekar J, Vaiyapuri Ramalingam P (2019) Biocontrol efficiency of native plant growth promoting rhizobacteria against rhizome rot disease of turmeric. Biol Control 129:55–64. https://doi.org/10.1016/j.biocontrol.2018.07.002

Chukwuneme CF, Babalola OO, Kutu FR, Ojuederie OB (2020) Characterization of actinomycetes isolates for plant growth promoting traits and their effects on drought tolerance in maize. J Plant Interact 15:93–105. https://doi.org/10.1080/17429145.2020.1752833

Cocq LK, Gurr SJ, Hirsch PR, Mauchline TH (2017) Exploitation of endophytes for sustainable agricultural intensification. Mol Plant Pathol 18:469–473. https://doi.org/10.1111/mpp.12483

Compant S, Duffy B, Nowak J, Clement C, Barka EA (2005) Use of plant growth-promoting bacteria for biocontrol of plant diseases: principles, mechanisms of action, and future prospects. Appl Environ Microbiol 71:4951–4959. https://doi.org/10.1128/AEM.71.9.4951-4959.2005

Compant S, Samad A, Faist H, Sessitsch A (2019) A review on the plant microbiome: ecology, functions, and emerging trends in microbial application. J Adv Res 19:29–37. https://doi.org/10.1016/j.jare.2019.03.004

Conn VM, Walker AR, Franco CMM (2008) Endophytic actinobacteria induce defense pathways in Arabidopsis thaliana. Mol Plant-Microbe Interact 21:208–218

Couillerot O, Loqman S, Toribio A, Hubert J, Gandner L, Nuzillard JM, Ouhdouch Y, Clément C, Barka EA, Renault JH (2014) Purification of antibiotics from the biocontrol agent Streptomyces anulatus S37 by centrifugal partition chromatography. J Chromatogr B Anal Technol Biomed Life Sci 944:30–34. https://doi.org/10.1016/j.jchromb.2013.11.008

Dede A, Güven K, Şahin N (2020) Isolation, plant growth-promoting traits, antagonistic effects on clinical and plant pathogenic organisms and identification of actinomycetes from olive rhizosphere. Microb Pathog 143:1–11. https://doi.org/10.1016/j.micpath.2020.104134

Dennis PG, Miller AJ, Hirsch PR (2010) Are root exudates more important than other sources of rhizodeposits in structuring rhizosphere bacterial communities? FEMS Microbiol Ecol 72:313–327. https://doi.org/10.1111/j.1574-6941.2010.00860.x

Díaz S, Grossi C, Zawoznik M, Daniela M (2016) Wheat seeds harbour bacterial endophytes with potential as plant growth promoters and biocontrol agents of Fusarium graminearum. Microbiol Res 186–187:37–43. https://doi.org/10.1016/j.micres.2016.03.002

Dimri AG, Chauhan A, Aggarwal ML (2019) Antibiotic potential of actinomycetes from different environments against human pathogens and microorganisms of industrial importance: a review. Sci Archiv 1:32–48

Edelvio d BG, Leo RLD, Rita de CM (2018) Actinomycetes bioactive compounds: biological control of fungi and phytopathogenic insect. Afr J Biotechnol 17:552–559. https://doi.org/10.5897/ajb2017.16323

Elliott AJ, Daniell TJ, Cameron DD, Field KJ (2020) A commercial arbuscular mycorrhizal inoculum increases root colonization across wheat cultivars but does not increase assimilation of mycorrhiza-acquired nutrients. Plants People Planet 1–12. https://doi.org/10.1002/ppp3.10094

Elsharkawy MM, El-Khateeb NMM (2019) Antifungal activity and resistance induction against Sclerotium cepivorum by plant growth-promoting fungi in onion plants. Egypt J Biol Pest Control 29:1–9. https://doi.org/10.1186/s41938-019-0178-9

El-Shatoury SA, Ameen F, Moussa H, Abdul Wahid O, Dewedar A, AlNadhari S (2020) Biocontrol of chocolate spot disease (Botrytis cinerea) in faba bean using endophytic actinomycetes Streptomyces : a field study to compare application techniques. PeerJ 8:2–14. https://doi.org/10.7717/peerj.8582

Emmanuel OC, Babalola OO (2020) Productivity and quality of horticultural crops through co-inoculation of arbuscular mycorrhizal fungi and plant growth promoting bacteria. Microbiol Res 239:1–11. https://doi.org/10.1016/j.micres.2020.126569

Ezawa T, Smith SE, Smith FA (2002) P metabolism and transport in AM fungi. Plant Soil 244:221–230. https://doi.org/10.1023/A:1020258325010

Fan B, Chen XH, Budiharjo A, Bleiss W, Vater J, Borriss R (2011) Efficient colonization of plant roots by the plant growth promoting bacterium Bacillus amyloliquefaciens FZB42, engineered

to express green fluorescent protein. J Biotechnol 151:303–311. https://doi.org/10.1016/j.jbiotec.2010.12.022
Fankem H, Nwaga D, Deubel A, Dieng L, Merbach W, Etoa FX (2006) Occurrence and functioning of phosphate solubilizing microorganisms from oil palm tree (Elaeis guineensis) rhizosphere in Cameroon. Afr J Biotechnol 5:2450–2460
Ferluga S, Steindler L, Venturi V (2008) N-acyl Homoserine lactone quorum sensing in gram-negative Rhizobacteria. In: Karlovsky P (ed) Secondary metabolites in soil ecology. Soil biology. Springer, Berlin, pp 69–90
Forchetti G, Masciarelli O, Alemano S, Alvarez D, Abdala G (2007) Endophytic bacteria in sunflower (Helianthus annuus L.): isolation, characterization, and production of jasmonates and abscisic acid in culture medium. Appl Microbiol Biotechnol 76:1145–1152. https://doi.org/10.1007/s00253-007-1077-7
Fuqua C, Parsek MR, Greenberg EP (2001) Regulation of gene expression by cell-to-cell communication: acyl-Homoserine lactone quorum sensing. Annu Rev Genet 35:439–468. https://doi.org/10.1146/annurev.genet.35.102401.090913
Gahan J, Schmalenberger A (2014) The role of bacteria and mycorrhiza in plant sulfur supply. Front Plant Sci 5:1–7. https://doi.org/10.3389/fpls.2014.00723
Glazebrook J (2005) Contrasting mechanisms of defense against biotrophic and necrotrophic pathogens. Annu Rev Phytopathol 43:205–227
Hacquard S (2016) Disentangling the factors shaping microbiota composition across the plant holobiont. New Phytol 209:454–457. https://doi.org/10.1111/nph.13760
Han X, Kahmann R (2019) Manipulation of phytohormone pathways by effectors of filamentous plant pathogens. Front Plant Sci 10. https://doi.org/10.3389/fpls.2019.00822
Hardoim PR, van Overbeek LS, Berg G, Pirttilä AM, Compant S, Campisano A, Döring M, Sessitsch A (2015) The hidden world within plants: ecological and evolutionary considerations for defining functioning of microbial endophytes. Microbiol Mol Biol Rev 79:293–320. https://doi.org/10.1128/MMBR.00050-14
Harley JL (1989) The significance of mycorrhiza. Mycol Res 92:129–139. https://doi.org/10.1016/S0953-7562(89)80001-2
Hashem A, Tabassum B, Fathi Abd Allah E (2019) Bacillus subtilis: a plant-growth promoting rhizobacterium that also impacts biotic stress. Saudi J Biol Sci 26:1291–1297. https://doi.org/10.1016/j.sjbs.2019.05.004
Henri F, Laurette NGON, Annette D, John Q (2008) Solubilization of inorganic phosphates and plant growth promotion by strains of Pseudomonas fluorescens isolated from acidic soils of Cameroon. Afr J Microbiol Res 2:171–178. https://doi.org/10.5897/AJMR.9000660
Hider RC, Kong X (2010) Chemistry and biology of siderophores. Nat Prod Rep 27:637. https://doi.org/10.1039/b906679a
Horstmann JL, Dias MP, Ortolan F, Medina-Silva R, Astarita LV, Santarém ER (2020) Streptomyces sp. CLV45 from Fabaceae rhizosphere benefits growth of soybean plants. Brazilian J Microbiol. https://doi.org/10.1007/s42770-020-00301-5
Hossain MM, Sultana F, Islam S (2017) Plant growth-promoting fungi (PGPF): Phytostimulation and induced systemic resistance. In: Plant-microbe interactions in agro-ecological perspectives. Springer, Singapore, pp 135–191
Hosseinzadeh MH, Ghalavand A, Mashhadi-Akbar M, Modarres-Sanavy SAM, Mokhtassi-Bidgoli A (2020) Communications in soil science and plant analysis increased medicinal contents of purslane by nitrogen and arbuscular mycorrhiza under drought stress increased medicinal contents of purslane by nitrogen and arbuscular mycorrhiza under drought stress. Commun Soil Sci Plant Anal 51:118–135. https://doi.org/10.1080/00103624.2019.1695828
Huang R, Zeng J, Zhao D, Cook KV, Hambright KD, Yu Z (2020) Sediment microbiomes associated with the rhizosphere of emergent macrophytes in a shallow, subtropical lake. Limnol Oceanogr 65:S38–S48. https://doi.org/10.1002/lno.11325
Hyakumachi M (1994) Plant-growth-promoting fungi from Turfgrass rhizosphere with potential for disease suppression. Soil Microorg 44:53–68. https://doi.org/10.18946/jssm.44.0_53

Igual JM, Valverde A, Cervantes E, Velázquez E (2001) Phosphate-solubilizing bacteria as inoculants for agriculture: use of updated molecular techniques in their study. Agronomie 21:561–568. https://doi.org/10.1051/agro:2001145

Islam S, Akanda AM, Prova A, Islam MT, Hossain MM (2016) Isolation and identification of plant growth promoting Rhizobacteria from cucumber rhizosphere and their effect on plant growth promotion and disease suppression. Front Microbiol 6:1–12. https://doi.org/10.3389/fmicb.2015.01360

Jing Y, He Z, Yang X (2007) Role of soil rhizobacteria in phytoremediation of heavy metal contaminated soils. J Zhejiang Univ 8:192–207. https://doi.org/10.1631/jzus.2007.B0192

Johnson NC, Graham JH, Smith FA (1997) Functioning of mycorrhizal associations along the mutualism-parasitism continuum. New Phytol 135:575–585. https://doi.org/10.1046/j.1469-8137.1997.00729.x

Jones JDG, Dangl JL (2006) The plant immune system. Nature 444:323–329. https://doi.org/10.1038/nature05286

Joseph S, Jisha MS (2009) Buffering reduces phosphate solubilizing ability of selected strains of bacteria. World J Agric Sci 5:135–137

Keswani C, Mishra S (2014) Unraveling the efficient applications of secondary metabolites of various Trichoderma spp. Appl Microbiol Biotechnol 98:533–544. https://doi.org/10.1007/s00253-013-5344-5

Keswani C, Bisen K, Chitara MK, Sarma BK, Singh HB (2017) Exploring the role of secondary metabolites of Trichoderma in tripartite interaction with plant and pathogens. In: Singh JS, Seneviratne G (eds) Agro-environmental sustainability. Springer, Cham, pp 63–79

Khan MAW, Bohannan BJM, Nüsslein K, Tiedje JM, Tringe SG, Parlade E, Barberán A, Rodrigues JLM (2019) Deforestation impacts network co-occurrence patterns of microbial communities in Amazon soils. FEMS Microbiol Ecol 95:1–12. https://doi.org/10.1093/femsec/fiy230

Khan MS, Zaidi A, Wani PA (2009) Role of phosphate solubilizing microorganisms in sustainable agriculture - a review. In: Lichtfouse E, Navarrete M, Debaeke P, Véronique S, Alberola C (eds) Sustainable agriculture. Springer, Dordrecht, pp 551–570

Kim JH, Choi JY, Park DH, Park D-J, Park MG, Kim SY, Ju YJ, Kim JY, Wang M, Kim C-J, Je YH (2020) Isolation and characterization of the insect growth regulatory substances from actinomycetes. Comp Biochem Physiol Part C Toxicol Pharmacol 228:1–9. https://doi.org/10.1016/j.cbpc.2019.108651

King AJ, Farrer EC, Suding KN, Schmidt SK (2012) Co-occurrence patterns of plants and soil bacteria in the high-alpine Subnival zone track environmental harshness. Front Microbiol 3:1–14. https://doi.org/10.3389/fmicb.2012.00347

Kirk PM, Cannon PF, David JC, Stalpers JA (2001) Ainsworth and Bisby's dictionary of the fungi, 9th edn. CABI publishing, Wallingford

Kiss L (2003) A review of fungal antagonists of powdery mildews and their potential as biocontrol agents. Pest Manag Sci 59:475–483. https://doi.org/10.1002/ps.689

Koornneef A, Pieterse CMJ (2008) Cross talk in defense signaling. Plant Physiol 146:839–844

Kumar SM, Reddy CG, Phogat M, Korav S (2018) Role of bio-fertilizers towards sustainable agricultural development: a review. J Pharmacogn Phytochem 7:1915–1921

Lai B, Lin M, Wu Z, Wang J (2020) Identification of an antagonistic actinomycetes and its biological control effect on banana fusarium wilt. J South Agric 51:836–843. https://doi.org/10.3969/j.issn.2095-1191.2020.04.014

Lemanceau P, Blouin M, Muller D, Moënne-Loccoz Y (2017) Let the Core microbiota be functional. Trends Plant Sci 22:583–595. https://doi.org/10.1016/j.tplants.2017.04.008

Leonetti P, Zonno MC, Molinari S, Altomare C (2017) Induction of SA-signaling pathway and ethylene biosynthesis in Trichoderma harzianum-treated tomato plants after infection of the root-knot nematode Meloidogyne incognita. Plant Cell Rep 36:621–631. https://doi.org/10.1007/s00299-017-2109-0

Lettice EP (2019) The rhizosphere: measuring the zone of interaction. In: Annual plant reviews online. Wiley, pp 219–236

Li X, Li B, Cai S, Zhang Y, Xu M, Zhang C, Yuan B, Xing K, Qin S (2020a) Identification of Rhizospheric Actinomycete Streptomyces lavendulae SPS-33 and the inhibitory effect of its volatile organic compounds against Ceratocystis fimbriata in postharvest sweet potato (Ipomoea batatas (L.) lam.). Microorganisms 8:1–13. https://doi.org/10.3390/microorganisms8030319

Li Y, Guo Q, He F, Li Y, Xue Q, Lai H (2020b) Biocontrol of root diseases and growth promotion of the tuberous plant Aconitum carmichaelii induced by actinomycetes are related to shifts in the rhizosphere microbiota. Microb Ecol 79:134–147. https://doi.org/10.1007/s00248-019-01388-6

Lugtenberg B (2015) Principles of plant-microbe interactions. Springer, Cham

Manzoor M, Gul I, Kallerhoff J, Arshad M (2019) Fungi-assisted phytoextraction of lead: tolerance, plant growth–promoting activities and phytoavailability. Environ Sci Pollut Res 26:23788–23797. https://doi.org/10.1007/s11356-019-05656-3

Markets and Markets Research Private Ltd (2020) Agricultural Microbials Market by Type (Bacteria, Fungi, Virus, Protozoa), Function (Soil Amendment and Crop Protection), Crop Type (Cereals & Grains, Oilseeds & Pulses, Fruits & Vegetables), Mode of Application, Formulation, and Region - Forecast to 2025. https://www.researchandmarkets.com/reports/5028320/agricultural-microbials-market-by-type-bacteria. Accessed 20 Sept 2020

Massalha H, Korenblum E, Malitsky S, Shapiro OH, Aharoni A (2017) Live imaging of root–bacteria interactions in a microfluidics setup. Proc Natl Acad Sci 114:4549–4554. https://doi.org/10.1073/pnas.1618584114

Mathew BT, Torky Y, Amin A, Mourad A-HI, Ayyash MM, El-Keblawy A, Hilal-Alnaqbi A, Abu Qamar SF, El-Tarabily KA (2020) Halotolerant marine rhizosphere-competent Actinobacteria promote Salicornia bigelovii growth and seed production using seawater irrigation. Front Microbiol 11:1–21. https://doi.org/10.3389/fmicb.2020.00552

Mayo S, Gutiérrez S, Malmierca MG, Lorenzana A, Campelo MP, Hermosa R, Casquero PA (2015) Influence of Rhizoctonia solani and Trichoderma spp. in growth of bean (Phaseolus vulgaris L.) and in the induction of plant defense-related genes. Front Plant Sci 6:1–11. https://doi.org/10.3389/fpls.2015.00685

Meghvansi MK, Varma A (2020) Biology of composts. Springer International Publishing, Cham

Mendes R, Garbeva P, Raaijmakers JM (2013) The rhizosphere microbiome: significance of plant beneficial, plant pathogenic, and human pathogenic microorganisms. FEMS Microbiol Rev 37:634–663. https://doi.org/10.1111/1574-6976.12028

Mishra V, Gupta A, Kaur P, Singh S, Gehlot P, Singh J (2016) Synergistic effects of arbuscular mycorrhizal fungi and plant growth promoting rhizobacteria in bioremediation of iron contaminated soils. Int J Phytoremediation 18:697–703. https://doi.org/10.1080/15226514.2015.1131231

Monard C, Gantner S, Bertilsson S, Hallin S, Stenlid J (2016) Habitat generalists and specialists in microbial communities across a terrestrial-freshwater gradient. Sci Rep 6:1–10. https://doi.org/10.1038/srep37719

Moyne AL, Shelby R, Clevel TE, Tuzun S (2000) Bacillomycin D: aniturin with antifungal activity against Aspergillus flavus. J Appl Microbiol 90:622–629. https://doi.org/10.1046/j.1365-2672.2001.01290.x

Mukherjee A, Kapoor A, Dutta S (2018) Organic food business in India: a survey of companies. Res Econ Manag 3:72. https://doi.org/10.22158/rem.v3n2p72

Mukhtar S, Zaheer A, Aiysha D, Abdulla Malik K, Mehnaz S (2017) Actinomycetes: a source of industrially important enzymes. J Proteomics Bioinform 10:316–319. https://doi.org/10.4172/jpb.1000456

Murphy B, Doohan F, Hodkinson T (2018) From concept to commerce: developing a successful fungal endophyte inoculant for agricultural crops. J Fungi 4:24. https://doi.org/10.3390/jof4010024

Nalini MS, Prakash HS (2017) Diversity and bioprospecting of actinomycete endophytes from the medicinal plants. Lett Appl Microbiol 64:261–270. https://doi.org/10.1111/lam.12718

Nejad KZ, Ghasemi M, Shamili M (2020) Effect of Mycorrhiza and Vermicompost on drought tolerance of lime seedlings (Citrus aurantifolia Cv. Mexican Lime) effect of Mycorrhiza and

Vermicompost on drought tolerance of lime seedlings (Citrus aurantifolia Cv. Mexican Lime). Int J Fruit Sci 20:646–657. https://doi.org/10.1080/15538362.2019.1678448
Ngamau NC (2012) Isolation and identification of endophytic bacteria of bananas (Musa spp.) in Kenya and their potential as biofertilizers for sustainable banana production. Afr J Microbiol Res 6:6414–6422. https://doi.org/10.5897/AJMR12.1170
Nishad R, Ahmed T, Rahman VJ, Kareem A (2020) Modulation of plant defense system in response to microbial interactions. Front Microbiol 11:1298. https://doi.org/10.3389/fmicb.2020.01298
Olanrewaju OS, Ayangbenro AS, Glick BR, Babalola OO (2019) Plant health: feedback effect of root exudates-rhizobiome interactions. Appl Microbiol Biotechnol 103:1155–1166. https://doi.org/10.1007/s00253-018-9556-6
Ossowicki A, Jafra S and and Garbeva P (2017) The antimicrobial volatile power of the rhizospheric isolate Pseudomonas donghuensis P482. PLoS One 12 (3):e0174362. https://doi.org/10.1371/journal.pone.0174362
Pandin C, Le Coq D, Briandet R (2017) Should the bio film mode of life be taken into consideration for microbial biocontrol agents ? Microb Biotechnol 10:719–734. https://doi.org/10.1111/1751-7915.12693
Patel KB, Thakker JN (2019) Growth promotion and biocontrol activity of Nocardiopsis dassonvillei strain YM12: an isolate from coastal agricultural land of Khambhat. Vegetos 32:571–582. https://doi.org/10.1007/s42535-019-00064-x
Pieterse CMJ, Van Loon LC (2004) NPR1: the spider in the web of induced resistance signaling pathways. Curr Opin Plant Biol 7:456–464
Pons C, Voß A, Schweiger R, Müller C (2020) Effects of drought and mycorrhiza on wheat and aphid infestation. Ecol Evol 10:10481–10491. https://doi.org/10.1002/ece3.6703
Potshangbam M, Devi SI, Sahoo D, Strobel GA (2017) Functional characterization of endophytic fungal community associated with Oryza sativa L. and Zea mays L. Front Microbiol 8:1–15. https://doi.org/10.3389/fmicb.2017.00325
Rajamanikyam M, Vadlapudi V, Amanchy R, Upadhyayula SM (2017) Endophytic fungi as novel resources of natural therapeutics. Brazilian Arch Biol Technol 60:451–454. https://doi.org/10.1590/1678-4324-2017160542
Rajesh RW, Rahul MS, Ambalal NS (2016) Trichoderma: a significant fungus for agriculture and environment. Afr J Agric Res 11:1952–1965. https://doi.org/10.5897/AJAR2015.10584
Rajulu GMB, Thirunavukkarasu N, Suryanarayanan TS, Ravishankar JP, El Gueddari NE, Moerschbacher BM (2011) Chitinolytic enzymes from endophytic fungi. Fungal Divers 47:43–53. https://doi.org/10.1007/s13225-010-0071-z
Ram RM, Keswani C, Bisen K, Tripathi R, Singh SP, Singh HB (2018) Biocontrol technology. Engineering. In: Omics Technologies and Bio-Elsevier, pp 177–190
Reimer-Michalski E, Conrath U (2016) Innate immune memory in plants. Semin Immunol 28:319–327. https://doi.org/10.1016/j.smim.2016.05.006
Ren XM, Guo SJ, Tian W, Chen Y, Han H, Chen E, Li BL, Li YY, Chen ZJ (2019) Effects of plant growth-promoting bacteria (PGPB) inoculation on the growth, antioxidant activity, cu uptake, and bacterial community structure of rape (Brassica napusL.) grown in cu-contaminated agricultural soil. Front Microbiol 10:1–12. https://doi.org/10.3389/fmicb.2019.01455
Romera FJ, García MJ, Lucena C, Martínez-Medina A, Aparicio MA, Ramos J, Alcántara E, Angulo M, Pérez-Vicente R (2019) Induced systemic resistance (ISR) and fe deficiency responses in dicot plants. Front Plant Sci 10:1–17. https://doi.org/10.3389/fpls.2019.00287
Ryan RP, An S, Allan JH, McCarthy Y, Dow JM (2015) The DSF family of cell–cell signals: an expanding class of bacterial virulence regulators. PLoS Pathog 11:e1004986. https://doi.org/10.1371/journal.ppat.1004986
Sahni S, Prasad BD (2020) Management of collar rot disease using vermicompost and a PGPR strain Pseudomonas sp. and their effect on defense-related enzymes in chickpea. Indian Phytopathol. https://doi.org/10.1007/s42360-020-00203-4
Saijo Y, Loo EP, Yasuda S (2018) Pattern recognition receptors and signaling in plant–microbe interactions. Plant J 93:592–613. https://doi.org/10.1111/tpj.13808

Saleem M, Arshad M, Hussain S, Bhatti AS (2007) Perspective of plant growth promoting rhizobacteria (PGPR) containing ACC deaminase in stress agriculture. J Ind Microbiol Biotechnol 34:635–648. https://doi.org/10.1007/s10295-007-0240-6

Samuel O, Bernard O, Olubukola RG, Babalola O (2017) Mechanisms of action of plant growth promoting bacteria. World J Microbiol Biotechnol 33:1–16. https://doi.org/10.1007/s11274-017-2364-9

Sasse J, Martinoia E, Northen T (2018) Feed your friends: do Plant exudates shape the root microbiome? Trends Plant Sci 23:25–41. https://doi.org/10.1016/j.tplants.2017.09.003

Shahabivand S, Maivan HZ, Goltapeh EM, Sharifi M, Aliloo AA (2012) The effects of root endophyte and arbuscular mycorrhizal fungi on growth and cadmium accumulation in wheat under cadmium toxicity. Plant Physiol Biochem 60:53–58. https://doi.org/10.1016/j.plaphy.2012.07.018

Shi S, Nuccio EE, Shi ZJ, He Z, Zhou J, Firestone MK (2016) The interconnected rhizosphere: high network complexity dominates rhizosphere assemblages. Ecol Lett 19:926–936. https://doi.org/10.1111/ele.12630

Singh DP, Patil HJ, Prabha R, Yandigeri MS, Prasad SR (2018) Actinomycetes as potential plant growth-promoting microbial communities. In: Prasad R, Gill SS, Tuteja N (eds) Crop improvement through microbial biotechnology. Elsevier, pp 27–38

Singh LS, Sharma H, Sahoo D (2019a) Actinomycetes: the unparallel potential source of antibiotics and natural products for drug discovery. Acta Sci Microbiol 2:15–17. https://doi.org/10.31080/ASMI.2019.02.0324

Singh LS, Sharma H, Sahoo D (2019b) Actinomycetes from soil of Lachung, a pristine high altitude region of Sikkim Himalaya, their antimicrobial potentiality and production of industrially important enzymes. Adv Microbiol 09:750–773. https://doi.org/10.4236/aim.2019.98046

Smith SE, Jakobsen I, Grønlund M, Smith FA (2011) Roles of arbuscular mycorrhizas in plant phosphorus nutrition: interactions between pathways of phosphorus uptake in arbuscular mycorrhizal roots have important implications for understanding and manipulating plant phosphorus acquisition. Plant Physiol 156:1050–1057. https://doi.org/10.1104/pp.111.174581

Song L, Jiang N, Wei S, Lan Z, Pan L (2020) Isolation, screening, and identification of actinomycetes with antifungal and enzyme activity assays against Colletotrichum dematium of Sarcandra glabra. Mycobiology 48:37–43. https://doi.org/10.1080/12298093.2020.1716604

Spagnoletti FN, Cornero M (2020) Arbuscular mycorrhiza protects soybean plants against Macrophomina phaseolina even under nitrogen fertilization. Eur J Plant Pathol 156:839–849. https://doi.org/10.1007/s10658-020-01934-w)

Sudheep NM, Marwal A, Lakra N, Anwar K, Mahmood S (2017) Plant-microbe interactions in agro-ecological perspectives. Springer, Singapore

Sujarit K, Pathom-aree W, Mori M, Dobashi K, Shiomi K, Lumyong S (2020) Streptomyces palmae CMU-AB204T, an antifungal producing-actinomycete, as a potential biocontrol agent to protect palm oil producing trees from basal stem rot disease fungus, Ganoderma boninense. Biol Control 148:1–12. https://doi.org/10.1016/j.biocontrol.2020.104307

Sunar K, Chakraborty U, Chakraborty BN (2014) Exploitation of native microflora of Darjeeling hills for sclerotial rot management and growth promotion in pulses. Indian Phytopathol 67:59–69

Sunar K, Dey P, Chakraborty U, Chakraborty B (2015) Biocontrol efficacy and plant growth promoting activity of Bacillus altitudinis isolated from Darjeeling hills, India. J Basic Microbiol 55:91–104. https://doi.org/10.1002/jobm.201300227

Sunar K, Chakraborty U, Chakraborty BN (2017) Influence of indigenous Bacilli isolated from Darjeeling hills on phosphate mobilization and induction of resistance against sclerotial blight disease of Tea cultivars. Int J Basic Appl Biol 4:128–134

Sunar K, Chakraborty U, Chakraborty BN (2020) Bacillus altitudinis and Bacilllus pumilus induce resistance in Brassica juncea L. against Thanatephorus cucumeris causing Seedling Blight. J Mycopathol Res 58:167–179

Tayung K, Barik B (2011) Endophytic fungi as potential sources of bioactive natural products: prospects and challenges. Annu Rev Plant Pathol 6:299–334

Thomashow LS (1996) Biological control of plant root pathogens. Curr Opin Biotechnol 7:343–347. https://doi.org/10.1016/S0958-1669(96)80042-5

Timmusk S, Behers L, Muthoni J, Muraya A, Aronsson A-C (2017) Perspectives and challenges of microbial application for crop improvement. Front Plant Sci 8:1–10. https://doi.org/10.3389/fpls.2017.00049

Tückmantel T, Leuschner C, Preusser S, Kandeler E, Angst G, Mueller CW, Meier IC (2017) Root exudation patterns in a beech forest: dependence on soil depth, root morphology, and environment. Soil Biol Biochem 107:188–197. https://doi.org/10.1016/j.soilbio.2017.01.006

Vallad GE, Goodman R (2004) Systemic acquired resistance and induced systemic resistance in conventional agriculture. Crop Sci 44:1920–1934. https://doi.org/10.2135/cropsci2004.1920

Van Wees SCM, Van der Ent S, Pieterse CMJ (2008) Plant immune responses triggered by beneficial microbes. Curr Opin Plant Biol 11:443–448

Velásquez A, Vega-Celedón P, Fiaschi G, Agnolucci M, Avio L, Giovannetti M, D'Onofrio C, Seeger M (2020) Responses of Vitis vinifera cv. Cabernet Sauvignon roots to the arbuscular mycorrhizal fungus Funneliformis mosseae and the plant growth-promoting rhizobacterium Ensifer meliloti include changes in volatile organic compounds. Mycorrhiza 30:161–170. https://doi.org/10.1007/s00572-020-00933-3

Venturi V, Fuqua C (2013) Chemical Signaling between plants and plant-pathogenic bacteria. Annu Rev Phytopathol 51:17–37. https://doi.org/10.1146/annurev-phyto-082712-102239

Venturi V, Keel C (2016) Signaling in the rhizosphere. Trends Plant Sci 21:187–198. https://doi.org/10.1016/j.tplants.2016.01.005

Vessey JK (2003) Plant growth promoting rhizobacteria as biofertilizers. Plant Soil 255:571–586. https://doi.org/10.1023/A:1026037216893

Vos CMF, De Cremer K, Cammue BPA, De Coninck B (2015) The toolbox of Trichoderma spp. in the biocontrol of Botrytis cinerea disease. Mol Plant Pathol 16:400–412. https://doi.org/10.1111/mpp.12189

Wang H, Hyde KD, Soytong K, Lin F (2008) Fungal diversity on fallen leaves of Ficus in northern Thailand. J Zhejiang Univ Sci B 9:835–841. https://doi.org/10.1631/jzus.B0860005

Wang M, Xue J, Ma J, Feng X, Ying H, Xu H (2020) Streptomyces lydicus M01 regulates soil microbial community and alleviates foliar disease caused by Alternaria alternata on cucumbers. Front Microbiol 11:1–13. https://doi.org/10.3389/fmicb.2020.00942

Waqas M, Khan AL, Kang S-M, Kim Y-H, Lee I-J (2014) Phytohormone-producing fungal endophytes and hardwood-derived biochar interact to ameliorate heavy metal stress in soybeans. Biol Fertil Soils 50:1155–1167. https://doi.org/10.1007/s00374-014-0937-4

Whitelaw MA (2000) Growth promotion of plants inoculated with phosphate-solubilizing fungi. In: Advances in agronomy. Academic Press, pp 99–151

Xie L, Lehvävirta S, Timonen S, Kasurinen J, Niemikapee J, Valkonen JPT (2018) Species-specific synergistic effects of two plant growth—promoting microbes on green roof plant biomass and photosynthetic efficiency. PLoS One 13:1–16. https://doi.org/10.1371/journal.pone.0209432

Xu Z, Zhang R, Wang D, Qiu M, Feng H, Zhang N, Shen Q (2014) Enhanced control of cucumber wilt disease by Bacillus amyloliquefaciens SQR9 by altering the regulation of its DegU phosphorylation. Appl Environ Microbiol 80:2941–2950. https://doi.org/10.1128/AEM.03943-13

Yamaji K, Watanabe Y, Masuya H, Shigeto A, Yui H, Haruma T (2016) Root fungal endophytes enhance heavy-metal stress tolerance of Clethra barbinervis growing naturally at mining sites via growth enhancement, promotion of nutrient uptake and decrease of heavy-metal concentration. PLoS One 11:1–15. https://doi.org/10.1371/journal.pone.0169089

Yanni YG, Dazzo FB (2010) Enhancement of rice production using endophytic strains of Rhizobium leguminosarum bv. Trifolii in extensive field inoculation trials within the Egypt Nile delta. Plant Soil 336:129–142. https://doi.org/10.1007/s11104-010-0454-7

Yimer D (2019) Components, mechanisms of action, success under greenhouse and field condition, market availability, formulation and inoculants development on biofertilizer. Biomed J Sci Technol Res 12:9366–9372. https://doi.org/10.26717/BJSTR.2019.12.002279

Yoo S-J, Sang MK (2017) Induced systemic tolerance to multiple stresses including biotic and abiotic factors by Rhizobacteria. Res Plant Dis 23:99–113. https://doi.org/10.5423/RPD.2017.23.2.99

Zak DR, Pregitzer KS, Burton AJ, Edwards HK (2011) Microbial responses to a changing environment: implications for the future functioning of terrestrial ecosystems. Fung Ecol 4: 386-395. https://doi.org/10.1016/j.funeco.2011.04.001

Zeriouh H, de Vicente A, Pérez-García A, Romero D (2014) Surfactin triggers biofilm formation of Bacillus subtilis in melon phylloplane and contributes to the biocontrol activity. Environ Microbiol 16:2196–2211. https://doi.org/10.1111/1462-2920.12271

Zhang S, Gan Y, Xu B (2016) Application of plant-growth-promoting fungi Trichoderma longibrachiatum T6 enhances tolerance of wheat to salt stress through improvement of antioxidative defense system and gene expression. Front Plant Sci 7:1–11. https://doi.org/10.3389/fpls.2016.01405

Zhang S, Guo X, Yun W, Xia Y, You Z, Rillig MC (2020a) Arbuscular mycorrhiza contributes to the control of phosphorus loss in paddy fields. Plant Soil 447:623–636. https://doi.org/10.1007/s11104-019-04394-2

Zhang SW, Ming Wu Z, Yang Y, Tai Li K (2020b) Antifungal action of Antifungalmycin N2 against Rhizoctonia solani by disrupting cell membrane and inhibiting succinate dehydrogenase. Curr Microbiol 77:254–260. https://doi.org/10.1007/s00284-019-01837-y

Zhang Y, Kang X, Liu H, Liu Y, Li Y, Yu X, Zhao K, Gu Y, Xu K, Chen C, Chen Q (2018) Endophytes isolated from ginger rhizome exhibit growth promoting potential for Zea mays. Arch Agron Soil Sci 64:1302–1314. https://doi.org/10.1080/03650340.2018.1430892

Zheng YK, Miao CP, Chen HH, Huang FF, Xia YM, Chen YW, Zhao LX (2017) Endophytic fungi harbored in Panax notoginseng: diversity and potential as biological control agents against host plant pathogens of root-rot disease. J Ginseng Res 41:353–360. https://doi.org/10.1016/j.jgr.2016.07.005

Zhou J, Deng Y, Luo F, He Z, Tu Q, Zhi X (2010) Functional molecular ecological networks. MBio 1:1–10. https://doi.org/10.1128/mBio.00169-10

Microbial Diversity of Different Agroecosystems: Current Research and Future Challenges

3

Keshani, Shivani Sharma, and Jasmine Kaur

Abstract

The presence of different environmental conditions on Earth affects the properties of soil which thereby affect the agricultural systems. The soils vary from place to place like mountain soil that is completely different from desert soil. These differences determine the diversity and community structure of native microorganisms, which indirectly affects the agriculture of that place. Microorganisms are an important component of the ecosystem and play a vital role in processes like nutrient cycling, soil formation by weathering of rocks, and waste recycling. The changes in climate due to extreme weather conditions like drought, heavy rainfall, frost, etc. are adversely affecting the soil properties. These changes also affect the species distribution, soil microbial community composition, and interactions. Also, this leads to changes in the interspecific relationship between competitors, host/parasite, and predator/prey which changes community structure and ecosystem functions. In this chapter, we will discuss the distribution of different microorganisms among different agroecosystems and the role of these microorganisms in maintaining the ecosystem processes.

Keywords

Agroecosystem · Soil biodiversity · Sustainable agriculture · Nutrient cycling

Keshani (✉) · S. Sharma · J. Kaur
Department of Microbiology, College of Basic Sciences and Humanities, Punjab Agricultural University, Ludhiana, India
e-mail: keshani@pau.edu

R. Soni et al. (eds.), *Microbiological Activity for Soil and Plant Health Management*,
https://doi.org/10.1007/978-981-16-2922-8_3

3.1 Introduction

Microorganisms represent an enormous component of Earth's biodiversity and play a vital role in sustaining all life forms due to their role in ecosystem processes (Falkowski et al. 2008; Prosser 2012). The relationship between microbial diversity and ecosystem functioning is an important factor for determining the ecosystem responses to a changing environment (Logue et al. 2015). Soil microbes constitute an important part of the microbial world and play a major role in the functioning of terrestrial ecosystems by catalyzing various conversions in the biogeochemical cycles (Whitman et al. 1998; van der Heijden et al. 2008).

The presence of diverse environmental stresses on earth like frost, heat waves, droughts, floods, etc. causes difficulties in the agriculture sector. Microbes can adapt and survive in extreme conditions varying from low water availability, high solar radiations, cold temperatures, and different pH and pressures (Azua Bustos and Gonzalez Silva 2014). Microorganisms being ubiquitous help to play a major role in overcoming these abiotic stresses and provide an increase in agricultural productivity. These microorganisms may help plants to overcome stress conditions like drought, salinity, heavy metals, etc. (Grover et al. 2011). Microorganisms have the ability to colonize the rhizosphere and phyllosphere and can also live inside plant tissues as endophytes. The plant–microbe interaction helps in nutrient acquisition, plant growth promotion under extreme environments, and elicits plant defense reactions against pests (Verma et al. 2017). The recent discovery of microorganisms from some extreme environments like glacial ice and their effectiveness in promoting plant growth has promoted the use of these microbes as biofertilizers (Yarzabal 2020).

The soil biodiversity is a significant asset that provides ecosystem processes essential to the functioning of natural and global systems. They are the biggest resource of the nature which forms the basis of important ecological processes like nutrient cycling, food chains and help to maintain relationships between different organisms (Onen et al. 2020).

3.2 Microbial Diversity of Agroecosystems

The microbial communities are being characterized through a wide variety of molecular tools by observing the differences in organismal assemblages or community characteristics (Gibbons and Gilbert 2015). Surveys at the global level for determining microbial diversity have been carried out within the study and have revealed prominent physicochemical drivers of microbial community structure (Ghiglione et al. 2012). The Earth Microbiome Project (http://www.earthmicrobiome.org/) carried out the biggest global survey for analyzing microbial diversity. Its results revealed the presence of 5.6 million OTUs in the first 15,000 samples, having a similarity of 97% at the V4 region of the 16S rRNA gene and without the singleton OTUs. This led to the setting of a new lower bound for the number of bacteria and archaeal phylotypes on Earth (Rideout et al. 2014).

Furthermore, with growth in sampling efforts, the tally of novel phylotypes discovered resumes to increase from the previously estimated numbers (Scheffers et al. 2012).

Microorganisms show varying metabolisms which allow them to use various energy sources ranging from sunlight to organic carbon and inorganic minerals. Due to the vast differences in their metabolism, microbes can survive in extreme conditions ranging from acid mines to thermal hot springs. Besides, their ability to survive on numerous metabolites like complex carbohydrates, peptides, antibiotics, and lipids grants them the ability to cross-feed one another, establish highly specific niches and distinct life history policies (Giovannoni et al. 2014; Morris et al. 2014). Furthermore, the diversity on ecological timescales can be affected by the continued evolution and speciation in microorganisms (Morris et al. 2014; Raeside et al. 2014). The soil microbial diversity has been shown to be improved by numerous agricultural activities. Practices like biological farming (Birkhofer et al. 2008; Verbruggen et al. 2010), tree-based intercropping (Lacombe et al. 2009; Bainard et al. 2012), crop rotation (Altieri 1999; Cavagnaro and Martin 2011), and extensification of land use (Postma-Blaauw et al. 2010; De Vries et al. 2012) have a positive effect on number and richness of specific groups of soil organisms including arbuscular mycorrhizal fungi and earthworms and the microbial diversity of soil. The environmental factors play a major role in governing the bacterial diversity in different agroecosystem and the bacterial diversity in turn play a role in maintaining the microbial interactions (Tan et al. 2020).

3.2.1 Temporal and Spatial Distribution

The diversity of microbes increases with the temporal and spatial distribution of bacteria. Soil microbes have greater diversity compared to other species. The functioning of microbial processes is determined by the microbial niches. The small size of microbes makes it difficult to evaluate them and due to their small size, their movement through small distances with the help of water and wind makes a major change. In addition, the soil texture and pore space influence community composition. The microenvironment of soil can vary from the adjacent soil due to differences in pore space and this leads to the creation of microenvironment-specific communities. This increases the overall diversity of soil. High microbial activity is observed in the macropores (Lee and Foster 1991), due to considerable root and micro- and macro-faunal activities occurring in these regions. The smaller pores and water filled microsites might increase the anaerobic activity, creating variability in the nutrient cycling processes (Tiedje et al. 1984). The diversity and distribution of microbes are also affected by the position of the landscape (Turco and Bezdicek 1987). At the greater depths in the soil, many diverse and metabolically active populations are found. However, the bacterial population declines with depth in the soil, due to lack of carbon which limits their proliferation (Fredrickson et al. 1991).

The density and functioning of microbes in the soil are affected by the minerals and organic matter content in the soil. Microbes play a major role in the functioning of plants by affecting their physiology and development. The number of microorganisms populating a plant can reach numbers much higher than the number of plant cells (Mendes et al. 2013). The narrow zone surrounding the roots, known as the rhizosphere is considered as a complex ecosystem as it is a hotspot for many microorganisms (Raaijmakers et al. 2009). The population of microbes in the rhizosphere region will depend on microbial responses to release exudates and substrates. Further, the diversity of the bacterial community can be affected by the plant community due to variability in the chemical composition of the exudates (Christensen 1989). The numerous rhizospheric organisms which have positive effects on plant include mycorrhizal fungi, nitrogen fixing bacteria, mycoparasitic fungi, biocontrol microorganisms, plant growth promoting rhizobacteria (PGPR), and protozoa. Some organisms present in the rhizosphere that are detrimental to plant health are oomycetes, pathogenic fungi, bacteria, and nematodes. Apart from these, some human pathogens can also be found in the rhizosphere (Mendes et al. 2013). The culture-independent approaches used for determining rhizosphere microbiomes and soil diversity have shown that the knowledge about microbial diversity is highly underestimated. A significant proportion of bacterial phyla recognized by the next-generation sequencing technologies have shown to have no cultured representative, indicating that only a small proportion of bacteria have been cultured with the help of current technologies.

The decomposition of plant remnants in the soil provides sites for microbial growth. These areas are favorable for fungal and bacterial growth and proliferation. Similar to the rhizosphere, they allow microbial proliferation and provide heterogeneity to the soil system. The chemical composition of these plant materials is responsible for forming a particular community structure (Wicklow et al. 1974; Kjoller and Struwe 1982). The bacterial diversity is also affected by the above ground plant diversity. Reduction in plant community due to overgrazing, pollutants, and tillage also decreases the bacterial population (Christensen 1989). Furthermore, the diversity of microbial communities is influenced by presence of aerobic and anaerobic conditions in the microsites (Lynch and Harper 1985; Aulakh et al. 1991).

The composition of the bacterial population affects the decomposition rate and nutrient cycling in conventionally tilled and untilled systems (Beare et al. 1993). The composition of some microbial groups and the bacterial population was altered with cultivation practices. In research from the prairie and cultivated soils, it was found that the diversity of microbes was higher in cultivated or disturbed soils compared to a prairie grassland (Kennedy and Smith 1995). Therefore, the diversity indices of a particular soil can be used to indicate the result of disturbance. Greater diversity does not mean a more stable system; rather, shifts in diversity may be more informative about the status of a soil bacterial population with management. Crop rotation is a critical component for maintaining sustainable systems; it breaks the cycle of pathogens, decreases weeds, and increases favorable species and interactions. Crop growth is also positively benefitted from crop rotation due to changes in bacterial community composition. The use of legumes in crop rotation helps provide fixed nitrogen to the plant, decreases pathogens, and helps in maintaining the water status of the system (Johnson et al. 1992).

3.2.2 Diversity in Different Agroecosystems

Microorganisms form the foundation of life on Earth. They have developed in every plausible niche on the planet. They redesign their environments (Falkowski et al. 2008) including oceans and atmospheres. It involves changes in oceanic weather patterns (Lovelock and Margulis 1974), the health of multi-cellular hosts (Turnbaugh et al. 2007), and oxidation of Earth's atmosphere.

Microbes are able to share genetic information easily, which causes a very rapid and ongoing diversification of species in natural habitat, making the genetic diversity of the entire soil microbiome to be considered huge (Monier et al. 2011). The microbial communities interact with each other to accomplish tasks like decomposition of plant waste, creation of food web structures, and processing of nutrient cycles. These processes are not a result of single organism (Aneja et al. 2006) and there are varied microbial species that exist in different agroecosystems (Table 3.1). Additionally, processes like legume rhizobia interactions are not accomplished by a

Table 3.1 Micro-biota found in different agroecosystems

S. no.	Agroecosystem	Microorganisms found
1.	Jhum agroecosystem	*Curtobacterium oceanosedimentum, Bacillus methylotrophicus, B. cereus, B. subtilis, B. licheniformis, B. clausii, B. megaterium, Pseudomonas stutzeri, P. aeruginosa, Trichoderma, Penicillium, Paecilomyces, Aspergillus, Cladosporium, Coniella, and Rhizoctonia*
2.	Sundarbans agroecosystem	*Streptomyces sundarbansensis* sp. nov., *Streptomyces euryhalinus* sp. nov., *Oxynema aestuarii* sp. nov
3.	North Western Himalayan agroecosystem	*Exiguobacterium acetylicum* strain 1P, *Pseudomonas* sp. strain PGERs 17, and bacteria belonging to group *Bacteroidetes, Acidobacteria, Proteobacteria, Firmicutes,* and *Actinobacteria*
4.	Thar agroecosystem	*Bacillus* spp., *Corynebacterium* spp., *Aeromonas* spp., *Acinetobacter* spp., *Staphylococcus* spp., *Ensifer, Bradyrhizobium, Phormidium, Oscillatoria, Lyngbya, Anabaena, Nostoc, Scytonema, and Calothrix*
5.	Coffee shade tree agroecosystem	*Glomus, Scutellospora, Gigaspora,* and *Acaulospora*
6.	Apatani wet rice agroecosystem	*Penicillium brevicompactum, Trichoderma* spp., *Aspergillus* spp., and phytoplanktons belonging to *Bacillariophyceae, Chrysophyceae, Chlorophyceae, Cyanophyceae, Dinophyceae, Euglenophyceae, Xanthophyceae,* and *Zygnemophyceae*
7.	Leh Ladakh agroecosystem	*Phormidium, Leptolyngbya, Microcoleus vaginatus,* Lyngbya, and microalgae belonging to Chlorophyceae and Xanthophyceae

Fig. 3.1 Agriculture system in different geographic terrains: (**a**) Step farming in hilly areas (**b**) Plain fields in flat landforms

single organism and involve the number of helper organisms for the creation of the infection phase (Frey-Klett et al. 2007). Figure 3.1 gives interpretation of how agroecosystem in different landforms exists, e.g. step farming in hilly areas and plain fields in flat terrain.

3.2.2.1 *Jhum* Agroecosystem

The northeast region of India is a diverse region in respect to the different indigenous practices and biological diversity still present there. It is home to diverse cultures and has fluctuating altitudes, which serve as an opportunity for the searching of biotechnologically useful microorganisms. The vast majority of microbes here is unknown and needs to be explored. The agricultural practice followed here commonly is "Jhum" farming which is a shifting type of cultivation. This type of cultivation greatly depends on the duration for which the land is cultivated and the duration during which it is fallowed. This impacts the overall nutrient loss and replenishment over time affecting the produce. Studies are been carried out to explore the microbial diversity of these shifting cultivated lands to excel in discovering microbes in different habitats. Due to the continuous tilt and shift in nutrients of the jhum fields, competitive conditions arise for the microorganisms in the soil (Joshi et al. 2019). The microbial diversity explored from the different fields of Jhum agroecosystem in North East India includes bacteria like *Curtobacterium oceanosedimentum, Bacillus methylotrophicus, B. cereus* (Banerjee et al. 2017), *B. subtilis, B. licheniformis, B. clausii, B. megaterium, Pseudomonas stutzeri,* and *P. aeruginosa* (Pandey et al. 2011). Also, fungi belonging to genera *Trichoderma, Penicillium, Paecilomyces, Aspergillus, Cladosporium, Coniella*, and *Rhizoctonia* have been isolated from jhum fields (Jain et al. 2016).

3.2.2.2 Microbial Diversity in Sundarbans

Sundarbans is a tidal mangrove forest which lies in the Bay of Bengal. It has enormous biodiversity and has a dynamic ecosystem. The rivers provide an inflow of waste materials and debris to this region which serves as a source of organic material for the microorganisms. Mangroves are a noticeable source of microbial diversity and are been explored for bioactive compounds for applications in pharmaceutical industries (Biswas and Mukherjee 2019). Two unique microbial species, *Streptomyces sundarbansensis* sp. nov. (Arumugam et al. 2011) and *Streptomyces euryhalinus* sp. nov. (Biswas et al. 2017) have been identified from this region. Also, from the many cyanobacteria isolated, *Oxynema aestuarii* sp. nov. was reported as a novel species. In addition, the bacteria isolated from Sundarbans provide industrially valuable enzymes like esterase, protease, and ribonuclease (Chakraborty et al. 2018). The unculturable bacteria and archaea have been discovered with the help of the metagenomic approach and bioinformatics tools (Basak et al. 2015). It has been reported that *Proteobacteria* form the dominant phyla in the mangrove forest, whereas *Firmicutes, Chloroflexi, Planctomycetes,* and *Actinobacteria* are found occasionally.

3.2.2.3 Agroecosystem of North Western Himalayas

The North Western Himalayas has diverse ecosystems including dense forests, mountain slopes, alpines, and agricultural lands. The North Western Himalayan region is a hotspot region for microbial diversity and different approaches have been utilized to explore the biodiversity of bacteria and fungi. Bacteria belonging to phyla *Bacteroidetes* and *Firmicutes* are dominant in low altitudes, whereas *Actinobacteria, Acidobacteria,* and *Proteobacteria* are dominant in high altitudes of North Western Himalayas. The hot springs are populated by *Firmicutes* and *Proteobacteria,* whereas alpine are populated by *Acidobacteria* and *Actinobacteria.* As microbes are under constant stresses of moisture and soil conditions in the mountain agroecosystems, a sustainable approach is required for conserving microbes for long term agricultural sustainability (Jaggi et al. 2020).

The microbial diversity of forest agroecosystem of North Western Himalayas is of particular interest due to the presence of adaptive enzymes in these microbes which enable them to inhabit the extreme temperature conditions. Although it has vast microbial diversity and richness, only a little has been explored (Pandey et al. 2006). Several new cold tolerant bacterial species have been isolated from the abundant biodiversity of NWH (Shivaji et al. 2005). A bacterial species *Exiguobacterium acetylicum* strain 1P was isolated from the rhizospheric soil of NWH at an altitude of 2200 m. This bacterium is shown to produce volatile compounds that act as a plant growth promoter as well as inhibit some phytopathogenic fungi (Selvakumar et al. 2011). *Pseudomonas* sp. strain PGERs 17 (MTCC 9000) isolated from sub-alpine region of NWH from garlic roots can also serve as an important inoculant for growing winter season crops in NWH (Mishra et al. 2008). The most dominant bacteria belonged to the group *Proteobacteria* followed by *Firmicutes* and *Actinobacteria* (Yadav et al. 2015).

3.2.2.4 Thar Agroecosystem

The deserts occupy one-seventh of the land area out of the different ecosystems. The lack of moisture makes the deserts dependent on the balance between rainfall and evaporation (Sharma and Mehra 2009). Although the environment is harsh in Thar Desert, the soils are rich in bacterial diversity. The different soil samples evaluated from sand dunes of Thar Desert revealed the presence of osmotolerant bacteria. Bacteria belonging to *Bacillus* spp., *Corynebacterium* spp., *Aeromonas* spp., *Acinetobacter* spp., and *Staphylococcus* spp. were isolated (Sharma et al. 2013).

The Thar Desert of India is also rich in rhizobium bacteria. The native leguminous plants of Thar Desert are nodulated with rhizobia bacteria in their roots. The traditional genera of rhizobia, i.e. *Ensifer* and *Bradyrhizobium* are found in the alkaline soils of Thar Desert. The strains of *Bradyrhizobium yuanmingenese* diversified to produce distinct strains which are microsymbionts of *Alysicarpus* spp., *Chamaecrista* spp., *Tephrosia* spp., *Crotalaria* spp., and *Ensifer* strains. In addition, wild legumes of the Thar agroecosystem also harbor *Tephrosia* and *Chamaecrista rhizobia* species. These bacteria can be used as a potential source for preparing consortia for agricultural fields (Tak and Gehlot 2019).

The Thar Desert also fosters diverse algal-cyanobacterial species. The soil samples of soils showed the presence of 79 morphotypes of cyanobacteria belonging to 21 genera and four morphotypes belonging to three genera of green algae. The unbranched cyanobacteria belonging to genera *Phormidium, Oscillatoria,* and *Lyngbya* were dominant. Diazotrophs like *Anabaena, Nostoc, Scytonema* and *Calothrix* were major contributors to tomorphotypic diversity. The Shannon diversity indices ranged from 0 to 0.6599 for algal and cyanobacteria morphotypes and 0–0.74 for diazotrophic cyanobacteria in the presence of nitrogen fortified medium (Bhatnagar et al. 2008).

3.2.2.5 Coffee Shade Tree Agroecosystem

The traditional method of growing coffee, i.e. growing coffee plants under mild shade of trees is a sustainable method of agriculture and helps increase the soil microbial diversity compared to non-shaded coffee plantations.

The soil microbes play a vital role in various processes like nutrient cycling and nitrogen fixation. A study was conducted by Evizal et al. (2012), to determine the bacterial diversity with respect to different agroecosystems of coffee plantations. The shade coffee agroecosystems were compared with the microbial diversity of unshaded coffee plantation. The results showed a higher bacterial diversity from soils of shaded agroecosystem compared to the unshaded system during a span of 15 years. A shaded coffee agroecosystem is even better at conserving the bacterial diversity compared to a no-shade coffee plantation.

The type of trees used for shade also affects the microbial community of the soil. The use of legume trees is shown to positively affect the diversity of nitrogen fixing bacteria in the soil. The diversity of legume nodulating bacteria is higher in soils shaded by legume trees in shade coffee agroecosystems than those having other trees for shade based on Shannon diversity indices (Evizal et al. 2013).

The roots of coffee plants are colonized by arbuscular mycorrhizal (AM) fungi. The different fungi found in soils of shade coffee agroecosystem include genera *Glomus, Scutellospora, Gigaspora,* and *Acaulospora* (Sewnet and Tuju 2013).

3.2.2.6 Apatani Wet Rice Agroecosystem

The Apatanis (community of Arunachal Pradesh) practice an advanced version of paddy cultivation. This practice involves paddy cum fish cultivation in both low and high elevations of North Eastern Himalayas. It is a highly efficient and economic agroecosystem which provides continuous yield year after year. The fields of paddy are kept submerged underwater during monsoon and fish are allowed to swim in these shallow waters. The fish get nutrients from the manure of the fields. Small pits are created on the terrace so the fish can stay in pits having water during a shortage of water supplies. Thus this system helps in the sustainable use of land for paddy and fish rearing (Rai 2005).

The various species of microbes found in the soils of paddy cum fish cultivation included *Penicillium brevicompactum, Trichoderma* spp., *Aspergillus* spp. (Majumder and Shukla 2012). A wetland rice fish agroecosystem in North East India revealed a diverse population of plankton species. Fifty-seven genera of phytoplankton from eight classes, i.e. *Bacillariophyceae, Chrysophyceae, Chlorophyceae, Cyanophyceae, Dinophyceae, Euglenophyceae, Xanthophyceae,* and *Zygnemophyceae* were recorded. Along with this, zooplankton species were also noticed (Das et al. 2014).

3.2.2.7 Agroecosystem of Leh Ladakh

Traditional agricultural practices are followed in Ladakh. It involves processes like manuring, plowing, sowing, watering, harvesting, etc. Organic manure prepared by a mixture of human and animal excreta is used. This manure serves as a source of nutrients like nitrogen, potassium, and phosphorus and helps maintain soil functioning (Pelliciardi 2012).

The analysis of soil microorganism from the cold deserts of Leh revealed the presence of cyanobacterial species. Cyanobacteria belonging to orders Chlorococcales, Nostocales, and Oscillatoriales were observed. The orders Oscillatoriales and Nostocales were more dominant. Species like *Phormidium, Leptolyngbya, Microcoleus vaginatus,* and *Lyngbya* like microorganisms were observed in order Oscillatoriales. The order Nostocales had two distinct taxonomic groups, *Nostoc* spp and other Nostocales. Apart from these, phototrophic microorganisms were found in the samples from mountain sites of Ladakh. The phototrophs included 14 morphotypes of cyanobacteria, ten green algae morphotypes, and four morphotypes of diatoms. Microalgae belonging to Chlorophyceae, Xanthophyceae, and diatoms formed a small proportion of phototrophic microbes (Rehakova et al. 2011).

3.2.2.8 Effect of Changing Environment on Microbial Diversity

Soil warming has been shown to alter the diversity and function of microbial communities (Mackelprang et al. 2015; Xiong et al. 2014). Soil carbon is being lost as methane and carbon dioxide from thawed permafrost soils due to microbial activity (McCalley et al. 2014).

3.3 Role of Microorganisms in Ecosystem Functioning

There are a number of mechanisms by which microbial diversity can support agroecosystem functioning and particular ecosystem functions such as plant productivity and decomposition. For instance, microbes can form "consortia" that enhance plant productivity (e.g. when different microbes provide different limiting resources to plants) or decomposition (e.g. when plant material is decomposed by specialized microbes with unique physiological properties that succeed each other). As a consequence, microbial diversity can promote ecosystem functioning. However, in other cases, the presence of keystone species (e.g. specific pathogens, nitrogen fixers) rather than diversity itself may determine agroecosystem functioning.

3.3.1 Nutrient Cycling

Microorganisms play a key role in the biogeochemical cycle of soil nutrients. They are essential for cycling of nitrogen, sulfur, phosphorus, and carbon on a global scale.

N_2 Microbes are an essential component of the nitrogen cycle because they carry out processes like nitrogen fixation, nitrification, anammox, and denitrification. The inability of plants to use atmospheric N_2 makes agriculture sometimes limited to N_2 availability. Microorganisms with the ability to reduce nitrogen to ammonia (plant accessible form) play an important role in making nitrogen available to the plants, The presence of the nitrogenase enzyme in these microorganisms makes them responsible for this conversion (Olivares et al. 2013; De Bruijn 2015). There are commonly three different forms of N_2 fixers which are free living, symbiotic, and associative (Olivares et al. 2013). The free living bacteria do not form associations with plants and are present freely in the soils. They use the fixed N_2 for their requirements with low direct transfer to plants. Conversely, symbiotic bacteria form a mutualistic symbiotic relationship with the plant roots. The fixed ammonia is transferred directly to the plant. Rhizobium fixes atmospheric N_2 by forming a mutualistic association with plants like legumes and in some non-legumes like *Parasponia*. The associative bacteria do not form specialized N_2 fixing structures but can invade the intracellular tissues and are present on the root surfaces (Olivares et al. 2013). Bacteria like *Azospirillum* (free living N2 fixer) form diazotrophic rhizocenosis and are closely associated with plant roots than other free living bacteria. The symbiotic association between legume and rhizobia is considered to be the most crucial N_2 fixing system (Azcon-Aguilar and Barea 2015). Nitrification is the conversion of ammonia to nitrite and then to nitrate. The nitrification in soils is an aerobic process and is carried out by some autotrophic bacteria and Crenarchaea. The first step is nitrification, i.e. conversion of ammonia to nitrite through biological oxidation by *Nitrosomonas* and *Nitrospira* and the conversion of nitrite to nitrate by *Nitrobacter* and *Nitrospira.* The next process is denitrification and consists of converting nitrate, nitrite, and nitric oxide to nitrous oxide or nitrogen gas. It is also a microbial process generally occurring in anaerobic, waterlogged soil. The denitrification is the last step that completes the nitrogen cycle, returning fixed nitrogen to the atmosphere (Aislabie and Deslippe 2013).

Phosphorus The phosphorus is made available to the plants by the microorganisms either from inorganic sources by solubilization or from organic source by mineralization. The inorganic phosphate is present as calcium, aluminum, or iron salts in the soil. Some species of bacteria and fungi belonging to genera *Bacillus, Rhizobium, Bradyrhizobium, Enterobacter, Erwinia, Pseudomonas; Aspergillus, Penicillium,* and *Trichoderma* possess the ability to solubilize phosphates (Marschner 2008). Mineralization of organic P which is the process of hydrolyzing organic P substrates results in the release of orthophosphate to the soil (Richardson et al. 2009). Bacteria and fungi responsible for P mineralization are *Bacillus* and *Pseudomonas* and *Aspergillus* and *Penicillium,* respectively (Marschner 2008). Mineralization is achieved by the various types of phosphatase enzymes produced by the microorganisms. The enzymes can be specific like phytases or nonspecific like acid and alkaline phosphatases (Jorquera et al. 2008).

Carbon The microbes play an important role in the cycling of carbon. The carbon dioxide is fixed by primary producers and converted to organic material. The primary source of organic material is plant in the terrestrial ecosystem. The carbon fixation is carried out by surface dwelling algae, cyanobacteria, and autotrophic microbes in the soil. The primary production leads to the production of organic materials which accumulate in living organisms and the non-living organic matter is derived from it. The non-living organic material is particularly recycled by the heterotrophic bacteria and fungi. They conclude the carbon cycle by producing carbon dioxide during respiration, thus converting the organic matter produced by primary producers. Fungi form the major component of soil biomass and carry out a large part of organic matter decomposition. Bacteria belonging to the genera Actinobacteria and Proteobacteria carry out degradation of organic molecules like amino acids, organic acids, and sugars (Eilers et al. 2010). The recalcitrant carbon compounds including cellulose, chitin, and lignin are degraded by bacteria such as Bacteroidetes (Treseder et al. 2011).

3.3.2 Soil Formation and Weathering

Microbial communities are an essential component in building up soils and other ecosystem processes linked to the terrestrial ecosystem. Soil properties including pH, redox conditions, organic matter content affect the microbial community structure and function. This complex relationship makes it evident that a fertile soil is a result of a hundred years of soil evolution. The microorganisms play an important role in the geological process of rock formation. The bacteria have been reported from structures as old as the Archean period indicating that bacteria help to activate crystallization and sedimentation of minerals and sediments. The paleo-weathering profiles in Fenno-Scandinavian shield (Karelian) dating back to 2.1–2.4 billion years revealed the presence of remains of fossil microorganisms. These included filaments or threads, fossil biofilms, and coccoidal forms. Rocks mostly contained destroyed cocci, shreds of filaments, and dumbbell like cells. The chemical composition of microfossils was similar to the rock matrix as well as the rock forming oxides of aluminum, silicon, potassium, iron, and magnesium.

A diverse range of fossil microbes were found in the paleo-weathering profiles of the Karelian Fenno-Scandinavian shield. It is 2.1–2.4 billion years old. The rocks consisted of destroyed cocci cells, dumbbell shaped cells, and filaments and threads of fossil biofilms. Also, the chemical composition of the rock matrix was similar to that of microfossils. It is likely that the microorganisms existing in the rock caused decay and decomposition of minerals present in the rock transforming them to clay like material (Schulz et al. 2013).

3.3.3 Waste Recycling

The microorganisms play an impressive role in the detoxification and biodegradation of dangerous substances. Many genera of bacteria—*Pseudomonas, Mycobacterium,* and *Sphingomonas*—help in oil degradation. The hydrocarbon degrading bacteria are ever-present in soil and need not be introduced from outside during oil spills and generally increase in numbers during oil spills (Aislabie and Deslippe 2013).

In addition, the degradation of pesticides is carried out by some bacteria and fungi. *Arthrobacter nicotinovorans* HIM (bacteria), isolated from New Zealand agricultural field was able to degrade atrazine by using it as the sole source of nitrogen and carbon. Similar triazine compounds like terbuthylazine, simazine, propazine, and cyanazine were also degraded by it (Aislabie et al. 2005). Some pesticides like DDT do not decompose easily and remain in the soil. Under aerobic conditions, DDT is converted to a dead end metabolite known as DDE. One of the microorganisms named *Terrabacter* sp. strain DDE-1 has been reported to metabolize DDE. It was isolated from Winchmore Research Station and degrades DDE when grown on biphenyl as an alternative source for growth (Aislabie et al. 1999).

Environmental pollutants like pentachlorophenol and dioxin are degraded by white rot fungi. The fungus *Phanerochaete chrysosporium* performs this degradation under co-metabolic conditions. This ability has been developed due to their ability to degrade lignin (Barr and Aust 1994).

To reduce the toxic effect of heavy metals, bacteria have evolved mechanisms for detoxification of these metals. The three different pathways followed by bacteria are: (1) metal sequestration in which the metals are bonded to the cell constituents like cell membrane and extracellular polymeric substances (EPS), thereby reducing the free ions concentration in the cytoplasm. Various metals including copper and lead can be sequestered in EPS (Harrison et al. 2007). (2) Reduction reactions—The intracellular ions can undergo reduction to cause detoxification. For example, the enzyme mercury reductase can reduce Hg^{+2} to Hg which further leaves the cell due to its low evaporation point (Nies 1999). (3) Ion extrusion—Involves the efflux of ions from the cell by efflux mechanism. The antiport Czc in *Alcaligenes eutrophus* expels metals like Cd^{+2} and Zn^{+2} outside through the cell membrane (Silver and Phung 1996). These properties of microbes are being exploited for use in the detoxification of heavy metals.

3.4 Effect of Changing Environment on Microbial Diversity

3.4.1 Soil Biodiversity, Resistance, and Resilience

The environmental changes occurring cause alterations in the soil ecosystem affecting their functioning. An in-depth study of soil and microorganism response to environmental changes is essential to help maintains soil functions. Changes in atmospheric conditions cause shifts in the functioning of soil microorganisms which can further make them vulnerable to transitions to different functional and taxonomic states. Press and pulse perturbations which include nutrient enrichment and gradual warming and drought and heat waves, respectively, can coincide in the ecosystem. So, the combined effects of these should be studied for analyzing their effect on soil microbial community resistance and resilience (Bardgett and Caruso 2020). The continuous nutrient enrichment causes a decrease in fungal population with respect to bacteria, which makes microbial population susceptible to transitions to different taxonomic and functional states (Gordon et al. 2008). In addition, the elevated CO_2 levels can also make microbes vulnerable to transitions due to changes in their microbial structure (Zhou et al. 2011).

The effect on microbial resistance and resilience due to soil compaction was studied. The compaction of soil due to logging significantly affects microbial diversity. Different microbes behave differently to compaction and also the effects vary from soil to soil. For example, clayey soils are less resistant and resilient compared to sandy soils. Also, fungi showed less resistance and resilience compared to bacteria. Soils having drastic changes in moisture conditions and air and water conductivities had the greatest effect on microbes. The impact was evident after 6–12 months of compaction. Resilience in microorganisms was observed after 4 years of disturbances only in lightly compacted soils. Anaerobic bacteria along with sulfur, sulfate, and metal reducers were associated with compacted soils. The saprobic and parasitic fungi increased relatively in compacted soil, whereas ectomycorrhizal species were affected deleteriously (Hartmann et al. 2014).

The influence of extrinsic factors on the intrinsic properties of soil microorganisms that make them susceptible to transitions to alternative states is activated by climate extremes. Though a lack of studies in this field persists, the evidence supporting this has emerged. For example, a decrease in soil moisture retention and an increase in carbon mineralization occurred as an effect of repeated droughts in healthy soil. Furthermore, this was followed by loss of soil moisture retention as an effect to extreme drought (Robinson et al. 2016). Continuous dry wet cycles can also cause shifts in the functionality of soils (Robinson et al. 2016) and the soils earlier having grasslands are stable to transitions compared to soils having a history of agricultural land (Todmam et al. 2018). A powerful and enduring shift in the soil microbial diversity and composition along with differences in microbial functioning with respect to the nitrogen cycle was observed as a result of severe drought (De vries et al. 2018). Thus, climate changes can cause shifts in soil properties and the diversity and functioning of soil microorganisms.

3.4.2 Nitrogen Deposition

The inputs of reactive nitrogen (N) due to human activities have increased significantly since the industrial revolution rising by twice in figure (Galloway et al. 2008; Gruber and Galloway 2008). The rise in N deposition helps improve primary productivity by overcoming N shortage (Thomas et al. 2010). However, it may also have negative impacts on the ecosystem (Erisman et al. 2013). The consequences of increasing N deposition on plants and soil nutrient cycle are mostly focused (Simkin et al. 2016; Deng et al. 2017). The changes in soil microorganisms due to fluctuating N availability have been less researched. Due to the major role of these microorganisms in the functioning of ecosystem processes (Brussaard et al. 2007; van der Heijden et al. 2008), they can have serious impacts on carbon and nitrogen cycles and climate changes due to changes in their responses to increasing N deposition.

Wang et al. (2018) conducted a study for observing the effect of N deposition on soil microbial diversity. An overall reduction in microbial diversity was recorded due to increasing N deposition which is related to decreasing microbial biomass. The decline in microbial diversity under N addition was related to the availability of soil nitrogen and soil carbon. The role of soil pH was rather insignificant. Changes in the microbial population were different for different phyla in the soil. A significant reduction in *Acidobacteria* and *Nitrospirae* was seen due to N addition but other bacterial groups were not affected. The fungal groups Ascomycota and Basidiomycota also showed little differences in their population due to N addition.

In the future, increasing N deposition rate may lead to significant reductions in microbial diversity. To minimize the damaging effects of N addition, it is better to study the background soil conditions before N addition. Also, further studies on fungal responses to N addition should be undertaken to provide relevant information.

3.4.3 Elevated Carbon Concentration

The increasing concentration of carbon dioxide is a major reason for climate change. Although elevated CO_2 (eCO_2) levels have positive effects on plant growth and primary productivity, the effects on microbial diversity are not well studied. A study was conducted for 10 years in a grassland experiment wherein atmospheric and elevated carbon concentrations were used to detect changes in soil microbial communities. The species richness was determined by detecting the number of operational taxonomic units (OTUs) with the help of phylogenetic microarrays (PhyloChip). A lower microbial biomass was observed at eCO_2 levels. The PhyloChip was able to detect 2269 OTUs comprising of 45 phyla, 55 classes, 99 orders, 164 families, and 190 subfamilies. The signal intensity of phyla like Crenarchaeota, Chloroflexi, Verrucomicrobia, Proteobacteria, Actinobacteria, Acidobacteria, and Bacteroidetes significantly decreased at eCO_2 concentrations (He et al. 2012).

An experiment to observe the effect of elevated CO_2 on bacterial species was conducted. Soil samples from a grassland experiment were taken which had been exposed to ambient and elevated carbon dioxide levels (600 ppm) for a period of 5 years. The analysis of bacterial community by temporal temperature gradient gel electrophoresis (TTGE) revealed that only little differences were present in the dominant bacterial populations in different soil samples. Further screening of samples by T-RFLP (terminal restriction fragment length polymorphism) also revealed that elevated CO_2 levels had no significant effect on soil bacterial composition and richness (Gruter et al. 2006).

3.5 Mitigation Strategies

3.5.1 Soil Biodiversity and Sustainable Agricultural Practices

Sustainable agriculture is the farming procedure which ensures full use of environmental resources and makes sure that no harm is done to it. The viable use of water and nutrients is the most important factor for increasing agricultural sustainability (Brussaard et al. 2007).

The response of plants to soil water availability and water use was studied in Mediterranean grasslands by Caldeira et al. (2001). The response to monocultures and species rich communities was evaluated on experimental plants by measuring the concentration of leaf N and soil water content. It was observed that species rich soil had more water availability in the top layer of soil where roots were present (Caldeira et al. 2001). This accumulation of water may be linked to the presence of mycorrhizas. Auge (2004) concluded that although the AM plants and non-AM plants had a similar amount of P content and same size, the AM plants often had higher stomatal conductance and greater water loss due to transpiration. Thus, mycorrhizal plants can effectively scavenge water of low activity. The study by van der Heijden et al. (1998) revealed an increase in vegetation biomass with an increase in mycorrhizal diversity. This is particularly due to increased water absorption efficiency. The mycorrhiza may indirectly affect soil structure which affects the soil moisture retention. Mycorrhiza can stabilize soil structure and affect the size and distribution of soil pores. Thus, it can benefit plant growth of non-mycorrhizal plants as well (Auge 2004).

Bender and van der Heijden (2015) studied the impact of soil microbial community on nutrient use efficiency, nutrient losses, and plant growth. The experiment was conducted using outdoor lysimeters, filled with soil. The soils were inoculated with reduced (microbial dominated) and enriched soil-life inoculum, along with arbuscular mycorrhiza fungi. The plants underwent crop rotation and were analyzed for nutrient content, nutrient losses, and plant growth for 2 years. The results showed an increase in N and P uptake by plants with a simultaneous decrease in leaching losses of N. Thus, soil biota (11 μm to 2 mm) along with AMF can have a great effect on nutrient leaching and crop biomass in agricultural practices.

3.5.2 Soil Biodiversity and Restoration Ecology

The biosphere is undergoing alterations due to human activities. To resolve this problem, a budget friendly solution is restoration ecology and its application (Suding et al. 2015). It follows a productive perspective between microbes and the ecosystem. The use of soil microorganisms for ecology restoration has been of interest recently. The microbes can be manipulated to help restore the biodiversity and functioning of degraded ecosystems (Young et al. 2005; Heneghan et al. 2008).

Various microbes can help in restoration ecology. The plant growth promoting rhizobia (PGPR) colonize the roots of plants and help protect the plant from diseases caused by pathogens (Kloepper et al. 2004). A PGPR strain *Achromobacter piechaudii* ARV8 produces1-aminocyclopropane-1-carboxylate (ACC) deaminase which confers stress tolerance in plants like tomato and pepper (Mayak et al. 2004). The oxidative damage caused by droughts can be diminished by co-inoculation of AMF (*Glomus mosseae* or *G. intraradices*) and *Pseudomonas mendocina* in lettuce by enhancing the production of catalase, which acts as an antioxidant (Kohler et al. 2008). Extremely disturbed areas like large scale mining can be restored by the use of AMF. They help in establishing plant species at a relatively low cost. The closeness of mycorrhizal networks and propagule sources to the restoration site is a prerequisite for developing new mycorrhizal associations. The addition of mycorrhizal symbionts has provided improvement in establishments in many desert ecosystems (Requena et al. 2001).

Bacteria also play a role to maintain a healthy ecosystem by decomposing dead matter. They release CO_2, water, nutrients, and energy for use by other organisms (Kling 2010). The decomposition process also helps water retention in the soil and prevents roots from the attack of parasites. The organic matter constitutes the humus which further helps improve soil fertility and health.

3.5.3 Agroecosystem Management with Core Microbiomes

Identification and modification of elements that drive ecosystem dynamics is a challenging process. Out of all the ecosystems, agroecosystems are the most difficult to manipulate. They consist of a large number of macro- and micro-organisms which have very strong webs of interaction among them (Busby et al. 2017). Due to the huge amount of damage caused by chemical fertilizers, scientists are now trying to maximize the use of microbial communities (microbiomes) as part of improving sustainable agriculture practices. The main aim is to find methods to convert the uncontrollable microbiomes into functionally rich sources for agriculture (Dangl et al. 2013).

The goal is not random colonization of microbes on plants, as this would lead to a large number of coexisting species. The microbiomes are structured so that the plant species can be classified into microbiome types. However, the shifting of microbial populations from original to alternative states requires the changes in environmental conditions. The microbiomes show resistance and resilience, returning to the original state after some time, thereby making it difficult to shift from a diseased to a healthy microbiome (Beisner et al. 2003).

Improved methods for making a suitable mix of microorganisms for inoculation in the host are required. In the agricultural plants, the creation of soil transplants or extracted soil microbiomes created using centrifugation is required (Dangl et al. 2013).

Although indigenous biota is seen as a source of disease in the agroecosystems; the plants in natural areas have developed mechanisms to gain maximum benefits from the microbiomes. The immune and signaling pathways of plants are revived with the help of core microbiome technologies (Foster et al. 2017). Though, the use of single culture has been favored previously, the use of heterogeneous cultures with different spatia, and temporal characteristics is gaining importance (King 2017). The optimization of both plant genetic varieties and core microbiomes is the secret for developing resource efficient and pathogen resistant agroecosystems (Toju et al. 2018).

3.6 Conclusion

Microorganisms are present all around the globe and their diversity varies from region to region based on different factors like altitude, temperature, and soil conditions. Apart from forming the basis of different processes like nutrient cycling, soil formation, and waste recycling, they are a prerequisite for carrying out sustainable agriculture in all the agroecosystems exist on this earth. Also, microbes from different terrains can be engineered for use in agricultural fields for different applications like biofertilizers. However, due to changing environmental conditions and other stresses the microbial diversity is under constant threat. Therefore, there is a need to develop strategies for overcoming the loss of microbial diversity and conserve the microbial biota from different agroecosystems.

References

Aislabie J, Deslippe JR (2013) Soil microbes and their contribution to soil services. In: Dymond JR (ed) Ecosystem services in New Zealand-conditions and trends. Manaaki Whenua Press, Lincoln, pp 143–161

Aislabie J, Davison AD, Boul HL, Franzmann PD, Jardine DR, Karuso P (1999) Isolation of *Terrabacter* sp. strain DDE-1 which metabolises DDE when induced with biphenyl. Appl Environ Microbiol 65:5607–5611. https://doi.org/10.1128/AEM.65.12.5607-5611.1999

Aislabie J, Bej AK, Ryburn J, Lloyd N, Wilkins A (2005) Characterization of *Arthrobacter nicotinovorans* HIM, an atrazine-degrading bacterium, from agricultural soil New Zealand. FEMS Microbiol Ecol 52:279–286. https://doi.org/10.1016/j.femsec.2004.11.012

Altieri MA (1999) The ecological role of biodiversity in agroecosystems. Agric Ecosyst Environ 74:19–31. https://doi.org/10.1016/B978-0-444-50019-9.50005-4

Aneja M, Sharma S, Schloter M, Munch JC (2006) Microbial degradation of beech litter – influence of soil type and litter quality on the structure and function of microbial populations involved in the turnover process. Microb Ecol 52:127–135

Arumugam M, Mitra A, Pramanik M (2011) *Streptomyces sundarbansensis* sp. nov., a novel actinomycete that produces 2-allyloxyphenol. Int J Syst Evol Microbiol 61:2664–2669. https://doi.org/10.1099/ijs.0.028258-0

Auge RM (2004) Arbuscular mycorrhizae and soil/plant water relations. Can J Soil Sci 84:373–381. https://doi.org/10.4141/S04-002

Aulakh MS, Doran JW, Walters DT, Mosier AR, Francis DD (1991) Crop residue type and placement effects on denitrification and mineralization. Soil Sci Soc Am J 55:1020–1025. https://doi.org/10.2136/sssaj1991.03615995005500040022x

Azcon-Aguilar C, Barea JM (2015) Nutrient cycling in the mycorrhizosphere. J Soil Sci Plant Nutr 15:372–396. https://doi.org/10.4067/S0718-95162015005000035

Azua Bustos A, Gonzalez Silva C (2014) Biotechnological applications derived from microorganisms of the Atacama Desert. Biomed Res Int 2014:1–7. https://doi.org/10.1155/2014/909312

Bainard LD, Koch AM, Gordon AM, Klironomos JN (2012) Temporal and compositional differences of arbuscular mycorrhizal fungal communities in conventional monocropping and tree-based intercropping systems. Soil Biol Biochem 45:172–180. https://doi.org/10.1016/j.soilbio.2011.10.008

Banerjee A, Bareh DA, Joshi SR (2017) Native microorganisms as potent bioinoculants for plant growth promotion in shifting agriculture (*Jhum*) systems. J Soil Sci Plant Nutr 17:127–140. https://doi.org/10.4067/S0718-95162017005000010

Bardgett RD, Caruso T (2020) Soil microbial community responses to climate extremes: resistance, resilience and transitions to alternative states. Philos Trans R Soc B 375:20190112. https://doi.org/10.1098/rstb.2019.0112

Barr DP, Aust SD (1994) Mechanisms white rot fungi use to degrade pollutants. Environ Sci Technol 28:79–87. https://doi.org/10.1021/es00051a002

Basak P, Pramanik A, Roy R (2015) Cataloguing the bacterial diversity of the Sundarbans mangrove, India in the light of metagenomics. Genom Data 4:90–92. https://doi.org/10.1016/j.gdata.2015.03.014

Beare MH, Pohlad BR, Wright DH, Coleman DC (1993) Residue placement and fungicide effects on fungal communities in conventional and no-tillage soils. Soil Sci Soc Am J 57:392–399. https://doi.org/10.2136/sssaj1993.03615995005700020018x

Beisner BE, Haydon DT, Cuddington K (2003) Alternative stable states in ecology. Front Ecol Environ 1:376–382. https://doi.org/10.1890/1540-9295(2003)001[0376:ASSIE]2.0.CO;2

Bender SF, van der Heijden MG (2015) Soil biota enhance agricultural sustainability by improving crop yield, nutrient uptake and reducing nitrogen leaching losses. J Appl Ecol 52:228–239. https://doi.org/10.1111/1365-2664.12351

Bhatnagar A, Makandar MB, Garg MK, Bhatnagar M (2008) Community structure and diversity of cyanobacteria and green algae in the soils of Thar Desert (India). J Arid Environ 72:73–83. https://doi.org/10.1016/j.jaridenv.2007.05.007

Birkhofer K, Bezemer TM, Bloem J, Bonkowski M, Christensen S, Dubois D, Ekelund F, Fliessbach A, Gunst L, Hedlund K, Mader P, Mikola J, Robin C, Setala H, Tatin-Froux F, Van der Putten WH, Scheu S (2008) Long-term organic farming fosters below and aboveground biota: implications for soil quality, biological control and productivity. Soil Biol Biochem 40:2297–2308. https://doi.org/10.1016/j.soilbio.2008.05.007

Biswas K, Mukherjee J (2019) Microbial diversity of the Sundarbans, the world's largest tidal mangrove forest, and its bioprospects. In: Satyanarayana T, Das S, Johri B (eds) Microbial diversity in ecosystem sustainability and biotechnological applications. Springer, Singapore, pp 231–256. https://doi.org/10.1007/978-981-13-8487-5_10

Biswas K, Chowdhury JD, Mahansaria R, Mukherjee J (2017) *Streptomyces euryhalinus* sp. Nov., a new actinomycete isolated from a mangrove forest. J Antibiot 70:747–753. https://doi.org/10.1038/ja.2017.3

Brussaard L, De Ruiter PC, Brown GG (2007) Soil biodiversity for agricultural sustainability. Agric Ecosyst Environ 121:233–244. https://doi.org/10.1016/j.agee.2006.12.013

Busby PE, Soman C, Wagner MR, Friesen ML, Kremer J, Bennett A et al (2017) Research priorities for harnessing plant microbiomes in sustainable agriculture. PLoS Biol 15:e2001793. https://doi.org/10.1371/journal.pbio.2001793

Caldeira MC, Ryel RR, Lawton JH, Pereira JS (2001) Mechanisms of positive biodiversity-production relationships: insights provided by $\delta^{13}C$ analysis in experimental Mediterranean grassland plots. Ecol Lett 4:439–443. https://doi.org/10.1046/j.1461-0248.2001.00238.x

Cavagnaro TR, Martin AW (2011) Arbuscular mycorrhizas in southeastern Australian processing tomato farm soils. Plant Soil 340:327–336. https://doi.org/10.1007/s11104-010-0603-z

Chakraborty S, Maruthanayagam V, Achari A (2018) *Oxynema aestuarii* sp. nov., (Microcoleaceae) isolated from an Indian mangrove forest. Phytotaxa 374:24–40. https://doi.org/10.11646/phytotaxa.374.1.2

Christensen M (1989) A view of fungal ecology. Mycologia 81:1–19. https://doi.org/10.1080/00275514.1989.12025620

Dangl JL, Horvath DM, Staskawicz BJ (2013) Pivoting the plant immune system from dissection to deployment. Science 341:746–751. https://doi.org/10.1126/science.1236011

Das T, Sarkar P, Prasad N (2014) Exploring the potential for concurrent rice-fish culture in wetlands of Assam, North East India. Intl Res J Biol Sci 3:60–69

De Bruijn FJ (2015) Biological nitrogen fixation. In: Lugtenberg B (ed) Principles of plant-microbe interactions. Springer, Berlin, pp 215–224

De Vries FT, Liiri ME, Bjornlund L, Bowker MA, Christensen S, Setala HM, Bardgett RD (2012) Land use alters the resistance and resilience of soil food webs to drought. Nat Clim Chang 2:276–280. https://doi.org/10.1038/nclimate1368

De Vries FT, Griffiths RI, Bailey M, Craig H, Girlanda M, Gweon HS, Hallin S, Kaisermann A, Keith AM, Kretzschmar M, Lemanceau P (2018) Soil bacterial networks are less stable under drought than fungal networks. Nat Commun 9:30–33. https://doi.org/10.1038/s41467-018-05516-7

Deng Q, Hui D, Dennis S, Reddy KC (2017) Responses of terrestrial ecosystem phosphorus cycling to nitrogen addition: a meta-analysis. Glob Ecol Biogeogr 26:713–728. https://doi.org/10.1111/geb.12576

Eilers KG, Lauber CL, Knight R, Fierer N (2010) Shifts in bacterial community structure associated with inputs of low molecular weight carbon compounds to soil. Soil Biol Biochem 42:896–903. https://doi.org/10.1016/j.soilbio.2010.02.003

Erisman JW, Galloway JN, Seitzinger S, Bleeker A, Dise NB, Petrescu AR, Leach AM, de Vries W (2013) Consequences of human modification of the global nitrogen cycle. Philos Trans R Soc B Biol Sci 368:20130116. https://doi.org/10.1098/rstb.2013.0116

Evizal R, Tohari T, Prijambada ID, Widada J, Widianto D (2012) Soil bacterial diversity and productivity of coffee-shade tree agro-ecosystems. J Trop Soils 17:181–187

Evizal R, Prijambada ID, Widada J, Widianto D (2013) Diversity of legume nodulating bacteria as key variable of coffee agroecosystem productivity. Int Res J Agric Sci Soil Sci 3:141–146

Falkowski PG, Fenchel T, Delong EF (2008) The microbial engines that drive earth's biogeochemical cycles. Science 320:1034–1039. https://doi.org/10.1126/science.1153213

Foster KR, Schluter J, Coyte KZ, Rakoff-Nahoum S (2017) The evolution of the host microbiome as an ecosystem on a leash. Nature 548:43–51. https://doi.org/10.1038/nature23292

Fredrickson JK, Balkwill DL, Zachara JM, Li SMW, Brockman FJ, Simmons MA (1991) Physiological diversity and distributions of heterotrophic bacteria in deep cretaceous sediments of the Atlantic coastal plain. Appl Environ Microbiol 57:402–411

Frey-Klett P, Garbaye J, Tarkka M (2007) The mycorrhiza helper bacteria revisited. New Phytol 176:22–36. https://doi.org/10.1111/j.1469-8137.2007.02191.x

Galloway JN, Townsend AR, Erisman JW, Bekunda M, Cai Z, Freney JR, Martinelli LA, Seitzinger SP, Sutton MA (2008) Transformation of the nitrogen cycle: recent trends, questions, and potential solutions. Science 320:889–892. https://doi.org/10.1126/science.1136674

Ghiglione JF, Galand PE, Pommier T, Pedros-Alio C, Maas EW, Bakker K, Bertilson S, Kirchman DL, Lovejoy C, Yager PL (2012) Pole-to-pole biogeography of surface and deep marine bacterial communities. Proc Natl Acad Sci 109:17633–17638. https://doi.org/10.1073/pnas.1208160109

Gibbons SM, Gilbert JA (2015) Microbial diversity- exploration of natural ecosystems and microbiomes. Curr Opin Genet Dev 35:66–72. https://doi.org/10.1016/j.gde.2015.10.003

Giovannoni SJ, Thrash JC, Temperton B (2014) Implications of streamlining theory for microbial ecology. ISME J8:1553–1565. https://doi.org/10.1038/ismej.2014.60

Gordon H, Haygarth PM, Bardgett RD (2008) Drying and rewetting effects on soil microbial community composition and nutrient leaching. Soil Biol Biochem 40:302–311. https://doi.org/10.1016/j.soilbio.2007.08.008
Grover M, Ali SZ, Sandhya V, Rasul A, Venkateswarlu B (2011) Role of microorganisms in adaptation of agriculture crops to abiotic stresses. World J Microbiol Biotechnol 27:1231–1240. https://doi.org/10.1007/s11274-010-0572-7
Gruber N, Galloway JN (2008) An earth-system perspective of the global nitrogen cycle. Nature 451:293–296. https://doi.org/10.1038/nature06592
Gruter D, Schmid B, Brandl H (2006) Influence of plant diversity and elevated atmospheric carbon dioxide levels on belowground bacterial diversity. BMC Microbiol 6:68. https://doi.org/10.1186/1471-2180-6-68
Harrison JJ, Ceri H, Turner R (2007) Mulitresistance and tolerance in microbial biofilms. Nat Rev Microbiol 5:928–938. https://doi.org/10.1038/nrmicro1774
Hartmann M, Niklaus PA, Zimmermann S, Schmutz S, Kremer J, Abarenkov K, Luscher P, Widmer F, Frey B (2014) Resistance and resilience of the forest soil microbiome to logging-associated compaction. ISME J 8:226–244. https://doi.org/10.1038/ismej.2013.141
He Z, Piceno Y, Deng Y, Xu M, Lu Z, De Santis T, Andersen G, Hobbie SE, Reich PB, Zhou J (2012) The phylogenetic composition and structure of soil microbial communities shifts in response to elevated carbon dioxide. ISME J 6:259–272. https://doi.org/10.1038/ismej.2011.99
Heneghan L, Miller SP, Baer S, Callaham MA Jr, Montgomery J, Pavao Zuckerman M (2008) Integrating soil ecological knowledge into restoration management. Restor Ecol 16:608–617. https://doi.org/10.1111/j.1526-100X.2008.00477.x
Jaggi V, Brindhaa NT, Sahgal M (2020) Microbial diversity in north western Himalayan agroecosystems: functions and applications. In: Microbiological advancements for higher altitude agro-ecosystems & sustainability. Springer, Berlin. https://doi.org/10.1007/978-981-15-1902-4_8
Jain R, Chaudhary D, Dhakar K, Pandey A (2016) A consortium of beneficial fungi survive the fire operations under shifting cultivation in northeast Himalaya, India. Natl Acad Sci Lett 39:343–346. https://doi.org/10.1007/s40009-016-0482-9
Johnson NC, Copeland PJ, Crookston BK, Pfleger FL (1992) Mycorrhizae: possible explanation for yield decline with continuous corn and soybean. Agron J 84:387–390. https://doi.org/10.2134/agronj1992.00021962008400030007x
Jorquera MA, Hernandez MT, Rengel Z, Marschner P, Mora ML (2008) Isolation of culturable phosphobacteria with both phytate-mineralization and phosphate-solubilization activity from the rhizosphere of plants grown in a volcanic soil. Biol Fertil Soil 44:1025–1034. https://doi.org/10.1007/s00374-008-0288-0
Joshi SR, Bareh D, Banerjee A (2019) Soil microbiota and sustainable jhum agroecosystem. In: Satyanarayana T, Das S, Johri B (eds) Microbial diversity in ecosystem sustainability and biotechnological applications. Springer, Singapore, pp 57–82. https://doi.org/10.1007/978-981-13-8487-5_3
Kennedy AC, Smith KL (1995) Soil bacterial diversity and ecosystem functioning. Plant Soil 170:75–86
King A (2017) The future of agriculture. Nature 544:21–23. https://doi.org/10.1038/544S21a
Kjoller A, Struwe S (1982) Microfungi in ecosystems: fungal occurrence and activity in litter and soil. Oikos 39:391–422. https://doi.org/10.2307/3544690
Kling G (2010) Microbes: transformers of matter and material. Ecology and evolutionary biology. University of Michigan, Ann Arbor, MI. Retrieved from global change. umich.edu/globalchange1
Kloepper JW, Ryu CM, Zhang S (2004) Induced systemic resistance and promotion of plant growth by Bacillus spp. Phytopathology 94:1259–1266. https://doi.org/10.1094/PHYTO.2004.94.11.1259

Kohler J, Hernandez JA, Caravaca F, Roldan A (2008) Plant-growth-promoting rhizobacteria and arbuscular mycorrhizal fungi modify alleviation biochemical mechanisms in water-stressed plants. Funct Plant Biol 35:141–151. https://doi.org/10.1071/FP07218

Lacombe S, Bradley RL, Hamel C, Beaulieu C (2009) Do tree-based intercropping systems increase the diversity and stability of soil microbial communities? Agric Ecosyst Environ 131:25–31. https://doi.org/10.1016/j.agee.2008.08.010

Lee KE, Foster RC (1991) Soil fauna and soil structure. Aust J Soil Res 29:745–775. https://doi.org/10.1071/SR9910745

Logue JB, Findlay SEG, Comte J (2015) Editorial: microbial responses to environmental changes. Front Microbiol 6:1364. https://doi.org/10.3389/fmicb.2015.01364

Lovelock JE, Margulis L (1974) Atmospheric homeostasis by and for the biosphere: the Gaia hypothesis. Tellus 26:2–10. https://doi.org/10.3402/tellusa.v26i1-2.9731

Lynch JM, Harper SHT (1985) The bacterial upgrading of straw for agricultural use. Phil Trans R Soc London 310:221–226. https://doi.org/10.1098/rstb.1985.0110

Mackelprang R, Jacobsen CS, Jansson J, Tas N (2015) Permafrost metagenomics and climate change. Annu Rev Earth Planet Sci 44:439–462. https://doi.org/10.1146/annurev-earth-060614-105126

Majumder M, Shukla AK (2012) Studies on micro fungal diversity under variable habitats in Arunachal Pradesh. In: Sati SC, Belwal M (eds) Microbes: diversity and biotechnology. Daya Publishing House, New Delhi, pp 137–149

Marschner P (2008) The role of rhizosphere microorganisms in relation to P uptake by plants. In: White PJ, Hammond J (eds) The ecophysiology of plant-phosphorus interactions. Series: plant ecophysiology, vol vol 7. Springer, Berlin, pp 165–176. https://doi.org/10.1007/978-1-4020-8435-5_8

Mayak S, Tirosh T, Glick BR (2004) Plant growth-promoting bacteria that confer resistance to water stress in tomatoes and peppers. Plant Sci 166:525–530. https://doi.org/10.1016/j.plantsci.2003.10.025

McCalley CK, Woodcroft BJ, Hodgkins SB, Wehr RA, Kim EH, Mondav R, Crill PM, Chanton JP, Rich VI, Tyson GW (2014) Methane dynamics regulated by microbial community response to permafrost thaw. Nature 514:478–481. https://doi.org/10.1038/nature13798

Mendes R, Garbeva P, Raaijmakers JM (2013) The rhizosphere microbiome: significance of plant beneficial, plant pathogenic, and human pathogenic microorganisms. FEMS Microbiol Rev 37:634–663. https://doi.org/10.1111/1574-6976.12028

Mishra PK, Mishra S, Selvakumar G, Bisht SC, Bisht JK, Kundu S, Gupta HS (2008) Characterization of a psychrotolerant plant growth promoting *Pseudomonas* sp. StrainPGERs17 (MTCC 9000) isolated from North Western Indian Himalayas. Ann Microbiol 58:561–568. https://doi.org/10.1007/BF03175558

Monier JM, Demaneche S, Delmont TO, Mathieu A, Vogel TM, Simonet P (2011) Metagenomic exploration of antibiotic resistance in soil. Curr Opin Microbiol 14:229–235. https://doi.org/10.1016/j.mib.2011.04.010

Morris JJ, Papoulis SE, Lenski RE (2014) Coexistence of evolving bacteria stabilized by a shared black queen function. Evolution 68:2960–2971. https://doi.org/10.1111/evo.12485

Nies DH (1999) Microbial heavy metal resistance. Appl Microbiol Biotechnol 51:730–750. https://doi.org/10.1007/s002530051457

Olivares J, Bedmar EJ, Sanjuan J (2013) Biological nitrogen fixation in the context of global change. Mol Plant-Microbe Interact 26:486–494. https://doi.org/10.1094/MPMI-12-12-0293-CR

Onen O, Aboh AA, Mfam AN, Akor MO, Nweke CN, Osuagwu AN (2020) Microbial diversity: values and roles in ecosystems. Asian J Biol 9:10–22. https://doi.org/10.9734/AJOB/2020/v9i130075

Pandey A, Trivedi P, Kumar B, Palni LMS (2006) Characterization of a phosphate solubilizing and antagonistic strain of Pseudomonas putida (B0) isolated from a sub-alpine location in the Indian central Himalaya. Curr Microbiol 53:102–107. https://doi.org/10.1007/s00284-006-4590-5

Pandey A, Chaudhry S, Sharma A, Choudhary VS, Malviya MK, Chamoli S, Rinu K, Trivedi P, Palni LMS (2011) Recovery of *Bacillus* and *Pseudomonas* spp. from the 'fired plots' under shifting cultivation in Northeast India. Curr Microbiol 62:273–280. https://doi.org/10.1007/s00284-010-9702-6

Pelliciardi V (2012) Nutrients (N, P, and K) recycling in traditional soil fertility practices in Leh district: a case study at small farm level. Int Assoc Ladakh Stud 28:27–35

Postma-Blaauw MB, de Goede RGM, Bloem J, Faber JH, Brussaard L (2010) Soil biota community structure and abundance under agricultural intensification and extensification. Ecology 91:460–473. https://doi.org/10.1890/09-0666.1

Prosser JI (2012) Ecosystem processes and interactions in a morass of diversity. FEMS Microbiol Ecol 81:507–519. https://doi.org/10.1111/j.1574-6941.2012.01435.x

Raaijmakers JM, Paulitz TC, Steinberg C, Alabouvette C, Moenne-Loccoz Y (2009) The rhizosphere: a playground and battlefield for soilborne pathogens and beneficial microorganisms. Plant Soil 321:341–361. https://doi.org/10.1007/s11104-008-9568-6

Raeside C, Gaffe J, Deatherage DE, Tenaillon O, Briska AM, Ptashkin RN, Cruveiller S, Medigue C, Lenski RE, Barrick JE (2014) Large chromosomal rearrangements during a long-term evolution experiment with *Escherichia coli*. MBio 5. https://doi.org/10.1128/mBio.01377-14

Rai SC (2005) Apatani paddy-cum-fish cultivation: an indigenous hill farming system of North East India. IJTK 4:65–71

Rehakova K, Chlumska Z, Dolezal J (2011) Soil cyanobacterial and microalgal diversity in dry mountains of Ladakh, NW Himalaya, as related to site, altitude, and vegetation. Microb Ecol 62:337–346. https://doi.org/10.1007/s00248-011-9878-8

Requena N, Perez-Solis E, Azcon-Aguilar C, Jeffries P, Barea JM (2001) Management of indigenous plant-microbe symbioses aids restoration of desertified ecosystems. Appl Environ Microbiol 67:495. https://doi.org/10.1128/AEM.67.2.495-498.2001

Richardson AE, Barea JM, McNeill AM, Prigent-Combaret C (2009) Acquisition of phosphorus and nitrogen in the rhizosphere and plant growth promotion by microorganisms. Plant Soil 321:305–339. https://doi.org/10.1007/s11104-009-9895-2

Rideout JR, He Y, Navas-Molina JA, Walters WA, Ursell LK, Gibbons SM, Chase J, McDonald D, Gonzalez A, Robbins-Pianka A (2014) Subsampled open-reference clustering creates consistent, comprehensive OTU definitions and scales to billions of sequences. PeerJ 2:545. https://doi.org/10.7717/peerj.545

Robinson DA, Jones SB, Lebron I, Reinsch S, Domínguez MT, Smith AR, Jones DL, Marshall MR, Emmett BA (2016) Experimental evidence for drought induced alternative stable states of soil moisture. Sci Rep 6:200–218. https://doi.org/10.1038/srep20018

Scheffers BR, Joppa LN, Pimm SL, Laurance WF (2012) What we know and don't know about Earth's missing biodiversity. Trends Ecol Evol 27:501–510. https://doi.org/10.1016/j.tree.2012.05.008

Schulz S, Brankatschk R, Dumig A, Kogel-Knabner I, Schloter M, Zeyer J (2013) The role of microorganisms at different stages of ecosystem development for soil formation. Biogeosciences 10:3983–3996. https://doi.org/10.3929/ethz-b-000070776

Selvakumar G, Joshi P, Suyal P, Mishra PK, Joshi GK, Bisht JK, Gupta HS (2011) Pseudomonas lurida M2RH3 (MTCC 9245), a psychrotolerant bacterium from the Uttarakhand Himalayas, solubilizes phosphate and promotes wheat seedling growth. World J Microbiol Biotechnol 27:1129–1135. https://doi.org/10.1007/s11274-010-0559-4

Sewnet TC, Tuju FA (2013) Arbuscular mycorrhizal fungi associated with shade trees and *Coffea arabica* L. in a coffee-based agroforestry system in Bonga, southwestern Ethiopia. Afrika Focus 26(2):4912. https://doi.org/10.21825/af.v26i2.4912

Sharma KK, Mehra SP (2009) The Thar of Rajasthan (India): ecology and conservation of a desert ecosystem. In: Faunal ecology and conservation of the great Indian desert. Springer, Berlin, Heidelberg, pp 1–11. https://doi.org/10.1007/978-3-540-87409-6_1

Sharma R, Manda R, Gupta S, Kumar S, Kumar V (2013) Isolation and characterization of osmotolerant bacteria from Thar Desert of Western Rajasthan (India). Rev Biol Trop 61:1551–1562

Shivaji S, Chaturvedi P, Reddy GSN, Suresh K (2005) Pedobacter himalayensis sp. nov., from Hamta glacier located in the Himalayan mountain range in India. Int J Syst Evol Microbiol 55:1083–1088. https://doi.org/10.1099/ijs.0.63532-0

Silver S, Phung LT (1996) Bacterial heavy metal resistance: new surprises. Annu Rev Microbiol 50:753–789. https://doi.org/10.1146/annurev.micro.50.1.753

Simkin SM, Allen EB, Bowman WD, Clark CM, Belnap J, Brooks ML, Cade BS, Collins SL, Geiser LH, Gilliam FS, Jovan SE (2016) Conditional vulnerability of plant diversity to atmospheric nitrogen deposition across the United States. PNAS 113:4086–4091. https://doi.org/10.1073/pnas.1515241113

Suding K, Higgs E, Palmer M, Callicott JB, Anderson CB, Baker M, Gutrich JJ, Hondula KL, LaFevor MC, Larson BMH, Randall A, Ruhi JB, Schwartz KZS (2015) Committing to ecological restoration. Science 348:638–640. https://doi.org/10.1126/science.aaa4216

Tak N, Gehlot HS (2019) Diversity of nitrogen-fixing symbiotic rhizobia with special reference to Indian Thar Desert. In: Satyanarayana T, Das S, Johri B (eds) Microbial diversity in ecosystem sustainability and biotechnological applications. Springer, Singapore. https://doi.org/10.1007/978-981-13-8487-5_2

Tan W, Wang J, Bai W, Qi J, Chen W (2020) Soil bacterial diversity correlates with precipitation and soil pH in long-term maize cropping systems. Sci Rep 10:6012. https://doi.org/10.1038/s41598-020-62919-7

Thomas RQ, Canham CD, Weathers KC, Goodale CL (2010) Increased tree carbon storage in response to nitrogen deposition in the US. Nat Geosci 3:13–17. https://doi.org/10.1038/ngeo721

Tiedje JM, Sextone AJ, Parkin TB, Revsbech NP, Shelton DR (1984) Anaerobic processes in soil. Plant Soil 76:197–212. https://doi.org/10.1007/BF02205580

Todmam LC, Fraser FC, Corstanje R, Harris JA, Pawlett M, Ritz K, Whitmore AP (2018) Evidence for functional state transitions in intensively-managed soil ecosystems. Sci Rep 8:1–10. https://doi.org/10.1038/s41598-018-29925-2

Toju H, Peay KG, Yamamichi M, Narisawa K, Hiruma K, NaitoK FS, Ushio M, Nakaoka S, Onoda Y, Yoshida K (2018) Core microbiomes for sustainable agroecosystems. Nat Plants 4:247–257. https://doi.org/10.1038/s41477-018-0139-4

Treseder KK, Kivlin SN, Hawkes CV (2011) Evolutionary trade-offs among decomposers determine responses to nitrogen enrichment. Ecol Lett 14:933–938. https://doi.org/10.1111/j.1461-0248.2011.01650.x

Turco RF, Bezdicek DF (1987) Diversity within two serogroups of *Rhizobium leguminosarum* native to soils in the Palouse of eastern Washington. Ann Appl Biol 111:103–114. https://doi.org/10.1111/j.1744-7348.1987.tb01437.x

Turnbaugh PJ, Ley RE, Hamady M, Fraser-Liggett CM, Knight R, Gordon JI (2007) The human microbiome project. Nature 449:804–810. https://doi.org/10.1038/nature06244

van der Heijden M, Klironomos JN, Ursic M, Moutoglis P, StreitwolfEngel R, Boller T, Wiemken A, Sanders IR (1998) Mycorrhizal fungal diversity determines plant biodiversity, ecosystem variability and productivity. Nature 396:69–72. https://doi.org/10.1038/23932

van der Heijden MGA, Bardgett RD, van Straalen NM (2008) The unseen majority: soil microbes as drivers of plant diversity and productivity in terrestrial ecosystems. Ecol Lett 11:296–310. https://doi.org/10.1111/j.1461-0248.2007.01139.x

Verbruggen E, Roling WFM, Gamper HA, Kowalchuk GA, Verhoef HA, van der Heijden MGA (2010) Positive effects of organic farming on below-ground mutualists: large-scale comparison of mycorrhizal fungal communities in agricultural soils. New Phytol 186:968–979. https://doi.org/10.1111/j.1469-8137.2010.03230.x

Verma P, Yadav AN, Kumar V, Singh DP, Saxena AK (2017) Beneficial plant-microbes interactions: biodiversity of microbes from diverse extreme environments and its impact for crop improvement. In: Plant-microbe interactions in agro-ecological perspectives. Springer, Singapore, pp 543–580. https://doi.org/10.1007/978-981-10-6593-4_22

Whitman WB, Coleman DC, Wiebe WJ (1998) Prokaryotes: the unseen majority. PNAS 95:6578–6583. https://doi.org/10.1073/pnas.95.12.6578

Wicklow MC, Bollen WB, Denison WC (1974) Comparison of soil microfungi in 40-year-old stands of pure alder, pure conifer, and alder–conifer mixtures. Soil Biol Biochem 6:73–78. https://doi.org/10.1016/0038-0717(74)90062-5

Xiong J, Sun H, Peng F, Zhang H, Xue X, Gibbons SM, Gilbert JA, Chu H (2014) Characterizing changes in soil bacterial community structure in response to short-term warming. FEMS Microbiol Ecol 89:281–292. https://doi.org/10.1111/1574-6941.12289

Yadav AN, Sachan SG, Verma P, Saxena AK (2015) Prospecting cold deserts of north western Himalayas for microbial diversity and plant growth promoting attributes. J Biosci Bioeng 119:683–693. https://doi.org/10.1016/j.jbiosc.2014.11.006

Yarzabal LA (2020) Perspectives for using glacial and periglacial microorganisms for plant growth promotion at low temperatures. Appl Microbiol Biotechnol 104:3267–3278. https://doi.org/10.1007/s00253-020-10468-4

Young TP, Petersen DA, Clary JJ (2005) The ecology of restoration: historical links, emerging issues and unexplored realms. Ecol Lett 8:662–673. https://doi.org/10.1111/j.1461-0248.2005.00764.x

Zhou J, Deng Y, Luo F, He Z, Yang Y (2011) Phylogenetic molecular ecological network of soil microbial communities in response to elevated CO2. MBio 2:e00122-11. https://doi.org/10.1128/mBio.00122-11

Soil Microbial Biomass as an Index of Soil Quality and Fertility in Different Land Use Systems of Northeast India

4

N. Bijayalaxmi Devi

Abstract

Soil microbes play an important role in various soil processes such as nutrient cycling, nitrogen fixation, litter decomposition, and soil remediation and help in improving the fertility of an ecosystem. Microbial biomass in the soil acts as a source of labile nutrients for the plants as well as serves as a temporary sink for nutrients and can be used as an index of soil fertility and quality. Critical information on the variation of soil microbial biomass due to change in land-use types, plant species composition, edaphic factors, climate, and land management practices is necessary for proper maintenance of soil. Therefore, the microbial biomass carbon (C), nitrogen (N), and phosphorus (P) concentration, stoichiometry, and microbial nutrient fractions in the soils of different land-use types in Northeast India was examined to understand the nutrient status of these systems. Microbial biomass was highest in Jhumland > forests > agroforests > grassland > agricultural lands > wastelands. Microbial C:N:P stoichiometry suggests N limitation in the soils of this region. Burning treatment improves soil nutrients while logging reduces nutrients in the soil. Conversion of forest to grassland and agricultural land will lead to loss of soil nutrients but the incorporation of trees in agricultural lands such as agroforestry systems will minimize the loss.

Keywords

Nutrient immobilization · Jhumland · Agricultural land · Forests · Grassland · Agroforestry · Wasteland

N. B. Devi (✉)
Department of Botany, Sikkim University, Gangtok, India

R. Soni et al. (eds.), *Microbiological Activity for Soil and Plant Health Management*,
https://doi.org/10.1007/978-981-16-2922-8_4

4.1 Introduction

Soil is one of the most important resources that provide us food and with an increase in population the importance of soil also increases. Anthropogenic activities such as mining, agriculture, land-use change and urbanization, deforestation, and natural calamities often lead to soil degradation. Soil degradation refers to the deterioration of soil quality due to the declination of productive and ecological function and is one of the important issues of the modern world. Soil microorganisms constitute an important part of the soil, and they have the potential to reclaim degraded soil. Nutrient fluxes of an ecosystem such as mineralization and immobilization are controlled by soil organisms (Van der Heijden et al. 2008; Mooshammer et al. 2012) through the process of decomposition. Soil microbes serve as an index of soil degradation, health, quality, and fertility because of their sensitive nature to environmental changes (Takoutsing et al. 2016). The health of all living organisms is directly linked to the soil quality and health as agricultural sustainability and environmental quality is determined by soil (Doran 2002), and minor variations in soil quality and health affect soil functions. Several scientific indices are used to understand soil quality and health such as physical and chemical properties, morphology, organic matter content, and microbial and macrofauna in soil (Velasquez et al. 2007). The diversity of microbes and microbial biomass in the soil is often used as an index of soil health as they can show early signs of changes in soil quality and degradation (Bhatia 2008). Most of the biological activity in soil is concentrated in the topsoil as they are rich in organic matter and humus. Microbial populations that thrive in soil consist of bacteria, fungi, actinomycetes, and protozoa. These microbes play an important role in the nutrient cycling of an ecosystem through the decomposition of organic matter and serve as a source and a sink for plant nutrients (Singh et al. 1989). The soil microbial biomass is the microbial communities present in soil and constitutes only about 1–5% of total soil nutrients, but forms an important pool of nutrients for the plants due to its fast turnover (Jenkinson and Ladd 1981). The capacity of the microbes to regulate nutrient cycling through immobilization and mineralization of nutrients varied across biomes (Martiny et al. 2006). Land use, environment, and vegetation change can shift the microbial composition (McKew et al. 2011; Griffiths and Philippot 2013) of soil. Besides the spatial change, soil microbes also exhibit temporal changes such as seasonal changes (Luo et al. 2020). The concentration of microbial nutrients and the ratio of microbial nutrients such as C:N, C:P, N:P, and C:N:P and microbial fractions can be used to understand the quality, fertility, and nutrient status of an ecosystem. Therefore, this chapter discusses about the fertility and nutrient status of important land-use types of Northeast India by using the microbial biomass C, N, and P and their corresponding ratios in soil.

4.2 Role of Soil Microbes in an Ecosystem

Soil host several microbes such as bacteria, fungi, algae, protists, and animals (Bonkowski et al. 2009; Müller et al. 2016), which release nutrients from the organic substrates and help in improving and enhancing the fertility of the soil. Microbes in soil secrete certain mucilaginous substances and polysaccharides to bind soil aggregates, thereby improving the structure of the soil. Also, fungi present in the soil acts as hairy threads binding the soil particles together. Microbes can transform nutrients by the metabolic depolymerization of the organic substrates and mineralized them directly in soil or released it through microbial cell lysis or predation (Bonkowski et al. 2009; Richardson et al. 2009). Hence, soil microbes can perform the function of nutrient immobilization and flux simultaneously.

Not only this, microorganisms in soil can be employed as an agent to improve contaminated and degraded soils because of their capacity to convert noxious and toxic substances into non-toxic products (Singh et al. 2019). This process is called bioremediation. Toxic substances such as organic pollutants present in soil, e.g. polychlorinated biphenyls, alkanes, crude oil, dyes, heavy metals, industrial effluents and slurry, pesticides, chemical fertilizers in soil can be successfully removed or reduced by using appropriate technique and microbes.

Nitrogen is another important essential nutrient required by the plants for growth and other metabolic activities. However, due to the inability of the plants to fix nitrogen directly from the atmosphere, certain nitrogen-fixing microorganisms present in the soil such as nitrifying bacteria, ammonifying bacteria, archaea are responsible for the supply of nitrogen to plants. Thus, microbes play a significant role in nitrogen fixation and N cycling in an ecosystem (Fig. 4.1). During the process of litter decomposition soil microbes and fauna break down easily soluble compounds such as starch, amino acids, etc., and tough plant components like cellulose, lignin released nutrients in the soil (Bani et al. 2018). The amount of global C released by the process of litter decomposition (60 Tg C y^{-1}) is about 11 times higher than that released by the burning of fossil fuels (Schlesinger and Bernhardt 2013). This suggests the substantial role of soil microbes in litter decomposition and nutrient cycling of an ecosystem.

4.3 Soil Microbial Biomass

Soil microbial biomass has been defined as the living part of soil organic matter and constitutes all organisms with a volume of less than $5 \times 10^3 \mu m^3$ in the soil (Brookes 2001). These organisms are responsible for breaking down the organic substrates into simple inorganic substances that can be readily utilized by the plants. Being the lived component of soil they respond fast to changes in soil conditions such as a change in organic matter content, soil moisture conditions, and land management practices and therefore can reflect the change in soil quality. Different microbes have specialized decomposing ability depending on the climate and type of substrate available to them and hence have bio-geographic restrictions, for example, tropical

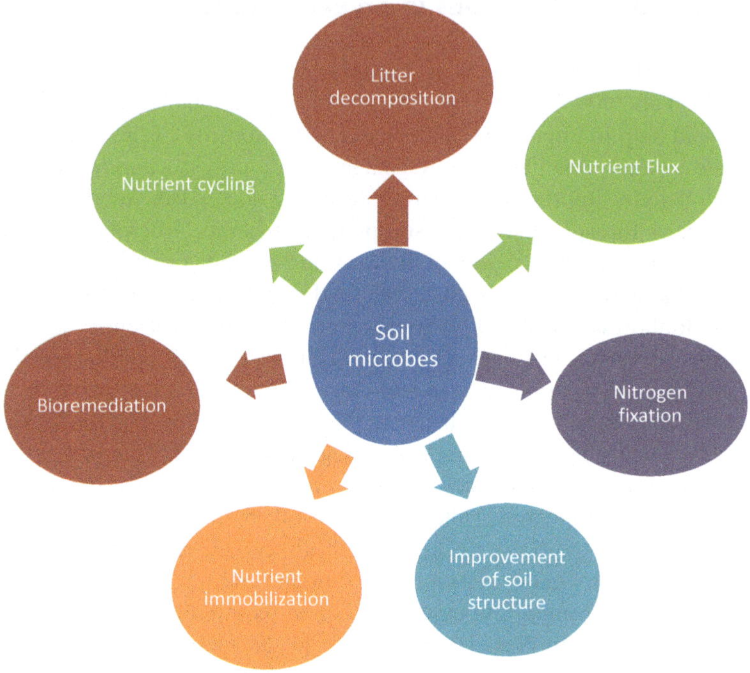

Fig. 4.1 Roles of soil microbes in an ecosystem

systems have nitrogen-fixing bacteria, while temperate boreal systems have mycorrhizae (Singh and Gupta 2018). Important regulators that significantly affect the soil microbial biomass are land-use types, soil factors, climate, plant species composition, quality, and quantity of substrates (Wardle 1992). It can act as a nutrient reservoir for the survival of plants in nutrient-depleted systems (Singh 2015; Vimal et al. 2017), but the release and immobilization of nutrients are controlled by the temporal pattern of the systems. The size of microbial nutrient fractions to total nutrients in an ecosystem indicates the nutrient status of an ecosystem and can successfully reveal the rehabilitation status of the degraded ecosystem (Singh and Gupta 2018).

4.4 Land-Use Types of Northeast India

The Northeastern region of India is located in the Eastern Himalayas and forms one of the important biodiversity hotspots of the world. This region has an area of 269,179 sq. km and comprises about 8% of the total geographical region of India and consists of eight states of India, namely Assam, Manipur, Nagaland, Mizoram, Meghalaya, Arunachal Pradesh, Tripura, and Sikkim. The presence of various ethnic groups in these states with diverse cultures and traditions led to the development of

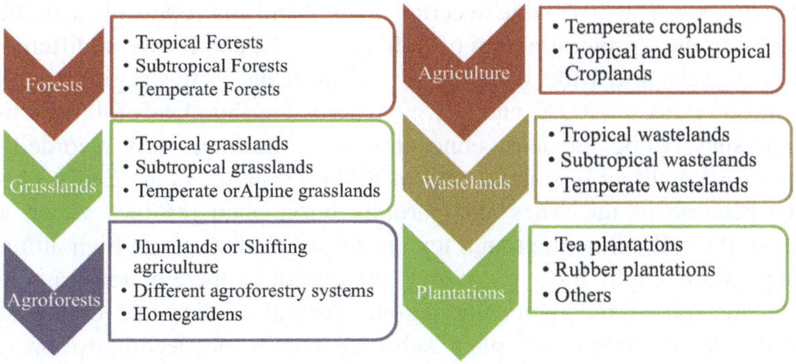

Fig. 4.2 Different land-use types of Northeast India

the various form of land-use types with different management practices in the region over and above the natural land uses. Besides the diversity in land use management practices and types, the northeast region also has a wide variation in terrain, altitude, and slope which directly or indirectly affects the nutrient status of soil resulting in a wide variation in soil fertility (Fig. 4.2). However, in the past few years, population growth has put pressure on land use of this region resulting in various activities such as agriculture expansion, deforestation, and mining leading to degradation of natural resources in this region (Lele and Joshi 2008). Important land-use types of the northeast region include forest, agroforestry, Jhumlands or shifting agriculture, cropland, grassland, and wastelands which have different soil fertility and quality. Among the agricultural land use of northeast India, jhuming is practiced in all the states of this region except Sikkim. About 3869 km^2 area is subjected to jhuming every year leading to a high rate of soil erosion (Saha et al. 2012). However, conventional farming and agroforestry including terraced farming are also practiced by the locals leading to the diverse farming pattern in this region. Forests of this region vary from dense to open type and tropical to the temperate forest with different tree species. The region has both tropical and temperate grasslands too.

4.5 Soil Nutrient Status of Different Land-Use Types of Northeast India

Northeast India, being a part of the Eastern Himalaya region has highly eroded soil due to heavy torrential rain and agricultural practices such as shifting cultivation (Saha et al. 2012). Soil productivity and sustainability are influenced by land-use types as it can change the physico-chemical and biotic properties of soil (Somasundaram et al. 2013). Different agricultural management practices influence soil quality and productivity (Di et al. 2013); therefore, sustainable use of soil resources is necessary for food production, decomposition of waste, carbon sequestration, heat storage, and gaseous exchange (Nanganoa et al. 2019). The most productive and fertile part of the soil is topsoil and about 24 billion tons of topsoil is lost (Nanganoa et al. 2019) due to certain natural and anthropogenic activities (Liu et al. 2006). In the Northeast region of India, soil C, N, and P vary in different land-use types suggesting a wide variation of soil nutrients in this region (Table 4.1). Maximum soil nutrients (C, N, and P) were recorded in Jhumlands followed by slash and burnt sites, while the least concentration of nutrients was recorded in the temperate zone and highly sloped areas of Sikkim (Sharma et al. 2004). This trend could be because of the ashes and partially burnt plant residues which add up nutrients to the soil, while the tough texture of plant substrates in high altitude and low temperature might have reduced decomposition, thereby reducing nutrient return in the soil in the temperate region. Several studies had established the influential role of climate and plant substrate quality on decomposition (Austin and Vitousek 2000; Devi and Yadava 2010). This suggests that the wide variation in soil nutrients of Northeast India has a direct link with the varying land-use types, elevation, slope, and soil management practices.

4.6 Changes in Microbial Biomass C, N, and P Due to Land-Use Types

Annual microbial biomass concentration trend in the different land uses of Northeast India follows Jhumlands > highland forests> other forest types> agroforests > grassland> agricultural lands >wastelands (Table 4.2). Increased organic C of the Jhumlands and burnt sites triggered the microbial activity in these sites leading to an increase in microbial biomass. Among the Jhumlands, the age of the Jhumland seems to be a regulating factor of microbial biomass rather than the elevation probably due to the recovery of soil nutrients from the effects of burning with time. Forests and other tree-based systems with high litter input have higher microbial nutrients as compared to other land-use types such as agriculture-based systems and wastelands. Subtropical forests and temperate forests have more microbial biomass and microbial activity than tropical forests. Disturbance in forests such as slash and burnt and burning decreases microbial nutrient in the initial years except for a *Dipterocarpus* forest but the microbial activity and nutrient sequestration increase tremendously with time. In contrast, systems with low organic matter input such as agriculture and wasteland have less microbial activity and biomass. Microbial biomass is more constrained to altitudinal gradient rather than species in agriculture-based systems. Agricultural lands of the temperate belt recorded the highest soil microbial nutrient than that of the subtropical and tropical belt. However, the microbial biomass value in the subtropical grassland is higher than agricultural lands of the tropical region and temperate agro-based system. The activity of the microbes and microbial biomass varies due to a change in plant species, management practices, and nutrient concentration of site, altitude, edaphic factors, climate, and topography. Conversion of natural forests to agricultural lands will lead to loss of microbial biomass nutrients; however, integration of trees- based agricultural systems such as agroforests will minimize this loss.

Table 4.1 Variation in soil C, N, and P in different land-use types of Northeast India

Forests	Soil C (%)	Soil N (%)	Soil P (%)	Soil depth (cm)	Altitude (m)	Location	References
Jhum-field lowland	4.20	0.40	Nil	0–15	300	Meghalaya	Arunachalam and Pandey (2003)
Jhum-field Highland	5.80	0.50	Nil	0–15	1700	Meghalaya	Arunachalam and Pandey (2003)
Jhum-fallows-1 year	1.90	0.20	Nil	0–15	300	Meghalaya	Arunachalam and Pandey (2003)
Jhum Fallow-16 year	3.60	0.70	Nil	0–15	300	Meghalaya	Arunachalam and Pandey (2003)
Jhum fallow Highland-1 year	2.40	0.30	Nil	0–15	1700	Meghalaya	Arunachalam and Pandey (2003)
Jhum fallow highland-16 year	4.80	0.60	Nil	0–15	1700	Meghalaya	Arunachalam and Pandey (2003)
Forest lowland	5.10	0.50	Nil	0–15	300	Meghalaya	Arunachalam and Pandey (2003)
Forest-Highland	5.70	0.60	Nil	0–15	1700	Meghalaya	Arunachalam and Pandey (2003)
Tropical wet evergreen	2.23	1.11	0.43	0–15	220	Assam	Barbhuiya et al. (2004)
Tropical wet evergreen disturbed Forest	1.23	0.56	0.39	0–15	220	Assam	Barbhuiya et al. (2004)
Tropical wet evergreen highly disturbed	0.89	0.39	0.22	0–15	220	Assam	Barbhuiya et al. (2004)
Agro-temperate	0.0027	0.000019	0.000075	0–15	2650	Sikkim	Sharma et al. (2004)
Agro-subtropical	0.0017	0.0002	0.000046	0–15	800	Sikkim	Sharma et al. (2004)
Temperate dense Forest	0.0032	0.00028	0.00008	0–15	2650	Sikkim	Sharma et al. (2004)
Temperate open Forest	0.0026	0.00024	0.000066	0–15	2650	Sikkim	Sharma et al. (2004)
Subtropical Forest	0.0002	0.00023	0.000076	0–15	800	Sikkim	Sharma et al. (2004)

(continued)

Table 4.1 (continued)

Forests	Soil C (%)	Soil N (%)	Soil P (%)	Soil depth (cm)	Altitude (m)	Location	References
Cardamom agroforestry	0.0028	0.00023	0.000076	0–15	300–2650	Sikkim	Sharma et al. (2004)
Madarin agroforestry	0.0016	0.00019	0.000075	0–15	300–2650	Sikkim	Sharma et al. (2004)
Wasteland temperate	0.0026	0.0002	0.000046	0–15	2650	Sikkim	Sharma et al. (2004)
Wasteland subtropical	0.0016	0.00015	0.000047	0–15	800	Sikkim	Sharma et al. (2004)
Protected *Dipterocarpus* Forest	0.69	0.14	0.039	0–15	300–360	Manipur	Yadava and Devi (2004)
Slash and burnt *Dipterocarpus* Forest	2.37	0.21	0.038	0–15	300–360	Manipur	Yadava and Devi (2004)
Subtropical mixed oak	3.50	0.47	0.076	0–15	780–910	Manipur	Devi and Yadava (2006)
Subtropical grassland	1.05	0.13	0.017	0–15	782	Manipur	Ibomcha Singh and Yadava (2006)
Agro-Paddy	0.95	0.12	0.016	0–15	782	Manipur	Ibomcha Singh and Yadava (2006)
Slash and burnt tropical forest	2.38	0.28	0.0411	0–15	902–944	Manipur	Binarani et al. (2009)
Protected tropical Forest	1.20	0.19	0.0404	0–15	902–944	Manipur	Binarani et al. (2009)
Agro-soybean	1.40	0.49	0.0042	0–15	130–752	Arunachal Pradesh	Bhuyan et al. (2013)
Agro-millet	1.80	0.46	0.0008	0–15	130–752	Arunachal Pradesh	Bhuyan et al. (2013)
Agro-maize	1.40	0.55	0.00085	0–15	130–752	Arunachal Pradesh	Bhuyan et al. (2013)
Agro-veg	0.86	0.42	0.0043	0–15	130–752	Arunachal Pradesh	Bhuyan et al. (2013)

Table 4.2 Microbial biomass C, N, and P variance in different land-use types of Northeast India ($\mu g\ g^{-1}$)

Land-use types	Mic C	Mic N	Mic P	Soil depth	Location	References
Jhum-field low	902.20	67.90	Nil	0–15	Meghalaya	Arunachalam and Pandey (2003)
Jhum-field high	1522.30	82.90	Nil	0–15	Meghalaya	Arunachalam and Pandey (2003)
Jhum-fallows-1 year low	801.40	32.30	Nil	0–15	Meghalaya	Arunachalam and Pandey (2003)
Jhum Fallow-16 year low	1431.90	98.20	Nil	0–15	Meghalaya	Arunachalam and Pandey (2003)
Jhum fallow High-1 year	651.40	61.30	Nil	0–15	Meghalaya	Arunachalam and Pandey (2003)
Jhum fallow high-16 year	1629.20	87.20	Nil	0–15	Meghalaya	Arunachalam and Pandey (2003)
Forest low	1287.60	90.20	Nil	0–15	Meghalaya	Arunachalam and Pandey (2003)
Forest-high	1620.40	93.10	Nil	0–15	Meghalaya	Arunachalam and Pandey (2003)
Tropical wet evergreen	809.45	74.12	36.85	0–15	Assam	Barbhuiya et al. (2004)
Tropical wet evergreen disturbed Forest	574.71	43.35	27.19	0–15	Assam	Barbhuiya et al. (2004)
Tropical wet Evergreen highly disturbed	368.07	31.34	18.90	0–15	Assam	Barbhuiya et al. (2004)
Agro-temperate	390.00	38.00	22.00	0–15	Sikkim	Sharma et al. (2004)
Agro-subtropical	291.00	34.00	16.00	0–15	Sikkim	Sharma et al. (2004)
Temperate dense Forest	864.00	142.00	43.00	0–15	Sikkim	Sharma et al. (2004)
Temperate open Forest	712.00	96.00	31.00	0–15	Sikkim	Sharma et al. (2004)
Subtropical Forest	764.00	71.00	31.00	0–15	Sikkim	Sharma et al. (2004)
Cardamom agroforestry	583.00	63.00	22.00	0–15	Sikkim	Sharma et al. (2004)
Madarin agroforestry	471.00	48.00	18.00	0–15	Sikkim	Sharma et al. (2004)
Wasteland temperate	259.00	33.00	13.00	0–15	Sikkim	Sharma et al. (2004)
Wasteland subtropical	219.00	30.00	16.00	0–15	Sikkim	Sharma et al. (2004)
Protected *Dipterocarpus* Forest	390.31	56.50	10.92	0–15	Manipur	Yadava and Devi (2004)

(continued)

Table 4.2 (continued)

Land-use types	Mic C	Mic N	Mic P	Soil depth	Location	References
Slash and burnt *Dipterocarpus* Forest	550.12	66.06	16.10	0–15	Manipur	Yadava and Devi (2004)
Subtropical mixed oak	839.39	82.31	37.33	0–15	Manipur	Devi and Yadava (2006)
Subtropical grassland	407.03	90.33	20.87	0–15	Manipur	Ibomcha Singh and Yadava (2006)
Agro-Paddy	330.30	38.60	18.00	0–15	Manipur	Ibomcha Singh and Yadava (2006)
Slash and burnt tropical forest	308.44	31.86	18.01	0–15	Manipur	Binarani et al. (2009)
Protected tropical Forest	363.03	40.08	29.90	0–15	Manipur	Binarani et al. (2009)
Agro-soybean	228.21	21.54	9.75	0–15	Arunachal Pradesh	Bhuyan et al. (2013)
Agro-millet	213.54	23.58	12.24	0–15	Arunachal Pradesh	Bhuyan et al. (2013)
Agro-maize	234.48	26.55	13.55	0–15	Arunachal Pradesh	Bhuyan et al. (2013)
Agro-veg	238.45	19.93	9.28	0–15	Arunachal Pradesh	Bhuyan et al. (2013)

4.7 Microbial C:N:P Stoichiometry in Different Land-Use Types of Northeast India

Microbial C:N:P stoichiometry has been widely used as a powerful biological tool to study ecosystem functions in terrestrial systems (Elser and Hamilton 2007; Sterner and Elser 2017) as it affects soil C, N, and P mineralization (Mooshammer et al. 2012). These microbial nutrient ratios can indicate nutrient status and help in the characterization of several important ecological processes (Reich and Oleksyn 2004; Cleveland and Liptzin 2007; Güsewell and Gessner 2009). Microbes can recover fast from stress and because of this strong capacity they are termed as resilient to environmental change (Guénon and Gros 2013; Phillips et al. 2015; Gao et al. 2016).

4.7.1 Microbial C:N Ratio

Forests have wide and high microbial C:N ratio depending on the type of forest which suggests spatial variability of microbial nutrient sequestration. Temperate forests and *Dipterocarpus* forests have low C:N ratio than that of tropical and subtropical forests indicating higher microbial N in the former two types of forests (Table 4.3). This suggests a difference in the microbial community between forest

Table 4.3 Microbial stoichiometry and proportion of microbial nutrients to total soil nutrient in different forests of Northeast India

Natural Forest	Mic C	Mic N	Mic P	C: N	C: P	N: P	Mic C/SOC	MicN/ TN	MicP/ P	Location	References
Undisturbed and protected forests											
Forest lowland	1287.60	90.20	Nil	14.27	–	–	2.52	1.80	–	Meghalaya	Arunachalam and Pandey (2003)
Forest-Highland	1620.40	93.10	Nil	17.40	–	–	2.84	1.55	–	Meghalaya	Arunachalam and Pandey, (2003)
Tropical wet evergreen	809.45	74.12	36.85	10.93	21.97	2.01	3.63	0.67	0.86	Assam	Barbhuiya et al. (2004)
Temperate dense Forest	864.00	142.00	43.00	6.08	20.09	3.30	2.70	4.9	7.8	Sikkim	Sharma et al. (2004)
Temperate open Forest	712.00	96.00	31.00	7.42	22.97	3.10	2.70	3.9	6.8	Sikkim	Sharma et al. (2004)
Subtropical Forest	764.00	71.00	31.00	10.77	24.66	2.29	5.20	3.0	6.5	Sikkim	Sharma et al. (2004)
Protected *Dipterocarpus* Forest	390.31	56.50	10.92	6.90	35.74	5.17	5.66	4.04	2.80	Manipur	Yadava and Devi (2004)
Subtropical mixed oak	839.39	82.31	37.33	10.20	22.49	2.20	2.40	1.75	4.91	Manipur	Devi and Yadava (2006)
Protected tropical Forest	363.03	40.08	29.90	9.06	12.14	1.34	3.03	2.11	7.40	Manipur	Binarani et al. (2009)
Disturbed and slash and burnt forests											
Tropical wet evergreen disturbed Forest	574.71	43.35	27.19	13.26	21.14	1.59	4.67	0.77	0.70	Assam	Barbhuiya et al. (2004)
Tropical wet Evergreen highly disturbed	368.07	31.34	18.90	11.74	19.48	1.66	4.14	0.80	0.86	Assam	Barbhuiya et al. (2004)
Slash and burnt *Dipterocarpus* Forest	550.12	66.06	16.10	8.33	34.17	4.10	2.32	3.15	4.24	Manipur	Yadava and Devi (2004)
Slash and burnt tropical forest	308.44	31.86	18.01	9.68	17.12	1.77	1.30	1.14	4.38	Manipur	Binarani et al. (2009)

types. C:N ratio of fungi ranges from 10–12, while bacteria have a range of 3–5 (Jenkinson and Ladd 1981). The microbial C:N ratios revealed that microbial communities in temperate forests and *Dipterocarpus* forest are mainly dominated by bacteria, while fungi form the main dominant microbial population of tropical and subtropical forests of Northeast India. Anthropogenic disturbances and burning treatment increase the microbial C:N ratio and change the microbial community of the area. The shift in microbial C:N:P stoichiometry is related to change in microbial communities (Fanin et al. 2013; Heuck et al. 2015). In agricultural land uses, there is no consistent trend of microbial C:N based on the agroclimatic zone but depends on the management practices and use of other chemical and organic manures in the fields (Table 4.4). Among the agroforests, Jhumlands especially younger ones have the highest C:N ratio than other Jhumlands and agroforests (Table 4.5). Subtropical grassland has the least microbial C:N ratio which suggests the dominance of bacteria in this land-use type. Both temperate and subtropical wastelands have a similar range of microbial C:N which is comparable with the values of temperate forest and tropical *Dipterocarpus* forests. The microbial C:N ratios of soils of the different land-use types of Northeast India are lower than their respective global range of microbial C:N reported by (Xu et al. 2013).

4.7.2 Microbial C:P Ratio

Microbial biomass C:P changes can be indicative of nutrient limitation in soil within a site (Griffiths et al. 2012). *Dipterocarpus* forest has the highest microbial C:P among the different forest types. This suggests high P limitation in this forest as compared to other forest types. Tree-based systems such as forests, agroforests recorded higher C: P ratios than other systems such as grassland, wasteland, and agricultural lands except for tropical soybean and vegetable agricultural fields (Tables 4.3, 4.4, and 4.5) where the values were exceptionally high. This is due to either less P demand or high P requirement for metabolic activities by the microbial community such as bacteria in these sites leading to a low P in microbes. Bacteria require more P than fungi for metabolic activity (Strickland and Rousk 2010; Fanin et al. 2013; Mouginot et al. 2014). Disturbances such as tree felling and slash and burnt treatment reduce C:P ratio in evergreen forests but not in the tropical forest. Tree felling and burning increase the soil pH and altered soil P availability by releasing occluded phosphorus in soil (DeBano and Klopatek 1988). Microbial C: P seems to be not affected by the vegetation type of the system.

4.7.3 Microbial N:P Ratio

The microbial N:P ratio is a powerful indicator of nutrient limitation in the tropics and N:P > 6.9 indicates P limitations, while N:P < 6.9 suggests N limitation (Cleveland and Liptzin 2007). All the land-use types of Northeast India showed N: P ratio less than 6.9 which indicates soil N limitations in this region (Table 4.3).

Table 4.4 Microbial stoichiometry and proportion of microbial nutrients to total soil nutrient in different agricultural and grassland of Northeast India

Agriculture	Mic C	Mic N	Mic P	C:N	C:P	N: P	Mic C/SOC	MicN/TN	MicP/P	Location	References
Agro-temperate	390.00	38.00	22.00	10.26	17.73	1.73	1.80	1.90	2.70	Sikkim	Sharma et al. (2004)
Agro-subtropical	291.00	34.00	16.00	8.56	18.19	2.13	2.20	1.70	2.90	Sikkim	Sharma et al. (2004)
Subtropical agro-Paddy	330.30	38.60	18.00	8.56	18.35	2.14	3.48	3.22	11.25	Manipur	Ibomcha Singh and Yadava (2006)
Tropical agro-soybean	228.21	21.54	9.75	10.59	23.41	2.21	1.63	0.44	23.21	Arunachal pradesh	Bhuyan et al. (2013)
Tropical agro-millet	213.54	23.58	12.24	9.06	17.45	1.93	1.19	0.51	153.00	Arunachal pradesh	Bhuyan et al. (2013)
Tropical agro-maize	234.48	26.55	13.55	8.83	17.30	1.96	1.67	0.48	159.41	Arunachal pradesh	Bhuyan et al. (2013)
Tropical agro-vegetable	238.45	19.93	9.28	11.96	25.70	2.15	2.77	0.47	21.58	Arunachal pradesh	Bhuyan et al. (2013)
Grassland											
Subtropical grassland	407.03	90.33	20.87	4.51	19.50	4.33	3.88	6.95	12.28	Manipur	Ibomcha Singh and Yadava (2006)

Table 4.5 Microbial stoichiometry and proportion of microbial nutrients to total soil nutrient in different agroforests and Jhumlands of Northeast India

Land-use types	Mic C	Mic N	Mic P	C:N	C:P	N: P	Mic C/SOC	MicN/ TN	MicP/ P	Location	References
Jhumlands											
Jhum-field low	902.20	67.90	Nil	13.29	–	–	2.15	1.70	–	Meghalaya	Arunachalam and Pandey (2003)
Jhum-field high	1522.30	82.90	Nil	18.36	–	–	2.62	1.66	–	Meghalaya	Arunachalam and Pandey (2003)
Jhum-fallows-1 year low	801.40	32.30	Nil	24.81	–	–	4.22	1.62	–	Meghalaya	Arunachalam and Pandey (2003)
Jhum Fallow-16 year Low	1431.90	98.20	Nil	14.58	–	–	3.98	1.40	–	Meghalaya	Arunachalam and Pandey (2003)
Jhum fallow High-1 year	651.40	61.30	Nil	10.63	–	–	2.72	2.04	–	Meghalaya	Arunachalam and Pandey (2003)
Jhum fallow high-16 year	1629.20	87.20	Nil	18.68	–	–	3.39	1.45	–	Meghalaya	Arunachalam and Pandey (2003)
Agroforests											
Cardamom agroforestry	583.00	63.00	22.00	9.25	26.50	2.86	2.60	2.70	3.70	Sikkim	Sharma et al. (2004)
Madarin agroforestry	471.00	48.00	18.00	9.81	26.17	2.67	3.30	2.50	3.10	Sikkim	Sharma et al. (2004)
Miscellaneous											
Wasteland temperate	0.0026	0.0002	0.000046	7.85	19.93	2.54	1.50	1.70	3.70	Sikkim	Sharma et al. (2004)
Wasteland subtropical	0.0016	0.00015	0.000047	7.30	13.69	1.88	1.90	1.90	3.30	Sikkim	Sharma et al. (2004)

Among the forests, *Dipterocarpus* forests have the highest microbial N:P ratio, while the least was observed in tropical forests which implies that microbial P concentration was high in tropical forests. A higher P in the microbes of the tropical forests could be due to P limitation in these forests as compared to other forests. Earlier studies have shown higher content of microbial P in P limited sites (Elser and Hamilton 2007; Xu et al. 2011). However, the variance of the microbial N:P ratio in the crop-based systems including the agroforests and wastelands is less but grassland exhibits a slight increase in N:P ratio (Tables 4.4 and 4.5).

4.8 Soil Nutrient Fractions in Microbial Biomass of Different Land Uses of Northeast India

The amount of soil nutrients present in the microbial biomass can also indicate the nutrient status of an ecosystem. Tropical forests showed high microbial soil nutrient fractions than other types of forests. This suggests nutrient limitations of tropical forests, as a higher nutrient fraction in microbial biomass indicates nutrient limitations to plants (Jonasson et al. 1999). Several studies from different nutrient-limited ecosystems reported high microbial nutrients, for example, elevated microbial N and P concentration in the boreal forest, a highly N limited (DeLuca et al. 2008), and a slightly P limiting system (Giesler et al. 2002). Anthropogenic disturbances such as logging increase soil nutrient fractions in microbial biomass but burning treatment reduce microbial nutrient accumulation. The felling of trees reduces plant litter or organic substrates ultimately reducing soil fertility and further stimulates the soil microbes to sequester more nutrients as a strategy to survive in stress. In contrast, the addition of the ashes and partially burnt plant materials due to burning enhance soil nutrients, which results in less microbial nutrient immobilization. Soil nutrient limitations may shift the decomposition ability of microbes to acquire N or P ultimately inducing high decomposition activity (Luo et al. 2013; Chen et al. 2014). Grassland has higher microbial nutrients than agricultural lands due to nutrient supplements in cropland in the form of organic manures and other forms of fertilizers. However, agroforestry systems and Jhumlands have comparable microbial nutrient fractions as that of forests. The relationship established between microbial nutrient fractions and soil nutrient contents using linear regression equations (Fig. 4.3) explained 2–62% variability of microbial biomass nutrients due to variance in soil nutrients concentrations of different land-use types.

4.9 Conclusion

The Northeastern part of India has a wide variation in land-use types, farming, and land management practices which therefore resulted in variation in soil fertility and quality. Change in land-use types and farming practices will have a significant impact on the soil of Northeast India. Slash and burn agriculture of Northeast India improves soil nutrient and fertility although it has a severe impact on

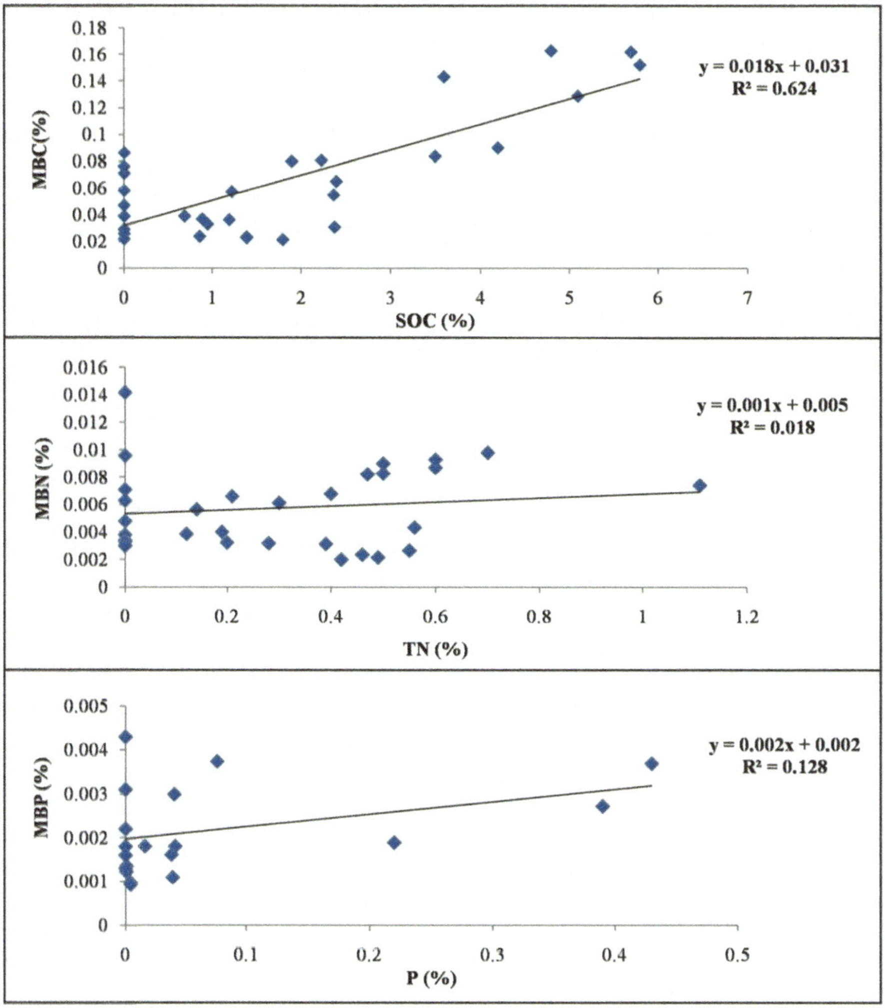

Fig. 4.3 Relationship between soil microbial nutrients and total soil nutrient concentration in different land-use types

vegetation especially trees due to deforestation. However, human-induced disturbances such as the felling of trees decrease microbial activity and nutrients in microbes. The trend of microbial biomass in different land-use types of Northeast India is Jhumlands > forests> agroforests > grassland > agricultural lands >wastelands. Tree-based systems such as agroforests exhibit higher microbial activity than purely agriculture-based systems. Microbial activity and nutrient sequestration changes due to variance in the site, plant species, climate, and land management. Microbial stoichiometry of different land-use types of Northeast India suggests N limitation in the soils of this region. Microbial C:N ratio of soils of this region indicates the dominance of fungi over bacteria which signifies nutrient mineralization in most of the sites leading to an increase in soil fertility. Deforestation and conversion of forests to grassland and further to agricultural land decline the microbial activity and biomass in soils which will deteriorate the soil fertility and quality of Northeast India. However, the incorporation of tree-based agriculture such as agroforestry can help in improving the soil quality and fertility. Transformation of natural forest to grassland will decrease microbial biomass by 46.26% and from grassland to agricultural land by 38.58% which will degrade the soil quality and fertility of Northeast India.

References

Arunachalam A, Pandey HN (2003) Ecosystem restoration of jhum fallows in Northeast India: microbial C and N along altitudinal and successional gradients. Restor Ecol 11:168–173. https://doi.org/10.1046/j.1526-100X.2003.00013.x

Austin AT, Vitousek PM (2000) Precipitation, decomposition and litter decomposability of Metrosideros polymorpha in native forests on Hawai'i. J Ecol 88:129–138. https://doi.org/10.1046/j.1365-2745.2000.00437.x

Bani A, Pioli S, Ventura M et al (2018) The role of microbial community in the decomposition of leaf litter and deadwood. Appl Soil Ecol. https://doi.org/10.1016/j.apsoil.2018.02.017

Barbhuiya AR, Arunachalam A, Pandey HN et al (2004) Dynamics of soil microbial biomass C, N and P in disturbed and undisturbed stands of a tropical wet-evergreen forest. Eur J Soil Biol 40:113–121. https://doi.org/10.1016/j.ejsobi.2005.02.003

Bhatia CR (2008) Role of microbial diversity for soil. Health and Plant Nutr:53–74. https://doi.org/10.1007/978-3-540-75575-3_2

Bhuyan SI, Tripathi OP, Khan ML (2013) Seasonal changes in soil microbial biomass under different agro-ecosystems of Arunachal Pradesh, north East India. J Agric Sci 8:142–152

Binarani RK, Yadava PS, Ibechoubi L (2009) Soil organic carbon and soil microbial biomass under the shifting cultivation systems of Kangchup Hills Manipur, north-East India. Int J Ecol Environ Sci 35:261–266

Bonkowski M, Villenave C, Griffiths B (2009) Rhizosphere fauna: the functional and structural diversity of intimate interactions of soil fauna with plant roots. Plant Soil 321:213–233. https://doi.org/10.1007/s11104-009-0013-2

Brookes P (2001) The soil microbial biomass: concept, measurement and applications in soil ecosystem research. Microbes Environ 16:131–140. https://doi.org/10.1264/jsme2.2001.131

Chen R, Senbayram M, Blagodatsky S et al (2014) Soil C and N availability determine the priming effect: microbial N mining and stoichiometric decomposition theories. Glob Chang Biol 20. https://doi.org/10.1111/gcb.12475

Cleveland CC, Liptzin D (2007) C:N:P stoichiometry in soil: is there a "Redfield ratio" for the microbial biomass? Biogeochemistry 85:235–252. https://doi.org/10.1007/s10533-007-9132-0

DeBano LF, Klopatek JM (1988) Phosphorus dynamics of pinyon-Juniper soils following simulated burning. Soil Sci Soc Am J 52:271–277. https://doi.org/10.2136/sssaj1988.03615995005200010048x

DeLuca TH, Zackrisson O, Gundale MJ, Nilsson MC (2008) Ecosystem feedbacks and nitrogen fixation in boreal forests. Science 80:320. https://doi.org/10.1126/science.1154836

Devi NB, Yadava PS (2010) Influence of climate and litter quality on litter decomposition and nutrient release in sub-tropical forest of Northeast India. J For Res 21:143–150. https://doi.org/10.1007/s11676-010-0023-1

Devi NB, Yadava PS (2006) Seasonal dynamics in soil microbial biomass C, N and P in a mixed-oak forest ecosystem of Manipur. North-east India Appl Soil Ecol 31. https://doi.org/10.1016/j.apsoil.2005.05.005

Di H, Cameron K, Shen J-P et al (2013) The role of bacteria and archaea in nitrification, nitrate leaching and nitrous oxide emissions in nitrogen-rich grassland soils. Mol Environ Soil Sci:79–89

Doran JW (2002) Soil health and global sustainability: translating science into practice. Agric Ecosyst Environ 88:119–127. https://doi.org/10.1016/S0167-8809(01)00246-8

Elser JJ, Hamilton A (2007) Stoichiometry and the new biology: the future is now. PLoS Biol 5: e181

Fanin N, Fromin N, Buatois B, Hättenschwiler S (2013) An experimental test of the hypothesis of non-homeostatic consumer stoichiometry in a plant litter-microbe system. Ecol Lett 16. https://doi.org/10.1111/ele.12108

Gao J, Feng J, Zhang X et al (2016) Drying-rewetting cycles alter carbon and nitrogen mineralization in litter-amended alpine wetland soil. Catena:145. https://doi.org/10.1016/j.catena.2016.06.026

Giesler R, Petersson T, Högberg P (2002) Phosphorus limitation in boreal forests: effects of aluminum and iron accumulation in the humus layer. Ecosystems 5. https://doi.org/10.1007/s10021-001-0073-5

Griffiths BS, Philippot L (2013) Insights into the resistance and resilience of the soil microbial community. FEMS Microbiol Rev 37:112–129

Griffiths BS, Spilles A, Bonkowski M (2012) C:N:P stoichiometry and nutrient limitation of the soil microbial biomass in a grazed grassland site under experimental P limitation or excess. Ecol Process 1:1–11. https://doi.org/10.1186/2192-1709-1-6

Guénon R, Gros R (2013) Frequent-wildfires with shortened time-since-fire affect soil microbial functional stability to drying and rewetting events. Soil Biol Biochem:57. https://doi.org/10.1016/j.soilbio.2012.07.006

Güsewell S, Gessner MO (2009) N:P ratios influence litter decomposition and colonization by fungi and bacteria in microcosms. Funct Ecol 23. https://doi.org/10.1111/j.1365-2435.2008.01478.x

Heuck C, Weig A, Spohn M (2015) Soil microbial biomass C:N:P stoichiometry and microbial use of organic phosphorus. Soil Biol Biochem 85. https://doi.org/10.1016/j.soilbio.2015.02.029

Ibomcha Singh L, Yadava PS (2006) Spatial distribution of microbial biomass in relation to land-use in subtropical systems of north-East India. Trop Ecol 47:63–70

Jenkinson DS, Ladd JN (1981) Microbial biomass in soil: measurement and turnover. Soil Biochem 5:415–471

Jonasson S, Michelsen A, Schmidt IK (1999) Coupling of nutrient cycling and carbon dynamics in the Arctic, integration of soil microbial and plant processes. Appl Soil Ecol 11:135–146

Lele N, Joshi PK (2008) Analyzing deforestation rates, spatial forest cover changes and identifying critical areas of forest cover changes in north-East India during 1972–1999. Environ Monit Assess 156:159. https://doi.org/10.1007/s10661-008-0472-6

Liu X, Herbert S, Hashemi M et al (2006) Effects of agricultural management on soil organic matter and carbon transformation - A review. Plant Soil Environ 52. https://doi.org/10.17221/3544-PSE

Luo G, Xue C, Zhang F, Guo S (2020) Soil carbon, nitrogen, and phosphorus cycling microbial populations and their resistance to global change depend on soil C:N:P stoichiometry. Appl Environ Sci. https://doi.org/10.1128/mSystems.00162-20

Luo Y, Durenkamp M, De Nobili M et al (2013) Microbial biomass growth, following incorporation of biochars produced at 350 °C or 700 °C, in a silty-clay loam soil of high and low pH. Soil Biol Biochem 57. https://doi.org/10.1016/j.soilbio.2012.10.033

Martiny J, Bohannan B, Brown J et al (2006) Microbial biogeography: putting microorganisms on the map. Nat Rev Microbiol 4:102–112. https://doi.org/10.1038/nrmicro1341

McKew BA, Taylor JD, McGenity TJ, Underwood GJC (2011) Resistance and resilience of benthic biofilm communities from a temperate saltmarsh to desiccation and rewetting. ISME J 5. https://doi.org/10.1038/ismej.2010.91

Mooshammer M, Wanek W, Schnecker J et al (2012) Stoichiometric controls of nitrogen and phosphorus cycling in decomposing beech leaf litter. Ecology 93:770–782. https://doi.org/10.1890/11-0721.1

Mouginot C, Kawamura R, Matulich KL et al (2014) Elemental stoichiometry of fungi and bacteria strains from grassland leaf litter. Soil Biol Biochem 76:278–285. https://doi.org/10.1016/j.soilbio.2014.05.011

Müller DB, Vogel C, Bai Y, Vorholt JA (2016) The plant microbiota: systems-level insights and perspectives. Annu Rev Genet 50:211–234. https://doi.org/10.1146/annurev-genet-120215-034952

Nanganoa LT, Okolle JN, Missi V et al (2019) Impact of different land-use systems on soil physicochemical properties and macrofauna abundance in the humid tropics of Cameroon. Appl Environ Soil Sci 2019:5701278. https://doi.org/10.1155/2019/5701278

Phillips LA, Schefe CR, Fridman M et al (2015) Organic nitrogen cycling microbial communities are abundant in a dry Australian agricultural soil. Soil Biol Biochem 86. https://doi.org/10.1016/j.soilbio.2015.04.004

Reich PB, Oleksyn J (2004) Global patterns of plant leaf N and P in relation to temperature and latitude. Proc Natl Acad Sci U S A 101. https://doi.org/10.1073/pnas.0403588101

Richardson AE, Barea J-M, McNeill AM, Prigent-Combaret C (2009) Acquisition of phosphorus and nitrogen in the rhizosphere and plant growth promotion by microorganisms. Plant Soil 321:305–339. https://doi.org/10.1007/s11104-009-9895-2

Saha R, Chaudhary RS, Somasundaram J (2012) Soil health management under hill agroecosystem of north East India. Appl Environ Soil Sci 2012:696174. https://doi.org/10.1155/2012/696174

Schlesinger WH, Bernhardt ES (2013) Chapter 11 - the global carbon cycle. In: Schlesinger WH, Bernhardt ES (eds) Biogeochemistry: an analysis of global change, 3rd edn. Academic Press, Boston, pp 419–444

Sharma P, Rai SC, Sharma R, Sharma E (2004) Effects of land-use change on soil microbial C, N and P in a Himalayan watershed. Pedobiologia (Jena) 48:83–92. https://doi.org/10.1016/j.pedobi.2003.09.002

Singh JS, Raghuvanshi AS, Singh RS, Srivastava S (1989) Microbial biomass acts as a source of plant nutrients in dry tropical forest and savanna. Nature 338:499–500

Singh JS (2015) Microbes: the chief ecological engineers in reinstating equilibrium in degraded ecosystems. Agric Ecosyst Environ 203:80–82. https://doi.org/10.1016/j.agee.2015.01.026

Singh JS, Gupta VK (2018) Soil microbial biomass: a key soil driver in management of ecosystem functioning. Sci Total Environ 634:497–500. https://doi.org/10.1016/j.scitotenv.2018.03.373

Singh P, Singh VK, Singh R et al (2019) Bioremediation: A sustainable approach for management of environmental contaminants. Elsevier Inc., Amsterdam

Somasundaram J, Singh RK, Parandiyal AK et al (2013) Soil properties under different land use systems in parts of Chambal region of Rajasthan. J Agric Phys 13:139–147

Sterner RW, Elser JJ (2017) 2. Biological chemistry: building cells from elements. In: Ecological stoichiometry. Princeton University Press, Princeton

Strickland MS, Rousk J (2010) Considering fungal:bacterial dominance in soils – methods, controls, and ecosystem implications. Soil Biol Biochem 42:1385–1395. https://doi.org/10.1016/j.soilbio.2010.05.007

Takoutsing B, Weber J, Aynekulu E et al (2016) Assessment of soil health indicators for sustainable production of maize in smallholder farming systems in the highlands of Cameroon. Geoderma 276:64–73. https://doi.org/10.1016/j.geoderma.2016.04.027

Van der Heijden M, Bardgett RD, van Straalen NM (2008) The unseen majority: soil microbes as drivers of plant diversity and productivity in terrestrial ecosystems. Ecol Lett 11(3):296–310. https://doi.org/10.1111/j.1461-0248.2008.01199.x

Velasquez E, Lavelle P, Andrade M (2007) GISQ, a multifunctional indicator of soil quality. Soil Biol Biochem 39:3066–3080. https://doi.org/10.1016/j.soilbio.2007.06.013

Vimal SR, Singh JS, Arora NK, Singh S (2017) Soil-plant-microbe interactions in stressed agriculture management: a review. Pedosphere 27:177–192. https://doi.org/10.1016/S1002-0160(17)60309-6

Wardle DA (1992) A comparative assessment of factors which influence microbial biomass carbon and nitrogen levels in soil. Biol Rev 67:321–358. https://doi.org/10.1111/j.1469-185X.1992.tb00728.x

Xu X, Thornton PE, Post WM (2013) A global analysis of soil microbial biomass carbon, nitrogen and phosphorus in terrestrial ecosystems. Glob Ecol Biogeogr 22:737–749. https://doi.org/10.1111/geb.12029

Xu X, Tian H, Pan Z, Thomas CR (2011) Modeling ecosystem responses to prescribed fires in a phosphorus-enriched Everglades wetland: II Phosphorus dynamics and community shift in response to hydrological and seasonal scenarios, Ecol Model:222. https://doi.org/10.1016/j.ecolmodel.2011.09.003

Yadava PS, Devi AS (2004) Impact of slash and burning on microbial biomass in semi-Evergreen tropical deciduous Forest of Manipur, north-East India. Korean J Ecol 27:225–230. https://doi.org/10.5141/jefb.2004.27.4.225

Microbes and Plant Mineral Nutrition 5

R. Rajkumar and C. Kurinjimalar

Abstract

Plants and microbes are inseparable constituent of ecosystem for plant mineral nutrition availability. The prime important factors for the plant productivity are macro and micronutrients next to water. A few plants can assimilate macronutrients and micronutrients such as nitrogen, phosphorus, potassium and iron, zinc, copper, manganese, respectively, from soil, while other cannot take up readily. The soil nutrient cycling and its availability to plants are invariantly caused by soil microbes and thus soil health is determined by soil microbes. The different modes of nutrients acquisition by plants are direct uptake from the soil or symbiotic association with soil microbes and mycorrhizal interaction with plants. The indirect way of soil nutrients availability to plants are biological nitrogen fixation, phosphate and sulfate solubilization. The important role of soil microbes in plant growth promotion includes improved plant health, plant growth, plant yield, and also antagonistic to plant pathogens. This chapter highlights different classes of plant nutrients and efficient microbes (cyanobacteria, bacteria, and fungi) helpful in plant nutrition availability are discussed.

Keywords

Bioavailability · Soil nutrients · Plant growth-promoting Rhizobacteria · Biofertilizer · Macro and micronutrients · Bioavailability of nutrients

R. Rajkumar (✉)
Department of Environmental Sciences, Bharathiar University, Coimbatore, India

C. Kurinjimalar
Centre for Advanced Studies in Botany, University of Madras, Chennai, India

R. Soni et al. (eds.), *Microbiological Activity for Soil and Plant Health Management*, https://doi.org/10.1007/978-981-16-2922-8_5

5.1 Introduction

The biological soil fertility is maintained by the soil microbial habitants, namely bacteria, fungi, actinomycetes, algae, and protozoa (Muller et al. 2016). The importance of microbes in soil is recycling of nutrient and thus plant nutrition has been realized for many years. The plants are not an individual entity as they co-habit with the microbes. The different plant–microbial interactions are symbiotic, pathogenic, epiphytic or endophytic (Iniguez et al. 2005). The beneficial plant-associated microbiota increases plant nutrient availability by different microbes including cyanobacteria, bacteria including N-fixing bacteria, actinomycetes, fungi (Arbuscular Mycorrhizal), and other Plant Growth-Promoting Rhizobacteria (PGPR) (Miransari 2010, 2011a). Commonly, soil nutrients are present as insoluble precipitates or bound to inorganic and organic components in the soil. Thus direct plant nutrients are not present in available forms. The soil microbes colonizing the plant root either decompose or mineralize organic matter and thus released nutrients in sustainable plant productivity (Hirel et al. 2011). Together with physical and chemical factors, biological factor improves the agricultural drylands. Thus plant–microbe interaction improves the plants growth under nutrient deprived soils.

The fundamental of life on the planet is maintained by soil microbes by nutrient recycling process. Bacteria are the fundamental decomposers of organic waste from inaccessible forms to usable forms of different nutrients. The *Pseudomonas* sp., *Streptomyces* sp. bacterial species and *Trichoderma harzianum,* and *Polyporus ostriformis* lignocellulolytic fungi after consumption for their nutrition release inorganic plant nutrients (Woo et al. 2014). The increase in soil fertility and thus improves plant productivity. In other ways, increasing world population necessitates the increased food production by the use of fertilizers and herbicides of chemical origin. This leads to irreversible environmental damages of ground water pollution and soil degradation. It is therefore need an alternate biological approach of plant nutrients availability without affecting the environment (Miransari 2011b). Therefore, soil inoculation with indigenous microflora is a bioprospecting approach for the necessary plant–microbe interaction. Gopal and Gupta (2016), shows plant dependence on plant-associated microbes including biofertilization, protection from diseases and tolerance to abiotic stresses.

Plant growth directly depends on soil, soil-borne organisms and its interaction with plants. When nutrients are not present in readily available forms for the plant growth, it causes reduction in crop yields or quality. This leads to reduction in overall biodiversity of living creatures, as plants being the primary producers of food web chain. A diverse range of ecological interactions between soil-borne organisms and plants is developed (competition, neutral, commensalism, and mutualism). The symbiotic association of mycorrhizal and legume-*Rhizobium* findings are the base of present studies on microbial role for plant growth. Recent biotechnological advances of next-generation 16S rRNA gene sequencing show major contribution of bacteria present in the core of plant microbiomes. Hawkes et al. (2007) reported as many as 1200 bacterial taxa associated with rhizosphere of 14 different plant species.

Nutrient deficiency can have a significant impact on agriculture, resulting in reduced crop yield or reduced plant quality. Plants require 17 essential nutrients which are derived from the soil. Macronutrients, required in large quantities includes carbon (C), nitrogen (N), oxygen (O_2), hydrogen (H), phosphorus (P), potassium (K), calcium (Ca), magnesium (Mg), and sulfur (S). Among these, carbon, hydrogen, and oxygen are the building block of organic macromolecules of the cell. Micronutrient includes iron (Fe), zinc (Zn), manganese (Mn), copper (Cu), chlorine (Cl), and boron (B) are the cofactors for enzyme activity required in very small amounts. These elements are required to plant growth for both structural and biochemical functions. The source of essential elements such as hydrogen, carbon, and oxygen are derived from carbon-di-oxide and water, while the others have to be absorbed from mineral nutrition in soil.

The soil microbes are an important component of ecosystem with essential function in plant nutrient nourishment and protecting plants. The use of soil-borne organism for promoting plant growth is biofortification (Miransari 2011b). These soil microbes metabolize soil-borne nutrients for their nutrition and thereby the elemental nutrients for plant use (Uroz et al. 2009). In other way N, P, S, and K are present in various organic molecules as complexes that are minimally bio-available to plants. This can be overcome by the conversion of organic to inorganic forms (N, P and S) through the process of depolymerization and mineralization (van der Heijden et al. 2008). Thus chemical fertilizer application can be minimized by the use of microbes and making easy availability of nutrients to plants. The microbe mediated nutrient availability is proved to be effective by keeping the environment clean and enhancing soil nutrient availability. Thus understanding the linkage between plant and microbes allows better nutrient management. The different mechanisms to increase the solubility of soil nutrients by symbiotic or non-symbiotic association of microbes with their host plant are: (1) Plant metabolism, (2) Increasing solubility and availability of soil nutrients, and (3) Interaction with other soil microbes (Miransari 2013). Therefore, sustainable agriculture can be developed by the application of plant growth-promoting bacteria as a new management strategy.

The different beneficial role of soil microbes by symbiotic or non-symbiotic association to boost growth of host plants are (1) Bioavailability of complex soil nutrients by solubility, (2) Microbial pathogenic strains control, (3) Manipulating the hormonal signaling of plants, (4) Nutrient cycling by mineralization of soil organic matter, (5) Production of different biochemicals such as plant hormones and enzymes, (6) Bioremediation of the toxic heavy metal species and degrading xenobiotic compounds in the polluted soils, (7) production of siderophores, (8) production of 1-aminocyclopropane-1-carboxylate (ACC) deaminase, (9) improving abiotic stress resistance, (10) drought stress tolerance (van der Heijden et al. 2008; Mendes et al. 2013; Verbon and Liberman 2016; Vurukonda et al. 2016) (Fig. 5.1).

Soil microbes perform essential functions of nourishing as well as protecting plants. The understanding of plant and soil microbe interactions allow better nutrient management for sustainable crop production. The niche of plant roots is habitat for microbial communities. The microbial conversion of organic complex material into

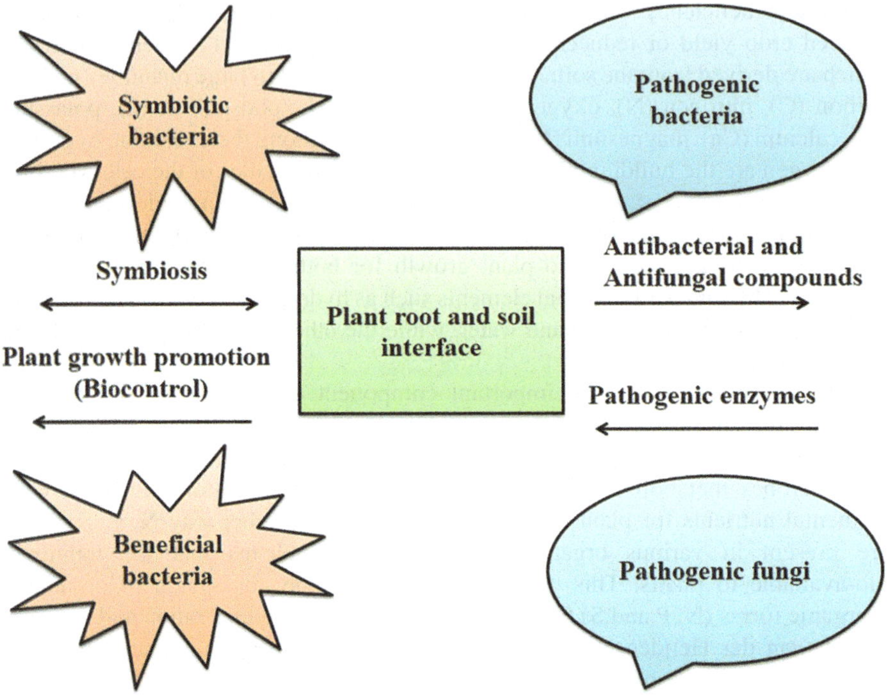

Fig. 5.1 Interaction of plant roots and microbes

Table 5.1 Different mechanisms of plant nutrient availability mediated by soil microbes

Nutrient	Mechanism of transformation by microbes
Nitrogen	Mineralization, Nitrification, Denitrification, N_2 fixation, extracellular enzyme activity (protease and chitinase)
Phosphorus	Mineralization, Solubilization, Extracellular phosphatase activity, organic acid mediated dissolution
Potassium	Mineralization, K solubilization
Iron	Chelation, production of siderophores, changing oxidation states
Zinc	Solubilization, Facilitated uptake by mycorrhizal fungi
Copper	Production of carboxylase and phenolic compounds, Facilitated uptake by mycorrhizal fungi
Manganese	Change in oxidation and reduction state

plant available forms as N, P, and S are carried over by soil microbes by various mechanisms (Table 5.1). This chapter will focus on the mechanism of increasing the soil-borne nutrients bioavailability to the plants.

5.2 An Overview of Soil Microorganisms for the Availability of Nutrients in Plants

The different roles of symbiotic and free living soil microbes in plant growth are nitrogen fixation, improving soil structure which promotes root growth and to control plant pathogens (Coyne and Mikkelsen 2015). Figure 5.2 shows the role of soil microbes involved in the major nutrients accessibility for plant growth. The nitrogen recovery process is carried out by mycorrhizal fungi either ecto-mycorrhizal or endo-mycorrhizal fungi. The process involves solubilization of organic matter to soluble forms by enzymatic process. Nitrogen fixation carried out by symbiotic bacteria, namely *Frankia* through infection in woody species. In root-associated asymbiotic bacteria, namely *Azospirillum* can provide nitrogen to the roots of grasses. Good soil structure enhances plant root growth and extraction of nutrients. The organic molecules like protein released by soil microbes bind soil particles and improve soil structure. The mechanism of pathogens control by plant rhizospheric organisms is competition between pathogens and beneficial microbes for essential nutrients. Many studies reported pathogen control by the application of antagonistic beneficial microbes in disease control.

Soil microbes in 1 g of root niche habitat estimates as 10^{11} microbe (Rolli et al. 2015). The root microbiome inhabiting soil microbes of food crops include maize (Peiffer et al. 2013), rice (Edwards et al. 2015), sugarcane (Yeoh et al. 2015), and sweet potato (Marques et al. 2014). Rhizosphere organisms are categorized into biofertilizers and biocontrol agents which enhance the nutrient availability to plants and prevention of plant diseases (Glick 2012). The rhizobacteria may be extracellular or intracellular, present in the rhizoplane and inside the root cells as nodules. The plant beneficial microbes are significant in agriculture due to sustainability, environmental safety, and multiple beneficiaries as improved nutrient acquisition, plant growth and tolerance to environmental stresses (Sharma et al. 2017). The direct soil nutrients availability to plants are the production of different compounds such as

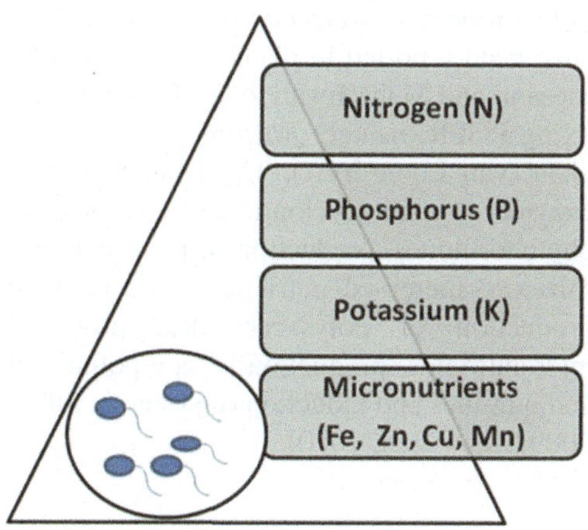

Fig. 5.2 Role of soil microbes in major nutrients accessibility for plant growth

organic (namely glucose, fructose, sucrose, ribose, maltose), organic acids (namely citric, lactic, pyruvic, malic, succinic, and oxalic acids), fatty acids, amino acids, vitamins, and putrescine by plant roots as signaling molecules which influences microbial population for the solubility and availability of nutrients (Ortíz-Castro et al. 2009; Johnson et al. 2010).

The different types of plant–microbe interrelations are neutral, negative, and positive interaction (Whipps 2001). The neutral interaction, namely commensalism have no significant effects on plant growth (Beattie 2007). In negative association, phytopathogenic microbes affect the plant growth by producing phytotoxic substances such as hydrogen cyanide (HCN) or ethylene, (Khalid et al. 2004). Therefore, in positive association the microbes can promote plant growth either directly or indirectly (Glick 2012). Therefore, different plant parts are colonized by different microbe's which are grouped into three groups such as rhizosphere microbes (resides on root surrounding area), rhizoplane microbes (resides on root surface), and endophytic microbes (resides within host tissues) (Andrews and Harris 2000).

A well-known endo-symbiotic association of plant and microbes are between plants and mycorrhizal fungi as well as nodulated bacteria in leguminous plants recognized date backs to the nineteenth century (Morton 1981). Mycorrhizal fungi have increased surface area with improved nutrient absorptive capacity of roots (Rillig 2004). A holistic approach to improve plant growth and yields are developed by crop seeds coating with bacterial cultures such as *Azotobacter* and *Bacillus megaterium* (Brown 1974). Rhizobacteria such as *Pseudomonas* and *Azospirillium* are commonly known as Plant Growth Promoting Rhizobacteria (PGPR) that enhances the plant growth (Burr et al. 1978). Thus plants access the recalcitrant soil-borne nutrients by the metabolic activities of soil microbes. Therefore, selection of soil microbes for biofertilization is based on the (1) inoculation potential, (2) persistence in soil and (3) able to survive under stress condition.

The major PGPRs genera include *Frankia*, *Azotobacter*, *Streptomyces*, *Arthrobacter*, *Rhizobium*, *Azospirillum*, *Pseudomonas*, *Bacillus*, *Pseudomonas*, *Flavobacterium*, *Thiobacillus*, *Enterobacter*, *Serratia*, *Chryseobacterium*, *Achromobacter*, *Aeromonas*, *Acetobacter*, *Bradyrhizobium*, and *Sinorhizobium* have been reported in plant growth promotion (Vessey 2003; Dimkpa et al. 2009; Etesami and Maheshwari 2018; Etesami and Beattie 2018) (Table 5.2). The nitrogen fixing PGPR, namely *Rhizobia* spp. develops symbiotic association with the host plant (van Loon 2007). The other beneficial benefits of PGPR are production of enzyme, 1-aminocyclopropane-1-carboxylate (ACC) deaminase, production of phytohormones, production of plant pathogens control products such as HCN, and rhizoxin, increased nutrient uptake in plants, induction of antioxidant enzymes, production of polysaccharides, production of carboxylates which chelate micronutrients, production of compounds like humic acid, riboflavin, phenazines, and quinines and induction of systemic tolerance (Hernandez et al. 2004; Uroz et al. 2009; Zhao et al. 2010).

Table 5.2 Important soil microbes in plant nutrient availability from soil

Rhizobacteria role	Mechanism of action	Microbes	References
Increase in plant macronutrients uptake	Solubilizing and mineralizing insoluble phosphate	*Rhizobium*, *Pseudomonas*, *Bacillus*, *Enterobacter*, and *Pantoea*	Sharpley et al. (1992); Sharma et al. (2013); Etesami and Maheshwari (2018)
Not mentioned	Solubilizing K-containing minerals by the synthesis of organic acids, inorganic acids, siderophores	*Bacillus mucilaginosus*, *Bacillus circulans*, *Bacillus edaphicus*, *Paenibacillus spp.*, *Pseudomonas*, *Acidithiobacillus ferrooxidans*, and *Burkholderia*	Botella et al. (1997); Ahmad et al. (2016); Sindhu et al. (2016); Etesami et al. (2017)
Not mentioned	Symbiotic and non-symbotic N_2 fixation by enzyme nitrogenase; Mineralizing organic forms of N	Not mentioned	Feigin (1985); Glick (2012); Santi et al. (2013); Etesami and Beattie (2017)
Increase in plant micronutrients uptake-Fe, Zn, Mn, Cu, Mo, Ni, Co	Generation of iron chelators, namely siderophores	*Bacillus*, *Pseudomonas*, and *Geobacter*	Zhuang et al. (2007); Iqbal et al. (2010); Etesami (2018a)
Decrease salinity stress	IAA-containing PGPR stimulate ACC levels	*P. putida*, *P. fluorescens*, *Variovorax paradoxus*, *Enterobacter* sp., *Arthrobacter* sp., *Bacillus* sp., and *Pantoea dispersa*	Bal et al. (2013); Glick (2014); Wang et al. (2016)
Production of phytohormones	PGPR synthesize IAA and cytokinin which induces root growth for positive effect on water acquisition	*Pseudomonas aurantiaca*, *Pseudomonas extremorientalis*, and *Bacillus subtilis*	Ilangumaran and Smith (2017); Barnawal et al. (2017); Etesami and Beattie (2018)
Salinity stress alleviation	Taking up K^+ within their cells and amassing compatible solutes such as polyols and derivatives, amino acids and their derivatives, sugars and derivatives, betaines, and ectoines	*Pseudomonas fluorescens*, *Serratia* sp., *Pseudomonas*, *Bacillus*, *Flavobacterium*, *Azospirillum*, *Chryseobacterium*, *Achromobacter*, *Sinorhizobium*, *Bradyrhizobium*, *Aeromonas*, and *Acetobacter*	Lugtenberg et al. (2013); Paul and Lade (2014); Qin et al. (2016); Etesami and Beattie (2017, 2018); Etesami and Maheshwari (2018)

(continued)

Table 5.2 (continued)

Rhizobacteria role	Mechanism of action	Microbes	References
Accumulation of osmolytes in plants	Production of proline and glycine betaine	*B. subtilis, Bacillus amyloliquefaciens, Bacillus aquimaris, Azospirillum brasilense, P. dispersa, Rhizobium tropici*, and *Paenibacillus polymyxa*	Ilangumaran and Smith (2017); Etesami and Maheshwari (2018)
Induction of antioxidant enzymes	Scavenging ROS by the antioxidant enzymes such as CAT, POD, SOD, polyphenol oxidase, phenylalanine ammonia-lyase, phenolics, and lipoxygenase	*Bacillus pumilus, B. cepacia, Promicromonospora* spp., *A. calcoaceticus, Bacillus* spp., *Exiguobacterium oxidotolerans, Pseudomonas pseudoalcaligenes*, , *B. subtilis*, and *Arthrobacter*	Jha et al. (2011); Upadhyay et al. (2012); Damodaran et al. (2014); Etesami (2018b)
Production of exopolysaccharides	Augmenting water and fertilizer availability to raise from seeds and plants	*Pseudomonas mendocina, Halomonas variabilis, Planococcus rifietoensis, Enterobacter* sp., *Bacillus* sp., *Bacillus amylolequifaciens, Bacillus insolitus, Microbacterium* spp., and *Pseudomonas syringae*	Tewari and Arora (2014); Sandhya and Ali (2015); Etesami and Maheshwari (2018)
Induction of systemic tolerance	Generation of volatile organic compounds	*B. subtilis*	Egamberdieva and Lugtenberg (2014); Etesami and Maheshwari (2018)

In nutrients assimilation, plant growth-promoting bacterial strains should possess successful colonization in the niche of rhizosphere. In nitrogen fixing symbiotic leguminous host plant nitrogen fixing diverse bacterial taxa possesses nitrogenase genes (Gyaneshwar et al. 2011; Mus et al. 2016). Rhizobia fixes free atmospheric nitrogen to the host plant and in turn depend on host for photosynthates (Hunter 2016). The Arbuscular mycorrhizal (AM) fungi develop symbiotic association with host plants called mycorrhizosphere with organelles called vesicles and arbuscules as storage organ and as hyphal branched structure, respectively (Smith and Read 2008). Endo-symbiotic Arbuscular mycorrhizal (AM) fungi establish increased water and nutrients uptake for their host plant in the exchange for carbon (Rogers and Oldroyd 2014).

5.3 Soil Microbes Induced Nitrogen Uptake by Plants

One of the most important plant growths limiting nitrogen element is not present in readily available form used by the plant. Despite nitrogen is the most abundant gaseous element which constitutes of about 78% in the atmosphere it cannot be directly assimilated by plants due to the presence of triple bonds between two N atoms. Green plants can only readily acquire inorganic nitrate (NO^{3-}) and ammonium (NH^{4+}) from the soil through the plant roots. In the absence of nitrate in the soil, it can be replenished using soil microbes (PGPR) by fixing atmospheric nitrogen (N_2) in the atmosphere.

Therefore plant species has evolved mechanisms of beneficial symbiotic and non-symbiotic process of nitrogen fixation with soil-borne microorganisms called diazotrophs. It can be classified based on the trait of N_2 fixation into three subgroups: symbiotic, free living, and associative. In symbiotic association plant species of Fabaceae family by the process of biological nitrogen fixation develops symbiotic relationship with a group of nitrogen fixing bacteria such as *Rhizobium* and other related genera. The interactions are based on chemical signaling between plant and microbial compounds, flavonoids and lipochitooligosaccharides, respectively. This cause root hair curling to envelop the bacteria through which infection thread grows into the cells of the root cortex and differentiate into structures called bacteroids. The subsequent division produces the nodules, which converts atmospheric nitrogen to ammonia catalyzed by nitrogenase enzyme complex, a readily available form to plants (Ferguson et al. 2010). As a result of this biological nitrogen fixing association between the host plant and microbes, both are benefited with productivity and survival, respectively. The deleterious approach of using nitrogen-rich fertilizers to combat nitrogen deficiency leads to ground water pollution with eutrophication in aquatic ecosystem. The root or legume associated bacteria fixes atmospheric nitrogen into usable N (176×10^{12} g year^{-1}).

The other types of non-symbiotic associative relationships of host plant with atmospheric nitrogen fixation bacteria are Cyanobacteria, *Azospirillum* spp., *Acetobacter* spp., *Azotobacter* spp., *Bacillus* spp., *Pseudomonas* spp., *Rhodospirillum* spp., *Corynebacterium* spp., *Beijerinckia* spp., and others (Saharan and Nehra 2011). Furthermore some bacteria involve mineralization of organic N compounds into inorganic forms (NH^{4+} and NO^{3-}) that are readily available for growth of plants. Non-symbiotic organisms degrade organic matter and fix the atmospheric nitrogen for plant use. For example, the plant parts that remain after harvest are the source of N to the soil upon decomposition. During the decomposition organic nitrogen is converted to inorganic ammonium into the soil and the process is called bio-mineralization. The different steps in bio-mineralization are aminization, ammonification, and nitrification, where the initial step involves break down of complex proteins to amino acids, amides, and amines, further converted to ammonium and finally nitrate formation occurs. The two steps of nitrification are carried out by different bacterial groups, namely *Nitrosomonas* and *Nitrobacter*, respectively. The common nitrogen fixing rhizobacteria either endophytic or free-living genera includes *Azotobacter, Azospirillum, Bacillus, Bradyrhizobium, Burkholderia,* and *Pseudomonas* are having positive effects on food crops (Igiehon and Babalola 2018).

5.4 Soil Microbes Induced Phosphate Uptake by Plants

The second most essential element phosphorus (P) in inorganic phosphate form is also a plant growth limiting nutrient. In natural plant habitat phosphorus is present in the form of rock phosphate, mineral salts, hydroxyapatite or organic compounds. This helps in root development, seed production, improves BNF and resistance to diseases (Murrell and Munson 1999). Phosphorus is an integral component of biochemicals such as nucleic acids, and phosphoproteins. Thus a large proportion of about 95–99% of phosphorus are present in unavailable forms as present in insoluble or precipitates, which causes phosphorus limitation. The two kinds of phosphate available to the plants are mineral phosphate and organic phosphate. Generally for phosphate fertilization triple super phosphate is applied to soil only 20% of soil phosphate is in available form and the remaining will precipitate. Soil pH affects the plant nutrient availability in the soil, in which solubility of nitrogen, potassium, and sulfur are not much affected as phosphate. At acidic and basic pH, phosphate ions react rapidly with Al, Fe, Ca, and Mg and they become less soluble forms. Therefore, phosphate availability is a pH-dependent. A number of studies reported different strains of both fungal and bacterial solubilize inorganic P, and also mineralize organic P (Ahemad and Kibret 2014).

The mechanism of inorganic and organic phosphate acquisition involves solubilization and mineralization by the release of microbial phosphatases (Illmer et al. 1995; Gouda et al. 2018). This bio-solubilization of P is most important in agricultural soils for enhanced utilization of P by plant. A wide range of soil bacteria especially PGPR and fungi could convert precipitated and organic phosphate in the form $H_2PO_4^-$ and HPO_4^{-2} ions (Whitelaw 2000). The lowering in pH of the medium by organic acids production in phosphate solubilizing microbes (PSB) dissolves organic phosphorus into inorganic form. The different organic acids are acetic acid, citric acid, fumaric acid, oxalic acid, lactic acid, propionic acid, malonic acid, succinic acid, 2-ketogluconic acid, glycolic acid, and gluconic acid are produced in PSB (Krishnaraj and Dahale 2014). Among the several soil bacterial communities, *Pseudomonas* and *Bacillus* spp. have been excellent phosphate solubilizers (Goswami et al. 2014). The common mechanisms of microbial action involves production of enzymes (phosphatases) and products such as low molecular weight organic products (carboxylic acids) pH, anions, and cations can readily increase phosphate availability (Houser and Richardson 2010; Salimpour et al. 2010). The PSB solubilizes inorganic soil phosphates of Ca, Fe, and Al via production of siderophores, several acids (organic), and hydroxyl and carboxyl groups, and chelating them to the bound phosphates and the available calcium (Sharma et al. 2013).

Among bacteria, the most efficient phosphate solubilizer belongs to genera such as *Rhizobium*, *Bacillus*, and *Pseudomonas*. The other two chickpea nodulating species, *Mesorhizobium ciceri* and *Mesorhizobium mediterraneum*, are reported as phosphate solubilizers (Rivas et al. 2006). A large number of phosphate solubilizing microorganisms has been reported by in vitro studies. The most commonly reported phosphate solubilizing organisms of bacteria such as *Azospirillum brasilense*, *Arthrobacter*, *Rhizobium*, *Pseudomonas*, *Arthrobacter*, *Beijerinckia*, *Erwinia*, *Thiobacillus ferrooxidans Bacillus*, and *Nitrobacter sp.* (Sharma et al. 2013); fungal genera are *Penicillium Aspergillus*, *Fusarium*, and Chaetomium spp. of the plant microbiome hydrolyze inorganic phosphorus into soluble forms (Uribe et al. 2010; Sharma et al. 2013). The different mechanisms involved in phosphorus availability by microbes are release of H^+, OH^-, organic acid, anions such as citrate, malate, and oxalate and also mineralization of organic P by release of various phosphatase enzymes (Marschner et al. 2010). Thus soil organisms can be applied as potent soil inoculants for plant growth.

Phosphate solubilizing bacteria produce organic acids helps in dissolution of glucose to gluconic acid. Organic acids have good chelation properties by the substitution of divalent cation of Ca^{2+} coupled with the release of phosphates from insoluble complexes (Behera et al. 2017). Thus most phosphate solubilizing microbes cause a reduction in the pH of the medium either by production of inorganic acids as H^+ extrusion (sulfuric and nitric acids) or by secretion of various organic acids like malic, citric, succinic, tartaric, and oxalic acids; and by the production of enzymes acting on fatty substrates (Alori et al. 2017). The organic acids secretion by microbes is mainly dependent on environmental properties and gene induced in bio-solubilization of phosphate (Zhen et al. 2016). In this study, revealed a clear relationship between profile of organic acids and source of phosphate. The other common P-mineralization process is governed by hydrolyzing enzymes, namely phytases and phosphatases produced by fungi and bacteria (Alori et al. 2017).

Other rhizospheric bacteria produce siderophores, phytohormones, including auxins, gibberellins, cytokinins, ethylene, and abscisic acid involved in plant growth and yield improvement through N_2 fixation. The phosphate solubilizing for P availability to plants occurs either by chelation with cations or in exchange of organic acid by ligands (Parker et al. 2005). A number of reports show plant growth promotion activities by the microbes besides making soluble P accessible for plants uptake (Sharma et al. 2013). Indole-3-acetic acid (IAA), produced by various PGPR are involved in plant growth and development, such as cell elongation, cell division, and tissue differentiation (Misra et al. 2012; Oves et al. 2013; Kaushal et al. 2017). The processes in plants like promotion of seed germination, nutritional signaling, expansion of leaf, and delay of senescence are also greatly influenced by cytokinins produced by PGPR (Wong et al. 2015).

Egamberdiyeva (2005) isolated and reported rhizospheric bacteria from the field, namely *Bacillus laevolacticus*, *B. amyloliquefaciens*, *Pseudomonas denitrificans*, *P. rathonis, and Arthrobacter simplex* from wheat, alfalfa, cotton, and tomato. *Acinetobacter* sp. and *Bacillus* sp. isolates from *Phyllanthus amarus* showed phosphate solubilizing property with promoted higher vigor index, phosphorus content, percentage of germination, plant biomass, phenolic content, and also antioxidative activity compared to uninoculated control. Therefore, many researchers reported the role of phosphate solubilizing bacteria towards enhancing plant growth and reducing the usuage of fertilizer with high salt tolerance (Alori et al. 2017). A large number of field-based studies evidenced a high uptake of phosphate that enhance the crops yield (Sawers et al. 2017).

5.5 Soil Microbes Induced Potassium Uptake by Plants

Potassium is the third essential macronutrient present in soil is absorbed by plants for its growth and yield. The different forms of potassium (K) in the soil are non-exchangeable K, exchangeable K, mineral non-exchangeable K, and K in soil solution (water-soluble K). The important functions of potassium within plant cell include control of stomatal opening/closing, enzyme activation, and balancing the charges of cellular anions. The deficiency of potassium leads to chlorosis (yellowing), browning of leaves, and curling of leaf tips with ultimate reduced growth and yield. The application of chemical and organic fertilization can make potassium availability for plant use. However, owing to soil erosion, leaching, intensive cropping, imbalanced fertilizer application, and presence of insoluble K sources, the availability of potassium to plants is decreasing (Zorb et al. 2014). As a result, deficiency of K in soils is reducing crop production. Only tightly bound mineral form of K constitute about 90–98% of soil K and it is unavailable for plant (Sparks 1987). Therefore K solubilizing microbes have been applied for sustainable production by mineralizing insoluble potassium into usable forms. Microbes solubilize K from insoluble sources of feldspar, mica, and others by extracellular production of organic acids, extracellular enzymes, siderophores, extracellular polysaccharides, organic ligands and formation of biofilms are the different key processes involved in the release of K by the dissolution of complex minerals (Keshavarz Zarjani et al. 2013; Meena et al. 2015; Das and Pradhan 2016).

Ullman (1996) reported that the bacterium, *Bacillus mucilaginosus* are able to solubilize potassium rock by the secretion of organic acids. Thus K-solubilizing microbes (KSB) play an effective role in K cycle to fulfill the K requirement of crops (Meena et al. 2014; Sindhu et al. 2014). The process involves conversion of mineral K into available K to plants. A number of genera reported to release K from K-bearing minerals, namely *Pseudomonas* spp., *Burkholderia* spp., *Acidithiobacillus ferrooxidans*, *Enterobacter hormaechei*, *Paenibacillus glucanolyticus*, *Arthrobacter* spp., *Paenibacillus mucilaginosus*, *P. glucanolyticus*, *Bacillus mucilaginosus*, *B. edaphicus*, and *B. circulans* (Lian et al. 2002; Keshavarz Zarjani et al. 2013; Zhang et al. 2013).

Overall the direct mechanisms used by KSB include: (i) acidolysis, a process of dissolving mineral K by organic acids (Shelobolina et al. 2012), (ii) chelation, a mechanism of forming complexes with reaction products with organic acids which can enhance dissolution (Ullman and Welch 2002), (iii) oxidation, involves breakdown of K-bearing minerals microbial Fe(II) oxidation in the rhizosphere, and (iv) production of CO_2 for the breakdown of K-bearing complex which result in carbonic acid formation (Barker et al. 1998).

5.6 Soil Microbes Mediated Micronutrient Acquisition in Plants

The plant nutrients that are required in significantly in lesser amounts in comparison to other macronutrients are micronutrients. The soil is a good reservoir for all plant nutrients including micronutrients. The inoculation of soil with rhizospheric microbes stimulates the micronutrient acquisition. The action mechanism of micronutrient availability to plants by soil microbes are organic and inorganic acids, chelating agents and also play a key role in disease control. Eight essential micronutrient elements required for plant growth are iron (Fe), zinc (Zn), manganese (Mn), chlorine (Cl), copper (Cu), boron (B), molybdenum (Mo), and nickel (Ni) (Kumar et al. 2016). As similar to macronutrients, micronutrient availability from soil to plants is highly dependent on interaction between plant roots and soil microbes. The process involves secretion of organic acids and chelating agents by the microbes for solubilization and mobilization of nutrients (Suri and Choudhary 2012, 2013). Thus, rhizospheric microbes play an important role towards micronutrient availability to plants such as iron, zinc, copper, and manganese.

5.6.1 Iron

Iron is relatively abundant element with an average concentration of 40 g kg^{-1} present in soil (Cornell and Schwertmann 2003). The iron deficiency in plant nutrient leads to chlorosis which ultimately leads to reduced agricultural productivity. The cellular processes of chlorophyll synthesis, mitochondrial respiration, oxygen transport, and as constituent of some enzymes and proteins in plants are dependent on iron (Jin et al. 2014). The mechanism of solubility of iron in soil by microbes and plants are carried out by the chelation of insoluble iron in producing siderophore (Sharma et al. 2003). The iron deficient conditions stimulate siderophore production in rhizobial strains to enhance bioavailability of the nutrient in the environment (Terpolilli et al. 2012). In leguminous plant roots, secretion of signalling phenolic compounds establish rhizobia nodulation in Fe-deficient soil. Further, microbes produce different types of siderophores such as ferrioxamines, enterobactin, pyoverdine and ferrichromes (Marschner et al. 2010) which plays key role in plant growth. The siderophore producing rhizobia are considered as the potential iron acquisition in leguminous plant. This process involves siderophore mediated chelation of ferric (Fe^{3+}) to ferrous (Fe^{2+}) ion in the cell surface (Mendes et al. 2013). Yaseen et al. (2018) show increased iron uptake in the wheat with co-inoculation with endophytic bacteria *Enterobacter* sp. and *Burkholderia phytofirmans*. The iron content was increased up to 10.14% with enhanced leaf area, height, and biomass of plant.

5.6.2 Zinc

Zinc is an essential micronutrient for plant growth. The concentration of zinc is generally very low (80 mg kg^{-1}) in soil. It is the most important nutrient required for the synthesis of auxin, proteins, carbohydrates, lipids, nucleic acid, and involved in chlorophyll and seed formation (Seilsepour 2006; Broadley et al. 2007). It also helps in catalytic roles in different enzyme classes such as isomerase, transferase, oxido-reductase, hydrolases, and ligases (Hafeez et al. 2013). In soil various mechanisms of zinc solubilization are achieved by soil pH reduction by the release of organic acids and chelation (Subramanian et al. 2009; Whiting et al. 2001). The acidification of medium by gluconic acid production also make Zn availability to plants by bacteria such as *Curtobacterium, Plantibacter, Pseudomonas, Stenotrophomonas, Streptomyces* reported by Costerousse et al. (2018). The other way includes beneficial symbiotic relationship by Arbuscular mycorrhizal fungi (AMF) with the crop plants for sustained crop productivity (Yadav et al. 2018). This can increase root surface area which helps in high Zn uptake from the soil. Jha (2019) shows zinc-solubilizing microbes, namely *Bacillus pumilus* and *Pseudomonas pseudoalcaligenes* protect plants from osmotic stress by enhancing antioxidant enzymes catalase (CAT), and peroxidase (PO).

Kamran et al. (2017) shows maximum shoot and root length with high zinc content in wheat plant tested with *E. cloacae*, similarly maximum zinc content in wheat plant observed with *Pseudomonas fragi* and *P. agglomerans*. The Zn-solubilizing bacteria such as *Pseudomonas aeruginosa*, *Ralstonia pickettii*, *Burkholderia cepacia*, and *Klebsiella pneumonia* aids in the Zn-biofortification in rice seedlings with potential for utilizing insoluble zinc compound with exopolysaccharide (EPS) production, 1-aminocycloproparane-1-carboxylic acid (ACC) utilization, and potassium and phosphate solubilization (Gontia-Mishra et al. 2017). Singh et al. (2017) show two endophytic bacteria *Bacillus subtilis* and *Arthrobacter* sp., enhanced two folds increased Zn content in the wheat plant. The Zn-solubilizing bacterium *Pseudomonas* sp. is also a plant growth-promoting rhizobacteria in wheat cultivation (Lasani-2008 and Faisalabad-2008). Foliar spray application of microbes significantly increase Zn-biofortification in wheat grains (Rehman et al. 2018). Together Zn-solubilizing rhizobacteria have the ability of phosphate solubilization and production of IAA, siderophore, HCN, ammonia, exopolysaccharides, with catalase, protease, chitinase and lipase activity which increase the plant growth of wheat (Mumtaz et al. 2017).

5.7 Copper

The other essential micronutrient for plant growth is copper. Comparatively lesser amount of copper is required as compared to other micronutrients. It plays important role in the process of respiration and enzyme activation in the plants. The copper improves plant growth as it is important cofactor for enzymes (Makoi and Ndakidemi 2007). The copper availability to plants is made by soil microbes by the release of carboxylase and phenolic compounds (Badri and Vivanco 2009). The role of copper is acting as a catalyst for respiration and activator of several enzymes.

5.7.1 Manganese

The other essential micronutrient for plant growth manganese concentration in the soil is in an average amount of 1000 mg kg^{-1}. Manganese is a part of multiple enzymes and is a catalyst of other enzymes and used in metabolism of nitrogen and inorganic acids, formation of vitamins, and other metabolic process. The microbial activities in the soil largely affecting enhanced plant growth activities (Dutta and Podile 2010).

5.8 Future Perspectives and Challenges in Plant Microbe Based Agro-Inputs

Growing world population and climate change causes a big challenge towards crop production. In this context, intensify agricultural production is required to combat abiotic stress agents, pathogens, and pests in a sustainable manner. Research efforts may be made at developing agricultural yields through the management of soil microbes. In fact, developments of microbial formulations are global interest which is very useful to increase the soil fertility, particularly in N- and P-deficient agro-systems. The combined usage of beneficial microbes and mineral nutrients can stimulate plant growth and protect the environmental health. This is considered as the novel approach and an emerging research area to study the various activities of beneficial soil microbes, especially in interaction with plants and mineral nutrients. Currently researches are directed towards finding individual organism and its gene in a community. However, in future it can be widened with molecular approach. The scientific knowledge on plant-associated bacteria is scarce and therefore interaction between plants and microbes has to be studied intensively.

5.9 Conclusion

The role of soil microbes on soil nutrient availability to plants is discussed in this chapter. In addition to that, the usage of mineral nutrients with beneficial microbes is greatly required to improve higher yield in a sustainable way. Indeed, appropriate combination of mineral and microbial resources with favorable climatic conditions are highly essential for enhancing crop productivity and soil fertility. Moreover, systemic approaches are required to maintain biodiversity and environmental health. This mainly shows way forward towards efficient soil microbes as biofertilizers in order to reduce chemical fertilizers and pesticides. Thereby protecting the niche of soil habitat by biological approach is the proven efficiency.

Acknowledgments The first author gratefully acknowledges the Tamil Nadu State Council for Science and Technology (TNSCST), India for providing the award of Young Scientist Fellowship (No. TNSCST/YSFS/VR/07/2018-2019/7120) during the course of this study.

References

Ahemad M, Kibret M (2014) Mechanisms and applications of plant growth promoting rhizobacteria: current perspective. J King Saud Univ Sci 26:1–20

Ahmad M, Nadeem SM, Naveed M, Zahir ZA (2016) Potassium-solubilizing bacteria and their application in agriculture. In: Meena VS, Maurya BR, Verma JP, Meena RS (eds) Potassium solubilizing microorganisms for sustainable agriculture. Springer, New Delhi, pp 293–313

Alori ET, Glick BR, Babalola OO (2017) Microbial phosphorus solubilization and its potential for use in sustainable agriculture. Front Microbiol 8:971. https://doi.org/10.3389/fmicb.2017.00971

Andrews JH, Harris RF (2000) The ecology and biogeography of microorganisms of plant surfaces. Annu Rev Phytopathol 38:145–180

Badri DV, Vivanco JM (2009) Regulation and function of root exudates. Plant Cell Environ 32:666–681

Bal HB, Nayak L, Das S, Adhya TK (2013) Isolation of ACC deaminase producing PGPR from rice rhizosphere and evaluating their plant growth promoting activity under salt stress. Plant Soil 366:93–105

Barker WW, Welch SA, Chu S, Banfield JF (1998) Experimental observations of the effects of bacteria on aluminosilicate weathering. Am Mineral 83:1551–1563

Barnawal D, Bharti N, Pandey SS, Pandey A, Chanotiya CS, Kalra A (2017) Plant growth-promoting rhizobacteria enhance wheat salt and drought stress tolerance by altering endogenous phytohormone levels and TaCTR1/TaDREB2 expression. Physiol Plant 161:502–514

Beattie GA (2007) Plant-associated bacteria: survey, molecular phylogeny, genomics and recent advances. In: Gnanamanickam SS (ed) Plant-associated bacteria. Springer, Dordrecht. https://doi.org/10.1007/978-1-4020-4538-7_1

Behera BC, Yadav H, Singh SK, Mishra RR, Sethid BK, Dutta SK, Thatoi HN (2017) Phosphate solubilization and acid phosphatase activity of Serratia sp. isolated from mangrove soil of Mahanadi river delta, Odisha, India. J Genet Eng Biotechnol 15:169–178

Botella M, Martinez V, Pardines J, Cerda A (1997) Salinity induced potassium deficiency in maize plants. J Plant Physiol 150:200–205

Broadley MR, White PJ, Hammond JP, Zelko I, Lux A (2007) Zinc in plants. New Phytol 173:677–702

Brown ME (1974) Seed and root bacterization. Annu Rev Phytopathol 12:181–197

Burr TJ, Schroth MN, Suslow T (1978) Increased potato yields by treatment of seed-pieces with specific strains of *Pseudomonas fluorescens* and *P. putida*. Phytopathology 68:1377–1383

Cornell RM, Schwertmann U (2003) The iron oxides, second ed.s. Wiley-VCH, Weinheim

Costerousse B, Schönholzer-Mauclaire L, Frossard E, Thonar C (2018) Identification of heterotrophic zinc mobilization processes among bacterial strains isolated from wheat rhizosphere (*Triticum aestivum* L.). Appl Environ Microbiol 84(1):01715–01717

Coyne MS, Mikkelsen R (2015) Soil microorganisms contribute to plant nutrition and root health. Better Crops 99:18–20

Damodaran T, Rai RB, Jha SK, Kannan R, Pandey BK, Sah V, Mishra VK, Sharma DK (2014) Rhizosphere and endophytic bacteria for induction of salt tolerance in gladiolus grown in sodic soils. J Plant Interact 9:577–584

Das I, Pradhan M (2016) Potassium-solubilizing microorganisms and their role in enhancing soil fertility and health. In: Potassium solubilizing microorganisms for sustainable agriculture. Springer, New Delhi, pp 281–291

Dimkpa C, Weinand T, Asch F (2009) Plant-rhizobacteria interactions alleviate abiotic stress conditions. Plant Cell Environ 32:1682–1694

Dutta S, Podile A (2010) Plant growth promoting rhizobacteria (PGPR): the bugs to debug the root zone. Crit Rev Microbiol 36:232–244

Edwards J, Johnson C, Santos-medellın C, Lurie E, Kumar N (2015) Structure, variation, and assembly of the root-associated microbiomes of rice. Proc Natl Acad Sci USA 112:911–920

Egamberdieva D, Lugtenberg B (2014) Use of plant growth-promoting rhizobacteria to alleviate salinity stress in plants, use of microbes for the alleviation of soil stresses, vol 1. Springer, pp 73–96

Egamberdiyeva D (2005) Plant-growth-promoting rhizobacteria isolated from a Calcisol in a semi-arid region of Uzbekistan: biochemical characterization and effectiveness. J Plant Nutr Soil Sci 168:94–99

Etesami H (2018a) Bacterial mediated alleviation of heavy metal stress and decreased accumulation of metals in plant tissues: mechanisms and future prospects. Ecotoxicol Environ Saf 147:175–191

Etesami H (2018b) Can interaction between silicon and plant growth promoting rhizobacteria benefit in alleviating abiotic and biotic stresses in crop plants? Agric Ecosyst Environ 253:98–112

Etesami H, Beattie GA (2017) Plant-microbe interactions in adaptation of agricultural crops to abiotic stress conditions. In: Probiotics and plant health. Springer, pp 163–200

Etesami H, Beattie GA (2018) Mining halophytes for plant growth-promoting halotolerant bacteria to enhance the salinity tolerance of non-halophytic crops. Front Microbiol 9:148

Etesami H, Maheshwari DK (2018) Use of plant growth promoting rhizobacteria (PGPRs) with multiple plant growth promoting traits in stress agriculture: action mechanisms and future prospects. Ecotoxicol Environ Saf 156:225–246

Etesami H, Emami S, Alikhani HA (2017) Potassium solubilizing bacteria (KSB): mechanisms, promotion of plant growth, and future prospects—a review. J Soil Sci Plant Nutr 17:897–911

Feigin A (1985) Fertilization management of crops irrigated with saline water. Biosalinity in action: bioproduction with saline water. Springer, pp 285–299

Ferguson BJ, Indrasumunar A, Hayashi S, Lin MH, Lin YH, Reid DE, Gresshoff PM (2010) Molecular analysis of legume nodule development and autoregulation. J Integr Plant Biol 52:61–76

Glick BR (2012) Plant growth-promoting bacteria: mechanisms and applications. Scientifica. https://doi.org/10.6064/2012/963401

Glick BR (2014) Bacteria with ACC deaminase can promote plant growth and help to feed the world. Microbiol Res 169:30–39

Gontia-Mishra I, Sapre S, Tiwari S (2017) Zinc solubilizing bacteria from the rhizosphere of rice as prospective modulator of zinc biofortification in rice. Rhizosphere 3:185–190

Gopal M, Gupta A (2016) Microbiome selection could spur next-generation plant breeding strategies. Front Microbiol 7:1971

Goswami D, Dhandhukia P, Patel P, Thakker JN (2014) Screening of PGPR from saline desert of Kutch: growth promotion in *Arachis hypogea* by *Bacillus licheniformis* A2. Microbiol Res 169 (1):66–75

Gouda S, Kerry RG, Das G, Paramithiotis S, Patra JK (2018) Revitalization of plant growth promoting rhizobacteria for sustainable development in agriculture. Microbiol Res 206:131–140
Gyaneshwar P, Hirsch AM, Moulin L, Chen WM, Elliott GN, Bontemps C, Santos PE, Gross E, Reis FBD, Sprent JI, Young JPW, James EK (2011) Legume-nodulating betaproteobacteria: diversity, host range, and future prospects. Mol Plant-Microbe Interact 24:1276–1288
Hafeez FY, Abaid-Ullah M, Hassan MN (2013) Plant growth-promoting rhizobacteria as zinc mobilizers: a promising approach for cereals biofortification. In: Bacteria in agrobiology: crop productivity. Springer, New York, pp 217–235
Hawkes CV, DeAngelis KM Firestone MK (2007) Root interactions with soil microbial communities and processes. In: Cardon Z, Whitbeck J (eds) The Rhizosphere. Elsevier, New York, pp 1–3
Hernandez M, Kappler A, Newman D (2004) Phenazines and other redox-active antibiotics promote microbial mineral reduction. Appl Environ Microbiol 70:921–928
Hirel B, Tétu T, Lea PJ, Dubois F (2011) Improving nitrogen use efficiency in crops for sustainable agriculture. Sustainability 3:1452–1485
Houser J, Richardson W (2010) Nitrogen and phosphorus in the upper Mississippi River: transport, processing, and effects on the river ecosystem. Hydrobiologia 640:71–88
Hunter P (2016) Plant microbiomes and sustainable agriculture. EMBO Rep 17(12):1696–1699
Igiehon NO, Babalola OO (2018) Rhizosphere microbiome modulators: contributions of nitrogen fixing bacteria towards sustainable agriculture. Int J Environ Res Public Health 15(4)pii: E574
Ilangumaran G, Smith DL (2017) Plant growth promoting rhizobacteria in amelioration of salinity stress: a systems biology perspective. Front Plant Sci 8:1768
Illmer P, Barbato A, Schinner F (1995) Solubilization of hardly-soluble AlPO 4 with P-solubilizing microorganisms. Soil Biol Biochem 27:265–270
Iniguez AL, Dong Y, Carter HD, Ahmer BM, Stone JM, Triplett EW (2005) Regulation of enteric endophytic bacterial colonization by plant defenses. Mol Plant-Microbe Interact 18:169–178
Iqbal U, Jamil N, Ali I, Hasnain S (2010) Effect of zinc-phosphate-solubilizing bacterial isolates on growth of *Vigna radiata*. Ann Microbiol 60:243–248
Jha Y (2019) The importance of zinc-mobilizing rhizosphere bacteria to the enhancement of physiology and growth parameters for paddy under salt-stress conditions. Jordan J Biol Sci 12:167–173
Jha Y, Subramanian RB, Patel S (2011) Combination of endophytic and rhizospheric plant growth promoting rhizobacteria in *Oryza sativa* shows higher accumulation of osmoprotectant against saline stress. Acta Physiol Plant 33:797–802
Jin CW, Ye YQ, Zheng SJ (2014) An underground tale: contribution of microbial activity toplant iron acquisition via ecological processes. Ann Bot 113:7–18
Johnson NC, Wilson GWT, Bowker MA, Wilson JA, Miller RM (2010) Resource limitation is a driver of local adaptation in mycorrhizal symbioses. Proc Natl Acad Sci 107:2093–2098
Kamran S, Shahid I, Baig DN, Rizwan M, Malik KA, Mehnaz S (2017) Contribution of zinc solubilizing bacteria in growth promotion and zinc content of wheat. Front Microbiol 8:2593
Kaushal M, Kumar A, Kaushal R (2017). *Bacillus pumilus strain* YSPMK11 as plant growth promoter and biocontrol agent against *Sclerotinia sclerotiorum*. 3 Biotechnol 7:90
Keshavarz Zarjani J, Aliasgharzad N, Oustan S, Emadi M, Ahmadi A (2013) Isolation and characterization of potassium solubilizing bacteria in some Iranian soils. Arch Agron Soil Sci 59:1713–1723
Khalid A, Arshad M, Zahir ZA (2004) Screening plant growth promoting rhizobacteria for improving growth and yield of wheat. J Appl Microbiol 96:473–480
Krishnaraj PU, Dahale S (2014) Mineral phosphate solubilization: concepts and prospects in sustainable agriculture. Proc Ind Natl Sci Acad 80:389–405
Kumar A, Choudhary AK, Pooniya V, Suri VK, Singh U (2016) Soil factors associated with micronutrient acquisition in crops- biofortification perspective. In: Chapter 13, U. Singh et al. (eds) Biofortification of food crops. https://doi.org/10.1007/978-81-322-2716-8_13

Lian B, Fu P, Mo D, Liu C (2002) A comprehensive review of the mechanism of potassium releasing by silicate bacteria. Acta Mineral Sin 22:179–183

Lugtenberg BJJ, Malfanova N, Kamilova F, Berg G (2013) Plant growth promotion by microbes. Mol Microb Ecol Rhizosph 1:559–573

Makoi JH, Ndakidemi PA (2007) Biological, ecological and agronomic significance of plant phenolic compounds in rhizosphere of the symbiotic legumes. Afr J Biotechnol 6:1358–1368

Marques JM, da Silva TF, Vollu RE, Blank AF, Ding GC, Seldin L, Smalla K (2014) Plant age and genotype affect the bacterial community composition in the tuber rhizosphere of field-grown sweet potato plants. FEMS Microbiol Ecol 88:424–435

Marschner P, Crowely D, Rengel Z (2010) Interaction between rhizosphere microorganisms and plants governing iron and phosphorous availability. 19th world congress of soil science, Brisbane, Australia, 52–55

Meena VS, Maurya BR, Verma JP (2014) Does a rhizospheric microorganism enhance K+ availability in agricultural soils? Microbiol Res 169:337–347

Meena VS, Maurya BR, Bahadur I (2015) Potassium solubilization by bacterial strain in waste mica. Bangladesh J Bot 43:235–237

Mendes R, Garbeva P, Raaijmakers JM (2013) The rhizosphere microbiome: significance of plant beneficial, plant pathogenic, and human pathogenic microorganisms. FEMS Microbiol Rev 37 (5):634–663

Miransari M (2010) Contribution of arbuscular mycorrhizal symbiosis to plant growth under different types of soil stress. Plant Biol 12(4):563–569

Miransari M (2011a) Arbuscular mycorrhizal fungi and nitrogen uptake. Arch Microbiol 193 (2):77–81

Miransari M (2011b) Soil microbes and plant fertilization. Appl Microbiol Biotechnol 92 (5):875–885

Miransari M (2013) Soil microbes and the availability of soil nutrients. Acta Physiol Plant 35. https://doi.org/10.1007/s11738-013-1338-2

Misra N, Gupta G, Jha PN (2012) Assessment of mineral phosphate-solubilizing properties and molecular characterization of zinc-tolerant bacteria. J Basic Microbiol 52:549–558

Morton AG (1981) History of botanical science. An account of the development of botany from ancient times to the present day. Academic Press, London

Muller DB, Vogel C, Bai Y, Vorholt JA (2016) The plant microbiota: systems-level insights and perspectives. In: Bonini NM (ed) Annual review of genetics, vol 50. (Annual Reviews), Palo Alto, pp 211–234

Mumtaz MZ, Ahmad M, Jamil M, Hussain T (2017) Zinc solubilizing *Bacillus* spp. potential candidates for biofortification in maize. Microbiol Res 202:51–60

Murrell TS, Munson RD (1999) Phosphorus and potassium economics in crop production: net returns. Better Crops 83:23–27

Mus F, Crook MB, Garcia K, Costas AG, Geddes BA, Kouri ED, Paramasivan P, Ryu MH, Oldroyd GED, Poole PS, Udvardi M, Voigt CA, Ané JM, Peters JW (2016) Symbiotic nitrogen fixation and the challenges to its extension to nonlegumes. Appl Environ Microbiol 82:3698–3710

Ortíz-Castro R, Contreras-Cornejo HA, Macías-Rodríguez L, López-Bucio J (2009) The role of microbial signals in plant growth and development. Plant Signal Behav 4(8):701–712

Oves M, Khan MS, Zaidi A (2013) Chromium reducing and plant growth promoting novel strain Pseudomonas aeruginosa OSG41 enhance chickpea growth in chromium amended soils. Eur J Soil Biol 56:72–83

Parker DR, Reichman SM, Crowley DE (2005) Metal chelation in the rhizosphere. Agronomy 48:57

Paul D, Lade H (2014) Plant-growth-promoting rhizobacteria to improve crop growth in saline soils: a review. Agron Sustain Dev 34:737–752

Peiffer JA, Spor A, Koren O, Jin Z, Tringe SG, Dangl JL, Buckler ES, Ley R (2013) Diversity and heritability of the maize rhizosphere microbiome under field conditions. Proc Natl Acad Sci U S A 110:6548–6553

Qin Y, Druzhinina IS, Pan X, Yuan Z (2016) Microbially mediated plant salt tolerance and microbiome-based solutions for saline agriculture. Biotechnol Adv 34:1245–1259

Rehman A, Farooq M, Naveed M, Ozturk L, Nawaz A (2018) *Pseudomonas*-aided zinc application improves the productivity and biofortification of bread wheat. Crop Pasture Sci 69:659–672

Rillig MC (2004) Arbuscular mycorrhizae, glomalin, and soil aggregation. Can J Soil Sci 84 (4):355–363

Rivas R, Peix A, Mateos PF, Trujillo ME, Martinez-Molina E, Velazquez E (2006) Biodiversity of population of phosphate solubilizing rhizobia that nodulates chickpea in different Spanish soils. Plant Soil 287:23–33

Rogers C, Oldroyd GED (2014) Synthetic biology approaches to engineering the nitrogen symbiosis in cereals. J Exp Bot 65(8):1939–1946

Rolli E, Marasco R, Vigani G, Ettoumi B, Mapelli F, Deangelis ML, Gandolfi C, Casati E, Previtali F, Gerbino R, Cei FP, Borin S, Sorlini C, Zocchi G, Daffonchio D (2015) Improved plant resistance to drought is promoted by the root-associated microbiome as a water stress-dependent trait. Environ Microbiol 17:316–331

Saharan BS, Nehra V (2011) Plant growth promoting rhizobacteria: a critical review. Life Sci Med Res 2011:LSMR-21

Salimpour S, Khavazi K, Nadian H, Besharati H, Miransari M (2010) Enhancing phosphorous availability to canola. *Brassica napus* L. using P solubilizing and sulfur oxidizing bacteria. Aust J Crop Sci 4:330–334

Sandhya V, Ali SZ (2015) He production of exopolysaccharide by *Pseudomonas putida* GAP-P45 under various abiotic stress conditions and its role in soil aggregation. Microbiology 84:512–519

Santi C, Bogusz D, Franche C (2013) Biological nitrogen fixation in non-legume plants. Ann Bot 111:743–767

Sawers RJH, Svane SF, Quan C, Grønlund M, Wozniak B, Gebreselassie MN, Munoz EG, Chavez-Montes RA, Baxter I, Goudet J, Jakobsen I, Paszkowski U (2017) Phosphorus acquisition efficiency in arbuscular mycorrhizal maize is correlated with the abundance of root-external hyphae and the accumulation of transcripts encoding PHT1 phosphate transporters. New Phytol 214:632–643

Seilsepour M (2006) Study of zinc effects on quantitative and qualitative traits of winter wheat in saline soil condition. Desert 11:17–23

Sharma A, Johri BN, Sharma AK, Glick BR (2003) Plant growth-promoting bacterium *Pseudomonas* sp. strain GRP3 influences iron acquisition in mung bean (*Vigna radiata* L. Wilzeck). Soil Biol Biochem 35:887–894

Sharma SB, Sayyed RZ, Trivedi MH, Gobi TA (2013) Phosphate solubilizing microbes: sustainable approach for managing phosphorus deficiency in agricultural soils. Springerplus 2:587

Sharma V, Salwan R, Sharma PN (2017) The comparative mechanistic aspects of *Trichoderma* and probiotics: scope for future research. Physiol Mol Plant Pathol 100:84–96

Sharpley A, Meisinger J, Power J, Suarez D (1992) Root extraction of nutrients associated with long-term soil management. Limitations to plant root growth. Springer, pp 151–217

Shelobolina E, Xu H, Konishi H, Kukkadapu R, Wu T, Blöthe M, Roden E (2012) Microbial lithotrophic oxidation of structural Fe(II) in biotite. Appl Environ Microbiol 78:5746–5752

Sindhu SS, Parmar P, Phour M (2014) Nutrient cycling: potassium solubilization by microorganisms and improvement of crop growth. In: Geomicrobiology and biogeochemistry. Springer, Berlin, pp 175–198

Sindhu SS, Parmar P, Phour M, Sehrawat A (2016) Potassium-solubilizing microorganisms (KSMs) and its effect on plant growth improvement, potassium solubilizing microorganisms for sustainable agriculture. Springer, pp 171–185

Singh D, Rajawat MVS, Kaushik R, Prasanna R, Saxena AK (2017) Beneficial role of endophytes in biofortification of Zn in wheat genotypes varying in nutrient use efficiency grown in soils sufficient and deficient in Zn. Plant Soil 416:107–116

Smith SE, Read DJ (2008) Mycorrhizal symbiosis. Academic Press, London

Sparks DL (1987) Potassium dynamics in soils. In: Advances in soil science. Springer, New York, pp 1–63

Subramanian K, Tenshia V, Jayalakshmi K, Ramach V (2009) Role of arbuscular mycorrhizal fungus (*Glomus intraradices*) (fungus aided) in zinc nutrition of maize. J Agric Biotechnol Sustain Dev 1:29–38

Suri VK, Choudhary AK (2012) Fertilizer economy through vesicular arbuscular mycorrhizal fungi under soil–test crop response targeted yield model in maize–wheat– maize crop sequence in Himalayan acid. Alfisol Commun Soil Sci Plant Anal 43(21):2735–2743

Suri VK, Choudhary AK (2013) Effects of vesicular arbuscular mycorrhizae and applied phosphorus through targeted yield precision model on root morphology, productivity and nutrient dynamics in soybean in an acid Alfisol. Commun Soil Sci Plant Anal 44(17):2587–2604

Terpolilli JJ, Hood GA, Poole PS (2012) What determines the efficiency of N-fixing *Rhizobium*-legume symbioses? Adv Microb Physiol 60:325–389

Tewari S, Arora NK (2014) Multifunctional exopolysaccharides from *Pseudomonas aeruginosa* PF23 involved in plant growth stimulation, biocontrol and stress amelioration in sunflower under saline conditions. Curr Microbiol 69:484–494

Ullman JB (1996) Structural equation modeling. In: Tabachnick BG, Fidell LS (eds) Using multivariate statistics, 3rd edn. HarperCollins College Publishers, New York, pp 709–819

Ullman WJ, Welch SA (2002) Organic ligands and feldspar dissolution water–rock interactions, ore deposits, and environmental geochemistry: a tribute to David Crerar. Special Publication 7:3–35

Upadhyay SK, Singh JS, Saxena AK, Singh DP (2012) Impact of PGPR inoculation on growth and antioxidant status of wheat under saline conditions. Plant Biol 14:605–611

Uribe D, Sánchez-Nieves J, Vanegas J (2010) Role of microbial biofertilizers in the development of a sustainable agriculture in the tropics. In: Dion P (ed) Soil biology and agriculture in the tropics. Soil biology, vol 21. Springer, Berlin

Uroz S, Calvaruso C, Turpault M-P, Frey-klett P (2009) Mineral weathering by bacteria: ecology, actors and mechanisms. Trends Microbiol 17:378–387. https://doi.org/10.1016/j.tim.2009.05.004

van der Heijden MGA, Bardgett RD, Van Straalen NM (2008) The unseen majority: soil microbes as drivers of plant diversity and productivity in terrestrial ecosystems. Ecol Lett 11:296–310

van Loon LC (2007) Plant responses to plant growth-promoting rhizobacteria. Eur J Plant Pathol 119:243–254

Verbon EH, Liberman LM (2016) Beneficial microbes affect endogenous mechanisms controlling root development. Trends Plant Sci 21. https://doi.org/10.1016/j.tplants.2016.01.013

Vessey JK (2003) Plant growth promoting rhizobacteria as biofertilizers. Plant Soil 255:571–586

Vurukonda SSKP, Vardharajula S, Shrivastava M, Skz A (2016) Enhancement of drought stress tolerance in crops by plant growth promoting rhizobacteria. Microbiol Res 184:13–24

Wang Q, Dodd IC, Belimov AA, Jiang F (2016) Rhizosphere bacteria containing 1-aminocyclopropane-1-carboxylate deaminase increase growth and photosynthesis of pea plants under salt stress by limiting Na+ accumulation. Funct Plant Biol 43:161–172

Whipps JM (2001) Microbial interactions and biocontrol in the rhizosphere. J Exp Bot 52:487–511

Whitelaw MA (2000) Growth promotion of plants inoculated with phosphate solubilizing fungi. Adv Agron 69:99–151

Whiting SN, de Souza MP, Terry N (2001) Rhizosphere bacteria mobilize Zn for hyperaccumulation by *Thlaspi caerulescens*. Environ Sci Technol 35:3144–3150

Wong WS, Tan SN, Ge L, Chen X, Yong JWH (2015) The importance of phytohormones and microbes in biofertilizers. In: Maheshwari DK (ed) Bacterial metabolites in sustainable agroecosystem. Springer, Cham, pp 105–158

Woo SL, Ruocco M, Vinale F, Nigro M, Marra R, Lombardi N, Pascale A, Lanzuise S, Manganiello G, Lorito M (2014) *Trichoderma*-based products and their widespread use in agriculture. Open Mycol J 8:71–126

Yadav AN, Kumar V, Prasad R, Saxena AK, Dhaliwal HS (2018) Microbiome in crops: diversity, distribution and potential role in crops improvements. In: Prasad R, Gill SS, Tuteja N (eds) Crop improvement through microbial biotechnology. Elsevier, USA, pp 305–332

Yaseen M, Abbas T, Aziz MZ, Wakeel A, Yasmeen H, Ahmed W, Ullah A, Naveed M (2018) Microbial assisted foliar feeding of micronutrients enhance growth, yield and biofotification of wheat. Int J Agric Biol 20:353–360

Yeoh YK, Paungfoo-Lonhienne C, Dennis PG, Robinson N, Ragan MA, Schmidt S, Hugenholtz P (2015) The core root microbiome of sugarcanes cultivated under varying nitrogen fertilizer application. Environ Microbiol 18:1338–1351

Zhang AM, Zhao GY, Gao TG, Wang W, Li J, Zhang S-F, Zhu BC (2013) Solubilization of insoluble potassium and phosphate by *Paenibacillus kribensis* CX-7: a soil microorganism with biological control potential. Afr J Microbiol Res 7:41–47

Zhao JL, Zhou LG, Wu JY (2010) Promotion of Salvia miltiorrhiza hairy root growth and tanshinone production by polysaccharideprotein fractions of plant growth-promoting rhizobacterium Bacillus cereus. Process Biochem 45:1517–1522

Zhen L, Bai T, Dai L, Wang F, Tao J, Meng S, Hu Y, Wang S, Hu S (2016) A study of organic acid production in contrasts between two phosphate solubilizing fungi: *Penicillium oxalicum* and *Aspergillus niger*. Sci Rep 6:25313

Zhuang X, Chen J, Shim H, Bai Z (2007) New advances in plant growth-promoting rhizobacteria for bioremediation. Environ Int 33:406–413

Zorb C, Senbayram M, Peiter E (2014) Potassium in agriculture–status and perspectives. J Plant Physiol 171:656–669

Drought Stress Alleviation in Plants by Soil Microbial Interactions

6

Yamini Tak, Vinod Kumar Yadav, Chirag Gautam, Rajendra Kumar, and Manpreet Kaur

Abstract

The biotic and abiotic stresses are major curtailment effects on crop production, food and its nutritional quality and most importantly global food security. A plant experiences various physiological, biochemical, and molecular changes under biotic and abiotic stresses that crashes overall plant development and growth. In order to increase crop productivity with doubling farmer income it becomes necessary to develop efficient low-cost technologies for biotic and abiotic stress management in field. Since microorganisms interaction is inherent part of ecosystems, so microbes are the native occupiers which exhibit extensive metabolic capacities to alleviate various stresses. Plant growth promoting rhizobacteria, arbuscular mycorrhiza, and actinomycetes increase plant growth under adverse conditions by regulating plant hormones, nutrition accession, induce antioxidative enzymes, and enhance the supply of minerals and nutrients. This chapter furnishes knowledge about the plant–microbial interactions under drought stress and the role of stress-adaptable microbes in relieving the same.

Y. Tak (✉)
Department of Biochemistry, Agriculture University, Kota, Rajasthan, India

V. K. Yadav
Department of Soil Science and Agricultural Chemistry, Agriculture University, Kota, Rajasthan, India

C. Gautam
Department of Plant Pathology, Agriculture University, Kota, Rajasthan, India

R. Kumar
Department of Entomology, Swami Keshwanand Rajasthan Agricultural University, Bikaner, Rajasthan, India

M. Kaur
Department of Biochemistry, Punjab Agricultural University, Ludhiana, Punjab, India

R. Soni et al. (eds.), *Microbiological Activity for Soil and Plant Health Management*, https://doi.org/10.1007/978-981-16-2922-8_6

Keywords

Drought · Microbial colonization · Mycorrhizae · PGPR · Stress · Tolerance

6.1 Introduction

Due to global climate change, plant faces many biotic and abiotic stresses leading to reduce in agricultural productivity of crops worldwide (Gull et al. 2019). The plant definitely faces some stress once in its life, but God also sent them with some intrinsic adaptive mechanisms to bear that adversity in plants to get out of those adverse environmental conditions. These stresses comprises high or low temperature, heavy rainfall, drought, flood, salt, metal toxicity, UV radiations, bacteria, fungi, viruses, insect and pests, wound, and sometimes weed stress (Koyro et al. 2011; Pandey et al. 2017). Due to unfavorable environmental conditions such as drought and temperature, causing a variety of negative losses in plants at the physiological, biochemical, and molecular levels. But among all of these, drought is a crucial restraint to plant growth and development, mainly for small or marginal farmers who have less irrigation water facility (Da Silva et al. 2013). Globally, every year about 55 million people are affected by drought directly or indirectly and nearly 42% of India's land area is cladding drought stress. Plant is totally dependent on soil for water and nutrition for growth, but in dry conditions due to water loss, soil loses its texture, structure, and soil carbon which negatively affects plant metabolic reactions (Schoonover and Crim 2015). Friends of the plant and soil, soil microorganisms assist them to reduce the burden of plant from various stress problems. Soil microbes such as plant growth promoting bacteria, fungi, viruses, actinomycetes play a crucial role in stress tolerance and help to maintain soil productivity (Glick 2012; Gouda et al. 2018). To cope up with drought stress the useful microbes colonize with the root of the host plant and stimulate developmental processes by direct and indirect mechanisms (Naylor and Coleman-Derr 2018). By secretion of root exudates most of the plants engage with soil microbes for nutrient and water absorption (Walker 2003). No doubt, water scarcity is not only injurious to plants, but also to advantageous fauna of soil. Under stress soil microorganisms improve root shoot growth and biomass, nutrient recycling, mobilization of some micro and macro nutrients, secondary metabolite phenol production, and make capable to plant fight against the drought stress (Backer et al. 2018; Meena et al. 2017a). Soil bacteria or fungi have symbiotic relationship with plants show wonderful influence on plant health, depend on exposure of stress, duration, developmental stage of plant, and individual plant genotype (Jacoby et al. 2017). ACC-deaminase-emitting soil bacteria benefitted to plants from harsh environmental conditions (Vurukonda et al. 2016). In these changing climatic conditions, the use of soil microbes as biofertilizers extends a ray of hope in agriculture field under drought conditions. In this book chapter, we have tried to explore the favorable consequences of soil microbes and their adaptive mechanism to cope up with drought conditions in the field.

6.2 Stresses, Soil Structure and Their Effect on Microbial Colonization

Plant stress is growing of the plant in adverse or in unhealthy conditions which negatively affects plant growth and development (Yadav et al. 2020). Throughout the life of the plant, it faces critical unfavorable conditions and various constraints which hinder the growth of the plant from the physiological to the molecular level. Sometimes the stress affects the plant beyond the limit so it can damage the tissues of the plant or maybe the plant will die (Fathi and Tari 2016). Plant stress can be categorized into biotic and abiotic stresses. Biotic stress can be due to bacteria, fungi, viruses, nematodes, insects and herbivores and weed (Tak and Kumar 2020), while the abiotic stress includes drought, temperature, salt, light, heavy metals, pH, or flood (Kumar et al. 2020). Plants itself have some adaptations to avoid stress, response of plant to particular stress also depends on the duration and severity of the particular adverse condition and growing stage of the plant when the stress is inflicted (Basu et al. 2016). Among the different abiotic stress conditions, water deficiency and temperature are the most demolition constraints.

Soil or media in which the plant spends its entire life is the major factor that most affects the health of the plant. Soil is a combination of inorganic and organic materials, which provides nutrients and water to plant during its entire life span (Ranjan et al. 2017). Plant roots absorb micro, macro nutrients and water from the soil, so soil texture plays an effective role in the uptake of nutrients in plants (Dhaliwal et al. 2019). Soil structure is the cue physical properties of soil that regulate organic matter, water absorption, nutrient uptake, and microbial community in the soil (Schoonover and Crim 2015). Hard soils with immense structures prohibit the availability of water and nutrients to the plants. Due to abiotic stresses, soil structure is altered, which affects the productivity of the plants, microbial colonization, and dynamics of available organic matter of that soil (Meena et al. 2017a, b). Microbial root colonization is a feasible phenomenon mainly depends on the characteristics of host plants and their linked microbial community (Kandel et al. 2017). Due to climate change, the temperature is constantly increasing, due to which the water evaporates and flies into the atmosphere and water is not available in sufficient quantity to the crop (Tkemaladze and Makhashvili 2016). Due to this rain storm happens in some area, while severe drought conditions in some regions. At the time of drought conditions, soil temperature is changed which negatively affects soil microbial activities, soil organic matter decomposition, decreasing the non-hydrolyzed carbon content and increasing carbon dioxide emission from the soil (Gorovtsov et al. 2019). Drought stress altered microbial community structure, colonization, reduces nutrient recycling and plant production (Naylor and Coleman-Derr 2018). Soil microbes changed their community structure under stress conditions but did not change their functions immediately (Jacoby et al. 2017). In loamy soil carbon transformation and bacterial colonization are affected under extreme dry conditions.

6.3 Microbes: As Protective Companion to Plants

Soil harbors many microorganisms like archaea, bacteria, fungi, actinomycetes, and protozoans which play a critical role in maintaining soil texture, structure, nutrient recycling, reduce uses of fertilizers in soil, and ultimately stimulate the growth and development of host plants (Sathya et al. 2016). In soil ecosystem microbes have a crucial role to maintain its productivity by decomposing complex organic material into simple (Jacoby et al. 2017). Plants have multiple interactions with soil microbes to maintain C and N ration in soil and organic C sequestration. Productive soil is a soil enriched with organic carbon and various soil microorganisms which help to maintain nutrients in the soil and make available for plants development (Gougoulias et al. 2014).

6.3.1 Bacteria

A soil bacterium plays a very important role in soil ecology and helps to maintain biogeochemical cycles from decades. Interaction of plant and bacteria from rhizosphere to phyllosphere directly and indirectly affects plant growth and productivity (Souza et al. 2015). Plant growth promoting bacteria (PGPR) are coupled with rhizosphere or colonize the plant rhizosphere and provide nutrients to the host plant (Backer et al. 2018). PGPR are mainly of two types: free living such as *Azospirillum*, *Klebsiella*, *Enterobacter*, and *Pseudomonas*, while symbiotic bacteria like *Rhizobium*, *Bradyrhizobium*, or *Azorhizobium* which colonize to plant root and prevent the deleterious consequences of biotic and abiotic stresses (Ahemad and Kibret 2014; Vejan et al. 2016). Microbes can be used in field as bio fertilizer supplementing surface area erupted by plant roots, solubilization of P, fixation of nitrogen fixation, iron sequestering, and formation of siderophore (Wang et al. 2020). Zn-mobilizing bacteria help the plant to absorb Zn and produce chelating agents and organic acids. Under stress conditions many root colonized bacteria produce phytohormone for root strengthening and activate stress related genes (Kamran et al. 2017). PGPR mainly produce auxin for elongation of roots and decreasing the stomata size to cope up with abiotic stresses. Cytokinin and ethylene deaminase secretion is also interlinked with soil bacterium colonization for protecting the plant from adverse conditions (Kandel et al. 2017). PGPR activate phytohormone mediated induced systemic resistance in plants, which triggers the activation of pathogenesis-related proteins against future biotic stresses (Olanrewaju et al. 2017).

6.3.2 AM Fungi

Arbuscular mycorrhizal fungi (AMF) present in natural environment of soil have a symbiotic relationship with the roots of plants and enhance uptake of nutrients from soil (Begum et al. 2019). AMF such as *Aphanomyces*, *Fusarium*, *Phytophthora*,

Pythium, Rhizoctonia, Sclerotinium, and *Verticillium* are obligate biotrophs and depend on photosynthetic product sugars and lipids to finish their life cycle (Luginbuehl et al. 2017). At the early stage root colonization is weak and transient but at later stages it becomes stronger (Pedrotti et al. 2013). Mycelial hyphae of fungi spread into the roots beyond the depletion zone and form a complicated network and make available N, P, K, Ca, Zn, and S nutrients to the plants. As a result for its symbiotic favors, the AM fungus gets fixed carbon in the form of sugars and lipids from the plant (Chen et al. 2018). Mycorrhizal fungi help plant to cope up with salt stress by limiting the uptake and increasing uptake of K^+, Ca^{2+}, and Mn^{2+} ions. AM fungal mycelium forms a three-dimensional cross-linked structure in soil which helps to bind the soil partials without compacting it (Wang et al. 2019). Glomalin is a glycoprotein produced by AM fungi which have an immuno-responsive property and helps in stabilization of soil aggregates. AMF protect the host plant from various stresses by increasing the lignification in roots and protect them from pathogen attack (Akhtar and Siddiqui 2008). *G. mosseae* colonization in host plant exhibited an increase in aspartic acid, glutamic acid, and serine amino acid which maintain pH of rhizosphere. AMF positively influence photosynthetic efficiency of plants, C and N acquisition, and enhanced efficiency of phytoremediation of heavy metals (Ma et al. 2016).

6.3.3 Actinomycetes

Actinomycetes are unicellular, aerobic or anaerobic, and Gram-positive bacteria belong to the actinomycetales order having an aerial mycelium. Actinomycetes play a crucial role in the organic matter and nutrient cycling in rhizosphere, act as scavengers or decompose complex organic material by enzymatic process, and cope up by biotic stress (Saini et al. 2015). They also ameliorate the soil quality by increasing the buffering capacity of soil and also play role in nitrogen fixation or enhance the synthesis of metabolite and plant growth regulators. Actinomycete isolates increase the production of IAA, siderophore and solubilize phosphate for maintaining the soil ecosystem (Anwar et al. 2016). In soil at the time of pathogen attack actinomycetes trigger the production of various antibiotics or metabolites which inhibit the pathogen growth or protect the plant from bacterial and fungal pathogens (Vurukonda et al. 2018). Biocontrol agents produce hydrolytic enzymes. *Streptomyces plicatus* enhanced the secretion of chitinase enzyme which dissolves bacteria and fungi cell wall and inhibits the plant from pathogens (Kobayashi et al. 2002).

6.3.4 Virus

Till date, we have seen that the virus only cause damage to the plant, it is difficult to believe that viruses can also benefit to the plants. Soil viruses can be a new hope for improvement in agricultural productivity because some soil viruses influence microbial mortality, nutrient recycling, and dynamics of food chain or web (Emerson 2019; Roy et al. 2020). Soil viruses are more abundantly present in soil and have ability to transfer the genes to host and maintain C:N ratio in soil (Winsley et al. 2014). Virus contamination in plants induces some biochemical metabolite to increase the survival capability of plants during abiotic stress conditions. During drought stress, rice plants infected with RTSV exhibited more stomatal conductance and high leaf water potential, showing RSTV-infected leaves have more hydrated and less expression of drought responsive gene OsNAC6 and OsDREB2a (Li et al. 2016) (Fig. 6.1).

6.4 Drought Stress Management

Drought stress is a major abiotic stress that continues to increase due to continual climate change so ultimately it negatively affects crop production and productivity (Yadav et al. 2020). Drought is expected to object consequential plant growth trouble for more than 50% of the arable lands by 2050 (Gong et al. 2014). As soon as the plant is affected by the drought stress, the metabolic activities of the plant such as photosynthesis, water transportation, respiration, synthesis of proteins and lipids start to be negatively affected (Da Silva et al. 2013). Due to water scarcity, uptake of water and nutrients is halted and if it will be for long time so cell membrane will be damage and electrolyte leakage will be there. At the time of drought stress condition metabolic flux and oxidative load on tissue are increased and ROS are generated in different cell organelles which in turn introduce oxidative stress (Abhinandan et al. 2018). Drought is a multidimensional stress, affects plant from gene to cellular level. Keeping in view the increasing population in world, if food production is to increase, then the drought tolerant varieties, agronomic management implications, and new technologies will have to be developed (Salehi-Lisar and Bakhshayeshan-Agdam 2016). Under drought stress, plant utilizes some alterations in gene expression to keep away from hazard effects of water stress. To increase plant growth and avoid drought conditions, soil microorganisms such as PGPR, AM fungi, and actinomycetes play a significant role (Sayyed et al. 2019). Under water deficit conditions, plants secrete root exudates, a type of communication between plants and microbes, to restore and enroll microorganisms with attributes that helps plants during a drought. Soil microbes improve soil structure, water holding capacity, and nutrient transport efficiency of soil (Gargallo-Garriga et al. 2018). Drought affects the plant growth and metabolism in different ways and soil microorganisms provide the ability to withstand the plant against drought by following strategies (Table 6.1).

6.4.1 Growth, Biomass, and Photosynthesis

Water scarcity inhibits cell enlargement, cell division, diminishes leaf expansion and hastens leaf senescence ultimately leading to death of leaf tissue. Root development

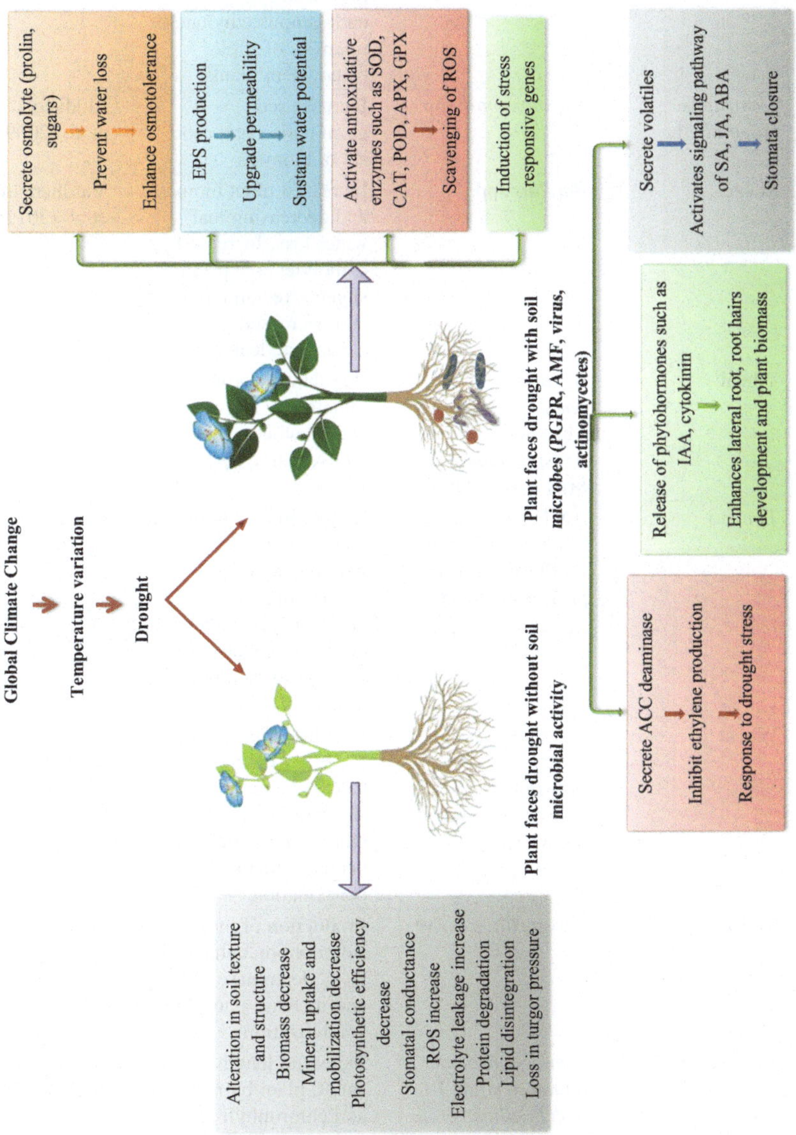

Fig. 6.1 Mechanisms to mitigation drought stress in plants by soil microbes

Table 6.1 Functions of soil microbes to diminish the drought stress in plants

S. no.	Crop/plant	Microbial strain	Function/mechanism	Reference
1.	Pea (*Pisum sativum*)	*Pseudomonas* spp.	Produce ACC deaminase enzyme and decrease production of endogenous ethylene or intensity of triple response in plants	Arshad et al. (2008)
2.	*Asparagus officinalis* L.	*Pseudomonas* spp.	Increase seed germination and root/ shoot biomass	Liddycoat et al. (2009)
3.	Maize	*Bacillus* spp.	Increased plant biomass and decreasing leaf water loss. Increased osmolytes like proline, sugars, free amino acids and decreased electrolyte leakage	Vardharajula et al. (2011)
4.	Wheat	*Bacillus amyloliquefaciens* 5113 and *Azospirillum brasilense* NO40	Regulate ascorbate–glutathione redox cycle and alleviate antioxidative defense system	Kasim et al. (2012)
5.	Tomato	AMF *Rhizophagus irregularis* and the PGPR *Variovorax paradoxus* 5C-2	Enhanced root hydraulic conductivity, CO_2 fixation capacity, photosynthetic efficiency, osmolyte and decrease oxidative damage to membrane lipids	Calvo-Polanco et al. (2016)
6.	Wheat	*Klebsiella* sp. IG 3, *Enterobacter ludwigii* IG 10, and *Flavobacterium* sp. IG 15	Enhance stress-related genes DREB2A and CAT1, increase root-shoot biomass, chlorophyll content and decrease membrane peroxidation	Gontia-Mishra et al. (2016)
7.	Maize	*Azospirillum* Az39	Production of indole and siderophore, enhanced ACC deaminase activity and increase phosphate solubilization	García et al. (2017)
8.	*Helianthus annuus*	*Planomicrobium chinense* strain P1 and *Bacillus cereus* strain P2	Increased length of root/ shoot, plant biomass, and chlorophyll content and decrease malondialdehyde content	Khan et al. (2018)

(continued)

Table 6.1 (continued)

S. no.	Crop/plant	Microbial strain	Function/mechanism	Reference
9.	Wheat	*Bacillus amyloliquefaciens* and *Agrobacterium fabrum*	Induces ACC deaminase activity and improve grain or straw yield	Zafar-ul-Hye et al. (2019)
10.	Pepper mint	(*Pseudomonas fluorescens* WCS417 r and *Bacillus amyloliquefaciens* GB03)	Decrease in membrane peroxidation and increase antioxidative enzymatic activities higher enzymatic activities were also observed in drought-stressed plants inoculated with PGPR	Chiappero et al. (2019)
11.	Strawberry	Endomycorrhizal fungi	Enhanced photosynthetic efficiency	Borkowska (2002)
12.	*Trigonella foenum-graecum*	*Bacillus subtilis* LDR2, *Ensifer meliloti,* and *Rhizophagus* irregularis	Reduce ACC and improve mycorrhizal colonization	Barnawal et al. (2013)
13.	*Robinia pseudoacacia*	AMF	Increase net photosynthetic rate, stomatal conductance, transpiration rate, leaf C, N, and P concentrations but decrease intercellular CO_2 concentration	Yang et al. (2014)
14.	*Cyclobalanopsis glauca*	*Glomus mosseae* and *Glomus intraradices*	Microbial colonization increases superoxide dismutase and peroxidase activity, maintains osmotic adjustment, and decreases proline content	Zhang et al. (2014)
15.	Soybean (*Glycine max*)	*Septoglomus constrictum* and *Glomus* sp.	Inhibit oxidative damage to membrane and increase chlorophyll content, osmotic metabolites	Grümberg et al. (2014)
16.	Maize (*Zea mays* L.)	*Rhizophagus intraradices*	Increase water uptake, nutrient uptake and relative water content	Zhao et al. (2015)
17.	*Zenia insignis*	*Funneliformis mosseae*, *Rhizoglomus intraradices*, and *Diversispora versiformis*	Enhance plant biomass, osmolytes, and antioxidative enzymatic system	Zhang et al. (2018)

(continued)

Table 6.1 (continued)

S. no.	Crop/plant	Microbial strain	Function/mechanism	Reference
18.	Wheat (*Triticum aestivum* L.)	AMF	Increased RWC in leaf and soil and photosynthetic efficiency	Mathur et al. (2018)
19.	Maize (Zea mays L.)	*Glomus versiforme*	Increase plant growth, photosynthetic efficiency, mineral uptake, antioxidant enzyme, and compatible solutes	Begum et al. (2019)
20.	Carob (*Ceratonia siliqua* L.)	AM fungus	Increasing the availability of inorganic solutes (P, K, Na, and Ca), water content uptake, soluble sugars, stomatal conductance and prevent oxidative damage	Boutasknit et al. (2020)
21.	*Glycyrrhiza uralensis*	*Acrocalymma vagum*, *Paraboeremia putaminum*, and *Fusarium acuminatum*	Increase photosynthetic efficiency and antioxidative defense system	He et al. (2019)

is inhibited under drought stress, so plant is not able to absorb water and minerals. Soil-associated microbiome plays a critical role in plant growth and productivity by increasing the availability of mineral and nutrients (Kim et al. 2020). He et al. (2019) exhibited that in water deficit stress, the inoculation of *P. putaminum* in licorice (*Glycyrrhiza uralensis*) plant enhanced the shoot and total biomass and lessens the root:shoot ratio. On the other side, inoculation of *F. acuminatum* diminished the total root and shoot biomass levels and root:shoot ratio as compared to the control. Under water deficient condition, plant growth promoting bacteria help to increase various growth parameters and biomass of the plant. Growth, nutrition, water status, and photosynthetic efficiency of the plant had a prominent possession on the mycorrhizal colonization under water deficit conditions (Vurukonda et al. 2016). AM fungi colonization increased growth, flower quality, and adaptation of rose plants under drought stress, mainly at maximum level of drought stress via improving their water use efficiency and photosynthetic status in arid and semi-arid regions (Abdel-Salam et al. 2017). PGPR ameliorate the plant cell membranes stability by triggering the antioxidant defense system, enhancing drought tolerance in plants.

AM fungal colonization remarkably enhances water uptake in different plants so plant is able to maintain leaf water content and water potential in leaves under drought stress. Mycorrhizal hyphae increase ground absorptive to leaf area ratio, so it is helpful for water uptake and transport (Begum et al. 2019). Under drought stress by hyphae elongation AMF have a strong ability to fix water uptake and improve WUE resulting in enhanced hydraulic conductivity of roots of plants (Bahadur et al. 2019). AMF colonization enhances root density and biomass resulted in more water uptake and WUE in entertainer plant. Microbes secrete exopolysaccharides which help the plant hold water and enhance soil water retention capacity (Posta and Hong Duc 2020). Soil microbes improve soil moisture under moderate drought conditions as measured by $\Psi_{TLP,}$ which had cataract effects on plant gas exchange during drought. Plant species have lower Ψ_{TLP} able to withstand against leaf dehydration and permit the plant to prolong photosynthesis under stress conditions (Bitterlich et al. 2018). Non-AM and AM soybean plants were grown under well-watered or drought-stressed conditions. AM plants were safeguard in opposition to drought conditions shown more shoot biomass and leaf water potential (−1.9 MPa) than in non-AM plants (−2.5 MPa) (Porcel 2004). AMF produce insoluble glue like material called as glomalin, which helps in modification of soil structure so plant water potential is increased under water deficit (Singh et al. 2012). *Bacillus licheniformis* (FMCH001) affects growth and physiology parameters of maize (*Zea mays* L. cv. Ronaldinho) under well-watered and drought-stressed conditions. Plants inoculated with *B. licheniformis* FMCH001 manifested enhanced root dry weight and plant water use efficiency as compared to uninoculated plants (Akhtar et al. 2020). When wheat genotypes were inoculated with rhizobacterial strains under drought increase in shoot (88%) and root dry weight (211%) compared to the control under drought conditions (Mutumba et al. 2018). *Azospirillum brasilense* Sp245 inoculation in wheat (*Triticum aestivum*) under drought stress exhibited more grain yield and better mineral quality (Mg, K, and Ca), with refined relative and absolute water content, water potential, apoplastic water fraction, and lower volumetric cell wall elasticity to increase drought resistance in plants (Creus et al. 2004). Soil microbes amended drought stress in *L. tulipifera* by limiting reductions in leaf water potential and by improving photosynthetic declines. But in *A. saccharum*, soil microbes minimize drought stress by attenuating leaf water potential and there was no significant change in photosynthetic efficiency (Kannenberg and Phillips 2016). Mycorrhizal symbioses increase gas exchange and transpiration flux capacity of the host plant. Under moderate drought stress, inoculation of *Acinetobacter calcoaceticus* X128 in *Sambucus williamsii* improves photosynthetic rate, stomatal conductance, intracellular CO_2 concentration, and total chlorophyll content (Liu et al. 2019). Maize inoculated with *B. subtilis* PRBS-1 exhibited reductions in transpiration rate under normal conditions when compared to non-inoculated with stress conditions. Under water stress, *B. subtilis* AP-3 or PRBS-1 inoculated maize had 40% lower transpiration rate as compared to non-inoculated plants (De Lima et al. 2019).

Under dry conditions due to loss in chlorophyll content and quantum yield, plant stomatal conductance, CO_2 assimilation, and photosynthetic efficiency are reduced. When plant is subjected to drought condition, the photosynthetic rate reduced by 40% as well as 61% reduction in stomatal conductance, as results of prolonged closure of stomata to prevent the loss of intracellular water content (Wang et al. 2018). AM colonization increases the rates of photosynthetic product and export by enhancing the number of photosynthetic units. Under drought stress concentrations of chlorophyll a, b and chlorophyll a + b of mycorrhizal plants were 18.6%, 27.5%, and 20.5%, respectively, higher as compared to non-mycorrhizal plants (Zhu et al. 2012). Water deficit demolishes the shape and role of PSII reaction center and deranges electron flow in photosynthetic apparatus, AM colonization helps to alleviate the unpropitious effects of drought stress on PSII reaction center and improves photosynthetic efficiency (Dalal and Tripathy 2018). In maize plant mycorrhizal colonization enhances Fv/Fm than in non-mycorrhizal plants under drought stress results in increased maximal fluorescence, maximum quantum efficiency of PSII photochemistry, and potential photochemical efficiency (Sheng et al. 2008). During water stress conditions when lettuce plant is co-inoculated with endophytic selenobacteria *Bacillus* sp., *Klebsiella* sp., or *Acinetobacter* sp. and the arbuscular mycorrhizal fungus showed tolerance to drought stress by enhancing relative water content, stomatal conductance and improve photosynthetic pigments such as chlorophyll and carotenoids (Durán et al. 2016). In *Arabidopsis* inoculation with *B. subtilis* improves photosynthesis by the modulation of plant endogenous sugar/abscisic acid signaling. PGPR help to enhance electron transport rate of PSII thereby keep away from photooxidative damage of photosystem (Zhang et al. 2008). *Lactuca sativa* plants inoculated with *Bacillus megaterium*, pigments like chlorophyll and carotenoids were decreased but consequently these pigments were increased when *B. megaterium* co-inoculated with an arbuscular mycorrhizal fungus (Marulanda-Aguirre et al. 2007).

6.4.2 Mineral Uptake and Mobilization

Drought stress causes soil moisture limitation and affects nutrient uptake and mobilization in plant. During dry period dead microbial organism helps in nutrient uptake and diffusion throughout the plant (Naylor and Coleman-Derr 2018). At the time of dry conditions N concentration is decreased and P concentration is increased in soil resulting to decoupling of N and P cycle. Most of the plants uptake nitrogen in the form of inorganic nitrate or ammonium by NRT1 and NRT 2 transporters and with a decline in moisture N and P uptake and transportation is decreased in soil (Zhang et al. 2019). As such, plant absorbs potassium from soil by roots. In plasma membrane proton pumps generate H^+ electrical gradient, it provides motive force to potassium ions to transport in plant cell (Ragel et al. 2019). Bacterial and fungal inoculation helps to absorb, mobilize, and translocate various micro and macro nutrients from dry soils. By the production and accumulation of organic acids and siderophores, AMF and bacteria encourage availability and mobilization of nutrients (Rashid et al. 2016). AMF help to boost up nutrient uptake and enhance availability of zinc and copper in plants under drought conditions which leads to improvement in photosynthate production and biomass accumulation (Coccina et al. 2019). PGPR help to increase uptake and solubilization of phosphate and promote root development in dry conditions. *Azospirillum lipoferum* inoculated with wheat seedlings under drought stress shows better growth, water potential, and N content as compared to non-stressed condition (Arzanesh et al. 2010). AMF construct fungal formations like arbuscules, which oblige in exchange of C, P, and various inorganic minerals, eventually conveying a remarkable vigor to plants (Begum et al. 2019). In *Ephedra foliata* Boiss AMF significantly enhanced the biosynthesis of phytohormones and uptake of essential nutrients like K, Mg, and Ca under drought stress, play role in cell division (Fahad Al-Arjani et al. 2019). Drought stress limits the uptake of N and P in sunflower leaves and seeds. When sunflower plant is inoculated with AMF *Glomus mosseae* and *Glomus hoi*, N and P concentration in leaves and seed was increased as well as plant biomass and growth were also increased (Gholamhoseini et al. 2013).

6.4.3 Redox Homeostasis and Membrane Stabilization

Under drought stress conditions reactive oxygen species (ROS) are produced in plants results from disabled electron transport in mitochondria, plasma membrane and chloroplast. H_2O_2 is produced from photorespiration and these ROS cause oxidative stress in plants (Sharma et al. 2012). At lower levels, ROS act as second messengers in stress signal transduction pathways in the cells, but in excess they cause damage to biomolecules such as denaturation of protein, lipid peroxidation, and mutation in nucleic acid (Tak and Asthir 2017). To cope up with oxidative stress plant synthesizes antioxidative enzyme for detoxification of these ROS. Antioxidants are categorized into enzymatic such as catalase, superoxide dismutase, peroxidase, glutathione reductase and non-enzymatic includes ascorbate, tocopherols, cysteine, and glutathione (Hasanuzzaman et al. 2019). Superoxide free radicals can damage membrane lipids by lipid peroxidation; SOD transforms superoxide free radicals into H_2O_2 and further catalase changes H_2O_2 into water and oxygen.

Peroxidases fester H_2O_2 by oxidation of phenolic compounds by utilizing guaiacol as electron donor in vitro (Birben et al. 2012). Inoculation with nitrogen fixing bacteria *Azotobacter* and *Azospirillum* in soybean plants enhanced the catalase, glutathione peroxidase activity, and superoxide dismutase activity during the pod-filling stage under drought stress. Under water deficit condition, when soybean plants were inoculated with bacteria, there were increase in Abscisic acid and proline content during flowering stage. This interrelationship between antioxidant enzymes and drought stress with bacteria revealed that antioxidant enzymes play a very important role in mitigating the deleterious effects of drought conditions (Zakikhani et al. 2012). Ascorbate and glutathione are powerful antioxidants and maintain cellular homeostasis in plant tissue (Kurutas 2015). ROS such as superoxide, peroxide, and hydroxyl radical are produced during drought stress and badly affect plant growth and development. Superoxide and hydrogen peroxide arise when molecular oxygen adventitiously oxidizes redox enzymes that normally transfer electrons to other substrates. Rice seeds were inoculated with *Bacillus methylotrophicus* RABA6 and yeast *Candida tropicalis* RAYN2. Plants inoculated with *Bacillus methylotrophicus* RABA6 manifested more cell membrane stability, improved antioxidative status, and decreased ROS content under moisture stress conditions (Punitha et al. 2019). Drought stress reduced the growth of rice plants but when these plants inoculated with *Trichoderma*, *Pseudomonas* and their mixture cut down the effect of watering mechanism. Microbial inoculated plant showed more antioxidant concentration and high values of DPPH, ABTS, Fe-ion reducing and chelation activity. Inoculated plants increased activity of peroxidase, ascorbate peroxidase, glutathione peroxidase, and glutathione reductase enzymes under drought stress (Singh et al. 2020). At water deficit condition *Cecropia pachystachya* and *Cariniana estrellensis* seedlings inoculated with PGPB (*Azospirillum brasilense* and *Bacillus* sp.) enhanced the antioxidant metabolism. *A. brasilense* mainly enhanced ascorbate peroxidase and superoxide dismutase activities and has significantly positive effect on non-enzymatic antioxidant compounds. Results revealed that *A. brasilense* and *Bacillus* sp. inoculation had high impact on the non-enzymatic pathway as compared to that in non-inoculated plants under drier conditions (Tiepo et al. 2020). In *Leymus chinensis* and *Hemarthria altissima* grasses, AMF limit the harsh influence of drought on plant growth and increased plant biomass, photosynthetic rate, and superoxide dismutase activity in *Leymus chinensis* and decreased malondialdehyde content under moderate drought (Li et al. 2019). Citrus (*Citrus tangerine*) was inoculated with arbuscular mycorrhizal fungus *Glomus versiforme* under well-watered and water stress conditions. Citrus roots with AM exhibited minimum malondialdehyde hydrogen peroxide and superoxide radicals whether superoxide dismutase, guaiacol peroxidase, and glutathione reductase were significantly more in AM than in non-AM roots. Water deficit AM roots indicated increased levels of ascorbate and GSSG which helps plant to withstand against dry conditions (Wu et al. 2006).

Oxidative stress damage membrane integrity of the cell causes peroxidation of polyunsaturated fatty acids and produced malondialdehyde (Ayala et al. 2014). Soybean plant was grown inoculated with AM fungi and well-watered or drought-stressed conditions. Water scares AM plants had 55% lower lipid peroxidation in shoots than in droughted non-AM plants. AM symbiosis maintained osmotic adjustment in roots and protect the plants against oxidative stress (Porcel 2004). Lipid peroxidation enhances permeability of membrane and electrolyte leakage of the ions ultimately damages cell membrane system. Mycorrhizal symbiosis alleviated the dismissive effects of electrolyte leakage resulted by drought stress in finger millet. AM fungi *Rhizophagus intraradices* and *Piriformospora indica* inoculation effective to mitigate water stress by reducing the electrolyte leakages, malondialdehyde and hydrogen peroxide content (Tyagi et al. 2017). *Gluconacetobacter diazotrophicus* Pal5 inoculation mitigates drought stress in red rice by decreasing the malondialdehyde content and causes less damage to plasma membrane phospholipids. Bacterial inoculation with *G. diazotrophicus* has a capacity to regulate the plant safeguarding against drought stress by increasing the expression of sod, gor, cat, P5CR, and BADH genes in leaves of rice plants (Filgueiras et al. 2019).

6.4.4 Osmolytes Regulation

Some plants have capability to cope up with adverse conditions mainly in dehydration. Under drought or temperature stress osmoregulation is the main adaptive mechanism, which decreasing cellular water potential by accumulation of different osmoprotectants such as sugars, proline, glycine betaine, and sugar alcohols (Chen and Jiang 2010). Osmolytes or osmoprotectants are small, electrically neutral, nontoxic molecules, and extremely soluble organic compounds that conserve osmotic pressure in plant tissue and stabilize proteins and lipid membranes under various stress conditions (Zulfiqar et al. 2019). During drought conditions osmolytes decrease the cellular water potential and inhibit the loss of water. Proline is the most common osmolyte which accumulates in cytoplasm, scavenge the free radicals and stabilize protein structure, integrity of membranes by regulate the cellular homeostasis in water dedicate conditions (Hayat et al. 2012). When green gram (*Vigna radiata*) subjected to drought and inoculated with *Pseudomonas fluorescens* Pf1 and *Bacillus subtilis* EPB5, EPB22, EPB 31 exhibit more proline accumulation in Pf1 inoculated plants contrasted to untreated plants (Saravanakumar et al. 2010). Tomato (*Lycopersicon esculentum* Mill) cv. Anakha inoculated with *Bacillus polymyxa* shows increase in proline accumulation to mitigate drought conditions (Shintu and Jayaram 2015). Maize is inoculated with group of *Azospirillum* isolates, Az19 strain increased 11.6 times the roots proline content as compared to uninoculated controls under water scarce conditions. Under drought plant inoculated with *Bacillus* strains enhanced proline content by upregulation of genes for P5CS, plays a chief role in proline biosynthesis (Oliveira et al. 2017). Soybean inoculated with AMF showed more proline accumulation in roots than in shoots under water scarce condition exhibit a positive symbiosis of mycorrhiza under stress (Begum et al. 2019). Higher content of proline, glucose, and total soluble protein in AMF-inoculated *E. foliate* plants furnishes ability to withstand against drought stress (Fahad Al-Arjani et al. 2019).

Glycine betaine is an electrically neutral, soluble, and nontoxic osmolyte which raises intracellular osmolality; maintains cellular integrity when cellular tissue imposed to stress conditions (Burg and Ferraris 2008). Glycine Betaine is synthesized by oxidation of choline by choline monooxygenase and betaine aldehyde dehydrogenase enzymatic reactions. Under water deficit conditions PGPR strains *Klebsiella variicola* F2 (KJ465989), *Raoultella planticola* YL2 (KJ465991), and *Pseudomonas fluorescens* YX2 (KJ465990) positively influence the accumulation of choline and glycinebetaine in maize leaves. These PGPR strains maintained the osmotic adjustment in the cells by accumulations of choline and subsequent GB, resulting in amelioration of water or nutrient mobilization and plant growth in maize under drought stress (Gou et al. 2015). When Arabidopsis inoculated with *Bacillus subtilis* (strain GB03) expression of PEAMT gene is induced and increased choline and glycine betaine synthesis to enhance plant tolerance under osmotic stress (Zhang et al. 2010). Fungal isolates, viz., UASDAMFS15, UASDAMFS25, and UASDAMFS36 increased glycine betaine with water stress conditions in soybean and revealed that glycine betaine as a constructive osmolyte increased under drought conditions and helps to stabilize and maintain membrane integrity under adverse conditions (Konnur et al. 2019). Polyamines are low molecular weight aliphatic nitrogenous bases consisting of two or more amino groups and play key roles in environmental stress retaliation. AMF also regulate polyamines metabolism to protect plant cells from the adverse effects of drought by increasing the antioxidant defense system, maintaining cellular pH and redox homeostasis under drought stress (Chen et al. 2019). Under restricted irrigation, *Rhizobium japonicum*, *Azotobacter chroococcum,* and *Azospirillum brasilense* and mixture of them showed increase in putrescine and spermidine polyamines to cope up with water stress conditions (Zahedi and Abbasi 2015).

Sugars act as osmolytes and accumulate in plants, maintain osmotic adjustment under stress conditions. Soluble sugars such as sucrose, hexose, trehalose, and sugar alcohols act as the osmoprotectants and help plant to maintain cellular structure, proficiency of photosynthesis and build up antioxidative system in the stress conditions (Gangola and Ramadoss 2018). Wheat seedlings treated with *Bacillus* were disclosed to drought stress, *Bacillus* inoculated wheat plants have more survival capacity in stress conditions by accumulation of amino acids, organic acids, and sugars in the cellular tissue (Abd El-Daim et al. 2019). Trehalose is a non-reducing sugar acts as osmolyte by maintain and stabilize membranes and enzymes. In maize plant increased trehalose content under drought conditions and mitigate the stress when inoculated with *A. brasilense* (Iordachescu and Imai 2011). AMF and plant growth promoting bacteria such as *Azotobacter chroococcum* and *Azospirillium lipofrum* help to mitigate the drought stress effects in walnut plants by increased osmoprotectant such as proline, total soluble sugar, and starch content (Behrooz et al. 2019). Transgenic rice was generated by fusion of E. coli gene for trehalose-6-phosphate synthase/phosphatase and ABA-inducible promoter. This transgenic rice showed better tolerance under drought, saline, and alkaline conditions and increasing relative water content, chlorophyll content, and photosynthetic efficiency (Joshi et al. 2019).

6.4.5 Hormonal Regulation and Volatiles

Phytohormones regulate plant growth, cell elongation and enlargement to escape the plant from stressful conditions (Takatsuka and Umeda 2014). During stress PGPR and AM fungi trigger the biosynthesis of metabolites and auxin, gibberellins, cytokinin, ABA, jasmonic acid, and salicylic acid plant growth regulators which affect plant survival, growth, and development (Egamberdieva et al. 2017). IAA, a naturally occurring auxin in plants performs an important role in tissue culture in root initiation. During drought conditions, soil microbes increase the biosynthesis of IAA, which increases the formation of adventitious roots and root hair for absorption of nutrients and water to mitigation of water loss from plants (Kim et al. 2020). *Azospirillum* symbiosis with wheat decreases leaf water potential due to biosynthesis of IAA plant hormone helps to boost up growth of root and growth of lateral roots to enhance water uptake and nutrients absorption under drier conditions (Arzanesh et al. 2010). *Bacillus* sp. (12D6) and *Enterobacter* sp. (16i) these two PGPR strains were inoculated with wheat and maize rhizosphere, help to reduce the negative effects of drought. Results revealed that both bacteria strains start biosynthesis of IAA and salicylic acid during drought, which alter the root system and increase root length, surface area, and branching of roots as compared to uninoculated plants (Jochum et al. 2019). Tomato seedlings are treated with A. brasilense, induce nitric oxide gas (NO) which enhanced IAA production. During IAA biosynthesis, NO acts as a signaling molecule and indirectly enhances the adventitious root emergence in tomato plants (Molina-Favero et al. 2008). *During drought stress, Ensifer meliloti* IAA-overproducing RD64 derivative strains (*Ms-RD64*) nodulated with *Medicago sativa* exhibited that nitrogenase enzymatic activity is increased in inoculated plant and they were less damaged as compared to control due to drought stress. Endogenous IAA level and osmolyte content were increased in *Ms-RD64* nodulated plants to cope up with drought stress (Defez et al. 2017). AM fungi inoculated with trifoliate orange stimulate IAA biosynthesis and root hair growth under drought conditions. AM treated seedlings showed increased root length and diameter, whereas PtYUC3 and PtYUC8 induced IAA biosynthesis and down regulate transcription factors of auxin efflux under drought stress (Liu et al. 2018).

PGPR inoculation which helps to increase biosynthesis of cytokinin under water stress conditions has a crucial role in cell division, proliferation, and delaying in senescence and abscission (Olanrewaju et al. 2017). During gain filling stage of plant if plant is exposed to water stress, cytokinin helps plant to stay green to exhibit against drought stress. Under drought stress *Bacillus subtilis* inoculated with seedling of *Platycladus orientalis* (oriental thuja) had more relative water content and leaf water potential compared with uninoculated plants. Bacillus inoculated plants raised cytokinin and ABA hormone in shoots which gives benefit to plants under drought stress (Liu et al. 2013). Lettuce seedling inoculated with cytokinin producing bacteria under water deficit conditions, plant showed more root and shoot biomass, stomatal conductance was hindered due to increases of ABA concentration (Arkhipova et al. 2007). ABA is known as a stress hormone and it provides the capability to plants to cope up with specially drought stress by stomatal regulation, reduce leaf expansion and root development. Under stress conditions ABA enhanced the synthesis of various osmolytes and maintains the turgor pressure of plant under dehydration (Fahad et al. 2017).

Ethylene is also known as ripening hormone; it regulates the leaf senescence or abscission and protects plant from harsh environmental conditions. 1-aminocyclopropane-1-carboxylate (ACC) acts as precursor in ethylene biosynthesis (Iqbal et al. 2017). At the time of drought stress soil microbes induce ACC deaminase enzyme which inhibits the ACC activity for ethylene synthesis. Under drought stress, endogenous ethylene controls the plant shoot and root length and maintains homeostasis of plant (Glick 2004). Rhizobacteria—*Ochrobactrum pseudogrignonense*RJ12, *Pseudomonas* sp.RJ15, and *Bacillus subtilis*RJ46 produce ACC deaminase enzyme in *Vigna mungo* L. and *Pisum sativum* L. under drought stress and enhanced germination percentage, root and shoot length. These strains decreased ACC accumulation and ACC oxidase enzyme in roots to improvement of plant under drought (Saikia et al. 2018). Integrated application of ACC deaminase generating PGPR and biochar decreased negative effects of drought in wheat and maintains moisture status in plants in its life cycle. Colonization of *E. cloacae* and *A. xylosoxidans* with maize decrease the ethylene biosynthesis by ACC deaminase results in better root elongation for uptake of nutrients and alleviate negative impact of drought stress. These bacteria strain increased yield, photosynthetic capacity, stomatal conductance, and pigment contents (Danish et al. 2020). Bacteria strain which produce ACC deaminase isolated from jujube (*Ziziphus jujuba*) enhanced plant biomass and relative water content in and make capable plant to survive in drought conditions (Zhang et al. 2020).

Late embryogenesis abundant proteins and microbial volatile organic compounds (VOCs) influence the plant growth and development under drought stress condition and provide tolerance to plant (Kaushal and Wani 2015). These VOCs act as signaling compounds for priming and systemic responses in the host and its neighboring plants. Some PGR like *Pseudomonas aeruginosa* strain Pa2 secreted exopolysaccharides which increase bacteria ability to perpetuate soil structure and moisture for drought tolerance (Ojuederie et al. 2019). PGPR such as *Bacillus*, *Fusarium Arthrobacter,* and *Pseudomonas* produce VOCs like 2-pentylfuran, 3-hydroxy-2-butanone, and 2,3-butanediol and help plant to cope up with drought in Arabidopsis and tobacco (Park et al. 2015). Some VOCs are also emitted by *Fusarium oxysporum* and *Verticillium dahliae* to protect the host plant from variety of environmental stresses (Brilli et al. 2019). Spermidine is a polyamine secreted by *B. megaterium* BOFC15 which enhanced biomass, photosynthetic efficiency in Arabidopsis plant under drier conditions (Zhou et al. 2016). *Pseudomonas chlororaphis* O6 colonizes in rhizosphere and produces 2R, 3R-butanediol VOCs which inhibit water loss from plant under drought stress (Vurukonda et al. 2016). PGPR secrete dimethylhexadecylamine and an indole volatile compounds which increase root length and hairs in *A. thaliana* plants to mitigate drought stress (Ojuederie et al. 2019).

6.5 Conclusion and Future Prospective

From last decade a lot of study had been done in laboratory and in the natural conditions, that soil microbe interactions with plant are beneficial for overall development of plant. Conventional plant breeding or genetic engineering techniques are being adopted to produce stress tolerant variety, but if the plant has to be saved from stress in a short time, then soil microbe is a better option; these are proving to be an easy and inexpensive medium to cope up with stress. Soil microbes such as PGPB, mycorrhizal fungi, and actinomycetes colonized with roots to give out the tolerance against abiotic stresses. PGPB and AMF can act as capable stress attenuators in plants and increase the sustainable abiotic stress management during these changing climatic conditions. Among the entire stresses drought is a severe environmental constraint under this changing climatic conditions and soil microorganism helps to provide tolerance and adaptation to plants by alterations in soil properties, increasing antioxidative system, by accumulation of osmolytes and cross talk of hormone and volatiles.

By synthesis of various hormones, production of ACC deaminase enzyme and secretion of EPS are important mechanisms that help plant to withstand against drought conditions. Various soil microbes increase defensive enzymes, metabolite like phenols and inhibit ethylene biosynthesis in plants to fight biotic and abiotic stresses. AM fungi colonization plays a key role under drought stress by osmolyte production, antioxidative enzymes, and enhanced nutrient uptake or plant biomass. Further research work should be carried out to create mycorrhizal field crops under different adverse conditions. This chapter describes various types of microorganisms that support plants in stress conditions and give the plant the courage to stand against the stress. In field condition, there is a need to explore diversity of novel microbes colonization with roots which gives strength to plant against biotic and abiotic stresses. There is also a need to pay attention on data generation on omics-based by following genomics, metagenomics, and proteomics approach mediated studies on particular plant–microbe abiotic stress will be demanded to sort out the reason beyond the stress tolerance mechanisms in plants.

References

Abd El-Daim I, Bejai S, Meijer J (2019) *Bacillus velezensis* 5113 induced metabolic and molecular reprogramming during abiotic stress tolerance in wheat. Sci Rep 9:16282. https://doi.org/10.1038/s41598-019-52567-x

Abdel-Salam E, Alatar A, El-Sheikh MA (2017) Inoculation with arbuscular mycorrhizal fungi alleviates harmful effects of drought stress on damask rose. Saudi J Biol Sci. https://doi.org/10.1016/j.sjbs.2017.10.015

Abhinandan K, Skori L, Stanic M, Hickerson NMN, Jamshed M, Samuel MA (2018) Abiotic stress signaling in wheat – an inclusive overview of hormonal interactions during abiotic stress responses in wheat. Front Plant Sci 9. https://doi.org/10.3389/fpls.2018.00734

Ahemad M, Kibret M (2014) Mechanisms and applications of plant growth promoting rhizobacteria: current perspective. J King Saud Univ Sci 26(1):1–20. https://doi.org/10.1016/j.jksus.2013.05.001

Akhtar MS, Siddiqui ZA (2008) Arbuscular mycorrhizal fungi as potential bioprotectants against plant pathogens. In: Siddiqui ZA (eds) Mycorrhizae: sustainable agriculture and forestry. Springer Science, pp 61–97

Akhtar SS, Amby DB, Hegelund JN, Fimognari L, Großkinsky DK, Westergaard JC, Muller R, Moelbak L, Liu F, Roitsch T (2020) *Bacillus licheniformis* FMCH001 increases water use efficiency via growth stimulation in both Normal and drought conditions. Front Plant Sci 11. https://doi.org/10.3389/fpls.2020.00297

Anwar S, Ali B, Sajid I (2016) Screening of rhizospheric actinomycetes for various in-vitro and in-vivo plant growth promoting (PGP) traits and for agroactive compounds. Front Microbiol 7. https://doi.org/10.3389/fmicb.2016.01334

Arkhipova TN, Prinsen E, Veselov SU, Martinenko EV, Melentiev AI, Kudoyarova GR (2007) Cytokinin producing bacteria enhance plant growth in drying soil. Plant Soil 292:305–315

Arshad M, Shaharoona B, Mahmood T (2008) Inoculation with Pseudomonas spp. containing ACC-deaminase partially eliminates the effects of drought stress on growth, yield, and ripening of pea (*Pisum sativum* L.). Pedosphere 18(5):611–620. https://doi.org/10.1016/s1002-0160(08)60055-7

Arzanesh MH, Alikhani HA, Khavazi K, Rahimian HA, Miransari M (2010) Wheat (*Triticum aestivum* L.) growth enhancement by *Azospirillum* sp. under drought stress. World J Microbiol Biotechnol 27(2):197–205. https://doi.org/10.1007/s11274-010-0444-1

Ayala A, Muñoz MF, Argüelles S (2014) Lipid peroxidation: production, metabolism, and signaling mechanisms of malondialdehyde and 4-hydroxy-2-nonenal. Ox Med Cell Longev 1–31. https://doi.org/10.1155/2014/360438

Backer R, Rokem JS, Ilangumaran G, Lamont J, Praslickova D, Ricci E, Smith DL (2018) Plant growth-promoting rhizobacteria: context, mechanisms of action, and roadmap to commercialization of biostimulants for sustainable agriculture. Front Plant Sci 9. https://doi.org/10.3389/fpls.2018.01473

Bahadur A, Batool A, Nasir F, Jiang S, Mingsen Q, Zhang Q, Pan J, Liu Y, Feng H (2019) Mechanistic insights into arbuscular mycorrhizal fungi-mediated drought stress tolerance in plants. Int J Mol Sci 20:4199. https://doi.org/10.3390/ijms20174199

Barnawal D, Maji D, Bharti N, Chanotiya CS, Kalra A (2013) ACC deaminase-containing Bacillus subtilis reduces stress ethylene-induced damage and improves mycorrhizal colonization and rhizobial nodulation in *Trigonella foenum-graecum* under drought stress. J Plant Growth Regul 32(4):809–822. https://doi.org/10.1007/s00344-013-9347-3

Basu S, Ramegowda V, Kumar A, Pereira A (2016) Plant adaptation to drought stress. F1000 Res 5:1554. https://doi.org/10.12688/f1000research.7678.1

Begum N, Ahanger MA, Su Y, Lei Y, Mustafa NSA, Ahmad P, Zhang L (2019) Improved drought tolerance by AMF inoculation in maize (*Zea mays*) involves physiological and biochemical implications. Plant 8(12):579. https://doi.org/10.3390/plants8120579

Behrooz A, Vahdati K, Rejali F, Lotfi M, Sarikhani S, Leslie C (2019) Arbuscular mycorrhiza and plant growth-promoting bacteria alleviate drought stress in walnut. J Am Soc Hortic Sci 54 (6):1087–1092

Birben E, Sahiner UM, Sackesen C, Erzurum S, Kalayci O (2012) Oxidative stress and antioxidant defense. World Allergy Organ J 5(1):9–19. https://doi.org/10.1097/wox.0b013e3182439613

Bitterlich M, Franken P, Graefe J (2018) Arbuscular mycorrhiza improves substrate hydraulic conductivity in the plant available moisture range under root growth exclusion. Front Plant Sci 9. https://doi.org/10.3389/fpls.2018.00301

Borkowska B (2002) Growth and photosynthetic activity of micropropagated strawberry plants inoculated with endomycorrhizal fungi (AMF) and growing under drought stress. Acta Physiol Plant 24(4):365–370. https://doi.org/10.1007/s11738-002-0031-7

Boutasknit A, Baslam M, Ait-El-Mokhtar M, Anli M, Ben-Laouane R, Douira A, Modafar CE, Mitsui T, Wahbi S, Meddich A (2020) Arbuscular mycorrhizal fungi mediate drought tolerance and recovery in two contrasting carob (*Ceratonia siliqua* L.) ecotypes by regulating stomatal, water relations, and (in) organic adjustments. Plant 9(1):80. https://doi.org/10.3390/plants9010080

Brilli F, Loreto F, Baccelli I (2019) Exploiting plant volatile organic compounds (VOCs) in agriculture to improve sustainable defense strategies and productivity of crops. Front Plant Sci 10. https://doi.org/10.3389/fpls.2019.00264

Burg MB, Ferraris JD (2008) Intracellular organic osmolytes: function and regulation. J Biol Chem 283(12):7309–7313. https://doi.org/10.1074/jbc.r700042200

Calvo-Polanco M, Sánchez-Romera B, Aroca R, Asins MJ, Declerck S, Dodd IC, Martínez-Andújar C, Albacete A, Ruiz-Lozano JM (2016) Exploring the use of recombinant inbred lines in combination with beneficial microbial inoculants (AM fungus and PGPR) to improve drought stress tolerance in tomato. Environ Exp Bot 131:47–57. https://doi.org/10.1016/j.envexpbot.2016.06.015

Chen H, Jiang JG (2010) Osmotic adjustment and plant adaptation to environmental changes related to drought and salinity. Environ Rev 18:309–319. https://doi.org/10.1139/a10-014

Chen M, Arato M, Borghi L, Nouri E, Reinhardt D (2018) Beneficial services of arbuscular mycorrhizal fungi - from ecology to application. Front Plant Sci 9. https://doi.org/10.3389/fpls.2018.01270

Chen D, Shao Q, Yin L, Younis A, Zheng B (2019) Polyamine function in plants: metabolism, regulation on development, and roles in abiotic stress responses. Front Plant Sci 9. https://doi.org/10.3389/fpls.2018.01945

Chiappero J, Cappellari L, Sosa Alderete LG, Palermo TB, Banchio E (2019) Plant growth promoting rhizobacteria improve the antioxidant status in *Mentha piperita* grown under drought stress leading to an enhancement of plant growth and total phenolic content. Ind Crops Prod 139:111553

Coccina A, Cavagnaro TR, Pellegrino E, Ercoli L, McLaughlin MJ, Watts-Williams SJ (2019) The mycorrhizal pathway of zinc uptake contributes to zinc accumulation in barley and wheat grain. BMC Plant Biol 19(1). https://doi.org/10.1186/s12870-019-1741-y

Creus CM, Sueldo RJ, Barassi CA (2004) Water relations and yield in Azospirillum-inoculated wheat exposed to drought in the field. Can J Bot 82(2):273–281. https://doi.org/10.1139/b03-119

Da Silva EC, de Albuquerque MB, de Azevedo Neto AD, da Silva Junior CD (2013) Drought and its consequences to plants – from individual to ecosystem. In: Responses of organisms to water stress. https://doi.org/10.5772/53833

Dalal VK, Tripathy BC (2018) Water-stress induced downsizing of light-harvesting antenna complex protects developing rice seedlings from photo-oxidative damage. Sci Rep 8(1). https://doi.org/10.1038/s41598-017-14419-4

Danish S, Zafar-ul-Hye M, Mohsin F, Hussain M (2020) ACC-deaminase producing plant growth promoting rhizobacteria and biochar mitigate adverse effects of drought stress on maize growth. PLoS One 15(4):e0230615. https://doi.org/10.1371/journal.pone.0230615

De Lima BC, Moro AL, Santos ACP, Bonifacio A, Araujo ASF, de Araujo FF (2019) *Bacillus subtilis* ameliorates water stress tolerance in maize and common bean. J Plant Int 14 (1):432–439. https://doi.org/10.1080/17429145.2019.1645896

Defez R, Andreozzi A, Dickinson M, Charlton A, Tadini L, Pesaresi P, Bianco C (2017) Improved drought stress response in Alfalfa plants nodulated by an IAA over-producing rhizobium strain. Front Microbiol:8. https://doi.org/10.3389/fmicb.2017.02466

Dhaliwal SS, Naresh RK, Mandal A, Singh R, Dhaliwal MK (2019) Dynamics and transformations of micronutrients in agricultural soils as influenced by organic matter build-up: a review. Environ Sustain Indicat 1-2:100007. https://doi.org/10.1016/j.indic.2019.100007

Durán P, Acuña J, Armada E, López-Castillo O, Cornejo P, Mora M, Azcón R (2016) Inoculation with selenobacteria and arbuscular mycorrhizal fungi to enhance selenium content in lettuce plants and improve tolerance against drought stress. J Soil Sci Plant Nutr:0–0. https://doi.org/10.4067/s0718-95162016005000017

Egamberdieva D, Wirth SJ, Alqarawi AA, Abd_Allah EF, Hashem A (2017) Phytohormones and beneficial microbes: essential components for plants to balance stress and fitness. Front Microbiol 8. https://doi.org/10.3389/fmicb.2017.02104

Emerson J B (2019) Soil viruses: a new hope mSystems 4(3). https://doi.org/10.1128/msystems.00120-19

Fahad Al-Arjani AB, Hashem A, Fathi Abd_Allah E (2019) Arbuscular mycorrhizal fungi modulates dynamics tolerance expression to mitigate drought stress in *Ephedra foliata* Boiss. Saudi J Biol Sci. https://doi.org/10.1016/j.sjbs.2019.10.008

Fahad S, Bajwa AA, Nazir U, Anjum SA, Farooq A, Zohaib A, Sadia S, Naim W, Adkins S, Saud S, Ihsan MZ, Alharby H, Wu C, Wang D, Huang J (2017) Crop production under drought and heat stress: plant responses and management options. Front Plant Sci 8. https://doi.org/10.3389/fpls.2017.01147

Fathi A, Tari DB (2016) Effect of drought stress and its mechanism in plants. Int J Life Sci 10(1):1. https://doi.org/10.3126/ijls.v10i1.14509

Filgueiras L, Silva R, Almeida I, Vidal M, Baldani JI, Meneses CHSG (2019) Gluconacetobacter diazotrophicus mitigates drought stress in *Oryza sativa* L. Plant Soil. https://doi.org/10.1007/s11104-019-04163-1

Gangola MP, Ramadoss BR (2018) Sugars play a critical role in abiotic stress tolerance in plants. In: Biochemical, physiological and molecular avenues for combating abiotic stress tolerance in plants, pp 17–38. https://doi.org/10.1016/b978-0-12-813066-7.00002-4

García JE, Maroniche G, Creus C, Suárez-Rodríguez R, Ramirez-Trujillo JA, Groppa MD (2017) In vitro PGPR properties and osmotic tolerance of different Azospirillum native strains and their effects on growth of maize under drought stress. Microbiol Res 202:21–29. https://doi.org/10.1016/j.micres.2017.04.007

Gargallo-Garriga A, Preece C, Sardans J, Oravec M, Urban O, Peñuelas J (2018) Root exudate metabolomes change under drought and show limited capacity for recovery. Sci Rep 8(1). https://doi.org/10.1038/s41598-018-30150-0

Gholamhoseini M, Ghalavand A, Dolatabadian A, Jamshidi E, Khodaei-Joghan A (2013) Effects of arbuscular mycorrhizal inoculation on growth, yield, nutrient uptake and irrigation water productivity of sunflowers grown under drought stress. Agric Water Manag 117:106–114. https://doi.org/10.1016/j.agwat.2012.11.007

Glick BR (2004) Bacterial ACC deaminase and the alleviation of plant stress. Adv Appl Microbiol:291–312. https://doi.org/10.1016/s0065-2164(04)56009-4

Glick BR (2012) Plant growth-promoting bacteria: mechanisms and applications. Scientifica 2012:1–15. https://doi.org/10.6064/2012/963401

Gong F, Yang L, Tai F, Hu X, Wang W (2014) "Omics" of maize stress response for sustainable food production: opportunities and challenges. OMICS: J Integr Biol 18(12):714–732. https://doi.org/10.1089/omi.2014.0125

Gontia-Mishra I, Sapre S, Sharma A, Tiwari S (2016) Amelioration of drought tolerance in wheat by the interaction of plant growth-promoting rhizobacteria. Plant Biol 18(6):992–1000. https://doi.org/10.1111/plb.12505

Gorovtsov AV, Minkina TM, Mandzhieva SS, Perelomov LV, Soja G, Zamulina IV, Rajput VD, Sushkova SN, Yao J (2019) The mechanisms of biochar interactions with microorganisms in soil. Environ Geochem Health 42:2495–2518. https://doi.org/10.1007/s10653-019-00412-5

Gou W, Zheng P, Chen F, Zhang L, Cui Z, Cao M, Zhang L, Hu J (2015) Accumulation of choline and glycinebetaine and drought stress tolerance induced in maize (*Zea mays*) by three plant growth promoting rhizobacteria (pgpr) strains. Pak J Bot 46(24):581–586

Gouda S, Kerry RG, Das G, Paramithiotis S, Shin HS, Patra JK (2018) Revitalization of plant growth promoting rhizobacteria for sustainable development in agriculture. Microbiol Res 206:131–140. https://doi.org/10.1016/j.micres.2017.08.016

Gougoulias C, Clark JM, Shaw LJ (2014) The role of soil microbes in the global carbon cycle: tracking the below-ground microbial processing of plant-derived carbon for manipulating carbon dynamics in agricultural systems. J Sci Food Agric 94(12):2362–2371. https://doi.org/10.1002/jsfa.6577

Grümberg BC, Urcelay C, Shroeder MA, Vargas-Gil S, Luna CM (2014) The role of inoculum identity in drought stress mitigation by arbuscular mycorrhizal fungi in soybean. Biol Fertil Soil 51(1):1–10. https://doi.org/10.1007/s00374-014-0942-7

Gull A, Ahmad Lone A, Ul Islam Wani N (2019) Biotic and abiotic stresses in plants. In: Abiotic and biotic stress in plants. https://doi.org/10.5772/intechopen.85832

Hasanuzzaman M, Bhuyan MHMB, Anee TI, Parvin K, Nahar K, Mahmud JA, Fujita M (2019) Regulation of ascorbate-glutathione pathway in mitigating oxidative damage in plants under abiotic stress. Antioxidants 8(9):384. https://doi.org/10.3390/antiox8090384

Hayat S, Hayat Q, Alyemeni MN, Wani AS, Pichtel J, Ahmad A (2012) Role of proline under changing environments. Plant Signal Behav 7(11):1456–1466. https://doi.org/10.4161/psb.21949

He C, Wang W, Hou J (2019) Plant growth and soil microbial impacts of enhancing licorice with inoculating dark septate endophytes under drought stress. Front Microbiol 10. https://doi.org/10.3389/fmicb.2019.02277

Iordachescu M, Imai R (2011) Trehalose and abiotic stress in biological systems. In: Abiotic stress in plants - mechanisms and adaptations https://doi.org/10.5772/22208

Iqbal N, Khan NA, Ferrante A, Trivellini A, Francini A, Khan MIR (2017) Ethylene role in plant growth, development and senescence: interaction with other phytohormones. Front Plant Sci 08. https://doi.org/10.3389/fpls.2017.00475

Jacoby R, Peukert M, Succurro A, Koprivova A, Kopriva S (2017) The role of soil microorganisms in plant mineral nutrition—current knowledge and future directions. Front Plant Sci 8. https://doi.org/10.3389/fpls.2017.01617

Jochum MD, McWilliams KL, Borrego EJ, Kolomiets MV, Niu G, Pierson EA, Jo YK (2019) Bioprospecting plant growth-promoting rhizobacteria that mitigate drought stress in grasses. Front Microbiol 10. https://doi.org/10.3389/fmicb.2019.02106

Joshi R, Sahoo KK, Singh AK, Anwar K, Gautam RK, Krishnamurthy SL, Sopory SK, Pareek A, Singla-Pareek SL (2019) Engineering trehalose biosynthesis pathway improves yield potential in rice under drought, saline and sodic conditions. J Exp Bot. https://doi.org/10.1093/jxb/erz462

Kamran S, Shahid I, Baig DN, Rizwan M, Malik KA, Mehnaz S (2017) Contribution of zinc solubilizing bacteria in growth promotion and zinc content of wheat. Front Microbiol 8. https://doi.org/10.3389/fmicb.2017.02593

Kandel S, Joubert P, Doty S (2017) Bacterial endophyte colonization and distribution within plants. Microorganism 5(4):77. https://doi.org/10.3390/microorganisms5040077

Kannenberg SA, Phillips RP (2016) Soil microbial communities buffer physiological responses to drought stress in three hardwood species. Oecologia 183(3):631–641. https://doi.org/10.1007/s00442-016-3783-2

Kasim WA, Osman ME, Omar MN, Abd El-Daim IA, Bejai S, Meijer J (2012) Control of drought stress in wheat using plant-growth-promoting bacteria. J Plant Growth Regul 32(1):122–130. https://doi.org/10.1007/s00344-012-9283-7

Kaushal M, Wani SP (2015) Plant-growth-promoting rhizobacteria: drought stress alleviators to ameliorate crop production in drylands. Ann Microbiol 66(1):35–42. https://doi.org/10.1007/s13213-015-1112-3

Khan N, Zandi P, Ali S, Mehmood A, Adnan Shahid M (2018) Impact of salicylic acid and PGPR on the drought tolerance and phytoremediation potential of *Helianthus annuus*. Front Microbiol 9. https://doi.org/10.3389/fmicb.2018.02507

Kim Y, Chung YS, Lee E, Tripathi P, Heo S, Kim KH (2020) Root response to drought stress in rice (Oryza sativa L.). Int J Mol Sci 21(4):1513. https://doi.org/10.3390/ijms21041513

Kobayashi DY, Reedy RM, Bick J, Oudemans PV (2002) Characterization of a Chitinase gene from Stenotrophomonas maltophilia strain 34S1 and its involvement in biological control. Appl Environ Microbiol 68(3):1047–1054. https://doi.org/10.1128/aem.68.3.1047-1054.2002

Konnur SR, Nirmalnath PJ, Jagadeesh KS (2019) Role of AM fungi in mitigating moisture stress in soybean (*Glycine max* L. Merill). Int J Curr Microbiol App Sci 8(7):1451–1462

Koyro HW, Ahmad P, Geissler N (2011) Abiotic stress responses in plants: an overview. In: Ahmad P, Prasad MNV (eds) Environmental adaptations and stress tolerance of plants in the era of climate change, pp 1–28. https://doi.org/10.1007/978-1-4614-0815-4_1

Kumar M, Tak Y, Potkule J, Choyal P, Tomar M, Meena NL, Kaur C (2020) Phenolics as plant protective companion against abiotic stress. In: Lone R (ed) Plant phenolics in sustainable agriculture. Springer, Singapore, pp 277–308

Kurutas EB (2015) The importance of antioxidants which play the role in cellular response against oxidative/nitrosative stress: current state. Nutr J 15(1). https://doi.org/10.1186/s12937-016-0186-5

Li J, Li Y, Yin Z, Jiang J, Zhang M, Guo X, Zhujia Y, Zhao Y, Xiong H, Zhang Z, Shao Y, Jiang C, Zhang H, An G, Paek NC, Ali J, Li Z (2016) OsASR5enhances drought tolerance through a stomatal closure pathway associated with ABA and H_2O_2 signalling in rice. Plant Biotechnol J15(2):183–196. https://doi.org/10.1111/pbi.12601

Li J, Meng B, Chai H, Yang X, Song W, Li S, Lu A, Zhang T, Sun W (2019) Arbuscular mycorrhizal fungi alleviate drought stress in C3 (*Leymus chinensis*) and C4 (*Hemarthria altissima*) grasses via altering antioxidant enzyme activities and photosynthesis. Front Plant Sci 10. https://doi.org/10.3389/fpls.2019.00499

Liddycoat SM, Greenberg BM, Wolyn DJ (2009) The effect of plant growth-promoting rhizobacteria on asparagus seedlings and germinating seeds subjected to water stress under greenhouse conditions. Can J Microbiol 55(4):388–394. https://doi.org/10.1139/w08-144

Liu F, Xing S, Ma H, Du Z, Ma B (2013) Cytokinin-producing, plant growth-promoting rhizobacteria that confer resistance to drought stress in *Platycladus orientalis* container seedlings. App Microbiol Biotechnol 97(20):9155–9164. https://doi.org/10.1007/s00253-013-5193-2

Liu CY, Zhang F, Zhang DJ, Srivastava A, Wu QS, Zou YN (2018) Mycorrhiza stimulates root-hair growth and IAA synthesis and transport in trifoliate orange under drought stress. Sci Rep 8(1). https://doi.org/10.1038/s41598-018-20456-4

Liu F, Ma H, Peng L, Du Z, Ma B, Liu X (2019) Effect of the inoculation of plant growth-promoting rhizobacteria on the photosynthetic characteristics of *Sambucus williamsii* Hance container seedlings under drought stress. AMB Express 9:169

Luginbuehl LH, Menard GN, Kurup S, Van Erp H, Radhakrishnan GV, Breakspear A, Oldroyd GED, Eastmond PJ (2017) Fatty acids in arbuscular mycorrhizal fungi are synthesized by the host plant. Sci 356(6343):1175–1178. https://doi.org/10.1126/science.aan0081

Ma Y, Oliveira RS, Freitas H, Zhang C (2016) Biochemical and molecular mechanisms of plant-microbe-metal interactions: relevance for phytoremediation. Front Plant Sci 7. https://doi.org/10.3389/fpls.2016.00918

Marulanda-Aguirre A, Azcón R, Ruiz-Lozano JM, Aroca R (2007) Differential effects of a Bacillus megaterium strain on Lactuca sativa plant growth depending on the origin of the arbuscular mycorrhizal fungus coinoculated: physiologic and biochemical traits. J Plant Growth Regul 27 (1):10–18. https://doi.org/10.1007/s00344-007-9024-5

Mathur S, Tomar RS, Jajoo A (2018) Arbuscular mycorrhizal fungi (AMF) protects photosynthetic apparatus of wheat under drought stress. Photosynth Res. https://doi.org/10.1007/s11120-018-0538-4

Meena KK, Sorty AM, Bitla UM, Choudhary K, Gupta P, Pareek A, Singh DP, Prabha R, Sahu PK, Gupta VK, Singh HB, Krishanani KK, Minhas PS (2017a) Abiotic stress responses and microbe-mediated mitigation in plants: the omics strategies. Front Plant Sci:8. https://doi.org/10.3389/fpls.2017.00172

Meena VS, Meena SK, Verma JP, Kumar A, Aeron A, Mishra PK, Bisht JP, Pattanayak A, Naveed M, Dotaniya ML (2017b) Plant beneficial rhizospheric microorganism (PBRM) strategies to improve nutrients use efficiency: a review. Ecol Eng 107:8–32. https://doi.org/10.1016/j.ecoleng.2017.06.058

Molina-Favero C, Creus CM, Simontacchi M, Puntarulo S, Lamattina L (2008) Aerobic nitric oxide production by Azospirillum brasilenseSp245 and its influence on root architecture in tomato. Mol Plant-Microbe Interact 21(7):1001–1009. https://doi.org/10.1094/mpmi-21-7-1001

Mutumba FA, Zagal E, Gerding M, Castillo-Rosales D, Paulino L, Schoebitz M (2018) Plant growth promoting rhizobacteria for improved water stress tolerance in wheat genotypes. J Soil Sci Plant Nutr:0–0. https://doi.org/10.4067/s0718-95162018005003003

Naylor D, Coleman-Derr D (2018) Drought stress and root-associated bacterial communities. Front Plant Sci:8. https://doi.org/10.3389/fpls.2017.02223

Ojuederie O, Olanrewaju O, Babalola O (2019) Plant growth promoting rhizobacterial mitigation of drought stress in crop plants: implications for sustainable agriculture. Agron 9(11):712. https://doi.org/10.3390/agronomy9110712

Olanrewaju OS, Glick BR, Babalola OO (2017) Mechanisms of action of plant growth promoting bacteria. World J Microbiol Biotechnol 33(11). https://doi.org/10.1007/s11274-017-2364-9

Oliveira ALM, Santos OJAP, Marcelino PRF, Milani KML, Zuluaga MYA, Zucareli C, Gonçalves LSA (2017) Maize inoculation with *Azospirillum brasilense* Ab-V5 cells enriched with exopolysaccharides and polyhydroxybutyrate results in high productivity under low N fertilizer input. Front Microbiol 8. https://doi.org/10.3389/fmicb.2017.01873

Pandey P, Irulappan V, Bagavathiannan MV, Senthil-Kumar M (2017) Impact of combined abiotic and biotic stresses on plant growth and avenues for crop improvement by exploiting physio-morphological traits. Front Plant Sci:8. https://doi.org/10.3389/fpls.2017.00537

Park YS, Dutta S, Ann M, Raaijmakers JM, Park K (2015) Promotion of plant growth by *Pseudomonas fluorescens* strain SS101 via novel volatile organic compounds. Biochem Biophy Res Commun 461(2):361–365. https://doi.org/10.1016/j.bbrc.2015.04.039

Pedrotti L, Mueller MJ, Waller F (2013) *Piriformospora indica* root colonization triggers local and systemic root responses and inhibits secondary colonization of distal roots. PLoS One 8(7): e69352. https://doi.org/10.1371/journal.pone.0069352

Porcel R (2004) Arbuscular mycorrhizal influence on leaf water potential, solute accumulation, and oxidative stress in soybean plants subjected to drought stress. J Exp Bot 55(403):1743–1750. https://doi.org/10.1093/jxb/erh18

Posta K, Hong Duc N (2020) Benefits of arbuscular mycorrhizal fungi application to crop production under water scarcity. Drought Detect Solut. https://doi.org/10.5772/intechopen.86595

Punitha S, Kalarani MK, Uthandi S (2019) Apoplast associated microbes influences antioxidant system of rice under drought stress. Madras Agric J 106(7–9)

Ragel P, Raddatz N, Leidi EO, Quintero FJ, Pardo JM (2019) Regulation of K^+ nutrition in plants front. Plant Sci. https://doi.org/10.3389/fpls.2019.00281

Ranjan P, Patle GT, Prem M, Solanke KR (2017) Organic mulching - a water saving technique to increase the production of fruits and vegetables. Curr Agri Res 5(3). https://doi.org/10.12944/CARJ.5.3.17

Rashid MI, Mujawar LH, Shahzad T, Almeelbi T, Ismail IMI, Oves M (2016) Bacteria and fungi can contribute to nutrients bioavailability and aggregate formation in degraded soils. Microbiol Res 183:26–41. https://doi.org/10.1016/j.micres.2015.11.007

Roy K, Ghosh D, DeBruyn JM, Dasgupta T, Wommack KE, Liang X, Wagner RE, Radosevich M (2020) Temporal dynamics of soil virus and bacterial populations in agricultural and early plant successional soils. Front Microbiol 11. https://doi.org/10.3389/fmicb.2020.01494

Saikia J, Sarma RK, Dhandia R, Yadav A, Bharali R, Gupta VK, Saikia R (2018) Alleviation of drought stress in pulse crops with ACC deaminase producing rhizobacteria isolated from acidic soil of Northeast India. Sci Rep 8(1). https://doi.org/10.1038/s41598-018-21921-w

Saini A, Aggarwal NK, Sharma A, Yadav A (2015) Actinomycetes: a source of lignocellulolytic enzymes. Enzyme Res:1–15. https://doi.org/10.1155/2015/279381

Salehi-Lisar SY, Bakhshayeshan-Agdam H (2016) Drought stress in plants: causes, consequences, and tolerance. In: Drought stress tolerance in plants, vol 1, pp 1–16. https://doi.org/10.1007/978-3-319-28899-4_1

Saravanakumar D, Kavino M, Raguchander T, Subbian P, Samiyappan R (2010) Plant growth promoting bacteria enhance water stress resistance in green gram plants. Acta Physiol Planta 33 (1):203–209. https://doi.org/10.1007/s11738-010-0539-1

Sathya A, Vijayabharathi R, Gopalakrishnan S (2016) Soil microbes: the invisible managers of soil fertility. In: Microbial inoculants in sustainable agricultural productivity, pp 1–16. https://doi.org/10.1007/978-81-322-2644-4_1

Sayyed RZ, Arora NK, Reddy MS (2019) Plant growth promoting rhizobacteria for sustainable stress management. Microorganisms Sustain. https://doi.org/10.1007/978-981-13-6536-2

Schoonover JE, Crim JF (2015) An introduction to soil concepts and the role of soils in watershed management. J Cont Water Res Edu 154(1):21–47. https://doi.org/10.1111/j.1936-704x.2015.03186.x

Sharma P, Jha AB, Dubey RS, Pessarakli M (2012) Reactive oxygen species, oxidative damage, and antioxidative defense mechanism in plants under stressful conditions. J Bot:1–26. https://doi.org/10.1155/2012/217037

Sheng M, Tang M, Chen H, Yang B, Zhang F, Huang Y (2008) Influence of arbuscular mycorrhizae on photosynthesis and water status of maize plants under salt stress. Mycorrhiza 18 (6–7):287–296. https://doi.org/10.1007/s00572-008-0180-7

Shintu PV, Jayaram KM (2015) Phosphate solubilising bacteria (*Bacillus polymyxa*) - an effective approach to mitigate drought in tomato (*Lycopersicon esculentum* Mill.). Trop Plant Res 2 (1):17–22

Singh PK, Singh M, Tripathi BN (2012) Glomalin: an arbuscular mycorrhizal fungal soil protein. Protoplasma 250(3):663–669. https://doi.org/10.1007/s00709-012-0453-z

Singh DP, Singh V, Gupta VK, Shukla R, Prabha R, Sarma BK, Patel JS (2020) Microbial inoculation in rice regulates antioxidative reactions and defense related genes to mitigate drought stress. Sci Rep 10(1). https://doi.org/10.1038/s41598-020-61140-w

Souza R, de Ambrosini A, Passaglia LMP (2015) Plant growth-promoting bacteria as inoculants in agricultural soils. Genet Mol Biol 38(4):401–419. https://doi.org/10.1590/s1415-475738420150053

Tak Y, Asthir B (2017) Fluoride-induced changes in the antioxidant defense system in two contrasting cultivars of *Triticum aestivum* L. Fluoride 50:324–333

Tak Y, Kumar M (2020) Phenolics: a key defensive secondary metabolite to counter biotic stress. In: Lone R (ed) Plant phenolics in sustainable agriculture. Springer, Singapore, pp 309–330

Takatsuka H, Umeda M (2014) Hormonal control of cell division and elongation along differentiation trajectories in roots. J Exp Bot 65(10):2633–2643. https://doi.org/10.1093/jxb/ert485

Tiepo AN, Constantino LV, Madeira TB, Gonçalves LSA, Pimenta JA, Bianchini E, Martinez de Oliveira AL, Oliveira HC, Stolf-Moreira R (2020) Plant growth-promoting bacteria improve leaf antioxidant metabolism of drought-stressed neotropical trees. Planta 251(4). https://doi.org/10.1007/s00425-020-03373-7

Tkemaladze GS, Makhashvili KA (2016) Climate changes and photosynthesis. Ann Agrar Sci 14 (2):119–126. https://doi.org/10.1016/j.aasci.2016.05.012

Tyagi J, Varma A, Pudake RN (2017) Evaluation of comparative effects of arbuscular mycorrhiza (*Rhizophagus intraradices*) and endophyte (*Piriformospora indica*) association with finger millet (*Eleusine coracana*) under drought stress. Eur J Soil Biol 81:1–10. https://doi.org/10.1016/j.ejsobi.2017.05.007

Vardharajula S, Zulfikar Ali S, Grover M, Reddy G, Bandi V (2011) Drought-tolerant plant growth promoting Bacillus spp.: effect on growth, osmolytes, and antioxidant status of maize under drought stress. J Plant Int 6(1):1–14. https://doi.org/10.1080/17429145.2010.535178

Vejan P, Abdullah R, Khadiran T, Ismail S, Nasrulhaq Boyce A (2016) Role of plant growth promoting rhizobacteria in agricultural sustainability-a review. Molecule 21(5):573. https://doi.org/10.3390/molecules21050573

Vurukonda SSKP, Vardharajula S, Shrivastava M, SkZ A (2016) Enhancement of drought stress tolerance in crops by plant growth promoting rhizobacteria. Microbiol Res 184:13–24. https://doi.org/10.1016/j.micres.2015.12.003

Vurukonda SSKP, Giovanardi D, Stefani E (2018) Plant growth promoting and biocontrol activity of *Streptomyces* spp. as endophytes. Int J Mol Sci 19(4):952. https://doi.org/10.3390/ijms19040952

Walker TS (2003) Root exudation and rhizosphere biology. Plant Physiol 132(1):44–51. https://doi.org/10.1104/pp.102.019661

Wang Z, Li G, Sun H, Ma L, Guo Y, Zhao Z, Gao H, Mei L (2018) Effects of drought stress on photosynthesis and photosynthetic electron transport chain in young apple tree leaves. Biol Open Bio 035279. https://doi.org/10.1242/bio.035279

Wang J, Fu Z, Ren Q, Zhu L, Lin J, Zhang J, Cheng X, Ma J, Yue J (2019) Effects of arbuscular mycorrhizal fungi on growth, photosynthesis, and nutrient uptake of *Zelkova serrata* (Thunb.) Makino seedlings under salt stress. Forest 10(2):186. https://doi.org/10.3390/f10020186

Wang J, Li R, Zhang H, Wei G, Li Z (2020) Beneficial bacteria activate nutrients and promote wheat growth under conditions of reduced fertilizer application. BMC Microbiol 20(1). https://doi.org/10.1186/s12866-020-1708-z

Winsley TJ, Snape I, McKinlay J, Stark J, van Dorst JM, Ji M, Ferrari BC, Siciliano SD (2014) The ecological controls on the prevalence of candidate division TM7 in polar regions. Front Microbiol 5. https://doi.org/10.3389/fmicb.2014.00355

Wu QS, Zou YN, Xia RX (2006) Effects of water stress and arbuscular mycorrhizal fungi on reactive oxygen metabolism and antioxidant production by citrus (*Citrus tangerine*) roots. Eur J Soil Biol 42(3):166–172. https://doi.org/10.1016/j.ejsobi.2005.12.006

Yadav S, Modi P, Dave A, Vijapura A, Patel D, Patel M (2020) Effect of abiotic stress on crops. In: Hasanuzzaman M (ed) Sustainable crop production https://doi.org/10.5772/intechopen.88434

Yang Y, Tang M, Sulpice R, Chen H, Tian S, Ban Y (2014) Arbuscular mycorrhizal fungi alter fractal dimension characteristics of *Robinia pseudoacacia* L. seedlings through regulating plant growth, leaf water status, photosynthesis, and nutrient concentration under drought stress. J Plant Growth Regul 33(3):612–625. https://doi.org/10.1007/s00344-013-9410-0

Zafar-ul-Hye M, Danish S, Abbas M, Ahmad M, Munir TM (2019) ACC deaminase producing PGPR Bacillus amyloliquefaciens and agrobacterium fabrum along with biochar improve wheat productivity under drought stress. Agron 9(7):343. https://doi.org/10.3390/agronomy9070343

Zahedi H, Abbasi S (2015) Effect of plant growth promoting rhizobacteria (PGPR) and water stress on phytohormones and polyamines of soybean. Ind J Agric Res 49(5). https://doi.org/10.18805/ijare.v49i5.5805

Zakikhani H, Ardakani MR, Rejali F, Gholamhoseini M, Joghan AK, Dolatabadian A (2012) Influence of diazotrophic bacteria on antioxidant enzymes and some biochemical characteristics of soybean subjected to water stress. J Int Agric 11(11):1828–1835. https://doi.org/10.1016/s2095-3119(12)60187-7

Zhang H, Xie X, Kim MS, Kornyeyev DA, Holaday S, Paré PW (2008) Soil bacteria augment Arabidopsis photosynthesis by decreasing glucose sensing and abscisic acid levels in planta. Plant J 56(2):264–273. https://doi.org/10.1111/j.1365-313x.2008.03593.x

Zhang H, Murzello C, Sun Y, Kim MS, Xie X, Jeter RM, Zak JC, Dowd SC, Paré PW (2010) Choline and osmotic-stress tolerance induced in Arabidopsis by the soil microbe *Bacillus subtilis* (GB03). Mol Plant-Microbe Interact 23(8):1097–1104. https://doi.org/10.1094/mpmi-23-8-1097

Zhang Z, Zhang J, Huang Y (2014) Effects of arbuscular mycorrhizal fungi on the drought tolerance of *Cyclobalanopsis glauca* seedlings under greenhouse conditions. New For 45(4):545–556. https://doi.org/10.1007/s11056-014-9417-9

Zhang Z, Zhang J, Xu G, Zhou L, Li Y (2018) Arbuscular mycorrhizal fungi improve the growth and drought tolerance of *Zenia insignis* seedlings under drought stress. New For. https://doi.org/10.1007/s11056-018-9681-1

Zhang S, Zheng Q, Noll L, Hu Y, Wanek W (2019) Environmental effects on soil microbial nitrogen use efficiency are controlled by allocation of organic nitrogen to microbial growth and regulate soil gross N mineralization. Soil Biol Biochem. https://doi.org/10.1016/j.soilbio.2019.05.019

Zhang M, Yang L, Hao R, Bai X, Wang Y, Yu X (2020) Drought-tolerant plant growth-promoting rhizobacteria isolated from jujube (*Ziziphus jujuba*) and their potential to enhance drought tolerance. Plant Soil. https://doi.org/10.1007/s11104-020-04582-5

Zhao R, Guo W, Bi N, Guo J, Wang L, Zhao J, Zhang J (2015) Arbuscular mycorrhizal fungi affect the growth, nutrient uptake and water status of maize (*Zea mays* L.) grown in two types of coal mine spoils under drought stress. Appl Soil Ecol 88:41–49. https://doi.org/10.1016/j.apsoil.2014.11.016

Zhou C, Ma Z, Zhu L, Xiao X, Xie Y, Zhu J, Wang J (2016) Rhizobacterial strain Bacillus megaterium BOFC15 induces cellular polyamine changes that improve plant growth and drought resistance. Int J Mol Sci 17(6):976. https://doi.org/10.3390/ijms17060976

Zhu XC, Song FB, Liu SQ, Liu TD, Zhou X (2012) Arbuscular mycorrhizae improves photosynthesis and water status of *Zea mays* L. under drought stress. Plant Soil Environ 58(4):186–191. https://doi.org/10.17221/23/2011-pse

Zulfiqar F, Akram NA, Ashraf M (2019) Osmoprotection in plants under abiotic stresses: new insights into a classical phenomenon. Planta 251(1). https://doi.org/10.1007/s00425-019-0329

Role of Nitrogen-Fixing Microorganisms for Plant and Soil Health

7

Jayani J. Wewalwela, Mihiri Seneviratne, and
Sara A. Shields-Menard

Abstract

With increasing global demand for both food and energy, the importance of microorganism based soil and plant health management has been widely considered. Nitrogen-fixing microorganisms, together with the contribution from plants and soil, can fulfill the nitrogen requirements of plants in different ecosystems. Such nutrient requirements are met either completely or partially, while maintaining soil sustainability. Symbiotic nitrogen fixation, associative and free-living nitrogen-fixing microorganisms can convert atmospheric nitrogen into ammonia, providing nitrogen to the plant while enhancing plant yield. Moreover, nitrogen fixation can indirectly affect the soil nitrogen pool and enrich microbial diversity in the soil to support soil health. These diverse nitrogen-fixing microorganisms have various mechanisms for fixing nitrogen, which contribute favorably towards the sustainability of plant and soil health.

Keywords

Nitrogen · Plant and soil health · Symbiotic nitrogen fixation · Associative nitrogen fixation · Free living · Nitrogen-fixing microorganisms

J. J. Wewalwela (✉)
Department of Agricultural Technology, Faculty of Technology, University of Colombo, Colombo, Sri Lanka
e-mail: jayani@at.cmb.ac.lk

M. Seneviratne
Department of Botany, Faculty of Natural Sciences, Open University of Sri Lanka, Colombo, Sri Lanka

S. A. Shields-Menard
Department of Biology, University of Texas at San Antonio, San Antonio, TX, USA

R. Soni et al. (eds.), *Microbiological Activity for Soil and Plant Health Management*,
https://doi.org/10.1007/978-981-16-2922-8_7

7.1 Introduction

Nitrogen is an essential element that contributes to plant development and growth. Although 78% of the atmosphere is comprised of nitrogen gas, plants are unable to capture it directly. However, the available nitrogen is taken up by plants, through their roots, in the form of ammonium and nitrates. Therefore, it can be seen that soil properties, especially chemical properties, directly influence plant growth. Evidently, the low bioavailability of nitrogen in soil is a limiting factor for plant growth and has negative effects on plant health. However, some prokaryotes, collectively named as diazotrophs, possess the ability to convert atmospheric nitrogen into more bioavailable forms (Franche et al. 2009), which fulfill the nitrogen requirement in plants, while enriching the soil nitrogen pool. Therefore, the diazotrophs, the plants, and the soil have a collective function toward sustainability (Fig. 7.1).

Nitrogen fixation rates of diazotrophs are different according to the several factors such as host type, physical, chemical, and biological characters of soil and host. Among biological nitrogen fixation (BNF), symbiotic nitrogen fixation can range up to 60% of the total annual *N* input. In contrast, associative nitrogen fixation only ranges up to 30–55% fixation (Wewalwela et al. 2020). This indicates that the contribution of BNF towards soil and plant health is significant.

Soil health is attributed to soil fertility and is considered a vital characteristic of good soil. A soil health can be defined using physical, chemical, and biological properties, and it is assumed that soil is a living system. Interactions that take place between different processes and properties of soil, along with the effect due to the activity of soil microbiota, are believed to influence the soil health immensely (Yang et al. 2020). Therefore, in microbiological aspect, soil health can be defined as the capacity of a living soil to function (Gothwal et al. 2008) and plant health is mainly influenced by the rhizosphere soil microbiome and endophytic microbiome (Puri et al. 2020). Moreover, soil health is the capacity of soil to promote animal, and plant health while maintaining the quality of the environment and sustaining biological productivity (Doran and Parkin 1994). Soil health and plant health are related to each

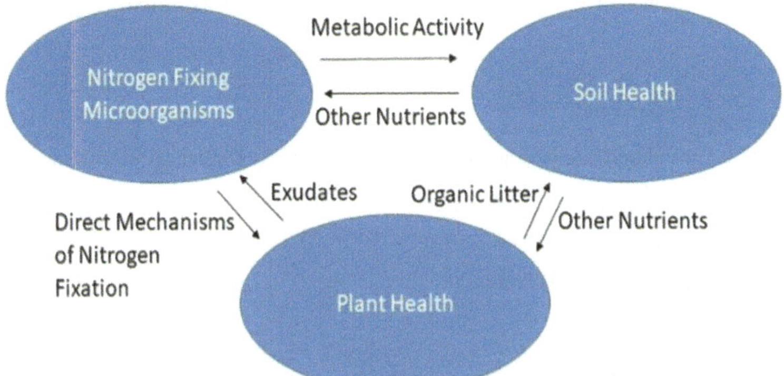

Fig. 7.1 Interactions between nitrogen fixing microorganisms for plants and soil health

other and plant health is the ability of a plant to exist in a physiologically stable manner and perform its function. Several factors of the soil environment effect for microbial community assembly around plant by forming diverse interactions between soil, plants and microbes. Plants can actively communicate with the surrounding environment of rhizosphere to horizontally transfer microbial community for sustained growth and plant health (Hardoim et al. 2015).

Both plants and nitrogen-fixing microorganisms obtain their nutrients from soil. They change the soil properties by secreting metabolites and via nitrogen enrichment, respectively. Nitrogen-fixing microorganisms in the form of symbiotic, associative and free living have a range of direct methods by which to fix nitrogen. Plants communicate with nitrogen-fixing microorganisms through metabolites exuded by the plant roots.

Plants are also influenced extensively by the rhizosphere microbiome, which is a complex assembly of organisms that result in marked changes in plant growth. Considering a more biological approach, such as nitrogen fixation, as an alternative to synthetic fertilizer is, therefore, a feasible option for improving soil and plant health. A variety of symbiotic (*Rhizobium* sp.) and non-symbiotic bacteria such as *Azotobacter*, *Azospirillum*, *Bacillus*, and *Klebsiella* sp., (Singh et al. 2017) are being closely examined and utilized to enhance plant and soil health (Matse et al. 2020; Ramasamy et al. 2020).

7.2 Biological Nitrogen Fixation

Biological nitrogen fixation is the process where atmospheric nitrogen is incorporated into plant tissues by the action of nitrogen-fixing organisms. In the pathway of biological nitrogen fixation, inert nitrogen gas is converted to ammonia (NH_3) under micro-aerobic conditions (Fig. 7.2). The process of BNF is driven by prokaryotes, which contains both eubacteria and archaea (Zehr et al. 2003). Nitrogen fixation microorganisms can be categorized as symbiotic and associative/free living (Fig. 7.3).

7.2.1 Symbiotic Nitrogen Fixation

Symbiotic nitrogen fixation is a property common to a limited number of bacterial groups, such as the genera *Rhizobium*, *Mesorhizobium*, *Sinorhizobium*, *Bradyrhizobium*, and *Azorhizobium* (collectively known as rhizobia) and *Frankia*. The plants, which are symbiotically associated with rhizobia (associate with legumes) and *Frankia* (with actinorhizal plants), have the potential to strive in

$$N_2 + 8H^+ + 8e^- + 16Mg\text{-}ATP \rightarrow 2NH_3 + H_2 + 16Mg\text{-}ADP + 16\,P$$

Fig. 7.2 The optimum stoichiometry of BNF

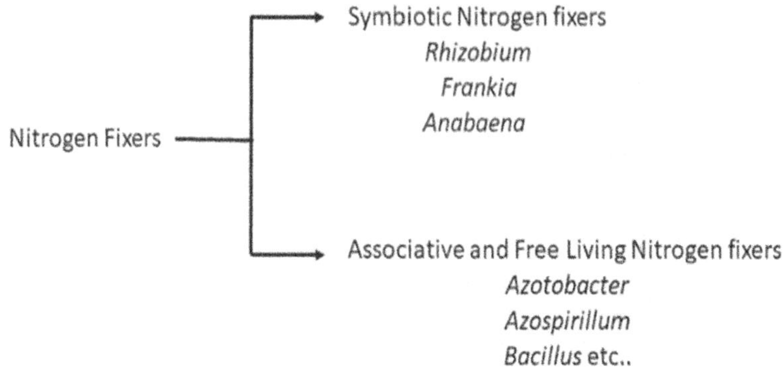

Fig. 7.3 Two groups of nitrogen fixing microorganisms of symbiotic nitrogen fixing species and associative and free-living microorganisms

marginally fertile soils. These plants can be considered as pioneer species of early plant community development. The process of BNF requires nitrogen gas, molybdenum-ferrous (Mo-Fe), and ferrous (Fe) proteins reductant (flavodoxin, ferredoxin, or sodium dithionite), magnesium-adenosine tri phosphate (MgATP plus an ATP-regenerating system) and an anaerobic environment or aerobic environment which is protected from oxygen damage.

7.2.2 Invasion and Infection

7.2.2.1 Release of Flavonoids

Legume rhizobium symbiosis initiates with the molecular dialog between rhizobia and the host plant. Flavonoids which are secreted by legume roots into the rhizosphere act as primary signals to rhizobia (Graham 1991; Dong and Song 2020). They are polyaromatic secondary metabolic products of central phenylpropanoid pathway (Ververidis et al. 2007). Central chalcone is the precursor of all the flavonoid derivatives (Selepe and Van Heerden 2013).

Flavonoids affect legume root nodulation via

- Inducing *nod* gene transcription (Firmin et al. 1986; Jiménez-Guerrero et al. 2018).
- Promoting movement of compatible rhizobia towards the legume roots—chemoattractant (Mathesius 2019).
- Enhancing the growth rate of bacterial cells (Coskun et al. 2017).

It has been reported that the release of flavonoids is specific to each legume species (Liu and Murray 2016).

7.2.2.2 Nod Factor

Flavonoid perception in rhizobia is facilitated via NodD protein (Moscatiello et al. 2010), where sensitivity of *Rhizobium* for flavonoids is associated with the increase in cellular calcium concentration, which induces the expression of NodD. NodD proteins activates bacterial nodulation genes (*nod, nol, noe*) which are located in pSyms (indigenous plasmid) in fast growing and on the chromosome of slow growing rhizobia (Banfalvi et al. 1981).

The role played by proteins encoded from nodulation genes is crucial for Nod Factor synthesis and secretion. Nod Factors are lipochitooligosaccharides (LCOs) signals (Fliegmann and Bono 2015; Kamst et al. 1998). The structure of Nod factors was first determined in 1990 for *Sinorhizobium meliloti* (Lerouge et al. 1990). Currently, different structures of Nod factors produced by most other rhizobia have been identified. Rhizobia produce a mixture of Nod factors and the environmental conditions affect the composition (Geurts et al. 2005).

The specificity of legume rhizobia interaction is due to structural variability of Nod factor and the nature of plant flavonoid signal. The Nod factors identified from different rhizobial species differed in the number of glucosamine residues, the length and saturation of the acyl chain, and the type of modification present on the basic backbone (Dénarié et al. 1996). The host specific modifications observed were due to the addition of different groups such as sulphuryl, methyl, carbamoyl, acetyl, fucosyl, and arabinosyl to different positions of the backbone (Mathesius 2019; Perret et al. 2000).

7.2.2.3 Nod Factor Perception

Differences among Nod factors and their specificity of perception by leguminous plants confer the host-bacterial specificity of legume rhizobium symbiosis (Dénarié et al. 1996). Nod factor receptors are suggested as being located in the plasma membrane (Kalsi and Etzler 2000). Literature supports that, at very low concentrations (picomolar concentrations) in the absence of *Rhizobium* bacteria, purified Nod factors are able to promote root hair depolarization (Ehrhardt and Atkinson 1992), root hair deformation (Heidstra et al. 1994), and primordium formation (Truchet et al. 1991). Moreover, it has suggested that Nod factors are recognized by a high affinity receptor (Heidstra and Bisseling 1996). Experiments have shown that Nod factors can bind to lectin nucleotide phosphohydrolase (LNP), and further analysis with specific antibodies against LNPs suggest that, LNPs have a role to play in the early steps of nodulation (Etzler et al. 1999). However, more research need to be done on understanding the function of LNP as a receptor on binding to Nod factor.

Epipolarization microscopy of radioactively tagged Nod factors (Heidstra et al. 1994), immunolocalization of fixed nodules (Timmers et al. 1998), fluorescence microscopy of fluorescent Nod factor derivatives (Gadella Jr et al. 1997), and fluorescence correlation spectroscopy (FCS) were used to study the behavior of Nod factors on legume root hairs. The diffusional behavior (Goedhart et al. 2000) of fluorescent Nod factor analogues have been examined on living *Vicia sativa* root hairs using FCS. The study has revealed that Nod factors are predominantly located in the cell wall.

7.2.2.4 Responses to Nod Factor

One of the initial responses observed in the plant to Nod factor (Ehrhardt and Atkinson 1992) are ionic fluctuations that take place across the plasma membrane of epidermal root cells. Changes in the levels of calcium present in the root hair cells, under the influence of Nod factors, have been reported in several studies (Kurkdjian 1995; Shaw and Long 2003). After the addition of Nod factors to legume roots, there were two phases of ionic changes that were observed in the root hair cells (Oldroyd and Downie 2004).

1. rapid influx of Calcium——immediately after the membrane depolarization.
2. calcium spiking—some minutes later.

It suggests that Ca^{+2} influx triggers activation of an anion channel which stops depolarization and initiates repolarization. The increase in Ca^{+2} concentration at the tip of a growing root hair creates a Ca^{+2} gradient along the root hair.

7.2.2.5 Root Hair Curling

After secreting the Nod factor, the next step is invasion of rhizobia. Developmental changes such as root hair deformation, membrane depolarization, intracellular calcium oscillations, and the initiation of cell division at the root cortex are induced by Nod factors (Verma 1992). As a result, under the influence of Nod factor, the cytoarchitecture and cytoplasmic streaming of the root hair change resulting in deformed root hairs (Heidstra et al. 1994).

At first a deformed root hair forms a bend which traps the bacteria attached to the root hair (Timmers et al. 1999). There are subsequent changes in actin and microtubule components present in the root hair cytoskeleton, after being exposed to compatible rhizobia. Nevertheless, the role that these bacteria play in infection thread growth is still under research. The presence of active cytoplasmic streaming at the tip region of extending threads supports the idea that there is a role that plant cytoskeleton plays in the growth of infection threads (Perrine-Walker et al. 2014).

7.2.2.6 Nodule Organogenesis

The root cortical cells start to divide, resulting in a cell mass that contains the nodule primordium. The cell division takes place before the infection thread reaches the base of the root hair cell and leads to the accumulation of a large amount of amyloplasts (Crespi and Gálvez 2000). The root cortical cells are the first to receive the intracellular bacteria that eventually differentiate into nitrogen-fixing bacteroids. Then the cells of the middle cortex divide, which contributes to form the nodule meristem (Gage 2004). The infection thread crosses the outer cell layers to reach the nodule primordium (Crespi and Gálvez 2000). The rhizobial populations in the tubule start to increase as the rhizobia inside the thread grow and divide. As the thread continues to branch, it enters the nodule primordium, leading to an increase in the number of sites which are infected by rhizobia. This ensures that most of the nodule cells are colonized by rhizobia (Haynes et al. 2004). Through endocytosis, the rhizobia are released from the tip of the infection thread into bacteroids (Hirsch 1992; Puppo et al. 2005). During the endocytosis, rhizobia are enclosed by a membrane derived from the plant (Hernández et al. 2004). It has been shown that the small GTPases of the Rab family and by phosphatidylinositide signaling molecules are responsible for regulating the movement of the vesicles (Jürgens 2004). However, further research has to be conducted in order to determine whether a plant or bacterium is responsible for the degradation. With the bacteria enclosed in a plant cell membrane, a new structure called a symbiosome is produced (Garg 2009; Roth et al. 1988).

There are two types of nodules, named determinate and indeterminate nodules. In determinate nodules, bacteroids are further divided within the symbiosome or individual symbiosomes fuse resulting in symbiosomes that contain several bacteroids. However, in indeterminate nodules, individual symbiosomes are further divided into single bacteroids which exist within the symbiosome (Prell and Poole 2006).

As the nodule and infection thread continues to grow, a highly branched network called an infection zone develops (Gage 2004). In-depth details related to the propagation of infection thread networks in the nodule is yet to be discovered. Symbiosome consists of the symbiosome membrane, bacteroid, and symbiosome space. A specialized membrane called the symbiome membrane forms a structural and functional interface between the legume and its bacterial counterpart (Lodwig et al. 2003). The proteins that are present in the symbiome membrane are involved in the transport, metabolism, nodule formation, and signaling processes. As the root nodule enlarges the symbiosome membrane is synthesized in order to accommodate proliferating bacteroids. Considering the energy requirement for symbiosome membrane synthesis, it is reported to be about 30 times greater than that required for plasma membrane synthesis (Catalano et al. 2004).

7.2.3 Regulation of Nitrogen Fixation

Nitrogen fixation is controlled by *nif* gene expression (Table 7.1). Nitrogen fixation is a highly, energy demanding process therefore the first regulatory factor is the fixed nitrogen status. The *nif* genes are expressed in response to changes in the environment via different regulatory mechanisms. These mechanisms differ for different genera and species of bacteria (Hübner et al. 1991; Singh et al. 2014).

Oxygen availability also plays an important role as nitrogenases are highly sensitive toward oxygen damage. One mechanism to maintain a low internal O_2 concentration is to respire at very high rates is, e.g. *Azotobacter vinelandii* (Dalton and Postgate 1968; Oelze 2000). Formation of transient protein aggregate with nitrogenase is another mechanism to provide protection against oxygen damage (Oelze 2000). This provides a conformational protection by formation of a complex between FeSII protein and nitrogenase under high intracellular oxygen levels

Table 7.1 *nif* gene and functions (Li and Chen 2020; Richau Kerstin et al. 2017; Rubio and Ludden 2005)

Gene	Structure	Function
nifH	Fe-protein subunit	FeMo biosynthesis
nifD	Mo-Fe α-subunit	Cofactor of holoprotein α 2β2 tetramer
nifK	Mo-Fe β-subunit	MoFe cofactor biosynthesis
nifB	Mo-Fe β-subunit	FeMo cofactor biosynthesis
nifE	Mo-Fe β-subunit	FeMO cofactor biosynthesis. Forms α 2β2 tetramer with *nifN* gene product
nifN	FeMo protein	Processing of Mo
	FeMo protein	Encodes a homocitrate
nifA	FeMo protein	Positively regulates the activity of *nif* transcription
nifL	FeMo protein	Negatively regulates the activity of *nif* transcriptase
nifF	Flavodoxin	Physiological reduction of Fe-protein
nifJ	Pyruvate flavodoxin oxidoreductase	Reduction of Flavodoxin
nifM	Pyruvate flavodoxin oxidoreductase	Activity of Fe-protein
nifS	Pyruvate flavodoxin oxidoreductase	Processing of MoFe protein
nifU	Pyruvate flavodoxin oxidoreductase	Processing of MoFe protein
nifY	Pyruvate flavodoxin oxidoreductase	Processing of MoFe protein but not required for diazotrophic growth
nifT	Unknown	Not required for diazotrophic growth
nifW	Unknown	Required for full activity of MoFe protein
nifZ	Unknown	Required for full activity of MoFe protein
nifX	Unknown	Involved in FeMo cofactor biosynthesis

(Moshiri et al. 1995; Ureta and Nordlund 2002). In some filamentous cyanobacteria (order Nostocales), nitrogenase is only synthesized in the specialized structures (heterocysts) that do not produce O_2, e.g. *Anabaena* sp., *Nostoc* sp. (Fay 1992; Flores et al. 2019; Neilson et al. 1971).

To control nitrogen fixation process, many diazotrophic bacteria have developed regulatory cascades. Rhizobia are members of the α-proteobacteria which comes under the order Rhizobiales. In proteobacteria, nitrogen fixation is regulated at three levels. At first, the status of the fixed nitrogen is sensed by the nitrogen regulation system (Ntr system). When the level of fixed nitrogen is low, it leads to the expression of the *nifA* gene, which codes for the central activator protein of all the other *nif* genes (Dixon and Kahn 2004). There are three proteins in the Ntr system; UTase/UR enzyme (GlnD), the P_{II} signal transduction protein, and the two-component regulatory system NtrB/NtrC. Under low nitrogen conditions NtrC phosphorylates and initiates the transcription of the *nifA* gene. As the central regulator of nitrogen fixation, the NifA protein activates transcription of all the other *nif* genes in proteobacteria (Dixon and Kahn 2004), e.g. structural genes of nitrogenase, *nifHDK,* and all the other *nif* genes.

The activity of the NifA is impaired in response to the increase in NH_4^+ and O_2. NifA proteins can be divided into two classes: O_2-sensitive NifA proteins, e.g. *Bradyrhizobium japonicum* NifA and O_2-tolerant NifA proteins, e.g. *Klebsiella. pneumoniae* NifA (Marchal and Vanderleyden 2000; Unden et al. 1995). Moreover, the regulatory systems of diazotrophic bacteria have the ability to rapidly and reversibly regulate nitrogenase activity in response to abrupt changes that take place in the environment (Halbleib and Ludden 2000).

7.2.4 Free Living and Associative Nitrogen Fixation

Other than by microbes in symbiotic associations, free-living microbes and microbes associated with non-nodulating plants, e.g. grass family (Poaceae) are also capable of fixing atmospheric nitrogen. In free-living nitrogen fixation, microbes fix nitrogen without a host, but in associative nitrogen fixation microbes enter to their host without forming external structures such as nodules. Diazotrophic bacteria has found in association with sugarcane, rice, maize, and tropical grasses (Monti 2012; Roley et al. 2018; Shrestha and Maskey 2005). When it comes to nitrogen fixation, free living and associative diazotrophs also have certain constrains to overcome. Some of these issues are, to optimize nitrogen fixation while managing the oxygen sensitivity of nitrogenase, to meet the energy demands of the fixation, to supply metals for the metal enzymes, and to utilize the remaining fixed nitrogen before fixing atmospheric nitrogen.

7.2.4.1 Free-Living Diazotrophs

7.2.4.1.1 *Azotobacter vinelandii*

Azotobacter vinelandii is considered as an important model system to study nitrogen fixation performed by free-living diazotrophs (Rubio and Ludden 2008). It fixes atmospheric nitrogen via oxygen sensitive Mo nitrogenase. Two mechanisms have been suggested to protect nitrogenase from oxygen inactivation. They are called respiratory protection and conformational protection. Respiratory protection is extremely higher respiration rates to maintain intracellular free oxygen at concentrations lower enough to so as nitrogenase is not affected (Thorneley and Ashby 1989). Protection of nitrogenase from oxygen damage is named conformational protection. In this state the protein components are inactive, but become active when the redox environment become favorable for nitrogenase function. It is suggested to achieve by the formation of a complex with FeSII protein (shethna protein) (Becking 1981; Maier and Moshiri 2000).

7.2.4.1.2 Cyanobacteria

Cyanobacteria are photosynthetic prokaryotes, which consist of unicellular forms to branched filamentous structures. They contribute greatly towards nitrogen economy in both aquatic and terrestrial ecosystems through their ability to fix atmospheric. Their habitats range from artic to tropics.

Cyanobacterial nitrogen fixation is mostly associated with heterocysts. Heterocystous cyanobacteria are capable of cell differencing. Under limited nitrogen conditions, a heterocyst will result after the vegetative cells undergo a series of changes. When a vegetative cell changes into a heterocyst, the photosystem II disappears preventing evolution of oxygen. But photosystem I remains intact harvesting light to generate ATP. Another strategy is formation of a thick extracellular glycolipid envelop to limit oxygen diffusion (Stal 2015; Walsby 1985). It has been reported that, heterocysts in cyanobacteria aerobically fix nitrogen forming micro-oxic environment by exhibiting high respiration rates (Ermakova et al. 2014; Valladares et al. 2003). Diffusion of oxygen into heterocysts can be regulated by pores at the vegetative cell-heterocyst junction (Wolk et al. 1994). However, certain other prokaryotes protect the nitrogenase from oxygen by performing this reaction inside vesicles in the cells which includes genera *Nostoc, Anabaena*, *Azetobactor, Klebsiella,* and *Clostridium.*

7.2.4.2 Associative Diazotrophs

Most associative diazotrophs belong to proteobacteria (Igiehon and Babalola 2018). However, there is no evidence that associative diazotrophs live inside host cells in an organelle. Plant associative diazotrophs that colonize inner tissues of plants without causing any cell disruptions (Fig. 7.4) are named as endophytic diazotrophs, e.g. *Gluconacetobacter*, *Azoarcus*, *Burkholderia* (Wewalwela 2020).

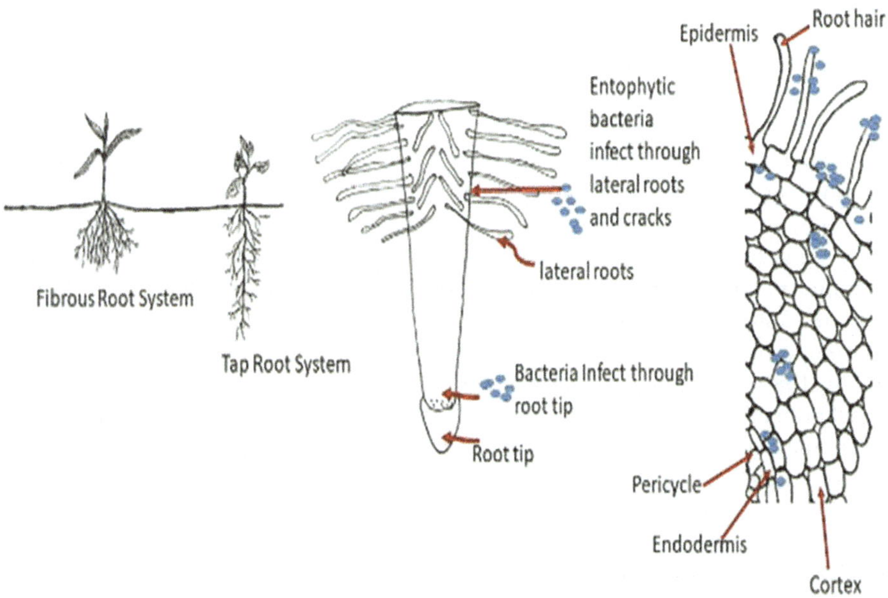

Fig. 7.4 Site of infection and colonization of roots in monocot and dicot plants

Azospirillum is an example of one of the earliest discovered associative nitrogen-fixing bacteria (Beijerinck 1925), which colonizes plants in the grass family. Several species has been identified such as *A. lipoferum, A. brasilense, A. amazonense, A. halopraeferans,* and *A. irakense* (Döbereiner 1989; Rodrigues et al. 2008; Volpon et al. 1981). *Azospirillum* has the ability to fix nitrogen under free-living conditions or when in association with grasses. It has been found in association with several cereal and forage grasses grown in temperature and tropical climates (Nosko et al. 1994). Nitrogen fixation of micro-aerobic *A.brasilense* is governed by the level of oxygen in the surrounding environment. *Azospirillum* has a unique regulatory system for nitrogen fixation. Not like in other diazotrophs *nifA* gene in *A. brasilense* is expressed under fixed nitrogen (Liang et al. 1991). It has been reported that the amount of fixed nitrogen and the level of oxygen will regulate the expression of the *nif* gene and activity of nitrogenase enzyme at the post-translation level (Zhang et al. 1997). However, more research studies are needed to carry out inorder to understand the oxygen regulation of nitrogen fixation in *A.brasilense*.

Gluconacetobacter diazotrophicus (formerly *Acetobacter diazotrophicus*) is a diazotrophic Gram-negative rod which is abundantly present endophytically in sugarcane and rice (Boddey et al. 1991; Parewa et al. 2018). It is estimated that *G. diazotrophicus* produces about 60% nitrogen of the sugarcane plant nitrogen content (Boddey et al. 2001; Subramanian et al. 2007). *G. diazotrophicus* has the ability to tolerate sucrose concentrations up to 30%, and this is an adaptation to live inside the sugar cane plant (Li and MacRae 1991). Several studies have shown poor survival of *G. diazotrophicus* in soil and this supports the endophytic nature of the bacterium (Baldani et al. 1997; Dong et al. 1994; Oliveira et al. 2004). Other than respiratory protection it has been revealed that *G. diazotrophicus* reduces reactive oxygen species production by increasing antioxidant transcript levels (Alquéres et al. 2010). Compared to other diazotrophs, *G. diazotrophicus* has the ability to fix nitrogen in the presence of ammonium as well (James et al. 1994).

The host plants of *G. diazotrophicus* contain high levels of asparagine, which enhance microbial growth but inhibit nitrogenase activity. Therefore, it has been suggested that *G. diazotrophicus* carries an alternative pathway to synthesize asparagine (Alquéres et al. 2012).

Among the 19 species of genus, *Burkholderia, B. vietnamiensis* were the first species found to fix nitrogen (Estrada-De Los Santos et al. 2001; Perin et al. 2006) while in association with roots of rice. Later it was found in association with other crops, such as maize and sugar cane as well (Luvizotto et al. 2010).

7.3 Application in Management Practices

Nitrogen in the atmosphere and nitrogen in soil has several fates and can exist as gaseous in the atmosphere (N_2, N_2O, NO), inorganic in soil (ammonium and nitrate) and organic in soil (amino acids). Denitrification, volatilization, immobilization, leaching and crop uptake decreases the availability of nitrogen in the soil while nitrification, nitrogen fixation and mineralization are all geochemical processes that increase available nitrogen in the soil. Therefore, above-mentioned processes imply that soil has a very dynamic environment and nitrogen fixation by above-mentioned mechanisms plays a major role in enhancing the soil nitrogen pool, which leads to a sustainable soil and plant ecosystem. In the rhizosphere, due to the role they play in nutrient uptake, symbiotic and endophytic bacteria are considered as extremely valuable members of the plant microbiome.

Examples of endophytic nitrogen-fixing microorganisms reported in natural forest ecosystems are with coniferous trees such as lodgepole pine (*Pinus contorta* var. *latifolia*) (Tang et al. 2017), limber pine (*Pinus flexilis*) (Moyes et al. 2016), and western red cedar (*Thuja Plicata*) (Anand and Chanway 2013). Additionally, deciduous trees like black cottonwood (*Populus trichocarpa*) and willow (*Salix sitchensis*) (Doty et al. 2009) are also reported to show such behavior. In these instances, endophytic bacteria have been reported to both colonize and provide nitrogen to host plants and influence the forest soil ecosystem.

Recently, Puri et al. (2020) reported that soils under spruce trees are lacking nutrients and have properties of marginal soil (Puri et al. 2020). However, trees can grow on soil due to the presence of plant nitrogen-fixing endophytic association, especially tress such as Engelmann spruce (*Picea engelmannii*) (Carrell and Frank 2014) and Norway Spruce (*Picea abies*) (Cankar et al. 2005) growing in the subalpine forests of Colorado, USA, and Slovenia. Moreover, they have isolated 18 species out of the 55 endophytic bacterial species from the tissues of hybrid white spruce trees in the West Chilcotin region. They showed nitrogen-fixing ability in vitro (i.e. nitrogenase enzyme activity) when analyzed using acetylene reduction assay (Matse et al. 2020). This evidence proves the effect of nitrogen-fixing microorganisms support for soil and plant health under extremely poor nutrient systems.

Soils affected by salt conditions are defined under infertile or unhealthy soils. However, some endophytic nitrogen-fixing bacteria can associate with plants growing in salty conditions and thereby, helps to sustain plant and soil health (Hrynkiewicz et al. 2019; Piernik et al. 2017; Yang et al. 2020). As an example, scientists have identified nitrogen-fixing endophytes classified under *Proteobacteria, Actinobacteria, Firmicutes* and *Bacteroidetes* associated with *Suaeda maritima* grown in salty soils of Iran (Alishahi et al. 2020) which have a higher nitrogen fixation ability. Moreover, alfalfa plant grown under salinity condition can associate with nitrogen-fixing *Klebsiella cowanii A37 and Klebsiella* sp. *A36* enhance the growth (Noori et al. 2018). In contrast, most of symbiotic nitrogen-fixing bacteria seem to reduce the ability of nitrogen fixation under salty conditions (Shao et al. 2020).

Introducing legumes in agricultural fields with extreme nitrogen deficient soil system is another approach to uplift soil and plant health. One example of such a study was conducted in an agricultural field in the mountainous areas of Azad Jammu and Kashmir (Abbasi and Khan 2004). Due to the sloppy conditions, the uppermost fertile layer of soil was depleted in Azad Jammu and Kashmir. Therefore, nitrogen becomes the extremely deficient nutrient in soil of mountainous areas of Azad Jammu and Kashmir. They have identified that growing white clover

(*Trifolium repens* L.) with perennial grasses had potential for fixing nitrogen in soil while providing other benefits as well (e.g., reduce soil erosion, fodder for animals, cover crop for soil, minimizing the use of fertilizers). The nitrogen-fixing potential of the white clover crop was around 77 kg/ha N and thereby introducing clove can increase the soil nitrogen pool and plant growth.

In some instances, co-inoculation of *Rhizobium* with plant growth-promoting rhizobacteria (PGPR) was observed to grow well in phosphorus deficient soil (Matse et al. 2020). Therefore, mutualism of PGPR with *Rhizobium* showed the ability to increase soil nitrogen pool while contributing for good plant health. Furthermore, they have done more co-inoculation studies of *Rhizobium* with salt tolerant non-rhizobial endophytic bacteria and boosted nitrogen fixation to achieve maximum productivity on slat affected infertile soil (Etesami and Adl 2020).

Excluding the bio-inoculation, biofertilizers that incorporated nitrogen-fixing microorganism is another technological application towards strengthening the soil and plant health in agricultural fields (Jain et al. 2020; Ramasamy et al. 2020; Singh 2018). Biofertilizers incorporated with symbiotic and associative nitrogen fixers can interact with rhizosphere of plants and infect through seed coats or root system can fix nitrogen in plant while sustaining the soil microbial composition. Microbes that exist in biofertilizers are used to enhance certain microbial processers (e.g., nitrogen fixation) to increase the availability of nutrients in a form which can be assimilated by plants and increase soil nutrient pools. Therefore, biofertilizers with nitrogen fixers has been proven as a safe and efficient method of plant and soil management.

Moreover, mixed cropping systems associated with legumes showed increasing nitrogen, in companion crops. The host legumes have identified to increase nitrogen pool in the soil and the fixed nitrogen can transfer to non-legumes in mixed cropping system. As an example soybean-wheat system have the potential to give higher yield and provide *N* to the soil (Mahmud et al. 2020).

7.4 Conclusions

As nitrogen fixed through BNF is directly absorbed by plant tissues, there is a minute tendency for it to undergo denitrification, leaching, and volatilization. As a result, it becomes important for plants and need to be optimize fixationation if to sustain both plant and soil health in nitrogen aspect. Therefore, it is important to conduct research to identify selected strains of nitrogen-fixing organisms, and include more legumes in agroecosystems or forests for efficient BNF. In this chapter, we provide a summary of the role of nitrogen fixation and areas that need focus if to successful adopt nitrogen fixation to improve plant and soil health. Therefore, the community can make a significant contribution to sustainable land management by transferring fixed nitrogen to improve soil and plant function with the aid of practical tools and approaches by which land managers can assess.

References

Abbasi MK, Khan MN (2004) Introduction of white clover for herbage production and nitrogen fixation in the hilly areas of Azad Jammu and Kashmir. Mt Res Dev 24(2):134–140

Alishahi F, Alikhani HA, Khoshkholgh-Sima NA, Etesami H (2020) Mining the roots of various species of the halophyte Suaeda for halotolerant nitrogen-fixing endophytic bacteria with the potential for promoting plant growth. Int Microbiol 23(3):415–427

Alquéres SM, Oliveira JHM, Nogueira EM, Guedes HV, Oliveira PL, Câmara F, Baldani JI, Martins OB (2010) Antioxidant pathways are up-regulated during biological nitrogen fixation to prevent ROS-induced nitrogenase inhibition in Gluconacetobacter diazotrophicus. Arch Microbiol 192(10):835–841

Alquéres SM, Cardoso AM, Brito-Moreira J, Baldani JI, Martins OB (2012) Transfer RNA-dependent asparagine biosynthesis in Gluconacetobacter diazotrophicus and its influence on biological nitrogen fixation. Plant Soil 356(1–2):209–216

Anand R, Chanway C (2013) N 2-fixation and growth promotion in cedar colonized by an endophytic strain of Paenibacillus polymyxa. Biol Fertil Soils 49(2):235–239

Baldani J, Caruso L, Baldani VL, Goi SR, Döbereiner J (1997) Recent advances in BNF with non-legume plants. Soil Biol Biochem 29(5–6):911–922

Banfalvi Z, Sakanyan V, Koncz C, Kiss A, Dusha I, Kondorosi A (1981) Location of nodulation and nitrogen fixation genes on a high molecular weight plasmid of R. meliloti. Mol Gen Genet MGG 184(2):318–325

Becking J-H (1981) The family Azotobacteraceae. In: The prokaryotes. Springer, pp 795–817

Beijerinck M (1925) Uber ein Spirillum, welches frei en Stick-stoff binden kann? Zentralbl Bakteriol 63:353–359

Boddey R, Urquiaga S, Reis V, Döbereiner J (1991) Biological nitrogen fixation associated with sugar cane. In: Nitrogen fixation. Springer, pp 105–111

Boddey RM, Polidoro JC, Resende AS, Alves BJ, Urquiaga S (2001) Use of the 15N natural abundance technique for the quantification of the contribution of N2 fixation to sugar cane and other grasses. Funct Plant Biol 28(9):889–895

Cankar K, Kraigher H, Ravnikar M, Rupnik M (2005) Bacterial endophytes from seeds of Norway spruce (Picea abies L. Karst). FEMS Microbiol Lett 244(2):341–345

Carrell AA, Frank AC (2014) Pinus flexilis and Picea engelmannii share a simple and consistent needle endophyte microbiota with a potential role in nitrogen fixation. Front Microbiol 5:333

Catalano CM, Lane WS, Sherrier DJ (2004) Biochemical characterization of symbiosome membrane proteins from Medicago truncatula root nodules. Electrophoresis 25(3):519–531

Coskun D, Britto DT, Shi W, Kronzucker HJ (2017) How plant root exudates shape the nitrogen cycle. Trends Plant Sci 22:661–673

Crespi M, Gálvez S (2000) Molecular mechanisms in root nodule development. J Plant Growth Regul 19(2):155–166

Dalton H, Postgate J (1968) Effect of oxygen on growth of Azotobacter chroococcum in batch and continuous cultures. Microbiology 54(3):463–473

Dénarié J, Debellé F, Promé J-C (1996) Rhizobium lipo-chitooligosaccharide nodulation factors: signaling molecules mediating recognition and morphogenesis. Annu Rev Biochem 65 (1):503–535

Dixon R, Kahn D (2004) Genetic regulation of biological nitrogen fixation. Nat Rev Microbiol 2 (8):621–631

Döbereiner J (1989) Isolation and identification of root associated diazotrophs. In: Nitrogen fixation with non-legumes. Springer, pp 103–108

Dong W, Song Y (2020) The significance of flavonoids in the process of biological nitrogen fixation. Int J Mol Sci 21(16):5926

Dong Z, Canny MJ, McCully ME, Roboredo MR, Cabadilla CF, Ortega E, Rodes R (1994) A nitrogen-fixing endophyte of sugarcane stems (a new role for the apoplast). Plant Physiol 105 (4):1139–1147

Doran JW, Parkin TB (1994) Defining and assessing soil quality. In: Defining soil quality for a sustainable environment, vol 35, pp 1–21

Doty SL, Oakley B, Xin G, Kang JW, Singleton G, Khan Z, Vajzovic A, Staley JT (2009) Diazotrophic endophytes of native black cottonwood and willow. Symbiosis 47(1):23–33

Ehrhardt DW, Atkinson EM (1992) Depolarization of alfalfa root hair membrane potential by Rhizobium meliloti Nod factors. Science 256(5059):998–1000

Ermakova M, Battchikova N, Richaud P, Leino H, Kosourov S, Isojärvi J, Peltier G, Flores E, Cournac L, Allahverdiyeva Y (2014) Heterocyst-specific flavodiiron protein Flv3B enables oxic diazotrophic growth of the filamentous cyanobacterium Anabaena sp. PCC 7120. Proc Natl Acad Sci 111(30):11205–11210

Estrada-De Los Santos P, Bustillos-Cristales RO, Caballero-Mellado J (2001) Burkholderia, a genus rich in plant-associated nitrogen fixers with wide environmental and geographic distribution. Appl Environ Microbiol 67(6):2790–2798

Etesami H, Adl SM (2020) Can interaction between silicon and non–rhizobial bacteria benefit in improving nodulation and nitrogen fixation in salinity–stressed legumes? A review. Rhizosphere, 100229

Etzler ME, Kalsi G, Ewing NN, Roberts NJ, Day RB, Murphy JB (1999) A nod factor binding lectin with apyrase activity from legume roots. Proc Natl Acad Sci 96(10):5856–5861

Fay P (1992) Oxygen relations of nitrogen fixation in cyanobacteria. Microbiol Mol Biol Rev 56 (2):340–373

Firmin J, Wilson K, Rossen L, Johnston A (1986) Flavonoid activation of nodulation genes in Rhizobium reversed by other compounds present in plants. Nature 324(6092):90–92

Fliegmann J, Bono J-J (2015) Lipo-chitooligosaccharidic nodulation factors and their perception by plant receptors. Glycoconj J 32(7):455–464

Flores E, Picossi S, Valladares A, Herrero A (2019) Transcriptional regulation of development in heterocyst-forming cyanobacteria. Biochim Biophys Acta 1862(7):673–684

Franche C, Lindström K, Elmerich C (2009) Nitrogen-fixing bacteria associated with leguminous and non-leguminous plants. Plant Soil 321(1–2):35–59

Gadella T Jr, Vereb G Jr, Hadri A-E, Röhrig H, Schmidt J, John M, Schell J, Bisseling T (1997) Microspectroscopic imaging of nodulation factor-binding sites on living Vicia sativa roots using a novel bioactive fluorescent nodulation factor. Biophys J 72(5):1986–1996

Gage DJ (2004) Infection and invasion of roots by symbiotic, nitrogen-fixing rhizobia during nodulation of temperate legumes. Microbiol Mol Biol Rev 68(2):280–300

Garg N (2009) Symbiotic nitrogen fixation in legume nodules: process and signaling: a review. In: Sustainable agriculture. Springer, pp 519–531

Geurts R, Fedorova E, Bisseling T (2005) Nod factor signaling genes and their function in the early stages of Rhizobium infection. Curr Opin Plant Biol 8(4):346–352

Goedhart J, Hink MA, Visser AJ, Bisseling T, Gadella TW Jr (2000) In vivo fluorescence correlation microscopy (FCM) reveals accumulation and immobilization of Nod factors in root hair cell walls. Plant J 21(1):109–119

Gothwal R, Nigam V, Mohan M, Sasmal D, Ghosh P (2008) Screening of nitrogen fixers from rhizospheric bacterial isolates associated with important desert plants. Appl Ecol Environ Res 6 (2):101–109

Graham TL (1991) Flavonoid and isoflavonoid distribution in developing soybean seedling tissues and in seed and root exudates. Plant Physiol 95(2):594–603

Halbleib CM, Ludden PW (2000) Regulation of biological nitrogen fixation. J Nutr 130 (5):1081–1084

Hardoim PR, Van Overbeek LS, Berg G, Pirttilä AM, Compant S, Campisano A, Döring M, Sessitsch A (2015) The hidden world within plants: ecological and evolutionary considerations for defining functioning of microbial endophytes. Microbiol Mol Biol Rev 79(3):293–320

Haynes JG, Czymmek KJ, Carlson CA, Veereshlingam H, Dickstein R, Sherrier DJ (2004) Rapid analysis of legume root nodule development using confocal microscopy. New Phytol 163 (3):661–668

Heidstra R, Bisseling T (1996) Nod factor-induced host responses and mechanisms of Nod factor perception. New Phytol 133(1):25–43

Heidstra R, Geurts R, Franssen H, Spaink HP, Van Kammen A, Bisseling T (1994) Root hair deformation activity of nodulation factors and their fate on Vicia sativa. Plant Physiol 105 (3):787–797

Hernández LE, Escobar C, Drøbak BK, Bisseling T, Brewin NJ (2004) Novel expression patterns of phosphatidylinositol 3-hydroxy kinase in nodulated Medicago spp. plants. J Exp Bot 55 (398):957–959

Hirsch AM (1992) Developmental biology of legume nodulation. New Phytol 122(2):211–237

Hrynkiewicz K, Patz S, Ruppel S (2019) Salicornia europaea L. as an underutilized saline-tolerant plant inhabited by endophytic diazotrophs. J Adv Res 19:49–56

Hübner P, Willison J, Vignais P, Bickle T (1991) Expression of regulatory nif genes in Rhodobacter capsulatus. J Bacteriol 173(9):2993–2999

Igiehon NO, Babalola OO (2018) Rhizosphere microbiome modulators: contributions of nitrogen fixing bacteria towards sustainable agriculture. Int J Environ Res Public Health 15(4):574

Jain S, Jain S, Tiwari S (2020) Biofertilizer: impact on soil health and plant growth agrinenv com, 80

James E, Reis V, Olivares F, Baldani J, Döbereiner J (1994) Infection of sugar cane by the nitrogen-fixing bacterium Acetobacter diazotrophicus. J Exp Bot 45(6):757–766

Jiménez-Guerrero I, Acosta-Jurado S, Del Cerro P, Navarro-Gómez P, López-Baena FJ, Ollero FJ, Vinardell JM, Pérez-Montaño F (2018) Transcriptomic studies of the effect of nod gene-inducing molecules in rhizobia: different weapons, one purpose. Genes 9(1):1

Jürgens G (2004) Membrane trafficking in plants. Annu Rev Cell Dev Biol 20:481–504

Kalsi G, Etzler ME (2000) Localization of a Nod factor-binding protein in legume roots and factors influencing its distribution and expression. Plant Physiol 124(3):1039–1048

Kamst E, Spaink HP, Kafetzopoulos D (1998) Biosynthesis and secretion of rhizobial lipochitin-oligosaccharide signal molecules. In: Plant-microbe interactions. Springer, pp 29–71

Kurkdjian AC (1995) Role of the differentiation of root epidermal cells in Nod factor (from Rhizobium meliloti)-induced root-hair depolarization of Medicago sativa. Plant Physiol 107 (3):783–790

Lerouge P, Roche P, Faucher C, Maillet F, Truchet G, Promé JC, Dénarié J (1990) Symbiotic host-specificity of Rhizobium meliloti is determined by a sulphated and acylated glucosamine oligosaccharide signal. Nature 344(6268):781–784

Li Q, Chen S (2020) Transfer of nitrogen fixation (nif) genes to non-diazotrophic hosts. Chembiochem

Li R, MacRae I (1991) Specific association of diazotrophic acetobacters with sugarcane. Soil Biol Biochem 23(10):999–1002

Liang Y, Kaminski P, Elmerich C (1991) Identification of a nifA-like regulatory gene of Azospirillum brasilense Sp7 expressed under conditions of nitrogen fixation and in the presence of air and ammonia. Mol Microbiol 5(11):2735–2744

Liu C-W, Murray JD (2016) The role of flavonoids in nodulation host-range specificity: an update. Plan Theory 5(3):33

Lodwig EM, Hosie AH, Bourdes A, Findlay K, Allaway D, Karunakaran R, Downie J, Poole PS (2003) Amino-acid cycling drives nitrogen fixation in the legume–Rhizobium symbiosis. Nature 422(6933):722–726

Luvizotto DM, Marcon J, Andreote FD, Dini-Andreote F, Neves AA, Araújo WL, Pizzirani-Kleiner AA (2010) Genetic diversity and plant-growth related features of Burkholderia spp. from sugarcane roots. World J Microbiol Biotechnol 26(10):1829–1836

Mahmud K, Makaju S, Ibrahim R, Missaoui A (2020) Current progress in nitrogen fixing plants and microbiome research. Plan Theory 9(1):97

Maier R, Moshiri F (2000) Role of the Azotobacter vinelandiinitrogenase-protective shethna protein in preventing oxygen-mediated cell death. J Bacteriol 182(13):3854–3857

Marchal K, Vanderleyden J (2000) The "oxygen paradox" of dinitrogen-fixing bacteria. Biol Fertil Soils 30(5–6):363–373

Mathesius U (2019) The role of the flavonoid pathway in Medicago truncatula in root nodule formation. A review. In: The model legume Medicago truncatula, pp 434–438

Matse DT, Huang C-H, Huang Y-M, Yen M-Y (2020) Effects of coinoculation of Rhizobium with plant growth promoting rhizobacteria on the nitrogen fixation and nutrient uptake of Trifolium repens in low phosphorus soil. J Plant Nutr 43(5):739–752

Monti A (2012) Switchgrass: a valuable biomass crop for energy. Springer Science & Business Media

Moscatiello R, Squartini A, Mariani P, Navazio L (2010) Flavonoid-induced calcium signalling in Rhizobium leguminosarum bv. viciae. New Phytol 188(3):814–823

Moshiri F, Crouse BR, Johnson MK, Maier RJ (1995) The "nitrogenase-protective" FeSII protein of Azotobacter vinelandii: overexpression, characterization, and crystallization. Biochemistry 34(40):12973–12982

Moyes AB, Kueppers LM, Pett-Ridge J, Carper DL, Vandehey N, O'Neil J, Frank AC (2016) Evidence for foliar endophytic nitrogen fixation in a widely distributed subalpine conifer. New Phytol 210(2):657–668

Neilson A, Rippka R, Kunisawa R (1971) Heterocyst formation and nitrogenase synthesis in Anabaena sp. Arch Mikrobiol 76(2):139–150

Noori F, Etesami H, Zarini HN, Khoshkholgh-Sima NA, Salekdeh GH, Alishahi F (2018) Mining alfalfa (Medicago sativa L.) nodules for salinity tolerant non-rhizobial bacteria to improve growth of alfalfa under salinity stress. Ecotoxicol Environ Saf 162:129–138

Nosko P, Bliss L, Cook F (1994) The association of free-living nitrogen-fixing bacteria with the roots of high arctic graminoids. Arct Alp Res 26(2):180–186

Oelze J (2000) Respiratory protection of nitrogenase in Azotobacter species: is a widely held hypothesis unequivocally supported by experimental evidence? FEMS Microbiol Rev 24 (4):321–333

Oldroyd GE, Downie JA (2004) Calcium, kinases and nodulation signalling in legumes. Nat Rev Mol Cell Biol 5(7):566–576

Oliveira AL, Canuto EL, Silva EE, Reis VM, Baldani JI (2004) Survival of endophytic diazotrophic bacteria in soil under different moisture levels. Braz J Microbiol 35(4):295–299

Parewa HP, Meena VS, Jain LK, Choudhary A (2018) Sustainable crop production and soil health management through plant growth-promoting Rhizobacteria. In: Role of rhizospheric microbes in soil. Springer, pp 299–329

Perin L, Martínez-Aguilar L, Castro-González R, Estrada-De Los Santos P, Cabellos-Avelar T, Guedes H, Reis V, Caballero-Mellado J (2006) Diazotrophic Burkholderia species associated with field-grown maize and sugarcane. Appl Environ Microbiol 72(5):3103–3110

Perret X, Staehelin C, Broughton WJ (2000) Molecular basis of symbiotic promiscuity. Microbiol Mol Biol Rev 64(1):180–201

Perrine-Walker F, Kouchi H, Ridge R (2014) Endoplasmic reticulum-targeted GFP reveals ER remodeling in Mesorhizobium-treated Lotus japonicus root hairs during root hair curling and infection thread formation. Protoplasma 251(4):817–826

Piernik A, Hrynkiewicz K, Wojciechowska A, Szymańska S, Lis MI, Muscolo A (2017) Effect of halotolerant endophytic bacteria isolated from Salicornia europaea L. on the growth of fodder beet (Beta vulgaris L.) under salt stress. Arch Agron Soil Sci 63(10):1404–1418

Prell J, Poole P (2006) Metabolic changes of rhizobia in legume nodules. Trends Microbiol 14 (4):161–168

Puppo A, Groten K, Bastian F, Carzaniga R, Soussi M, Lucas MM, De Felipe MR, Harrison J, Vanacker H, Foyer CH (2005) Legume nodule senescence: roles for redox and hormone signalling in the orchestration of the natural aging process. New Phytol 165(3):683–701

Puri A, Padda KP, Chanway CP (2020) Can naturally-occurring endophytic nitrogen-fixing bacteria of hybrid white spruce sustain boreal forest tree growth on extremely nutrient-poor soils? Soil Biol Biochem 140:107642

Ramasamy M, Geetha T, Yuvaraj M (2020) Role of biofertilizers in plant growth and soil health. In: Nitrogen fixation. IntechOpen

Richau Kerstin H, Petar P, Philippe N, Anita S (2017) Nitrogenase and hydrogenase of the Actinobacteria Frankia: from gene expression to proteins function

Rodrigues EP, Rodrigues LS, de Oliveira ALM, Baldani VLD, dos Santos Teixeira KR, Urquiaga S, Reis VM (2008) Azospirillum amazonense inoculation: effects on growth, yield and N 2 fixation of rice (Oryza sativa L.). Plant Soil 302(1–2):249–261

Roley SS, Duncan DS, Liang D, Garoutte A, Jackson RD, Tiedje JM, Robertson GP (2018) Associative nitrogen fixation (ANF) in switchgrass (Panicum virgatum) across a nitrogen input gradient. PLoS One 13(6):e0197320

Roth E, Jeon K, Stacey G (1988) Molecular genetics of plant-microbe interactions. In: Homology in endosymbiotic systems: the term "Symbiosome", pp 220–225

Rubio L, Ludden P (2005) Maturation of nitrogenase: a biochemical puzzle. J Bacteriol 187:405–414. https://doi.org/10.1128/JB.187.2.405-414.2005

Rubio LM, Ludden PW (2008) Biosynthesis of the iron-molybdenum cofactor of nitrogenase. Annu Rev Microbiol 62

Selepe MA, Van Heerden FR (2013) Application of the Suzuki-Miyaura reaction in the synthesis of flavonoids. Molecules 18(4):4739–4765

Shao J, Markham J, Renault S (2020) Nitrogen fixation symbiosis and salt tolerance of the boreal woody species Elaeagnus commutata. Acta Physiol Plant 42(6)

Shaw SL, Long SR (2003) Nod factor elicits two separable calcium responses inMedicago truncatula root hair cells. Plant Physiol 131(3):976–984

Shrestha RK, Maskey SL (2005) Associative nitrogen fixation in lowland rice. Nepal Agric Res J 6:112–121

Singh B (2018) Are nitrogen fertilizers deleterious to soil health? Agronomy 8(4):48

Singh S, Shrivastava A, Singh V (2014) Arsenic and cadmium are inhibitors of cyanobacterial dinitrogenase reductase (nifH1) gene. Funct Integr Genomics 14(3):571–580

Singh RK, Singh P, Li H-B, Yang L-T, Li Y-R (2017) Soil–plant–microbe interactions: use of nitrogen-fixing bacteria for plant growth and development in sugarcane. In: Plant-microbe interactions in agro-ecological perspectives. Springer, pp 35–59

Stal LJ (2015) Nitrogen fixation in cyanobacteria. eLS:1–9

Subramanian S, Stacey G, Yu O (2007) Distinct, crucial roles of flavonoids during legume nodulation. Trends Plant Sci 12(7):282–285

Tang Q, Puri A, Padda KP, Chanway CP (2017) Biological nitrogen fixation and plant growth promotion of lodgepole pine by an endophytic diazotroph Paenibacillus polymyxa and its GFP-tagged derivative. Botany 95(6):611–619

Thorneley RN, Ashby GA (1989) Oxidation of nitrogenase iron protein by dioxygen without inactivation could contribute to high respiration rates of Azotobacter species and facilitate nitrogen fixation in other aerobic environments. Biochem J 261(1):181–187

Timmers A, Auriac M-C, de Billy F, Truchet G (1998) Nod factor internalization and microtubular cytoskeleton changes occur concomitantly during nodule differentiation in alfalfa. Development 125(3):339–349

Timmers A, Auriac M-C, Truchet G (1999) Refined analysis of early symbiotic steps of the Rhizobium-Medicago interaction in relationship with microtubular cytoskeleton rearrangements. Development 126(16):3617–3628

Truchet G, Roche P, Lerouge P, Vasse J, Camut S, de Billy F, Promé J-C, Dénarié J (1991) Sulphated lipo-oligosaccharide signals of Rhizobium meliloti elicit root nodule organogenesis in alfalfa. Nature 351(6328):670–673

Unden G, Becker S, Bongaerts J, Holighaus G, Schirawski J, Six S (1995) O2-sensing and O2-dependent gene regulation in facultatively anaerobic bacteria. Arch Microbiol 164(2):81–90

Ureta A, Nordlund S (2002) Evidence for conformational protection of nitrogenase against oxygen in Gluconacetobacter diazotrophicus by a putative FeSII protein. J Bacteriol 184 (20):5805–5809

Valladares A, Herrero A, Pils D, Schmetterer G, Flores E (2003) Cytochrome c oxidase genes required for nitrogenase activity and diazotrophic growth in Anabaena sp. PCC 7120. Mol Microbiol 47(5):1239–1249

Verma D (1992) Signals in root nodule organogenesis and endocytosis of Rhizobium. Plant Cell 4 (4):373

Ververidis F, Trantas E, Douglas C, Vollmer G, Kretzschmar G, Panopoulos N (2007) Biotechnology of flavonoids and other phenylpropanoid-derived natural products. Part I: Chemical diversity, impacts on plant biology and human health. Biotechnol J 2(10):1214–1234

Volpon AG, De-Polli H, Döbereiner J (1981) Physiology of nitrogen fixation in Azospirillum lipoferum Br 17 (ATCC 29 709). Arch Microbiol 128(4):371–375

Walsby AE (1985) The permeability of heterocysts to the gases nitrogen and oxygen. Proc R Soc Lond Ser B Biol Sci 226(1244):345–366

Wewalwela JJ (2020) Associative nitrogen-fixing bacteria and their potential to support the growth of bioenergy grasses on marginal lands. Mississippi State University

Wewalwela JJ, Tian Y, Donaldson JR, Baldwin BS, Varco JJ, Rushing B, Lu H, Williams MA (2020) Associative nitrogen fixation linked with three perennial bioenergy grasses in field and greenhouse experiments. GCB Bioenergy. https://doi.org/10.1111/gcbb.12744

Wolk CP, Ernst A, Elhai J (1994) Heterocyst metabolism and development. In: The molecular biology of cyanobacteria. Springer, pp 769–823

Yang T, Siddique KH, Liu K (2020) Cropping systems in agriculture and their impact on soil health – a review. Global Ecol Conservation e01118

Zehr JP, Jenkins BD, Short SM, Steward GF (2003) Nitrogenase gene diversity and microbial community structure: a cross-system comparison. Environ Microbiol 5(7):539–554

Zhang Y, Burris RH, Ludden PW, Roberts GP (1997) Regulation of nitrogen fixation in Azospirillum brasilense. FEMS Microbiol Lett 152(2):195–204

Serendipita indica Mediated Drought and Heavy Metal Stress Tolerance in Plants

8

Surbhi Dabral, Ajit Varma, Deepesh Bhatt, and Manoj Nath

Abstract

Serendipita indica, a root colonizing fungus, is well known to promote plant growth. *S. indica* belongs to the order Sebacinales and can be grown axenically. In addition, this fungus is also reported to alleviate the stressful environment in different plants. Notably, *S. indica* is well reported in improving the plant growth under abiotic stress conditions. Recent studies revealed that *S. indica* has the ability to upgrade the defense system of plants. Moreover under heavy metal and drought stress *S. indica* is known to initiate antioxidative mechanisms, thus alleviating the harmful effect of these stress conditions. Here, in this chapter we have focused majorly on the involvement of *S. indica* in various plants under drought and heavy metal stress.

Keywords

Plant–microbe interaction · Drought stress · Heavy metal stress

S. Dabral · A. Varma
Amity Institute of Microbial Technology, Amity University, Noida, Uttar Pradesh, India

D. Bhatt
Faculty of Science, Sarvajanik University, Surat, Gujarat, India

M. Nath (✉)
ICAR-Directorate of Mushroom Research, Chambaghat, Solan, Himachal Pradesh, India
e-mail: manoj.nath@icar.gov.in

R. Soni et al. (eds.), *Microbiological Activity for Soil and Plant Health Management*,
https://doi.org/10.1007/978-981-16-2922-8_8

8.1 Introduction

The world population is expected to increase from about 7 billion to 9.6 billion by the end of 2050; therefore, food security is going to be a major challenge in future (Mahajan and Tuteja 2005). Additionally, sufficient food supply required to feed the growing population with limited agriculture land will further exacerbate the situation. Therefore, key attention is required to develop high yielding varieties in different crops in order to address food security and climate change related issues. Stress affects plant growth, which ultimately reduces the crop yield. Among stresses, abiotic stress is one of the major factors responsible for this decline. Presently developing abiotic stress tolerant crops has become a core area of crop production where abiotic stress can be specifically categorized into various types, namely salinity, cold, freezing, drought, water logging, UV radiation, high temperature, nutrient deficiencies, heavy metal toxicity, and nutrient imbalances. All the mentioned stress types negatively affect the essential cellular metabolic processes resulting in excessive generation of reactive oxygen species (ROS), leading to cell damage. Prolonged abiotic stress environment severely affects the plant growth and ultimately results in cell death (Tuteja 2010). Alternately, extensive use of chemical fertilizers also deteriorates the soil structure, soil health as well as the overall environment. Repeated application of these fertilizers leads to accumulation and toxic buildup of chemicals like arsenic and cadmium in soil. Recent studies indicate the importance of the microbial counterpart to promote the plant growth during plant–microbial interactions (Gill et al. 2016). Utilization of the microbial consortium as a potential biofertilizer can overcome the problems of chemical fertilizers. Plant–microbe interaction can be positive, negative, or neutral. However, focusing on the positive interactions is one of the most beneficial for attainment of sustainable agriculture (Bandyopadhyay et al. 2016; Finkel et al. 2017).

Microbial associations with plants can be below or above-ground. Though, depending on their habitat on plants, these microbes can be classified as rhizospheric (inhabiting soil closely associated with roots), epiphyte (present on the surface), phyllospheric (growing on leaf surface), and endophyte (found inside the tissue). Notably, microbes residing in rhizosphere are the most active and have maximum impact on plant growth and nutrition (Kowalchuk et al. 2010; Lakshmanan et al. 2014; Berendsen et al. 2012; Bakker et al. 2013; Mendes et al. 2013). Plant root system interacts with numerous beneficial microbes and further this association improves plant growth development (Sanders 2011; Mine et al. 2014) and alleviates the stressful environment (Goh et al. 2013; Schouteden et al. 2015; Doty 2016). However, several beneficial microbes of the root microbiome still need to be explored for better utilization towards sustainable agriculture (Mendes et al. 2011; Philippot et al. 2013). Plants accompanied by microbes like rhizobacteria and mycorrhizal fungi are believed to function together and have evolved as holobiont. Conversely, interactions among plants and mycorrhizal fungi are greatly influenced by rhizospheric bacteria and vice versa, but the associated mechanisms are still unanswered. Host and symbionts are dependent on each other, for instance, *Rhizobium* is involved in fixing atmospheric nitrogen in association with the host, while acquisition of phosphate from the soil is due to the mycorrhizal connotation (Udvardi and Poole 2013). Role of plant growth promoting rhizobacteria (PGPR) and plant growth promoting fungi (PGPF) in acquirement of nutrition is also well reported. Particularly, arbuscular mycorrhizal (AMF) interactions are known to play a major role in stress tolerance as well as for growth development of plants (Muthukumar and Udaiyan 2010; Porcel et al. 2012; Tahat and Sijam 2012). AMF also acts as a biofertilizer, bioprotectant, and biodegrader (Xavier and Boyetchko 2004).

Studies suggest that AMF has huge impact on plant stress tolerance mechanisms, for which one of the effective contenders is *Serendipita indica* (formerly known as *Piriformospora indica*), a root colonizing fungus. It belongs to the order Sebacinales and is capable of growing axenically. Alleviation of the stress and growth promotory activities are some of the important characteristics attributed to this fungus (Waller et al. 2005; Deshmukh and Kogel 2007; Serfling et al. 2007; Sarma et al. 2011; Dolatabadi et al. 2012; Varma et al. 2012; Jogawat et al. 2013; Gill et al. 2016; Zhang et al. 2017; Dabral et al. 2019). *S. indica* has a broad plant host spectrum and helps in stress tolerance, disease resistance, nutrient uptake, and plant growth promotion activities (Unnikumar et al. 2013). *S. indica* acts a powerful weapon for enhancement of various crops and its usage has been established in seed germination, plant development, and biomass production (Ansari et al. 2014). Apart from this the fungus has the capability to modulate the generation and scavenging of ROS under stressful regimes (Beneventi et al. 2013; Goh et al. 2013; Hashem et al. 2016; Mo et al. 2016). Role of *S. indica* in alleviating drought (Tsai et al. 2020; Sherameti et al. 2008; Sun et al. 2010) and heavy metal (Mohd et al. 2017; Nanda and Agrawal 2018; Dabral et al. 2019) stress conditions has been reported in several crops. Here, in this chapter we have emphasized the role of *S. indica* in mitigating drought and heavy metal stress in plants.

8.2 Role of *S. indica* in Heavy Metal Stress Tolerance

Elements having atomic mass over 20 and density higher than 5 g/cm^{-3} are categorized under heavy metals. They are non-biodegradable and mutagenic in nature which have genotoxic and cytotoxic effects in plants, humans, and animals (Flora et al. 2008; Rascio and Navari-Izzo 2011; Wuana and Okieimen 2011). There are two types of metals which are available in soil, viz. essential micronutrients like Fe, Mn, Zn, Cu, Mg, Mo, and Ni which are required for normal plant growth and nonessential elements like Cd, Pb, Cr, Co, Ag, Hg, etc. which are lesser known to have any physiological or biological function (Rascio and Navari-Izzo 2011; Schutzendubel and Polle 2002; Tangahu et al. 2011; Zhou et al. 2014). Concentration of both essential and nonessential elements in plant system is of great importance as a little variation can disrupt the normal growth and development (Zengin and Munzuroglu 2005). Heavy metals are supplemented in the atmosphere due to various anthropogenic deeds which leads to their deposition in the environment by air and water (Dean et al. 1972). Due to various human activities accumulation of heavy metals in soil occurs through municipal and industrial wastes, smelting, mining, contaminated irrigation water, use of fertilizers and pesticides, and ejection of waste from leather industries (Al-Hobaib et al. 2012; Arunakumara et al. 2013; Gallego et al. 2012). Heavy metals act as toxic agents for plants, e.g. alter the imperative cell molecules (Hossain et al. 2012) and enzymes (Ali et al. 2013) involved in normal functioning of plant cells, undesirably affecting the membrane of plant cell (Farid et al. 2013). These heavy metals can also hamper the metabolic process of plants in many ways like degradation and displacement of plant proteins (Hall 2002) and affecting the vital processes like respiration and photosynthesis in plants (Hossain et al. 2012). Apart from this heavy metal toxicity is also responsible for ROS overproduction, namely superoxides, hydroxyl free radicals, and non-free radical species like singlet oxygen, hydrogen peroxide (H_2O_2), etc. Thus it can be related that imbalanced production of ROS is responsible for causing oxidative stress in plants (Sytar et al. 2013; Hossain et al. 2012) which have various after effects like redox imbalance, DNA degradation, protein and lipid oxidation, ion leakage followed by denaturing of cell structure which results in apoptosis (Hatata and Abdel-Aal 2008; Rellan-Alvarez et al. 2006; Sharma et al. 2012).

Heavy metal accumulation in soil can lead to yield loss in crops and also causes numerous health constraints as they enter the food chain (Li et al. 2016). One of the most common heavy metals which is making its way to the agricultural lands is cadmium (Cd) (Adriano 2001). Plants are highly affected due to Cd stress as it inhibits the photosynthetic activity (Paunov et al. 2018; Bączek-Kwinta et al. 2019; Song et al. 2019). In addition Cd also reduces the plant growth, delay seed germination, necrosis and chlorosis in leaves (Hall 2002; Xu et al. 2009; Dias et al. 2013). Moreover translocation of Cd from roots to the aerial parts of the plants can cause serious implications to the consumers who feed them (Du et al. 2013; Aziz et al. 2015; Song et al. 2015). It was observed that spike in heavy metal concentrations is the main reason for cell death as ROS production is elevated which leads to oxidative burst, thus resulting in cell death (Guo et al. 2009; Tripathi et al. 2012). Interestingly, *S. indica* has been reported for extenuating heavy metal stress in plants (Shahabivand et al. 2012; Hui et al. 2015; Shahabivand et al. 2017; Mohd et al. 2017; Nanda and Agrawal 2018). Dabral et al. (2019) reported the ability of the *S. indica* to grow under various Cd concentrations (0.01–0.1 mM) and further explored its role in mitigating the stress tolerance in rice. Interestingly, several reports suggest that process of acquisition of nutrients from soil responsible for plant growth promotion is assisted by *S. indica* (Hartley and Gange 2009) which in turn activates various biochemical pathways required to achieve optimum growth (Sirrenberg et al. 2007; Vadassery et al. 2009a, 2009b). Studies in tobacco (Hui et al. 2015) and sunflower (Shahabivand et al. 2017) demonstrated *S. indica* inoculation under Cd stress tends to store more Cd in roots to restrict heavy metal movement to aerial parts of the plants. Recent study conducted in rice varieties N22 and IR64 with the help of confocal microscopy gave an insight to the possible mechanism of *S. indica* in mitigating the stress. Higher stress survivability of the *S. indica* colonized plants demonstrated the sequestration ability of the fungus (Dabral et al. 2019). Moreover, Stein et al. (2008) reported that the colonization of plant roots by the fungus encourages ROS signaling pathway followed by various other defense mechanisms like instigation of jasmonic acid and salicylic acid, which in turn activates antioxidant enzymes that protect the plants against high level of Cd. Another study conducted in wheat signified that *S. indica* inoculation had a positive effect on growth and physiology of wheat plants, the results revealed the decrease in the concentrations of leaf MDA and Cd contents of shoot and increased activity of antioxidant enzymes. The beneficial effects of the root endophytic fungus suggest the promotion of this symbiotic association to mitigate the Cd stress (Shahabivand and Aliloo 2016). In tobacco *S. indica* was able to sequester Cd in roots of tobacco, thereby increasing the concentration of antioxidant enzymes and proline contents in roots. Moreover expression of phytochelatin, a stress related gene in fungus inoculated plants was also recorded higher (Hui et al. 2015). Studies conducted in sunflower indicated increased chlorophyll a and b content, augmented proline level, and an overall increased growth, thus suggesting property of this fungus to alleviate metal toxicity in host plants and also phyto-stabilization of Cd-polluted soils. Several studies thereby suggest that *S. indica* can act well to impart Cd tolerance (Shahabivand et al. 2017).

Arsenic (As) is also one of the heavy metals involved in reduction of shoots, roots, and biomass of plants. Colonization of *S. indica* is also known to alleviate arsenic toxicity in plants as axenic culture of the fungus able to tolerate toxicity up to 1 mM. Studies in rice revealed the role of *S. indica* as a bioprotectant against toxic nature of arsenic. *S. indica* inoculated rice plants revealed increased biomass and better growth in comparison to non-inoculated plants. The study also highlighted the detoxification capacity of *S. indica* as the fungus helps to reduce As translocation from root to shoot by immobilizing As into the roots itself. This study also explored the different mechanisms performed by *S. indica*, first to condense As availability in plant atmosphere, second conversion of available As to insoluble particulate matter, and third by activating the antioxidative system of the plant (Mohd et al. 2017).

Copper (Cu) is an essential micronutrient which is required by plants for normal growth and functioning. However, increased concentration of Cu in soil also leads to the negative consequences in plants (Lombardi and Sebastiani 2005). Toxic effects of Cu include necrosis, discoloration symptoms in leaves, chlorosis, lipid peroxidation in cell membranes, and production of free radicals in different parts of the plant (Yruela 2009; Chen et al. 2000; Panda et al. 2016). Nanda and Agrawal (2018) reported that *S. indica* ameliorates Cu phytotoxicty in *C. angustifolia.* A similar study indicated that *S. indica* colonization helped in alleviating stress conditions by restricting the heavy metal in roots. Moreover, the fungus also helped the plant by strengthening the defense mechanism by enhancing the enzymatic activities. Low cell death and H_2O_2 content and significant increase in the proline content were also observed in *S. indica* inoculated plants.

8.3 Drought Stress Tolerance Mediated by *S. indica*

Drought is also known as water deficit stress which is responsible for enormous agricultural loss. Various crop growth models predict that this condition will be more severe in the coming times (Farooq et al. 2012). Decrease of water levels, stomatal closure, decreased and delayed cell growth and enlargement, turgor loss, and reduced leaf water potential are some of the characteristics related to drought stress (Jaleel et al. 2008). Abiotic stress damages the normal functioning of plants and causes distressed enzyme activity and metabolism of the cell, affects the membrane proteins, and also reduces photosynthetic activity (Tuteja 2010; Farooq et al. 2012). Apparently, closing of stomata occurring during water stress causes a series of changes in normal plant functioning, it reduces availability of carbon dioxide which in turn affects photosynthesis (Ramachandra Reddy et al. 2004). Acceleration of photo-inhibition results in excessive ROS generation, which is responsible for oxidative damage in plant cells (Takahashi and Murata 2008; Tripathy and Oelmüller 2012). In response to stress, plants also practice various defense mechanisms by activating various anti-oxidation pathways comprising of superoxide dismutase (SOD), ascorbate peroxidase (APX), catalase (CAT) (Miller et al. 2010; Osakabe et al. 2014; Blokhina et al. 2003; Foyer and Noctor 2011). Furthermore, drought stress related signaling involves re-establishment of the ionic/osmotic equilibrium in the cell, to repair the damage caused by stress.

Drought stress severely affects the initial stages of plant development (Anjum et al. 2003; Bhatt and Rao 2005; Kusaka et al. 2005; Shao et al. 2008). An array of the responses can be observed during drought stress, for instance, decreased shoot length in soybean (Zhang et al. 2004; Specht et al. 2001), *Abelmoschus esculentus* (Sankar et al. 2007); *Vigna unguiculata* (Manivannan et al. 2007), and potato (Heuer and Nadler 1995) were observed during stress condition. Notably, shoot and root fresh weights of rice were significantly reduced under drought conditions (Tsai et al. 2020). Apart from this, drought stress is also responsible for delayed leaf growth in many plants like sunflower (Manivannan et al. 2008) and *Vigna unguiculata* (Manivannan et al. 2007). Seed yield is worst affected as reported in soybean (Specht et al. 2001), common bean and green gram (Webber et al. 2006), *parsley* (Petropoulos et al. 2008), and maize (Monneveux et al. 2006). In addition, studies revealed that drought stress has huge impact on chlorophyll content, decrease in chlorophyll content in crops like cotton (Massacci et al. 2008) and sunflower (Kiani et al. 2008).

Plant–microbe interactions play an important role and mutualistic symbiosis is the basis of beneficial establishment during this interaction (Kogel et al. 2006; Johnson and Oelmüller 2009; Singh et al. 2011). *S. indica* was also reported to improve the plant growth under stress conditions (Varma et al. 1999; Waller et al. 2005; Lee et al. 2011). Drought related studies revealed improved survival ability of *S. indica* colonized plants (Arabidopsis, Chinese cabbage, and Maize) under stress (Sherameti et al. 2008; Sun et al. 2010; Xu et al. 2017). Notably, exposure of seedling stage of Arabidopsis to drought stress (84 h) revealed that 50% of *S. indica* colonized plants survived and produced seeds at later stage, while no seed was observed in non-colonized plants. PEG-mediated drought stress in Chinese cabbage revealed that *S. indica* promotes shoot-root growth and activates various enzymes like SOD, catalases, and peroxidases within 24 h of onset of drought. The fungus reversed the impact of drought, i.e. degeneration in the photosynthetic activity and chlorophyll. Various drought related genes like CBL1, ANAC072, DREB2A, and RD29A were analyzed and their expression was upregulated with onset of drought stress in *S. indica* colonized plants (Sun et al. 2010). In response to 6 h and 12 h of drought stress, enhanced root length and leaf size were observed in *S. indica* colonized maize plants as compared with non-colonized one. Furthermore, fungal responsive genes also stimulated the hormone production, viz. auxin, salicylic acid, cytokinins, and abscisic acid (Zhang et al. 2017). A recent study in rice revealed augmented photosynthetic efficiency and grain yield in colonized plants than non-colonized one. Under drought stress, *S. indica* promote stomatal closure and upregulation of ROS scavenging enzymes, viz. glutathione reductase and catalase (Tsai et al. 2020). Vadassery et al. (2009a, 2009b) reported that *S. indica* induced monodehydroascorbate reductase 2 (MDHAR2) and dehydroascorbate reductase 5 (DHAR5) in the colonized plant.

Drought stress affects the photosynthetic ability of the plant. *S. indica* on the other side helps the plant in overcoming stress environment by activating antioxidative molecules/enzyme (Scandalios 2005). The photosynthetic efficiency of the plant is measured by F variable/F maximum (Fv/Fm) values (Bjorkman and Demming 1987; Maxwell and Johnson 2000). Drought stress primarily affects this value and *S. indica* combat with stress environment by maintaining photosynthetic efficiency (Sherameti et al. 2008; Sun et al. 2010). In addition, enhanced photosynthetic efficiency, mineral uptake, and improved morphological traits were observed in *S. indica* inoculated rice plants under osmotic stress (Saddique et al. 2018).

Moreover, an array of genes is associated with *S. indica* mediated drought stress tolerance. Some of the genes are specifically induced under drought stress, viz. PLDd gene (related to plasma membrane). This gene is triggered in response to hydrogen peroxide and phosphatidic acid and diminishes hydrogen peroxide induced programmed cell death (Zhang et al. 2003). While DREB2A is one of the key genes involved in drought stress tolerance (Sakuma et al. 2006). Upregulation of SDIR1 activates ABA-dependent pathway leading to drought tolerance (Zhang et al. 2007). Notably, drought associated genes DREB2A, CBL1, ANAC072, and RD29A were upregulated in *S. indica* inoculated stressed plants (Sun et al. 2010). Conversely, stress also induces the calcium signaling pathway (Harper and Harmo 2005, Vadassery et al. (2009a, 2009b)). Table 8.1 provides an overview of the *S. indica* involvement under stress condition.

8.4 Conclusion

S. indica can colonize a wide range of plants and mediate stress tolerance. An axenic culture ability of this fungus has opened wide doors for more effective research in the field of plant–microbe interaction. *S. indica* is capable to combat the stressful

Table 8.1 *S. indica* mediated drought and heavy metal tolerance in different plants

S. no.	Crop	Stress	Key parameters observed in *S. indica* inoculated plants	References
1	Rice	Heavy metal (Cd)	Acquisition of Cd by rice roots	Dabral et al. (2019)
2	Wheat	Heavy metal (Cd)	Decreased Cd in shoots, increased activity of antioxidant enzymes	Shahabivand and Aliloo (2016)
3	Tobacco	Heavy metal (Cd)	Cd sequestration in roots	Hui et al. (2015)
4	Sunflower	Heavy metal (Cd)	Increased growth and chlorophyll content	Shahabivand et al. (2017)
5	Rice	Heavy metal (As)	As immobilization to roots	Mohd et al. (2017)
6	*C. angustifolia*	Heavy metal (Cu)	Ameliorates Cu phytotoxicty	Nanda and Agrawal (2018)
7	Arabidopsis	Drought	Improved survival ability under stress	Sherameti et al. (2008)
8.	Chinese cabbage	Drought	Drought related genes upregulated	Sun et al. (2010)
9.	Maize	Drought	Stimulated production of hormones	Zhang et al. (2017)
10.	Rice	Drought	Increased grain yield and photosynthetic efficiency	Tsai et al. (2020)

environment and further can be utilized to attain maximum crop yield under abiotic stress conditions.

References

Adriano DC (2001) Cadmium in trace elements in terrestrial environments: biogeochemistry, bioavailability, and risks of metals, 2nd edn. Springer, New York, pp 264–314

Al-Hobaib AS, Al-Jaseem KQ, Baioumy HM, Ahmed AH (2012) Environmental impact assessment inside and around Mahd Adh Dhahab gold mine, Saudi Arabia. Arab J Geosci 4:199–205

Ali H, Khan E, Sajad MA (2013) Phytoremediation of heavy metals—concepts and applications. Chemosphere 91(7):869–881

Anjum F, Yaseen M, Rasul E, Wahid A, Anjum S (2003) Water stress in barley (*Hordeum vulgare* L.). I. Effect on morphological characters. Pak J Agric Sci 40:43–44

Ansari MW, Gill SS, Tuteja N (2014) Piriformospora indica a powerful tool for crop improvement. Proc Indian Natl Sci Acad 80:317–324. https://doi.org/10.16943/ptinsa/2014/v80i2/55109

Arunakumara KKIU, Walpola BC, Yoon MH (2013) Current status of heavy metal contamination in Asia's rice lands. Environ Sci Technol. https://doi.org/10.1007/s11157-013-9323-1

Aziz R, Rafiq MT, Li T, Liu D, He Z, Stoffella PJ, Sun KW, Xiaoe Y (2015) Uptake of cadmium by rice grown on contaminated soils and its bioavailability/toxicity in human cell lines (Caco-2/HL-7702). J Agric Food Chem 63:3599–3608

Bączek-Kwinta R, Juzon K, Borek M, Antonkiewicz J (2019) Photosynthetic response of cabbage in cadmium-spiked soil. Photosynthetica 57(3):731–739. https://doi.org/10.32615/ps.2019.070

Bakker PAHM, Berendsen RL, Doornbos RF, Wintermans PCA, Pieterse CMJ (2013) The rhizosphere revisited: root microbiomics. Front Plant Sci 4:165. https://doi.org/10.3389/fpls.2013.00165

Bandyopadhyay P, Bhuyan SK, Yadava PK, Varma A, Tuteja N (2016) Emergence of plant and rhizospheric microbiota as stable interactomes. Protoplasma 254:617–626. https://doi.org/10.1007/s00709-016-1003-x

Beneventi MA, da Silva OB, de Sa MEL, Firmino AAP, de Amorim RMS, Albuquerque EVS et al (2013) Transcription profile of soybean–root-knot nematode interaction reveals a key role of phythormones in the resistance reaction. BMC Genomics 10:322. https://doi.org/10.1186/1471-2164-14-322

Berendsen RL, Pieterse CMJ, Bakker PAHM (2012) The rhizosphere microbiome and plant health. Trends Plant Sci 17:478–486

Bhatt RM, Rao S (2005) Influence of pod load response of okra to water stress. Indian J Plant Physiol 10:54–59

Bjorkman O, Demming B (1987) Photon yield of O2 evolution and chlorophyll fluorescence at 77K among vascular plants of diverse origins. Planta 170:489–504

Blokhina O, Virolainen E, Fagerstedt KV (2003) Antioxidants, oxidative damage and oxygen deprivation stress: a review. Ann Bot 91:179–194. https://doi.org/10.1093/aob/mcf118

Chen LM, Lin CC, Kao CH (2000) Copper toxicity in rice seedlings: changes in antioxidative enzyme activities, H2O2 level, and cell wall peroxidase activity in roots. Bot Bull Acad Sin 41 (2):99–103

Dabral S, Yashaswee, Varma A, Choudhary DK, Bahuguna RN, Nath M (2019) Biopriming with Piriformospora indica ameliorates cadmium stress in rice by lowering oxidative stress and cell death in root cells. Ecotoxicol Environ Saf 186:109741. https://doi.org/10.1016/j.ecoenv.2019.109741

Dean JG, Bosqui FL, Lanouette KH (1972) Removing heavy metals from wastewater. Environ Sci Technol 6:518–524.a

Deshmukh SD, Kogel KH (2007) Piriformospora indica protects barley from root rot caused by Fusarium graminearum. J Plant Dis Prot 114:263–268

Dias MC, Monteiro C, Moutinho-Pereira J, Correia C, Gonçalves B, Santos C (2013) Cadmium toxicity affects photosynthesis and plant growth at different levels. Acta Physiol Plant 35:1281–1289

Dolatabadi HK, Goltapeh EM, Mohammadi N, Rabiey M, Rohani N, Varma A (2012) Biocontrol potential of root endophytic fungi and Trichoderma species against Fusarium wilt of lentil under in vitro and greenhouse conditions. J Agric Sci Technol 14:407–420

Doty SL (2016) Plant–microbe symbiotic interactions. Plant Mol Biol 90:535. https://doi.org/10.1007/s11103-016-0470-y

Du Y, Hu XF, Wu XH, Shu Y, Jiang Y, Yan XJ (2013) Affects of mining activities on Cd pollution to the paddy soils and rice grain in Hunan province, Central South China. Environ Monit Assess 185:9843–9856

Farid M, Shakoor MB, Ehsan A, Ali S, Zubair M, Hanif MS (2013) Morphological, physiological and biochemical responses of different plant species to Cd stress. Int J Chem Biochem Sci 3:53–60

Farooq M, Hussain M, Wahid A, Siddique KHM (2012) Drought stress in plants: an overview. In: Aroca R (ed) Plant responses to drought stress. Springer, Berlin-Heidelberg

Finkel OM, Castrillo G, Herrera Paredes S, Salas González I, Dangl JL (2017) Understanding and exploiting plant beneficial microbes. Curr Opin Plant Biol 38:155–163. https://doi.org/10.1016/j.pbi.2017.04.018

Flora SJS, Mittal M, Mehta A (2008) Heavy metal induced oxidative stress & its possible reversal by chelation therapy. Indian J Med Res 128(4):501–523

Foyer CH, Noctor G (2011) Ascorbate and glutathione: the heart of the redox hub. Plant Physiol 155:2–18. https://doi.org/10.1104/pp.110.167569

Gallego SM, Pena LB, Barcia RA, Azpilicueta CE, Iannonea MF, Rosales EP, Zawoznika MS, Groppaa MD, Benavides MP (2012) Unravelling cadmium toxicity and tolerance in plants: insight into regulatory mechanisms. Environ Exp Bot 83:33–46

Gill SS, Gill R, Trivedi DK, Anjum NA, Sharma KK (2016) Piriformospora indica: potential and significance in plant stress tolerance. Front Microbiol 7:332. https://doi.org/10.3389/fmicb.2016.00332

Goh C, Vallejos DFV, Nicotra AB, Mathesius U (2013) The impact of beneficial plant-associated microbes on plant phenotypic plasticity. J Chem Ecol 39:826–839. https://doi.org/10.1007/s10886-013-0326-8

Guo B, Liang Y, Zhu Y (2009) Does salicylic acid regulate antioxidant defense system, cell death, cadmium uptake and partitioning to acquire cadmium tolerance in rice? J Plant Physiol 166:20–31. https://doi.org/10.1016/j.jplph.2008.01.002

Hall JL (2002) Cellular mechanisms for heavy metal detoxification and tolerance. J Exp Bot 53 (366):1–11

Harper JF, Harmo A (2005) Plants, symbiosis and parasites: a calcium signalling connection. Nat Rev 6:555–567

Hartley SE, Gange AC (2009) Impacts of plant symbiotic fungi on insect herbivores: mutualism in a multitrophic context. Annu Rev Entomol 54:323–342

Hashem A, Abd Allah EF, Alqarawi AA, Al Huqail AA, Egamberdieva D, Wirth S (2016) Alleviation of cadmium stress in Solanum lycopersicum L. by arbuscular mycorrhizal fungi via induction of acquired systemic tolerance. Saudi J Biol Sci 23:272–281. https://doi.org/10.1016/j.sjbs.2015.11.002

Hatata MM, Abdel-Aal EA (2008) Oxidative stress and antioxidant defense mechanisms in response to cadmium treatments. Am Eurasian J Agric Environ Sci 4(6):655–669

Heuer B, Nadler A (1995) Growth, development and yield of potatoes under salinity and water deficit. Aust J Agric Res 46:1477–1486

Hossain MA, Piyatida P, da Silva JAT, Fujita M (2012) Molecular mechanism of heavy metal toxicity and tolerance in plants: central role of glutathione in detoxification of reactive oxygen species and methylglyoxal and in heavy metal chelation. J Bot 2012:872875

Hui F, Liu J, Gao Q, Lou B (2015) Piriformospora indica confers cadmium tolerance in Nicotiana tabacum. J Environ Sci 37:184–191. https://doi.org/10.1016/j.jes.2015.06.005

Jaleel CA, Manivannan P, Lakshmanan GMA, Gomathinayagam M, Panneerselvam R (2008) Alterations in morphological parameters and photosynthetic pigment responses of *Catharanthus roseus* under soil water deficits. Colloids Surf B Biointerfaces 61:298–303

Jogawat A, Saha S, Bakshi M, Dayaman V, Kumar M, Dua M, Varma A, Oelmuller R, Tuteja N, Johri AK (2013) Piriformospora indica rescues growth diminution of rice seedlings during high salt stress. Plant Signal Behav 8(10):e26891. https://doi.org/10.4161/psb.26891

Johnson J, Oelmüller R (2009) Mutualism or parasitism: life in an unstable continuum. What can we learn from the mutualistic interaction between Piriformospora indica and Arabidopsis thaliana. Endocyt Cell Res 19:81–111

Kiani SP, Maury P, Sarrafi A, Grieu P (2008) QTL analysis of chlorophyll fluorescence parameters in sunflower (*Helianthus annuus* L.) under well-watered and water-stressed conditions. Plant Sci 175:565–573

Kogel KH, Franken P, Huckelhoven R (2006) Endophyte or parasite-what decides? Curr Opin Plant Biol 9:358–363. https://doi.org/10.1016/j.pbi.2006.05.001

Kowalchuk GA, Yergeau E, Leveau JHJ, Sessitch A, Bailey M (2010) Plant-associated microbial communities. In: Liu W-T, Jansson JK (eds) Environmental molecular microbiology. Caister Academic Press, Norwich, FC, pp 133–147

Kusaka M, Ohta M, Fujimura T (2005) Contribution of inorganic components to osmotic adjustment and leaf folding for drought tolerance in pearl millet. Physiol Plant 125:474–489

Lakshmanan V, Selvaraj G, Bais HP (2014) Functional soil microbiome: below ground solutions to an aboveground problem. Plant Physiol 166:689–700

Lee YC, Johnson JM, Chien CT, Sun C, Cai D, Lou B, Oelmüller R, Yeh KW (2011) Growth promotion of Chinese cabbage and Arabidopsis by Piriformospora indica is not stimulated by mycelium-synthesized auxin. Mol Plant-Microbe Interact 24:421–431. https://doi.org/10.1094/mpmi-05-10-0110

Li H, Luo N, Zhang LJ, Zhao HM, Li YW, Cai QY, Wong MH, Mo CH (2016) Do arbuscular mycorrhizal fungi affect cadmium uptake kinetics, subcellular distribution and chemical forms in rice? Sci Total Environ. https://doi.org/10.1016/j.scitotenv.2016.07.124

Lombardi L, Sebastiani L (2005) Copper toxicity in *Prunus cerasifera*: growth and antioxidant enzymes responses of in vitro grown plants. Plant Sci 168(3):797–802

Mahajan S, Tuteja N (2005) Cold, salinity and drought stresses: an overview. Arch Biochem Biophys 444:139–158

Manivannan P, Jaleel CA, Kishorekumar A, Sankar B, Somasundaram R, Sridharan R, Panneerselvam R (2007) Changes in antioxidant metabolism of *Vigna unguiculata* (L.) Walp. by propiconazole under water deficit stress. Colloids Surf B Biointerfaces 57:69–74

Manivannan P, Jaleel CA, Somasundaram R, Panneerselvam R (2008) Osmoregulation and antioxidant metabolism in drought stressed Helianthus annuus under triadimefon drenching. Comp Rend Biol 331:418–425

Massacci A, Nabiev SM, Pietrosanti L, Nematov SK, Chernikova TN, Thor K, Leipner J (2008) Response of the photosynthetic apparatus of cotton (*Gossypium hirsutum*) to the onset of drought stress under field conditions studied by gas-exchange analysis and chlorophyll fluorescence imaging. Plant Physiol Biochem 46:189–195

Maxwell K, Johnson GN (2000) Chlorophyll fluorescence – a practical guide. J Exp Bot 51:659–668

Mendes R, Kruijt M, de Bruijn I, Dekkers E, van der Voort M et al (2011) Deciphering the rhizosphere microbiome for disease-suppressive bacteria. Science 332:1097–1100

Mendes R, Garbeva P, Raaijmakers JM (2013) The rhizosphere microbiome: significance of plant beneficial, plant pathogenic, and human pathogenic microorganisms. FEMS Microbiol Rev 37:634–663

Miller G, Suzuki N, Ciftci-Yilmaz S, Mittler R (2010) Reactive oxygen species homeostasis and signalling during drought and salinity stresses. Plant Cell Environ 33:453–467. https://doi.org/10.1111/j.1365-3040.2009.02041.x

Mine A, Sato M, Tsuda K (2014) Toward a systems understanding of plant–microbe interactions. Front Plant Sci 5:423. https://doi.org/10.3389/fpls.2014.00423

Mo Y, Wang Y, Yang R, Zheng J, Liu C, Li H et al (2016) Regulation of plant growth, photosynthesis, antioxidation and osmosis by an arbuscular mycorrhizal fungus in watermelon seedlings under well-watered and drought conditions. Front Plant Sci 7:644. https://doi.org/10.3389/fpls.2016.00644

Mohd S, Shukla J, Kushwaha AS, Mandrah K, Shankar J, Arjaria N, Saxena PN, Narayan R, Roy SK, Kumar M (2017) Endophytic fungi Piriformospora indica mediated protection of host from arsenic toxicity. Front Microbiol 6:754

Monneveux P, Sánchez C, Beck D, Edmeades GO (2006) Drought tolerance improvement in tropical maize source populations: evidence of progress. Crop Sci 46:180–191

Muthukumar T, Udaiyan K (2010) Growth response and nutrient utilization of Casuarina equisetifolia seedlings inoculated with bioinoculants under tropical nursery conditions. New For 40:101–118. https://doi.org/10.1007/s11056-009-9186-z

Nanda R, Agrawal V (2018) Piriformospora indica, an excellent system for heavy metal sequestration and amelioration of oxidative stress and DNA damage in Cassia angustifolia Vahl under copper stress. Ecotoxicol Environ Saf 156:409–419

Osakabe Y, Osakabe K, Shinozaki K, Tran LS (2014) Response of plants to water stress. Front Plant Sci 5:86. https://doi.org/10.3389/fpls.2014.00086

Panda SK, Choudhury S, Patra HK (2016) Heavy-metal-induced oxidative stress in plants: physiological and molecular perspectives. In: Tuteja N, Gill SS (eds) Abiotic stress response in plants. Wiley-VCH Verlag GmbH & Co. KGaA, Weinheim, pp 219–232

Paunov M, Koleva L, Vassilev A, Vangronsveld J, Goltsev V (2018) Effects of different metals on photosynthesis: cadmium and zinc affect chlorophyll fluorescence in Durum Wheat. Int J Mol Sci 19(3):787

Petropoulos SA, Daferera D, Passam HC, Polissiou MG (2008) The effect of water deficit stress on the growth, yield and composition of essential oils of parsley. Sci Hortic 115:393–397

Philippot L, Raaijmakers JM, Lemanceau P, van der Putten WH (2013) Going back to the roots: the microbial ecology of the rhizosphere. Nat Rev Microbiol 11:789–799

Porcel R, Aroca R, Ruiz-Lozano JM (2012) Salinity stress alleviation using arbuscular mycorrhizal fungi. A review. Agron Sust Dev 32:181–200. https://doi.org/10.1007/s13593-011-0029-x

Ramachandra Reddy A, Chaitanya KV, Vivekanandan M (2004) Drought-induced responses of photosynthesis and antioxidant metabolism in higher plants. J Plant Physiol 161:1189–1202. https://doi.org/10.1016/j.jplph.2004.01.013

Rascio N, Navari-Izzo F (2011) Heavymetal hyperaccumulating plants: how and why do they do it? And what makes them so interesting? Plant Sci 180(2):169–181

Rellan-Alvarez RC, Ortega-Villasante A, Alvarez-Fernandez A, Campo FFD, Hernandez LE (2006) Stress responses of *Zea mays* to cadmium and mercury. Plant Soil 279(1–2):41–50

Saddique MAB, Ali Z, Khan AS et al (2018) Inoculation with the endophyte *Piriformospora indica* significantly affects mechanisms involved in osmotic stress in rice. Rice 11:34. https://doi.org/10.1186/s12284-018-0226-1

Sakuma Y, Maruyama K, Osakabe Y, Qin F, Seki M, Shinozaki K, Yamaguchi-Shinozaki K (2006) Functional analysis of an Arabidopsis transcription factor, DREB2A, involved in drought-responsive gene expression. Plant Cell 18:1292–1309

Sanders IR (2011) Mycorrhizal symbioses: how to be seen as a good fungus. Curr Biol 21:R550–R552. https://doi.org/10.1016/j.cub.2011.06.022

Sankar B, Jaleel CA, Manivannan P, Kishorekumar A, Somasundaram R, Panneerselvam R (2007) Effect of paclobutrazol on water stress amelioration through antioxidants and free radical scavenging enzymes in *Arachis hypogaea* L. Colloids Surf B: Biointerfaces 60:229–235

Sarma MVRK, Kumar V, Saharan K, Srivastava R, Sharma AK, Prakash A, Sahai V, Bisaria VS (2011) Application of inorganic carrier-based formulations of fluorescent pseudomonads and Piriformospora indica on tomato plants and evaluation of their efficacy. J Appl Microbiol 111:456–466. https://doi.org/10.1111/j.1365-2672.2011.05062.x

Scandalios JG (2005) Oxidative stress: molecular perception and transduction of signals triggering antioxidant gene defenses. Braz J Med Biol Res 38:995–1014

Schouteden N, De Waele D, Panis B, Vos CM (2015) Arbuscular mycorrhizal fungi for the biocontrol of plant-parasitic nematodes: a review of the mechanisms involved. Front Microbiol 6:1280. https://doi.org/10.3389/fmicb.2015.01280

Schutzendubel A, Polle A (2002) Plant responses to abiotic stresses: heavy metal-induced oxidative stress and protection by mycorrhization. J Exp Bot 53(372):1351–1365

Serfling A, Wirsel SGR, Lind V, Deising HB (2007) Performance of the biocontrol fungus Piriformospora indica on wheat under greenhouse and field conditions. Phytopathology 97(4):523–531

Shahabivand S, Aliloo AA (2016) Piriformospora indica promotes growth and antioxidant activities of wheat plant under cadmium stress. Yuzuncu Yil Univ J Agric Sci 26:333–340

Shahabivand S, Maivan HZ, Goltapeh EM, Sharifi M, Aliloo AA (2012) The effects of root endophyte and arbuscular mycorrhizal fungi on growth and cadmium accumulation in wheat under cadmium toxicity. Plant Physiol Biochem 60:53–58. https://doi.org/10.1016/j.plaphy.2012.07.018

Shahabivand S, Parvaneh A, Aliloo AA (2017) Root endophytic fungus Piriformospora indica affected growth, cadmium partitioning and chlorophyll fluorescence of sunflower under cadmium toxicity. Ecotoxicol Environ Saf 145:496–502

Shao HB, Chu LY, Shao MA, Abdul Jaleel C, Hong-Mei M (2008) Higher plant antioxidants and redox signaling under environmental stresses. C R Biol 331:433–441

Sharma P, Jha AB, Dubey RS, Pessarakli M (2012) Reactive oxygen species, oxidative damage, and antioxidative defense mechanism in plants under stressful conditions. J Bot 2012:217037

Sherameti I, Tripathi S, Varma A, Oelmuller R (2008) The root colonizing endophyte Pirifomospora indica confers drought tolerance in Arabidopsis by stimulating the expression of drought stress-related genes in leaves. Mol Plant-Microbe Interact 21:799–807

Singh LP, Gill SS, Tuteja N (2011) Unraveling the role of fungal symbionts in plant abiotic stress tolerance. Plant Signal Behav 6:175–191. https://doi.org/10.4161/psb.6.2.14146

Song WE, Chen SB, Liu JF, Li CH, Song NN, Ning LI, Bin LI (2015) Variation of Cd concentration in various rice cultivars and derivation of cadmium toxicity thresholds for paddy soil by species-sensitivity distribution. J Integr Agric 14:1845–1854

Song X, Yue X, Chen W, Jiang H, Han Y, Li X (2019) Detection of cadmium risk to the photosynthetic performance of hybrid pennisetum. Front Plant Sci 10:798. https://doi.org/10.3389/fpls.2019.00798

Specht JE, Chase K, Macrander M, Graef GL, Chung J, Markwell JP, Germann M, Orf JH, Lark KG (2001) Soybean response to water. A QTL analysis of drought tolerance. Crop Sci 41:493–509

Stein E, Molitor A, Kogel KH, Waller F (2008) Systemic resistance in Arabidopsis conferred by the mycorrhizal fungus Piriformospora indica requires jasmonic acid signaling and the cytoplasmic function of NPR1. Plant Cell Physiol 49:1747–1751

Sun C, Johnson MJ, Cai D, Sherameti I, Oelmullera R, Lou B (2010) Piriformospora indica confers drought tolerance in Chinese cabbage leaves by stimulating antioxidant enzymes, the expression of drought related genes and the plastid-localized CAS protein. J Plant Physiol 167:1009–1017

Sytar O, Kumar A, Latowski D, Kuczynska P, Strzałka K, Prasad MNV (2013) Heavymetal-induced oxidative damage, defense reactions, and detoxification mechanisms in plants. Acta Physiol Plant 35(4):985–999

Tahat MM, Sijam K (2012) Mycorrhizal fungi and abiotic environmental conditions relationship. Res J Environ Sci 6:125–133. https://doi.org/10.3923/rjes.2012.125.133

Takahashi S, Murata N (2008) How do environmental stresses accelerate photoinhibition? Trends Plant Sci 13:178–182. https://doi.org/10.1016/j.tplants.2008.01.005

Tangahu BV, Abdullah SRS, Basri H, Idris M, Anuar N, Mukhlisin M (2011) A review on heavy metals (As, Pb, and Hg) uptake by plants through phytoremediation. Int J Chem Eng 2011:939161

Tripathi DK, Singh VP, Kumar D, Chauhan DK (2012) Rice seedlings under cadmium stress: effect of silicon on growth, cadmium uptake, oxidative stress, antioxidant capacity and root and leaf structures. Chem Ecol 28:281–291. https://doi.org/10.1080/02757540.2011.644789

Tripathy BC, Oelmüller R (2012) Reactive oxygen species generation and signaling in plants. Plant Signal Behav 7:1621–1633. https://doi.org/10.4161/psb.22455

Tsai H-J, Shao K-H, Chan M-T, Cheng C-P, Yeh K-W, Oelmüller R, Wang S-J (2020) *Piriformospora indica* symbiosis improves water stress tolerance of rice through regulating stomata behavior and ROS scavenging systems. Plant Signal Behav. https://doi.org/10.1080/15592324.2020.1722447

Tuteja N (2010) Cold, salinity, and drought stress. In: Hirt H (ed) Plant stress biology: from genomics to systems biology. WILEY-VCH Verlag Gmbh & Co. KGaA, Weinheim, pp 137–160

Udvardi M, Poole PS (2013) Transport and metabolism in legume-rhizobia symbioses. Annu Rev Plant Biol 64:781–805

Unnikumar KR, Sowjanya SK, Varma A (2013) Piriformospora indica: a versatile root endophytic symbiont. Symbiosis 60:107–113. https://doi.org/10.1007/s13199-013-0246-y

Vadassery J, Tripathi S, Prasad R, Varma A, Oelmuller R (2009a) Monodehydroascorbate reductase 2 and dehydroascorbate reductase 5 are crucial for a mutualistic interaction between Piriformospora indica and Arabidopsis. J Plant Physiol 166:1263–1274

Vadassery J, Ranf S, Drzewiecki C, Mithofer A, Mazars C, Scheel D, Lee J, Oelmuller R (2009b) A cell wall extract from the endophytic fungus Piriformospora indica promotes growth of Arabidopsis seedlings and induces intracellular calcium elevation in roots. Plant J 59:193–206

Varma A, Verma S, Sahay N, Butehorn B, Franken P (1999) Piriformospora indica, a cultivable plant-growth-promoting root endophyte. Appl Environ Microbiol 65:2741–2744. https://doi.org/10.1128/AEM.65.6.2741-2744.1999

Varma A, Bakshi M, Lou B, Hartmann A, Oelmueller R (2012) Piriformospora indica: a novel plant growth-promoting mycorrhizal fungus. Agric Res 1:117–131. https://doi.org/10.1007/s40003-012-0019-5

Waller F, Achatz B, Baltruschat H, Fodor J, Becker K, Fischer M, Heier T, Hückelhoven R, Neumann C, Wettstein D, Franken P, Kogel KH (2005) The endophytic fungus Piriformospora indica reprograms barley to salt-stress tolerance, disease resistance, and higher yield. Proc Natl Acad Sci U S A 102(38):13386–13391. https://doi.org/10.1073/pnas.0504423102

Webber M, Barnett J, Finlayson B, Wang M (2006) Pricing China's irrigation water. working paper, School of Anthropology, Geography and Environmental Studies, The University of Melbourne, Victoria, Australia

Wuana RA, Okieimen FE (2011) Heavy metals in contaminated soils: a review of sources, chemistry, risks and best available strategies for remediation. ISRN Ecol 2011:402647

Xavier LJC, Boyetchko SM (2004) Arbuscular mycorrhizal fungi in plants disease control. In: Arora DK (ed) Fungal biotechnology in agricultural, food, and environmental applications. Marcel Dekker, New York, pp 183–194

Xu J, Yin HX, Li X (2009) Protective effects of proline against cadmium toxicity in micropropagated hyperaccumulator, Solanum nigrum L. Plant Cell Rep 28:325–333

Xu L, Wang A, Wang J, Wei Q, Zhang W (2017) Piriformospora indica confers drought tolerance on Zea mays L. through enhancement of antioxidant activity and expression of drought-related genes. Crop J 5:251–258. https://doi.org/10.1016/j.cj.2016.10.002

Yruela I (2009) Copper in plants: acquisition, transport and interactions. Funct Plant Biol 36 (5):409–430

Zengin FK, Munzuroglu O (2005) Effects of some heavy metals on content of chlorophyll, proline and some antioxidant chemicals in bean (*Phaseolus vulgaris* L.) seedlings. Acta Biol Cracov Ser Bot 47(2):157–164

Zhang W, Wang C, Qin C, Wood T, Olafsdottir G, Welti R, Wang X (2003) The oleate-stimulated phospholipase D, PLDd, and phosphatidic acid decrease H2O2-induced cell death in Arabidopsis. Plant Cell 15:2285–2295

Zhang M, Duan L, Zhai Z, Li J, Tian X, Wang B, He Z, Li Z (2004) Effects of plant growth regulators on water deficit-induced yield loss in soybean. In: Proceedings of the 4th international crop science congress, Brisbane, Australia

Zhang Y, Yang C, Li Y, Zhang N, Chen H, Zhao Q, Gao T, Guo H, Xie Q (2007) SDIR1 is a RING finger E3 ligase that positively regulates stress responsive abscisic acid signaling in Arabidopsis. Plant Cell 19:1912–1929

Zhang W, Wang J, Xu L, Wang A, Huang L, Du H, Qiu L, Oelmüller R (2017) Drought stress responses in maize are diminished by Piriformospora indica. Plant Signal Behav 13(1): e1414121. https://doi.org/10.1080/15592324.2017.1414121

Zhou B, Yao W, Wang S, Wang X, Jiang T (2014) The metallothionein gene, *TaMT3*, from *Tamarix androssowii* confers Cd2+ tolerance in Tobacco. Int J Mol Sci 15(6):10398–10409

Role of Rhizosphere and Endophytic Microbes in Alleviation of Biotic and Abiotic Stress in Plants

9

N. D. Sooriyaarachchi, M. C. M. Zakeel, M. I. S. Safeena, and K. M. R. D. Abhayapala

Abstract

The role of rhizosphere and endophytic microbes in agriculture is substantial in many ways such as adapting to unfavourable environmental conditions (biotic and abiotic stresses), enhancing the efficiency of phytoremediation, promoting plant growth, alleviating metal stress, and reducing metal phytotoxicity. The molecular mechanisms behind stress alleviation in host plants through the process of beneficial interaction by the rhizosphere and endophytic microorganisms are crucial in the management of stress conditions. At present, many biotechnological approaches are in use to enhance the effectiveness of rhizosphere and endophytic microbes.

N. D. Sooriyaarachchi
Flemings Nurseries QLD, Woombye, QLD, Australia

M. C. M. Zakeel (✉)
Department of Plant Science, Faculty of Agriculture, Rajarata University of Sri Lanka, Puliyankulama, Anuradhapura, Sri Lanka

Centre for Horticultural Science, Queensland Alliance for Agriculture and Food Innovation, The University of Queensland, Brisbane, QLD, Australia
e-mail: zakeel@agri.rjt.ac.lk

M. I. S. Safeena
Department of Biological Science, Faculty of Applied Science, South Eastern University of Sri Lanka, Sammanthurai, Sri Lanka

K. M. R. D. Abhayapala
Department of Biosystems Technology, Faculty of Technology, University of Sri Jayewardenepura, Homagama, Sri Lanka

R. Soni et al. (eds.), *Microbiological Activity for Soil and Plant Health Management*, https://doi.org/10.1007/978-981-16-2922-8_9

Keywords

Endophytes · Plant–microbe interactions · Induced systemic resistance · System acquired resistance · Microbe-induced plant growth · Stress alleviation

9.1 Introduction

For many million years, plants have been playing a vital role in the ecosystem. When plants perform their ecological services, they have to overcome many challenges. Among many of these challenges, biotic and abiotic stress conditions that plants have to withstand are confronting. For the survival in these challenging environments, plants have to adjust their growth and development to alleviate the impact of stress. Therefore, plants, in addition to their own tolerance mechanisms against stress conditions, can develop alliances with members of their ecosystem to flourish in their natural environment (Afzal et al. 2019). Plants always show interactions with a wide range of microorganisms. It has been well researched and proved that plants are supported by a wide range of microorganisms living in and around plants (Zakeel 2015; Santoyo et al. 2016).

A diverse range of microorganisms in their natural ecosystem supports almost every plant on the earth and these microbes colonize inside tissues and on the surfaces of different plant parts such as roots, stem, leaves, seeds and fruits (Zakeel 2015). Beneficial microbes associated with plants play a vital role in alleviating biotic and abiotic stresses and help plants in growth, development and survival under adverse environmental conditions (Abedinzadeh et al. 2019). Interactions between plants and microbes occur through symbiotic, endophytic or associative processes (Souza et al. 2015). Microorganisms colonized in rhizosphere are known as rhizosphere microbes and those colonized in plant internal tissues are endophytic microbes (Fig. 9.1A; Etesami and Maheshwari 2018). The diversity of these microbes is governed by various environmental factors and the host plant they are associated with (Dastogeer et al. 2020). Both endophytic and rhizosphere microbes share numerous characteristics that are vital for host plant growth promotion (Afzal et al. 2019).

One of the challenging problems in agricultural production in the recent history is water shortage (Abedinzadeh et al. 2019). In this context, the use of microbes to reduce water stress has become an important focus for many scientists around the world and thus far, many plant-microbe associations have been revealed to have beneficial effects in reducing water stress. For example, mycorrhizal associations can be used in crop production to alleviate water stress and drought conditions as they are more effective in scavenging for water in micropores and the fungal hyphae can retain water efficiently in addition to various hormonal and nutritional mechanisms (Augé 2001; Lakshmanan et al. 2017). Heavy metal toxicity is another problem that has serious impacts on agriculture (Ma et al. 2016) and some researches proved that rhizosphere and endophytic microbes, are helpful in alleviating the heavy metal toxicity stresses in plants (Etesami and Maheshwari 2018). Moreover, some

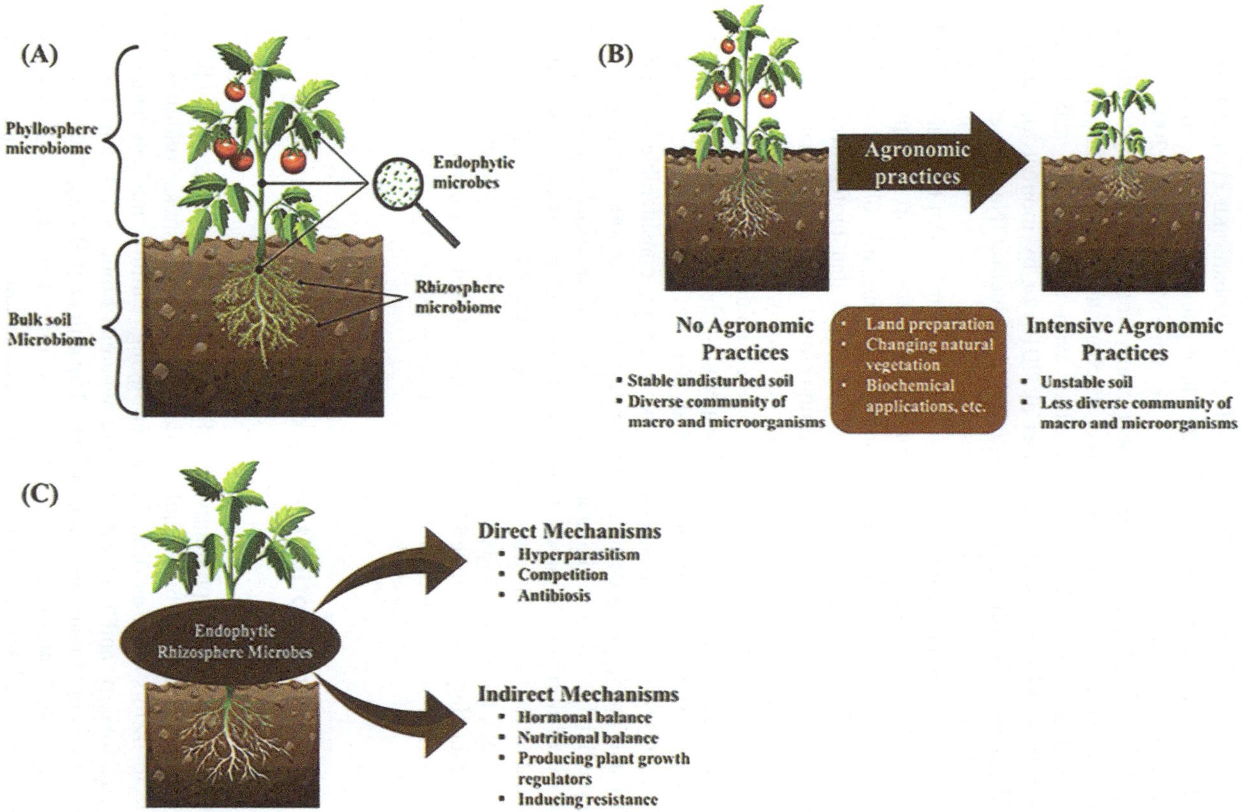

Fig. 9.1 Plant-associated microbiomes and their interactions with hosts. (**a**) Colonization of rhizosphere and endophytic microbes; (**b**) Dynamics of composition and diversity of rhizosphere and endophytic microbes with changes of environmental conditions; (**c**) Stress alleviation by rhizosphere and endophytic microbes via direct and indirect methods

beneficial microorganisms have shown their ability to transform nutrients into plant available forms and assist in boosting plant tolerance to environmental stress (Bardi and Malusà 2012). Some microorganisms are capable of producing of phytohormones and make direct influence of plant (Abbamondi et al. 2016). For example, *Pseudomonas spp.* associated with wheat plants are capable of secreting indole acetic acid (IAA) and induce plant growth (Mishra et al. 2008). Some rhizosphere and endophytic microbes protect plants from phytopathogens (Kumar and Verma 2018). It can be by using direct mechanisms such as competition for nutrients, production of secondary metabolites with antimicrobial activity and induced systemic resistance (Abbamondi et al. 2016) or by indirect mechanisms such as siderophore, phytohormone and antibiotics production (Nadeem et al. 2015; Złoch et al. 2016).

The quality of agricultural produce is another important aspect that needs significant consideration particularly when the products are for export or supermarkets. Recent studies have underlined the importance of using microbes in improving product quality. For example, a recent study showed that biofertilizer enriched with *Trichoderma* enhances production and nutritional quality of tomatoes (Molla et al. 2012). Some *Trichoderma* species are known to produce secondary metabolites, which are important for plant growth regulation. Further, they produce organic acids that decrease soil pH which permits phosphate solubilization and increase micronutrient availability (Molla et al. 2012).

Land degradation and climate change have received considerable attention in recent past and the importance of increasing agricultural production in poor land under unpredictable climate is challenging. In consequence, maintaining the sustainability of agricultural production systems is vital in order to meet future food demand to feed the ever-increasing global population (Yadav et al. 2018). To achieve this goal, the understanding of plant–microbe relationship is indispensable. Proper understanding of the underlying mechanisms that these microorganisms are using to associate with plants is important to get maximum benefits from these associations.

9.2 Biotic and Abiotic Stress and Their Impacts on Crop Production

Plant growth, development, productivity and their distribution across deferent types of environments affects by biotic and abiotic stress. Abiotic stress conditions may occur due to factors such as drought, waterlogging, salinity, nutrient deficiency, heavy metal toxicity and herbicide toxicity whereas biotic stress conditions are mainly caused by phytopathogens and insects (Hussain et al. 2018). Plant can suffer from more than one stress conditions at a given time and more often one stress condition can make plants susceptible to other stress conditions. For example, a plant suffering from abiotic stress can easily subject to biotic stress compared to normal healthy plant and occasionally two stress conditions can impair and share same properties (Pandey et al. 2012). All stress conditions pose serious threat to plant health, which in turn decreases agricultural production. Stress conditions can also affect soil physical, chemical and biological properties and disturb ecological balance.

Out of abiotic stress conditions, drought is the most serious factor that is associated with significant crop loss worldwide and experienced in many parts of the world (Lakshmanan et al. 2017). Low water availability can lead to a series of physiological and molecular changes in plants, causing severe loss in crop production in both quality and quantity. For example, water deficit can induce ethylene production in plants, causing chlorophyll degradation and subsequently inhibition of photosynthesis (Kumar and Verma 2018). Further, limited water availability can affect leaf water potential and stomatal opening, reduce cell size, decrease seed number, size and viability delay flowering and fruiting and promote leaf senescence (Lata et al. 2018). Moreover, drought can also induce the accumulation of free radicals that induce changes in membrane function, protein confirmation and lipid peroxidation leading to cell death (Kumar and Verma 2018).

Salinity is another important abiotic stress that occurs mainly due to accumulation of salt that is present in soil, water or addition of inorganic fertilizers. Insufficient precipitation can intensify salt accumulation around the root zones of plants causing severe problems to plant physiology. Soil salinity can also have complications on seed germination, embryo viability and damage cellular structure that restrains plant growth and inhibits photosynthesis by reducing leaf area, chlorophyll content and stomatal conductance, and also by increasing premature senescence (Pandey et al. 2012). Salinity can also impose ion toxicity and nutrient deficiency. It removes K^+, Ca^{2+}, NO_3^-, Fe^{2+}, $Zn^{2+}$and Mg^{2+} from rhizosphere and replaces them with Na^+ and Cl^-, thus making those ions unavailable for plants. Saline soils reduce phosphorus uptake by plants because phosphate ions are precipitated with calcium ions in saline soils (Shrivastava and Kumar 2015). At present, over 20% of agricultural soils are affected by soil salinity problem and this figure is estimated to reach approximately 50% by year 2050 (Pandey et al. 2012).

Heavy metal toxicity is another abiotic stress to plants. In fact, plants may need heavy metals in trace amounts for their growth (Emamverdian et al. 2015; Shahid et al. 2015), however, excessive amount can be toxic to plants and this toxicity is severe in acidic soils (Shahid et al. 2015; Pandey et al. 2016). Primarily, heavy metal toxicity can reduce seed production, viability and germination, cause stunted growth, chlorosis and wilting, reduce leaf area, fruit size and dry matter production, inhibit root growth, reduce antioxidant enzyme activities and reduce plant protein, amino acids, starch and sugar content (Wuana and Okieimen 2011; Shahid et al. 2015). Generation of reactive oxygen species (ROS), enzyme activation, damage to lipid and DNA, chromosomal aberration, cell injury and death are secondary effects of heavy metal toxicity in plants (Shahid et al. 2015). Moreover, heavy metals can replace essential cations from cation exchange sites and inhibit soil microbial activities that are imperative to organic matter decomposition, all of which lead to severe reduction in soil nutrients (Wuana and Okieimen 2011). Nutrient deficiency on its own is an abiotic stress condition that can affect plants in every stage of their life cycle. Urbanization, industrialization and intensive agriculture are the main reasons for the accumulation of heavy metals that are quite non-degradable in the natural environment (Ma et al. 2016).

Crop loss due to biotic stress caused by pests and diseases is also substantial. Among the biotic stresses, plant diseases caused by bacteria, fungi, viruses and nematodes as well as insect pest damages cause most extensive loss in crop productivity. These biological agents directly or indirectly depend on host plants for their habitat and nutritional needs and these plant–pathogen/insect interactions may sometimes end up with the death of the host plants. For example, *Phytophthora cinnamomi*, known as the biggest threat in avocado farming, is a fungus that depends on avocado roots for its nutrient (Anon 2020). This fungus makes the roots rot leading the tree to wilt, causing leaf drop, branch dieback and fruit drop, and ultimately the tree slowly dies (Anon 2020).

Despite being immobile and not having adaptive immune system, plants can resist both biotic and abiotic stresses by adapting efficient strategies. Among those defence mechanisms, their interactions with rhizosphere and endophytic microbes are vital. Application of organic matters, microbes and green remediation promised improvement of plant health and performance under biotic and abiotic stress conditions (Zakeel and Safeena 2019).

9.3 Diversity and Consortium of Rhizosphere and Endophytic Microbes

Microbes are an important group of living entities that have shown to develop beneficial interactions with a diverse range of plant species (Santoyo et al. 2016). These plant-associated microbes primarily benefit the host plants in various ways that mainly include plant growth promotion and tolerance to various stress conditions (Afzal et al. 2019). Microorganisms that colonize outside the host plants are known as epiphytes (on plant surfaces) or rhizosphere (the zone of soil or substrate surrounding a plant root) microbes (Compant et al. 2010) and that live inside plant tissues are known as endophytes (Hardoim et al. 2008).

Soil is known as a biologically active powerhouse as it is a rich source of microflora that comprise a far-fetched diversity of microbes including algae, actinomycetes, fungi, bacteria and cyanobacteria (Zakeel and Safeena 2019). Soil microflora contain a many fungal and bacterial species that can promote plant growth, thus collectively known as plant growth promoting microorganisms (PGPM) (Zakeel and Safeena 2019). Although soil is generally known as biological hotspot, rhizosphere is the interface where the soil, plant roots, microbes and soil fauna exhibit strong interactions (Bardi and Malusà 2012). The term "rhizosphere", originating from a Greek word "rhiza" (meaning root), was first coined in 1904 by the German agronomist and plant physiologist Lorenz Hiltner to designate the plant–root interface (Hiltner 1904, as cited in Hartmann et al. 2008). Rhizosphere can be classified into three zones, i.e. ectorhizosphere, rhizoplane and endorhizosphere, based on its relative proximity to the root (McNear Jr. 2013). Ectorhizosphere is the outermost zone extending from the root surface out into the soil around root, rhizoplane is the middle zone that includes root surface, particularly root epidermis and mucilage, and endorhizosphere is the innermost zone which includes the cortex and endodermis of the root (McNear 2013) (Bardi and Malusà 2012). It has been proven that the population density of microbes in rhizosphere is greater than that in the adjacent soil (Berendsen et al. 2012). Rhizosphere is the most dynamic interface on the earth where each gram of root contains about 10^{11} microbial cells that includes over 30,000 prokaryotic species (Si et al. 2018). The rhizosphere is composed of soil fraction that is found in close proximity to plant roots (Meena et al. 2017) and the thickness of the rhizosphere soil may range between less than one millimetre and several millimetres depending on the presence of root exudates (Nadeem et al. 2014).

The main reason for microbes being attracted toward rhizosphere is the plant root exudates that are rich in nutrients, minerals and various metabolites, and that accumulate in rhizosphere (Dimkpa et al. 2009; Meena et al. 2017). The interactions between microbes and host plants occur via complex exchanges between microorganisms and roots that are regulated by complex molecular signalling (Nadeem et al. 2014). These useful interactions have positive effects on crop production by means of significantly influence on plant growth and development (Nadeem et al. 2014). Root exudates contain readily available carbon source and energy particularly sugars, vitamins, organic compounds, auxins and flavonoids that support microbial growth, development and physiology (Somers et al. 2004; Nadeem et al. 2014).

Endophytes are also equally important for plants as they provide a variety of benefits to the host. Such benefits are highly indispensable when plants are challenged by stress conditions (Afzal et al. 2019). Endophytic microbes can be found in in different plant parts such as roots, stems, leaves, flowers, fruits and seeds (Miliute et al. 2015; Tamosiune et al. 2017). However, higher density of endophyte population is colonized in plant roots and other belowground tissues (Chi et al. 2005). Plant roots are the main entry point of the endophytic microbes from soil and serve as a base camp for colonization of other plant organs (Tamosiune et al. 2017). The interactions between host plants and endophytes are governed by compounds that are produced by the microorganisms and the host cells (Miliute et al. 2015). The endophytes help host plants for their growth and development by reshaping plant hormone production and increasing the availability of nutrients such as nitrogen and phosphorus (Glick 2012; Miliute et al. 2015; Afzal et al. 2019). Endophytic bacteria secrete a wide range of compounds including antibiotics and chitinase enzyme that can inhibit plant pathogens, thus act as biocontrol agents (Miliute et al. 2015). Evidence suggests that plants devoid of endophytic microbes are highly susceptible to environmental stress conditions and would be less able to deal with pathogens (Santoyo et al. 2016). Compared to rhizosphere bacteria, endophytes have the opportunity to always be in direct contact with plant cells as they live within plant tissues and therefore readily exert direct benefits to host plants (Santoyo et al. 2016).

9.4 Environmental and Host Influence on the Rhizosphere and Endophytic Microbes

Although rhizosphere and endophytic microbes are adapted to associate with specific plant genotypes, many studies have indicated that the community structure of these microbes is influenced by many factors and interactions including environment conditions, microbe–microbe interactions, plant–microbe interactions and agronomic practices (Fig. 9.1B; Dastogeer et al. 2020). Changes in the above factors and interactions lead to changes in plant phenotype and ultimately changes in the composition and diversity of microbial population associated with plants (Tardieu 2013).

9.4.1 Environmental Effects

The composition and diversity of microbes found in rhizosphere and endophytic associations of a given plant genotype vary depending on the hosts environment (Peiffer et al. 2013). Several studies have shown that environmental factors including soil parameters, temperature and rainfall may influence the microbial composition in both associations (Dastogeer et al. 2020). Among these factors, soil parameters such as soil type, pH, C:N ratio, soil carbon, available P and K, and water content are the major factor determining the rhizosphere community (Peiffer et al. 2013; Dastogeer et al. 2020).

9.4.2 Effects of Agronomic Practices

Human influences on ecosystem, which include land use and cultivation practices, are the most important causes leading to the loss of biodiversity in the environment. Land use intensity has significant effects on a range of environmental factors such as soil structure, the quality and quantity of nutrients and general biodiversity pattern (Birkhofer et al. 2012). Agronomic practices including land preparation, changing natural vegetation, application of agro-chemicals and irrigation influence the diversity and structure of soil microbes and that influence can be either positive or negative (Dastogeer et al. 2020). Tillage and irrigation practices generally change soil properties via soil compaction and mechanical destruction that may lead to the decline of soil microbial diversity (Tamosiune et al. 2017). Indiscriminate use of agro-chemicals also has serious impacts on the function and structure of soil microbial populations by inhibiting microbial growth and changing the structure of agricultural ecosystems (Pampulha and Oliveira 2006). Excessive use of synthetic fertilizers can inhibit the colonization of some endophytic microbes. For example, application of high levels of nitrogen fertilizers to sugarcane reduces the colonization ability of *Acetobacter diazotrophicus*, a nitrogen fixing endophytic bacterium (Fuentes-Ramírez et al. 1999). However, the use of balanced organic fertilizers may have positive effects on the diversity and metabolic activity of soil microbial community (Zhong et al. 2010). Intensive agricultural practices strongly influence the soil structure and nutrient availability that ultimately change the plant phenotype and may modulate plant-associated microbial communities (Estendorfer et al. 2017).

9.4.3 Influence of Host Plants

The contribution of plants to facilitate and maintain a biological system at the associative interface is crucial. The genotypic diversity of plants has significant influence on the composition and diversity of microbiomes at this interface (Meena et al. 2017). The rhizosphere consists of a soil microenvironment in close proximity to root region where the average count of microorganisms is higher than the rest of the bulk soil, which is an indicative of the influence of plants on rhizosphere microbes (Berendsen et al. 2012). Plant roots that are generally rich in various nutrients and metabolites is a key factor that attracts microorganisms to associate with plant tissues (Meena et al. 2017). Different plant species growing adjacent to one another can harbour distinct microbiomes, as their metabolite and nutritional composition are different from each other (Bouffaud et al. 2014). Rhizosphere microbial communities of different plant species growing on the same soil may often differ from each other (Viebahn et al. 2005; Berg et al. 2006; Berendsen et al. 2012), whereas same plant species can create similar communities in different soils (Berendsen et al. 2012). Sometimes even within the same species, different plant genotypes can develop distinct microbial communities in the rhizosphere, indicating that plants are able to determine the composition of the microbiome that they harbour in their rhizosphere (Micallef et al. 2009; Berendsen et al. 2012). This ability of plants is attributed to the active secretions of compounds that specifically stimulate or repress the individual species of the microbial community (Berendsen et al. 2012).

9.5 Role of Rhizosphere and Endophytic Microbes in Agriculture

There is plenty of evidence to support that rhizosphere and endophytic microbes support host plants to adapt to unfavourable environmental conditions, enhance the efficiency of phytoremediation, promote plant growth, alleviate metal stress and reduce metal phytotoxicity (Ma et al. 2016). The mechanisms by which rhizosphere and endophytic microbes promote plant growth and development depend on the microbes and host plants and a better understanding of these mechanisms is indispensable in sustainable agriculture in order to maximize the efficacy of agroecosystem. These mechanisms can be either direct or indirect and are summarized in Table 9.1.

By increasing root surface area, the rhizosphere and endophytic microbes have the ability to improve soil characteristics and encourage plant development under normal as well as stress conditions (Sosa-Hernández et al. 2019). For instance,

Table 9.1 Details of plant growth promotion by plant-associated microbes in selected plant species

Microbe	Host plant	Plant growth promoting mechanism	Reference
Pseudomonas spp.	Peanut	production of IAA and ACC deaminase, siderophore production and phosphate solubilization	Sharma et al. (2016)
Azospirillum lipoferu	Rice, maize and wheat	nitrogen fixation and secretion of phytohormone	Wisniewski-Dye et al. (2011)
Pseudomonas fragi	Wheat	phosphate solubilization, and IAA and HCN production	Selvakumar et al. (2009)
Pseudomonas spp.	Wheat	IAA production, tricalcium phosphate solubilization, HCN and siderophore production and inhibition of several phytopathogenic fungi	Mishra et al. (2008)
Pseudomonas saponiphila	Capsicum	production of IAA, solubilizing phosphate, siderophore formation and production of HCN antibiotics	Wu et al. (2016)
Azospirillum spp.	Rice	nitrogen fixation and phytohormone secretion	Kaneko et al. (2010)
Gluconacetobacter diazotrophicus	Sugarcane, rice, coffee and tea	nitrogen fixation and auxin synthesis	Bertalan et al. (2009)
Burkholderia *phytofirmans*	Potato, tomato, maize, barley, onion, canola and grapevine	IAA synthesis and ACC deaminase	Weilharter et al. (2011)

mycorrhizal fungi can extend their hyphae deep into areas of soil which roots cannot reach and promote nutrient uptake, water availability and withstand stress conditions like salinity (Lakshmanan et al. 2017). Moreover, mycorrhizal fungi can grow in subsoils that are characterized by higher bulk densities mainly due to compaction, reduced pore spaces and thus lower oxygen levels, and contribute to the reclamation of the soil to support plant growth (Lynch and Wojciechowski 2015; Weil and Brady 2016).

9.5.1 Plant Growth Promotion by Increasing Nutrient Availability

Nutrient availability is crucial for plant growth and development. Rhizosphere and endophytic microbes increase nutrient availability via direct or indirect means (Shameer and Prasad 2018). Direct contribution to nutrient availability includes nitrogen (N) fixation, phosphorus (P) and potassium (K) solubilisation, etc., and indirect methods include displacement of sorption equilibrium that leads to increased net transfer of nutrients into soil solution and transformation of nutrients among different pools (Bardi and Malusà 2012). Many rhizosphere and endophytic microbes are known to fix atmospheric N, solubilise P and K, and make many other macro and microelements available for associated host plants (Sosa-Hernández et al. 2019).

Nitrogen is one of the key nutrients required for plant growth and therefore the contribution of rhizosphere and endophytic microbes in supplying N for host plant is remarkable. A variety of N fixing bacteria such as *Arthrobacter, Azoarcus, Azospirillum, Azotobacter, Bacillus, Rhizobium*, etc. have been isolated from the rhizosphere of various crops (Dimkpa et al. 2009). Some of the above microbes establish endophytic or symbiotic associations with host plants while others are free-living or non-symbiotic (Gouda et al. 2018). The relationship between legumes and *Rhizobium* spp. is a well-known symbiosis association that fixes atmospheric N to support the growth of legumes in N poor environments (Franche et al. 2009).

Phosphate is another important nutrient for plant growth and the role of mycorrhizal symbiotic association in relation to P absorption is highly researched. Although soil is generally rich in P, it is often unavailable for absorption by plants because they are mostly present in the form of precipitates with iron, aluminium and calcium or fixed in soil as organic or insoluble mineral complex (Rengel and Marschner 2005). Plant root association with mycorrhizal fungi generally known as arbuscular mycorrhizae (AM) converts the above unavailable P into available forms for plants to absorb either by expanding fungal hyphae into subsoils and absorbing precipitated phosphorus (Wang et al. 2017) or by secreting exudates that can solubilise insoluble mineral form or organic complex of P (Tawaraya et al. 2006). In addition to mycorrhizal symbionts, many other symbiotic and non-symbiotic microbes play vital roles in making P available for plant growth. For example, bacterial species such as *Pseudomonas* and *Bacillus* are very effective in solubilising P from precipitated inorganic sources by producing several different organic acids (Bardi and Malusà 2012). Other macro and microelements are also important for plant growth and development. The contribution of rhizosphere and endophytic microbes to increase the availability of nutrients such as K, Ca, Mg, Mn, Cu and Zn is highly appreciated in agriculture (Mirza et al. 2001; Smith and Smith 2011; Miliute et al. 2015; Sosa-Hernández et al. 2019).

9.5.2 Plant Growth Promotion by Hormone Production

Phytohormones produced by plants play an important role in their growth and development (Porcel et al. 2014). In addition, rhizosphere and endophytic microbes, particularly plant growth promoting rhizobacteria (PGPR) have the ability to synthesize plant hormones that stimulate plant growth and development under stress conditions (Kumar and Verma 2018). Examples for hormones produced by PGPR include IAA, abscisic acid (ABA), 1-aminocyclopropane-1-carboxylic acid (ACC) deaminase, chitinase and glucanase. During drought stress, IAA regulates cell division, shoot growth, adventitious and lateral root differentiation, and vascular tissue differentiation (Goswami et al. 2015). The ABA in plants helps transcription of drought related genes and increases root hydraulic conductivity (Jiang et al. 2013).

9.5.3 Defend Plants Against Biotic Stress

Rhizosphere and endophytic microbes are able to alleviate biotic stress by either direct or indirect antagonistic methods (Fig. 9.1C; Shinwari et al. 2019). The direct mechanism is by producing certain metabolites which reduce pathogen population around host plants, whereas the indirect mechanism helps improve crop resistance against phytopathogens (Pundir and Jain 2015). Direct mechanisms can be classified into three as hyperparasitism, competition and antibiosis (Shinwari et al. 2019).

Hyperparasitism involves the killing of the disease-causing microbes or their propagules by directly attacking them. *Trichoderma* is one of the most popular examples for hyperparasitism in which the fungus can release toxic endochitinases, which can degrade the cell wall of the target fungal pathogen (Gautam and Avasthi 2019). Competition for resources (nutrient, water and space) is another mechanism that beneficial microbes use to defend against pathogens (Köhl et al. 2019). Soil and surfaces of living plants or plant tissues are usually the environments with limited nutrients for microbes (Shinwari et al. 2019). Beneficial microbes compete against pathogens for nutrients and thereby suppress diseases. For example, siderophore produced by these beneficial microbes in rhizosphere reduce iron availability to certain pathogens and thus reduce their growth (Santoro et al. 2015; Złoch et al. 2016). The third mechanisms, antibiosis, is by producing some secondary metabolites known as antibiotics that are detrimental to the growth or metabolic activities of plant pathogens. Antibiotics can also inhibit spore germination and hyphal growth of pathogens (Köhl et al. 2019).

9.5.4 Increase Abiotic Stress Tolerance in Plants

Abiotic stress is the stress condition experienced by plants from certain environmental factors such as water, temperature, salts, pH and essential nutrients. Plant growth, development and survival under stress conditions can be improved by the application of certain stress tolerant plant growth promoting microbes (PGPM) and AM fungi (Nadeem et al. 2014). Rhizosphere and endophytic microbes use direct or indirect approaches for the promotion of plant growth and development under stress conditions (Kumar and Verma 2018). For example, PGPM promotes plant growth by regulating hormonal and nutritional balance and producing plant growth (Kumar and Verma 2018).

Drought, being a major challenge to global agriculture, limits plant growth and productivity. Some drought tolerant microbes are able to enhance plant growth and development under water deficit conditions using direct or indirect mechanisms (Miliute et al. 2015; Lakshmanan et al. 2017). For example, PGPR regulate physiology of plants to withstand drought stress by producing important growth regulators such as ABA, which enhances drought stress resistance of plants via regulating root hydraulic conductivity and transcription of drought related genes (Jiang et al. 2013). As an indirect mechanism, mycorrhizal fungi extend their hyphae to deeper levels of soil where plant roots hardly reach and scavenge for water to make it available for plants (Lakshmanan et al. 2017; Wang et al. 2017).

Salinity, heavy metal toxicity and toxic exudates from weeds are other important stress conditions that challenge agricultural production. Endophytic and rhizosphere bacteria have also been associated with alleviating salinity stress of many plant species via a variety of mechanisms including phytohormones production, nitrogen fixation, phosphate solubilization, siderophore production and enhanced nutrient uptake (Shinwari et al. 2019). Some PGPR that are capable of metal biosorption, a phenomenon that binds metals to negatively charged groups of bacteria, can alleviate metal toxicity (Syed and Chinthala 2015; Shameer and Prasad 2018). Bacterial genera such as *Bacillus* and *Pseudomonas* are well known for alleviation of metal toxicity (Shameer 2016). Some microorganisms use various mechanisms such as volatilization, exopolysaccharide (EPS) sequestration, enzymatic detoxification and metal complexation that make host plants tolerate heavy metal toxicity (Kumar and Verma 2018). Stress due to the presence of weeds and their exudates is minimized by rhizobacteria through the production of antibiotics and phytotoxins that predominantly inhibit weed growth (Mishra et al. 2013).

9.6 Molecular Mechanisms of Stress Alleviation

Plant-associated endophytic and rhizosphere microbes help plant growth, fitness and health under different biotic and abiotic stress conditions (Vimal et al. 2017). Plant growth and development are enhanced by these microbes via an array of molecular and biochemical mechanisms (Kumar and Verma 2018). For instance, PGPM regulate nutritional and hormonal balance, produce plant growth regulators and induce resistance to plant pathogens in order to promote plant growth (Spence and Bais 2015). Many molecular interactions occur between these microbes and host plants. These interactions begin with the recognition of plant exudates by the microbes and once the microbes enter the host tissues cross talks of signal molecules which make the mode of communication between plants and microbes commence (Khare et al. 2018; Zhang et al. 2019). Many plants produce different flavonoids that function as chemical signals that are recognized by microbes. For example, flavonoids produced by plant roots attract arbuscular mycorrhizal fungi (AMF), their spore formation, hyphal growth, differentiation and root colonization (Steinkellner et al. 2007; Badri et al. 2009). Various signal molecules including Nod factor and strigolactone are well known to activate plant–microbe interactions (Gough and Cullimore 2011; Rozpądek et al. 2018). To overcome harmful effects to plants under adverse conditions, certain bacterial species possess sigma factors that can alter gene expressions (Gupta et al. 2013).

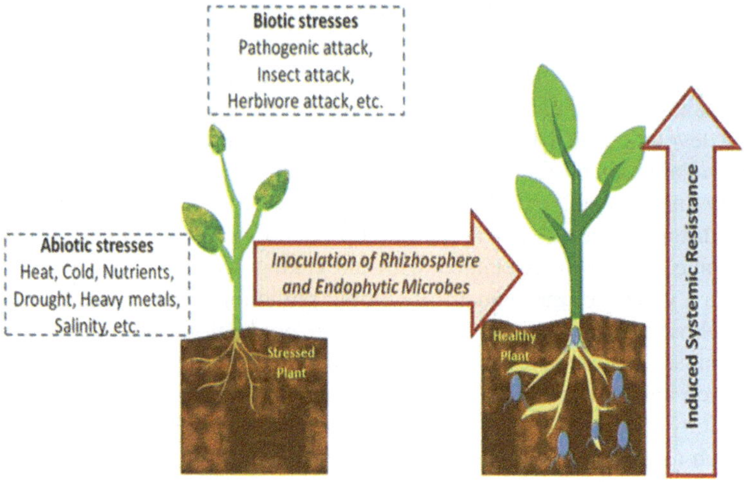

Fig. 9.2 Schematic diagram of microbe-mediated induced systemic resistance to biotic and abiotic stresses in plants

9.6.1 Microbe-Mediated Induced Systemic Tolerance to Abiotic Stress

Plants that are exposed to different environmental stress conditions including drought, flooding, depletion of nutrients and the presence of toxic metals in substantial quantities show adaptation mechanisms exhibited in various forms including changes in root morphology and production of phytohormones (Potters et al. 2007). Majority of root-associated microbes help plant growth under water stress by producing IAA which increases root growth and enhances the formation of lateral roots and root hairs (Fig. 9.2; Patten and Glick 2002; Rajkumar et al. 2005; Chakraborty et al. 2006; Long et al. 2008). Enhancement of root growth increases root surface, which improves the nutrient and water uptake by plants. Further, endophytic and rhizosphere microbes are involved in reducing abiotic stress through ACC deaminase activity and regulation of ACC mechanism (Saleem et al. 2007). Reduction in ethylene levels in plants that is catalysed by the hydrolysis of ACC in bacteria can also enhance root growth (Belimov et al. 2007; Long et al. 2008). Nonetheless, the stress status of plants changes with the decline in ethylene levels (Glick 2005). Further, both biotic and abiotic stress conditions trigger plants to synthesize proline, an amino acid that can function as a defence molecule in plants (Barka et al. 2006; Hayat et al. 2012). Different microbes mediated induced systems have developed to alleviate abiotic stresses of plants. In the presence of bacteria such as *Burkholderia*, *Arthrobactor* and *Bacillus*, plants synthesize proline that alleviate abiotic stress in plants (Barka et al. 2006; Sziderics et al. 2007).

9.6.1.1 Amelioration of Nutrient Deficiency

Plants mostly require macronutrients (C, H, O, N, P, K, Ca, Mg, S) and micronutrients (Fe, Zn, Cu, Mn, Mo, Cl, B) for their growth and development. Among these essential nutrients, C, H, O and N are freely available in soil in the form of complex compounds. The farmers usually utilize chemical fertilizers to increase crop production. The utilization of these synthetic fertilizers in high doses cause nitrates and various other chemical compounds to leach and subsequently contaminate surface and underground water, in addition to disturbing soil fertility. Ingredients of synthetic chemical fertilizers may pose toxic effects to human and various animals, thus leading to serious health consequences. Therefore, application of biofertilizers or microbes that have potential to promote plant growth is a sustainable option to increase crop productivity (Garima and Nath 2015). Many recent studies have shown that bacterial and fungal endophytes can be used as biofertilizers or bioformulations in agriculture to protect and improve crop production (Garima and Nath 2015). Rhizosphere and endophytic microbes can solubilize macro and microelements for plants and facilitate the mobilization and uptake of macronutrients and micronutrients to their host plants (Dobbelaere et al. 2001).

Nitrogen is one of the important macroelements required for increasing the crop growth and yield. About 80% of nitrogen gas in atmosphere is in the form of di-nitrogen and cannot be taken up by plants because plants can absorb nitrogen only in the form of ammonium and nitrate (Steenhoudt and Vanderleyden 2000). Therefore, N-based fertilizers are used to increase N bioavailability for plants, and overuse of N-based fertilizers directly contributes to soil contamination which has negative impacts on human health. However, endophytic and rhizosphere microbes are capable of fixing more atmospheric nitrogen and thus, enhance the bioavailability to plants. Microorganisms form symbiotic interactions with legumes and these associations provide fixed N that helps growth and development of plants. Researchers found, *Diazotrophic* bacteria could fix atmospheric nitrogen by nitrogenase enzyme in root nodules of legume plants (Dobbelaere and Okon 2007). Some non-legume plants in associations with endophytic bacteria such as *Azospirillum spp.*, *Azoarcus spp.* and *Herbaspirillum* are involved in biological nitrogen fixation (Doty et al. 2009; Desbrosses and Stougaard 2011). Nitrogen fixing bacterium, *Gluconacetobacter diazotrophicus* was reported for colonization of crops like maize, rice, wheat and other major non-legume crops, such as oilseed rape and tomato (Cocking et al. 2005). Plant growth promoting rhizosphere microbes perform the process of associative N_2-fixation even in different non-legume crops such as maize, sugar beet, wheat, rice and sugar cane (Şahin et al. 2004; Jha and Saraf 2015).

The phosphorus present as inorganic phosphate remains unavailable for the plants due to its non-soluble state in soil. Phosphorous is essential for metabolic processes of plants including N fixation in legumes, respiration, photosynthesis, signal transduction and energy transfer (Kouas et al. 2005; Khan et al. 2010). Endophytes involve converting the inorganic phosphate into a soluble form of phosphate with secretion of organic acid or by mineralizing the organic phosphorus, and produces different phosphatases thereby solubilizing the phosphorus (Ramachandran et al. 2007). Among other endophytes *Bacillus, Pseudomonas, Erwinia, Agrobacterium* and *Flavobacterium spp.,* are involved in solubilizing the inorganic phosphate compounds by phosphatases (Rodríguez and Fraga 1999). *Escherichia, Salmonella, Enterobacter, Azotobacter, Arthrobacter, Streptomyces, Micrococcus, Serratia, Thiobacillus, Alcaligenes, Chromobacterium and Bradyrhizobium* are some endophytic bacteria that secrete organic acids to solubilize insoluble phosphorus (Zhu et al. 2011). Moreover, soil can be rich in phosphorus from organic and inorganic material reserves, but it may not be available to plants due to fixation with Mg, Ca in alkaline soil and Fe, Al in acidic soils. Phosphorous solubilizing microbes like bacteria can enhance phosphorous availability by lowering the pH through release of organic acids (citric, oxalic, lactic, tartaric, malic and gluconic acid) or H^+ (Jha and Saraf 2015). Meantime, rhizobacterial species (*Bacillus thuringiensis)* are involved in increasing the availability of nutrients such as Ca, Mg, K, Cu, Zn and Mn (Armada et al. 2014).

9.6.1.2 Water, Temperature and Salinity Stress Tolerance

Drought or water stress is a key environmental factor affecting plant physiological and biochemical processes and it directly limits the growth and development of plants. It disturbs carbon assimilation and protein synthesis, triggers photorespiration, changes cell homeostasis, enhances the accumulation of ROS in cells and affects plant hormone balance (Cohen et al. 2015; Talaat 2015; Talaat and Shawky 2016). Microbial inoculants have capacity to change negative responses of different plant physiological and biochemical processes under water stress conditions. Inoculation of endophytic bacteria such as *Enterobacter sp.* FD 17 and *Burkholderia phytofirmans* PsJN enhanced drought resistance capacity in maize plants by colonizing internal portion of roots, shoots and leaves (Naveed et al. 2014). *Azospirillum* inoculated wheat seedlings contained more phospholipids in its roots than non-inoculated seedling roots under water stress condition and this bacterial inoculation led to change the root cell membrane elasticity and it cause to build up tolerance to water stress (Pereyra et al. 2006). Further, water stress tolerance of wheat plant observed after association with *B. phytofirmans* strain PsJN through modulation of metabolism and improving the ionic balance (Naveed et al. 2014). Plant root colonization with rhizobacteria *Pseudomonas chlororaphis* 06 induces tolerance to drought stress (Cho et al. 2008).

Maize plants reduced its shoot and root growth under low temperature condition (Baek and Skinner 2012; Saeidnejad et al. 2012). However, the inoculation of *Pseudomonas spp., Bacillus amyloliquefaciens* subsp. *plantarum* and *Bacillus* complex strain R41 with micronutrients (Zn/Mn) showed beneficial effects on maize under low temperature (Bradáčová et al. 2016). Chilling resistance of grapevine plantlets were enhanced after inoculation of plant growth promoting rhizobacteria, *Burkholderia phytofirmans* strain PsJN (Barka et al. 2006). Inoculation of temperature-tolerant *Pseudomonas putida* strain AKMP7 improved growth and development of wheat plant under heat stressed condition (Ali et al. 2011). Moreover, the strain AKM-P6 of this bacterium helped sorghum seedlings to overcome heat stress by producing proline and heat shock proteins (HSP) which reduced membrane injury and elevated protein and chlorophyll content in leaves (Ali et al.

2009). The association of a rhizosphere fungus *Paraphaeosphaeria quadriseptata* with *Arabidopsis thaliana* showed improved tolerance to heat stress, which was induced by HSP101 and HSP70, two conserved HSP components (McLellan et al. 2007). An endophytic fungus, *Paecilomyces formosus* strain LWL1 has shown to alleviate heat stress by enhancing plant growth in Japonica rice varieties (Waqas et al. 2015).

Soil salinization is a crucial problem for agricultural practices. Accumulation of salt in the soil negatively affects the growth and development of plants. Talaat and Shawky (2013) and Talaat and Shawky (2014) reported that inhibition of respiration, carbon assimilation and protein synthesis, disruption in nutrient accumulation, metabolic toxicity and enzyme destruction, decline in the translocation of assimilates from source to sink, disturbance in water and osmotic potential, increased ROS accumulation in chloroplast and toxicity of excessive sodium and chloride ions occur due to salt stress. Endophytic and rhizosphere microbes facilitate plant growth in saline soil by triggering various biochemical and physical changes in plants. Salt tolerance of *A. thaliana* was enhanced by *Bacillus subtilis* strain GB03 via a tissue specific regulation of high affinity K^+ transporter 1 (HKT1) that lowered the accumulation of Na^+ in the plant (Zhang et al. 2008). Salinity resistant microbes such as *Pseudomonas stutzeri, P. aeruginosa and P. fluorescens* showed amelioration of salt stress in *Solanum lycopersicum* (tomato) plants and increased shoot and root growth (Tank and Saraf 2010). Endophytic bacteria such as *Pseudomonas pseudoalcaligenes* induce the buildup of glycine betaine compounds at higher concentrations that helps rice plants to tolerate salt stress (Jha et al. 2011). Nadeem et al. (2007) reported high K^+/Na^+ ratio, relative water content, chlorophyll level and low proline content were observed in salt stressed maize plants associated with *Pseudomonas syringae, Enterobacter aerogenes* and *P. fluorescens* containing ACC deaminase activity. Further, wheat seedlings inoculated with bacteria that produce EPS have shown halt of Na^+ uptake and stimulation of plant growth in saline environments (Grover et al. 2011). Salt tolerance ability of paddy seedlings enhanced through inoculation of *P. pseudoalcaligenes* and *Bacillus pumilus* bacteria (Jha and Subramanian 2014). Further, *Azospirillum* alleviated salt stress in barely seedlings (Zawoznik et al. 2011). Bacilio et al. (2004) observed inoculation of genetically modified *A. lipoferum* could reduce negative effects of salt in wheat seedlings.

9.6.1.3 Tolerance of Stress Due to Heavy Metal and Herbicide Toxicity

Accumulation of heavy metals in soil affects agricultural activities and human health. However, some heavy metals are micronutrients that are required for plant growth and development. If used in excess, heavy metals may cause serious damages to plant metabolism by changing the plant hormonal balance, photosynthesis process, uptake and distribution of macro and micronutrients, permeability and function of plasma membranes and enzymes function (Garg and Singla 2012; Islam et al. 2016). Therefore, microbial inoculants could protect plants from the negative impact of heavy metals. For example, microorgansims such as *Proteobacteria, Firmicutes* and *Actinobacteria* are involved in the removal heavy metals including manganese

(Mn), lead (Pb) and arsenic (As) from metal polluted soil (Zhang et al. 2015). *P. aeruginosa* OSG 41 enhanced chickpea growth under chromium (Cr) stress (Oves et al. 2013). *Hordeum vulgare* (barley) plants grown in Cd contaminated soil and inoculated with *Klebsiella mobilis* CIAM 880 have shown higher grain yield and decrease in grain Cd content as free Cd ions are bound to bacteria making the plant uptake of the ion difficult (Pishchik et al. 2002). *Burkholderia spp.* and *Methylobacterium oryzae* reduce stress due to the accumulation of cadmium (Cd) and nickel (Ni) by lowering the uptake and translocation in tomato (Madhaiyan et al. 2007). Endophytic bacteria *Sphingomonas* SaMR 12 influenced the content of root exudates and decreased chelating Cd ions, thus alleviated toxic metal stress in *Sedum alfredii* (Chen et al. 2014). An Endophytic bacterium *Sphingomonas* SaMR12 has shown to reduce Cd toxicity by chelating the metal via influencing the exudation of tartaric acid, malic acid and oxalic acid (Chen et al. 2014). AM fungi alleviated effects of Cd stress by reducing malonaldehyde and hydrogen peroxide (Hashem et al. 2016). Jing et al. (2014) have shown the removal of Cd, Pb and zinc (Zn) from contaminated soil by *Enterobacter* spp. and *Klebsiella* spp.

9.6.2 Microbe-Mediated Induced Systemic Resistance to Biotic Stresses

Biotic stresses in plants occur due to damage caused by other living creatures such as insects, weeds, nematodes, fungi, bacteria and viruses. Rhizosphere and endophytic microbes have shown to protect plants from biotic stress by competing with those stress-causing organisms for space and nutrients, via processes such as antibiosis and mycoparasitism, and by inducing plant defence mechanisms (Fig. 9.2; Barea et al. 2013). Applications of endophytes and rhizosphere microbes have shown to trigger plant defence mechanisms (Ma 2017). These microbes regulate genes, which are involved in ethylene and jasmonic acid synthesis pathways (Pangesti et al. 2016). Further, Samain et al. (2017) suggested changes of biochemical responses such as increased synthesis of ROS and phenolic metabolic compounds that are mediated by these microbes. Similarly, anatomical modifications like the deposition of cellulose and lignin in the endophytic colonized tissues also occur (Constantin et al. 2019). Further, microbes induced resistance against biotic stress through the production of allopathic compounds such as siderophores and antibiotics showed to effectively protect plants from pathogens and to inhibit the growth microbial pathogens (Choudhary and Johri 2009; Jain et al. 2013).

Microbes are also involved in developing induced systemic resistance against biotic stress in plants. For example, bacteria such as *Acinetobactor lwoffii*, *P. fluorescens*, *P. putida*, *Paenibacillus alvei*, *Serratia marcescens*, *Bacillus pumilus*, *Chryseobacterium balustinum* and *Azospirillum brasilense* colonize roots and protect plants from foliar diseases under field and greenhouse conditions (van Loon 2007). Association of the endophytic *Burkholderia cenocepacia* 869T2 with banana showed decrease in Fusarium wilt disease (Ho et al. 2015). Pyrrolnitrin, an antibiotic produced by *Pseudomonas fluorescens* BL915 could prevent the damage from *Rhizoctonia solani* that causes damping off of cotton plants (Hill et al. 1994). Pavlo et al. (2011) found that *Pseudomonas* sp. IM BG 294 reduced symptoms of soft rot diseases caused by bacterial pathogen *Pectobacterium atrosepticum* in potato. Moreover, endophytic *Serratia plymuthica, Methylobacterium* sp. and *Streptomyces* spp. have shown to mediate induced systematic resistance in many plants (Benhamou et al. 2000; Conn et al. 2008; Pavlo et al. 2011). Moreover, cucumber anthracnose could be prevented by *P. fluorescens* strain 89B-61 (Kloepper and Ryu 2006). Bacterial strains *Bacillus licheniformis* AE6 and *Streptomyces* spp. Zapt 10 have shown to induce systemic resistance against downy mildew caused by *Pseudoperonospora cubensis* in *Cucumis sativus* (cucumber) (Sen et al. 2014). Black rot disease of cabbage caused by *Xanthomonas campestris* was effectively controlled by a biological control agent *Paenibacillus* sp. P16 that activated induced systemic resistance in the host plant (Ghazalibiglar et al. 2016). Chithrashree et al. (2011) have observed *Bacillus* strains to cause induced resistance in rice against bacterial leaf blight caused by *Xanthomonas oryzae* pv *oryzae*. An endospore forming bacterium *Brevibacillus laterosporus* Laubach has shown to effectively inhibit insects of orders Coleopteran and Lepidoptera (Boets et al. 2004; De Oliveira et al. 2004), nematodes (Singer 1996) and plant pathogenic fungi (Saikia et al. 2011).

9.6.3 Defence Mechanisms of Rhizosphere and Endophytic Microbes Against Biotic Stresses

Rhizospheres and endophytes are considered as effective group of microbes, which take a significant part in promoting growth, development and protection of plants against biotic stress. These microbes significantly stimulate host performance or innate resistance during the stress conditions, especially under biotic stress (Lugtenberg et al. 2016; White et al. 2019). There are two systems of resistance mechanisms, namely induced systemic resistance (ISR) or system acquired resistance (SAR) that are coordinated and performed by these two categories of microbes against biotic stress (Mengistu 2020). Many useful microorganisms associated with host plants can regulate resistance systems. Among them, facilitation and enhancement of immune system are important methods of protection against biotic stress in plant in which structurally conserved molecules that are associated with microbes or pathogen (MAMPs/PAMPs: Microbe—Associated Molecular Pattern/Pathogen—Associated Molecular Pattern respectively) are recognized (Pozo et al. 2013; Eid et al. 2019). Biological control of pest and diseases is gaining importance in the field of plant protection for maintaining a threshold level below to reduce the environmental and economic impact for enabling the plant to reclaim the ability to tolerate the biotic stress (Bale et al. 2008). The interaction among rhizospheres, endophytes and host plants have made the "coevolution" of colonization and symbiotic relationship for the ultimate benefit to the plants to protect from biotic stress caused by phytopathogens, phytophagous pests, etc.

9.6.4 Defence Against Phytopathogens

Defence against plant pathogens and their competitions with the host for resources has a pivotal role in crop protection in order to uphold crop productivity (Oerke and Dehne 2004). Rhizosphere and endophytic microbes are distinct and functional alternatives to the traditional chemical-based crop protection methods. These microbes follow the similar pattern of potential colonization by overcoming phases relevant to plant and other competitive organisms for holding a sustainable niche for them. At the initial stage, these organisms manipulate the host immune system for colonization and later they enhance the system to withstand or overcome the conquering phytopathogens (Reinhold-Hurek and Hurek 2011; Mitter et al. 2013). Successful colonization of endophytes is determined by several factors which are relevant to plant (age, tissue type, genotype, species, etc.), the number and density of inoculum, strains of microorganisms and environmental conditions (biotic stress, abiotic stress and growing media) (Hardoim et al. 2015; Bamisile et al. 2018). Bamisile et al. (2020) have revealed that the microbial colonization of rhizospheres and endophytes carry out colonization systematically from root → shoot → flowers →fruit → seeds. The colonization is almost similar to the pattern followed by phytopathogens and it takes place through planting materials (vegetative parts of plant), seeds, rhizosphere soil and phyllosphere (Rosenblueth and Martínez-Romero 2004).

The mechanisms expressed by rhizosphere and endophytes can be antibiosis or competition or parasitism or induced systemic resistance or increased growth response or can be a combination of more than one mechanism to control over the phytopathogens. However, both rhizosphere and endophytes involve directly in dealing with the phytopathogens for their either reduction or suppression of disease-causing ability and survival (Ownley et al. 2010). In many instances, the process of anti-phytopathogenic activity is established by the rhizosphere and endophytic microbes by synthesizing a number of enzymes, antibiotics and volatile organic compounds (VOCs which are bioactive) (Ownley and Windham 2007). Strobel et al. (2001) have proven that Ascomycota fungi (*Muscodor albus*) synthesize a mixture of VOCs with a wide spectrum of toxicity against a plethora of microorganisms.

Endophytic fungi produce various kind of secondary metabolites and hydrolases against the pathogens and thereby strongly affect the growth and development of pathogenic microorganisms due to the antifungal and antibacterial activities (Berg and Hallmann 2006; Gunatilaka 2006). Enzymatic biocontrol activities of endophytes fungi are different from usual way of mycoparasitism towards plant pathogens. Some endophytic fungi like *Talaromyces flavus* produce glucose oxidase which has an ability to kill *Verticillium* spp. by catalyzing the production of hydrogen peroxide that has the biological control activity (Fravel 1988). An effective competition can be observed between endophytes and other phytopathogens especially for niches and sources of nutrients. These kinds of competitions are mostly prevailing in the rhizosphere, phyllosphere and in the intercellular regions of the plant. The success of rhizospheres and endophytes depend on their ability for initial colonization in the host plants. Similarly, *Trichoderma* spp. produces low level of exochitinase before having the contact made with phytopathogens during the colonization. The enzyme diffuses and catalyses the cell wall oligomolecules released by the target pathogenic fungi. This process activates the *Trichoderma* to release a fungal toxic endochitinase to abolish the cell wall components of pathogenic fungi completely (Harman et al. 2004). There are several kinds of lytic enzymes involve in degradation of the cell walls of plant pathogens and that include lipophytic enzymes (lipase), glucanases, proteolytic enzymes (Protease), exo and endochitinases, etc. and secondary metabolites such as alkaloids, terpenoids, polypeptides, aromatic compounds, etc. For example, *Alternaria* sp. shows antibacterial activities through signal interference antagonism method against several pathogenic bacteria by secreting a new secondary metabolite "altersetin" (Hellwig et al. 2002). However, the biocontrol activities of rhizospheres and endophytes are; some of them may apply multiple mechanisms to control a specific pathogen or some may apply different mechanisms against for different pathogenic fungi (Korolev et al. 2008; Ownley et al. 2010). Nevertheless, "biopesticides" produced by these beneficial organisms to control the phytopathogens are an effective, alternative and safe crop management procedure in modern agriculture systems for the future and as well as for current use (Syed-Ab-Rahman et al. 2018).

The ideal beneficiary microbes should be able to colonize easily in rhizospheres in order to provide a combat zone of protection to the root area from various pathogenic attacks or create a suppressive environment to the pathogen for weaken them. The beneficial microbes in the rhizosphere deliver toxic substances/ antagonizing metabolites to the vicinity of roots and conventionally affect the pathogenic bacteria (Shoda 2000). Beneficial rhizosphere microbes or endophyte–host plant mutual relationships generally exist in nature to either improve the nutrient level of host plants or augment the host plant's ability to withstand the stress situation induced by the biotic or abiotic factors. Although, in any situation of this natural mechanism leads to the improvement and proliferation of plant growth (Haney et al. 2015). A wide range of endophytic and rhizosphere microorganisms utilize the nutrients released by the host plant and mutually secrete metabolites in the rhizosphere zone for overcoming the effects of disease caused by bacteria and fungi for optimizing the health (Gray and Smith 2005; Kiely et al. 2006; Knief 2014). It has been demonstrated by Castillo et al. (2002) that an entophytic bacterium *Streptomyces* sp. (strain NRRL 30562) secrets a toxic substance munumbicins which can inhibit the growth of phytopathogenic fungi, *Pseudomonas ultimum*, and *Fusarium oxysporum* under *in vitro* conditions. Similarly, the application of several species of *Bacillus*: *B. subtilis* (IN937b), *B. pumilus* (SE34) and *B. amyloliquefaciens* (IN937a) for treating tomato seeds to induce systemic resistance against CMV has expressively improved the performance of tomato plants while reducing the severity of diseases (Zehnder et al. 2000).

However, this defensive mechanism is not unanimous for every combination of endophyte–host plant–phytopathogen interaction (Rabiey et al. 2019). For example, the protective functions of an ectomycorrhizal fungus *Suillus luteus* were investigated against other fungal pathogens (*Heterobasidion irregular* and

Heterobasidion annosum) of Scots pine (*Pinus sylvestris*) and was found that it only reduced host tree susceptibility to *H. annosum,* not both pathogens (Gonthier et al. 2019). In a similar manner, some tree endophytes have shown a promising effect to inhibit the growth of phytopathogens using antifungal and antibacterial compounds (Fadiji and Babalola 2020). A *Bacillus pumilus* strain (JK-SX001) produces a combination of substances which have some extracellular enzymes such as cellulases and proteases and secondary metabolites to retard the infection rate and growth of canker causing pathogens; *Cytospora chrysosperma*, *Phomopsis macrospora* and *Fusicoccum aesculi* (Ren et al. 2013). An endophyte associated with black pepper; *P. putida* (strain BP25) inhibits the growth of several other pathogenic fungi and nematode such as *P. capsici, P. myriotylum, G. moniliformis, R. solani, A. rolfsii, C. gloeosporioides* and plant parasitic nematode, *R. similis* using VOCs (Sheoran et al. 2015).

In recent years, a significant challenge is faced in biocontrol of phytopathogens using rhizosphere and endophytes especially due to the rapid fluctuations of climate which alter the ranges of pathogens and challenging the plant health or by subjecting plants to stress (Shaw and Osborne 2011). The climatic changes will have diverse degrees of effects depending on the type of pathosystems and geographical region. Climate change can influence especially the dispersal of a key symbiont mycorrhizal fungus and may restrict the migration and colonization of them in new-fangled environments (Pickles et al. 2015). Wilson et al. (2016) have revealed that warming due to extreme temperatures, heat waves may induce a decrease in arbuscular mycorrhizal fungal colonization. On the other hand, there is a possibility of transition of "intimate relationship" between plant and endophytes in to "pathogenic relationship" when plants are exposed to stress conditions evolved due to environment and physiological nature. The similar process has been observed in healthy *Quercus cerris* in which endophytic fungi *Discula quercina* which lives in trees becomes pathogenic nature during the drought stress and thereby causing causes damage to host structure and function (Ragazzi et al. 2001; Moricca and Ragazzi 2008).

9.6.5 Defence Against Phytophagous Insects

There are many biological control strategies based on natural enemies (predatory insects, antagonizing microorganisms or entomopathogenic microorganisms, parasitoids, etc.) that are used as an alternative to traditional control to manage phytophagous insects in agriculture. Similarly, there is a new trend of using microorganisms such as entomopathogenic fungi, bacteria, nematodes and viruses in rhizosphere, phyllosphere and endosphere to increase crop yield and also to enhance the defence mechanisms of the host plants against phytophagous insects or other herbivores at both above and below ground levels by retardation of their feeding habit or via antibiosis in diverse environments (Ruiu 2018; Francis et al. 2020).

The use of endophytes to protect plants from phytophagous insects is not a new concept and has been well studied and used since early 1980s (Azevedo et al. 2000). A first report on control of phytophagous insect was probably elaborated by (Webber 1981), in which the larvae of beetle *Physocnemum brevilineum* (vector of *Ceratocystis ulmi* which is the causal agent of elm Dutch disease in trees) was killed by toxic compounds produced by an endophytic fungus; *Phomopsis oblonga*.

The beneficial microorganisms that are considered to be natural "biocontrol" agents against phytophagous insects can prime the plant to prompt on the insect pest by triggering some specific molecular biological pathways like jasmonic and salicylic acids and/or ethylene induced systemic resistance (ISR) (Pieterse et al. 2012). In addition, the ISR leads to the initiation of hypersensitive induced oxidative stress and the process is considered to be a crucial component for the plants to exhibit defence system against insects and other herbivores (Rashid and Chung 2017). However, a limited understanding is prevailing about the endophytes as how they interact with plants through a diverse biochemical and molecular pathway and how are those pathways actually applied to manage and mediate plant–microbes–insects interaction for the betterment of plant health. It is believed that the interaction is very much correlated with the degree of specificity of the insect to the plant. For example, generally various kinds of insects are negatively affected by toxic metabolites secreted by certain plant species, while some specific insects use those compounds, which do not affect the insects, to distinguish the host plants. Van Oosten et al. (2008) have demonstrated the differential behaviour of insect larvae against the ISR produced by *Pseudomonas fluorescens*, in which the specific caterpillar *Pieris rapae* was not disturbed whereas *Spodoptera exigua* (a generalist insect) was harmfully affected.

Beneficial soil microbes (arbuscular mycorrhizal, non-mycorrhizal fungi, PGPR, endophytes, etc.) can play a major role in plant–insect interactions in three different ways.

1. Contribute to increase the size, vigour, nutritional status and creating a favourable environment in the host plant in order to attract the insects for feeding (Bukovinszky et al. 2008),
2. Induce systemic resistance in plants or enhance pest resistance ability through the biochemical and molecular pathways of production of enzymes, toxic secondary metabolites, etc. (van Lenteren et al. 2018),
3. Modify the VOCs that are released from the host plant and thereby interfere the plant signalling. The ultimate impact to inhibit the performance of the insects widely depends on the balance situation between the enhanced plant growth and the negative impact of ISR (Rasmann et al. 2017).

Entophytic influenced metabolites enhance herbivore resistance because of the feeding discouragement or antibiosis (Vega 2008). Many endophytic and entomopathogenic fungi like *Beauveria bassiana*, *Metarhizium anisopliae* and *Lecanicillium lecanii* secrete some metabolites that can enhance plant fitness and show resistance to herbivores (Saikkonen et al. 2006; Vega 2008). Remarkably, these fungi have ability to upload nitrogen from above ground insect remains to the plant through fungal mycelia (Behie et al. 2012). A screening for identifying further entomopathogenic fungi and bacteria may help in agriculture to improve crop production by increasing the resistance of plants against herbivores including plant-feeding insects.

9.7 Influence of Rhizosphere and Endophytic Microbes on Product Quality

The endophytes and host plants show the symbiotic relationship of interaction in which the plants provide niches to the microorganisms and in turn they produce metabolites for enhancing the plant protection as well as for the improvement of growth and development of the plant (Xia et al. 2015). The interaction between both endophytes and host plants evolved through coevolution that was directed through colonization and motivated due to the factors associated with plants, microbes and environment. The coevolution process leads to the establishment of close association and causing effective changes in cellar and molecular level activities for the growth and development of the plant (Aravind et al. 2010; Costa and de Melo 2012). The entophytic bacteria and fungi have direct and indirect impact on crop improvement. These influence or impact of them for improving nutrient and water uptake, increase nitrogen fixation, facilitate nutrients, mineral solubilization and absorption and production of metabolites such as phytohormones for conquer the biotic and abiotic stress conditions and act as biological controlling agents (Chebotar et al. 2015; Xia et al. 2015).

Interactions between host plants and beneficial bacteria eventually have great effects on the growth and development of plants, crop productivity, plant health and soil quality (Compant et al. 2008). This can be witnessed in sugarcane where *Azospirillum amazonense*, *Burkholderia tropica, Herbaspirillum seropedicae, Herbaspirillum rubrisubalbicans* and *Gluconacetobacter diazotrophicus* increase and accelerate nitrogen fixation, biomass accumulation, IAA and siderophores production, and improve phosphate solubilization that enhance budding process (Beneduzi et al. 2013). Similarly, *Azospirillum lipoferum, Azospirillum brasilense, Bacillus lentimorbus*, *B. subtilis*, *Burkholderia cepacia* and *Streptomyces* sp. have shown to influence growth promotion, IAA synthesis, nitrogen fixation, reduction of toluene evapotranspiration, antagonistic effect to pathogenic fungi and increase of maize yield (He et al. 2009; Ferreira et al. 2013).

Phytohormones like auxins, cytokinin and gibberellin involve in regulation of plant growth and development. These hormones are synthesized through phytostimulation and they modulate the morphological characters of root cells for efficient absorption (Santos et al. 2018). The direct ethylene (ET) precursor, ACC is a member of auxin group that reduces the effect of high level of ethylene which can regulate and inhibit cell division, molecular DNA synthesis, cell division and elongation of roots and aerial parts of plants (Vanderstraeten and Van Der Straeten 2017). Endophytes retard and regulate the high ethylene production by altering and reprogramming the signalling pathways using ACC and IAA and thereby promoting the root growth and seed germination (Gaiero et al. 2013). For example, some entophytic bacteria such as *Bacillus* sp., *Pseudomonas* sp., *Azotobacter* sp., *Azospirillum* sp. and *Acinetobacter* sp. synthesize phytohormones like IAA, IBA, cytokinin, gibberellin and some other compounds, which show resemblance to the activity of jasmonates (Egamberdieva et al. 2017; Santos et al. 2018). The products of these endophytes help plants for healthy survival by assisting them in maintaining proper osmotic nature, modification of root cells for efficient absorption of water and minerals, natural nitrogen fixation, etc. (Chebotar et al. 2015). The introduction of *Bacillus* sp. in barley and *Acinetobacter johnsonii* in beet have increased the minerals and ions (Manganese, Zinc and Copper) and carbohydrate content due to the efficient photosynthetic activity, respectively (Canbolat et al. 2006; Shi et al. 2010).

Several endophytic bacteria and fungi involve in biofertilization of plants for promoting growth and product quality. It is considered that rhizosphere and endophytic are efficient and alternative natural nitrogen fixing organisms since they have capacity for easy colonization not only in the rhizosphere area but also in the stems and leaves (Balachandar et al. 2006). The inoculation of bacterial mixtures (that contain selective entophytic bacteria *Burkholderia* sp., *Azospirillum amazonense*, *Gluconacetobacter diazotrophicus*, *Herbaspirillum seropedicae* and *H. rubrisubalbicans*) to sugarcane has revealed 30% increase in nitrogen content by converting atmospheric nitrogen into ammonia for direct utilization by the plant (Oliveira et al. 2002). Likewise, *Burkholderia* sp. is an endophytic bacteria entophytic bacterium having ability to produce organic acids such as oxalic, citric and tartaric acids, which makes phosphorus available for easy absorption by plants (Gaiero et al. 2013).

It is evidence from the previous studies on *Vitis vinifera* (grape) that the interaction process between the host plant and endophytic fungi leads to the accumulation of new metabolites in the host plants to improve the characteristics and quality of grapes (Huang et al. 2018; Pacifico et al. 2019). Yang et al. (2016) also observed that the inoculation of endophytic fungal isolates to field grown grape vines modified the physio-chemical status of grapes. Moreover, grapevines have revealed to produce various metabolites in the flesh cells of grapes berries when inoculated with different endophytes in dual cultures (Huang et al. 2018).

9.8 Biotechnological Approaches for Enhancing the Effectiveness of Rhizosphere and Endophytic Microbes

The success of plant to gain the most benefits out of rhizosphere and endophytic microbes virtuously depend on how resourcefully the organisms have colonized the plant at the rhizosphere, phyllosphere and in the intercellular regions of the plant. Endophytes such as *Enterobacter* sp. are able to synthesize various antioxidant enzymes including superoxide dismutase, (SOD), catalase (CAT) and hydroperoxide reductase to overcome oxidative stress in *Populus trichocarpa* (Poplar tree) during the colonization of endophytes (Taghavi et al. 2010). Similarly, the colonization and the density of endophyte *Azoarcus olearius* in rice roots was limited due to the constant activity of differentially expressed genes (DEGs) of JA signalling pathway than SA pathway (Chen et al. 2020). A gene product of BSn5 of endophyte bacterium *B. subtilis* binds with flagellin that is the main forerunner substance in flg22 induced plant defence system and thereby easily colonizing *Arabidopsis thaliana* (Deng et al. 2019).

A few endophytic microbes are able to modify the features that are relevant to the effective colonization of the host plant. For example, swimming motility and biofilm formation are stimulated by *Serratia plymuthica* using G3 and QS genes. The downstream product of shr5 gene of *Gluconacetobacter diazotropicus* and antioxidant genes expression of diazotrophic beneficial bacteria like *Klebsiella pneumonia* play a key role in regulation and deactivation of ROS during endophytic colonization (Hardoim and van Elsas 2013; Malfanova et al. 2013; Kandel et al. 2017). Similarly, microbial VOCs influence the signal eliciting defence mechanisms (systemic resistance) by enhanced expression of encoding genes and also VOCs facilitate the successful colonization of endophytes (Chung et al. 2016). Trichdiene, which is a VOC produced by *Trichoderma arundinaceum*, affects *Botrytis cinerea* by inducing the expression of defence related genes that encode for salicylic acid (SA) and jasmonate (JA) pathways in host plant (Malmierca et al. 2012). In addition, genetically engineered endophytic *P. putida* with antifungal phz gene (obtained from *P. fluorescens*) has exhibited the reduction of pathogenic other fungi in the soil (Glandorf et al. 2001). It is apparent that there are many promising genes associated with the colonization of microorganisms in rhizosphere and endophyte and those can be modified or reprogrammed the gene expression to increase the colonization pattern of microorganisms of rhizosphere and endophytes as well as to improve plant defence mechanisms and enrich growth and development.

The microbiome engineering is a new approach in plant breeding for sustaining beneficial plant–microbes interactions and to produce microbe-optimized plants for attracting and maintaining valuable biocontrol microorganisms (Syed-Ab-Rahman et al. 2018). It is interesting to note that plants have their own strategies to create a favourable environment to attract beneficial microbes and to have effective interaction with them (Trivedi et al. 2017). In addition, plant growth promoters, which are secreted by beneficial microbiomes, also help protect the environment from the use of conventional chemical substances (Timmusk et al. 2017). Hence, it is imperative to understand the mechanisms that naturally occurring microbes or artificially inoculated microbes to plants use to maintain positive plant–microbe interactions. Different molecular techniques can be used to unravel these mechanisms, which can be manipulated to enhance those interactions of beneficial microbes with crop plants. Therefore, the generation of microbe-optimized plants would be possible through genetic engineering and plant breeding if we can find the answers for the above.

Similarly, a new encouraging approach has been already introduced to genetic engineering of the microorganisms in endophytes to synthesize proteins (like lectins) to control pest insects. For example, PtA gene product from endophytic fungi and bacteria has been used as bioinsecticidal substance to control sap-sucking pest, white backed plant hopper, etc. (Zhao et al. 2010; Zhang et al. 2011).

A wide range of bioactive secondary metabolites (e.g. Alkaloids, phenolic compounds, steroids, quinone, terpenoids, flavonoids, etc.) are synthesized by endophytic microorganisms and they are extensively used as medicine, aromatics agents, agro-chemicals, anticancer compounds, antitoxins, antioxidants, antiparasitic drugs, recreational drugs, etc. (Rana et al. 2016a, b; Yadav 2017; Yadav et al. 2017). However, the nature, quality and quantity of the secondary metabolites may vary depending on the biotype of the microbes, environmental factors and geographical locations (Firáková et al. 2007; Mohiuddin 2019). Nevertheless, the elucidation of plant–microbe interaction, biochemical and molecular investigation of pathways of secondary metabolites production will enlighten the extensive application of these microorganisms in rhizosphere and endophytes in biotechnological aspects.

9.9 Conclusion and Future Perspective

Rhizosphere and endophytic microbes are considered as effective group of microbes which take a significant part in promoting growth, development and protection of plants against biotic and abiotic stress conditions. These microbes significantly stimulate host performance or innate resistance during the stress conditions. It is believed that endophytes represent an eco-friendly option for the resources of novel bioactive natural products. Although extensive research is taking place in the view of exploring novel genes associated with endophytic characters, only a few terrestrial plant species (1%) have been studied to find effective associations between rhizosphere and endophytic microbes and host plants. Several important bioactive compounds, hormones and enzymes produced by endophyte–plant interactions are exploited in the management of biotic and abiotic stress conditions. It is therefore imperative to explore and elucidate the biochemical and physiological processes that are triggered by plant-associated beneficial microbes to mitigate the stress conditions. Many studies have identified different endophytic microbes that are suitable for bioformulations or biofertilizers that are very effective for stress alleviation and reclamation of contaminated soil in agriculture. Thus, detailed studies of rhizosphere and endophytic microorganisms associated with agriculturally important crops at genomics and metabolomics levels are essential to maximize the agricultural productivity by exploiting the benefits of these microbes.

References

Abbamondi GR, Tommonaro G, Weyens N, Thijs S, Sillen W, Gkorezis P et al (2016) Plant growth-promoting effects of rhizospheric and endophytic bacteria associated with different tomato cultivars and new tomato hybrids. Chem Biol Technol Agricult 3(1):1

Abedinzadeh M, Etesami H, Alikhani HA (2019). Characterization of rhizosphere and endophytic bacteria from roots of maize (*Zea mays* L.) plant irrigated with wastewater with biotechnological potential in agriculture. Biotechnol Rep 21:e00305

Afzal I, Shinwari ZK, Sikandar S, Shahzad S (2019) Plant beneficial endophytic bacteria: mechanisms, diversity, host range and genetic determinants. Microbiol Res 221:36–49

Ali SZ, Sandhya V, Grover M, Kishore N, Rao LV, Venkateswarlu B (2009) *Pseudomonas* sp. strain AKM-P6 enhances tolerance of sorghum seedlings to elevated temperatures. Biol Fertil Soils 46(1):45–55

Ali SZ, Sandhya V, Grover M, Linga VR, Bandi V (2011) Effect of inoculation with a thermotolerant plant growth promoting *Pseudomonas putida* strain AKMP7 on growth of wheat (*Triticum* spp.) under heat stress. J Plant Interact 6(4):239–246

Anon (2020). Phytophthora root rot. Queensland Government Department of Agriculture and Fisheries

Aravind R, Eapen S, Kumar A, Dinu A, Ramana K (2010) Screening of endophytic bacteria and evaluation of selected isolates for suppression of burrowing nematode (*Radopholus similis* Thorne) using three varieties of black pepper (*Piper nigrum* L.). Crop Prot 29(4):318–324

Armada E, Roldán A, Azcon R (2014) Differential activity of autochthonous bacteria in controlling drought stress in native *Lavandula* and *Salvia* plants species under drought conditions in natural arid soil. Microb Ecol 67(2):410–420

Augé RM (2001) Water relations, drought and vesicular-arbuscular mycorrhizal symbiosis. Mycorrhiza 11(1):3–42

Azevedo JL, Maccheroni W Jr, Pereira JO, de Araújo WL (2000) Endophytic microorganisms: a review on insect control and recent advances on tropical plants. Electron J Biotechnol 3 (1):15–16

Bacilio M, Rodriguez H, Moreno M, Hernandez J-P, Bashan Y (2004) Mitigation of salt stress in wheat seedlings by a gfp-tagged *Azospirillum lipoferum*. Biol Fertil Soils 40(3):188–193

Badri DV, Weir TL, van der Lelie D, Vivanco JM (2009) Rhizosphere chemical dialogues: plant–microbe interactions. Curr Opin Biotechnol 20(6):642–650

Baek K-H, Skinner DZ (2012) Production of reactive oxygen species by freezing stress and the protective roles of antioxidant enzymes in plants. J Agricult Chem Environ 1:34–40

Balachandar D, Sandhiya G, Sugitha T, Kumar K (2006) Flavonoids and growth hormones influence endophytic colonization and in planta nitrogen fixation by a diazotrophic *Serratia* sp. in rice. World J Microbiol Biotechnol 22(7):707–712

Bale J, Van Lenteren J, Bigler F (2008) Biological control and sustainable food production. Philosophical Transact R Soc B Biol Sci 363(1492):761–776

Bamisile BS, Dash CK, Akutse KS, Keppanan R, Afolabi OG, Hussain M et al (2018) Prospects of endophytic fungal entomopathogens as biocontrol and plant growth promoting agents: an insight on how artificial inoculation methods affect endophytic colonization of host plants. Microbiol Res 217:34–50

Bamisile BS, Senyo Akutse K, Dash CK, Qasim M, Ramos Aguila LC, Ashraf HJ et al (2020) Effects of seedling age on colonization patterns of citrus limon plants by endophytic *Beauveria bassiana* and *Metarhizium anisopliae* and their influence on seedlings growth. J Fungi 6(1):29

Bardi L, Malusà E (2012) Drought and nutritional stresses in plant: alleviating role of rhizospheric microorganisms. In: Haryana N, Punj S (eds) Abiotic stress: new research. Nova Science Publishers Inc, Hauppauge, pp 1–57

Barea JM, Pozo MJ, Azcón R, Azcón-Aguilar C (2013) Microbial interactions in the rhizosphere. In: de Bruijn F (ed) Molecular microbial ecology of the rhizosphere, vol 1. Wiley Blackwell, Hoboken, pp 29–44

Barka EA, Nowak J, Clément C (2006) Enhancement of chilling resistance of inoculated grapevine plantlets with a plant growth-promoting rhizobacterium, *Burkholderia phytofirmans* strain PsJN. Appl Environ Microbiol 72(11):7246–7252

Behie S, Zelisko P, Bidochka M (2012) Endophytic insect-parasitic fungi translocate nitrogen directly from insects to plants. Science 336(6088):1576–1577

Belimov A, Dodd I, Safronova V, Hontzeas N, Davies W (2007) *Pseudomonas brassicacearum* strain Am3 containing 1-aminocyclopropane-1-carboxylate deaminase can show both pathogenic and growth-promoting properties in its interaction with tomato. J Exp Bot 58 (6):1485–1495

Beneduzi A, Moreira F, Costa PB, Vargas LK, Lisboa BB, Favreto R et al (2013) Diversity and plant growth promoting evaluation abilities of bacteria isolated from sugarcane cultivated in the south of Brazil. Appl Soil Ecol 63:94–104

Benhamou N, Gagné S, Le Quéré D, Dehbi L (2000) Bacterial-mediated induced resistance in cucumber: beneficial effect of the endophytic bacterium *Serratia plymuthica* on the protection against infection by *Pythium ultimum*. Phytopathology 90(1):45–56

Berendsen RL, Pieterse CM, Bakker PA (2012) The rhizosphere microbiome and plant health. Trends Plant Sci 17(8):478–486

Berg G, Hallmann J (2006) Control of plant pathogenic fungi with bacterial endophytes. In: Schulz BJE, Boyle CJC, Sieber TN (eds) Microbial root endophytes. Springer, Berlin, pp 53–69

Berg G, Opelt K, Zachow C, Lottmann J, Götz M, Costa R et al (2006) The rhizosphere effect on bacteria antagonistic towards the pathogenic fungus *Verticillium* differs depending on plant species and site. FEMS Microbiol Ecol 56(2):250–261

Bertalan M, Albano R, de Pádua V, Rouws L, Rojas C, Hemerly A et al (2009) Complete genome sequence of the sugarcane nitrogen-fixing endophyte *Gluconacetobacter diazotrophicus* Pal5. BMC Genomics 10(1):1–17

Birkhofer K, Schöning I, Alt F, Herold N, Klarner B, Maraun M et al (2012) General relationships between abiotic soil properties and soil biota across spatial scales and different land-use types. PLoS One 7(8):e43292

Boets A, Arnaut G, Van Rie J, Damme N (2004) Toxins. Google Patents (Vol. U.S. Patent 6,706,860): Bayer BioScience NV

Bouffaud ML, Poirier MA, Muller D, Moënne-Loccoz Y (2014) Root microbiome relates to plant host evolution in maize and other Poaceae. Environ Microbiol 16(9):2804–2814

Bradáčová K, Weber NF, Morad-Talab N, Asim M, Imran M, Weinmann M et al (2016) Micronutrients (Zn/Mn), seaweed extracts, and plant growth-promoting bacteria as cold-stress protectants in maize. Chem Biol Technol Agricult 3(1):19

Bukovinszky T, van Veen FF, Jongema Y, Dicke M (2008) Direct and indirect effects of resource quality on food web structure. Science 319(5864):804–807

Canbolat MY, Bilen S, Çakmakçı R, Şahin F, Aydın A (2006) Effect of plant growth-promoting bacteria and soil compaction on barley seedling growth, nutrient uptake, soil properties and rhizosphere microflora. Biol Fertil Soils 42(4):350–357

Castillo UF, Strobel GA, Ford EJ, Hess WM, Porter H, Jensen JB et al (2002) Munumbicins, wide-spectrum antibiotics produced by *Streptomyces* NRRL 30562, endophytic on *Kennedia nigriscans*. Microbiology 148(9):2675–2685

Chakraborty U, Chakraborty B, Basnet M (2006) Plant growth promotion and induction of resistance in *Camellia sinensis* by *Bacillus megaterium*. J Basic Microbiol 46(3):186–195

Chebotar V, Malfanova N, Shcherbakov A, Ahtemova G, Borisov AY, Lugtenberg B et al (2015) Endophytic bacteria in microbial preparations that improve plant development. Appl Biochem Microbiol 51(3):271–277

Chen B, Zhang Y, Rafiq MT, Khan KY, Pan F, Yang X et al (2014) Improvement of cadmium uptake and accumulation in *Sedum alfredii* by endophytic bacteria *Sphingomonas* SaMR12: effects on plant growth and root exudates. Chemosphere 117:367–373

Chen X, Marszałkowska M, Reinhold-Hurek B (2020) Jasmonic acid, not salicyclic acid restricts endophytic root colonization of rice. Front Plant Sci 10:1758

Chi F, Shen S-H, Cheng H-P, Jing Y-X, Yanni YG, Dazzo FB (2005) Ascending migration of endophytic rhizobia, from roots to leaves, inside rice plants and assessment of benefits to rice growth physiology. Appl Environ Microbiol 71(11):7271–7278

Chithrashree, Udayashankar AC, Chandra Nayaka S, Reddy MS, Srinivas C (2011) Plant growth-promoting rhizobacteria mediate induced systemic resistance in rice against bacterial leaf blight caused by *Xanthomonas oryzae* pv. *oryzae*. Biol Control 59(2):114–122. https://doi.org/10.1016/j.biocontrol.2011.06.010

Cho SM, Kang BR, Han SH, Anderson AJ, Park J-Y, Lee Y-H et al (2008) 2R, 3R-butanediol, a bacterial volatile produced by *Pseudomonas chlororaphis* O6, is involved in induction of systemic tolerance to drought in *Arabidopsis thaliana*. Mol Plant-Microbe Interact 21 (8):1067–1075

Choudhary DK, Johri BN (2009) Interactions of *Bacillus* spp. and plants–with special reference to induced systemic resistance (ISR). Microbiol Res 164(5):493–513

Chung J-H, Song GC, Ryu C-M (2016) Sweet scents from good bacteria: case studies on bacterial volatile compounds for plant growth and immunity. Plant Mol Biol 90(6):677–687

Cocking EC, Stone PJ, Davey MR (2005) Symbiosome-like intracellular colonization of cereals and other crop plants by nitrogen-fixing bacteria for reduced inputs of synthetic nitrogen fertilizers. Sci China Ser C Life Sci 48(2):888–896

Cohen AC, Bottini R, Pontin M, Berli FJ, Moreno D, Boccanlandro H et al (2015) *Azospirillum brasilense* ameliorates the response of *Arabidopsis thaliana* to drought mainly via enhancement of ABA levels. Physiol Plant 153(1):79–90

Compant S, Kaplan H, Sessitsch A, Nowak J, Ait Barka E, Clément C (2008) Endophytic colonization of *Vitis vinifera* L. by *Burkholderia phytofirmans* strain PsJN: from the rhizosphere to inflorescence tissues. FEMS Microbiol Ecol 63(1):84–93

Compant S, Clément C, Sessitsch A (2010) Plant growth-promoting bacteria in the rhizo-and endosphere of plants: their role, colonization, mechanisms involved and prospects for utilization. Soil Biol Biochem 42(5):669–678

Conn V, Walker A, Franco C (2008) Endophytic actinobacteria induce defense pathways in *Arabidopsis thaliana*. Mol Plant-Microbe Interact 21(2):208–218

Constantin ME, de Lamo FJ, Vlieger BV, Rep M, Takken FL (2019) Endophyte-mediated resistance in tomato to *Fusarium oxysporum* is independent of ET, JA, and SA. Front Plant Sci 10:979

Costa FDC, de Melo IS (2012) Endophytic and rhizospheric bacteria from *Opuntia ficus-indica* mill and their ability to promote plant growth in cowpea, *Vigna unguiculata* (L.) Walp. Afr J Microbiol Res 6(6):1345–1353

Dastogeer KMG, Tumpa FH, Sultana A, Akter MA, Chakraborty A (2020) Plant microbiome–an account of the factors that shape community composition and diversity. Curr Plant Biol 23:100161. https://doi.org/10.1016/j.cpb.2020.100161

De Oliveira EJ, Rabinovitch L, Monnerat RG, Passos LKJ, Zahner V (2004) Molecular characterization of *Brevibacillus laterosporus* and its potential use in biological control. Appl Environ Microbiol 70(11):6657–6664

Deng Y, Chen H, Li C, Xu J, Qi Q, Xu Y et al (2019) Endophyte *Bacillus subtilis* evade plant defense by producing lantibiotic subtilomycin to mask self-produced flagellin. Communicat Biol 2(1):1–12

Desbrosses GJ, Stougaard J (2011) Root nodulation: a paradigm for how plant-microbe symbiosis influences host developmental pathways. Cell Host Microbe 10(4):348–358

Dimkpa C, Weinand T, Asch F (2009) Plant–rhizobacteria interactions alleviate abiotic stress conditions. Plant Cell Environ 32(12):1682–1694

Dobbelaere S, Okon Y (2007) The plant growth-promoting effect and plant responses. In: Elmerich C, Newton W (eds) Associative and endophytic nitrogen-fixing bacteria and cyanobacterial associations. Nitrogen Fixation: Origins, Applications, and Research Progress), vol 5. Springer, Dordrecht, pp 145–170

Dobbelaere S, Croonenborghs A, Thys A, Ptacek D, Vanderleyden J, Dutto P et al (2001) Responses of agronomically important crops to inoculation with *Azospirillum*. Funct Plant Biol 28(9):871–879. https://doi.org/10.1071/PP01074

Doty SL, Oakley B, Xin G, Kang JW, Singleton G, Khan Z et al (2009) Diazotrophic endophytes of native black cottonwood and willow. Symbiosis 47(1):23–33

Egamberdieva D, Wirth SJ, Alqarawi AA, Abd Allah EF, Hashem A (2017) Phytohormones and beneficial microbes: essential components for plants to balance stress and fitness. Front Microbiol 8:2104–2104. https://doi.org/10.3389/fmicb.2017.02104

Eid AM, Salim SS, Hassan SE-D, Ismail MA, Fouda A (2019) Role of endophytes in plant health and abiotic stress management. In: Kumar V, Prasad R, Kumar M, Choudhary DK (eds) Microbiome in plant health and disease: challenges and opportunities. Springer, Singapore, pp 119–144

Emamverdian A, Ding Y, Mokhberdoran F, Xie Y (2015). Heavy metal stress and some mechanisms of plant defense response. Sci World J 2015:Article ID 756120

Estendorfer J, Stempfhuber B, Haury P, Vestergaard G, Rillig MC, Joshi J et al (2017) The influence of land use intensity on the plant-associated microbiome of *Dactylis glomerata* L. Front Plant Sci 8:930

Etesami H, Maheshwari DK (2018) Use of plant growth promoting rhizobacteria (PGPRs) with multiple plant growth promoting traits in stress agriculture: action mechanisms and future prospects. Ecotoxicol Environ Saf 156:225–246. https://doi.org/10.1016/j.ecoenv.2018.03.013

Fadiji AE, Babalola OO (2020) Exploring the potentialities of beneficial endophytes for improved plant growth. Saudi J Biol Sci, In Press. https://doi.org/10.1016/j.sjbs.2020.08.002

Ferreira A, Pires R, Rabelo P, Oliveira R, Luz J, Brito C (2013) Implications of *Azospirillum brasilense* inoculation and nutrient addition on maize in soils of the Brazilian Cerrado under greenhouse and field conditions. Appl Soil Ecol 72:103–108

Firáková S, Šturdíková M, Múčková M (2007) Bioactive secondary metabolites produced by microorganisms associated with plants. Biologia 62(3):251–257

Franche C, Lindström K, Elmerich C (2009) Nitrogen-fixing bacteria associated with leguminous and non-leguminous plants. Plant Soil 321(1–2):35–59

Francis F, Jacquemyn H, Delvigne F, Lievens B (2020) From diverse origins to specific targets: role of microorganisms in indirect pest biological control. Insects 11(8):533

Fravel DR (1988) Role of antibiosis in the biocontrol of plant diseases. Annu Rev Phytopathol 26 (1):75–91

Fuentes-Ramírez LE, Caballero-Mellado J, Sepúlveda J, Martínez-Romero E (1999) Colonization of sugarcane by *Acetobacter diazotrophicus* is inhibited by high N-fertilization. FEMS Microbiol Ecol 29(2):117–128

Gaiero JR, McCall CA, Thompson KA, Day NJ, Best AS, Dunfield KE (2013) Inside the root microbiome: bacterial root endophytes and plant growth promotion. Am J Bot 100 (9):1738–1750

Garg N, Singla P (2012) The role of *Glomus mosseae* on key physiological and biochemical parameters of pea plants grown in arsenic contaminated soil. Sci Hortic 143:92–101

Garima G, Nath JP (2015) Screening of potential PGPR candidates as future biofertilizers-a strategic approach from lab to field. Res J Biotechnol 10(11):48–62

Gautam AK, Avasthi S (2019) Fungal endophytes: potential biocontrol agents in agriculture. In: Role of plant growth promoting microorganisms in sustainable agriculture and nanotechnology. Elsevier, pp 241–283

Ghazalibiglar H, Hampton JG, de Jong EVZ, Holyoake A (2016) Is induced systemic resistance the mechanism for control of black rot in *Brassica oleracea* by a *Paenibacillus* sp.? Biol Control 92:195–201

Glandorf DC, Verheggen P, Jansen T, Jorritsma J-W, Smit E, Leeflang P et al (2001) Effect of genetically modified *Pseudomonas putida* WCS358r on the fungal rhizosphere microflora of field-grown wheat. Appl Environ Microbiol 67(8):3371–3378

Glick BR (2005) Modulation of plant ethylene levels by the bacterial enzyme ACC deaminase. FEMS Microbiol Lett 251(1):1–7

Glick BR (2012) Plant growth-promoting bacteria: mechanisms and applications. Scientifica 2012: Article ID 963401

Gonthier P, Giordano L, Zampieri E, Lione G, Vizzini A, Colpaert JV et al (2019) An ectomycorrhizal symbiosis differently affects host susceptibility to two congeneric fungal pathogens. Fungal Ecol 39:250–256

Goswami D, Thakker JN, Dhandhukia PC (2015) Simultaneous detection and quantification of indole-3-acetic acid (IAA) and indole-3-butyric acid (IBA) produced by rhizobacteria from l-tryptophan (Trp) using HPTLC. J Microbiol Methods 110:7–14

Gouda S, Kerry RG, Das G, Paramithiotis S, Shin H-S, Patra JK (2018) Revitalization of plant growth promoting rhizobacteria for sustainable development in agriculture. Microbiol Res 206:131–140. https://doi.org/10.1016/j.micres.2017.08.016

Gough C, Cullimore J (2011) Lipo-chitooligosaccharide signaling in endosymbiotic plant-microbe interactions. Mol Plant-Microbe Interact 24(8):867–878. https://doi.org/10.1094/mpmi-01-11-0019

Gray EJ, Smith DL (2005) Intracellular and extracellular PGPR: commonalities and distinctions in the plant–bacterium signaling processes. Soil Biol Biochem 37(3):395–412. https://doi.org/10.1016/j.soilbio.2004.08.030

Grover M, Ali SZ, Sandhya V, Rasul A, Venkateswarlu B (2011) Role of microorganisms in adaptation of agriculture crops to abiotic stresses. World J Microbiol Biotechnol 27 (5):1231–1240. https://doi.org/10.1007/s11274-010-0572-7

Gunatilaka AAL (2006) Natural products from plant-associated microorganisms: distribution, structural diversity, bioactivity, and implications of their occurrence. J Nat Prod 69 (3):509–526. https://doi.org/10.1021/np058128n

Gupta G, Panwar J, Jha PN (2013) Natural occurrence of *Pseudomonas aeruginosa*, a dominant cultivable diazotrophic endophytic bacterium colonizing *Pennisetum glaucum* (L.) R. Br. Appl Soil Ecol 64:252–261

Haney CH, Samuel BS, Bush J, Ausubel FM (2015) Associations with rhizosphere bacteria can confer an adaptive advantage to plants. Nat Plants 1(6):15051. https://doi.org/10.1038/nplants.2015.51

Hardoim PR, van Elsas JD (2013) Properties of bacterial endophytes leading to maximized host fitness. In: de Bruijn FJ (ed) Molecular microbial ecology of the rhizosphere, vol 1. John Wiley & Sons, pp 405–411

Hardoim PR, van Overbeek LS, van Elsas JD (2008) Properties of bacterial endophytes and their proposed role in plant growth. Trends Microbiol 16(10):463–471

Hardoim PR, Van Overbeek LS, Berg G, Pirttilä AM, Compant S, Campisano A et al (2015) The hidden world within plants: ecological and evolutionary considerations for defining functioning of microbial endophytes. Microbiol Mol Biol Rev 79(3):293–320

Harman GE, Howell CR, Viterbo A, Chet I, Lorito M (2004) *Trichoderma* species—opportunistic, avirulent plant symbionts. Nat Rev Microbiol 2(1):43–56

Hartmann A, Rothballer M, Schmid M (2008) Lorenz Hiltner, a pioneer in rhizosphere microbial ecology and soil bacteriology research. Plant Soil 312(1–2):7–14

Hashem A, Abd Allah E, Alqarawi A, Al Huqail AA, Egamberdieva D, Wirth S (2016) Alleviation of cadmium stress in *Solanum lycopersicum* L. by arbuscular mycorrhizal fungi via induction of acquired systemic tolerance. Saudi J Biol Sci 23(2):272–281

Hayat S, Hayat Q, Alyemeni MN, Wani AS, Pichtel J, Ahmad A (2012) Role of proline under changing environments: a review. Plant Signal Behav 7(11):1456–1466

He R-L, Wang G-P, Liu X, Zhang C-L, Lin F-C (2009) Antagonistic bioactivity of an endophytic bacterium isolated from *Epimedium brevicornu* Maxim. Afr J Biotechnol 8(2):191–195

Hellwig V, Grothe T, Mayer-Bartschmid A, Endermann R, Geschke F-U, Henkel T et al (2002) Altersetin, a new antibiotic from cultures of endophytic *Alternaria* spp. J Antibiot 55 (10):881–892

Hill D, Stein J, Torkewitz N, Morse A, Howell C, Pachlatko J et al (1994) Cloning of genes involved in the synthesis of pyrrolnitrin from *Pseudomonas fluorescens* and role of pyrrolnitrin synthesis in biological control of plant disease. Appl Environ Microbiol 60(1):78–85

Ho Y-N, Chiang H-M, Chao C-P, Su C-C, Hsu H-F, Guo C-T et al (2015) In planta biocontrol of soilborne *Fusarium* wilt of banana through a plant endophytic bacterium, *Burkholderia cenocepacia* 869T2. Plant Soil 387(1–2):295–306

Huang L-H, Yuan M-Q, Ao X-J, Ren A-Y, Zhang H-B, Yang M-Z (2018) Endophytic fungi specifically introduce novel metabolites into grape flesh cells *in vitro*. PLoS One 13(5): e0196996

Hussain SS, Mehnaz S, Siddique KH (2018) Harnessing the plant microbiome for improved abiotic stress tolerance. In: Egamberdieva D, Ahmad P (eds) Plant microbiome: stress response. Springer, Singapore, pp 21–43

Islam F, Yasmeen T, Ali Q, Mubin M, Ali S, Arif MS et al (2016) Copper-resistant bacteria reduces oxidative stress and uptake of copper in lentil plants: potential for bacterial bioremediation. Environ Sci Pollut Res 23(1):220–233

Jain S, Vaishnav A, Kasotia A, Kumari S, Gaur RK, Choudhary DK (2013) Bacteria-induced systemic resistance and growth promotion in *Glycine max* L. Merrill upon challenge inoculation with *Fusarium oxysporum*. Proc Natl Acad Sci India Sect B Biol Sci 83(4):561–567

Jha CK, Saraf M (2015) Plant growth promoting rhizobacteria (PGPR): a review. J Agricult Res Dev 5(2):108–119

Jha Y, Subramanian R (2014) PGPR regulate caspase-like activity, programmed cell death, and antioxidant enzyme activity in paddy under salinity. Physiol Mol Biol Plants 20(2):201–207

Jha Y, Subramanian R, Patel S (2011) Combination of endophytic and rhizospheric plant growth promoting rhizobacteria in *Oryza sativa* shows higher accumulation of osmoprotectant against saline stress. Acta Physiol Plant 33(3):797–802

Jiang S, Zhang D, Wang L, Pan J, Liu Y, Kong X et al (2013) A maize calcium-dependent protein kinase gene, ZmCPK4, positively regulated abscisic acid signaling and enhanced drought stress tolerance in transgenic *Arabidopsis*. Plant Physiol Biochem 71:112–120

Jing YX, Yan JL, He HD, Yang DJ, Xiao L, Zhong T et al (2014) Characterization of bacteria in the rhizosphere soils of *Polygonum pubescens* and their potential in promoting growth and cd, Pb, Zn uptake by *Brassica napus*. Int J Phytoremediation 16(4):321–333

Kandel SL, Joubert PM, Doty SL (2017) Bacterial endophyte colonization and distribution within plants. Microorganisms 5(4):77

Kaneko T, Minamisawa K, Isawa T, Nakatsukasa H, Mitsui H, Kawaharada Y et al (2010) Complete genomic structure of the cultivated rice endophyte *Azospirillum* sp. B510. DNA Res 17(1):37–50. https://doi.org/10.1093/dnares/dsp026

Khan MS, Zaidi A, Ahemad M, Oves M, Wani PA (2010) Plant growth promotion by phosphate solubilizing fungi–current perspective. Arch Agron Soil Sci 56(1):73–98

Khare E, Mishra J, Arora NK (2018) Multifaceted interactions between endophytes and plant: developments and prospects. Front Microbiol 9:2732

Kiely PD, Haynes J, Higgins C, Franks A, Mark GL, Morrissey JP et al (2006) Exploiting new systems-based strategies to elucidate plant-bacterial interactions in the rhizosphere. Microb Ecol 51(3):257–266

Kloepper JW, Ryu C-M (2006) Bacterial endophytes as elicitors of induced systemic resistance. In: Schulz BJE, Boyle CJC, Sieber TN (eds) Microbial root endophytes, vol 9. Springer, Berlin, pp 33–52

Knief C (2014) Analysis of plant microbe interactions in the era of next generation sequencing technologies. Front Plant Sci 5:216

Köhl J, Kolnaar R, Ravensberg WJ (2019) Mode of action of microbial biological control agents against plant diseases: relevance beyond efficacy. Front Plant Sci 10:845

Korolev N, David DR, Elad Y (2008) The role of phytohormones in basal resistance and *Trichoderma*-induced systemic resistance to *Botrytis cinerea* in *Arabidopsis thaliana*. BioControl 53(4):667–683

Kouas S, Labidi N, Debez A, Abdelly C (2005) Effect of P on nodule formation and N fixation in bean. Agron Sustain Dev 25(3):389–393

Kumar A, Verma JP (2018) Does plant—microbe interaction confer stress tolerance in plants: a review? Microbiol Res 207:41–52

Lakshmanan V, Ray P, Craven KD (2017) Toward a resilient, functional microbiome: drought tolerance-alleviating microbes for sustainable agriculture. In: Sunkar R (ed) Methods in molecular biology: plant stress tolerance. Humana Press, Springer, New York, pp 69–84
Lata R, Chowdhury S, Gond SK, White JF Jr (2018) Induction of abiotic stress tolerance in plants by endophytic microbes. Lett Appl Microbiol 66(4):268–276
Long HH, Schmidt DD, Baldwin IT (2008) Native bacterial endophytes promote host growth in a species-specific manner; phytohormone manipulations do not result in common growth responses. PLoS One 3(7):e2702
Lugtenberg BJ, Caradus JR, Johnson LJ (2016) Fungal endophytes for sustainable crop production. FEMS Microbiol Ecol 92(12)
Lynch JP, Wojciechowski T (2015) Opportunities and challenges in the subsoil: pathways to deeper rooted crops. J Exp Bot 66(8):2199–2210
Ma Y (2017) Beneficial bacteria for disease suppression and plant growth promotion. In: Singh D, Singh H, Prabha R (eds) Plant-microbe interactions in agro-ecological perspectives. Springer, Singapore, pp 513–529
Ma Y, Rajkumar M, Zhang C, Freitas H (2016) Beneficial role of bacterial endophytes in heavy metal phytoremediation. J Environ Manag 174:14–25
Madhaiyan M, Poonguzhali S, Sa T (2007) Metal tolerating methylotrophic bacteria reduces nickel and cadmium toxicity and promotes plant growth of tomato (*Lycopersicon esculentum* L.). Chemosphere 69(2):220–228
Malfanova N, Lugtenberg B, Berg G (2013) Bacterial endophytes: who and where, and what are they doing there? In: de Bruijn FJ (ed) Molecular microbial ecology of the rhizosphere, vol 1. John Wile & Sons, pp 393–403
Malmierca M, Cardoza R, Alexander N, McCormick S, Hermosa R, Monte E et al (2012) Involvement of *Trichoderma trichothecenes* in the biocontrol activity and induction of plant defense-related genes. Appl Environ Microbiol 78(14):4856–4868
McLellan CA, Turbyville TJ, Wijeratne EMK, Kerschen A, Vierling E, Queitsch C et al (2007) A rhizosphere fungus enhances *Arabidopsis* thermotolerance through production of an HSP90 inhibitor. Plant Physiol 145(1):174–182. https://doi.org/10.1104/pp.107.101808
McNear DH Jr (2013) The rhizosphere - roots, soil and everything in between. Nat Educ Knowl 4 (3):1
Meena KK, Sorty AM, Bitla UM, Choudhary K, Gupta P, Pareek A et al (2017) Abiotic stress responses and microbe-mediated mitigation in plants: the omics strategies. Front Plant Sci 8:172
Mengistu AA (2020) Endophytes: colonization, behaviour, and their role in defense mechanism. Int J Microbiol 2020:6927219. https://doi.org/10.1155/2020/6927219
Micallef SA, Shiaris MP, Colón-Carmona A (2009) Influence of *Arabidopsis thaliana* accessions on rhizobacterial communities and natural variation in root exudates. J Exp Bot 60 (6):1729–1742. https://doi.org/10.1093/jxb/erp053
Miliute I, Buzaite O, Baniulis D, Stanys V (2015) Bacterial endophytes in agricultural crops and their role in stress tolerance: a review. Zemdirbyste-Agriculture 102(4):465–478
Mirza MS, Ahmad W, Latif F, Haurat J, Bally R, Normand P et al (2001) Isolation, partial characterization, and the effect of plant growth-promoting bacteria (PGPB) on micro-propagated sugarcane *in vitro*. Plant Soil 237(1):47–54
Mishra PK, Mishra S, Selvakumar G, Bisht SC, Bisht JK, Kundu S et al (2008) Characterisation of a psychrotolerant plant growth promoting *Pseudomonas* sp. strain PGERs17 (MTCC 9000) isolated from North Western Indian Himalayas. Ann Microbiol 58(4):561–568
Mishra S, Upadhyay RS, Nautiyal CS (2013) Unravelling the beneficial role of microbial contributors in reducing the allelopathic effects of weeds. Appl Microbiol Biotechnol 97 (13):5659–5668. https://doi.org/10.1007/s00253-013-4885-y
Mitter B, Petric A, SG Chain P, Trognitz F, Nowak J, Compant S et al (2013) Genome analysis, ecology, and plant growth promotion of the endophyte *Burkholderia phytofirmans* strain PsJN. Mol Microbial Ecol Rhizosphere 1:865–874
Mohiuddin A (2019) Chemistry of secondary metabolites. Ann Clin Toxicol 2(1):1014

Molla AH, Haque MM, Haque MA, Ilias G (2012) Trichoderma-enriched biofertilizer enhances production and nutritional quality of tomato (*Lycopersicon esculentum* Mill.) and minimizes NPK fertilizer use. Agricult Res 1(3):265–272

Moricca S, Ragazzi A (2008) Fungal endophytes in Mediterranean oak forests: a lesson from *Discula quercina*. Phytopathology 98(4):380–386

Nadeem SM, Zahir ZA, Naveed M, Arshad M (2007) Preliminary investigations on inducing salt tolerance in maize through inoculation with rhizobacteria containing ACC deaminase activity. Can J Microbiol 53(10):1141–1149

Nadeem SM, Ahmad M, Zahir ZA, Javaid A, Ashraf M (2014) The role of mycorrhizae and plant growth promoting rhizobacteria (PGPR) in improving crop productivity under stressful environments. Biotechnol Adv 32(2):429–448

Nadeem SM, Naveed M, Ahmad M, Zahir ZA (2015) Rhizosphere bacteria for crop production and improvement of stress tolerance: mechanisms of action, applications, and future prospects. In: Arora N (ed) Plant microbes symbiosis: applied facets. Springer, New Delhi, pp 1–36

Naveed M, Hussain MB, Zahir ZA, Mitter B, Sessitsch A (2014) Drought stress amelioration in wheat through inoculation with *Burkholderia phytofirmans* strain PsJN. Plant Growth Regul 73 (2):121–131

Oerke E-C, Dehne H-W (2004) Safeguarding production—losses in major crops and the role of crop protection. Crop Prot 23(4):275–285

Oliveira A d, Urquiaga S, Döbereiner J, Baldani J (2002) The effect of inoculating endophytic N 2-fixing bacteria on micropropagated sugarcane plants. Plant Soil 242(2):205–215

Oves M, Khan MS, Zaidi A (2013) Chromium reducing and plant growth promoting novel strain *Pseudomonas aeruginosa* OSG41 enhance chickpea growth in chromium amended soils. Eur J Soil Biol 56:72–83

Ownley BH, Windham MT (2007) Biological control of plant pathogens. In: Trigiano RN, Windham MT, Windham AS (eds) Plant pathology concepts and laboratory exercises, 2nd edn. CRC Press, New York, pp 423–436

Ownley BH, Gwinn KD, Vega FE (2010) Endophytic fungal entomopathogens with activity against plant pathogens: ecology and evolution. BioControl 55(1):113–128. https://doi.org/10.1007/s10526-009-9241-x

Pacifico D, Squartini A, Crucitti D, Barizza E, Lo Schiavo F, Muresu R et al (2019) The role of the endophytic microbiome in the grapevine response to environmental triggers. [Review]. Front Plant Sci 10:1256. https://doi.org/10.3389/fpls.2019.01256

Pampulha ME, Oliveira A (2006) Impact of an herbicide combination of *Bromoxynil* and *Prosulfuron* on soil microorganisms. Curr Microbiol 53(3):238–243. https://doi.org/10.1007/s00284-006-0116-4

Pandey PK, Yadav SK, Singh A, Sarma BK, Mishra A, Singh HB (2012) Cross-species alleviation of biotic and abiotic stresses by the endophyte *Pseudomonas aeruginosa* PW09. J Phytopathol 160(10):532–539. https://doi.org/10.1111/j.1439-0434.2012.01941.x

Pandey PK, Singh S, Singh AK, Samanta R, Yadav RNS, Singh MC (2016) Inside the plant: bacterial endophytes and abiotic stress alleviation. J Appl Nat Sci 8(4):1899–1904

Pangesti N, Reichelt M, van de Mortel JE, Kapsomenou E, Gershenzon J, van Loon JJA et al (2016) Jasmonic acid and ethylene signaling pathways regulate glucosinolate levels in plants during rhizobacteria-induced systemic resistance against a leaf-chewing herbivore. J Chem Ecol 42 (12):1212–1225. https://doi.org/10.1007/s10886-016-0787-7

Patten CL, Glick BR (2002) Role of *Pseudomonas putida* indoleacetic acid in development of the host plant root system. Appl Environ Microbiol 68(8):3795–3801

Pavlo A, Leonid O, Iryna Z, Natalia K, Maria PA (2011) Endophytic bacteria enhancing growth and disease resistance of potato (*Solanum tuberosum* L.). Biol Control 56(1):43–49

Peiffer JA, Spor A, Koren O, Jin Z, Tringe SG, Dangl JL et al (2013) Diversity and heritability of the maize rhizosphere microbiome under field conditions. Proc Natl Acad Sci 110 (16):6548–6553. https://doi.org/10.1073/pnas.1302837110

Pereyra MA, Zalazar CA, Barassi CA (2006) Root phospholipids in *Azospirillum*-inoculated wheat seedlings exposed to water stress. Plant Physiol Biochem 44(11):873–879. https://doi.org/10.1016/j.plaphy.2006.10.020

Pickles BJ, Twieg BD, O'Neill GA, Mohn WW, Simard SW (2015) Local adaptation in migrated interior Douglas-fir seedlings is mediated by ectomycorrhizas and other soil factors. New Phytol 207(3):858–871. https://doi.org/10.1111/nph.13360

Pieterse CMJ, Does DVD, Zamioudis C, Leon-Reyes A, Wees SCMV (2012) Hormonal modulation of plant immunity. Annu Rev Cell Dev Biol 28(1):489–521. https://doi.org/10.1146/annurev-cellbio-092910-154055

Pishchik VN, Vorobyev NI, Chernyaeva II, Timofeeva SV, Kozhemyakov AP, Alexeev YV et al (2002) Experimental and mathematical simulation of plant growth promoting rhizobacteria and plant interaction under cadmium stress. Plant Soil 243(2):173–186. https://doi.org/10.1023/A:1019941525758

Porcel R, Zamarreño ÁM, García-Mina JM, Aroca R (2014) Involvement of plant endogenous ABA in *Bacillus megaterium* PGPR activity in tomato plants. BMC Plant Biol 14(1):36. https://doi.org/10.1186/1471-2229-14-36

Potters G, Pasternak TP, Guisez Y, Palme KJ, Jansen MA (2007) Stress-induced morphogenic responses: growing out of trouble? Trends Plant Sci 12(3):98–105

Pozo MJ, Jung SC, Martínez-Medina A, López-Ráez JA, Azcón-Aguilar C, Barea J-M (2013) Root allies: arbuscular mycorrhizal fungi help plants to cope with biotic stresses. In: Aroca R (ed) Symbiotic endophytes. Springer, Berlin, pp 289–307

Pundir RK, Jain P (2015) Mechanism of prevention and control of medicinal plant-associated diseases. In: Egamberdieva D, Shrivastava S, Varma A (eds) Plant-growth-promoting Rhizobacteria (PGPR) and medicinal plants. Springer, Cham, pp 231–246

Rabiey M, Hailey LE, Roy SR, Grenz K, Al-Zadjali MAS, Barrett GA et al (2019) Endophytes vs tree pathogens and pests: can they be used as biological control agents to improve tree health? Eur J Plant Pathol 155(3):711–729. https://doi.org/10.1007/s10658-019-01814-y

Ragazzi A, Moricca S, Capretti P, Dellavalle I, Mancini F, Turco E (2001) Endophytic fungi in *Quercus cerris*: isolation frequency in relation to phenological phase, tree health and the organ affected. Phytopathol Mediterr 40(2):165–171

Rajkumar M, Lee KJ, Lee WH, Banu J (2005) Growth of *Brassica juncea* under chromium stress: influence of siderophores and indole 3 acetic acid producing rhizosphere bacteria. J Environ Biol 26(4):693–699

Ramachandran K, Srinivasan V, Hamza S, Anandaraj M (2007) Phosphate solubilizing bacteria isolated from the rhizosphere soil and its growth promotion on black pepper (*Piper nigrum* L.) cuttings. In: Velázquez E, Rodríguez-Barrueco C (eds) First International Meeting on Microbial Phosphate Solubilization, Dordrecht, 2007, (Developments in Plant and Soil Sciences), vol 102. Springer, Netherlands, pp 325–331

Rana K, Kour D, Yadav A, Kumar V, Dhaliwal H (2016a) Biotechnological applications of endophytic microbes associated with barley (Hordeum vulgare L.) growing in Indian Himalayan regions. In: Proceeding of 86th annual session of NASI & symposium on "science, technology and entrepreneurship for human welfare in the Himalayan Region", 2016a (p 80)

Rana K, Kour D, Yadav A, Kumar V, Dhaliwal H (2016b) Endophytic microbes from wheat: diversity and biotechnological applications for sustainable agriculture. In: Proceeding of 57th association of microbiologist of India & International symposium on "microbes and biosphere: What's new What's next", 2016b, p 453

Rashid MH-O, Chung YR (2017) Induction of systemic resistance against insect herbivores in plants by beneficial soil microbes. [Review]. Front Plant Sci 8:1816, https://doi.org/10.3389/fpls.2017.01816

Rasmann S, Bennett A, Biere A, Karley A, Guerrieri E (2017) Root symbionts: powerful drivers of plant above and belowground indirect defenses. Insect Sci 24(6):947–960. https://doi.org/10.1111/1744-7917.12464

Reinhold-Hurek B, Hurek T (2011) Living inside plants: bacterial endophytes. Curr Opin Plant Biol 14(4):435–443. https://doi.org/10.1016/j.pbi.2011.04.004

Ren J-H, Li H, Wang Y-F, Ye J-R, Yan A-Q, Wu X-Q (2013) Biocontrol potential of an endophytic *Bacillus pumilus* JK-SX001 against poplar canker. Biol Control 67(3):421–430. https://doi.org/10.1016/j.biocontrol.2013.09.012

Rengel Z, Marschner P (2005) Nutrient availability and management in the rhizosphere: exploiting genotypic differences. New Phytol 168(2):305–312. https://doi.org/10.1111/j.1469-8137.2005.01558.x

Rodrı́guez H, Fraga R (1999) Phosphate solubilizing bacteria and their role in plant growth promotion. Biotechnol Adv 17(4):319–339. https://doi.org/10.1016/S0734-9750(99)00014-2

Rosenblueth M, Martínez-Romero E (2004) *Rhizobium etli* maize populations and their competitiveness for root colonization. Arch Microbiol 181(5):337–344. https://doi.org/10.1007/s00203-004-0661-9

Rozpądek P, Domka AM, Nosek M, Ważny R, Jędrzejczyk RJ, Wiciarz M et al (2018) The role of strigolactone in the cross-talk between *Arabidopsis thaliana* and the endophytic fungus *Mucor* sp. [Original Research]. Front Microbiol 9:441. https://doi.org/10.3389/fmicb.2018.00441

Ruiu L (2018) Microbial biopesticides in agroecosystems. Agronomy 8:235

Saeidnejad AH, Pouramir F, Naghizadeh M (2012) Improving chilling tolerance of maize seedlings under cold conditions by spermine application. Notulae Scientia Biologicae 4(3):110–117

Şahin F, Çakmakçi R, Kantar F (2004) Sugar beet and barley yields in relation to inoculation with N2-fixing and phosphate solubilizing bacteria. Plant Soil 265(1):123–129. https://doi.org/10.1007/s11104-005-0334-8

Saikia R, Gogoi DK, Mazumder S, Yadav A, Sarma RK, Bora TC et al (2011) *Brevibacillus laterosporus* strain BPM3, a potential biocontrol agent isolated from a natural hot water spring of Assam, India. Microbiol Res 166(3):216–225. https://doi.org/10.1016/j.micres.2010.03.002

Saikkonen K, Lehtonen P, Helander M, Koricheva J, Faeth SH (2006) Model systems in ecology: dissecting the endophyte–grass literature. Trends Plant Sci 11(9):428–433. https://doi.org/10.1016/j.tplants.2006.07.001

Saleem M, Arshad M, Hussain S, Bhatti AS (2007) Perspective of plant growth promoting rhizobacteria (PGPR) containing ACC deaminase in stress agriculture. J Ind Microbiol Biotechnol 34(10):635–648

Samain E, van Tuinen D, Jeandet P, Aussenac T, Selim S (2017) Biological control of septoria leaf blotch and growth promotion in wheat by *Paenibacillus* sp. strain B2 and *Curtobacterium plantarum* strain EDS. Biol Control 114:87–96. https://doi.org/10.1016/j.biocontrol.2017.07.012

Santoro MV, Cappellari L, Giordano W, Banchio E (2015) Systemic induction of secondary metabolite biosynthesis in medicinal aromatic plants mediated by rhizobacteria. In: Egamberdieva D, Shrivastava S, Varma A (eds) Plant-growth-promoting Rhizobacteria (PGPR) and medicinal plants. Springer, Cham, pp 263–285

Santos MLD, Berlitz DL, Wiest SLF, Schünemann R, Knaak N, Fiuza LM (2018) Benefits associated with the interaction of endophytic bacteria and plants. Braz Arch Biol Technol 61: e18160431

Santoyo G, Moreno-Hagelsieb G, del Carmen Orozco-Mosqueda M, Glick BR (2016) Plant growth-promoting bacterial endophytes. Microbiol Res 183:92–99. https://doi.org/10.1016/j.micres.2015.11.008

Selvakumar G, Joshi P, Nazim S, Mishra P, Bisht J, Gupta H (2009) Phosphate solubilization and growth promotion by *Pseudomonas fragi* CS11RH1 (MTCC 8984), a psychrotolerant bacterium isolated from a high altitude Himalayan rhizosphere. Biologia 64(2):239–245. https://doi.org/10.2478/s11756-009-0041-7

Sen K, Sengupta C, Saha J (2014) PGPR consortium in alleviating downy mildew of cucumber. Int J Plant Animal Environ Sci 4:150–159

Shahid M, Khalid S, Abbas G, Shahid N, Nadeem M, Sabir M et al (2015) Heavy metal stress and crop productivity. In: Hakeem KR (ed) Crop production and global environmental issues. Springer, Cham, pp 1–25

Shameer S (2016) Haloalkaliphilic *Bacillus* species from solar salterns: an ideal prokaryote for bioprospecting studies. Ann Microbiol 66(3):1315–1327. https://doi.org/10.1007/s13213-016-1221-7

Shameer S, Prasad TNVKV (2018) Plant growth promoting rhizobacteria for sustainable agricultural practices with special reference to biotic and abiotic stresses. Plant Growth Regul 84 (3):603–615. https://doi.org/10.1007/s10725-017-0365-1

Sharma S, Kulkarni J, Jha B (2016) Halotolerant rhizobacteria promote growth and enhance salinity tolerance in peanut. [original research]. Front Microbiol 7:1600. https://doi.org/10.3389/fmicb.2016.01600

Shaw MW, Osborne TM (2011) Geographic distribution of plant pathogens in response to climate change. Plant Pathol 60(1):31–43. https://doi.org/10.1111/j.1365-3059.2010.02407.x

Sheoran N, Valiya Nadakkakath A, Munjal V, Kundu A, Subaharan K, Venugopal V et al (2015) Genetic analysis of plant endophytic *Pseudomonas putida* BP25 and chemo-profiling of its antimicrobial volatile organic compounds. Microbiol Res 173:66–78. https://doi.org/10.1016/j.micres.2015.02.001

Shi Y, Lou K, Li C (2010) Growth and photosynthetic efficiency promotion of sugar beet (*Beta vulgaris* L.) by endophytic bacteria. Photosynth Res 105(1):5–13. https://doi.org/10.1007/s11120-010-9547-7

Shinwari ZK, Tanveer F, Iqrar I (2019) Role of microbes in plant health, disease management, and abiotic stress management. In: Kumar V, Prasad R, Kumar M, Choudhary DK (eds) Microbiome in plant health and disease: challenges and opportunities. Springer Singapore, Singapore, pp 231–250

Shoda M (2000) Bacterial control of plant diseases. J Biosci Bioeng 89(6):515–521. https://doi.org/10.1016/S1389-1723(00)80049-3

Shrivastava P, Kumar R (2015) Soil salinity: a serious environmental issue and plant growth promoting bacteria as one of the tools for its alleviation. Saudi J Biol Sci 22(2):123–131. https://doi.org/10.1016/j.sjbs.2014.12.001

Si P, Shao W, Yu H, Yang X, Gao D, Qiao X, et al (2018) Rhizosphere microenvironments of eight common deciduous fruit trees were shaped by microbes in northern China. [Original Research]. Front Microbiol 9:3147. https://doi.org/10.3389/fmicb.2018.03147

Singer S (1996) The utility of strains of morphological group II Bacillus. In: Neidleman SL, Laskin AI (eds) Advances in applied microbiology, vol 42. Academic Press, pp 219–261

Smith SE, Smith FA (2011) Roles of arbuscular mycorrhizas in plant nutrition and growth: new paradigms from cellular to ecosystem scales. Annu Rev Plant Biol 62(1):227–250. https://doi.org/10.1146/annurev-arplant-042110-103846

Somers E, Vanderleyden J, Srinivasan M (2004) Rhizosphere bacterial signalling: a love parade beneath our feet. Crit Rev Microbiol 30(4):205–240. https://doi.org/10.1080/10408410490468786

Sosa-Hernández MA, Leifheit EF, Ingraffia R, Rillig MC (2019) Subsoil arbuscular mycorrhizal fungi for sustainability and climate-smart agriculture: a solution right under our feet? Front Microbiol 10:744. https://doi.org/10.3389/fmicb.2019.00744

Souza R d, Ambrosini A, Passaglia LMP (2015) Plant growth-promoting bacteria as inoculants in agricultural soils. Genet Mol Biol 38:401–419

Spence C, Bais H (2015) Role of plant growth regulators as chemical signals in plant–microbe interactions: a double edged sword. Curr Opin Plant Biol 27:52–58

Steenhoudt O, Vanderleyden J (2000) *Azospirillum*, a free-living nitrogen-fixing bacterium closely associated with grasses: genetic, biochemical and ecological aspects. FEMS Microbiol Rev 24 (4):487–506. https://doi.org/10.1111/j.1574-6976.2000.tb00552.x

Steinkellner S, Lendzemo V, Langer I, Schweiger P, Khaosaad T, Toussaint J-P et al (2007) Flavonoids and strigolactones in root exudates as signals in symbiotic and pathogenic plant-fungus interactions. Molecules 12(7):1290–1306

Strobel GA, Dirkse E, Sears J, Markworth C (2001) Volatile antimicrobials from *Muscodor albus*, a novel endophytic fungus. Microbiology 147(11):2943–2950. https://doi.org/10.1099/00221287-147-11-2943

Syed S, Chinthala P (2015) Heavy metal detoxification by different *Bacillus* species isolated from solar salterns. Scientifica 2015:319760. https://doi.org/10.1155/2015/319760

Syed-Ab-Rahman SF, Carvalhais LC, Chua E, Xiao Y, Wass TJ, Schenk PM (2018) Identification of soil bacterial isolates suppressing different *Phytophthora* spp. and promoting plant growth. [Original Research]. Front Plant Sci 9(1502). https://doi.org/10.3389/fpls.2018.01502

Sziderics A, Rasche F, Trognitz F, Sessitsch A, Wilhelm E (2007) Bacterial endophytes contribute to abiotic stress adaptation in pepper plants (*Capsicum annuum* L.). Can J Microbiol 53 (11):1195–1202

Taghavi S, van der Lelie D, Hoffman A, Zhang Y-B, Walla MD, Vangronsveld J et al (2010) Genome sequence of the plant growth promoting endophytic bacterium *Enterobacter* sp. 638. PLOS Genet 6(5):e1000943. https://doi.org/10.1371/journal.pgen.1000943

Talaat NB (2015) Effective microorganisms improve growth performance and modulate the ROS-scavenging system in common bean (*Phaseolus vulgaris* L.) plants exposed to salinity stress. J Plant Growth Regul 34(1):35–46. https://doi.org/10.1007/s00344-014-9440-2

Talaat NB, Shawky BT (2013) Modulation of nutrient acquisition and polyamine pool in salt-stressed wheat (*Triticum aestivum* L.) plants inoculated with arbuscular mycorrhizal fungi. Acta Physiol Plant 35(8):2601–2610

Talaat NB, Shawky BT (2014) Protective effects of arbuscular mycorrhizal fungi on wheat (*Triticum aestivum* L.) plants exposed to salinity. Environ Exp Bot 98:20–31. https://doi.org/10.1016/j.envexpbot.2013.10.005

Talaat NB, Shawky BT (2016) Dual application of 24-epibrassinolide and spermine confers drought stress tolerance in maize (*Zea mays* L.) by modulating polyamine and protein metabolism. J Plant Growth Regul 35(2):518–533. https://doi.org/10.1007/s00344-015-9557-y

Tamosiune I, Baniulis D, Stanys V (2017) Role of endophytic bacteria in stress tolerance of agricultural plants: diversity of microorganisms and molecular mechanisms. In: Kumar V, Kumar M, Sharma S, Prasad R (eds) Probiotics in agroecosystem. Springer, Singapore, pp 1–29

Tank N, Saraf M (2010) Salinity-resistant plant growth promoting rhizobacteria ameliorates sodium chloride stress on tomato plants. J Plant Interact 5(1):51–58. https://doi.org/10.1080/17429140903125848

Tardieu F (2013) Plant response to environmental conditions: assessing potential production, water demand, and negative effects of water deficit. [Methods]. Front Physiol 4:17. https://doi.org/10.3389/fphys.2013.00017

Tawaraya K, Naito M, Wagatsuma T (2006) Solubilization of insoluble inorganic phosphate by hyphal exudates of arbuscular mycorrhizal fungi. J Plant Nutr 29(4):657–665. https://doi.org/10.1080/01904160600564428

Timmusk S, Behers L, Muthoni J, Muraya A, Aronsson A-C (2017) Perspectives and challenges of microbial application for crop improvement. [Review]. Front Plant Sci 8:49. https://doi.org/10.3389/fpls.2017.00049

Trivedi P, Schenk PM, Wallenstein MD, Singh BK (2017) Tiny microbes, big yields: enhancing food crop production with biological solutions. Microb Biotechnol 10(5):999–1003. https://doi.org/10.1111/1751-7915.12804

van Lenteren JC, Bolckmans K, Köhl J, Ravensberg WJ, Urbaneja A (2018) Biological control using invertebrates and microorganisms: plenty of new opportunities. BioControl 63(1):39–59. https://doi.org/10.1007/s10526-017-9801-4

van Loon LC (2007) Plant responses to plant growth-promoting rhizobacteria. In: Bakker PAHM, Raaijmakers JM, Bloemberg G, Höfte M, Lemanceau P, Cooke BM (eds) New perspectives and approaches in plant growth-promoting Rhizobacteria research. Springer Netherlands, Dordrecht, pp 243–254

Van Oosten VR, Bodenhausen N, Reymond P, Van Pelt JA, Van Loon LC, Dicke M et al (2008) Differential effectiveness of microbially induced resistance against herbivorous insects in *Arabidopsis*. Mol Plant-Microbe Interact 21(7):919–930. https://doi.org/10.1094/mpmi-21-7-0919

Vanderstraeten L, Van Der Straeten D (2017) Accumulation and transport of 1-aminocyclopropane-1-carboxylic acid (ACC) in plants: current status, considerations for future research and agronomic applications. [Review]. Front Plant Sci 8(38). https://doi.org/10.3389/fpls.2017.00038

Vega FE (2008) Insect pathology and fungal endophytes. J Invertebr Pathol 98(3):277–279

Viebahn M, Veenman C, Wernars K, van Loon LC, Smit E, Bakker PA (2005) Assessment of differences in ascomycete communities in the rhizosphere of field-grown wheat and potato. FEMS Microbiol Ecol 53(2):245–253

Vimal SR, Singh JS, Arora NK, Singh S (2017) Soil-plant-microbe interactions in stressed agriculture management: a review. Pedosphere 27(2):177–192

Wang C, White PJ, Li C (2017) Colonization and community structure of arbuscular mycorrhizal fungi in maize roots at different depths in the soil profile respond differently to phosphorus inputs on a long-term experimental site. Mycorrhiza 27(4):369–381

Waqas M, Khan AL, Shahzad R, Ullah I, Khan AR, Lee I-J (2015) Mutualistic fungal endophytes produce phytohormones and organic acids that promote japonica rice plant growth under prolonged heat stress. J Zhejiang Univ-Sci B 16(12):1011–1018

Webber J (1981) A natural biological control of Dutch elm disease. Nature 292(5822):449–451

Weil RR, Brady NC (2016) The nature and properties of soils. Pearson, Columbus

Weilharter A, Mitter B, Shin MV, Chain PSG, Nowak J, Sessitsch A (2011) Complete genome sequence of the plant growth-promoting endophyte *Burkholderia phytofirmans* strain PsJN. J Bacteriol 193(13):3383–3384. https://doi.org/10.1128/jb.05055-11

White JF, Kingsley KL, Zhang Q, Verma R, Obi N, Dvinskikh S et al (2019) Endophytic microbes and their potential applications in crop management. Pest Manag Sci 75(10):2558–2565

Wilson H, Johnson BR, Bohannan B, Pfeifer-Meister L, Mueller R, Bridgham SD (2016) Experimental warming decreases arbuscular mycorrhizal fungal colonization in prairie plants along a Mediterranean climate gradient. PeerJ 4:e2083

Wisniewski-Dye F, Borziak K, Khalsa-Moyers G, Alexandre G, Sukharnikov LO, Wuichet K et al (2011) *Azospirillum* genomes reveal transition of bacteria from aquatic to terrestrial environments. PLoS Genet 7(12):e1002430

Wu L, Shang H, Wang Q, Gu H, Liu G, Yang S (2016) Isolation and characterization of antagonistic endophytes from *Dendrobium candidum* wall ex Lindl., and the biofertilizing potential of a novel *Pseudomonas saponiphila* strain. Appl Soil Ecol 105:101–108

Wuana RA, Okieimen FE (2011) Heavy metals in contaminated soils: a review of sources, chemistry, risks and best available strategies for remediation. Isrn Ecol 2011:402647

Xia Y, DeBolt S, Dreyer J, Scott D, Williams MA (2015) Characterization of culturable bacterial endophytes and their capacity to promote plant growth from plants grown using organic or conventional practices. Front Plant Sci 6:490

Yadav AN (2017) Agriculturally important microbiomes: biodiversity and multifarious PGP attributes for amelioration of diverse abiotic stresses in crops for sustainable agriculture. Biomed J Sci Technical Res 1:1–4

Yadav AN, Verma P, Kour D, Rana KL, Kumar V, Singh B et al (2017) Plant microbiomes and its beneficial multifunctional plant growth promoting attributes. Int J Environ Sci Nat Res 3(1):1–8

Yadav AN, Kumar V, Dhaliwal HS, Prasad R, Saxena AK (2018) Microbiome in crops: diversity, distribution, and potential role in crop improvement. In: Prasad R, Gill SS, Tuteja N (eds) Crop improvement through microbial biotechnology. Elsevier, pp 305–332

Yang M-Z, Ma M-D, Yuan M-Q, Huang Z-Y, Yang W-X, Zhang H-B et al (2016) Fungal endophytes as a metabolic fine-tuning regulator for wine grape. PLoS One 11(9):e0163186

Zakeel M (2015 Bio-filmed biofertilizers for sustainable agriculture and environment. SOBA Environment Magazine, pp 49–51

Zakeel M, Safeena M (2019) Biofilmed biofertilizer for sustainable agriculture. In: Ansari R, Mahmood I (eds) Plant health under biotic stress. Springer, Singapore, pp 65–82

Zawoznik MS, Ameneiros M, Benavides MP, Vázquez S, Groppa MD (2011) Response to saline stress and aquaporin expression in *Azospirillum*-inoculated barley seedlings. Appl Microbiol Biotechnol 90(4):1389–1397

Zehnder GW, Yao C, Murphy JF, Sikora ER, Kloepper JW (2000) Induction of resistance in tomato against cucumber mosaic cucumovirus by plant growth-promoting rhizobacteria. BioControl 45 (1):127–137

Zhang H, Kim M-S, Sun Y, Dowd SE, Shi H, Paré PW (2008) Soil bacteria confer plant salt tolerance by tissue-specific regulation of the sodium transporter HKT1. Mol Plant-Microbe Interact 21(6):737–744

Zhang J, Wang L-H, Yang J-C, Liu H, Dai J-L (2015) Health risk to residents and stimulation to inherent bacteria of various heavy metals in soil. Sci Total Environ 508:29–36

Zhang X, Li J, Qi G, Wen K, Lu J, Zhao X (2011) Insecticidal effect of recombinant endophytic bacterium containing *Pinellia ternata* agglutinin against white backed planthopper, *Sogatella furcifera*. Crop Prot 30(11):1478–1484

Zhang Y, Yu X, Zhang W, Lang D, Zhang X, Cui G et al (2019) Interactions between endophytes and plants: beneficial effect of endophytes to ameliorate biotic and abiotic stresses in plants. J Plant Biol 62(1):1–13

Zhao X, Qi G, Zhang X, Lan N, Ma X (2010) Controlling sap-sucking insect pests with recombinant endophytes expressing plant lectin. Nat Preced:1–1

Zhong W, Gu T, Wang W, Zhang B, Lin X, Huang Q et al (2010) The effects of mineral fertilizer and organic manure on soil microbial community and diversity. Plant Soil 326(1–2):511–522

Zhu F, Qu L, Hong X, Sun X (2011) Isolation and characterization of a phosphate-solubilizing halophilic bacterium *Kushneria* sp. YCWA18 from Daqiao Saltern on the coast of Yellow Sea of China. Evid-Based Complement Alternat Med 2011:615032

Złoch M, Thiem D, Gadzała-Kopciuch R, Hrynkiewicz K (2016) Synthesis of siderophores by plant-associated metallotolerant bacteria under exposure to Cd2+. Chemosphere 156:312–325

Augmentation of Plant Salt Stress Tolerance by Microorganisms

10

Kamal Pandey, Ankur Adhikari, Vinita Pant, and Tara Singh Bisht

Abstract

Salinity stress is one of the major abiotic stress, which has drastically resulted in a global reduction of agricultural productivity. Around 20% of the global irrigated land falls under the category of saline soil. The elevated amount of soluble salts present in the cultivable land poses severe problems for crop survival and affects various physiological and metabolic processes of the plants. Enhancing the quality of saline soils by chemical methods have shown limited achievements as they also disturb the balance of natural soil ecosystem. Thus, an alternative strategy to compensate for the excessive amounts of salts present in the soil with no harm to the environment is required. Salt-tolerant microorganisms or halophiles can survive in the harsh saline environment and are also important in plant growth and survival. Lately, the plant growth-promoting rhizobacteria, mycorrhiza, and fungal endophytes have been extensively studied for the reclamation of extremely saline habitats. Thus, understanding of the molecular mechanisms and identification of salt-tolerant microorganisms which can benefit crop survivability under extreme saline conditions without harming the ecosystem are necessary. Here, we provide an overview of our current understanding of salt stress and various microorganisms including bacteria, fungi, and endophytes useful to enhance the production of crops under saline environment.

K. Pandey
Kumaun University, Department of Biotechnology, Bhimtal, Uttarakhand, India

A. Adhikari · V. Pant
Govind Ballabh Pant University of Agriculture and Technology, Pantnagar, Uttarakhand, India

T. S. Bisht (✉)
Patanjali Research Institute, Haridwar, Uttarakhand, India

R. Soni et al. (eds.), *Microbiological Activity for Soil and Plant Health Management*,
https://doi.org/10.1007/978-981-16-2922-8_10

Keywords

Stress · Microorganims · Salinity · Soil · Endophytes

10.1 Introduction

The dawn of twenty-first century is marked by global increase in demand vs. supply ratio of food owing to the escalating human population and shortage of fertile agricultural land. Moreover, in the past few decades there have been drastic changes in global climatic pattern leading to abiotic stresses like shift in temperature, drought, and salinity which are considered as major restraints in plant growth, adding to the decreased agricultural yield (Shrivastava and Kumar 2015). Salinity is considered one of the major environmental stresses that trigger other secondary stresses such as hyperosmotic, oxidative, and hyperionic stress that alter the molecular, physiological, and biochemical processes in plants resulting in significant loss of soil fertility and crop productivity (Tang et al. 2015).

Salinization is one of the major detrimental factors of soil degradation causing a steady fall in agricultural productivity. Salt stress is defined as the osmotic force exercised on plants, while they grow under excessively saline habitats. Soil salinity is the amount of water soluble minerals and salts in the soil and the process of their accumulation in the soil is known as salinization (Kalev and Toor 2018). Salts are intrinsically available in the form of charged ions which may root from primary and/or secondary (also known as anthropogenic) causes. The primary source involves weathering of rocks, insufficient rainfall, and entry of salts from ocean/brackish water through rain and wind. The secondary source includes excessive use of synthetic fertilizers, seepage of effluent from industries in agricultural lands, irrigation practices with improper drainage system, and utilizing low grade ground water for irrigation (Bhise and Dandge 2019). The following dissolved ions are found in most aqueous soil extract, listed in order of importance: sodium (Na^+), chloride (Cl^-), calcium (Ca^{2+}), sulfate ($SO_4{}^{2-}$), bicarbonate (HCO_3), potassium (K^+), magnesium (Mg^{2+}), and nitrate ($NO_3{}^-$). The soil salinity can be determined by measuring the electrical conductivity of the saturation extract from the root zone. If it surpasses 4 dSm^{-1} ($\approx$40 mM NaCl) at 25 °C and possess 15% of the exchangeable sodium, the soil is reckoned to be salt afflicted (Stavridou et al. 2017; Rath et al. 2019). A substantial amount of Na^+ ions in soil create an imbalance in the ratio of monovalent cations to divalent cations resulting in sodicity measured by the exchangeable sodium percent.

Elevated levels of these soluble salts greatly reduce the pH of the soil and can therefore limit plant growth, even by more than 90% in some cases (Kalev and Toor 2018). In large number of salts, sodium chloride is predominantly available and the chloride anions are believed to be destructive for plants and at high levels cause plant growth to retard (Bhatt and Nailwal 2018). Soil salinity induces plant stress in the following ways: (1) affecting root density, root turgor pressure, and water absorption, eventually leading to osmotic stress and (2) inducing toxicity through acquisition of high salt concentrations in the plant (Kumar et al. 2019; Bhatt and Barh 2018; Bhatt 2018) .

As per FAO (2016), greater than 6% of the total agricultural land particularly in arid and semiarid zones across the world is salt afflicted posing a serious hazard to agriculture and food abundance. Soil is typically saline in these dry areas and has low agricultural potential, so crops are usually grown under irrigation whose inadequate management thereby causes secondary salinization which affects 20% of irrigated land worldwide (Mayak et al. 2004). Moreover, reduced rainfall, higher land evaporation, weathering, saline water irrigation, inadequate drainage during irrigation and poor cultural exercises have led to a 10% annual increase rate of salinized areas (Shrivastava and Kumar 2015).

As predicted, 60% more food will be needed by 2050 to feed the 9.5 billion people. Thus, it is necessary to explore the methods to mitigate the harmful effects of salt stress, increase salt tolerance of plants, and ultimately enhance the agricultural yield in saline soils in order to feed the constantly expanding population. Soil and water management practices along with the traditional perspective of selectively breeding salt-tolerant plant varieties have been unsuccessful to efficiently resolve the issue. Moreover, there is an urgent need to increase the pace of crop productivity/ yield to resolve the problem of food security. Since customizing genotypes of all the crop plants for the diverse range of threats is equally time-consuming, costly, and unlikely. Alternatively, a myriad of microbes harbored by plants rhizosphere have shown to remarkably improve crop productivity, gaining the attention of researchers to experiment further. The primary challenge in the field of agriculture is the evolution of technologies, particularly for salt affected soils, and promotes sustainable agriculture for enhancing crop yields. Research on alleviation of salinity stress and soil pollution is of primary importance for increasing crop productivity in order to facilitate the demands of growing populations. In many developing countries, agriculture contributes to a massive share of national income and export earnings while also ensuring food security and employment. The capability of plant-based microbes is recognized globally for environmentally friendly, cost-cutting, and sustainable agriculture, given the increasing costs and negative impact of synthetic fertilizers on the environment (Singh 2013). PGPB's potential for strengthening plant growth and development, whether rhizospheric or endophytic, has been known for ages and thus presents an alluring way of replacing the use of chemical fertilizers, pesticides, and other supplements. Not only is the application of PGPB confined to biofertilizers or as a biocontrol agent, but it is now being used to improve soil salinity (Bhattacharyya and Jha 2012).

10.1.1 Impact of Soil Salinization on Plants

Various factors impede plant growth, like imbalanced hormone and nutrition levels, accumulation of toxic ions, reduced water potential, increased reactive oxygen species (ROS), and enhanced susceptibility towards diseases, which all root from salt stress. Plants subjected to salinity stress show decreased growth and productivity because they suffer from

1. High osmotic and oxidative stress.
2. Nutritional imbalance, since high Na^+ and Cl^- concentration reduces the uptake of K^+, NO^-, $PO_4{}^{3-}$, and other vital elements (Nawaz et al. 2010).
3. Ions toxicity.
4. Decreased photosynthetic rates, CO_2 assimilation rate, and electron transport chain (Stepien and Klobus 2005).

Generally, crops show growth under saline condition, although the overall productivity is reduced. The most important phases of the plant growth cycle including germination of seed and growth of seedling are highly affected by the level of salt present in the soil, both extremely low and higher salt concentrations are harmful for plant growth, lower salt concentration elicits quiescence and decreases germination, whereas a considerably high salt concentration increases the germination time needed and lowers the germination percentage. High concentrations of Na^+ and Cl^- impact the existence of other vital elements and can decline the ability of plants to access and absorb essential minerals, nutrients, and their dissemination in plants, leading to nutritional imbalance that decreases physiological activity and plant growth. Their accumulation in chlorophyll and chloroplast results in inhibition of photosynthesis (Zhang et al. 2004). The photosystem-II is specifically sensitive to salinity stress. Accumulated Na^+ in plant tissue stimulates the amount of superoxide (O^{2-}), single oxygen ($^1O_2$), hydroxyl radical (OH^-) and hydrogen peroxide (H_2O_2), ROS (Ahmad et al. 2011). Although ROS, at lower levels, function as important signal transduction molecule that modulates plant response to stress and its normal growth, but at higher levels pose significant threat to plant by causing irreversible DNA damage and cell death. ROS can also lead to oxidative degradation of lipids, covalent protein modifications, enzyme inactivation (Islam et al. 2015), and degradation of chlorophyll (Verma and Mishra 2005).

10.1.2 Plant Growth-Promoting Bacteria

A healthy soil is rich in bacterial population with their concentrations higher around the roots, owing to the release of root exudates, which acts as chemical stimuli to attract a variety of microbial organisms existing in the soil. These plant growth-promoting rhizobacteria are root-associated microbial community which are in symbiotic association with host plant. The microbial-based approach has been found more effective in sustainable plant growth than any other techniques like plant breeding, genetic modification approaches, etc. Whereas bacterial endophytes are highly specialized microorganisms inhabiting tissues of roots, stem, leaves, flowers, fruits, and seeds of the plants. Proteobacteria (~50%) is the most easily available endophytic bacteria, along with Firmicutes (~10%), Actinobacteria (~10%), and Bacteroidetes (~10%). Under stress conditions PGPB are known to promote plant growth by improving osmotic balance, altering root size and morphology, increasing nutrients uptake, and adjusting the nitrogen metabolism. In exchange, the plants support and protect the microbes via rhizodeposition, provide carbon and nitrogen sources for growth, and affect the functioning and organization of the microbial communities (Qin et al. 2016).

PGPB relieves salinity stress by replenishing nutrients, preserving a high potassium and sodium ratio, surge the accumulation of osmoprotectants, namely betaine, proline, and non-reducing sugars (Nahar et al. 2016), improving photosynthesis and antioxidant enzyme activity. The activity of 1-aminocyclopropane 1 carboxylate (ACC) deaminase, therefore reducing the level of ethylene (Mayak et al. 2004), the synthesis of plant hormones including indole acetic acid (IAA), gibberellic acid (GA), abscisic acid (ABA) (Shahzad et al. 2017), cytokinin and exopolysaccharides, are some of the different plant growth promotion characteristics exhibited by PGPB. These bacteria induced physical and chemical changes in plants can cause elicitation of plant's defense against pathogens, described as induced systemic resistance (ISR) and/or it can enhance plant's tolerance towards abiotic stress. Various research conducted with application of PGPB to plants grown under salinity resulted in an overall improved growth and yield in maize (Bano and Fatima 2009), wheat (Tiwari et al. 2011), and white clover (Han et al. 2017).

The use of organic methods for cultivation of medicinal plants rather than chemical based pesticides and fertilizers is more favored. Considering the importance of rhizospheric microbiome, Joe et al. (2016) studied two halotolerant endophytes, *Acinetobacter* sp. and *Bacillus* sp. and found that both exhibited various plant growth enhancing properties such as solubilization of phosphate, production of IAA, siderophore production, and enzyme activity of ACC deaminase. Also, in the presence of salt stress they were able to promote higher germination (%), plant biomass, and improved various other biochemical aspects of *Phyllanthus amarus*. Singh and Jha (2015) demonstrated that treating wheat plants under salt stress with *Klebsiella* sp. increased plants biomass, chlorophyll content, decreased Na^+ accumulation, and also triggered IST through its high ACCD activity. *Bacillus amyloliquefaciens* was also found to help tolerate salt stress by eliciting IST, maintaining rate of photosynthesis, and protecting against osmotic stress and Na^+ toxicity in maize (Verma et al. 2021) and in medicinal plant *Codonopsis pilosula* (Han et al. 2017). *B. amyloliquefaciens* is also reported to trigger ISR in strawberries against *Botrytis cinerea* and in pepper against *cucumber mosaic virus*. A number of studies demonstrated that PGPB can be equally beneficial in restraining the effect of metal toxicity along with salinity. The isolates of *Pseudomonas* strains improved germination percentage, growth of roots, and promoted formation of roots exposed to salt and aluminum stress in maize plants (Zerrouk et al. 2016) and *Beta vulgaris* exposed to excessive salt amounts (Piernik et al. 2017).

As a response to salt stress, leading to excessive ethylene production, plants accumulated the stress hormone ethylene (Khan et al. 2017). However, a decrease in the level of ethylene is needed to maintain normal plant growth. In order to alleviate the increased levels of ethylene, PGPB containing ACC deaminase activity reduce the ACC concentration in plants (Bhise and Dandge 2019). The activity of ACC deaminase is widely studied in soil bacteria belonging to the *Alcaligenes*, *Variovorax*, *Rhodococcus*, and *Bacillus* genera and various *Pseudomonas* species (Belimov et al. 2005). Earlier reports have also shown that PGPB producing ACC deaminase can improve rice growth (Bal et al. 2013), tomato (Ali et al. 2014), canola, groundnut, mung bean (Ahmad et al. 2011), musli (Barnawal et al. 2016), and pepper, under salt stress conditions.

Part of the rhizospheric bacterial community is endophytic bacteria. Various mechanisms of endophytes have been shown in numerous studies to enhance plant growth under various abiotic stresses (Brader et al. 2014; Santoyo et al. 2016). Bacterial endophytes are being used adeptly to counter the negative effects of salts and to improve the growth of plants under conditions of stress. Osmotic adaptation is linked with these compounds, and cell components and free radical scavengers are stabilized. *Pseudomonas pseudoalcaligenes* bacterization has been known to stimulate glycine betaine aggregation, a compound that has improved salinity tolerance in rice plants (Jha et al. 2011). The impact of the endophytic bacterium Bacillus subtilis BERA 71 on chickpea plants under saline conditions has previously been explored by Abd-Allah et al. (2018). *Bacillus*, *Pseudomonas*, *Klebsiella*, *Serratia*, *Arthrobacter*, *Streptomyces*, *Isoptericola*, and *Microbacterium* were recovered from the halophyte plant *Limonium sinense* in a study of ACC deaminase-producing endophytic bacteria of different genera (Qin et al. 2014). In connexion to high salt concentration, such bacteria were able to demonstrate their ACC deaminase properties. A salt-tolerant bacterial endophyte, *Brachybacterium paraconglomeratum*, was isolated from the roots of *Chlorophytum borivilianum* by Barnawal et atl. (2016). Oxidative and osmotic damage done by salinity in the host plants was whittled down by bacterial ACC deaminase activity. The bacterial endophyte, *Bacillus amyloliquefaciens* RWL-1, produced ABA and had the ability to strengthen salt tolerance in rice plants in a study conducted (Shahzad et al. 2017). Inoculation of RWL-1 substantially increased expression essential amino acids and triggered endogenous salicylic acid production in plants, helping rice plants to survive under salinity stress. Halotolerant endophytes have been isolated and analyzed for their plant growth-promoting activity in wheat from a weed, *Psoralea corylifolia* L. The cell-free extract of these isolates optimized wheat seed germination under saline stress through the production of IAA.

10.1.3 Mycorrhizal and Endophytic Fungi

Fungi can be associated with host plant by either residing entirely inside plant tissues (endophytic fungi) or residing only in roots and rhizosphere regions (mycorrhizal fungi). Endophytic fungi and mycorrhizae, under extreme environmental conditions, uphold plant fitness and health. These fungi form symbiotic union with the host plant and help in alleviating salt stress through multiple mechanisms. Among them, due to their ubiquitous distribution among large taxa of terrestrial plants, arbuscular mycorrhizal fungi (AMF) are deemed many of the most common associations. AMF establishes a direct physical link between plant roots and soils and facilitates the acquisition of mineral nutrients from soils by the host plant, notably under conditions of nutrient stress, and modifies the rhizosphere environment, thus alleviating the detrimental effects of salinity stress (Jahromi et al. 2008; Evelin et al. 2009). In various plant species, such as tomatoes, cucumbers, maize, lettuce, clover, fenugreek, sesbania, and acacia, AMF has been thought to promote salt tolerance (Al-Karaki 2000; Feng et al. 2002; Giri and Mukerji 2004; Giri et al. 2007; Evelin et al. 2012, 2013). In order to counteract the deleterious effects of toxic ions under saline conditions, AMF modulates several physiological and biochemical processes and regulates the expression of salt-related genes (Evelin et al. 2009; Porcel et al. 2012). It has been broadly acknowledged that AMF improves the efficiency of water use and plant nutrient absorption under saline conditions, thereby assisting to reduce the negative effects of salt stress. AMF reduces the deleterious effects of toxic ions on the permeability of membranes and cell organelles, maintains the standard of compatible organic solutes, stimulates the amount of antioxidants (both enzymatic and non-enzymatic), and positively controls salt-related gene expression. Several physiological, biochemical, and molecular approaches were portrayed by researchers by which AM plants could alleviate salt stress (Evelin et al. 2009; Ruiz-Lozano et al. 2012).

10.2 Molecular Mechanism Involved in Salt Tolerance

10.2.1 General Mechanisms of Augmenting Salt Tolerance in Plants

The process of salinity tolerance is very intricate. Research has revealed that elements of various pathways are involved in the development of salinity tolerance in plants. A study involving cDNA microarray of 7000 Arabidopsis genes revealed that 194 genes were overexpressed under elevated salt stress), indicating that in stress signal transduction pathways, several transcriptional regulatory mechanisms are involved. Proline, ROS, ABA, and numerous ion pumps and small molecules like calcium play a prominent role in this process. Studies have demonstrated that certain calcium-binding proteins, transcription factors, and enzymes such as helicases and protein kinases also play important role in the salinity stress tolerance (Tuteja 2007). High salt stress induces a Na^+ homeostasis imbalance that is preserved mostly by concerted regulation of different downstream pumps, ions, Ca^{2+} sensors, and their associated partners, eventually leading to the efflux of surplus Na^+ ions. Some channels display K^+ over Na^+ with more selectivity. This included the K^+ inward rectifying channel mediating the influx of K^+ following cell membrane hyperpolarization and selectively accumulating K^+ over Na^+ ions. The histidine kinase transporter (HKT) is a main determinant of the tolerance of plant salinity and it is a low affinity Na^+ ion transporter which prevents Na^+ ions from entering the cytosol (Platten et al. 2006). The non-specific cation channel (NSCC) is a voltage-independent channel that continues to serve as a gate to the plant cells for Na^+ input. In addition, there is the K^+ outward-rectifying channel that further opened up during the depolarization of the plasma membrane and enables the efflux of K^+ and the

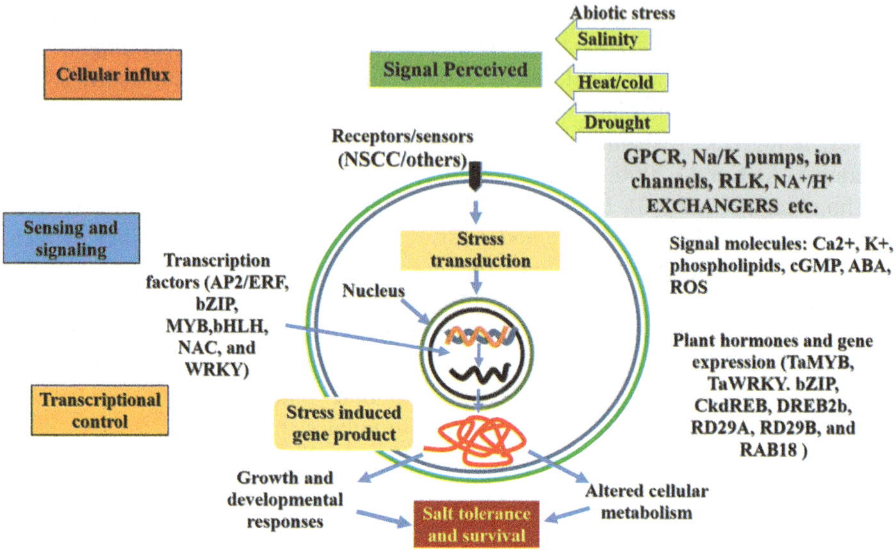

Fig. 10.1 Plant response pathway during abiotic stress. Extracellular signal received by the receptors with the help of membrane receptor which lead to the signal transduction under the signaling molecules. The signaling cascade resulted in the expression of multiple genes that respond to stress, the products of which may stress tolerance. It includes coordinated action of various transcription factors and many genes, which may cross talk with each other. GPCR G-protein-coupled receptor, RLK receptor-like kinase, ABA abscisic acid, ROS reactive oxygen species, cGMP 3′,5′-cyclic guanosine monophosphate, NSCC non-selective cation channels, AP2/ERF APETALA2/ethylene response factor, MYB myeloblastosis, bZIP basic leucine zipper domain, bHLH basic helix-loop-helix

inflow of Na^+ ions, due to the accumulation of Na^+ cytosols. The Na^+/H^+ vacuolar exchanger (NHX) assists in pushing surplus Na^+ ions into vacuoles. Na + extrusion from plant cells is driven by the electrochemical gradient triggered by H^+-ATPases, enabling the NHX to integrate the passive movement of H^+ inside all along electrochemical gradient with the filtration of Na^+ from the cytosol. Another pump, the H^+/Ca^{2+} antiporter (CAX1), assists with Ca^{2+} homeostasis (Mahajan et al. 2006; Zhang et al. 2004; Zhu 2002) (Fig. 10.1).

Advances at the cellular level have concentrated upon salt-induced initial signaling over the past decade, illustrating a vital function for the both calcium waves and reactive oxygen (ROS) species and their associated functions, whereas the cell wall is also involved only as amplifier of cell expansion throughout salt stress. Furthermore, gene expression, mRNA stability, and translational regulation are modified to alter protein abundance under salt stress. At the cellular level, we analyze how distinct ion channels and transporters influence Na^+/K^+ homeostasis, with implications for their function in plants. Roots are the forefront part of a plant that are closely associated with the soil's salt that will need to evolve to support the development and absorption of nutrients and water.

In particular, salt stress helps in decreasing root mass and alters the proportion of various components and the framework of the root system distinctively affects the rate of growth in the main and lateral roots. Until now, it is not clear how salt is regulated by the shoot apical meristem and shoot architecture. A recent report demonstrated that reduced soil NaCl levels, through BR and ABA signaling pathways, reduced hypocotyl elongation (Hayes et al. 2019).

The finest cellular responses to salt, sodium import, and sodium sensing are the least acknowledged and are in a black box in salt-induced signaling pathways. Through NSCCs, which carry sodium throughout the plasma membrane, salt can enter the root (Demidchik and Maathuis 2007; Demidchik and Tester 2002). Various salt-induced signals, such as ROS, cGMP, Ca^{2+} regulate NSCCs.

Rapid salt-specific responses in roots, along with sodium-specific calcium waves, have been recognized (Choi et al. 2014). Moreover, the rapid and sodium-specific impact of salt on the growth of the root direction (halotropism) foresees the existence of a root-based sodium sensor (Galvan-Ampudia et al. 2013). Sodium could be perceived intercellularly, extracellularly, as well as by ion transporters within the plasma membrane (Van Zelm et al. 2020).

The likely functioning of MOCA1 (MONOCATION-INDUCED [Ca^{2+}] INCREASES 1) in extracellular salt sensing, but are not limited to Na^+ ions, has actually become significantly advanced (Jiang et al. 2019). Few of the primary salt stress responses described are ROS production (Miller et al. 2010), rise in cytosolic Ca^{2+}) concentration and cGMP (Donaldson et al. 2004). Three forms of calcium fluxes are characterized in effect to sodium: cellular calcium surges and fast as well as late-response calcium waves (Choi et al. 2014; Feng et al. 2018). And subsequently, the regional application of salt indicated that the treatment of high salt, but not osmotic stress, induced the spread calcium waves of long-range. These waves appear to be almost instantaneous, start 10 s upon application of salt, or could promulgate all through the root within 30 s and perhaps even reach the leaves (Choi et al. 2014). All in all, calcium waves and spikes are not a sodium-specific phenomenon, as diverse then other stimuli, such as touch, cold, and osmotic stress, are sometimes regarded to stimulate effects on cellular calcium concentrations, but the peak oscillation pattern, peak amplitude, and wave propagation show salt specificity (Van Zelm et al. 2020).

Increased levels of apoplastic ROS molecules like H_2O_2, $^1O_2$ and $_2O^-$ and OH^- are quickly generated by salt stress, which disrupts redox homeostasis and triggers plant cell oxidative damage (Miller et al. 2010). NADPH oxidases generated by plant ROS are RESPIRATORY BURST OXIDASE HOMOLOGs (RBOHs). In *Arabidopsis*, the expression patterns of 10 RBOH genes change dynamically, and they begin generating ROS waves after salt stress over a 24-h period (Xie et al. 2011), this indicates also that complex production network for ROS is constantly in motion and plays an essential role in the early reaction to salt. Cross-talking between ROS and Ca^{2+} signaling is needed to expand the Ca^{2+} signal between cells (Evans et al. 2016).

The sodium and potassium cellular balance is extremely necessary for crop growth in saline soils. Multiple channels, transporters, and antiporters help in sustaining Na^+/K^+ equilibrium during the salt stress (Almeida et al. 2017), the first being the NA^+/H^+ EXCHANGERS (NHXs) family. The NHX7/SOS1, which is localized in the plasma membrane transports sodium primarily and is absolutely vital for the segregation of sodium from the root (Ji et al. 2013).

The other members of the NHX family are situated in intercellular compartments in addition to the NHX7 and NHX8 plasma membrane and are capable of transporting both K^+ and Na^+ that has been shown to be appropriate for the tolerance of salt (Almeida et al. 2017). These data together indicate the key role played by NHX7/SOS1 in Na exports, protein sorting, intercellular potassium carry, and the upkeep of the endomembrane pH, endomembrane and tonoplast localized NEXs appear to be beneficial under salt stress. Furthermore, almost all of our knowledge of transporters which have a function in the homeostasis of Na^+/K^+ is at either the cell or the organ level but rarely integrated between these levels.

The growth rate at the main root is significantly lowered by salt and a temporary pause in growth is called the dormant phase (Geng et al. 2013). The intensity and amplitude of salt-instigated dormancy are governed by hormone levels. The amount of abscisic acid and signaling transcripts was highly correlated with the quiescent phase. Thus, abscisic acid concentrations were whittled down mostly during recovery phase, while brassinosteroids (BR), gibberellic acid (GA), and jasmonic acid (JA) levels raised and transcriptional programs have been triggered downstream. The production of GA was observed between 5 and 8 h upon salt treatment, while it was able to recover from growth, accompanied by a reduction of 24 h (Geng et al. 2013), consistent with initial findings of bioactive GA reduction after several days of salt treatment (Achard et al. 2006). Salt stress triggers JA biosynthesis and signaling and therefore inhibits elongation of plant cells and primary root growth (Valenzuela et al. 2016). Growth rate modulation correlates with changes in the management of cellular water and solutions, photosynthesis, elongation of cells, and division of cells. Disruption of the ion homeostatic mechanisms of the plant is known to be one of the most adverse effects of salt stress. In specific, the analogous radii of Na^+ and K^+ make it challenging to differentiate in between two ions for transport proteins. Thereby, it is substantial uptake of Na^+ through K^+ transporters or channels under influence of high external Na^+ (Blumwald et al. 2000). In order to establish a strong reaction to deal with salt stress, plants have developed the ability to comprehend both the hyperosmotic component and the ionic Na^+ component of stress. These sensory modes are evident because certain NaCl reactions are different from purely osmotic stress reactions (Deinlein et al. 2014). Transcription factors intrinsically link salt sensory pathways to other tolerance responses. Particular genes encoding for transcription factors are essentially expressed in response to increasing environmental salinity, which includes basic leucine zipper, WRKY, APETALA2/ETHYLENE RESPONSE FACTOR (AP2/ERF), MYB, basic helix-loop-helix, and NAC families (Golldack et al. 2011; Yang et al. 2009a, b; Jiang and Deyholos 2009; Kasuga et al. 1999; Cui et al. 2013; Jiang et al. 2009; Tran et al. 2004). These transcription factors in turn regulate the levels of expression of various genes that may deeply influence the level of salt tolerance of plants. In order to avoid a detrimental decline in the K^+/Na^+ ratio, there are many possible tactics that plants could use: reduce the cell Na^+

entry, remove Na^+ from the cell, or divide Na^+ into the vacuole where cellular function cannot be interrupted.

10.2.2 Specific Mechanisms in Regulating Salt Tolerance by Microorganisms

The stress of soil salinity adversely affects the microbial abundance within and nearby plant roots. In environments where the salt concentration exceeds 200 mM of NaCl, halophytes can be classified as plants that can live and thrive (Flowers and Colmer 2008). The need for salinity-affected ecosystems to be revegetated and remediated has concentrated the research interest in understanding salt tolerance mechanisms in halophytes (Shabala 2013). Salt tolerance is an intricate phenomenon involving different mechanisms of biochemistry and physiology that are also particular to species. Together, these mechanisms entail enhancing the efficacy of extraneous and intrinsic Na^+ sequestration in trichomes and vacuoles, via trichome shape manipulation or modulation of ion channels and transporters, tonoplast antiporters, or by the formation of large vacuoles, frequently with altered composition of lipid, etc. (Shabala 2013; Hasanuzzaman et al. 2014). In addition, halophytes have the capacity to produce and gather compatible osmolytes that make a significant contribution to their osmoregulation while precluding prolonged transport of Na^+ to the shoots and have demonstrated great stomatal leverage that allows them to optimize the efficiency of water use under stressful conditions of salinity (Redondo-Gómez et al. 2010; Shabala 2013). While halophyte microbes may add value to plant survival and salt tolerance, studies with halophyte-inhabiting bacteria are still rare and most of them are primarily descriptive, i.e. studies that have attempted to categorize and enumerate microbial species (Ruppel et al. 2013). General mechanisms underlying those positive impacts include, among others, phosphorus solubilization, nitrogen fixation, iron sequestration, phytohormone (mainly auxins), and ACC (1-aminocyclopropane-1-carboxylic acid) deaminase synthesis and biofilm formation (Ullah et al. 2015). Several studies observed the use of PGPB in this perspective to improve salt tolerance and plant productivity under saline conditions (Ruppel et al. 2013).

A plant experiences several morphological, physiological, and molecular alterations under salinity stress, which hamper its growth and development (Fig. 10.2). These various modifications demand modulation of the various genes associated with stress that are implicated in regulatory and signal transduction (Bharti et al. 2016). Understanding the mechanisms associated in signaling salinity stress in plants and also between plants and bacteria is integral. The transduction of plant signals initiates with the receiving of receptor communication on the surface, preceded by the inception of secondary messenger molecules such as inositol phosphates and ROS. Secondary messengers exploit proteins such as CDPK (calcium-dependent protein kinase), MAPK (microtubule-associated protein kinase), and protein phosphatase responsible for controlling gene expression by modulating the concentration of Ca^+. Due to its exceptional roles in modulating

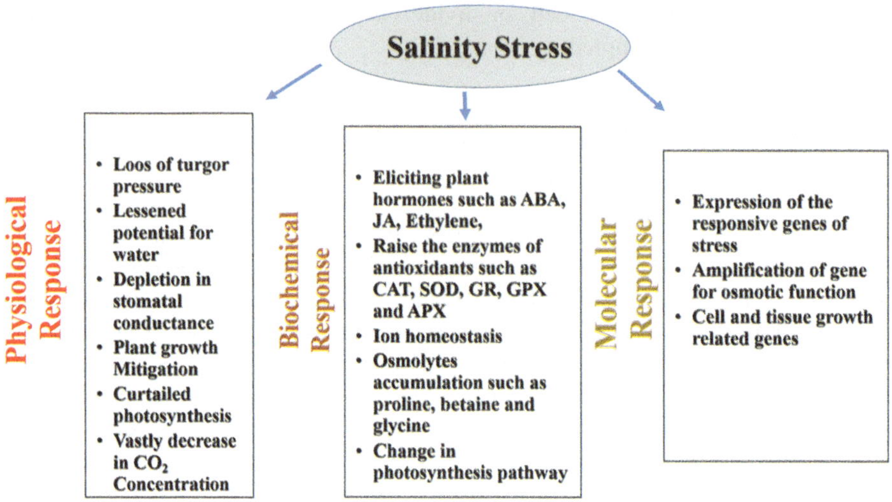

Fig. 10.2 Alterations in various pathways during salt stress

different stress-responsive genes, transcription factors are quite prominent in modulating salt stress response in plants. Some substantial families, such as AP2/ERF, bZIP, MYB, NAC, and WRKY, that regulate the expression and function of multiple genes, include these transcription factors (TFs). The modulation of the genes associated with salinity stress also relies on the post-transcriptional modulation of TFs. Resistance to salinity stress was found in the overexpression of bZIP and CkdREB genes in *Tamarix hispida* and *Caragana korshinskii* (Wang et al. 2010, 2011).

PGPB such as *Arthrobacter protophormiae* (SA3) and *Dietzia natronolimnaea* (STR1) improved the tolerance of salinity stress in wheat plants by regulating the expression of the ethylene signaling pathway regulatory component CTR1 (constitutive triple response1) and DREB2 TF (Barnawal et al. 2017). Under salinity stress, *Enterobacter spp.* induces the levels of genes responsible for salt stress such as DREB2b, RD29A, RD29B, and RAB18 in *Arabidopsis* and *D. natronolimnaea* induce the expression of TaMYB and TaWRKY genes in wheat (Bharti et al. 2016; Kim et al. 2014). Furthermore, epigenetic phenomenon like methylation of DNA and post-translational modifications in histones affect the process of gene expression triggered by stress under salinity stress (Dietz et al. 2010; Golldack et al. 2011). In the future, advanced tools like genomics, transcriptome analysis using microarray strategy, and proteomics will allow us to better understand plant stress signaling in full depth in salinity tolerance. In addition, a high concentration of salt influences the activities of enzymes, stomatal conductance, and photosynthesis rate. The overabundance of Na^+ and Cl^- ions also contributes to hypertonic stress due to salinity. With the utilization of transgenic methods and traditional breeding techniques, many salt-tolerant crop varieties have been produced but these approaches are labor-intensive and time-consuming. In view of the future challenges, the concurrent use of alternative technologies to encourage sustainable agriculture, such as the use of PGPB, seems to have become essential. The microbial diversity of the rhizosphere is challenging, which can be deemed to be the natural relationship between plants and microbes (Vacheron et al. 2013; Daliakopoulos et al. 2016; Tiwari et al. 2016). PGPB also promotes plant nutrient absorption and homeostasis and increases the activity of antioxidants during salt stress. In healthy plant tissues, plant growth-promoting bacterial endophytes live without causing them any disease. Such endophytic PGPB can also facilitate tolerance of salinity stress and plant growth (Ali et al. 2014). Kumar et al. (2020) revealed that by applying PGPB consortia against salt stress, the growth and yield of *Phaseolus vulgaris* are optimized. Due to the simultaneous scrummaging of salt stress effects and the improvement of accessible iron in saline soils, siderophore-producing rhizobacteria may provide a potential alternative to chemical fertilizers (Ferreira et al. 2019). In order to have a sustainable agricultural solution and cope with salinity stress, it is therefore essential to determine the diversity of microbes in order to recognize their physiological and functional characteristics and saddle their potential.

The phytomicrobiome is the microbial communities that live inside or even on the plant surface. In evaluating the survival of the plant under stressful conditions, such as salinity and drought stresses, the root microbiome of the plant is extremely important. *P. fluorescens* was isolated from Sahara region rhizosphere soil and revealed a PGPB property in maize under salinity stress (Zerrouk et al. 2016). A substantial adaptation of halophilic microbes is that, even at high salt levels, they preserve the protein structure and enzymatic activity for distinct metabolic processes (Ruppel et al. 2013). The common principle through which salinity-tolerant microbes flourish in saline habitats is by avoiding potential concentrations of salt within the cytoplasm. This is achieved by modifying the cell wall structure in which specific membrane proteins, lipids, and exopolysaccharides are produced. Certain survival adaptations under high salinity conditions also include development of proteins and enzymes worthy of performing metabolic functions. A few other microbes also develop organic osmolytes that also, under high salinity stress, accumulate in the cytoplasm to make them resistant to osmotic pressure. Organic osmolytes are also called compatible solutes because, by maintaining an appropriate molar concentration in the cytoplasm, they offer resistance to various molar concentrations of salinity stress (Kunte 2006).

10.3 Microbial Stimulation of Salt Tolerance

10.3.1 Salt Tolerance by Bacteria

The rhizosphere is inhabited by plant growth-promoting bacteria (PGPB) that stimulate plant growth directly or indirectly. While endophytic bacteria reside inside plants and help in promoting plant growth in challenging conditions. The direct mechanism exerted by PGPB involve production of phytohormones, nitrogen fixation, and facilitating nutrients intake such as iron and phosphorus, whereas the indirect mechanism involves protection against disease caused by pathogens (Hayat et al. 2010; Satyaprakash et al. 2017). Often plants are exposed to various environment stress such as temperature, salinity, drought, pesticides, and heavy metal contamination. Such abiotic stress hinders the plant growth and productivity resulting in crop loss worldwide. Amidst these stresses, salt stress is regarded as one of the major abiotic stresses as it can reduce the rate photosynthesis, respiration, protein synthesis, and trigger deleterious effect on various plants physiological processes such as nitrification, denitrification, soil biodiversity, and microbial activity (Schirawski and Perlin 2018).

In the last few years, reports have suggested the role of various PGPB in saline stress which proved to be an asset in augmenting plant growth and productivity. According to studies, PGPB produce variety of phytohormones such as auxins, gibberellins, and cytokinins that are beneficial for plant to function. IAA, a naturally occurring auxin, produced by many salt-tolerant PGPB such as *Azotobacter, Arthrobacter*, *Azospirillum*, *Pseudomonas*, and *Stenotrophomonas* proved to stimulate plant cell growth and proliferation under saline stress (Egamberdieva et al. 2016). One of the studies illustrated that IAA synthesized by *Pseudomonas putida* enhanced the growth of cotton under salt stress (Yao et al. 2010). *Streptomyces* sp. isolates produced IAA which ameliorated the root system and growth of wheat under saline stress (Sadeghi et al. 2012). Phytohormones other than auxin are also known to stimulate plant growth and development under salt stress conditions. When tomatoes treated with gibberellin (phytohormone responsible for stem elongation, seed germination, and induction of flowering) was subjected to salt stress it leads to acceleration in the water uptake and pruning of the stomatal resistance (Maggio et al. 2010). In another study it was observed that gibberellin leads to modification of osmotic stress and tissue water content in wheat (Manjili et al. 2012) and maize (Tuna et al. 2008). In addition, gibberellic acid slashed the levels of reactive oxygen species (ROS) that in turn intensified the antioxidant enzyme activity and augmented the growth of plant under saline stress. Under the abiotic stress conditions including salt, induce the production of ethylene, a plant hormone which regulates plant homeostasis and results in reduced growth of plants. However, ACC (1-aminocyclopropane-1-carboxylic) which is a precursor of ethylene can be degraded by ACC deaminase activity of bacteria, thus helping plants to grow normally under salt stress environment. Sarkar et al. showed that halotolerant *Enterobacter* species could effectively counter salt stress and help in rescuing the normal growth of rice seedling through ACC deaminase activity of bacteria (Sarkar et al. 2018). A detailed review on the ACC deaminase activity of bacteria to enhance the tolerance of plants in salt contaminated soil has been recently published (del Carmen Orozco-Mosqueda et al. 2020). Salt affected soil contains high amount of sodium (Na^+) which also effects the availability and uptake of essential nutrients (potassium, nitrogen, phosphorus, etc.) by plants. Inoculation of beneficial bacteria in salt affected soil has shown to enhance the uptake of nutrients and better growth of plants (Vejan et al. 2016).

Another plant hormone cytokinin is important for root callus differentiation and formation of shoot. Bucio et al. showed that endophytic bacterium *Bacillus megaterium* promoted plant growth and improved the biomass production in bean and Arabidopsis through cytokinin signaling (López-Bucio et al. 2007). Recently, *B. megaterium* was also found to alleviate salt stress and promote growth in alfalfa plants (Chinnaswamy et al. 2018). The bacterium produced IAA and showed ACC deaminase activity in vitro which explains the growth-promoting activity exhibited by the endophyte. Endophyte *Bacillus subtilis* (BERA71) inoculation also increased biomass, decreased the reactive oxygen species, and lipid peroxidation levels in chickpea plants under salt stress conditions. The bacterium also reduced Na^+ accumulation and increased the potassium, nitrogen, and magnesium levels in the plants (Abd-Allah et al. 2018). Another halotolerant bacterium *Glutamicibacter halophytocola* promoted the root growth, length, osmolyte content, and salt tolerance of tomato seedlings (Xiong et al. 2019). The bacterium affected the expression of multiple genes involved in nitrogen fixation, biosynthesis of siderophore, and genes related to high salt tolerance. Looking into the benefits plant growth-promoting halotolerant bacteria can be utilized as biofertilizers for sustainable agricultural practice. Table 10.1 depicts plant growth-promoting bacteria in salt stress alleviation and their effects on the different plants.

10.3.2 Salt Tolerance by Fungi

Similar to rhizobacterial interactions with plants, multiple studies have shown the involvement of fungi in survival of host plants under various abiotic stress conditions, including salt stress. Plants under salt stress produce low amount of plant growth hormones like gibberellins while produce high amounts of abscisic and jasmonic acid responsible for limited growth of crops (Sah et al. 2016). Endophytic fungi have been known to produce gibberellins and indole acetic acid that promote the growth of host plants under stress conditions (Khan et al. 2012; Waqas et al. 2012). Moreover, application of endophytic fungi *Phoma glomerata* and *Penicillium species* also increased the assimilation of essential nutrients like potassium, magnesium, and calcium and in the treated plants. Hamayun et al. also showed novel endophtyic fungi *Porostereum spadiceum AGH786* could boost the salt tolerance ability of soybean plant through inducing gibberellin production and inhibition of abscisic and jasmonic acid (Hamayun et al. 2017). Researchers have shown that plants colonized by mycorrhizal fungi grow better under salt stress and generate more biomass than non-mycorrhizal plants (Latef and Chaoxing 2011; Porcel et al. 2016). Under saline conditions, AM symbiosis impacts different morphological parameters, like plant height, leaf area, root density, and fresh and dry plant weight (Campanelli et al. 2013). The important chunk of the mycorrhizal fungi to improving plant growth under salt stress is thought to be improved absorption of phosphorus; however, other metabolic processes facilitated by nutrients such as N, K, and Mg have to be engaged in reducing deleterious salinity effects (Giri and Mukerji 2004; Evelin et al. 2009, 2012). The impact of arbuscular mycorrhiza *Glomus fasciculatum* on the growth of plant *Acacia nilotica* under various concentrations of salts was examined by Giri et al. (2007). They reported that soil salinity reduced root and dry

Table 10.1 Studies on growth promoting bacteria in salt stress alleviation

PGPB	Crop/plant	Plant growth-promoting traits	Effects	References
Achromobacter piechaudii.A	Tomato (*Lycopersicon esculentum*)	ACC deaminase activity	Increased the fresh and dry weights of tomato seedlings and uptake of phosphorus and potassium; reduced the production of ethylene	Mayak et al. (2004)
Serratia marcescens	Wheat (*Triticum aestivum*)	ACC deaminase activity, phosphate solubilization, nitrogen fixation, production of siderophore, indole acetic acid, and ammonia	Reduced ethylene levels and antioxidative enzymes activities; increased concentration of osmoprotectants and chlorophyll content	Singh and Jha (2015)
Pseudomonas fluorescens	Maize	Siderophores and IAA production	Improved root growth and promoted root formation	Zerrouk et al. (2016)
	Sunflower (*Helianthus annuus*)		Increased plant biomass and favored the K^{+}/Na^{+} ratio	Shilev et al. (2012)
Streptomyces	Stevia (*Stevia rebaudiana*)	ACCD activity; auxin and siderophore production	All growth parameters were improved; increased protein expression (RuBisCo, EF-Tu)	Tolba et al. (2019)
	Wheat		Increase in germination rate, percentage, and uniformity, shoot length, and dry weight	Sadeghi et al. (2012)
Azospirillum brasilense	Barley (*Hordeum vulgare*)	Enhanced the production of photoprotective pigments; phytohormone synthesis; decreased Na^{+} availability	Promoted overall plant growth; increased chlorophyll content; reduced oxidative stress	Omar et al. (2009)
	White clover (*Trifolium repens*)		Enhanced plant biomass, leaf area, chlorophyll content, and K^{+}/Na^{+} ratio; declined MDA content	Khalid et al. (2017)

B. Halodenitrificans, Halobacillus	Wheat	Exopolysaccharides, IAA, and siderophore production, ACCD activity, and P-solubilization	Reduction in stress ethylene levels, stimulate seed germination, root elongation; increase in root biomass and water uptake	Ramadoss et al. (2013)
Bacillus licheniformis	Wheat (*Triticum aestivum*)	ACC deaminase activity; exopolysaccharide production	Decrease in Na^+ content, and increased K^+ and Ca^{2+} content; increase in osmolytes	Singh and Jha (2015)
Acinetobacter	*Phyllanthus amarus*	IAA production, siderophore production, mineral phosphate solubilization	Germination (%), plant biomass, P content, phenolic content, radical scavenging, and antioxidative activity were improved	Joe et al. (2016)
Klebsiella sp.	Wheat (*Triticum aestivum*)	ACCD activity, production of IAA and GA	Increased plant biomass and chlorophyll content; reduction in plant growth inhibition; Na^+ exclusion and increased uptake of K^+	Singh and Jha (2015)
Curtobacterium oceanosedimentum	Soybean (*Glycine max*)	Abscisic acid, jasmonic acid, and ACC deaminase enzyme production	Reduced oxidative stress, ethylene stress; increased water use efficiency of plants	Khan et al. (2019)
Bacillus aquimaris	Maize (*Zea mays*)	IAA synthesis	Improved chlorophyll content, leaf relative water content; accumulation of proline and total phenolic compound; decreased lipid peroxidation levels and Na^+ content	Li and Jiang (2017)
Azotobacter chroococcum	Maize (*Zea mays*)	Improved antioxidative capacity, mineral nutrition; phytohormones synthesis; regulation of ions transporters	Improved K^+/Na^+ ratio, chlorophyll content	Rojas-Tapias et al. (2012)
Bacillus amyloliquefaciens	Rice (*Oryza sativa*)	ACCD activity and ABA and auxin-producing properties; alteration in rhizosphere microbial community	Enhanced water and nutrient uptake; enrichment of osmoprotectant utilizing rhizo-microflora	Nautiyal et al. (2013) and Shahzad et al. (2017)

(continued)

Table 10.1 (continued)

PGPB	Crop/plant	Plant growth-promoting traits	Effects	References
Pseudomonas syringae, *Pseudomonas fluorescens*, *Pseudomonas fluorescens*, *Rhizobium phaseoli*	Mung bean (*Vigna radiata*)	ACCD activity, exopolysaccharide production	Improving growth and nodule formation	Ahmad et al. (2011)
Bacillus thuringiensis	Sweet pepper (*Capsicum annuum*)	Reduced lipid peroxidation, electrolyte leakage, and improved proline accumulation and enzyme activity	Increase in relative water content, chlorophyll content, and fruit yield; reduced osmotic stress	ALKahtani et al. (2020)

shoot biomass, but mycorrhizal inoculation of acacia plants exhibited significant positive impact on plant growth and accretion of biomass regardless of salt concentrations. Inoculation of rice plants procured from saline soils with AM fungal strain *Claroideoglomus etunicatum* (Estrada et al. 2013) increased the yield of shoot dry biomass at all salinity levels (Porcel et al. 2016). Porras-Soriano et al. (2009) have applied *Glomus mosseae*, *Glomus intraradices*, or *Glomus claroideum* inoculum to olive plants grown in both non-saline and saline media. A major rise in plant growth and uptake of N, P, and K has been observed. Experimental results on mycorrhizal fungi and citrus seedlings by Khalil et al. (2011) exhibited a higher plant growth under salinity stress. Table 10.2 shows the salt tolerance in plants inoculated with arbuscular mycorrhizal fungi which affected the plant metabolism in several ways.

10.4 Combinatorial Benefits of PGPB and Mycorrhizal Fungi

Interactions among bacteria and fungi can be synergistic or antagonistic, which can be either beneficial or harmful towards crops. Thus, identification of synergistic interactions that co-operate to produce combined results much efficient than individual inoculants can be very useful in mitigating salt stress affected soil. Recent reports have demonstrated better results post co-inoculation with the two microorganisms. Hashem et al. showed the interaction between arbuscular mycorrhiza and endophytic bacteria increased *Acacia gerrardii* growth under salt stress conditions. Co-inoculation of *Bacillus subtilis* (BERA 71) and *Rhizophagus intraradices* not only enhanced nutrient availability but also reduced in the concentrations of Na and Cl in the soil (Hashem et al. 2016a, b). Furthermore, another study by Hashem et al. also demonstrated the increased levels of osmoprotectants like proline, glycine, and betaine and enhanced amounts of antioxidants (superoxide dismutase, catalase, etc.) after inoculation (Hashem et al. 2016a, b). Similar results of enhanced enzymatic and non-enzymatic antioxidants were observed when *Piriformospora indica* and *Azotobacter* species were applied to salt affected soil (Arora et al. 2020). The dual treatment resulted in better root, shoot growth and had enhanced survivability than uninoculated plants. Furthermore, ACC deaminase activity of both bacteria (*Pseudomonas putida)* and fungi (*Gigaspora rosea*) synergistically enhanced the tolerance towards salt in cucumber plants (Gamalero et al. 2010). Thus, dual treatment of synergistic microorganisms can induce significant affects in growth of plants under salt stress as compared to mono-treatment approach.

10.5 Conclusion and Future Perspective

Continuously changing environmental conditions and abiotic stresses have substantially affected the plant growth and subsequently result in low agricultural yield. Latest available sophisticated molecular techniques, omics technology such as metagenomics, metabolomics, and proteomics have been conducive in decoding

Table 10.2 Studies on salt tolerance in plants inoculated with beneficial fungi

Beneficial fungi		Mechanisms inducing salt tolerance	Effects	References
Glomus fasciculatum	Acacia (*Acacia auriculiformis*)	Protects root and alters its morphogenetic, lowered electrical conductivity of the soil, and improved phosphorus (P) nutrition; reduced Na^+ uptake	Improved root and shoot growth, nutrient acquisition, water uptake, and chlorophyll content	Giri and Mukerji (2004) and Giri et al. (2007)
	Lettuce (*Lactuca sativa*)	Better P nutrition, proline production, transpiration, stomatal conductance, carbon exchange rate, and water use efficiency	Greater plant biomass and photosynthetic rates	Ruiz-Lozano et al. (1996)
Diversispora versiformis	Chrysanthemum (*Chrysanthemum morifolium*)	Enhanced root nitrogen uptake	Higher root length, total dry weight, and root nitrogen concentration	Wang et al. (2018)
Funneliformis mosseae (syn. Glomus mosseae)	Cucumber (*Cucumis sativus*)	Lowered lipid peroxidation, electrolyte leakage, and Na^+ uptake; increased production of phenols and proline	Enhanced the biomass, synthesis of pigments; mineral elements, salicylates, and jasmonates accumulation; reduced activity of antioxidant enzymes	Hashem et al. (2018)
Glomus macrocarpum	Acacia (*Acacia auriculiformis*)	Protects root and alters its morphogenetic, lowered electrical conductivity of the soil, and improved P nutrition; reduced Na^+ uptake	Improved root and shoot growth, nutrient acquisition, water uptake, and chlorophyll content	Giri and Mukerji (2004)
Glomus intraradices	Tomato (*Solanum lycopersicum*)	Elevates water uptake, stomatal conductance, proline production, P nutrition; protects photochemical processes of PSII; lowered lipid peroxidation	Elevated net assimilation rates, reduced ROS scavenging enzymes activity	Ouziad et al. (2006) and Hajiboland et al. (2010)
	Pepper plants (*Capsicum annuum*)		Improved nutrient uptake, membranes stability, and plant growth	Beltrano et al. (2013)
Paraglomus occultum	Tangerine (*Citrus tangerine*)	Altered root morphology; reduced Na^+ uptake	Greater total plant biomass, photosynthetic capacity, transpiration rate, and stomatal conductance	Wu et al. (2010)
Acaulospora mellea	Sweet sorghum (*Sorghum bicolor*)		Improved plant biomass and mineral uptake, as well as higher K^+/Na^+ ratio	

the intricacies of plant–microbe interactions. Large number of studies both in the laboratory and greenhouse done in the recent years have shown promising use of plant associated microorganisms to overcome various abiotic stresses including salt stress. Moreover, utilization microbes to overcome salt stress is an eco-friendly approach to achieve superior crop growth and yield. Salt-tolerant plant growth-promoting bacteria, endophytic fungi as well as mycorrhiza have been extensively studied regarding their beneficial effects in saline environments. Recent research has broadened our understanding of the complex interaction between plant and microbes, yet, a lot remains to be learned about the mechanisms and changes at the molecular level induced by microorganisms for their purposeful utilization under saline conditions. Identification of the genes responsible for regulating various mechanisms and signaling pathways under salt stress environment can provide a better insight into the microbe mediated tolerance of salt stress in plants. Moreover, a better understanding of how different plants behave while treatment with different microorganisms and conditions can be useful in replacing the chemical driven mitigation of salt stress which is often detrimental.

Majority of research experiments conducted are limited to laboratory or greenhouse chambers, which are closely monitored for each requirement, which is not the case in the open agricultural land prone to various environmental factors. Thus, more focus on the field experiments could be beneficial to study the outcome of growth-promoting bacteria and fungi. Furthermore, the combinatorial treatment with synergistic bacteria and endophytic or mycorrhizal fungi can potentially generate better results for sustainable agriculture.

References

Abd-Allah EF, Alqarawi AA, Hashem A, Radhakrishnan R, Al-Huqail AA, Al-Otibi FON, Egamberdieva D (2018) Endophytic bacterium Bacillus subtilis (BERA 71) improves salt tolerance in chickpea plants by regulating the plant defense mechanisms. J Plant Interact 13 (1):37–44

Achard P, Cheng H, De Grauwe L, Decat J, Schoutteten H, Moritz T, Harberd NP (2006) Integration of plant responses to environmentally activated phytohormonal signals. Science 311(5757):91–94

Ahmad M, Zahir ZA, Asghar HN, Asghar M (2011) Inducing salt tolerance in mung bean through coinoculation with rhizobia and plant-growth-promoting rhizobacteria containing 1-aminocyclopropane-1-carboxylate deaminase. Can J Microbiol 57:578–589

Ali S, Charles TC, Glick BR (2014) Amelioration of high salinity stress damage by plant growth-promoting bacterial endophytes that contain ACC deaminase. Plant Physiol Biochem 80:160–167

ALKahtani MD, Attia KA, Hafez YM, Khan N, Eid AM, Ali MA, Abdelaal KA (2020) Chlorophyll fluorescence parameters and antioxidant defense system can display salt tolerance of salt acclimated sweet pepper plants treated with chitosan and plant growth promoting rhizobacteria. Agronomy 10:1180

Al-Karaki GN (2000) Growth of mycorrhizal tomato and mineral acquisition under salt stress. Mycorrhiza 10(2):51–54

Almeida DM, Oliveira MM, Saibo NJ (2017) Regulation of Na+ and K+ homeostasis in plants: towards improved salt stress tolerance in crop plants. Genet Mol Biol 40(1):326–345

Arora M et al (2020) Interaction between Piriformospora indica and Azotobacter chroococcum diminish the effect of salt stress in Artemisia annua L. by enhancing enzymatic and non-enzymatic antioxidants. Symbiosis 80(1):61–73

Bal HB, Nayak L, Das S, Adhya TK (2013) Isolation of ACC deaminase producing PGPR from rice rhizosphere and evaluating their plant growth promoting activity under salt stress. Plant Soil 366:93–105

Bano A, Fatima M (2009) Salt tolerance in Zea mays (L). following inoculation with Rhizobium and Pseudomonas. Biol Fertil Soils 45(4):405–413

Barnawal D, Bharti N, Tripathi A, Pandey SS, Chanotiya CS, Kalra A (2016) ACC-deaminase-producing endophyte Brachybacterium paraconglomeratum strain SMR20 ameliorates Chlorophytum salinity stress via altering phytohormone generation. J Plant Growth Regul 35 (2):553–564

Barnawal D, Bharti N, Pandey SS, Pandey A, Chanotiya CS, Kalra A (2017) Plant growth promoting rhizobacteria enhance wheat salt and drought stress tolerance by altering endogenous phytohormone levels and TaCTR1/TaDREB2 expression. Physiol Plant 161:502–514. https://doi.org/10.1111/ppl.12614

Belimov AA, Hontzeas N, Safronova VI, Demchinskaya SV, Piluzza G, Bullitta S, Glick BR (2005) Cadmium-tolerant plant growth-promoting bacteria associated with the roots of Indian mustard (Brassica juncea L. Czern.). Soil Biol Biochem 37:241–250

Beltrano J, Ruscitti M, Arango MC, Ronco M (2013) Effects of arbuscular mycorrhiza inoculation on plant growth, biological and physiological parameters and mineral nutrition in pepper grown under different salinity and P levels. J Soil Sci Plant Nutr 13:123–141

Bharti N, Pandey SS, Barnawal D, Patel VK, Kalra A (2016) Plant growth promoting rhizobacteria Dietzia natronolimnaea modulates the expression of stress responsive genes providing protection of wheat from salinity stress. Sci Rep 6(1):1–16

Bhatt P (2018) Insilico tools to study the bioremediation in microorganisms. In: Pathak V, Navneet (eds) Handbook of research on microbial tools for environmental waste management. IGI Global, Hershey, PA, pp 389–395. https://doi.org/10.4018/978-1-5225-3540-9.ch018

Bhatt P, Barh A (2018) Bioinformatic tools to study the soil microorganisms: an in silico approach for sustainable agriculture. In: Choudhary D, Kumar M, Prasad R, Kumar V (eds) In silico approach for sustainable agriculture. Springer, Singapore. https://doi.org/10.1007/978-981-13-0347-0_10

Bhatt P, Nailwal TK (2018) Crop improvement through microbial technology: a step toward sustainable agriculture. In: Crop improvement through microbial biotechnology. Ch-11. Elsevier, Amsterdam, pp 245–253. https://doi.org/10.1016/B978-0-444-63987-5.00011-6

Bhattacharyya PN, Jha DK (2012) Plant growth-promoting rhizobacteria (PGPR): emergence in agriculture. World J Microbiol Biotechnol 28(4):1327–1350

Bhise KK, Dandge PB (2019) Mitigation of salinity stress in plants using plant growth promoting bacteria. Symbiosis 79(3):191–204

Blumwald E, Aharon GS, Apse MP (2000) Sodium transport in plant cells. Biochim Biophys Acta 1465(1–2):140–151

Brader G, Compant S, Mitter B, Trognitz F, Sessitsch A (2014) Metabolic potential of endophytic bacteria. Curr Opin Biotechnol 27:30–37

Campanelli A, Ruta C, De Mastro G, Morone-Fortunato I (2013) The role of arbuscular mycorrhizal fungi in alleviating salt stress in Medicago sativa L. var. icon. Symbiosis 59(2):65–76

Chinnaswamy A et al (2018) A nodule endophytic Bacillus megaterium strain isolated from Medicago polymorpha enhances growth, promotes nodulation by Ensifer medicae and alleviates salt stress in alfalfa plants. Ann Appl Biol 172(3):295–308

Choi WG, Toyota M, Kim SH, Hilleary R, Gilroy S (2014) Salt stress-induced Ca^{2+} waves are associated with rapid, long-distance root-to-shoot signaling in plants. Proc Natl Acad Sci 111 (17):6497–6502

Cui MH, Yoo KS, Hyoung S, Nguyen HTK, Kim YY, Kim HJ, Shin JS (2013) An Arabidopsis R2R3-MYB transcription factor, AtMYB20, negatively regulates type 2C serine/threonine protein phosphatases to enhance salt tolerance. FEBS Lett 587(12):1773–1778

Daliakopoulos IN, Tsanis IK, Koutroulis A, Kourgialas NN, Varouchakis AE, Karatzas GP, Ritsema CJ (2016) The threat of soil salinity: a European scale review. Sci Total Environ 573:727–739

Deinlein U, Stephan AB, Horie T, Luo W, Xu G, Schroeder JI (2014) Plant salt-tolerance mechanisms. Trends Plant Sci 19(6):371–379

del Carmen Orozco-Mosqueda M, Glick BR, Santoyo G (2020) ACC deaminase in plant growth-promoting bacteria (PGPB): an efficient mechanism to counter salt stress in crops. Microbiol Res 235:126439

Demidchik V, Maathuis FJ (2007) Physiological roles of nonselective cation channels in plants: from salt stress to signalling and development. New Phytol 175(3):387–404

Demidchik V, Tester M (2002) Sodium fluxes through nonselective cation channels in the plasma membrane of protoplasts from Arabidopsis roots. Plant Physiol 128(2):379–387

Dietz KJ, Vogel MO, Viehhauser A (2010) AP2/EREBP transcription factors are part of gene regulatory networks and integrate metabolic, hormonal and environmental signals in stress acclimation and retrograde signalling. Protoplasma 245(1–4):3–14

Donaldson L, Ludidi N, Knight MR, Gehring C, Denby K (2004) Salt and osmotic stress cause rapid increases in Arabidopsis thaliana cGMP levels. FEBS Lett 569(1–3):317–320

Egamberdieva D et al (2016) A synergistic interaction between salt-tolerant Pseudomonas and Mesorhizobium strains improves growth and symbiotic performance of liquorice (Glycyrrhiza uralensis fish.) under salt stress. Appl Microbiol Biotechnol 100(6):2829–2841

Estrada B, Aroca R, Maathuis FJ, Barea JM, Ruiz-Lozano JM (2013) Arbuscular mycorrhizal fungi native from a Mediterranean saline area enhance maize tolerance to salinity through improved ion homeostasis. Plant Cell Environ 36(10):1771–1782

Evans MJ, Choi WG, Gilroy S, Morris RJ (2016) A ROS-assisted calcium wave dependent on the AtRBOHD NADPH oxidase and TPC1 cation channel propagates the systemic response to salt stress. Plant Physiol 171(3):1771–1784

Evelin H, Kapoor R, Giri B (2009) Arbuscular mycorrhizal fungi in alleviation of salt stress: a review. Ann Bot 104(7):1263–1280

Evelin H, Giri B, Kapoor R (2012) Contribution of Glomus intraradices inoculation to nutrient acquisition and mitigation of ionic imbalance in NaCl-stressed Trigonella foenum-graecum. Mycorrhiza 22(3):203–217

Evelin H, Giri B, Kapoor R (2013) Ultrastructural evidence for AMF mediated salt stress mitigation in Trigonella foenum-graecum. Mycorrhiza 23(1):71–86

FAO (2016) Saline soils and their management. Food and Agriculture Organization of the United Nations. http://www.fao.org/3/x5871e/x5871e04.htm. (cross reference: Litalien, A. & Zeeb, B. (2020). Curing the earth: A review of anthropogenic soil salinization and plant-based strategies for sustainable mitigation. Science of the Total Environment, 698, 134235)

Feng G, Zhang F, Li X, Tian C, Tang C, Rengel Z (2002) Improved tolerance of maize plants to salt stress by arbuscular mycorrhiza is related to higher accumulation of soluble sugars in roots. Mycorrhiza 12(4):185–190

Feng W, Kita D, Peaucelle A, Cartwright HN, Doan V, Duan Q, Yvon R (2018) The FERONIA receptor kinase maintains cell-wall integrity during salt stress through Ca^{2+} signaling. Curr Biol 28(5):666–675

Ferreira MJ, Silva H, Cunha A (2019) Siderophore-producing rhizobacteria as a promising tool for empowering plants to cope with Iron limitation in saline soils: a review. Pedosphere 29 (4):409–420

Flowers TJ, Colmer TD (2008) Salinity tolerance in halophytes. New Phytol:945–963

Galvan-Ampudia CS, Julkowska MM, Darwish E, Gandullo J, Korver RA, Brunoud G, Testerink C (2013) Halotropism is a response of plant roots to avoid a saline environment. Curr Biol 23 (20):2044–2050

Gamalero E et al (2010) Interactions between Pseudomonas putida UW4 and Gigaspora rosea BEG9 and their consequences for the growth of cucumber under salt-stress conditions. J Appl Microbiol 108(1):236–245

Geng Y, Wu R, Wee CW, Xie F, Wei X, Chan PMY, Dinneny JR (2013) A spatio-temporal understanding of growth regulation during the salt stress response in Arabidopsis. Plant Cell 25 (6):2132–2154
Giri B, Mukerji KG (2004) Mycorrhizal inoculant alleviates salt stress in Sesbania aegyptiaca and Sesbania grandiflora under field conditions: evidence for reduced sodium and improved magnesium uptake. Mycorrhiza 14(5):307–312
Giri B, Kapoor R, Mukerji KG (2007) Improved tolerance of Acacia nilotica to salt stress by arbuscular mycorrhiza, Glomus fasciculatum may be partly related to elevated K/Na ratios in root and shoot tissues. Microb Ecol 54:753–760
Golldack D, Lüking I, Yang O (2011) Plant tolerance to drought and salinity: stress regulating transcription factors and their functional significance in the cellular transcriptional network. Plant Cell Rep 30(8):1383–1391
Hajiboland R, Aliasgharzadeh N, Laiegh SF, Poschenrieder C (2010) Colonization with arbuscular mycorrhizal fungi improves salinity tolerance of tomato (Solanum lycopersicum L.) plants. Plant Soil 331:313–327
Hamayun M et al (2017) Gibberellins producing endophytic fungus Porostereum spadiceum AGH786 rescues growth of salt affected soybean. Front Microbiol 8:686
Han QQ, Wu YN, Gao HJ, Xu R, Paré PW, Shi H, Zhao Q, Li HR, Khan SA, Wang YQ, Wang SM (2017) Improved salt tolerance of medicinal plant Codonopsis pilosula by Bacillus amyloliquefaciens GB03. Acta Physiol Plant 39(1):–35
Hasanuzzaman M, Nahar K, Alam M, Bhowmik PC, Hossain M, Rahman MM, Fujita M (2014) Potential use of halophytes to remediate saline soils. BioMed Res Int 2014:589341
Hashem A et al (2016a) Induction of osmoregulation and modulation of salt stress in Acacia gerrardii Benth. by arbuscular mycorrhizal fungi and Bacillus subtilis (BERA 71). Biomed Res Int:2016
Hashem A et al (2016b) The interaction between arbuscular mycorrhizal fungi and endophytic bacteria enhances plant growth of Acacia gerrardii under salt stress. Front Microbiol 7:1089
Hashem A, Alqarawi AA, Radhakrishnan R, Al-Arjani ABF, Aldehaish HA, Egamberdieva D, Abd-Allah EF (2018) Arbuscular mycorrhizal fungi regulate the oxidative system, hormones and ionic equilibrium to trigger salt stress tolerance in Cucumis sativus L. Saudi J Biol Sci 25: 1102–1114
Hayat R et al (2010) Soil beneficial bacteria and their role in plant growth promotion: a review. Ann Microbiol 60(4):579–598
Hayes S, Pantazopoulou CK, van Gelderen K, Reinen E, Tween AL, Sharma A, Pierik R (2019) Soil salinity limits plant shade avoidance. Curr Biol 29(10):1669–1676
Islam F, Yasmeen T, Ali S, Ali B, Farooq MA, Gill RA (2015) Priming-induced antioxidative responses in two wheat cultivars under saline stress. Acta Physiol Plant 37:153
Jahromi F, Aroca R, Porcel R, Ruiz-Lozano JM (2008) Influence of salinity on the in vitro development of Glomus intraradices and on the in vivo physiological and molecular responses of mycorrhizal lettuce plants. Microb Ecol 55(1):45
Jha Y, Subramanian RB, Patel S (2011) Combination of endophytic and rhizospheric plant growth promoting rhizobacteria in Oryza sativa shows higher accumulation of osmoprotectant against saline stress. Acta Physiol Plant 33(3):797–802
Ji H, Pardo JM, Batelli G, Van Oosten MJ, Bressan RA, Li X (2013) The salt overly sensitive (SOS) pathway: established and emerging roles. Mol Plant 6(2):275–286
Jiang Y, Deyholos MK (2009) Functional characterization of Arabidopsis NaCl-inducible WRKY25 and WRKY33 transcription factors in abiotic stresses. Plant Mol Biol 69 (1–2):91–105
Jiang Y, Yang B, Deyholos MK (2009) Functional characterization of the Arabidopsis bHLH92 transcription factor in abiotic stress. Mol Gen Genomics 282(5):503–516
Jiang Z, Zhou X, Tao M, Yuan F, Liu L, Wu F, Li C (2019) Plant cell-surface GIPC sphingolipids sense salt to trigger Ca^{2+} influx. Nature 572(7769):341–346

Joe MM, Devaraj S, Benson A, Sa T (2016) Isolation of phosphate solubilizing endophytic bacteria from Phyllanthus amarus Schum & Thonn: evaluation of plant growth promotion and antioxidant activity under salt stress. J Appl Res Med Aromatic Plants 3:71–77

Kalev SD, Toor GS (2018) The composition of soils and sediments. In: Green chemistry. Elsevier, pp 339–357

Kasuga M, Liu Q, Miura S, Yamaguchi-Shinozaki K, Shinozaki K (1999) Improving plant drought, salt, and freezing tolerance by gene transfer of a single stress-inducible transcription factor. Nat Biotechnol 17(3):287–291

Khalid M, Bilal M, Hassani D, Iqbal HM, Wang H, Huang D (2017) Mitigation of salt stress in white clover (Trifolium repens) by Azospirillum brasilense and its inoculation effect. Bot Stud 58:1–7

Khalil HA, Eissa AM, El-Shazly SM, Nasr AMA (2011) Improved growth of salinity-stressed citrus after inoculation with mycorrhizal fungi. Sci Hortic 130(3):624–632

Khan AL et al (2012) Endophytic fungal association via gibberellins and indole acetic acid can improve plant growth under abiotic stress: an example of Paecilomyces formosus LHL10. BMC Microbiol 12(1):3

Khan NA, Khan MIR, Ferrante A, Poor P (2017) Ethylene: a key regulatory molecule in plants. Front Plant Sci 8:1782

Khan MA, Asaf S, Khan AL, Ullah I, Ali S, Kang SM, Lee IJ (2019) Alleviation of salt stress response in soybean plants with the endophytic bacterial isolate Curtobacterium sp. SAK1. Ann Microbiol 69:797–808

Kim K, Jang YJ, Lee SM, Oh BT, Chae JC, Lee KJ (2014) Alleviation of salt stress by Enterobacter sp. EJ01 in tomato and Arabidopsis is accompanied by up-regulation of conserved salinity responsive factors in plants. Mol Cells 37(2):109

Kumar N, Gangola S, Bhatt P, Jeena N, Khwairakpam R (2019) Soil genesis, survey and classification. In: Varma A, Choudhary D (eds) Mycorrhizosphere and pedogenesis. Springer, Singapore. https://doi.org/10.1007/978-981-13-6480-8_8

Kumar V, Kumar P, Khan A (2020) Optimization of PGPR and silicon fertilization using response surface methodology for enhanced growth, yield and biochemical parameters of French bean (Phaseolus vulgaris L.) under saline stress. Biocatal Agric Biotechnol 23:101463

Kunte HJ (2006) Osmoregulation in bacteria: compatible solute accumulation and osmosensing. Environ Chem 3(2):94–99

Latef AAHA, Chaoxing H (2011) Effect of arbuscular mycorrhizal fungi on growth, mineral nutrition, antioxidant enzymes activity and fruit yield of tomato grown under salinity stress. Sci Hortic 127(3):228–233

Li HQ, Jiang XW (2017) Inoculation with plant growth-promoting bacteria (PGPB) improves salt tolerance of maize seedling. Russ J Plant Physiol 64:235–241

López-Bucio J et al (2007) Bacillus megaterium rhizobacteria promote growth and alter root-system architecture through an auxin-and ethylene-independent signaling mechanism in Arabidopsis thaliana. Mol Plant-Microbe Interact 20(2):207–217

Maggio A et al (2010) Contrasting effects of GA 3 treatments on tomato plants exposed to increasing salinity. J Plant Growth Regul 29(1):63–72

Mahajan S, Sopory S, Tuteja N, Asaf A, Marg A (2006) CBL-CIPK paradigm: role in calcium and stress signaling in plants. Proc Indian Natl Sci Acad 72(2):63

Manjili FA, Sedghi M, Pessarakli M (2012) Effects of phytohormones on proline content and antioxidant enzymes of various wheat cultivars under salinity stress. J Plant Nutr 35 (7):1098–1111

Mayak S, Tirosh T, Glick BR (2004) Plant growth-promoting bacteria confer resistance in tomato plants to salt stress. Plant Physiol Biochem 42:565–572

Miller GAD, Suzuki N, Ciftci-Yilmaz S, Mittler RON (2010) Reactive oxygen species homeostasis and signalling during drought and salinity stresses. Plant Cell Environ 33(4):453–467

Nahar K, Hasanuzzaman M, Fujita M (2016) Roles of osmolytes in plant adaptation to drought and salinity. In: Osmolytes and plants acclimation to changing environment: emerging omics technologies. Springer, New Delhi

Nautiyal CS, Srivastava S, Chauhan PS, Seem K, Mishra A, Sopory SK (2013) Plant growth-promoting bacteria Bacillus amyloliquefaciens NBRISN13 modulates gene expression profile of leaf and rhizosphere community in rice during salt stress. Plant Physiol Biochem 66:1–9
Nawaz K, Hussain K, Majeed A, Khan F, Afghan S, Ali K (2010) Fatality of salt stress to plants: morphological, physiological and biochemical aspects. Afr J Biotechnol 9(34)
Omar MNA, Osman MEH, Kasim WA, Abd El-Daim IA (2009) Improvement of salt tolerance mechanisms of barley cultivated under salt stress using Azospirillum brasilense. In: Salinity and water stress. Springer, Dordrecht
Ouziad F, Wilde P, Schmelzer E, Hildebrandt U, Bothe H (2006) Analysis of expression of aquaporins and Na+/H+ transporters in tomato colonized by arbuscular mycorrhizal fungi and affected by salt stress. Environ Exp Bot 57:177–186
Piernik A, Hrynkiewicz K, Wojciechowska A, Szymańska S, Lis MI, Muscolo A (2017) Effect of halotolerant endophytic bacteria isolated from Salicornia europaea L. on the growth of fodder beet (Beta vulgaris L.) under salt stress. Arch Agron Soil Sci 63:1404–1418
Platten JD, Cotsaftis O, Berthomieu P, Bohnert H, Davenport RJ, Fairbairn DJ, Mäser P (2006) Nomenclature for HKT transporters, key determinants of plant salinity tolerance. Trends Plant Sci 11(8):372–374
Porcel R, Aroca R, Ruiz-Lozano JM (2012) Salinity stress alleviation using arbuscular mycorrhizal fungi. A review. Agron Sustain Dev 32(1):181–200
Porcel R, Aroca R, Azcon R, Ruiz-Lozano JM (2016) Regulation of cation transporter genes by the arbuscular mycorrhizal symbiosis in rice plants subjected to salinity suggests improved salt tolerance due to reduced Na^+ root-to-shoot distribution. Mycorrhiza 26(7):673–684
Porras-Soriano A, Soriano-Martín ML, Porras-Piedra A, Azcón R (2009) Arbuscular mycorrhizal fungi increased growth, nutrient uptake and tolerance to salinity in olive trees under nursery conditions. J Plant Physiol 166(13):1350–1359
Qin S, Zhang YJ, Yuan B, Xu PY, Xing K, Wang J, Jiang JH (2014) Isolation of ACC deaminase-producing habitat-adapted symbiotic bacteria associated with halophyte Limonium sinense (Girard) Kuntze and evaluating their plant growth-promoting activity under salt stress. Plant Soil 374(1–2):753–766
Qin Y, Druzhinina IS, Pan X, Yuan Z (2016) Microbially mediated plant salt tolerance and microbiome-based solutions for saline agriculture. Biotechnol Adv 34(7):1245–1259
Ramadoss D, Lakkineni VK, Bose P, Ali S, Annapurna K (2013) Mitigation of salt stress in wheat seedlings by halotolerant bacteria isolated from saline habitats. Springerplus 2:6
Rath KM, Fierer N, Murphy DV, Rousk J (2019) Linking bacterial community composition to soil salinity along environmental gradients. ISME J 13(3):836–846
Redondo-Gómez S, Mateos-Naranjo E, Figueroa ME, Davy AJ (2010) Salt stimulation of growth and photosynthesis in an extreme halophyte, Arthrocnemum macrostachyum. Plant Biol 12 (1):79–87
Rojas-Tapias D, Moreno-Galván A, Pardo-Díaz S, Obando M, Rivera D, Bonilla R (2012) Effect of inoculation with plant growth-promoting bacteria (PGPB) on amelioration of saline stress in maize (Zea mays). Appl Soil Ecol 61:264–272
Ruiz-Lozano JM, Azcon R, Gomez M (1996) Alleviation of salt stress by arbuscular-mycorrhizal Glomus species in Lactuca sativa plants. Physiol Plant 98:767–772
Ruiz-Lozano JM, Porcel R, Azcón C, Aroca R (2012) Regulation by arbuscular mycorrhizae of the integrated physiological response to salinity in plants: new challenges in physiological and molecular studies. J Exp Bot 63(11):4033–4044
Ruppel S, Franken P, Witzel K (2013) Properties of the halophyte microbiome and their implications for plant salt tolerance. Funct Plant Biol 40(9):940–951
Sadeghi A, Karimi E, Dahaji PA, Javid MG, Dalvand Y, Askari H (2012) Plant growth promoting activity of an auxin and siderophore producing isolate of Streptomyces under saline soil conditions. World J Microbiol Biotechnol 28:1503–1509
Sah SK, Reddy KR, Li J (2016) Abscisic acid and abiotic stress tolerance in crop plants. Front Plant Sci 7:571

Santoyo G, Moreno-Hagelsieb G, del Carmen Orozco-Mosqueda M, Glick BR (2016) Plant growth-promoting bacterial endophytes. Microbiol Res 183:92–99

Sarkar A et al (2018) A halotolerant Enterobacter sp. displaying ACC deaminase activity promotes rice seedling growth under salt stress. Res Microbiol 169(1):20–32

Satyaprakash M et al (2017) Phosphorous and phosphate solubilising bacteria and their role in plant nutrition. Int J Curr Microbiol App Sci 6(4):2133–2144

Schirawski J, Perlin MH (2018) Plant–microbe interaction 2017—the good, the bad and the diverse. Multidisciplinary Digital Publishing Institute

Shabala S (2013) Learning from halophytes: physiological basis and strategies to improve abiotic stress tolerance in crops. Ann Bot 112(7):1209–1221

Shahzad R, Khan AL, Bilal S, Waqas M, Kang SM, Lee IJ (2017) Inoculation of abscisic acid-producing endophytic bacteria enhances salinity stress tolerance in Oryza sativa. Environ Exp Bot 136:68–77

Shilev S, Sancho ED, Benlloch-González M (2012) Rhizospheric bacteria alleviate salt-produced stress in sunflower. J Environ Manag 95:S37–S41

Shrivastava P, Kumar R (2015) Soil salinity: a serious environmental issue and plant growth promoting bacteria as one of the tools for its alleviation. Saudi J Biol Sci 22:123–131

Singh JS (2013) Plant growth promoting rhizobacteria. Resonance 18:275–281

Singh RP, Jha PN (2015) The plant-growth-promoting bacterium Klebsiella sp. SBP-8 confers induced systemic tolerance in wheat (Triticum aestivum) under salt stress. J Plant Physiol 184: 57–67

Stavridou E, Hastings A, Webster RJ, Robson PR (2017) The impact of soil salinity on the yield, composition and physiology of the bioenergy grass Miscanthus × giganteus. GCB Bioenergy 9 (1):92–104

Stepien P, Klobus G (2005) Antioxidant defense in the leaves of C3 and C4 plants under salinity stress. Physiol Plant 125:31–40

Tang X, Mu X, Shao H, Wang H, Brestic M (2015) Global plant-responding mechanisms to salt stress: physiological and molecular levels and implications in biotechnology. Crit Rev Biotechnol 35:425–437

Tiwari S, Singh P, Tiwari R, Meena KK, Yandigeri M, Singh DP, Arora DK (2011) Salt-tolerant rhizobacteria-mediated induced tolerance in wheat (Triticum aestivum) and chemical diversity in rhizosphere enhance plant growth. Biol Fertil Soils 47(8):907

Tiwari S, Lata C, Chauhan PS, Nautiyal CS (2016) Pseudomonas putida attunes morphophysiological, biochemical and molecular responses in Cicer arietinum L. during drought stress and recovery. Plant Physiol Biochem 99:108–117

Tolba ST, Ibrahim M, Amer EA, Ahmed DA (2019) First insights into salt tolerance improvement of Stevia by plant growth-promoting Streptomyces species. Arch Microbiol 201:1295–1306

Tran LSP, Nakashima K, Sakuma Y, Simpson SD, Fujita Y, Maruyama K, Yamaguchi-Shinozaki K (2004) Isolation and functional analysis of Arabidopsis stress-inducible NAC transcription factors that bind to a drought-responsive cis-element in the early responsive to dehydration stress 1 promoter. Plant Cell 16(9):2481–2498

Tuna AL et al (2008) The combined effects of gibberellic acid and salinity on some antioxidant enzyme activities, plant growth parameters and nutritional status in maize plants. Environ Exp Bot 62(1):1–9

Tuteja N (2007) Mechanisms of high salinity tolerance in plants. In: Methods in enzymology, vol 428. Academic Press, pp 419–438

Ullah A, Heng S, Munis MFH, Fahad S, Yang X (2015) Phytoremediation of heavy metals assisted by plant growth promoting (PGP) bacteria: a review. Environ Exp Bot 117:28–40

Vacheron J, Desbrosses G, Bouffaud ML, Touraine B, Moënne-Loccoz Y, Muller D, Prigent-Combaret C (2013) Plant growth-promoting rhizobacteria and root system functioning. Front Plant Sci 4:356

Valenzuela CE, Acevedo-Acevedo O, Miranda GS, Vergara-Barros P, Holuigue L, Figueroa CR, Figueroa PM (2016) Salt stress response triggers activation of the jasmonate signaling pathway

leading to inhibition of cell elongation in Arabidopsis primary root. J Exp Bot 67 (14):4209–4220
Van Zelm E, Zhang Y, Testerink C (2020) Salt tolerance mechanisms of plants. Annu Rev Plant Biol 71
Vejan P et al (2016) Role of plant growth promoting rhizobacteria in agricultural sustainability—a review. Molecules 21(5):573
Verma S, Mishra SN (2005) Putrescine alleviation of growth in salt stressed Brassica juncea by inducing antioxidative defense system. J Plant Physiol 162:669–677
Verma S, Bhatt P, Verma A et al (2021) Microbial technologies for heavy metal remediation: effect of process conditions and current practices. Clean Techn Environ Policy. https://doi.org/10.1007/s10098-021-02029-8
Wang Y, Gao C, Liang Y, Wang C, Yang C, Liu G (2010) A novel bZIP gene from Tamarix hispida mediates physiological responses to salt stress in tobacco plants. J Plant Physiol 167(3):222–230
Wang X, Chen X, Liu Y, Gao H, Wang Z, Sun G (2011) CkDREB gene in Caragana korshinskii is involved in the regulation of stress response to multiple abiotic stresses as an AP2/EREBP transcription factor. Mol Biol Rep 38(4):2801–2811
Wang Y, Wang M, Li Y, Wu A, Huang J (2018) Effects of arbuscular mycorrhizal fungi on growth and nitrogen uptake of Chrysanthemum morifolium under salt stress. PLoS One 13:e0196408
Waqas M et al (2012) Endophytic fungi produce gibberellins and indoleacetic acid and promotes host-plant growth during stress. Molecules 17(9):10754–10773
Wu QS, Zou YN, He XH (2010) Contributions of arbuscular mycorrhizal fungi to growth, photosynthesis, root morphology and ionic balance of citrus seedlings under salt stress. Acta Physiol Plant 32:297–304
Xie YJ, Xu S, Han B, Wu MZ, Yuan XX, Han Y, Shen WB (2011) Evidence of Arabidopsis salt acclimation induced by up-regulation of HY1 and the regulatory role of RbohD-derived reactive oxygen species synthesis. Plant J 66(2):280–292
Xiong Y-W et al (2019) Enhancement of growth and salt tolerance of tomato seedlings by a natural halotolerant actinobacterium Glutamicibacter halophytocola KLBMP 5180 isolated from a coastal halophyte. Plant Soil 445(1–2):307–322
Yang J, Kloepper JW, Ryu CM (2009a) Rhizosphere bacteria help plants tolerate abiotic stress. Trends Plant Sci 14(1):1–4
Yang O, Popova OV, Süthoff U, Lüking I, Dietz KJ, Golldack D (2009b) The Arabidopsis basic leucine zipper transcription factor AtbZIP24 regulates complex transcriptional networks involved in abiotic stress resistance. Gene 436(1–2):45–55
Yao L et al (2010) Growth promotion and protection against salt stress by Pseudomonas putida Rs-198 on cotton. Eur J Soil Biol 46(1):49–54
Zerrouk IZ, Benchabane M, Khelifi L, Yokawa K, Ludwig-Müller J, Baluska F (2016) A Pseudomonas strain isolated from date-palm rhizospheres improves root growth and promotes root formation in maize exposed to salt and aluminum stress. J Plant Physiol 191:111–119
Zhang JZ, Creelman RA, Zhu JK (2004) From laboratory to field. Using information from Arabidopsis to engineer salt, cold, and drought tolerance in crops. Plant Physiol 135(2):615–621
Zhu JK (2002) Salt and drought stress signal transduction in plants. Annu Rev Plant Biol 53 (1):247–273

Impact of Plant Exudates on Soil Microbiomes

11

Yekti Maryani and Rohlan Rogomulyo

Abstract

The plant exits root exudate into the rhizosphere. This exudate acts as a regulator of soil microbial communities around the roots and supports a favorable symbiosis between plants and microbes. This symbiotic mechanism occurs because the root exudate is a source of carbon and energy for microbes. Population and microbial activity in the rhizosphere are higher than non-rhizosphere due to the availability of substrate in root exudate.

Keywords

Rhizosphere · Root exudate · Soil · Microorganisms

11.1 Rhizosphere

The rhizosphere is an ecological system in soil that is very dynamic. Conceptually, the root area of the plant is divided into two zones, namely the rhizosphere which includes the closest area which is the place where the roots of the plant directly contact the soil. The second zone is the part of the soil that is not close to the root. It is not the place where the roots of the plant are directly in contact with the soil. It is called bulk soil. In this concept, the rhizosphere is a more dynamic part of the ecosystem because of the interaction between plant roots and soil microbes. The plant root system release exudates from the roots and it is used by microbes as

Y. Maryani (✉)
Agriculture Faculty, Sarjanawiyata Tamansiswa University, Yogyakarta, Indonesia

R. Rogomulyo
Agriculture Faculty, Gadjah Mada University, Yogyakarta, Indonesia

R. Soni et al. (eds.), *Microbiological Activity for Soil and Plant Health Management*,
https://doi.org/10.1007/978-981-16-2922-8_11

nutrients for their growth (Yang et al. 2020). The presence of microbes can have a direct or indirect effect on plant growth (Yueling et al. 2020). In the rhizosphere, microbes obtain a source of nutrition from plant root exudates (Omotayo and Babalola 2021). The availability of nutrients and the highest number of microbes are found in the rhizosphere. In the part of the soil that is far from the rhizosphere (bulk soil), the availability of root exudate nutrients and number of microbes is less (Bakker et al. 2013).

The rhizosphere is defined as the zone of soil where plant roots and microbes interact effectively. The soil where the plant root system is associated has a different condition from the soil without a root system. Association of plant root systems with soil microbes can occur directly or indirectly (Bakker et al. 2013; Xing-Feng et al. 2014). In the rhizosphere, plant roots influence the physical, chemical, and biological properties of the soil. Microbial activity at the place where the roots meet the soil triggers and spurs biochemical reactions that play an important role in providing nutrients for plants. The effect of the association of root systems with microbes on the rhizosphere can be done in two ways. First, the microbes degrade the root cells which slough off and die when the roots penetrate the soil material. Second, microbes take advantage of root exudates in the form of organic compounds for their growth and development.

The intensity of plant root exudate utilization by microbes is influenced by the distance and time span of exudate release to the environment in the rhizosphere. It is called the rhizosphere effect. The effect of the rhizosphere is defined as the influence of the association of plant roots with the rhizosphere on microbes. The rhizosphere effect is influenced by soil type, soil moisture, pH, temperature, age, and plant character. The real rhizosphere effect can be seen based on the abundant number of microbes in the area (Paul et al. 2010; Brunel et al. 2020; Yang et al. 2020). The rate of microbial metabolism in the rhizosphere is different from the rate of microbial metabolism in non-rhizosphere soils. In the land of the rhizosphere, it was in the area around plant roots. It is called the transition area. There are 100 times more microbes than in non-rhizosphere areas.

Exudates of plant roots in the rhizosphere contain organic compounds that promote microbial growth and development. This is the reason why the number of microbes is very high in the rhizosphere. This condition can be used to assess soil fertility by determining the R/S ratio. The R notation is the number of microbes in the rhizosphere, while S is the number of microbes in non-rhizosphere soil (bulk soil). The small R/S index means that the source of microbial nutrition does not come from root exudates, but more from non-rhizosphere soil organic compounds. This shows that the lower the R/S index, the higher the soil fertility.

Conversely, if the R/S index is greater, the nutrients for microbial growth and development come from exudates in rhizosphere soil. This indicates that the soil is poor in organic compounds or nutrients for microbial growth. Thus, it shows that the soil is not fertile. The R/S ratio value, namely the rhizosphere (R) to soil (S) ratio, can be used to predict changes in the microbial population in the soil caused by plant growth. The R/S ratio is calculated based on the calculation of the number of microbes in the rhizosphere divided by the number of microbes in non-rhizosphere soil. Non-rhizosphere soil is soil that is free from the influence of plant root growth. The R/S ratio value is influenced by the type of microbe. In general, the order of the R/S ratio from the largest to the smallest index is bacteria, actinomycetes, fungi, protozoa, and algae.

The rhizosphere is divided into the endorhizosphere, rhizoplane, and ectorhizosphere. Endorhizosphere, namely the proliferation of microbes that occur in plant roots. Rhizoplane, namely the proliferation of microbes that occur on the surface of the roots. The ectorhizosphere is the proliferation of microbes that occur outside of plant roots.

The physical, chemical, and biological properties of soil are influenced by the interactions between plant roots, soil, and microbes in the rhizosphere. So this interaction also affects the microbial population in the rhizosphere. Interactions between plant roots and microbial communities are also mediated by root exudates in the rhizosphere (Xing-Feng et al. 2014; Yang et al. 2020). 5–21% of the carbon fixed in plant photosynthesis is released by plant roots in the form of a solution of sugars, amino acids, and secondary metabolites (Paul et al. 2010; Dayakar et al. 2013a, b; Xing-Feng et al. 2014; Brunel et al. 2020). The compounds from the exudates are used for the growth and development of microbes in the rhizosphere. There are two groups of root exudates, namely compounds with low molecular weight, for example, amino acids, organic acids, sugars, phenolic compounds, and other secondary metabolites. These two are compounds with high molecular weight, for example, polysaccharides and proteins. The qualitative and quantitative composition of root exudates is determined by the type of cultivar, plant growth phase, soil type, soil pH, temperature, and microbial location (Sugiyama 2019; Khare et al. 2020). It affects the microbial community in the rhizosphere. Each microbial community in the rhizosphere has a certain degree of specificity for each type of plant.

Root exudate compounds are secreted and distributed by means of transport mechanisms into the rhizosphere (Xing-Feng et al. 2014). In general, root exudates are released by diffusion or passive mechanisms, especially for organic compounds with low molecular weight and small polar compounds that are not charged. Diffusion is a process that is influenced by the permeability of the cell membrane. Root exudates such as secondary metabolites, polysaccharides, and proteins for the role of proteins can be bound to different membranes (Xing-Feng et al. 2014).

The interaction of plants with plants, plants with microbes, and plants with fauna is also mediated by phytochemicals secreted by plant roots. The degree of interaction can be neutral, beneficial, and detrimental. In certain cases, it is found that harmful microbes (pathogens) can switch beneficial functions (symbiosis) for plants. For example, N-binding bacteria, in interacting with plants, can be beneficial to neutral. The range of interaction properties is influenced by the level of N availability in the soil (Chaparro et al. 2014). If the availability of N in the soil is limited, legume plants will release flavones and flavonols to bind rhizobia. There was an initiation of symbiosis between legume plants and rhizobia (Chaparro et al. 2014; Xing-Feng et al. 2014). The availability of nutrients also determines the nature of interactions, for example, the relationship between mycorrhizal and *Medicago truncatula* plants. Mycorrhizae get a lot of carbon from these plants, whereas plants get phosphorus from mycorrhizae (Haris 2010; Mendes et al. 2013; Chaparro et al. 2014; Xing-Feng et al. 2014).

The results of tests on eight plant species turned out that only 10% of microbes colonized the rhizoplane. The distribution of microbes in rhizoplane is not randomly distributed but forms a colony pattern, namely the first microcolony pattern and the second the macro colony pattern. The formation of microbial colony patterns is by the place of exudation (Paul et al. 2010; Bakker et al. 2013).

In general, the rhizosphere environment has very high acidity, low oxygen content, and high CO_2 concentration. Soil chemical properties in the rhizosphere are influenced by plant root exudates that are absorbed by the soil. The type of exudate released by plant roots will determine the soil to become more acidic or alkaline. The kinds of root exudates are determined by the kinds of nutrients absorbed by the plant. For example, if plants absorb element N, in the form of ammonium, the roots will release hydrogen ions, so that the rhizosphere becomes more acidic. When plants absorb N elements in the form of nitrates, the roots release hydroxyl ions, so that the rhizosphere becomes more alkaline. However, these dynamics in the rhizosphere will not affect the overall pH of the soil. It is the microbial population that is affected because the microbes in the rhizosphere are not able to move far (Paul et al. 2010; Enny 2017).

The increase in carbonic or bicarbonic acid is caused by the respiration of plant roots and microbes in the rhizosphere. Root and microbial respiration release CO_2, resulting in increased concentrations of carbonic or bicarbonic acid, thus lowering the pH of the rhizosphere. The pH of the rhizosphere is higher than the pH of the soil as a whole because in the rhizosphere there is a process of root respiration and microbial colonies that are larger than in the non-rhizosphere (Berendsen et al. 2012).

11.2 Root Exudate

Naturally, the root system of plants in the rhizosphere releases exudate, and the part of the root that is actively doing exudation is the root tip and the point of growth of the lateral root. This means that the amount of exudate is more influenced by the number of lateral roots than the length of the lateral roots (Paul et al. 2010). This is proven through research on the *Avena barbata*. The results showed that the accumulation of sucrose was more around the root tips. Tryptophan exudate accumulates around the lateral roots (Paul et al. 2010; Haichar et al. 2014). In the area of accumulation of the exudate, there were also more bacterial colonies than in other parts.

11.2.1 Rhizodeposition

Root exudate is part of rhizodeposit, the result of the rhizodeposition process, based on the root exudate content is the main source of soil organic carbon compounds.

Based on molecular weight, rhizodeposit compounds are divided into two, namely (1) root exudates with low molecular weight, such as amino acids, organic acids, sugars, and other secondary metabolites and (2) root exudates with high molecular weight, such as polysaccharides and protein lenders. In the rhizosphere, the concentration of root exudate is higher than the surrounding soil solution, because exudation from plant roots is continuous (Haichar et al. 2014).

Rhizodeposition results in the accumulation of metabolite compounds called rhizodeposits. The accumulation of rhizodeposit can change the chemical, physical, and biological characteristics in the rhizosphere. The dynamics of changes in the nature of the rhizosphere are influenced by many factors related to the composition and content of rhizodeposit compounds. The quantity and quality of rhizodeposit are influenced by the type of plant, plant cultivar, plant growth phase, and the age of the plant that produced the root exudate. Other factors that affect the quantity and quality of plant root exudates are soil and climatic properties. Soil characteristics, especially soil type, soil texture, soil structure, solum thickness, depth of water table, soil fertility. The dominant climatic elements that influence are light intensity, rainfall, rainfall distribution, temperature, humidity. The factors of plants and the environment in which they grow vary in their influence on the rhizosphere, as they interact. The nature of the interaction between plant factors and the environment in which they grow are influenced by the applied plant cultivation techniques.

Plant rhizodeposits contain five types of metabolite compounds (Paul et al. 2010). The five types of compounds are exudate, secret, mucilage, mucigel, lysate. Exudate compounds are metabolite compounds resulting from plant root education. The secret is a metabolite compound that results from plant root secretions. Mucilage is a metabolite compound in the form of mucus, sap, adhesive liquid. Mucilage properties are thick and sticky, produced by almost all plants and microbes. Microbes use the mucus for their movement, the direction of microbial movement is opposite to the direction of the release of the mucilage. Mucilage is a polar glucoprotein and an esopoly of saccharides. Mucigel is also a secondary metabolite compound in the form of mucus that is thicker than the root tip. It is a hydrated polysaccharide. It is thought to be pectin secreted by the outer cells of the root caps, which are formed in the Golgi body of these cells through an exocytosis process. The lysate is a metabolite compound resulting from the disintegration of cell parts consisting of plasmids, receptor tests, proteins, DNA, RNA. Lysate can occur due to damaged root cells, exfoliated when the roots penetrate the soil particles.

Root exudates released by root cells into the space between the cells and soil particles consist of low molecular weight compounds and high molecular weight exudates. Low molecular weight compounds are sugars, amino acids, aromatic compounds.

Secretion is a bioproduct of physiological activity through the process of plant metabolism. The secret is a secretion metabolite that is actively released into the rhizosphere. Secret includes both low and high molecular weight compounds.

Mucilage is a metabolite compound derived from root tip epidermal cells that fall off when the roots grow, extending through the soil particles of the rhizosphere. In sandy soils, the physical properties of the soil have loose, porous, loose particles, so that the root tips of the plants are easier to penetrate. In contrast to clay or clay soils, the physical properties of the soil are denser, the soil particles are soft, dense, not porous, making it more difficult for plant roots to penetrate. There is a greater chance of root friction, especially when soil moisture is minimal or dry. This is due to clay, when dry it becomes hard and solidifies. The fall of plant root cells is determined by the abrasion force of the roots when penetrating or opposing the forces of soil particles. These cells contain cellulose, pectin, starch, and lignin (Paul et al. 2010).

It is a metabolic compound that forms a thin layer covering the root surface. This mucigel becomes a bridge connecting plant roots with soil particles and microbes in the rhizosphere (Paul et al. 2010). Mucigel functions as a liaison for the mediator between the roots and soil particles in the rhizosphere or areas of plant expansion. Mucigel accumulates more commonly in the center of the roots and hairs than at the tips of the plant roots.

The lysate is a type of plant rhizodeposit that is released by root epidermal cells when the root tip breaks or dies. Root epidermal cells are broken, releasing metabolite compounds that can invite microbes. Microbes that approach the tip of the root respond positively to lysate with its aroma and taste. Lysates released from the tips of plant roots to the rhizosphere, where microbes can use them as a source of food or nutrition (Paul et al. 2010; Haichar et al. 2014).

Besides using the above mentioned five metabolite compounds as a source of nutrition, microbes also release compounds including antibiotics, hormones, vitamins, and other molecules that can be beneficial for plant growth and development (Paul et al. 2010; Haichar et al. 2014). There are three forms of rhizodeposit in the rhizosphere, namely: liquid (exudate), mucus, and gas (ethylene, CO_2, H_2). All three can be a source of food and energy for microbes in the rhizosphere. Based on the level of diffusivity, exudates only include metabolites of low molecular weight (Haichar et al. 2014). Root exudates are liquid chemical compounds released by plant roots into the soil around plant roots. Root exudate functions include general functions and special functions. The function of root exudates is as: (a) to support the mechanical system of plants, (b) to support the absorption of water and nutrients, (c) to support the process of distributing photosynthate. The special functions of root exudates are: (a) synthesizing chemical compounds, (b) accumulating chemical compounds, (c) secreting chemical compounds. Root exudate also maintains the stability of soil micro-aggregates, absorbs and stores ions, ions resulting from the selection. Besides that, root exudates are also able to regulate soil microbial community in the rhizosphere, prevent and control microbial pests, support mutualism symbiosis, rhizosphere, inhibit the growth of competitor plants. Thus it can be concluded that the chemical compounds of root exudates can act as an attractant (Haichar et al. 2014).

Mucus is an organic compound root rhizodeposit in the form of a root layer, covering the tip or root cap. This mucus is a type of redeposit which is clearly visible at the root tips. The components of mucus are polysaccharides, proteins, and phospholipids (Haichar et al. 2014). The benefits of mucus redeposit for plants: (1) protect plant root meristems from toxic metal elements, (2) increase soil aggregate stability, (3) maintain soil aeration stability, (4) maintain plant root growth,

(5) prevent soil erosion, due to soil particles forming larger soil aggregates, so that it is not easily carried away by surface water flows, (6) maintaining water flow to the rhizoplane. The border cells between the root tips and rhizosphere soil particles actively release root rhizodeposit. Border cells also sort and select root redeposit compounds to be released according to plant needs and conditions in the rhizosphere (Haichar et al. 2014).

Plants through the process of root respiration release CO_2 gas to the rhizosphere by 17.5%. This will increase the dissolution of CO_2 CO_3 in the soil to produce Ca ++ that plants need (Haichar et al. 2014). Hydrogen gas (H_2) is a byproduct of N_2 fixation in legume plants (Fisk et al. 2015; Teng et al. 2019). H_2 gas released into the rhizosphere reaches 5% of net photosynthesis (Kalam et al. 2020). Some rhizomes or microbes in the rhizosphere have the coding genes to absorb the enzyme hydrogenase (Hup). This causes the H_2 to be oxidized by bacteria to produce more energy. However, many symbioses lack the hydrogenase enzyme, so that H_2 is produced by nitrogenase, diffuses out of the nodule, and releases into the rhizosphere (Haichar et al. 2014). Accumulation of H_2 into the rhizosphere activates the soil-oxidizing bacterial community. This creates soil biological properties that support plant growth (Haichar et al. 2014).

11.2.2 Root Exudate and Organic Acid

The exudate released by the plant roots is influenced by the variety of plant growth media. Plants grown in the soil secrete exudates that contain carbon compounds as much as 20% of the dry amount of the plant. If the plant is grown hydroponically, it turns out that only 0.1–0.4% of the exudate containing carbon compounds is released. The quantity and quality of root exudates may vary, influenced by factors of plant species and the environment in which the plant grows. Variations in the quantity and quality of root exudates include the number of molecular, molecular weight, and kinds of influence on biochemical activity in the rhizosphere.

The compounds that are dominant and always present in root exudates are simple carbohydrates, compounds, sugars, amino acids, and organic acids (Xing et al. 2020). For example, wheat plants are grown on fertile soil, nutrient adequacy, root exudates are dominated by carbohydrate compounds (Paul et al. 2010; Haichar et al. 2014). The composition of plant root exudate compounds determines the type and number of microbes that dominate. The microbial community in the rhizosphere can change according to changes in the quality of root exudates due to changes in nutritional status (Paul et al. 2010; Enny 2017).

Organic acid of root exudate compounds plays a role in inducing microbial growth in the rhizosphere. It is as a chemoattractant or chemical attractive aroma and taste, thus inducing a number of microbes to move to the root surface (Shukla et al. 2011; Xing-Feng et al. 2014).

The soil environment around plant roots is affected by exudate that is released by the roots continuously. It continuously interacts with microbes effectively (Enny 2017). The structure and composition of microbes in the rhizosphere vary depending on plant species so that interactions between plants and microbes also vary (Paul et al. 2010; Haichar et al. 2014).

Organic compounds of root exudate compounds are continuously released by the roots into the rhizosphere intensively. This causes microbes to be attracted to the root area of plants because of the attractive aroma and taste of exudate compounds (Shukla et al. 2011; Xing-Feng et al. 2014). The structure and composition of microbes in the rhizosphere are determined by the concentration and composition of root exudates. Chemical attractants are called chemoattractants. It is a source of energy for microbial life in the rhizosphere (Shukla et al. 2011; Xing-Feng et al. 2014).

The microbial community in the rhizosphere degrades plant debris into nutrients that support plant growth and other organic compounds to the rhizosphere (Venant et al. 2011). The microbial population is also affected by the abiotic environment such as the intensity of sunlight (Enny 2017). The intensity of sunlight affects the rate of photosynthesis. The resulting photosynthate differs in number and variety. This results in root exudates containing different amounts and kinds of organic compounds.

Indirectly, root exudates of plant species determine the composition of microbial communities in the rhizosphere. The root exudates of certain plant species are also of interest to certain microbes. Thus, the concentration and composition of plant root exudates determine the structure and components of the microbial community (Koranda et al. 2011; Cesco et al. 2012). Intensively microbes in the rhizosphere are selected by root exudates (Paul et al. 2010). Thus, the composition and diversity of the microbial community are diverse, due to differences in the concentration and composition of root exudates.

In the rhizosphere, interactions occur between microbes and plants, in this case, the proteins in root exudates. The protein composition of root exudates is influenced by microbial dynamics in the plant area (Xing-Feng et al. 2014). The association of plants with microbes like symbiotic mutualism or mutually beneficial are mycorrhizae, rhizobium, and PGPR (Agustian and Lusi 2012; Abedinzadeh et al. 2019).

The extracellular compound is a component of root exudates. These extracellular compounds are recognized by microbes (Xing-Feng et al. 2014). Therefore, the composition of the exudate compounds determines the type of microbes that dominate in the rhizosphere.

Plant roots have the ability to release exudate into the rhizosphere as much as 20% of photosynthesis results (Xing et al. 2020). The quality and quantity of root exudates are determined by plant species, plant age, biotic environment, and abiotic environment (Mendes et al. 2013).

The microbial community in the rhizosphere is controlled by plant root exudates. Root exudates also regulate herbivores, inhibiting the growth of other competing plant species (Enny 2013). The presence of microbes in the rhizosphere can be detected by plants, through the mechanism of exudation, the secretion of metabolite compounds into the rhizosphere.

Through the mediation of root exudates, plants can respond positively and negatively to existence of microbes in the rhizosphere (Venant et al. 2011). The positive response of plants to microbes, for example, is the association of plant roots with rhizobium which results in symbiotic or mutually beneficial interactions.

The release of root exudates into the rhizosphere causes microbial proliferation in plant roots. Proliferation is an event of a repetition of the cell cycle without a hitch. Proliferation is different from mitosis. The proliferation process results in an increase in the number of cells with the same shape in a fast time.

The content of the rhizosphere consists of various microbes with different density levels, namely: bacteria with a density level of 10^6–10^9, protozoa with a density level of 10^4, nematodes with a density level of 10^1–10^2, fungi with a density level of 10^5–10^6 in every gram of rhizosphere soil (Haichar et al. 2014). Carbon compounds are released by plant roots into the rhizosphere and become a source of microbial nutrition.

Plant root exudates and compounds released by microbes can increase nutrient content in the rhizosphere. This can be interpreted that the rhizodeposit can increase soil fertility, especially in the area of plant roots or the rhizosphere. Nutrient elements contained in the rhizosphere with the presence of root exudation activity and microbial response individually or in the form of their interactions can be increasingly available to plants because microbes are able to convert nutrients into nutrient elements in a form that is easily absorbed by plant roots.

For example, the mechanism of symbiosis mutualism is as follows: legume plants release flavonoids in the rhizosphere. These flavonoid compounds activate the performance of nitrogen-fixing bacteria, namely Rhizobium meliloti bacteria. These bacteria will bind free nitrogen into an available form so that it can be absorbed by plant roots. Thus the plant needs for the element N can be sufficient.

The exudate which is the result of the photosynthesis process is released by plant roots to the rhizosphere by 20% of the total photosynthetic. For microbes, root exudate can also be used for growth. For example, root exudates are utilized by mycorrhizae to get a source of nutrition (Xing-Feng et al. 2014).

Microbes in the rhizosphere also play a direct role in the mineralization of organic compounds into inorganic compounds that can be absorbed by plant roots (Paul et al. 2010). Microbes also secrete indole compounds into the rhizosphere (Yekti et al. 2019). Based on this phenomenon, it is evident that root exudates play an important role in determining the various dynamics of interactions that occur in the rhizosphere. The diversity of the quantity and quality of root exudates is a reflection of the dynamics of the interaction between roots and soil in the rhizosphere. The root tips are lubricated by chemical compounds of root exudates so that the root tips are protected from desiccation (dryness) so that the roots become easier to penetrate into the soil in the rhizosphere.

Root exudate also maintains the stability of soil micro-aggregates, absorbs and stores the selected ions. Besides that, root exudates are also able to regulate soil microbial communities in rhizospheres, prevent and control microbial pests, support mutualism symbiosis, inhibit the growth of competitor plants. Thus it can be concluded that the chemical compounds of root exudates can act as attractants.

Data on the quantity and quality properties of root exudates are limited. This is constrained by the correct methodology. Several studies through hydroponic cultivation of plants in a sterile environment have been carried out, but the research data is still unable to predict the real conditions in nature (Hooper et al. 2015). In this study, the root exudate was quantified by measuring the CO_2 formed in the rhizosphere. Plant roots release 1–20% of photosynthetic plant photosynthesis products into the rhizosphere (Yang et al. 2020). The composition of the compounds contained in root exudates consists of water, dissolved sugars, organic acids, amino acids, hormones, vitamins, phenolics, and phosphate sugar esters (Haichar et al. 2014; Shukla et al. 2011; Koranda et al. 2011; Xing-Feng et al. 2014).

The quantity and quality of the root exudate composition are influenced by various factors. Broadly speaking, three main factors are determining the quantity and quality of root exudates, namely: internal factors, external factors, and plant cultivation techniques. Internal factors include all characteristics of a plant, such as type of plant, plant cultivar, plant growth phase, plant life cycle (annual crops or annual plant), harvest age. External characters include all the characteristics possessed by the environment in which plants grow including soil and climate characteristics. The character of plant cultivation techniques includes the applied science and technology. The three interacting factors consist of: the interaction of plants with the environment in which they are grown, plants with cultivation techniques, and cultivation techniques with the environment in which they are grown. Environmental characteristics are thought to have a greater impact on the quantity and quality of root exudates (Enny 2017). Environmental factors that dominate this role are soil type, soil pH, oxygen status, soil porosity, light intensity, soil temperature, nutrients, the presence of microbes, and microbial positions in the rhizosphere.

Carbon compounds put on by the roots of young plants that are still active in vegetative growth reach 50% of the total photosynthetic results of photosynthesis (Paul et al. 2010). In mature plants going into the sense phase or aging, the quantity of root exudate released into the rhizosphere is less (Paul et al. 2010). The nature of root exudates also varies depending on the phase of plant growth. During the early vegetative growth phase, the plant has six leaves or less, the root exudates are mostly mucus and carboxylate compounds.

The organic acids released by the roots into the rhizosphere are considered the most important of the various other compounds. This is due to the fact that organic acids play an important role in the metabolic processes of plants or microbes in the rhizosphere. In plant mechanical processes, absorption of water and nutrients, secretions are some good examples of this process. In the presence of root exudates, the stability of the interaction between roots and soil can be maintained.

Among the various organic compounds in root exudates, organic acids are considered the most important to be studied. Organic acids play an important role in providing substrate as nutrients and a source of energy for microbial metabolism. Organic acids are also able to mediate biochemical reactions in the rhizosphere. The main organic acids in the micro rhizosphere zone, where the soil contacts the root tips, are organic acids with low molecular weight (Haichar et al. 2014).

Low molecular weight organic acids play an important role in biochemical processes in the rhizosphere (Anandyawati et al. 2017), among others, making nutrients in available forms, so it can be absorbed by plant roots. These organic acids helps by increasing the solubility of P nutrient, increasing chemotaxis response and by forming symbiotic associations, detoxify Al in the rhizosphere so that plant roots are protected from Al elemental poisoning (Haichar et al. 2014).

In a condition where the plant is severely deficient in nutrients, the plant is still able to survive and develop normally. This is due to the release of organic acids into the rhizosphere which is converted by microbes into nutrients available to plants (Zhang et al. 2014). Although the soil in which the plants are cultivated is poor in nutrients, thanks to the performance of microbes it can produce organic material that is useful for plant growth (Haichar et al. 2014).

11.3 Plant Interaction with Microbes

The interactions between plants and microbes are mediated by root exudates. The microbes in the rhizosphere are influenced by the plant species that produce root exudates. In the rhizosphere, there is an effective and continuous interaction of plants with microbes. The microbes are captivated by the aroma and taste of organic root exudates (Shukla et al. 2011; Xing-Feng et al. 2014; Enny 2017).

Plants' interaction with microbes is mediated by the exchange of chemical compounds. These chemical compounds are a means of interacting with plants and microbes (Xing-Feng et al. 2014). Plants do not have the ability to move places and have no brain activity. However, plants have the character of growth and development that is in harmony with and equal to the performance of the nervous system (Venturi and Keel 2016).

The root tip is a growing point that produces auxin compounds (Paul et al. 2010). Auxin compounds play a role in root growth activity. The root tips of plants accumulate root exudate organic compounds which are attractive to microbes for growth so that the degree of plant interaction with microbes will increase. In addition, the nature of plant interactions with microbes can change, for example, initially being harmful can be mutually beneficial. So the nature of plant interactions with microbes is dynamic (Paul et al. 2010; Xing-Feng et al. 2014).

The community of microbial diversity is determined by plant species (Shukla et al. 2011; Enny 2013; Xing-Feng et al. 2014; Pérez-Jaramillo et al. 2016). Therefore, the interaction between plants and microbes is determined by two main factors, namely internal and external factors. Internal factors include characters related to plant genetic traits. External factors concern the biotic and abiotic environment. The biotic environment is related to living components, including microbial activity and other organisms, both beneficial and harmful. The abiotic environment includes soil and climate. Soil factors include chemical properties, physical properties, and soil biological properties. Climatic factors include light intensity, temperature, air humidity, rainfall, wind, and gases in the air.

The presence of root exudates attracts microbes and colonizes the rhizosphere. The microbial community is able to modify the physical, chemical, and biological properties of the rhizosphere. Chemical changes in the rhizosphere are caused by the humification process of organic matter, mineralization of N, P, and S elements into available forms, so they can be absorbed by plant roots. In the rhizosphere, there are also microbes capable of fixing free nitrogen into a form that can be absorbed by plant roots. This kind of interaction has the advantage of both. Microbes benefit from getting food from root exudates, while plants are also made easy to absorb nutrients due to microbial activity.

The interaction mechanism between plants and microbes is through the chemotaxis process. Chemotaxis is a microbial mechanism in responding to changes in the composition of the root exudate compound (chemo-effector) around it effectively, efficiently, and unique (Ofek et al. 2014). Microbes will approach preferred root exudate compounds and away from harmful compounds (chemorepellent) (Paul et al. 2010; Haichar et al. 2014). Chemotaxis is a mechanism of plant interaction with microbes that are activated by secondary metabolite compounds. Microbes have basic properties both in situ and in vitro. This shows the ability of microbes to respond to nutrients such as amino acids, simple carbohydrates, and organic compounds.

It can be stated that chemotaxis is the initial stage of the microbial colonization process. Chemotaxis is also the basis for explaining biological phenomena, including the process of biofilm formation, pathogen infection, an association of diazotrophic bacteria, mobility in water and soil, bioremediation in situ (Haichar et al. 2014).

Interactions between plants and microbes can be mutually beneficial, neutral, or mutually detrimental or competitive. There are times when plants interact with microbes and compete for water and nutrients (Shukla et al. 2011; Enny 2017). This happens when water and nutrients are available in limited quantities, unable to meet the minimum needs for plant and microbial life.

The dynamics of plant–microbial interactions are also caused by the diversity of plant root exudates in the rhizosphere. If the root exudate aroma and taste are interested in the microbes, so the microbes will approach the root area of the plant. The number of microbes increases and forms colonies, especially at the root tips and lateral root growth points. However, if the compounds released by the roots are considered dangerous or the smell and taste are not favored by microbes, so the microbes will stay away from the rhizosphere area.

Plants have the ability to control the nature of interactions with microbes. A type of plant is the most dominant determining factor in determining the nature of a rhizosphere. Other determinants, such as soil characteristics and microbial colonies, are not as strong as plant species. Root exudate character is determined by plant genetic characteristics. Root exudates have the ability to invite or repel specific microbial populations. Various plants have the genetic trait to be tolerant in interacting with harmful microbes in the rhizosphere. Thus plant variety determines the diversity of microbial communities in the rhizosphere. The age and health level of plants also determine the dynamics of the microbial community in the rhizosphere. The structure and composition of the microbial community in the rhizosphere can also be controlled by plants. In addition, root exudates can also determine the dynamics of the physical properties of the rhizosphere. During the day, the plant's transpiration rate increases more than at night. When the transpiration flow strengthens, the water potential in the rhizosphere changes rapidly. This causes physical changes in the rhizosphere soils in the form of fluctuations in water potential. Meanwhile, in non-rhizosphere soils, this does not happen (Venant et al. 2011; Enny 2017). The results showed that the soil in the rhizosphere is wetter than the non-rhizosphere soil. With such a mechanism, plant roots can still absorb water and nutrients from the rhizosphere, so that plants can avoid nutrient deficiencies.

At night, the transpiration flow in the plant weakens. The osmotic pressure at the root tip decreases so that the absorption of the roots is also reduced. This allows for the release of root exudate into the rhizosphere. The interaction through contact of root exudates with microbes in the rhizosphere is strengthened, root exudates also make contact with soil particles in the rhizosphere. During the day, the transpiration flow in the plant body strengthens again, and more water is absorbed by the plant roots. This causes the water in the rhizosphere soil particles to decrease, the condition of the soil solution in the rhizosphere becomes more concentrated. Root exudate thickens and coagulates, adhering to the soil particles of the rhizosphere. When the rhizosphere begins to dry out and the potential for soil hydrolysis decreases, root exudates will release into the soil (Venant et al. 2011; Anandyawati et al. 2017; Enny 2017). This causes the soil in the rhizosphere to remain moist, wetter than the soil non-rhizosphere. These events explain that plants through root exudates can affect the physical properties of soil in the rhizosphere.

The interaction of plant roots with rhizosphere soil particles can affect soil porosity. In clay soil, the mobility of root exudate compounds and microbial colonies is slower. This is because clay has finer soil particles so that the pore space is smaller than sandy soil. Sandy soils have larger particles, so the space between the particles or the pore space is smaller. Thus, on sandy soil exudate and microbial colonies in the rhizosphere are faster than in clay soil. This shows that the interaction between plants and microbes in sandy soil can be faster because the exudate released by the roots into the rhizosphere moves more freely and quickly reaches soil particles around the root area. This accelerates contacts and expands the contact area of plants with microbes (Anandyawati et al. 2017). Plant root exudate is continuously secreted by the root cap so that the particles in the rhizosphere will be covered with root exudate to form aggregate. In clay, the particles bind with the root exudate to form the aggregate, and its size becomes larger. This causes the space between the aggregate to become larger. This condition causes the pore space of the clay in the rhizosphere to become bigger too so that the clay has a greater porosity, the soil is looser. This condition is created continuously, when the root cells are damaged, exfoliated, die and experience decay, it will change the clay to be more porous, loose, and fertile.

The interactions of plants with microbes can affect the chemical properties of the soil in the rhizosphere. Plant roots release a wide variety of compounds into the rhizosphere. It is through rhizodeposits that various interactions are mediated. In the rhizosphere, there are interactions between plant roots in the same species, interactions between plant roots in different species, interactions between plants and soil particles, interactions between plants and microbes, interactions between microbes and microbes, and interactions between microbes and soil particles. The interaction of these various components affects the dynamics of the ecological system in the rhizosphere. Environmental factors, both abiotic and biotic, will influence the dynamics of these interactions, especially the interactions between plants and microbes.

Rhizodeposits formation is influenced by the type of plant, plant growth phase, and biotic and abiotic environmental stress (Xuliang et al. 2013; Vanesa and Mary 2014). In the rhizosphere, there are various compounds including exudates of plant roots, secondary metabolites produced by microbes such as bacteria, fungi, lichens, invertebrates, and plants. These various compounds affect various processes in the soil ecosystem, especially the dynamics of the interaction of various components of the soil ecosystem. One of the important interactions is the interaction of plants with microbes. These reactions have an impact on the physical, chemical, and biological properties of the rhizosphere. These various compounds influence various processes in the soil ecosystem, especially the dynamics of the interaction of various components of the soil ecosystem. One of the important interactions that are affected is the interaction of plants with microbes which have an impact on the physical, chemical, and biological properties of the rhizosphere soil (Shukla et al. 2011; Xuliang et al. 2013; Haichar et al. 2014; Anandyawati et al. 2017).

Changes in the interaction of internal and external factors affect the nature of the interactions that occur between plants and microbes. Initially, neutral plant and microbial interactions can turn into mutually beneficial or mutually harmful. Plant cultivation techniques always try to create positive strategies between plant factors and the environment in which they grow, including in the rhizosphere. Thus it is hoped that the interaction between plants and microbes is mutually beneficial, at least neutral, not mutually harmful. The application of plant cultivation technology can affect the nature and degree of interaction between plants and microbes in the rhizosphere. The application of appropriate plant cultivation technology can synergize plant factors with the environment in which plants grow, both biotic and abiotic environments. The synergy of plant factors with their environment can result in mutually beneficial interactions between plants and microbes, not mutually harmful ones. One example of technical culture action in plant cultivation that results in mutually beneficial interactions between plants and microbes is the selection of healthy, high-yielding, and broadly adaptable plant species. Healthy plants can grow rapidly and produce large amounts of root exudate continuously into the rhizosphere. Root exudate compounds that are of interest to microbes because of the aroma and taste favored by microbes will cause the movement of microbes to the rhizosphere.

This can occur because the plant root exudate compounds are considered not to endanger the activities of microbial life. When plant roots come into contact with soil particles, friction occurs which makes the plant root cells peel off. The exfoliated cells will die and be degraded by microbes. The result of the degradation of the accumulated dead root cells of the plant is in the form of organic compounds available to plants as a source of nutrients. Plant roots can absorb nutrients dissolved in soil colloids following the rate of transpiration flow. Thus plants benefit from the presence of microbes in the rhizosphere so that plants can grow well and can produce maximum biomass and biochemistry.

Conversely, if the cultivated plant is unhealthy, low quality, then the release of root exudates into the rhizosphere is not as much and less intensive as healthy plants. This is because plants are easily exposed to unfavorable environmental conditions where they grow. This results in a limited number and diversity of microbial communities in the rhizosphere. The degradation activity of the remains of dead plant root cells is not intensive, so the availability of nutrients resulting from degradation is also limited. The subsequent effect of soil conditions on the rhizosphere limits nutrients to plants. In this case, the plant does not benefit from plant–microbial interactions. Microbes too, because they do not get enough nutrients from plant root exudates.

In the event of interaction between plants and microbes, it is known as chemotaxis. Chemotaxis is a mechanism that microbes have in responding to changes in the composition of compounds in their environment effectively and efficiently. The principle of the chemotaxis mechanism is as follows, the microbes will quickly approach the preferred compound. This can work the other way around, the microbes will stay away from harmful compounds themselves (Paul et al. 2010; Haichar et al. 2014). Chemotaxis is a mechanism that is activated by secondary metabolites.

In the event of plant interactions with microbes, plant roots secrete exudate compounds. The root exudate compound can be a bioactive compound. For the benefit of plants, bioactive compounds can have multiple functions. First, if the bioactive compounds released into the rhizosphere are aromatic compounds. For example, from the benzoate group that microbes are interested in. The microbes will respond positively and get closer to the plant root area. Microbes benefit from getting nutrients from root exudates. Likewise, plants, with the presence of microbes in the rhizosphere, the remains of damaged root cell organs, peel off when penetrating soil particles, are degraded by microbes into nutrients that can be absorbed by plant roots. In this event, microbes and plants interact with each other which is called symbiotic mutualism.

Second, if the secondary metabolites released by plant roots harm microbes, the microbes will quickly move away from the plant root area. In this case, the bioactive compounds contained in plant root exudates can serve as self-defense against pathogenic microbial attacks called antimicrobial (Baetz and Martinoia 2014). In this event, the interrelationship between plants and microbes is antagonistic. Based on this case, it can be understood that the interactions between plants and microbes are dynamic. The intensity of the interaction is also influenced by plant genetic characteristics, microbial genetic characteristics, physical properties of the rhizosphere, chemical properties of the rhizosphere, biological characteristics of the rhizosphere, and climate. In addition, plant cultivation techniques can also affect the nature of plant interactions with microbes. The influence of plant cultivation techniques on plant interactions with microbes can be carried out through the stages of using planting materials, land management.

The use of planting material will determine the type of plant, cultivar, plant life cycle, and crop harvest age. Soil cultivation techniques will affect the physical, chemical, and biological properties of the rhizosphere. Planting techniques can determine the depth and area of the contact area of plant roots with soil particles. Maintenance techniques in the form of fertilization and irrigation also affect the physical, chemical, and biological conditions of the rhizosphere. Harvest age determines the duration of plant interactions with microbes in the rhizosphere.

11.4 Root Exudate Impact

The impact of root exudate is divided into two, namely internally and externally. Internal impact, namely the impact of exudate on plants that release exudate. External impact, namely the impact of exudates on the environment where plants grow, including biotic and abiotic environments. The biotic environment includes a variety of microbes and other plant roots in the rhizosphere. The abiotic environment includes soil and climate.

The effects of root exudates in the rhizosphere can be beneficial or detrimental to the interacting components. The beneficial effects of root exudates can be explained as follows: root exudates released into the rhizosphere serve as a source of nutrition for compatible microbes. This benefits the microbes because microbes can grow and develop well in the rhizosphere. Microbes in the rhizosphere carry out activities that make rhizosphere land more fertile, crumbly, more porous. This rhizosphere condition has a beneficial impact on plants to growth and development.

The response of plants to changes in microbial composition and structure in the rhizosphere is through root exudates. Plants can attract compatible microbes and repel pathogenic microbes in the rhizosphere. For example, ABC Arabidopsis secretes more phenolics than wild-type Arabidopsis (Jin et al. 2019). This is supported by the statement that phenolic compounds have an impact on the structure and composition of the microbial community in the rhizosphere (Michalet et al. 2013; Jin et al. 2019). According to Fang et al. (2013), a decrease in the quantity and quality of phenolic compounds in transgenic has an impact on the microbial community in the rhizosphere, resulting in an increase in the microbial community compared to wild-type plants. This proves that there is a change in the microbial community naturally (Xing-Feng et al. 2014). Badri et al. (2013a) stated that the relationship between exudate compounds and microbes was more compatible, the higher the level of microbial density in the places where root exudate accumulated in the rhizosphere.

Changes in the composition of root exudates are related to beneficial bacterial communities as PGPR, N_2-binding agents, and metal remediation bacteria (Xing-Feng et al. 2014). The addition of various mixtures of natural chemicals derived from Arabidopsis exudates to the rhizosphere resulted in colonies of different microbial communities. This is evidence that the type and composition of root exudates can have a specific impact on the type and composition of microbes in the rhizosphere (Badri et al. 2013b).

Various research results show that plants through root exudates can mobilize, form, and select microbial communities in the rhizosphere (Chaparro et al. 2014). For example, cucumber seed root exudates contain p-coumaric acid which harms bacterial and fungal colonies. Changes in the structure and composition of bacterial and fungal communities in the rhizosphere have increased the population of cucumber pathogenic microbes that are carried by soil particles (Xingang and Fengzhi 2012).

In addition, the impact of plant root exudates also functions as a plant self-defense system against attack by pathogenic microbes, namely jasmonic acid. This acid, which is a root exudate, can change the microbial presence in the rhizosphere (Carvalhais et al., 2013). The microbial community in the rhizosphere is also influenced by plant species and plant growth phases (Chaparro et al. 2014). A study reported that different plant species resulted in different quantity and quality of root exudate chemical compounds. This change in root exudate affects the dynamics of the microbial community. Research on 27 maize lines has proven that these 27 lines affect the variety and diversity of microbes in the rhizosphere (Peiffer et al. 2012). Other studies have shown that rhizosphere soils originating from pine and maize plantations actually contain different microbes (Zolla et al. 2013).

High ethylene levels can exacerbate roots against environmental stress, such as drought stress, so that the physiological activity of the roots is disrupted, eventually plant growth is inhibited (Stearns et al. 2012). Many rhizosphere soil microbes are able to reduce plant response to ethylene production, by catalyzing ACC (aminocyclopropane-carboxylic acid) which is a direct precursor to ethylene, into ketobutyrate and ammonia (Haichar et al. 2014). This means that a decrease in ethylene levels can increase plant resistance from biotic or abiotic stress. ACC deaminase activity has been shown to help increase the ability of plants to withstand drought stress, likewise for water stress and salinity (Xing-Feng et al. 2014), so that plant growth continues to be normal (Yang et al. 2020). For example, the soil bacterium *Achromobacter piechaudii* ARV8 which has ACC deaminase activity can increase the biomass of tomato and pepper seedlings with environmental stress (Xing-Feng et al. 2014). Thus the presence of bacteria containing the ACC deaminase enzyme has a positive impact because it can increase plant resistance to environmental stress. This is because the decreased levels of ethylene inhibit the performance of auxins so that auxins can still support plant growth.

References

Abedinzadeh M, Etesami H, Alikhani HA (2019) Characterization of rhizosphere and endophytic bacteria from roots of maize (Zea mays L.) plant irrigated with wastewater with biotechnological potential in agriculture. Biotechnol Rep 20:1–12. https://doi.org/10.1016/j.btre.2019.e00305

Agustian AD, Lusi M (2012) Effect of various root extracts in the stimulation of arbuscular mycorrhizal fungi (AMF) infection in *Tithonia diversifolia*. J Solum 91:36–43

Anandyawati SA, Budi N, Rahayu W, Supiandi S (2017) Study of root exudate organic acids and microbial population in the rhizosphere of oil palm seedling. J Trop Soils 22(1):29–36. https://doi.org/10.5400/jts.2017.v22il.29-36

Badri DV, Chaparro JM, Zhang R, Shen Q, Vivanco JM (2013a) Application of natural blends of phytochemicals derived from the root exudates of Arabidopsis to the soil reveal that phenolic-related compounds predominantly modulate the soil microbiome. J Biol Chem 2887:4502–4512. https://doi.org/10.1074/jbc.M112.433300

Badri DV, Zolla G, Bakker MG, Manter DK, Vivanco JM (2013b) Potential impact of soil microbiomes on the leaf metabolome and on herbivore feeding behavior. New Phytol 1981:264–273. https://doi.org/10.1111/nph.12124

Baetz U, Martinoia E (2014) Root exudates: the hidden part of plant defense. Trends Plant Sci 192:90–98. https://doi.org/10.1016/j.tplants.2013.11.006

Bakker PAHM, Berendsen RL, Doornbos RF, Wintermans PCA, Pieterse CMJ (2013) The rhizosphere revisited: root microbiomics. Front Plant Sci 4:165. https://doi.org/10.3389/fpls.2013.00165

Berendsen RL, Pieterse CMJ, Bakker PAHM (2012) The rhizosphere microbiome and plant health. Trends Plant Sci 178:478–486. https://doi.org/10.1016/j.tplants.2012.04.001

Brunel C, Pouteau R, Dawson W, Pester M, Ramirez KS, van Kleunen M (2020) Towards unraveling macroecological patterns in rhizosphere microbiomes. Trends Plant Sci 2510:1017–1029. https://doi.org/10.1016/j.tplants.2020.04.015

Carvalhais LC, Dennis PG, Badri DV, Tyson GW, Vivanco JM, Schenk PM (2013) Activation of the jasmonic acid plant defence pathway alters the composition of rhizosphere bacterial communities. PLoS One 8(2):e56457. https://doi.org/10.1371/journal.pone.0056457

Cesco S, Mimmo T, Tonon G, Tomasi N, Pinton R, Terzano R, Neumann G, Weisskopf L, Renella G, Landi L, Nannipieri P (2012) Plant-borne flavonoids released into the rhizosphere: impact on soil bioactivities related to plants nutrition. Rev Biol Fertil Soil 48:123–149. https://doi.org/10.1007/s00374-011-0653-2

Chaparro J, Badri D, Vivanco J (2014) Rhizosphere microbiome assemblage is affected by plant development. ISME J 8:790–803. https://doi.org/10.1038/ismej.2013.196

Enny W (2013) Understanding on plants-microbes interaction. Tekno Hutan Tanaman 61:13–20

Enny W (2017) Understanding underground-plants communication to optimize land management. Jurnal Sumberdaya Lahan 111:33–42

Fang C, Zhuang Y, Xu T, Li Y, Li Y, Lin W (2013) Changes in rice allelopathy and rhizosphere microflora by inhibiting rice phenylalanine ammonia-lyase gene expression. J Chem Ecol 39 (2):204–212. https://doi.org/10.1007/s10886-013-0249-4

Fisk LM, Barton L, Jones DL, Glanville HC, Murphy DV (2015) Root exudate carbon mitigates nitrogen loss in a semi-arid soil. Soil Biol Biochem 88:380–389. https://doi.org/10.1016/j.soilbio.2015.06.011

Haichar FZ, Santaella C, Heulin T, Achouak WA (2014) Root exudates mediated interactions belowground. Soil Biol Biochem 77:69–80. https://doi.org/10.1016/j.soilbio.2014.06.017

Haris T (2010) Mycorrhizal vesicular-arbuscular (MVA) in plant. Proceeding National Cereal Week ISSN: 978-979-89-40-29-3

Hooper AM, Caulfield JC, Hao B, Pickett JA, Midega CAO, Khan ZR (2015) Isolation and identification of Desmodium root exudates from drought tolerant species used as intercrops against *Striga hermonthica*. Phytochemistry 117:380–387

Jin J, Wang M, Lu W, Zhang L, Jiang Q, Jin Y, Lu K, Sun S, Cao Q, Wang Y, Xiao M (2019) Effect of plants and their root exudate on bacterial activities during rhizobacterium–plant remediation of phenol from water. Environ Int 127:114–124. https://doi.org/10.1016/j.envint.2019.03.015

Kalam S, Basu A, Podile AR (2020) Functional and molecular characterization of plant growth promoting Bacillus isolates from tomato rhizosphere. Heliyon 6(8):e04734. https://doi.org/10.1016/j.heliyon.2020.e04734

Khare E, Tyagi S, Patil KS (2020) Language of plant-microbe-microbe interactions in rhizospheric ecosystems. In: Molecular aspects of plant beneficial microbes in agriculture. Elsevier, India, pp 59–76

Koranda M, Schnecker J, Kaiser C, Fuchslueger L, Kitzler B, Stange CF, Sessitsch A, Zechmeister-Boltenstern S, Richter A (2011) Microbial processes and community composition in the rhizosphere of European beech - the influence of plant C exudates. Soil Biol Biochem 43:551–558. https://doi.org/10.1016/j.soilbio.2010.11.022

Mendes R, Garbeva P, Raaijmakers JM (2013) The rhizosphere microbiome: significance of plant beneficial, plant pathogenic, and human pathogenic microorganisms. FEMS Microbiol Rev 37 (50):634–663. https://doi.org/10.1111/1574-6976.12028

Michalet S, Rohr J, Denis W, Clemen B, Jean-Christoher R, Domenach A-M, Sonia C, Thomas P, Bruno C, Nadine G, Froiant B, Gillles C, Franck P (2013) Phytochemical analysis of mature tree root exudates in situ and their role in shaping soil microbial communities in relation to tree N-acquisition strategy. Plant Physiol Biochem 72:169–177. https://doi.org/10.1016/j.plaphy.2013.05.003

Ofek M, Voronov-Goldman M, Hadar Y, Minz D (2014 Jul) Host signature effect on plant root-associated microbiomes revealed through analyses of resident vs. active communities. Environ Microbiol 16(7):2157–2167. https://doi.org/10.1111/1462-2920.12228

Omotayo OP, Babalola OO (2021) Resident rhizosphere microbiome's ecological dynamics and conservation: towards achieving the envisioned sustainable development goals. Int Soil Water Conserv Res 9:127–142. https://doi.org/10.1016/j.iswer.2020.08.002

Paul GD, Anthony JM, Penny RH (2010) Are root exudates more important than other sources of rhizodeposits in structuring rhizosphere bacterial communities? FEMS Microbiol Ecol 72:313–327. https://doi.org/10.1111/j.1574-6941.2010.00860.x

Peiffer JA, Spor A, Koren O, Jin Z, Tringe SG, Dangl JL, Buckler ES, Ley RE (2012) Diversity and heritability of the maize rhizosphere microbiome under field conditions. PNAS 110 (16):6548–6553. https://doi.org/10.1073/pnas.1302837110

Pérez-Jaramillo JE, Mendes R, Raaijmakers JM (2016) Impact of plant domestication on rhizosphere microbiome assembly and functions. Plant Mol Biol 90:635–644. https://doi.org/10.1007/s11103-015-0337-7

Shukla KP, Sharma S, Singh NK, Singh V, Tiwari K, Singh S (2011) Nature and role of root exudates: efficacy in bioremediation. Afr J Biotechnol 1048:9717–9724. http://www.academicjournals.org/AJB. https://doi.org/10.5897/AJB10.2552

Stearns JC, Woody OZ, McConkey BJ, Glick BR (2012) Effects of bacterial ACC deaminase on Brassica napus gene expression. Mol Plant-Microbe Interact 25(5):668–676. https://doi.org/10.1094/MPMI-08-11-0213

Sugiyama A (2019) The soybean rhizosphere: metabolites, microbes, and beyond—a review. J Adv Res 19:67–73. https://doi.org/10.1016/j.jare.2019.03.005

Teng Y, Xu Y, Wang X, Christie P (2019) Function of biohydrogen metabolism and related microbial communities in environmental bioremediation. Front Microbiol 10:106. https://doi.org/10.3389/fmicb.2019.00106

Vanesa CC, Mary ER (2014) The rhizosphere micobiota of plant invaders: an overview of recent advances in the microbiomic of invasive plants. Front Microbiol 5(368):1–10. https://doi.org/10.3389/fmicb.2014.00368

Venant N, Marc O, Maïté S, Philippe T (2011) Beneficial effect of the rhizosphere microbial community for plant growth and health. Biotechnol Agron Soc Environ 152:327–337

Venturi V, Keel C (2016) Signaling in the rhizosphere. Trends Plant Sci 21(3):187–198. https://doi.org/10.1016/j.tplants.2016.01.005

Xing S, Fan Y, Chunwang X, Yong Z (2020) Increased contribution of root exudates to soil carbon input during grassland degradation. Soil Biol Biochem 146(2020):107817

Xingang Z, Fengzhi W (2012) p-Coumaric acid influenced cucumber rhizosphere soil microbial communities and the growth of Fusarium oxysporum f.sp. cucumerinum Owen. PLos One 7 (10):e48288. https://doi.org/10.1371/journal.pone.0048288

Xing-Feng H, Jacqueline MC, Kenneth FR, Ruifu Z, Qirong S, Jorge MV (2014) Rhizosphere interactions: root exudates, microbes, and microbial communities. Botany 92:267–275

Xuliang Z, Jie G, Anzhou M, Shenglei F, Guoqiang Z (2013) Review bioactive molecules in soil ecosystems: masters of the underground. Int J Mol Sci 2013(14):8841–8868. https://doi.org/10.3390/ijms14058841

Yang X, Huang X, Wu W, Xiang Y, Du L, Zhang L, Liu Y (2020) Effects of different rotation patterns on the occurrence of clubroot disease and diversity of rhizosphere microbes. J Integr Agric 199:2265–2273. https://doi.org/10.1016/S2095-3119(20)63186-0

Yekti M, Sudadi, Widyatmani SD, Ahamad Y (2019) Isolation and screening of calcareous and non calcareous soil rhizobacteria producing osmoprotectant and indol acetic acid in Gunung Kidul, Yogyakarta, Indonesia. Bulgarian J Agric Sci 25(1):36–41

Yueling Q, Adam O, Xiaomei Y, Esperanza HL, Francisco D, Violette G, Paolina G (2020) Effects of plastic mulch film residues on wheat rhizosphere and soil properties. J Hazard Mater 387:1–7. https://doi.org/10.1016/j.jhazmat.2019.121711

Zhang N, Wang D, Liu Y et al (2014) Effects of different plant root exudates and their organic acid components on chemotaxis, biofilm formation and colonization by beneficial rhizosphere-associated bacterial strains. Plant Soil 374:689–700. https://doi.org/10.1007/s11104-013-1915-6

Zolla G, Badri DV, Bakker MG, Manter DK, Vivanco JM (2013) Soil microbiomes vary in their ability to confer drought tolerance to Arabidopsis. Appl Soil Ecol 68:1–9. https://doi.org/10.1016/j.apsoil.2013.03.007

Global Climate Change and Microbial Ecology: Current Scenario and Management

12

Zia Ur Rahman Farooqi, Muhammad Ashir Hameed, Waqas Mohy-Ud-Din, Muhammad Hayder Ali, Abdul Qadir, and Muhammad Mahroz Hussain

Abstract

Microbes are part of life and they support every life-sustaining activity on the earth right from food crops production through increasing the soil fertility, helping in food processing via fermentation, and decomposition of dead animals to save environmental nuisance. Microbes support one of the biggest industries in the world (agriculture) through nutrients cycling, optimizing soil properties for better crop production, retain nutrients for plant supply and fertility, and remediate soil pollutants through bioremediation. Climate change has aggravated and disturbed various processes undertaken by soil microbes, i.e. microbial populations, diversity, processes undertaken by them, and nutrient cycles by killing them via increased temperature and soil salinity and associated problems. This chapter is an effort to comprehensively describe the benefits of microbes in the life, effects of climate change on population, associated processes, and ultimate effects on the environment.

Keywords

Soil microbiota · Nutrient cycling · Carbon sequestration · Greenhouse gasses

12.1 Introduction

Variabilities in climatic conditions are changing the distribution of living species and simultaneously affecting the interactions between them (Bardgett et al. 2014; Van der Putten 2012). Such communities are naturally diverse and consist of various

Z. U. R. Farooqi · M. A. Hameed · W. Mohy-Ud-Din (✉) · M. H. Ali · A. Qadir · M. M. Hussain
Institute of Soil and Environmental Sciences, University of Agriculture, Faisalabad, Pakistan
e-mail: 2015ag559@uaf.edu.pk

R. Soni et al. (eds.), *Microbiological Activity for Soil and Plant Health Management*,
https://doi.org/10.1007/978-981-16-2922-8_12

species with distinct features of life history, resistance, and capabilities. In addition, interactions between different communities may be helpful, dangerous, or may have functional influence. These interfaces may be altered with the change in ecological balance (Agler et al. 2016). Several experiments have shown that species dynamics influence the functions of the terrestrial environment and their biodiversity (Hoegh-Guldberg and Bruno 2010; Sharma et al. 2018; Langley and Hungate 2014), and soil community (Cantón et al. 2012; De Vries et al. 2012). Soil microbes influence each other in many ways and interact with plants to shape and maintain the characteristics of ecosystems. In fact, the interaction between soil microorganisms and plants can alter the landscape pattern, diversity, and composition of plants and animals (Pecl et al. 2017).

Soil microbial species comprise pathogenic, symbionts, and decomposers. Some microbial species can reduce plant growth (pathogenic microbes), while some species can enhance plant productivity, i.e. plant growth-promoting rhizobacteria (PGPR) and nitrogen-fixing symbiotic bacteria. So, the relationship between plants and microorganisms is considered positive when the soil microbial communities enhance plant growth. When the microbial communities like nematodes and pathogenic bacteria spread disease in plants, the interaction is considered negative. Therefore, it is imperative to study their importance, mutual interactions, and ecosystem characteristics in response to climate change. One of the major functions of the ecosystem is the storage of carbon contents in it (Pecl et al. 2017; McCormack et al. 2015; van der Putten et al. 2016).

The annual carbon fluxes of 120 Gt into and out of the terrestrial environment are due to the burning of fossil fuels (Jenouvrier et al. 2015). It induces climatic changes and disturbs the normal functioning of the microbes present in the soil. In general, microbes function as a pool of nutrients for crops, recycle the nutrients, reduce the risks of soil pollution by bioremediation as well as work together to stabilize the soil properties and structure via carbon sequestration (Farooqi et al. 2018, 2020). The change in carbon concentrations in the atmosphere can cause global warming. Therefore, climate change has a significant influence on the microbial communities and carbon concentration in the soil and atmosphere in the future (Classen et al. 2015).

During the events of global warming and climate change, microbial degradation and soil salinity due to higher temperatures will eventually control the life-sustaining processes. However, activities of different microbes affect plants, the carbon feedback between the atmosphere and soil is ambiguous (Todd-Brown et al. 2013; Lehmann and Kleber 2015). If the soil community boosts its activity with respect to animal and plant input during higher temperatures, then the amount of soil carbon will decrease from the atmospheric (Doetterl et al. 2015; Leff et al. 2015). In adjunct to directly controlling the decomposition process, the soil microbial community can also affect plant characteristics, such as food production and quality (Feng et al. 2002; Coleman et al. 2017), which can synchronize the flux in the carbon cycle during carbon sequestration from the atmosphere.

12.2 Microbial Functions in the Environment

Environment-friendly technologies and environmental protection are very important in human life sustainability. Scientists have been studying technologies that can be used to enhance agriculture produce while minimum waste production and pollutant remediation. The technologies based on native microbes (bioremediation and biofertilizers) are such an eccentric, which are widely used in the environmental remediation and soil productivity enhancement. Indigenous microbes are a cluster of inherent microbial communities. They live on the surface of the soil and all internal and external organisms. They can convert toxic substances to less/non-toxic through biodegradation and give a boost to soil fertility through bio-composting products nitrogen fixation and producing plant-growth-promoting hormones. Without these microbes, on this lively planet, human existence would have been impossible. Therefore, the focus of this chapter is to restore and protect the climate through indigenous microorganisms and to convert useless and unhelpful waste into productive biological resources.

12.3 Applications in Agriculture

The soil microorganisms can increase agricultural production. Scientists have used microbes for the enhancement of soil fertility as biological fertilizers for help in plant growth (Rawat et al. 2019; Joshi et al. 2019). The microorganisms living in the soil can help plants to take up the required types of nutrients. Plants together with these microbes participate in nutrients recycling. Microbes also facilitate plants to absorb the necessary nutrients. In return, the plant donates its waste by-products to microorganisms for their use as food (Fig. 12.1).

12.3.1 Nutrient Recycling

Healthy soil can fulfill the growing demands of food for the growing population of the world. It can do this by maintaining soil microbiota and growth of the plant, recycling nutrients, degrading soil organic matter (SOM), remediating toxic substances, inhibiting disease-causing agents, and maintaining soil and water quality. At the same time, they keep the balance of each nutrient to enhance the sustainability of the growth of plants (Meena et al. 2017; Meena et al. 2016; Shrivastava et al. 2016). Deterioration of soil and water quality is an indicator of poor soil conditions (Biswas et al. 2017; Abhilash et al. 2016). Soil health is the result of the process of protection and decomposition. It is extremely dependent on the biological components of the soil environment and affects plant growth, ecological health, and quality and safety of food (Nath et al. 2017; Takoutsing et al. 2016). It is also a fact that multiple layers of soil have a distinct microbial population and their numbers (Table 12.1) to support soil and plant stainability. Healthy soil can protect nutrients, pollutants, and other solutes by adsorbing or combining with

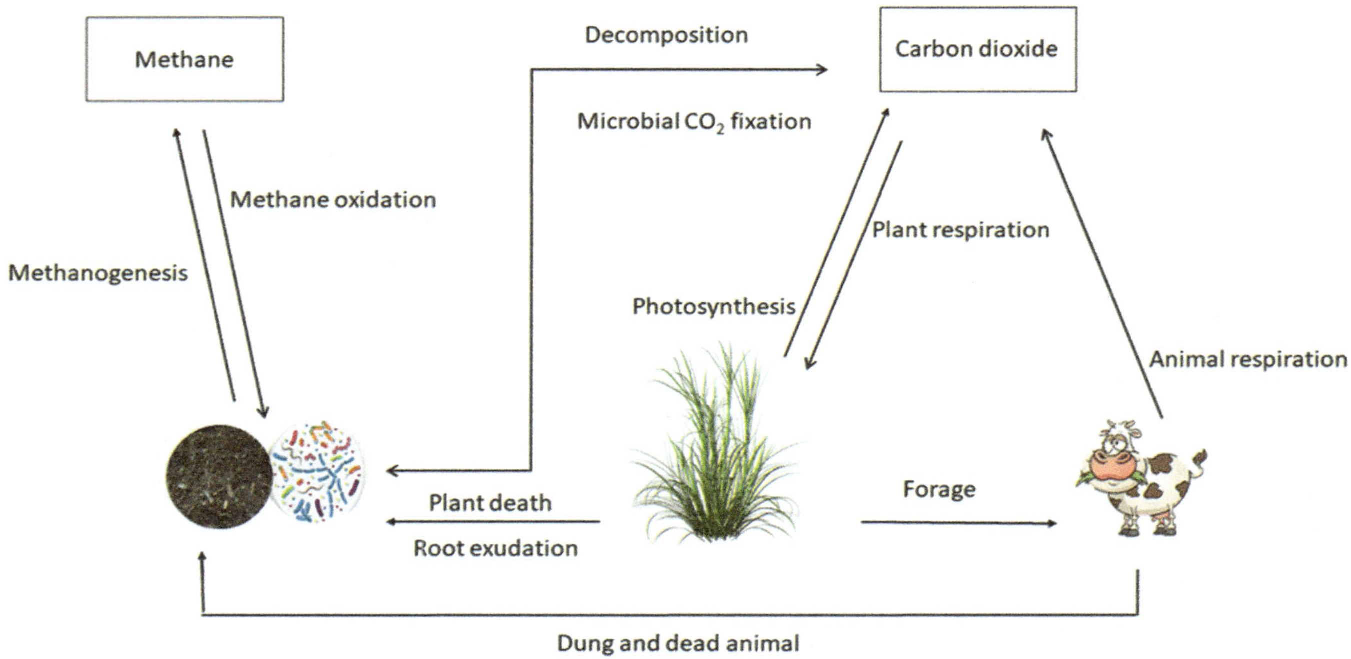

Fig. 12.1 Microbial functions in the environment

Table 12.1 Soil layers and microbial communities in it (Adopted from Sahu et al. 2017b)

	Soil microbial communities			
Depth (cm)	Aerobic bacteria	Anaerobic bacteria	Actinomyces	Fungi
3-8	7,800,000	1,950,000	2,080,000	119,000
20-25	1,800,000	379,000	245,000	50,000
35-40	472,000	98,000	49,000	14,000
65-75	10,000	1000	5000	3000
135	100	400		3000

particles of clay and SOM. Soil acts as a filter to remove detrimental components from (Bastida et al. 2008; Allen et al. 2011). These highly effective communities of microbes can be regarded as soil designers (Deiner et al. 2015; Sharma et al. 2013; Vayssier-Taussat et al. 2014; Pii et al. 2015). Microbial communities generate their own environment by complexing with soil particles in the presence of other biochemical nutrients (Teotia et al. 2016; Finley et al. 2013). Various microbial communities have a close relationship with each other and share the mutual trait of degradation which aids to degrade the plants litter as well as developing food web structures and helps to recycle the nutrients (Schulz et al. 2013; Bünemann et al. 2018).

12.3.2 Sustaining Optimal Soil Structure for Agriculture

The soil organic matter serves as a reservoir for phosphorus (P), nitrogen (N), and sulfur (S). Inorganic (N) levels are very low in the soil, while most of its part is present in the atmosphere or in the form of the organic compounds. Soil microbial communities produce the hydrolytic enzymes which help to degrade the most of organic substance into simpler ones, yielding N, P, K, and S. During the degradation of the organic compound, soil enzymes participate during the biochemical process. The vital soil enzymes are cellulase, amylase, chitinase, arylsulfatase, phosphatase, urease, and dehydrogenase which are produced by microbes and plants (Das and Varma 2010). The amount of SOM degradation is affected by various environmental aspects such as the presence of soil moisture, temperature, and diversity of microbial colonies (Umar et al. 2008). According to the decomposition pattern, there are four stages. First stage: organic matter is easily decomposed into protein, starch, and sugar-rich compounds by microbial communities. In the second stage, those compounds which take 2–10 years to decompose, such as plant cellulose and lignin. In the third stage, those compounds that take more than 10 years to degrade, such as certain fat-like compounds (waxes and phenols). While in the fourth stage, those compounds which take 10,000 years, including humus-like substances, these humus-like substances are the integration of plant decomposition products and microbial decomposition products (Sahu et al. 2017a). The decomposition of organic compounds is a process in which soil microbial communities are involved to degrade organic compounds into a simpler form of substances or mineralized nutrients.

Consequently, the existence of SOM in soil affects the amount of N available for the plants, and the high C/N ratio of organic substance might cause nitrogen insufficiency in crops at a minimum in the small period (Marín-Spiotta et al. 2014; Schjonning et al. 2015).

12.3.3 Mineralization and Humification

The decomposition of various types of organic material in the soil is one of the key roles of soil microbes. It is mainly composed of the remnants of deceased animals and plants and biological excrement. These carbon-containing components ought to be altered into simple forms such as minerals to get them usable for autotrophic living organisms. This process of converting carbon-containing compounds into simple mineral forms is known as mineralization, and it is primarily achieved by microbes present in the soil (Six et al. 2006; Liu et al. 2017). The organic material of the soil can be divided into three categories: simply decomposable, medium-decomposable, and difficult to decompose. Microbial mineralization carries by attaching of the microbe to the organic compounds. In the process of organic mineralization, the oxidation rate of significant biological elements is very important for plant bodies. Microorganisms perform a crucial role in transforming complex substances into usable forms of plants. Therefore, the processes of mineralization and humification coexist naturally (Sardans et al. 2012).

12.4 Role of Soil Enzymes in Decomposition of Organic Matter

Soil enzymes are a cluster of enzymes that usually present in the soil. They continue to play an important role in sustaining ecology and soil health, fertility, chemical, and physical properties and are responsible for the entire degradability of carbon-containing compounds in the soil system (Makoi and Ndakidemi 2008). They are essential for catalyzing the living processes of microorganisms in the soil and the soil structure stability, the decomposition of organic waste, the formation of organic matter, nutrient cycling, as well as play a significant role in agriculture. Changes in levels of enzymes in soil systems are mainly due to the different amounts of organic matter processes in each soil type. In practice, biochemical reactions are mainly produced by the catalysis of enzymes and variable substrates, and the substrates are the energy source of numerous microbes (Kumar et al. 2011; Griffiths and Philippot 2013).

These soil enzymes immediately mediate the biological SOM and mineral ingredient catabolism. Soil enzyme production is connected to the organic matter of the soil and its quality. These variations are much earlier than other factors and can therefore provide early signs of variations in soil sustainability (Gomez et al. 2006). These enzymes might comprise of β-glucosidase, amylase, cellulase, arylsulfatase, dehydrogenase, chitinase, protease, and phosphatase excrete from microorganisms and plants present soil (Xu et al. 2015). These factors distressing the activity of the soil enzymes and can adsorb on the inorganic and organic soil particles. These complex substances are released into the soil by microbial decomposition.

12.4.1 Amylase

It is known that starch hydrolase amylase consists of α-amylase and β-amylase. Alpha-amylase is produced by microbes, plants, and animals, while β-amylase is mainly produced by plants. It performs an important part in the degradation of starch, which transforms the amyloid compound into β-amylase and glucose, which transforms starch into maltose (En et al. 2019). However, plant bodies can impact soil amylase activity by immediately providing enzymes from its remains or emitted substances or implicitly offering compound for the plants-microbial system.

12.4.2 Arylsulfatase

Arylsulfatase is commonly distributed in nature and soil and is liable for sulfate hydrolysis (Kumar et al. 2011). Rhizosphere bacteria secrete them into the outer environment in response to sulfur constraint, and their existence is usually related to the biomass of microbes and the rate of sulfur sequestration. The enzyme plays a part in the breakdown of aromatic sulfates ($R–O–SO_3$) into phenol (R–OH) and sulfate or sulfate sulfur (SO_4^{2-} or SO_4-S) in the presence of water. Microbial genera and species which play a key role in the organic soil sulfur cycle, among which arylsulfatase is a key enzyme, are still rarely seen (Sahu et al. 2017a).

12.4.3 β-Glucosidase

Glucosidase is an important hydrolytic enzyme commonly found in the soil. It performs an essential role in biodegradation and catalyzing with the help of water. β-glucosidase enzymes are produced during decomposition of plant residues and added into the soil after their decomposition (Sahu et al. 2017a). Its end-product is glucose, which is a significant source of carbon for soil microorganisms. This can usually be used as an indication of soil quality, reflect previous biological activity, and the ability of the soil to alleviate SOM.

12.4.4 Cellulose

Cellulose, which makes up about 50% of the biomass created by photosynthesis, is a rich organic compound. In most agricultural soils, the development and endurance of microorganisms are important, depending on the carbon source confined in plant material in the soil (Hu et al. 2011). Cellulase is a cluster of enzymes that breakdown the cellulose during the process of degradation. Polysaccharides are composed of β-1,4-linked glucose components. Several researchers report that the cellulose in the soil primarily comes from plant residues mixed into the soil, and a small quantity is also derived from soil microbes and fungi. Ever since cellulase plays a crucial role in the large-scale recycling of cellulose, the highly plentiful polymer in nature, it is important to better understand this enzyme so that it can be used more often in our soil fertility program (Piazza et al. 2019; Muter et al. 2017).

12.4.5 Chitinase

Chitinase enzymes are essential components that are responsible for hydrolysis and chitin (poly-β-1-4-(2-acetamido-2-deoxy)-D-glucoside) decomposition. They are often known to be the principal structural feature in numerous fungal cells, used in combination with pest/pathogen destruction by hyper-parasitism processes. These biological elements often decrease disease-production components by means of other mechanisms, including competition or antibiosis approaches, and different living organisms containing microbes and plants produce or release them.

12.4.6 Dehydrogenase

Dehydrogenase action is usually utilized as a sign of soil biological activity. The enzyme is an important component of intact cells, but it does not gather outside the cell in the soil (Zeng et al. 2010). A method of oxidizing soil organic material by transporting protons and electrons from the substrate to the acceptor is also known. Soil moisture and temperature indirectly affect the dehydrogenase activity by influencing the redox state of the soil (Maharudrappa et al. 2000; Xu et al. 2015).

12.4.7 Phosphatase

In the geosphere, these enzymes are known to play a vital role in the phosphorus cycle, since they contribute to phosphorus toxicity and plant growth, and are also excellent soil fertility indicators. For instance, when P deficiency occurs in the soil, the acid phosphatase secretion increases from the plant roots. This increase in phosphatase secretion enhances the dissolution and migration of P, thereby enhancing the ability of plants to survive with P stress situations (Sahu et al. 2017a).

12.4.8 Proteases

The protease in the soil plays a significant part in N-mineralization, controlling the content of accessible nitrogen in plants and the development of plants. They are related to organic and inorganic colloids (Sahu et al. 2017a). The extracellular enzyme activities not only indicate the biological ability of soil to enzymatically alter substrates but also play an essential part in the microbial ecosystem.

12.5 Pollutants Mitigation

Numerous anthropogenic activities, such as fisheries, agriculture, and production units, will generate an unusual accumulation of distinct substances. These substances cause environmental hazards due to their concentration and/or characteristics, and occasional ecological disasters, like that Exxon Valdez event, Cher Nobel, the Gulf of Mexico, and the recent Aznarcollar issue in Spain. Although most pollutants are biodegradable and contain pollutants just like natural materials (agricultural or animal residues), they cannot be eliminated in a short time due to the large output. Considering that microorganisms use nutrients to degrade pollutants in aqueous solutions, and oil lacks these nutrients such as nitrogen and phosphorus, the degradation rate of other natural carbon-containing substances (like hydrocarbons) may be exceptionally slow. Engineering activities will generate many chemical substances that do not exist in the atmosphere and on soil, such as distinct types of plastic, biocides, etc. These chemical substances come into the ecosystem in huge quantities and are difficult to biodegrade.

Since most of the synthetic chemicals have distinct types of structure or subunits that present in the ecosystem. So, the methods used to characterize them as degradable and non-biodegradable may be incorrect or at least too much fierce. A further problem is that pollutants that are hazardous to microorganisms, such as heavy metals or synthetic chemicals that hinder the bioremediation of organic pollutants in the short term (He and Yang 2007; Harms et al. 2011; Gadd 2007). The contagion of soil and residues of petroleum is international anxiety. A novel microbial strain *Dechloromonas agitate* CKB has been isolated, which can degrade hydrocarbons in the dearth of oxygen (Cao et al. 2005). *Agrobacterium radiobacter* J14a can use atrazine (Cao et al. 2005) which is currently one of the most used herbicides. Microbial biomass is also efficient to degrade plastics in the soil (Sudesh and Iwata 2008).

12.6 Bioremediation

The increase in environmental pollution leads to the gradual deterioration of the quality of the ecosystem, which may disturb plants and animals, and therefore, immediately disturb human health (Pandey et al. 2009). This situation confronts society to discover effective remedies (Mohy-Ud-Din et al. 2020; Hussain et al. 2021). Traditional treatment methods for dangerous pollutants, such as immobilization, incineration, and volatilization of dangerous pollutants, only convert the pollution and generate new waste, but fail to eradicate the problem. Bioremediation skills can lead to decontamination and mineralization of dangerous pollutants and is an alluring unconventional method that can produce economic benefits in some cases. As mentioned above, effective bioremediation varies on the accessibility of appropriate microorganisms (Juwarkar et al. 2010; Debbarma et al. 2017; Giri et al. 2017).

This consideration, and the prospective use of engineered microbes, provides an extended time scale technique (Nicolaou et al. 2010). There are numerous ways to perform bioremediation on site. If local microorganisms can metabolize pollutants due to the adaptation process, non-native microorganisms can be selected in other locations to inoculate the environment (biofortification) or regular nutritional supplements (biostimulation). The combination of these basic technologies represents the constant supply of nutrients and aerobic environments to microbes being used for bioremediation. Bioaugmentation and biostimulation may also be used as ex-situ techniques to invigorate the biodegradation of dangerous pollutants. Bioreactors exemplify an extremely controllable technique of treating distinct pollutants, such as pH, nutrient content, temperature, and agitation can be monitored in batch or continuous feed bioreactors. If a reactor is based on compost or slurry, the microbial activity can be optimized to the greatest extent, thereby optimizing the degradation of pollutants. Solid-phase remediation is usually utilized as an ex-situ remediation method, which can be processed in a pile or in a treatment pond. Many oil companies also use land farming techniques to treat contaminated soil (Okoh 2006).

12.7 Impact of Climate Change on the World's Agriculture

Climate change may directly affect global food production. An upsurge in average seasonal temperature will decrease the period of various crops, thereby reducing the ultimate yield. In fields where the temperature is detrimental for crop survival, further warming adversely affects crop yield (Knutson et al. 2010). Due to global warming, world agriculture is facing a severe decline in this century (Table 12.2). Overall, by 2080, the world's agricultural productivity is expected to drop by 3–16%. Emerging countries (several of which have average temperatures close to or higher than crop forbearance levels) are expected to decrease by a norm of 10–25%. In developed countries where average temperatures are generally lower, the average impact will be milder or even positive, from a rise in productivity of 8% to a decrease of 6%. Developing countries face a greater decline. For example, India may drop by 30–40%.

12.7.1 Effects of Temperature

Although many aspects (such as parent material, climate, topography, soil texture, vegetation cover type, and soil microbial community structure) determine the soil's capability to sequester carbon, while microbial decomposition due to higher temperatures eventually determines the amount of carbon decomposition and its loss to atmosphere (Kaisermann et al. 2017). Global warming causes physiological

Table 12.2 Effects of climate change on crops yield (Adopted from Dubey et al. 2019)

Crops	Scientific name of crops	Percentage decrease
Extreme rise in temperature		
Kharif crops		
Groundnut	*Arachis hypogaea*	4
Pearl millet	*Pennisetum glaucum*	9
Rice	*Oryza sativa*	5
Sorghum	*Sorghum bicolor*	7
Soya	*Glycine max*	13
Rabi crops		
Chickpea	*Cicer arietinum*	10
Linseed	*Linum usitatissimum*	7
Wheat	*Triticum aestivum*	3
Extreme decline in rainfall		
Kharif crops		
Groundnut	*Arachis hypogaea*	13
Pearl millet	*Pennisetum glaucum*	16
Pulses		17
Rice	*Oryza sativa*	14
Sorghum	*Sorghum bicolor*	14
Rabi crops		
Barley	*Hordeum vulgare*	4
Chickpea	*Cicer arietinum*	10
Linseed	*Linum usitatissimum*	8
Rape and mustard seed	*Brassica napus* and *Brassica nigra*, *B. juncea*, *B. hirta*	3
Wheat	*Triticum aestivum*	5

changes in the decomposers and affects carbon dioxide emissions in the soil (Zechmeister-Boltenstern et al. 2015). Increased temperature may accelerate the decomposition of fungi, leading to increased carbon dioxide emissions in soil. Though, elevated temperatures also increase the N content in the soil, thereby inhibiting the decomposition rate of fungi. In fact, higher available N content harms the activity of different microorganisms (Dutta and Dutta 2016). On the other side, the efficiency of microbial biochemical reactions below the pressure of global warming is low. Therefore, these microorganisms release more carbon in the form of carbon dioxide as compared to transforming it into waste biomass (Dutta and Dutta 2016).

The intake of higher consumed CO_2 levels and more approaches accelerate nitrous oxide and methane gas emissions from plants (Alshaal et al. 2017). However, the ultimate microbial reaction of soil decomposition depends on the temperature sensitivity of decomposition materials, the supply of substrates, associations with soil processes, and the environment such as moisture and possible physiological adaptability. It must be emphasized here that due to decomposition, the effect of high temperature on the release of carbon dioxide is not equal between distinct soils. Microbial species in controlled agricultural soil can reduce the impacts of temperature rise due to carbon dioxide release. Though, the ultimate stimulus occurred in the Arctic and northern ecosystems where the global warming effects are prominent than anyother region (Azzaro et al. 2019). The efficiency of carbon usage is a key factor in determining the long-term constancy of carbon in the soil, because the microorganism biomass, instead of stubborn plants, is the best resilient to decay (Paustian et al. 2016; Cotrufo et al. 2013).

12.7.2 Moisture Fluctuations

Climate change also impacts soil environments through inducing precipitation variations as well as temperature fluctuations. Water is the primary component and is a major influence in many terrestrial habitat's soil breathing patterns (Aanderud et al. 2011). Microbial decomposition and their activities are disturbed by many components, which change with the changes of moisture level and water, like the flow of water, solute and gas diffusion, and microbial endurance and movement (Rodrigo et al. 1997). Water may also inhibit microbial activity in numerous environments (like saline soil and water). The small amount of available water reduces the potential of intracellular water, thereby reducing the activity of enzymes and hydration (Stark and Firestone 1995). Moisture content in the soil also has a strong impact on the dynamic changes and emissions of carbon dioxide (Aanderud et al. 2011).

12.7.3 Significance of Terrestrial and Aquatic Ecosystems

The soil is a major and dynamic carbon pool and a central element in climatic changes related to activities and climate change. Around 2/3 of all terrestrial carbon is deposited into the Earth (Smith et al. 2008; Schmidt et al. 2011). Some other scientists modify the statement saying that terrestrial soils store extra carbon than the plant and atmosphere (Davidson and Janssens 2006). The soil-atmosphere substitution of CO_2 is an essential part of the global carbon (C) cycle (Raich et al. 2002). Approximately 10% of ambient CO_2 passes into the earth every year (Luo et al. 2001). Every year, carbon dioxide flux produces 50–75 Pg of carbon emissions and is a significant contributor to the global carbon cycle. It must be noted that the annual primary production absorption of autotrophic soil microorganisms is about 120 billion tons of carbon, and the cumulative emission of soil heterotrophic microorganisms is around 119 billion tons of carbon (Ciais et al. 2014).

Aquatic environments are as important as climate change issues. The ocean is where carbon dioxide enters, so it is called a biological pump (Prichard and Granek 2016). The upper layer of the ocean is the periphery for the accumulation of carbon dioxide and other greenhouse gases for the years, and the subterranean ocean is the greatest terrifying carbon pool in the world (Bindoff et al. 2007). In marine ecosystems, microorganisms process approximately 60 gigatons of carbon per year (Myers and Worm 2003). Together, the ocean and soil ecosystems constitute a global carbon sink, which absorbs approximately 39,1012 kilograms of carbon and absorbs 40% of the carbon dioxide emissions from the burning of fossil fuels (Singh et al. 2010).

12.8 The Relevance of the Microbial World to the Problem

In ascertaining the history of the earth, the activities of microorganisms are critical to determining the intensity of greenhouse gases (GHGs) in the atmosphere like methane (CH_4), carbon dioxide (CO_2), and nitrous oxide (N_2O) (Singh et al. 2010). Since more than 3.5 billion years ago, microbes developed and extracted GHGs in the marine field and later moved to the ground approximately two billion years ago (Zimmer 2010). In detail, these are the major procedures that cause greenhouse gas instabilities among the soil and the atmosphere (Faust and Raes 2012), and greatly affect ground carbon dioxide productions (Brevik et al. 2015). Here, it should be noted that the world's average concentration of CO2 in 2005 was around 380 ppm, which in the last 650,000 years has been around 80 ppm above the maximum concentration (Change 2007). Microbes have caused numerous alterations in the world's climate, and these changes have in turn affected them (Alshaal et al. 2017). In reality, numerous microorganisms may be altered by climate change, which may cause adverse economic, social, and environmental impacts (Pautasso et al. 2012).

12.9 Role of Terrestrial Microbes

12.9.1 Production of Carbon Dioxide and Methane

The volume of carbon released by a microbial soil disintegration in the environment is 7.5–9 times that of carbon generated worldwide through anthropogenic discharges annually (Crowther et al. 2016). Every year, microbes release 5.59×10^{12} kg of carbon dioxide into the soil by decomposing organic matter in the farm, 8 times more than that of humans. This has had a major influence on carbon dioxide feedback and has also been expected by the Intergovernmental Protocol on Climate Change (IPCC). Soil biological interactions are also highly significant because they are essential for the input on climate change from soil microorganisms (Crowther et al. 2016). However, the breakdown of microbial organic matter (OM) is exceptionally sensitive to climate trends (Crowther et al. 2016). Compared with CO_2, CH_4 emissions worldwide are more precisely dependent on microbes (Singh et al. 2010). In fact, microorganisms account for approximately 85% of the world's methane emissions (Abatenh et al. 2018). In this case, special mention should be made of spongy soil, where many microbial genes and proteins are liable for CH_4 emissions.

This can be demonstrated by methyl-Coenzyme M reductase, which is liable for converting carbon dioxide into methane. The increasing temperature trend may prompt several of these native microorganisms to release huge volumes of methane. Microorganisms use approximately 60% of the world's methane (Abatenh et al. 2018). In point, CH_4-consuming microbes can remove methane from the atmosphere yet at very minimal concentrations, and it exists both on soil and in the ocean (Abatenh et al. 2018). In this sense, microorganisms can help control methane emissions, thereby regulating climate change. In this respect, we can also mention the microorganisms of natural grassland, which play an important role in alleviating carbon flux. Though, changes in precipitation forms owing to climate change will disturb these microorganisms. In particular, the carbon balance in this setting might be determined considerably by the soil carbon storage (Bhattacharjee and Prakash 2019).

12.9.2 The Issue of Agricultural Lands

Due to microbial nitrification and denitrification in farmland, a large amount of nitrous oxide (N_2O) is produced, and farmland is mainly a supply of N_2O. More than a third of GHGs pollution sources are responsible for this (Change 2007). In statistics, for every 1000 kg of active N (especially fertilizers) deposited, 10–50 kg of N_2O will be emitted (Singh et al. 2010). In 2000, there were 1.59×10^6 square kilometers of arable land, 3.4799×10^6 square kilometers of pasture, and 2.8099×10^5 square kilometers of irrigated land. It is predictable that by 2050, these numbers will increase by 18.5%, 13.4%, and 47.1%, respectively. Agricultural soils are the core of the overall carbon and nitrogen cycle, and it has been noticed that under traditional farming methods, cultivated soils discharge roughly ten times additional nitrous oxide than forest soils (Robertson et al. 2000). He et al. (2007) stated that as temperature increases, the amount of N_2O related to nitrification declines, though, the temperature does not alter denitrifying enzyme activity (Cheng et al. 2012). It is also the situation that microbial soil activities release greenhouse gases (GHGs) (such as N_2O, CH_4, and CO_2), soil also behaves like a carbon sink on farmed land (Houghton 2001).

12.10 Role of Aquatic Microbes

Almost 93% of the global carbon dioxide is deposited in the ocean. The ocean circulates about 90 billion tons of carbon dioxide each year, while six billion tons are manmade sources (Dutta and Dutta 2016). The mechanism of the ocean carbon cycle is mainly controlled via microplankton, nanoplankton, and picoplankton, containing bacteria and archaea (Dutta and Dutta 2016). The proliferation of photosynthetic microbes occurs in the ocean, particularly in polar areas (such as Arctic sea ice). These microorganisms eradicate large amounts of carbon present in the atmosphere. In point, since 1960, oceanic microbes have isolated about a quarter of anthropogenic carbon dioxide (Weiman 2015). Also, seawater contains many viruses, which can destroy/dissolve nearly 50% of marine bacteria in every single day. In this sense, they have a major impact on worldwide geochemistry by affecting the accumulation of OM and changes in respiration, and OM is an essential factor in climate change (Jiao et al. 2010). Methanotrophic microbes can cushion the effects of a large quantity of CH_4 released in certain circumstances (Singh et al. 2010). For example, they utilize large quantities of CH_4, not only from sediments of marine but also from sudden discharge such as Deepwater Horizon oil spills (Zimmerman and Labonte 2015).

The marine ecosystem is a powerful resource of CH_4 released into the environment because this GHG is continuously leaking from the pouches on the seafloor. But each methane leak has its own unique CH_4 culture using microorganisms as these sites have no popular species in the deep sea. Before the gas enters the atmosphere, these microbial communities eliminate approximately 75% of the CH_4 production. Therefore, these microbes give a crucial service for sustaining the climate by decreasing GHGs emissions (Dutta and Dutta 2016). The CH_4 production archaea in wetlands, oceans, rumen, and termite intestines is mainly due to the natural emission of CH_4 (250 million tons per year) (Singh et al. 2010). In this respect, landscape attributes are a vital determinant of microbial greenhouse gas emissions and carbon storage. For example, the microbial communities in saline-alkali wetlands release less methane compared to areas with active water flow (Mangodo et al. 2020). Another key factor is the state of the water body. The restored wetlands are colonized by microbial communities, which generate methane at a huge quantity than undisturbed methane. The greater growth of plants in the restored wetland may accelerate the biological process of methane emission (Mangodo et al. 2020).

12.11 Microbial Adaptations to Cope with Climate Change

To overcome the worlds feeding issue, intensive agricultural practices from the past few decades decreased the soil fertility and impair long-term farming and stability due to the excessive use of inorganic fertilizers (Agegnehu et al. 2016; Xia et al. 2017). Adaptation of standard procedures to maintain good quality soil and enhanced production should be the main concern (Li et al. 2015). There are several organic materials like poultry manure, farm-yard manure, rice straws, wheat straws, sugarcane bagasse, compost, and a mixture of organic waste that can be used as instead of inorganic fertilization (Steffen et al. 2015). These amendments improve the quality along with the long-term fertility status of the soil (Ling et al. 2016).

Mulch is an English word that is derived from the German word "molsch," which means easy or soft to decay (Mohsin 2019). Instead of other organic amendments that are incorporated into the soil profile, mulches are applied to the upper soil surface (Mohsin 2019). Mulching is the best technique due to several benefits like covering of soil surface, water conservation, increased temperature, reduced the chances of soil erosion and surface runoff. It can also reduce weed growth, minimize salinity stress, and improve crop yields. Furthermore, mulching can considerably impair the soil structure and composition and affecting the community and metabolic activity of microorganisms. Alleviating soil temperature under mulching material promotes microbial activities and catalyzes the decomposition of organic material in the soil (Mohsin 2019).

Mulching is of two types, organic and inorganic. Organic mulching contains organic materials like wood industrial wastes (sawdust), agriculture wastes (stalks, straw), processing industry residues (rice husks), and animal wastes (poultry and farm-yard manure), while the inorganic mulching includes polyethylene plastic films and synthetic polymers (Mohsin 2019). Other materials used for mulching include special materials like sand and concrete, but these materials are expensive and reduce soil fertility. New kinds of mulching techniques include flat mulching, plastic mulching with holes, ridge shape mulching, ridge-furrow mulching, and double mulching system. Microbes are the key element that controls the decomposition procedure. Similarly, both the microbial communities and microbial biomass adaptive diversity play a significant role in plant litter decomposition and rhizospheric-carbon cycling. Mulching also provides food to soil microorganisms like algae, bacteria, mosses, fungi, and many other entities like earthworms and arthropods (Mohsin 2019).

12.12 Plant–Microbe Interactions in Managing Stressed Agriculture

Rhizospheric functioning is influenced by different environmental stresses that ultimately affect agricultural outputs (Singh 2015). A healthy rhizosphere provides better support to the plant in terms of nutrients availability and water facilitation, besides also delivers long-term advantages to native microbial diversity that mutually improves plant health (Vimal et al. 2017). The attributes and composition of native beneficial microbial diversity near the root zones enhance plant growth promotion and includes mycorrhizal fungi, rhizobacteria, and miscellaneous microbiota (Rouphael et al. 2015; Etesami and Beattie 2017). Beneficial aspects of these symbiotic associations include control of pathogens, the release of plant growth-promoting hormones, metabolites, enzymes, and organic volatiles that increased resistance against stresses and ultimately promote the plant health and soil fertility (Glick 2014). It is also noted that beneficial hyphae networks around the rhizosphere protect plants against numerous pathogens, provide support to act as a bio-control agent, and enhance the uptake of nutrients and water under drought conditions (Barnawal et al. 2014; Spagnoletti et al. 2016).

12.13 Climate Change Impacts on the Soil Microbiome

The disturbances associated with climate change can significantly change the native microbes and functional profiles (Naylor et al. 2020). Climate change affects the soil carbon and nitrogen cycles through positive or negative feedbacks that mean the production of GHGs and carbon assimilation into the plant or microbial biomass, respectively (Bradford et al. 2016). To make a better understanding of the interaction of climate change and microbes, we must consider other factors that contribute to these changes.

12.13.1 Soil Warming

The present scenario of climate modeling predicts a 3.7 °C rise in global temperature by the end of the twenty-first century that indicates an unavoidable impact of these changes on microbial communities (Ramanathan and Carmichael 2008). Warming of soil indicates two types of phases, first include the rise in microbial biomass due to increased decomposition of OM that in some cases increased about 40–150%, while the second phase reduces the microbial biomass due to the elimination of labile carbon sources (Melillo et al. 2017). It has been studied that changes in microbial communities, composition, and physiology were observed under warming conditions and microbes shift their metabolism to remaining recalcitrant carbon sources (Rocca et al. 2019). Warming enhances the microbial community of *Actinobacteria* and *Acidobacteria* phyla and class alphaproteobacteria (DeAngelis et al. 2015). These microbial shifts overlap with functional attributes like Oligotrophic taxa (slow-growing microbes in low nutrient soil) are promoted over Copiotrophic taxa (fast-growing microbes in the high nutrient environment) that adapt themselves in response to soil warming (Li et al. 2019). Assessable differences in functional corporation responsible for ammonia oxidation (Li et al. 2020) or diazotrophy (Carrell et al. 2019) have also been noticed under soil warming.

Soil warming affects the microbial function in two ways, i.e. accelerating the enzymatic reactions and stimulate the microbes through rhizodeposition by altering the soil ecosphere. Phosphorus (P) and sulfur (S) cycling were stimulated under high temperature (Frey et al. 2013). Increased soil temperature influences the N-cycle (N-fixation, N-mineralization, nitrification, and denitrification) and disturbs the carbon cycle (Frey et al. 2013). Soil warming increases soil inorganic nitrogen pool in soil (Terrer et al. 2016) and ultimately minimizing the microbial decomposition rate and N-cycling (Lladó et al. 2017). Instead of that, carbon cycling is initially activated by soil warming because of the sufficient bioavailable carbon composition (DeAngelis et al. 2015).

12.13.2 Elevated Carbon Dioxide

Increased soil warming directly or indirectly affects the soil microbiota. Short-term studies indicate that carbon dioxide increased microbial numbers, respiration rate, and genetic signals for carbon cycling. Carbon dioxide stimulates the copiotrophs in the rhizosphere to stimulate plant growth by decomposing labile and then recalcitrant carbon. Taxonomic trends under increased carbon dioxide affect the soil microbiota, e.g. depletion of Acidobacteria, and increment in the number of oligotrophs at the same time and this might be happened due to the depletion of labile carbon and decreased soil moisture (Dunbar et al. 2012; Yang et al. 2019). Similarly, a higher carbon dioxide rate activates the soil enzymes that speed up the P-cycling, along with enhanced the N-cycling because of microbial denitrification and plant assimilation of N depletes the inorganic N from the soil, while enzymatic rates are often declined. These activities introduce a higher abundance of nitrogen fixers (Rhizobiales and $NH4^+$ oxidizers) in most cases.

12.13.3 Microbial Biochemical Pathways and Climate Change

The biochemical pathway of microbes is inevitably affected by climate change and their interaction with plants and microbes. Recent literature indicates that microbial communities take up root exudates as their carbon source under stressed (salinity and drought) environments (von Rein et al. 2016). Due to environmental stresses, the species of native and exotic origins thrive due to a shift in root exudates (Castro et al. 2019). Additionally, climate change may alter the plant communities or plant cover and increased C_4: C_3 ratio (Zhou et al. 2012), which may affect the rhizodeposition under stressed conditions (Lladó et al. 2017; Wang et al. 2017). Likewise, their C_4 plants show more root exudation relative to C_3 plants under higher carbon dioxide concentration (Zhou et al. 2012). It is estimated that climate change affects the ecosystem by altering its native community's composition and stimulate the different microbial metabolisms and pathways.

12.14 Plant, Microbe, and Climate Change

The shift in climatic conditions such as solar radiation, precipitation, and the temperature is quite likely to impact on crop production (van der Putten et al. 2016). Soil microbes have complicated association with each and others and with the plants because of thermal tolerance, life traits, and dispersal ability base. These microbial communities may act as pathogen or symbiont, due to climatic disturbances (Durán et al. 2018; Van der Putten 2013).

12.14.1 Direct Impacts of Climate Change on Soil Communities and Plants

Climatic change has direct impacts on the working and abundance of the microorganism communities existing in the soil. Microbial species also exhibit greater variations based on morphology, development, and temperature stress sensitivities (Briones et al. 2014; Zhang et al. 2018). Continuing warming of soil headed towards changing microbial diversities in temperate soils (Webb et al. 2017). In temperate forest soils, the increase in 5 °C temperature leads towards disturbance of native microbial communities such as changing the ratio of bacteria to fungi (Bintanja 2018). Global weather changes straightly affect the rates of microbial respiration as the processes they moderate are temperature sensitive as described by many researchers in the last decade (Ediriweera et al. 2016; Gao et al. 2018).

12.14.2 Indirect Effects of Climate Change on Plants and Soil Microbiome

Plant distribution in a community is directly affected by changing climate because of the alteration of microbial ecosphere as well as plant phenology (Kaisermann et al. 2017). Plants that propagate effectively in stressed conditions produce a higher defense system against different xenobiotic compounds (Lal 2016). Increase in global terrestrial temperature or predicted impact on climate change for crop yield, insect, weeds, and plants diseases, e.g.

- Higher temperature increases the number of insect pests, weeds as well as pathogens population as it fastens the rate of their life cycles.
- More pest population in a single season due to the long growing season.
- Temperature and humidity stresses are directly influenced by climate change and it increases the chances of disease attacks.

Effective application of microbiota helps in maintaining soil health by enhancing root growth, water holding capacity, nutrient cycling, and carbon storage along with reduced bioavailability of pollutants and biodiversity conservation (Keesstra et al. 2016; Burns et al. 2013). Excessive use of synthetic fertilizers is triggering a balance between soil's biogeochemical cycles and is blameworthy for soil degradation, GHGs emissions, and eutrophication (Meerow et al. 2016). Moreover, the manufacturing of inorganic N-fertilizer relies upon higher energy which is extracted from fossil fuels burning and it ultimately causes resource depletion and climate change because of higher emission of GHGs (Erisman et al. 2013). Over the globe, it was thoroughly investigated that microbial diversity is significantly correlated with multiple functions in the biosphere. Climate changes have a great influence on N-cycle in the ecosphere. In a common aspect, microbiota and vegetation responses to climate change vary among populations be contingent on their functional traits.

12.14.3 Interactions Between Climate Change and Microbial Ecosystems in Terrestrial Regions

Soil warming presents increased microbial metabolism because of the conversion of OM to carbon dioxide or CH_4 and fluctuates N-cycle which disturbs the biosphere (Caldeira and Wickett 2005; Schuur et al. 2015). Methane is produced by archaea and eukaryotes that all play an important part in CH_4_cycle (Nobu et al. 2016). Climate change specifically impacts the global CH_4_cycle as bacteria use or remove organic matter easily as temperatures are higher (Saunois et al. 2016). Carbon dioxide to CH_4 ratio is very important because CH_4 can trap more amount of heat as compared to CO_2 but had a shorter residence time in the atmosphere instead of carbon dioxide. Methanotrophs (CH_4 consuming bacteria) are those microbes that are produced in the environment where excess CH_4 is present and these microbes are potential candidates for the mitigation of climate change impacts.

12.14.4 Soil and Agriculture

Water stress, increased carbon dioxide, and N levels affect soil processes by disturbing terrestrial food webs and the interfaces between microbiota like viruses, protists, and nematodes that feed on these microbes. Rainfall patterns can also make huge impacts on soil structure, moisture contents, and flora–microbiota interactions, all of these can change the native inhabitant populations and alters the food chain (Mariotte et al. 2015). These interactions can change carbon and N fluxes, and the microbes like protozoa and other microorganisms require higher water contents to make their metabolism activate (Sharma and Gobi 2016), that is why prolonged water stress eliminates these microbiotas and disturbs the soil microclimate. Moisture contents of soil also cause the spread of diseases, like most of the viruses infects 40% of the rhizospheric bacterial population (Zablocki et al. 2016). Climate change leads towards the spread of diseases at a much faster rate and pathogenic fungal and viruses proliferate in the same manner that disturbs the micro and macroclimate of the biosphere.

Symbiotic microbes (mycorrhizae) enhance the plant growth by up-taking of nutrients and water from the deep horizons and mitigate the drought and temperature stresses. So, these microbes are used as a probiotic to overcome the effects of climate change. Changes in agricultural practices like tilled, non-tilled soil, same and changed cropping patterns influence the nutrient balance by affecting the native microbial communities (Wright and Wimberly 2013).

12.14.5 Freshwater

Climate change enhances heavy precipitation that flushes nutrients from the fields to freshwater bodies thus, causes eutrophication. When microbes use these nutrient fluxes, some dominate the ecosystem and produce algal blooms. Cyanobacteria and dinoflagellates are the common microbes that cause these types of blooms in water bodies. An increase in temperature and precipitation patterns supports algal blooms frequency (Glibert 2017). Extreme precipitations also affect recreational water security and waste-water security.

12.15 Conclusion

Since the climate is changing continuously, understanding of the possible reactions from the soil to the climate system is much more important. Microorganisms can promote plant growth and increase the resistance and abiotic control of diseases. Climate change impacts, including high carbon dioxide levels, drought, and hot weather are gradually being examined on the interaction of microorganisms. Living insects and thousands of other animals live in connection with some advantageous animals and pathogens impacting diverse populations. High carbon dioxide has a beneficial impact on the proliferation of plants, shrubs, trees, and ectomycorrhizal fungi among the various causes associated with climate change. While the effects on bacteria that sustain plant growth and endophytic fungi vary considerably. Temperature changes are extremely variable, positive, neutral, and negative on clean plant-related microorganisms, which were similarly normal and variable across temperatures. Likewise, plants under drought stress conditions positively impact microorganisms that encourage the growth of plants (e.g., fungi and bacteria).

References

Aanderud ZT, Schoolmaster DR, Lennon JT (2011) Plants mediate the sensitivity of soil respiration to rainfall variability. Ecosystems 14(1):156–167

Abatenh E, Gizaw B, Tsegaye Z, Tefera G (2018) Microbial function on climate change-a review. Environ Pollut Clim Change 2(1):1–6

Abhilash P, Dubey RK, Tripathi V, Gupta VK, Singh HB (2016) Plant growth-promoting microorganisms for environmental sustainability. Trends Biotechnol 34(11):847–850

Agegnehu G, Nelson PN, Bird MI (2016) Crop yield, plant nutrient uptake and soil physicochemical properties under organic soil amendments and nitrogen fertilization on Nitisols. Soil Tillage Res 160:1–13

Agler MT, Ruhe J, Kroll S, Morhenn C, Kim S-T, Weigel D, Kemen EM (2016) Microbial hub taxa link host and abiotic factors to plant microbiome variation. PLoS Biol 14(1):e1002352

Allen RG, Pereira LS, Howell TA, Jensen ME (2011) Evapotranspiration information reporting: I. factors governing measurement accuracy. Agric Water Manag 98(6):899–920

Alshaal T, El-Ramady H, Al-Saeedi AH, Shalaby T, Elsakhawy T, Omara AE-D, Gad A, Hamad E, El-Ghamry A, Mosa A (2017) The rhizosphere and plant nutrition under climate change. In: Naeem M, Ansari A, Gill S (eds) Essential plant nutrients. Springer, Cham, pp 275–308

Azzaro M, Packard TT, Monticelli LS, Maimone G, Rappazzo AC, Azzaro F, Grilli F, Crisafi E, La Ferla R (2019) Microbial metabolic rates in the Ross Sea: the ABIOCLEAR project. Nat Conserv 34:441

Bardgett RD, Mommer L, De Vries FT (2014) Going underground: root traits as drivers of ecosystem processes. Trends Ecol Evol 29(12):692–699

Barnawal D, Bharti N, Maji D, Chanotiya CS, Kalra A (2014) ACC deaminase-containing Arthrobacter protophormiae induces NaCl stress tolerance through reduced ACC oxidase activity and ethylene production resulting in improved nodulation and mycorrhization in Pisum sativum. J Plant Physiol 171(11):884–894

Bastida F, Zsolnay A, Hernández T, García C (2008) Past, present and future of soil quality indices: a biological perspective. Geoderma 147(3-4):159–171

Bhattacharjee S, Prakash T (2019) Microbial genomics in carbon management and energy production. In: Tripathi V, Kumar P, Tripathi P, Kishore A (eds) Microbial genomics in sustainable agroecosystems. Springer, Cham, pp 161–173

Bindoff NL, Willebrand J, Artale V, Cazenave A, Gregory JM, Gulev S, Hanawa K, Le Quere C, Levitus S, Nojiri Y (2007) Observations: oceanic climate change and sea level. In: Solomon S, Qin D, Manning M, Chen Z, Marquis M, Averyt KB, Tignor M, Miller HL (eds) Climate change 2007: the physical science basis. Contribution of working group I to the fourth assessment report of the intergovernmental panel on climate change. Cambridge University Press, New York, pp 387–429

Bintanja R (2018) The impact of Arctic warming on increased rainfall. Sci Rep 8(1):1–6

Biswas S, Hazra G, Purakayastha T, Saha N, Mitran T, Roy SS, Basak N, Mandal B (2017) Establishment of critical limits of indicators and indices of soil quality in rice-rice cropping systems under different soil orders. Geoderma 292:34–48

Bradford MA, Wieder WR, Bonan GB, Fierer N, Raymond PA, Crowther TW (2016) Managing uncertainty in soil carbon feedbacks to climate change. Nat Clim Chang 6(8):751–758

Brevik E, Cerdà A, Mataix-Solera J, Pereg L, Quinton J, Six J, Van Oost K (2015) The interdisciplinary nature of soil. Soil 1(1):117–129

Briones MJI, McNamara NP, Poskitt J, Crow SE, Ostle NJ (2014) Interactive biotic and abiotic regulators of soil carbon cycling: evidence from controlled climate experiments on peatland and boreal soils. Glob Chang Biol 20(9):2971–2982

Bünemann EK, Bongiorno G, Bai Z, Creamer RE, De Deyn G, de Goede R, Fleskens L, Geissen V, Kuyper TW, Mäder P (2018) Soil quality–a critical review. Soil Biol Biochem 120:105–125

Burns RG, DeForest JL, Marxsen J, Sinsabaugh RL, Stromberger ME, Wallenstein MD, Weintraub MN, Zoppini A (2013) Soil enzymes in a changing environment: current knowledge and future directions. Soil Biol Biochem 58:216–234

Caldeira K, Wickett ME (2005) Ocean model predictions of chemistry changes from carbon dioxide emissions to the atmosphere and ocean. J Geophys Res Oceans 110(C9):C002671

Cantón R, González-Alba JM, Galán JC (2012) CTX-M enzymes: origin and diffusion. Front Microbiol 3:110

Cao J, Elliott D, W-x Z (2005) Perchlorate reduction by nanoscale iron particles. J Nanopart Res 7 (4-5):499–506

Carrell AA, Kolton M, Glass JB, Pelletier DA, Warren MJ, Kostka JE, Iversen CM, Hanson PJ, Weston DJ (2019) Experimental warming alters the community composition, diversity, and N2 fixation activity of peat moss (Sphagnum fallax) microbiomes. Glob Chang Biol 25 (9):2993–3004

Castro SP, Cleland EE, Wagner R, Al Sawad R, Lipson DA (2019) Soil microbial responses to drought and exotic plants shift carbon metabolism. ISME J 13(7):1776–1787

Change IC (2007) The physical science basis. Cambridge Univ Press, Cambridge

Cheng L, Booker FL, Tu C, Burkey KO, Zhou L, Shew HD, Rufty TW, Hu S (2012) Arbuscular mycorrhizal fungi increase organic carbon decomposition under elevated CO_2. Science 337 (6098):1084–1087

Ciais P, Sabine C, Bala G, Bopp L, Brovkin V, Canadell J, Chhabra A, DeFries R, Galloway J, Heimann M (2014) Carbon and other biogeochemical cycles. In: Stocker TF, Qin D, Plattner GK, Tignor M, Allen SK, Boschung J, Nauels A, Xia Y, Bex V, Midgley PM (eds) Climate change 2013: the physical science basis. Contribution of working group I to the fifth assessment report of the intergovernmental panel on climate change. Cambridge University Press, New York, pp 465–570

Classen AT, Sundqvist MK, Henning JA, Newman GS, Moore JA, Cregger MA, Moorhead LC, Patterson CM (2015) Direct and indirect effects of climate change on soil microbial and soil microbial-plant interactions: what lies ahead? Ecosphere 6(8):1–21

Coleman DC, Callaham MA, Crossley D Jr (2017) Fundamentals of soil ecology. Academic Press, Cambridge, MA

Cotrufo MF, Wallenstein MD, Boot CM, Denef K, Paul E (2013) The M icrobial E fficiency-M atrix S tabilization (MEMS) framework integrates plant litter decomposition with soil organic matter stabilization: do labile plant inputs form stable soil organic matter? Glob Chang Biol 19 (4):988–995

Crowther TW, Todd-Brown KE, Rowe CW, Wieder WR, Carey JC, Machmuller MB, Snoek B, Fang S, Zhou G, Allison SD (2016) Quantifying global soil carbon losses in response to warming. Nature 540(7631):104–108

Das SK, Varma A (2010) Role of enzymes in maintaining soil health. In: Shukla G, Varma A (eds) Soil enzymology. Springer, Cham, pp 25–42

Davidson EA, Janssens IA (2006) Temperature sensitivity of soil carbon decomposition and feedbacks to climate change. Nature 440(7081):165–173

De Vries FT, Liiri ME, Bjørnlund L, Bowker MA, Christensen S, Setälä HM, Bardgett RD (2012) Land use alters the resistance and resilience of soil food webs to drought. Nat Clim Chang 2 (4):276–280

DeAngelis KM, Pold G, Topçuoğlu BD, van Diepen LT, Varney RM, Blanchard JL, Melillo J, Frey SD (2015) Long-term forest soil warming alters microbial communities in temperate forest soils. Front Microbiol 6:104

Debbarma P, Raghuwanshi S, Singh J, Suyal DC, Zaidi MGH, Goel R (2017) Comparative in situ biodegradation studies of polyhydroxybutyrate film composites. 3Biotech 7(178):1–9

Deiner K, Walser J-C, Mächler E, Altermatt F (2015) Choice of capture and extraction methods affect detection of freshwater biodiversity from environmental DNA. Biol Conserv 183:53–63

Doetterl S, Stevens A, Six J, Merckx R, Van Oost K, Pinto MC, Casanova-Katny A, Muñoz C, Boudin M, Venegas EZ (2015) Soil carbon storage controlled by interactions between geochemistry and climate. Nat Geosci 8(10):780–783

Dubey A, Malla MA, Khan F, Chowdhary K, Yadav S, Kumar A, Sharma S, Khare PK, Khan ML (2019) Soil microbiome: a key player for conservation of soil health under changing climate. Biodivers Conserv 28(8):2405–2429

Dunbar J, Eichorst SA, Gallegos-Graves LV, Silva S, Xie G, Hengartner N, Evans RD, Hungate BA, Jackson RB, Megonigal JP (2012) Common bacterial responses in six ecosystems exposed to 10 years of elevated atmospheric carbon dioxide. Environ Microbiol 14(5):1145–1158

Durán P, Thiergart T, Garrido-Oter R, Agler M, Kemen E, Schulze-Lefert P, Hacquard S (2018) Microbial interkingdom interactions in roots promote Arabidopsis survival. Cell 175 (4):973–983.e914

Dutta H, Dutta A (2016) The microbial aspect of climate change. Energy Ecol Environ 1 (4):209–232

Ediriweera DS, Kasturiratne A, Pathmeswaran A, Gunawardena NK, Wijayawickrama BA, Jayamanne SF, Isbister GK, Dawson A, Giorgi E, Diggle PJ (2016) Mapping the risk of snakebite in Sri Lanka-a national survey with geospatial analysis. PLoS Neg Trop Dis 10(7): e0004813

En H, Gupta RK, Lou F, Moon SH (2019) Role of microbes in environmental sustainability and food preservation. In: Chandra R, Sobti RC (eds) Microbes for sustainable development and bioremediation. CRC Press, Boca Raton, pp 325–336

Erisman J, Galloway J, Seitzinger S, Bleeker A, Dise NB, Petrescu AR, Leach AM, de Vries W (2013) Consequences of human modification of the global nitrogen cycle. Philos Trans R Soc Lond Ser B Biol Sci 368:20130116

Etesami H, Beattie GA (2017) Plant-microbe interactions in adaptation of agricultural crops to abiotic stress conditions. In: Kumar V, Kumar M, Sharma S, Prasad R (eds) Probiotics and plant health. Springer, Singapore, pp 163–200

Farooqi ZUR, Sabir M, Zeeshan N, Naveed K, Hussain MM (2018) Enhancing carbon sequestration using organic amendments and agricultural practices. In: Agarwal RK (ed) Carbon capture, utilization and sequestration. IntechOpen, London, pp 17–35

Farooqi ZUR, Sabir M, ZiaurRehman M, Hussain MM (2020) Mitigation of climate change through carbon sequestration in agricultural soils. In: Raj A, Jhariya MK, Yadav DK, Banerjee A (eds) Climate change and agroforestry systems: adaptation and mitigation strategies. CRC Press, Boca Raton, pp 87–118

Faust K, Raes J (2012) Microbial interactions: from networks to models. Nature Rev Microbiol 10 (8):538–550

Feng G, Zhang F, Li X, Tian C, Tang C, Rengel Z (2002) Improved tolerance of maize plants to salt stress by arbuscular mycorrhiza is related to higher accumulation of soluble sugars in roots. Mycorrhiza 12(4):185–190

Finley RL, Collignon P, Larsson DJ, McEwen SA, Li X-Z, Gaze WH, Reid-Smith R, Timinouni M, Graham DW, Topp E (2013) The scourge of antibiotic resistance: the important role of the environment. Clin Infect Dis 57(5):704–710

Frey SD, Lee J, Melillo JM, Six J (2013) The temperature response of soil microbial efficiency and its feedback to climate. Nat Clim Change 3(4):395–398

Gadd GM (2007) Geomycology: biogeochemical transformations of rocks, minerals, metals and radionuclides by fungi, bioweathering and bioremediation. Myco Res 111(1):3–49

Gao D, Hagedorn F, Zhang L, Liu J, Qu G, Sun J, Peng B, Fan Z, Zheng J, Jiang P (2018) Small and transient response of winter soil respiration and microbial communities to altered snow depth in a mid-temperate forest. Appl Soil Ecol 130:40–49

Giri K, Rai JPN, Pandey S, Mishra G, Kumar R, Suyal DC (2017) Performance evaluation of isoproturon-degrading indigenous bacterial isolates in soil microcosm. Chem Ecol 33 (9):817–825

Glibert PM (2017) Eutrophication, harmful algae and biodiversity—challenging paradigms in a world of complex nutrient changes. Marine Poll Bull 124(2):591–606

Glick BR (2014) Bacteria with ACC deaminase can promote plant growth and help to feed the world. Microbiol Res 169(1):30–39

Gomez E, Ferreras L, Toresani S (2006) Soil bacterial functional diversity as influenced by organic amendment application. Bioresour Technol 97(13):1484–1489

Griffiths BS, Philippot L (2013) Insights into the resistance and resilience of the soil microbial community. FEMS Microbiol Rev 37(2):112–129

Harms H, Schlosser D, Wick LY (2011) Untapped potential: exploiting fungi in bioremediation of hazardous chemicals. Nat Rev Microbiol 9(3):177–192

He Z-l, Yang X-e (2007) Role of soil rhizobacteria in phytoremediation of heavy metal contaminated soils. J Zhejiang Univ Sci B 8(3):192–207

He J, Shen J, Lm Z, Zhu Y, Zheng Y, Xu M, Di H (2007) Quantitative analyses of the abundance and composition of ammonia-oxidizing bacteria and ammonia-oxidizing archaea of a Chinese upland red soil under long-term fertilization practices. Environ Microbiol 9(9):2364–2374

Hoegh-Guldberg O, Bruno JF (2010) The impact of climate change on the world's marine ecosystems. Science 328(5985):1523–1528

Houghton J (2001) Climate change 2001: the scientific basis. Cambridge Univ Press, Cambridge

Hu J, Lin X, Wang J, Dai J, Chen R, Zhang J, Wong MH (2011) Microbial functional diversity, metabolic quotient, and invertase activity of a sandy loam soil as affected by long-term application of organic amendment and mineral fertilizer. J Soils Sediment 11(2):271–280

Hussain MM, Farooqi ZUR, Latif J, Mubarak MU, Younas F (2021) Bioremediation: a green solution to avoid pollution of the environment. In: Soil bioremediation: an approach towards sustainable technology. Willey and Sons, Hoboken, NJ, pp 15–40

Jenouvrier S, Péron C, Weimerskirch H (2015) Extreme climate events and individual heterogeneity shape life-history traits and population dynamics. Ecol Monogr 85(4):605–624

Jiao N, Herndl GJ, Hansell DA, Benner R, Kattner G, Wilhelm SW, Kirchman DL, Weinbauer MG, Luo T, Chen F (2010) Microbial production of recalcitrant dissolved organic matter: long-term carbon storage in the global ocean. Nat Rev Microbiol 8(8):593–599

Joshi D, Chandra R, Suyal DC, Kumar S, Goel R (2019) Impact of bioinoculants Pseudomonas jesenii MP1 and Rhodococcus qingshengii S10107 on Cicer arietinum yield and soil nitrogen status. Pedosphere 29(3):388–399

Juwarkar AA, Singh SK, Mudhoo A (2010) A comprehensive overview of elements in bioremediation. Rev Environ Sci Bio/Technol 9(3):215–288

Kaisermann A, de Vries FT, Griffiths RI, Bardgett RD (2017) Legacy effects of drought on plant–soil feedbacks and plant–plant interactions. New Phytol 215(4):1413–1424

Keesstra SD, Bouma J, Wallinga J, Tittonell P, Smith P, Cerdà A, Montanarella L, Quinton JN, Pachepsky Y, Van Der Putten WH (2016) The significance of soils and soil science towards realization of the United Nations sustainable development goals. Soil 2(2):111–128

Knutson TR, McBride JL, Chan J, Emanuel K, Holland G, Landsea C, Held I, Kossin JP, Srivastava A, Sugi M (2010) Tropical cyclones and climate change. Nat Geosci 3(3):157–163

Kumar A, Pandey AK, Singh SS, Shanker R, Dhawan A (2011) Engineered ZnO and TiO2 nanoparticles induce oxidative stress and DNA damage leading to reduced viability of Escherichia coli. Free Radic Biol Med 51(10):1872–1881

Lal R (2016) Soil health and carbon management. Food Energy Secur 5(4):212–222

Langley JA, Hungate BA (2014) Plant community feedbacks and long-term ecosystem responses to multi-factored global change. AoB Plants 6:plu035

Leff JW, Jones SE, Prober SM, Barberán A, Borer ET, Firn JL, Harpole WS, Hobbie SE, Hofmockel KS, Knops JM (2015) Consistent responses of soil microbial communities to elevated nutrient inputs in grasslands across the globe. Proc Nat Acad Sci 112(35):10967–10972

Lehmann J, Kleber M (2015) The contentious nature of soil organic matter. Nature 528 (7580):60–68

Li H, Guevara N, Herndon N, Caragea D, Neppalli K, Caragea C, Tapia AH (2015) Twitter mining for disaster response: a domain adaptation approach. In ISCRAM

Li Y, Lv W, Jiang L, Zhang L, Wang S, Wang Q, Xue K, Li B, Liu P, Hong H (2019) Microbial community responses reduce soil carbon loss in Tibetan alpine grasslands under short-term warming. Glob Chang Biol 25(10):3438–3449

Li Z, Zeng Z, Tian D, Wang J, Fu Z, Zhang F, Zhang R, Chen W, Luo Y, Niu S (2020) Global patterns and controlling factors of soil nitrification rate. Glob Chang Biol 26(7):4147–4157. https://doi.org/10.1111/gcb.15119

Ling N, Zhu C, Xue C, Chen H, Duan Y, Peng C, Guo S, Shen Q (2016) Insight into how organic amendments can shape the soil microbiome in long-term field experiments as revealed by network analysis. Soil Biol Biochem 99:137–149

Liu X-JA, Sun J, Mau RL, Finley BK, Compson ZG, van Gestel N, Brown JR, Schwartz E, Dijkstra P, Hungate BA (2017) Labile carbon input determines the direction and magnitude of the priming effect. Appl Soil Ecol 109:7–13

Lladó S, López-Mondéjar R, Baldrian P (2017) Forest soil bacteria: diversity, involvement in ecosystem processes, and response to global change. Microbiol Mol Biol Rev 81(2):e00063–e00016

Luo Y, Wan S, Hui D, Wallace LL (2001) Acclimatization of soil respiration to warming in a tall grass prairie. Nature 413(6856):622–625

Maharudrappa A, Srinivasamurthy C, Nagaraja M, Siddaramappa R, Anand H (2000) Decomposition rates of litter and nutrient release pattern in a tropical soil. J Indian Soc Soil Sci 48(1):92–97

Makoi JH, Ndakidemi PA (2008) Selected soil enzymes: examples of their potential roles in the ecosystem. Afr J Biotechnol 7(3):181–191

Mangodo C, Adeyemi T, Bakpolor V, Adegboyega D (2020) Impact of microorganisms on climate change: a review. World News Nat Sci 31:36–47

Marín-Spiotta E, Gruley K, Crawford J, Atkinson E, Miesel J, Greene S, Cardona-Correa C, Spencer R (2014) Paradigm shifts in soil organic matter research affect interpretations of aquatic carbon cycling: transcending disciplinary and ecosystem boundaries. Biogeochemistry 117 (2-3):279–297

Mariotte P, Robroek BJ, Jassey VE, Buttler A (2015) Subordinate plants mitigate drought effects on soil ecosystem processes by stimulating fungi. Funct Ecol 29(12):1578–1586

McCormack ML, Dickie IA, Eissenstat DM, Fahey TJ, Fernandez CW, Guo D, Helmisaari HS, Hobbie EA, Iversen CM, Jackson RB (2015) Redefining fine roots improves understanding of below-ground contributions to terrestrial biosphere processes. New Phytol 207(3):505–518

Meena SK, Rakshit A, Meena VS (2016) Effect of seed bio-priming and N doses under varied soil type on nitrogen use efficiency (NUE) of wheat (*Triticum aestivum* L.) under greenhouse conditions. Biocatal Agric Biotechnol 6:68–75

Meena VS, Meena SK, Verma JP, Kumar A, Aeron A, Mishra PK, Bisht JK, Pattanayak A, Naveed M, Dotaniya M (2017) Plant beneficial rhizospheric microorganism (PBRM) strategies to improve nutrients use efficiency: a review. Ecol Eng 107:8–32

Meerow S, Newell JP, Stults M (2016) Defining urban resilience: a review. Landsc Urban Plan 147: 38–49

Melillo JM, Frey SD, DeAngelis KM, Werner WJ, Bernard MJ, Bowles FP, Pold G, Knorr MA, Grandy AS (2017) Long-term pattern and magnitude of soil carbon feedback to the climate system in a warming world. Science 358(6359):101–105

Mohsin RM (2019) Landscape performance of Callistemon citrinus under environmental stress conditions. Mississippi State University, Starkville, MS

Mohy-Ud-Din W, Irfan M, Younas F, Hussain MM, Shareef A, Muntaha S, Sonia A (2020) Recent advances in rhizoremediation of glyphosate: a review. Plant Environ 1(01):24–34

Muter O, Ingus P, Selga T, Berzins A, Gudra D, Radovica-Spalvina I, Fridmanis D, Bartkevics V (2017) Removal of pharmaceuticals from municipal wastewaters at laboratory scale by treatment with activated sludge and biostimulation. Sci Total Environ 584:402–413

Myers RA, Worm B (2003) Rapid worldwide depletion of predatory fish communities. Nature 423 (6937):280–283

Nath D, Maurya BR, Meena VS (2017) Documentation of five potassium-and phosphorus-solubilizing bacteria for their K and P-solubilization ability from various minerals. Biocata Agric Biotechnol 10:174–181

Naylor D, Sadler N, Bhattacharjee A, Graham EB, Anderton CR, McClure R, Lipton M, Hofmockel KS, Jansson JK (2020) Soil microbiomes under climate change and implications for carbon cycling. Annu Rev Environ Res 45:29–59. https://doi.org/10.1146/annurev-environ-012320-082720

Nicolaou SA, Gaida SM, Papoutsakis ET (2010) A comparative view of metabolite and substrate stress and tolerance in microbial bioprocessing: from biofuels and chemicals, to biocatalysis and bioremediation. Metab Eng 12(4):307–331

Nobu MK, Narihiro T, Kuroda K, Mei R, Liu W-T (2016) Chasing the elusive Euryarchaeota class WSA2: genomes reveal a uniquely fastidious methyl-reducing methanogen. ISME J 10 (10):2478–2487

Okoh AI (2006) Biodegradation alternative in the cleanup of petroleum hydrocarbon pollutants. Biotechnol Mol Biol Rev 1(2):38–50

Pandey J, Chauhan A, Jain RK (2009) Integrative approaches for assessing the ecological sustainability of in situ bioremediation. FEMS Microbiol Rev 33(2):324–375

Paustian K, Lehmann J, Ogle S, Reay D, Robertson GP, Smith P (2016) Climate-smart soils. Nature 532(7597):49–57

Pautasso M, Döring TF, Garbelotto M, Pellis L, Jeger MJ (2012) Impacts of climate change on plant diseases—opinions and trends. Eur J Plant Pathol 133(1):295–313

Pecl GT, Araújo MB, Bell JD, Blanchard J, Bonebrake TC, Chen I-C, Clark TD, Colwell RK, Danielsen F, Evengård B (2017) Biodiversity redistribution under climate change: impacts on ecosystems and human well-being. Science 355(6332):eaai9214

Piazza G, Ercoli L, Nuti M, Pellegrino E (2019) Interaction between conservation tillage and nitrogen fertilization shapes prokaryotic and fungal diversity at different soil depths: evidence from a 23-year field experiment in the Mediterranean area. Front Microbiol 10:2047

Pii Y, Mimmo T, Tomasi N, Terzano R, Cesco S, Crecchio C (2015) Microbial interactions in the rhizosphere: beneficial influences of plant growth-promoting rhizobacteria on nutrient acquisition process. A review. Biol Fertil Soils 51(4):403–415

Prichard E, Granek EF (2016) Effects of pharmaceuticals and personal care products on marine organisms: from single-species studies to an ecosystem-based approach. Environ Sci Pollut Res 23(22):22365–22384

Raich JW, Potter CS, Bhagawati D (2002) Interannual variability in global soil respiration, 1980–94. Glob Chang Biol 8(8):800–812

Ramanathan V, Carmichael G (2008) Global and regional climate changes due to black carbon. Nat Geosci 1(4):221–227

Rawat N, Sharma M, Suyal DC, Singh DK, Joshi D, Singh P, Goel R (2019) Psyhcrotolerant bio-inoculants and their co-inoculation to improve Cicer arietinum growth and soil nutrient status for sustainable mountain agriculture. J Soil Sci Plant Nutr 19(3):639–647

Rocca JD, Simonin M, Blaszczak JR, Ernakovich JG, Gibbons SM, Midani FS, Washburne AD (2019) The microbiome stress project: toward a global meta-analysis of environmental stressors and their effects on microbial communities. Front Microbiol 9:3272

Robertson GP, Paul EA, Harwood RR (2000) Greenhouse gases in intensive agriculture: contributions of individual gases to the radiative forcing of the atmosphere. Science 289 (5486):1922–1925

Rodrigo A, Recous S, Neel C, Mary B (1997) Modelling temperature and moisture effects on C–N transformations in soils: comparison of nine models. Ecol Model 102(2–3):325–339

Rouphael Y, Franken P, Schneider C, Schwarz D, Giovannetti M, Agnolucci M, De Pascale S, Bonini P, Colla G (2015) Arbuscular mycorrhizal fungi act as biostimulants in horticultural crops. Sci Hortic 196:91–108

Sahu N, Vasu D, Sahu A, Lal N, Singh S (2017a) Strength of microbes in nutrient cycling: a key to soil health. In: Meena V, Mishra P, Bisht J, Pattanayak A (eds) Agriculturally important microbes for sustainable agriculture. Springer, Singapore, pp 69–86

Sahu N, Vasu D, Sahu A, Lal N, Singh SK (2017b) Strength of microbes in nutrient cycling: a key to soil health. In: Meena VS, Mishra PK, Bisht JK, Pattanayak A (eds) Agriculturally important microbes for sustainable agriculture, Plant-soil-microbe nexus, vol I. Springer, Singapore, pp 69–86

Sardans J, Rivas-Ubach A, Peñuelas J (2012) The C: N: P stoichiometry of organisms and ecosystems in a changing world: a review and perspectives. Perspect Plant Ecol Evol Syst 14 (1):33–47

Saunois M, Bousquet P, Poulter B, Peregon A, Ciais P, Canadell JG, Dlugokencky EJ, Etiope G, Bastviken D, Houweling S (2016) The global methane budget 2000–2012. Earth Syst Sci Data 8 (2):697–751

Schjonning P, van den Akker JJ, Keller T, Greve MH, Lamande M, Simojoki A, Stettler M, Arvidsson J, Breuning-Madsen H (2015) Driver-pressure-state-impact-response (DPSIR) analysis and risk assessment for soil compaction—a European perspective. In: Sparks DL (ed) Advances in agronomy, vol 133. Elsevier, Amsterdam, pp 183–237

Schmidt MW, Torn MS, Abiven S, Dittmar T, Guggenberger G, Janssens IA, Kleber M, Kögel-Knabner I, Lehmann J, Manning DA (2011) Persistence of soil organic matter as an ecosystem property. Nature 478(7367):49–56

Schulz S, Brankatschk R, Dümig A, Kögel-Knabner I, Schloter M, Zeyer J (2013) The role of microorganisms at different stages of ecosystem development for soil formation. Biogeosciences 10(6):3983–3996

Schuur EA, McGuire AD, Schädel C, Grosse G, Harden J, Hayes DJ, Hugelius G, Koven CD, Kuhry P, Lawrence DM (2015) Climate change and the permafrost carbon feedback. Nature 520 (7546):171–179

Sharma SB, Gobi TA (2016) Impact of drought on soil and microbial diversity in different agroecosystems of the semiarid zones. In: Hakeem K, Akhtar M, Abdullah S (eds) Plant, soil and microbes. Springer, Cham, pp 149–162

Sharma SB, Sayyed RZ, Trivedi MH, Gobi TA (2013) Phosphate solubilizing microbes: sustainable approach for managing phosphorus deficiency in agricultural soils. Springerplus 2(1):587

Sharma G, Singh S, Singh T, Kaur A, Malhotra S, Sharma S (2018) A brief insight in to impact of temperature rise due to climate change on soil microflora. Int J Theor Appl Sci 10(1):41–45

Shrivastava M, Srivastava P, D'Souza S (2016) KSM soil diversity and mineral solubilization, in relation to crop production and molecular mechanism. In: Meena V, Maurya B, Verma J, Meena R (eds) Potassium solubilizing microorganisms for sustainable agriculture. Springer, New Delhi, pp 221–234

Singh JS (2015) Microbes: the chief ecological engineers in reinstating equilibrium in degraded ecosystems. Agric Ecosyst Environ 203:80–82

Singh BK, Bardgett RD, Smith P, Reay DS (2010) Microorganisms and climate change: terrestrial feedbacks and mitigation options. Nat Rev Microbiol 8(11):779–790

Six J, Frey S, Thiet R, Batten K (2006) Bacterial and fungal contributions to carbon sequestration in agroecosystems. Soil Sci Soc Am J 70(2):555–569

Smith P, Martino D, Cai Z, Gwary D, Janzen H, Kumar P, McCarl B, Ogle S, O'Mara F, Rice C (2008) Greenhouse gas mitigation in agriculture. Philos Trans R Soc B Biol Sci 363 (1492):789–813

Spagnoletti FN, Balestrasse K, Lavado RS, Giacometti R (2016) Arbuscular mycorrhiza detoxifying response against arsenic and pathogenic fungus in soybean. Ecotoxicol Environ Saf 133:47–56

Stark JM, Firestone MK (1995) Mechanisms for soil moisture effects on activity of nitrifying bacteria. Appl Environ Microbiol 61(1):218–221

Steffen W, Richardson K, Rockström J, Cornell SE, Fetzer I, Bennett EM, Biggs R, Carpenter SR, De Vries W, De Wit CA (2015) Planetary boundaries: guiding human development on a changing planet. Science 347(6223):1259855

Sudesh K, Iwata T (2008) Sustainability of biobased and biodegradable plastics. CLEAN Soil Air Water 36(5/6):433–442

Takoutsing B, Weber J, Aynekulu E, Martín JAR, Shepherd K, Sila A, Tchoundjeu Z, Diby L (2016) Assessment of soil health indicators for sustainable production of maize in smallholder farming systems in the highlands of Cameroon. Geoderma 276:64–73

Teotia P, Kumar V, Kumar M, Shrivastava N, Varma A (2016) Rhizosphere microbes: potassium solubilization and crop productivity–present and future aspects. In: Meena V, Maurya B, Verma J, Meena R (eds) Potassium solubilizing microorganisms for sustainable agriculture. Springer, New Delhi, pp 315–325

Terrer C, Vicca S, Hungate BA, Phillips RP, Prentice IC (2016) Mycorrhizal association as a primary control of the CO2 fertilization effect. Science 353(6294):72–74

Todd-Brown K, Randerson J, Post W, Hoffman F, Tarnocai C, Schuur E, Allison S (2013) Causes of variation in soil carbon simulations from CMIP5 earth system models and comparison with observations. Biogeosciences 10(3):1717–1736

Umar I, Wali V, Kher R, Sharma A (2008) Impact of integrated nutrient management on strawberry yield and soil nutrient status. Appl Biol Res 10(1 and 2):22–25

Van der Putten WH (2012) Climate change, aboveground-belowground interactions, and species' range shifts. Annu Rev Ecol Evol Syst 43:365–383

Van der Putten P (2013) The European Union and decreasing regional stability in East and South-East Asia. The Norwegian Peacebuilding Resource Centre. https://www.clingendael.org/sites/default/files/pdfs/20130326_noref_putten.pdf

van der Putten WH, Bradford MA, Pernilla Brinkman E, van de Voorde TF, Veen G (2016) Where, when and how plant–soil feedback matters in a changing world. Fun Ecol 30(7):1109–1121

Vayssier-Taussat M, Albina E, Citti C, Cosson JF, Jacques M-A, Lebrun M-H, Le Loir Y, Ogliastro M, Petit M-A, Roumagnac P (2014) Shifting the paradigm from pathogens to pathobiome: new concepts in the light of meta-omics. Front Cell Infect Microbiol 4:29

Vimal SR, Singh JS, Arora NK, Singh S (2017) Soil-plant-microbe interactions in stressed agriculture management: a review. Pedosphere 27(2):177–192

von Rein I, Gessler A, Premke K, Keitel C, Ulrich A, Kayler ZE (2016) Forest understory plant and soil microbial response to an experimentally induced drought and heat-pulse event: the importance of maintaining the continuum. Glob Chang Biol 22(8):2861–2874

Wang P, Marsh EL, Ainsworth EA, Leakey AD, Sheflin AM, Schachtman DP (2017) Shifts in microbial communities in soil, rhizosphere and roots of two major crop systems under elevated CO 2 and O 3. Sci Rep 7(1):1–12

Webb HE, Brichta-Harhay DM, Brashears MM, Nightingale KK, Arthur TM, Bosilevac JM, Kalchayanand N, Schmidt JW, Wang R, Granier SA (2017) Salmonella in peripheral lymph nodes of healthy cattle at slaughter. Front Microbiol 8:2214

Weiman S (2015) Microbes help to drive global carbon cycling and climate change. Microbe Mag 10(6):233–238

Wright CK, Wimberly MC (2013) Recent land use change in the Western Corn Belt threatens grasslands and wetlands. Proc Natl Acad Sci 110(10):4134–4139

Xia L, Lam SK, Yan X, Chen D (2017) How does recycling of livestock manure in agroecosystems affect crop productivity, reactive nitrogen losses, and soil carbon balance? Environ Sci Technol 51(13):7450–7457

Xu Z, Yu G, Zhang X, Ge J, He N, Wang Q, Wang D (2015) The variations in soil microbial communities, enzyme activities and their relationships with soil organic matter decomposition along the northern slope of Changbai Mountain. Appl Soil Ecol 86:19–29

Yang S, Zheng Q, Yuan M, Shi Z, Chiariello NR, Docherty KM, Dong S, Field CB, Gu Y, Gutknecht J (2019) Long-term elevated CO2 shifts composition of soil microbial communities in a Californian annual grassland, reducing growth and N utilization potentials. Sci Total Environ 652:1474–1481

Zablocki O, Adriaenssens EM, Cowan D (2016) Diversity and ecology of viruses in hyperarid desert soils. Appl Environ Microbiol 82(3):770–777

Zechmeister-Boltenstern S, Keiblinger KM, Mooshammer M, Peñuelas J, Richter A, Sardans J, Wanek W (2015) The application of ecological stoichiometry to plant–microbial–soil organic matter transformations. Ecol Monogr 85(2):133–155

Zeng G, Yu M, Chen Y, Huang D, Zhang J, Huang H, Jiang R, Yu Z (2010) Effects of inoculation with Phanerochaete chrysosporium at various time points on enzyme activities during agricultural waste composting. Bioresour Technol 101(1):222–227

Zhang L, Xie Z, Zhao R, Zhang Y (2018) Plant, microbial community and soil property responses to an experimental precipitation gradient in a desert grassland. Appl Soil Ecol 127:87–95

Zhou J, Xue K, Xie J, Deng Y, Wu L, Cheng X, Fei S, Deng S, He Z, Van Nostrand JD (2012) Microbial mediation of carbon-cycle feedbacks to climate warming. Nat Clim Chang 2 (2):106–110

Zimmer C (2010) The microbe factor and its role in our climate future. Retrieved 16 Oct 2020

Zimmerman L, Labonte B (2015) Climate change and the microbial methane banquet. Clim Alert 27:1–6

Biotic Stress Management by Microbial Interactions in Soils

13

S. Raghu, M. S. Baite, M. K. Yadav, S. R. Prabhukarthikeyan, U. Keerthana, and P. Pati

Abstract

Global agriculture is facing numerous challenges under changing climatic conditions. The production and productivity of a nation directly indicate its food security and sustainability. There are numerous constraints such as biotic and abiotic stresses continuously acting as major obstacles for productivity and quality of food. Post-green revolution era has witnessed tremendous success in food production and at the same time it also brought some serious negative effects such as increased insect, diseases, nematodes, soil salinity, alkalinity, and other related problems. Since the day we started using inorganic fertilizers and intensive use of chemical pesticides, the soil fertility started deteriorating and causing environmental pollution along with residual effect in soil and water. As a result, the soil we grow food, the food we eat daily is contaminated and the residues of pesticides are entering human and animal body through food chain and causing serious health issues. Hence, it is a high-time to wake up and start extensive adoption of beneficial and ecofriendly agriculture production. In this situation, microbes become potential agents which can bring positive change in food productions. Microbes can be used in a number of ways for a number of processes including biotic stress management in crops. Application of plant growth promoting microbes (PGPMs) will enhance the growth by producing plant hormones, helping in mineral nutrition cycling and other mechanisms. They

S. Raghu (✉) · M. S. Baite · M. K. Yadav · S. R. Prabhukarthikeyan · U. Keerthana
Crop Protection Division, ICAR-National Rice Research Institute (NRRI), Cuttack, Odisha, India
e-mail: s.raghu@icar.gov.in

P. Pati
Crop Protection Division, ICAR-National Rice Research Institute (NRRI), Cuttack, Odisha, India

Department of Agricultural Entomology, Institute of Agricultural Sciences (IAS), Siksha 'O' Anusandhan (SOA) Deemed to be University, Bhubaneswar, Odisha, India

R. Soni et al. (eds.), *Microbiological Activity for Soil and Plant Health Management*, https://doi.org/10.1007/978-981-16-2922-8_13

also induce systemic acquired resistance (SAR) and induced systemic resistance (ISR) against various plant pathogenic microbes. The sustainable exploration of plant–microbe interactions will help in solving problems of agriculture and ultimately sustainable agricultural production. In this chapter, we will discuss the source of biocontrol agents, their potential role in mitigating biotic and abiotic stresses and enhancing soil health.

Keywords

Plant–microbe interaction · Biotic and abiotic stresses · Soil health · Sustainable production

13.1 Introduction

The production and productivity of the crops have increased to many folds over last few decades. The food consumption pattern has been changed over the years and accordingly the choice of crop grown in a particular area is also changed. India, being an agrarian country has achieved self-sufficiency in food production and food security due to many folds increase in food production. Agriculture contributes to the nation's GDP is quite significant. More than 65% people in India are directly or indirectly involved in agriculture which means it provides employment to people directly and indirectly depending on it. India achieved a record food production of 284 MT during 2019–20 (www.indiaagristat.com). Thanks to the research and innovations being carried out on various aspects of agriculture and allied sectors with an aim of increase in food production. Intensive agricultural practices like introduction of high yielding input use efficient varieties increased the production and productivity to many folds. But, this intensification of agriculture has the potential of both blessing and curse, depending on how it is done at the farmer's level. Intensive agriculture brought various new challenges like biotic and abiotic stresses, reduction in flora and fauna diversity, water scarcity, degraded lands, etc. Among the biotic constraints, pest, weeds, and disease are most important barriers to production. They are known to reduce the productivity of food crops, leading to low efficiencies in input use, quality losses, suppressed crop growth, and ultimately reduced food production and productivity (Strange and Scott 2005; Gregory et al. 2005; Lal 2009; Waddington et al. 2010; Knox et al. 2012). They can cause 20–35% of losses based on environmental conditions, varietal susceptibility, and other factors (Savary et al. 2012; Kumar 2014). History of agriculture has witnessed a number of disease epidemics starting with "*The Great Irish Famine*" between 1845 and 1849 in Ireland which was due to late blight disease of potato caused by *Phytopthora infestans* which destroyed potato crop resulting in death of over one million people due to starvation. Another classical example from India is "*The Great Bengal Famine*" in rice during 1943–1944. This was due to the brown spot disease of rice caused by *Helminthosporium oryzae* which devastated rice crop leading to the death of over 2.1–3 million people. Similarly, number of diseases caused epidemics in past with significant crop losses thereby affected country's economy. Reducing crop losses from biotic stress is becoming of increasing importance as in the present situation of challenges to meet growing world population. Management of the pest and diseases should be prioritized considering the magnitude of losses they cause (Raghu et al. 2020). There are numbers of strategies being employed to manage the pest and diseases starting from varietal resistance, cultural management, optimization of plant density, nutrient management, use of biological control agents, and finally synthetic chemicals. Biological control plays a pivotal role in sustainable pest management in agricultural ecosystem in interaction with other components of ecosystem. This chapter discusses microbe mediated biotic stress management with special emphasis on source of isolation, characterization, host–microbe interaction, evaluation of potential agents, commercial production, and management of soilborne diseases.

13.2 Important SoilBorne Diseases and Losses in Different Crops

Crop plants continuously cope up with the rapid fluctuations in environmental adversities which in turn impair their metabolic capabilities (Simontacchi et al. 2015). Any serious/observable changes in environmental conditions can cause infection of number of pest and diseases. Some of the diseases can emerge as more severe and many minor pest and diseases can emerge as major problems (Raghu et al. 2018). These pest and diseases reduce crop productivity in various ways either directly or indirectly. Some of the diseases causing agents are fungi, bacteria, viruses, plant parasitic nematodes, mycoplasm, spiroplasma, viroids, and actinomycetes (Fig. 13.1; Agriose 2005). On the other hand, crops also infested by number of insect pests like chewing, sucking pests, borers, defoliators, etc., (Vasantaraj and Ramamurthy 2016). Crop losses due to pests and diseases pose a major threat to rural families and food security worldwide (Cerda et al. 2017). Considering an average crop loss of 20% due to pest and diseases, the loss comes up to 1,40,000 crore (7 lakh crore of gross income by the agriculture) which is colossal (Kumar and Gupta 2012). At least 50% of the loss if we minimize, we can save 70,000 crore annually. Table 13.1 gives picture of extent of losses due to major soilborne plant diseases.

13.3 Role of Soil Health in Plant Disease Incidence

The quality of an agricultural soil is most important in performing any cropping activities. Healthy soil provides support for optimum and sustainable plant productivity (Abawi and Widmer 2000; Frac et al. 2018). The sick soil will cause plants that grow on that soil various diseases which significantly decline crop productivity (Huang et al. 2006). The sick soils are not only unfertile but also have an abundance of soil pathogen population which may result in outbreaks of many diseases (Huang

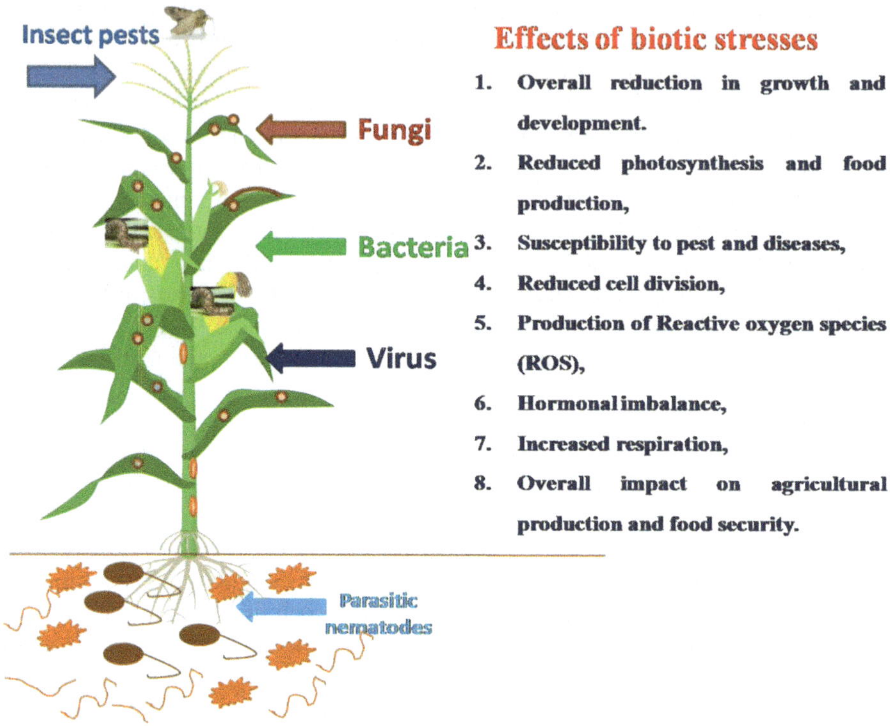

Fig. 13.1 Various biotic stresses of crop plant and their effect on physiological processes

et al. 2006). The physical characters such as texture, structure, chemical characters like pH, organic matter content, soil temperature, and nutrient status will strongly influence the activity of soilborne pathogens and soil health (Fiers et al. 2012). The sick soils will help thc pathogens for their multiplication, survival and causing epidemics in the subsequent seasons. The distribution of pathogens in soil is either horizontal or vertical based on the production practices, cropping history, and other related factors. Along a vertical axis, the pathogen inoculums of most root pathogens lie within the top 10 in. of the soil profile, the layers where the host roots and tissues and other organic substrates found. Management of soilborne diseases should also give great importance to reclamation of problematic soils so that, they become healthy in the way to encourage good crop growth and development (Fig. 13.2).

Table 13.1 Major soilborne diseases of crop plants

S. no.	Crop	Name of the disease	Causal organism	Extent of losses	Reference
1.	Pigeon pea	Fusarium wilt	*Fusarium udum*	22.60%	Kannaiyan et al. (1981)
2.	Chili	Fusarium wilt	*Fusarium solani*	13.40%	Madhukar (2001)
3.	Chick pea	Wilt complex	*Fusarium oxysporum f. sp. ciceris*	12.26%	Nikahm et al. (2008)
4.	Chili	Wilt complex	*Fusarium solani*	5–57%	Raghu (2014)
5.	Chili	Bacterial wilt	*Ralstonia solanacearum*	26–32%	Umesha et al. (2005)
6.	Vegetables	Damping off	*Pythium spp.*		
7.	Ginger	Wilt complex	*Fusarium solani, Ralstonia solanacearum, Sclerotium rolfsii, Meloidogyne incognita*	22–55%	Raghu (2011)
8.	Tomato	Fusarium wilt	*Fusarium oxysporum sp. lycopersici.*	10–90%	Singh and Kamal (2012)
9.	Rice	Sheath blight	*Rhizoctonia solani*	50%	Wu et al. (2013)
10.	Potato	Bacterial wilt	*Ralstonia solanacearum*	33–99%	Karim et al. (2018)
11.	Potato	Sclerotium wilt	*Sclerotium rolfsii*	20–45%	Basamma (2008)
12.	Groundnut	Root rot	*Sclerotium rolfsii*	20–34%	Ranganathswamy et al. (2014)
13.	Cotton	Verticillium wilt	*Verticillium dahliae*	20–45%	Wheeler et al. (2010)
14.	Soybean	Wilt complex	*Fusarium solani, Ralstonia solanacearum, Sclerotium rolfsii, Rhizoctonia bataticola*	15–40%	Sangeetha (2011).
15.	Capsicum	Damping off	*Pythium* spp. and *Rhizoctonia* spp.	62–90%	Majeed et al. (2018)
16.	Capsicum	Verticillium wilt	*Verticillium dahliae*	–	–
17.	Tomato	Phytopthora root rot	Phytopthora spp.	100%	Nowicki et al. (2012)
18.	Brinjal	Fusarium wilt		75–81%	Pandey and Dubey (2017)

(continued)

Table 13.1 (continued)

S. no.	Crop	Name of the disease	Causal organism	Extent of losses	Reference
			Fusarium oxysporum f. sp. melonginae		
19.	Coconut	Coconut wilt	*Ganoderma* spp.	5–25%	Snehalatharani et al. (2016)
20.	Pomegranate	Wilt complex	*Ceratocystis fimbriata, Meloidogyne incognita, Fusarium spp.*	45%	Sonyal et al. (2016)

Authors own compilation

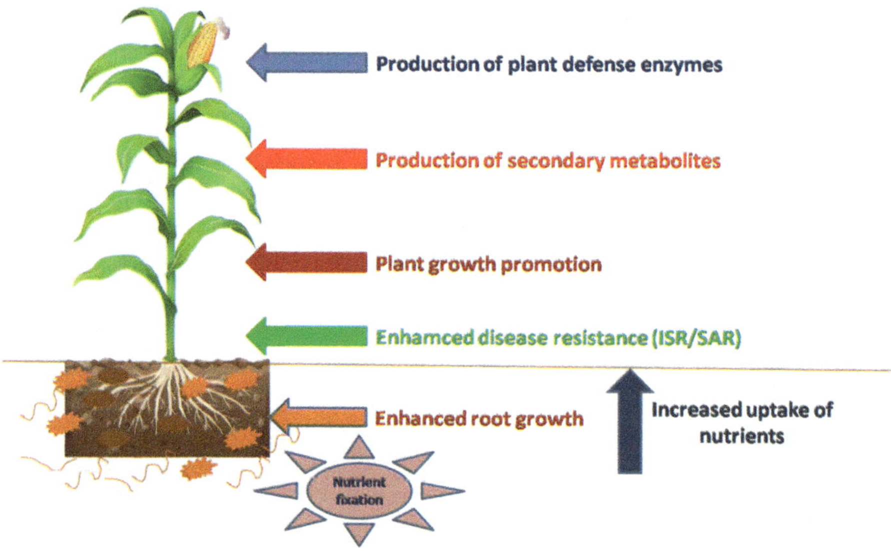

Fig. 13.2 Beneficial role of plant growth promoting microorganisms in soil on enhancing physiological activities in crop plants

13.4 Potential of Biological Agents for Disease Management

The increased reflection on environmental concern over pesticide use has been instrumented in a large upsurge of the biological pest control (Kumar 2014). The strategy of biological control agents usage in plant disease management involves the use of antagonistic microorganisms before or after the infection by insect pest or disease. Based on the type of infection and parts infected the biological control agents can be applied and manage the stress. Biological control agent may be a fungi, bacteria, actinomycetes, or virus which can kill or reduce the severity of pathogenic microbes by one or different mode of actions. Number of bacterial and fungal biocontrol agents is exploited for their beneficial role in plant growth promotion and disease management. The bacterial genera such as *Acetobacter, Achromobacter, Anabaena, Arthrobacter, Azoarcus, Azospirillum, Azotobacter, Bacillus, Burkholderia, Clostridium, Enterobacter, Flavobacterium, Frankia, Hydrogenophaga, Kluyvera, Microcoleus, Phyllobacterium, Pseudomonas, Serratia, Staphylococcus, Streptomyces, and Vibrio* have proven their potential through plant growth promotion and disease reduction. In recent years, fungal biocontrol agents proved their tremendous potential for their beneficial role. They are contributing significantly to the enhancement of crop yield and food production. The fungal BCAs do not cause any harm to the environment. There is no report of development of resistance against fungal biocontrol agents in any of the target pest such as insect, weeds, and pathogens. This is because of their complex mode of action. Because of this reason, they have been proved to be alternative against undesirable use of chemical pesticides (Savita 2019). Fungal biocontrol agents like *Trichoderma, Metarhizium anisopliae, Beauveria bassiana, Aspergillus spp. Culicinomyces, Hirsutella, Metarhizium, Nomuraea, Paecilomyces, Tolypocladium,* and *Verticillium* have proven their efficacy in controlling various pest and diseases (Jyoti and Singh 2016).

13.5 Mechanism of Action of Biocontrol Agents and Host–BCAs Interaction

The plant–microbe association in rhizosphere involves molecular recognition between both plant and microbe through signaling network mediated by the plant's hormones such as salicylic acid (SA), Jasmonic acid (JA), and ethylene (ET). The beneficial microbes induce systemic resistance against pathogens due to the involvement of signal transduction pathways where jasmonic acid and ethylene are involved. The signal transduction pathway through salicylic acid accumulation is found in the systemic acquired resistance (SAR) induced due to the attack by the pathogenic microbes (Van Wees et al. 2008). Strains belonging to the genera *Azospirillum* (Dobbelaere et al. 2001), *Bacillus* (Kokalis-Burelle et al. 2006), and *Pseudomonas* (Meyer et al. 2010) have been used in experimental tests on a wide range of economically important crops. Numarous antiviral compounds will produced when the host plant is naturally infected or artificially inoculated with viral pathogens. There are many antimicrobial compounds, including, but not limited to, antimicrobial peptides (AMP), antifungal proteins, enzymes that breakdown pathogen cell wall and infection structure, and phytoalexins (low molecular weight antimicrobial compounds) (Stuiver and Custers 2001). In order to enhance disease resistance, genes encoding those antimicrobial compounds are expressed in host plant. Two of the most popular genes to express in transgenic plants were chitinase and glucanase as they can breakdown the fungal cell wall components chitin and glucan, respectively. Either of the two genes isolated from different plant or non-plant sources has been expressed in plants like rice (Datta et al. 2001), tomato (Jabeen et al. 2015), cotton (Emani et al. 2003), and brinjal (Singh et al. 2014) showing enhanced resistance to various pathogens. Antimicrobial peptides, positively charged, cysteine-rich, and thermostable small peptides having antimicrobial properties owing to their ability to damage pathogen membrane, have also been used to develop transgenic crop plants with enhanced disease resistance. For example, transgenic rice (Iwai et al. 2002), tomato (Chan et al. 2005), papaya (Zhu et al. 2007), and wheat (Roy-Barman et al. 2006) plants expressing AMPs from different sources exhibited increased resistance to bacteria *Burkholderia plantari, Ralstonia solanacearum*, fungus *Phytophthora palmivora* and *Neovossia indica,* respectively. In a similar note, transformation of phtoalexin biosynthetic gene like grapevine *Vst1* improved resistance of rice to *Pyricularia oryzae* (Stark-Lorenzen et al. 1997), of wheat to *Blumeria graminis* (Liang et al. 2000), papaya to *Phytophthora palmivora* (Zhu et al. 2007).

Economically important concept for plant–pathogen interaction is gene-for-gene concept which is widely studied in different crops for disease resistance (Simmonds and Smartt 1997). Plants defense through effector-triggered immunity (ETI) based on the highly specific interaction between plant pathogen products called avirulent genes (AVr) and products from host resistant genes (R) produced according to the gene-for-gene hypothesis (Flor 1971), which suggests that, plants have many R-genes and pathogens have many Avr genes and plant disease resistance is observed if the product of any particular R gene has recognition specificity for a compound produced due to a particular pathogen Avr gene. We will see below that many Avr gene products contribute to pathogen virulence. R proteins can recognize pathogen effectors directly or indirectly through their effects on host cells. Indirect recognition of plant occurs through R protein mediated monitoring of effector disturbances in distinct host cellular targets of an effector, consistent with the "guard hypothesis" (Dangl and Jones 2001). Two variations of this model have been proposed currently. In first, R receptor is continuously associated with the host intermediate factor, whereas in the second the pathogen effector first associates with a host target and the complex formed is then recognized by the immune receptor (Elmore et al. 2011). Guard hypothesis was tested R/Avr system between Arabidopsis thaliana and Pseudomonas syringe pv tomato. Here modification of the host factor RIN4 by the bacterial Avr gene product activates the R protein RPM1 which results in plants resistance to bacteria (Mackey et al. 2002). R-genes are structurally present at a central nucleotide-binding site (NBS) domain, a C-terminal LRR region to mediate pathogen recognition, and N-terminal variable domain mainly identified as TIR (Toll/Interleucina-1) or CC (Coiledcoil) (Elmore et al. 2011). Besides TIRNBS-LRR and CC-NBS-LRR, other major classes of R-genes include the RLKs (containing an extracellular LRR, a transmembrane domain, and a cytoplasmic kinase domain), RLPs (which are similar to the RLKs but lack the kinase domain), and cytoplasmic enzymatic R-genes that contain neither LRR nor NBS groups (Gururani et al. 2012).

13.6 Expression of Plant Defense Genes During Host–BCAs Interaction

Unlike animal, plants are unable to move from one place to other compelling them to endure any kind of stress at a standstill condition (Molla et al. 2015). They depend on their innate immune system which relies on activation of an array of defense systems including production of phytoalexins, cell wall modification, synthesis of PR genes, generation of reactive oxygen species (ROS), callose deposition, activation of different defense signal transduction pathway, etc. In order to enhance resistance to pathogen by boosting plants own defense system, the endogenous defense genes or well-known defense genes from other plants are over-expressed in transgenic plants. Phenylalanine ammonia lyase (PAL) plays an important role in plant defense by mediating an important step in the shikimic acid pathway for the synthesis of many phenolic compounds, including lignin, which play a role in the defense mechanism (Molla et al. 2013). PAL gene over-expressing transgenic tobacco plants exhibited reduced susceptibility to the fungal pathogen *Cercospora nicotianae* (Shadle et al. 2003). Cao et al. (2008) showed that overexpression of a rice defense-related F-box protein gene *OsDRF1* improves disease resistance through enhanced defense gene expression in transgenic tobacco. Elevated expression of PR genes in host plant in response of pathogen attack is a common phenomenon in many pathosystems. There are more than ten different types of well-characterized PR proteins found in plants (Van Loon and Van Strien 1999). Two well-known plant PR genes are β-1,3-glucanase (PR2) and chitinase (PR3) which have been widely utilized to generate transgenic plants for developing resistance against various pathogens. Thaumatin like protein (TLP), a PR5 protein, exhibited enhanced resistance to sheath blight disease when expressed in transgenic rice (Datta et al. 1999) and to fusarium wilt in transgenic banana (Mahdavi et al. 2012). Plant genome harbors different master controller genes which act in different defense signaling pathways.

13.7 Modification of Soil Environment to Manage Plant Diseases

Many efforts have been made by several researchers to reduce soilborne disease by effective management of the soil environment which is quite successful. The management measure that augments the time frame of living cover and diversity of carbon inputs is associated with improved activity and presence of soil microbes. Large-scale establishment of non-pathogenic microbes in soil will tend to suppress soilborne pathogens through various mechanisms of actions such as competition for the resources and rhizosphere habitats (Whipps 2001), production of antagonistic compounds by beneficial microbes (Robleto et al. 1998), degradation of nutrients and making them available for the plants enhancing the growth and development, also they help in degradation of pathogen cell walls (Snapp et al. 2003) and induction of systemic resistance in the crop plant against an invading pathogen (Van Wees et al. 1997). The soilborne phytopathogens generally encounter antagonism from the beneficial microbes in the rhizosphere before, during, and after primary infections and subsequent secondary spread within host roots. The classical example is of the Take-all disease of wheat which is a highly soilborne devastating disease caused by *Gaeumannomyces graminis* var. *tritici* (Cook 1993). Monoculture of wheat caused severe epidemics of the disease worldwide. The efforts have been made to reduce the disease through altered dynamics of rhizosphere bacteria which are consistently associated with suppression of take-all disease. Very recently Pseudomonads that secrete 2,4-diacetylphloroglucinol (2,4-DAPG), a compound produced by Pseudomonads has been found effective in suppression of *Gaeumannomyces graminis* var. *tritici* (Mazzola 2004). This is one of the classical examples of modification of rhizosphere ecology for effective management of plant disease by managing complex-microbial and host interactions. From quite a long time inoculation of legumes with symbiotic *Rhizobium* has been practiced which is a tremendous success in not only reducing soilborne diseases, but also plant growth promotion. The suppressive soils can also be managed by cultural practices like addition of organic compounds, specific crop rotations, growing of resistant varieties (Raajimakers et al. 2009). For example, Fusarium wilt can be suppressed by continuous cropping of partially resistant cultivars. Inoculation of non-pathogenic *Fusarium oxysporum* strains is reported to induce resistance against pathogenic *Fusarium oxysporum* in tomato.

13.8 Commercial Formulation of Biocontrol Agents in Market

The research and application of biological control is gaining much more attention and interest. Historically, the success of biological control agents is still limited when they have applied on a large scale. They are mostly successful under controlled environment. Although the number of registered biological control agents has been increased over a period of time, only a small number of strains could able to make larger impact when they have been taken into field. The knowledge on the biological control system and obtaining successful formulations are the key to success of biological control agents (Emmert and Handelsman 1999). Increased market demand for biological pesticides has driven acceleration in research efforts from both industry and academia. The goal of achieving effective BCAs formulation must be competitive with chemical pesticide (Schisler et al. 2004). The microbial formulations should stabilize and encourage organism during production, storage and application. The formulation should also consider the efficacy on target pathogen, and interaction with other beneficial flora and fauna. The mode of application and ease of handling by the customers are very important points to be considered. Finally, the formulation must be safe and acceptable to regulatory agencies before its commercial release (Table 13.2).

Table 13.2 Important biological agents and their mode of action

S. no.	Name of the biocontrol agent	Pathogen species	Crop plants
1.	*Actinomycetes*	*Pythium ultimum*	Beet
2.	*Bacillus subtilis*	*Rhizoctonia solani; Gaeumannomyces graminis var tritici*	Wheat
3.	*Comamonas acidovorans*	*Magnoporthe*	Kentuki blue grass
4.	*Burkholderia cepacia*	*Fusarium graminearum*	Wheat
5.	*Enterobacter spp.*	*Magnoporthe*	Kentukigrass
6.	*Pseudomonas spp.*	*Fusarium oxysporum f. sp. radicis-lycopersici*	Tomato
7.	*Pseudomonas chlororaphis*	*Drechslera graminea*	Barley
8.	*Pseudomonas aureofaciens*	*Pythium ultimum*	Tomato
9.	*Pseudomonas fluorescens*	*Gaeumannomyces graminis var tritici*	Wheat
10.	*Pseudomonas fluorescens*	*Rhizoctonia solani*	Rice
11.	*Trichoderma harzianum Th-87*	*Rhizoctonia solani*	Brinjal
12.	*Trichoderma harzianum BAFC 742*	*Sclerotinia sclerotiorum*	Soybean
13.	*Trichoderma viride*	*Rhizoctonia solani*	Brinjal
14.	*Trichoderma harzianum*	*Fusarium solani*	Ginger
15.	*Trichoderma virens GL-1, GL-21, GL-23*	*Rhizoctonia solani*	Pea
16.	*Pseudomonas spp.*	*Pythium spp.*	Ginger
17.	*Trichoderma virens GL-3*	*Fusarium graminearum, Pythium arrhenomanes, Pythium ultimum*	Maize
18.	*Chaetomium spp.*	*Helminthosporium oryzae*	Rice
19.	*Bacillus spp.*	*Helminthosporium oryzae*	Rice
20.	*Pseudomonas spp.*	*Meloidogyne graminicola*	Rice

Modified from Whipps (2001)

13.9 Conclusion

Soilborne diseases are the most important and destructive obstacles for the crop production. A number of crop plants including cereals, high value vegetable crops, flowers, fruits, and ornamental crops are vulnerable for these destructive pathogens. They reduce the yield directly and quality and marketability indirectly. Most of the soilborne plant pathogens survive in soil for one or more years by producing overwintering structures like sclerotial bodies, mycelia, and other structures. Though a number of management options are available for the management of these diseases, chemical control has overtaken all other measures. This has led to the serious effects on soil, water, and human life. Many of the non-chemical options like soil solarization, growing resistant varieties, application of organic amendments, soil nutrition are available, and biological control agent provides an effective and environmentally safe management of the soilborne diseases. Exploring new strains of these biological agents should be taken up in priority for their antipathogenicity properties, growth promotion, and disease management. Finally, preparation of commercial formulation and large-scale application of these BCAs will manage soilborne diseases in an ecofriendly and sustainable manner.

References

Abawi GS, Widmer TL (2000) Impact of soil health management practices on soilborne pathogens, nematodes and root diseases of vegetable crops. Appl Soil Ecol 15:37–47

Agriose GN (2005) Plant pathology, 5th edn. Academic Press, New York

Basamma (2008) Integrated management of Sclerotium wilt of potato caused by *Sclerotium rolfsii* Sacc. M.Sc (Agri) thesis, Univ. Agric Sci. Dharwad, Karnataka (India)

Cao Y, Yang Y, Zhang H, Li D, Zheng Z, Song F (2008) Overexpression of a rice defense-related F-box protein gene OsDRF1 in tobacco improves disease resistance through potentiation of defense gene expression. Physiol Plant 134:440–452

Cerda R, Avelino J, Gary C, Tixier P, Lechevallier E, Allinne C (2017) Primary and secondary yield losses caused by pests and diseases: assessment and modeling in coffee. PLoS One 12(1): e0169133. https://doi.org/10.1371/journal.pone.0169133

Chan Y, Prasad V, Sanjaya, Chen KH, Liu PC, Chan M, Cheng C (2005) Transgenic tomato plants expressing an Arabidopsis thionin (Thi2.1) driven by fruit-inactive promoter battle against phytopathogenic attack. Planta 221:386–393

Cook RJ (1993) making greater use of introduced microorganisms for biological control of plant pathogens. Annu Rev Phytopathol 31:53–80

Dangl JL, Jones JDG (2001) Defence responses to infection. Nature 411:826–833

Datta K, Velazhahan R, Oliva N, Ona I, Mew T, Khush G, Muthukrishnan S et al (1999) Overexpression of the cloned rice thaumatin-like protein (PR-5) gene in transgenic rice plants enhances environmental friendly resistance to Rhizoctonia solani causing sheath blight disease. Theor Appl Genet 98:1138–1145

Datta K, Tu J, Oliva N, Ona I, Velazhahan R, Mew TW et al (2001) Enhanced resistance to sheath blight by constitutive expression of infection related rice chitinase in transgenic elite indica rice cultivars. Plant Sci 160:405–414. https://doi.org/10.1016/S0168-9452(00)00413-1

Dobbelaere S, Croonenborghs A, Thys A, Ptacek D, Vanderleyden J, Dutto P, Labandera-Gonzalez C, Caballero-Mellado J, Aguirre JF, Kapulnik Y, Brener S, Burdman S, Kadouri D, Sarig S, Okon Y (2001) Responses of agronomically important crops to inoculation with *Azospirillum*. Aust J Plant Physiol 28:871–879

Elmore JM, Lin ZJD, Coaker G (2011) Plant NB-LRR signaling: upstreams and downstreams. Curr Opin Plant Biol 14:365–371

Emani C, Garcia JM, Lopata-Finch E, Pozo MJ, Uribe P, Kim DJ, Sunilkumar G, Cook DR, Kenerley CM, Rathore KS (2003) Enhanced fungal resistance in transgenic cotton expressing an endochitinase gene from Trichoderma virens. Plant Biotechnol J 1:321–336

Emmert EAB, Handelsman J (1999) Biocontrol of plant disease: a gram-positive perspective. FEMS Microbiol Lett 171:1–9

Fiers, Edel-Hermann, Chatot, Hingrat, Alabouvette et al (2012) Potato soil-borne diseases. A review. Agron Sustain Dev 32(1):93–132

Flor HH (1971) Current status of the gene-for-gene concept. Annu Rev Phytopathol 9:275–296. https://doi.org/10.1146/annurev.py.09.090171.001423
Frac M, Hannula SE, Belka M, Jedryczka M (2018) Fungal biodiversity and their role in soil health. Front Microbiol 9:707. https://doi.org/10.3389/fmicb.2018.00707
Gregory PJ, Ingram JSI, Brklacich M (2005) Climate change and food security. Philos Trans R Soc B 360:2139–2148
Gururani MA, Venkatesh J, Upadhyaya CP, Nookaraju A, Pandey SK, Park SW (2012) Plant disease resistance genes: current status and future directions. Physiol Mol Plant Pathol 78:51–65
Huang HC, Chou CH, Erickson RS (2006) Soil sickness and its control. Allelopath J 18(1):1–21
Iwai T, Kaku H, Honkura R, Nakamura S, Ochiai H, Sasaki T, Ohashi Y (2002) Enhanced resistance to seed-transmitted bacterial diseases in transgenic rice plants overproducing an oat cell-wall-bound thionin. Mol Plant-Microbe Interact 15:515–521
Jabeen N, Chaudhary Z, Gulfraz M, Rashid H, Mirza B (2015) Expression of rice chitinase gene in genetically engineered tomato confers enhanced resistance to fusarium wilt and early blight. Plant Pathol J 31:252–258. https://doi.org/10.5423/PPJ.OA.03.2015.0026
Jyoti S, Singh DP (2016) Fungi as biocontrol agents in sustainable agriculture. In: Microbes and environmental management. Studium Press, New Delhi, pp 172–194
Kannaiyan MV, Reddy MV, Nene YL, Raju TN (1981) Prevalence of pigeon pea wilt and sterility mosaic in India (1975-1980). Int Pigeonpea Newsl 17:25
Karim Z, Sakhawat M, Hossain S, Begum MM (2018) Ralstonia solanacearum: a threat to potato production in Bangladesh. Fundam Appl Agric 3(1):407–421. https://doi.org/10.5455/faa.280361
Knox J, Hess T, Daccache A, Wheeler T (2012) Climate change impacts on crop productivity in Africa and South Asia. Environ Res Lett 7(3):034032
Kokalis-Burelle N, Kloepper JW, Reddy MS (2006) Plant growth-promoting rhizobacteria as transplant amendments and their effects on indigenous rhizosphere microorganisms. Appl Soil Ecol 31(1–2):91–100
Kumar S (2014) Plant disease management in India: advances and challenges. Afr J Agric Res 9 (15):1207–1217. https://doi.org/10.5897/AJAR2014.7311
Kumar S, Gupta O (2012) Expanding dimensions of plant pathology. JNKVV ResJ 46(3):286–293
Lal R (2009) Soil degradation as a reason for inadequate human nutrition. Food Secur 1(1):45–57
Liang H, Zheng J, Shuange JIA, Wang D, Ouyang J, Li J, Li L, Tian W, Jia X, Duan X, Sheng B, Hain R (2000) A transgenic wheat with a stilbene synthase gene resistant to powdery mildew obtained by biolistic method. Chin Sci Bull 45:634–638
Mackey D, Holt BF, Wiig A, Dangl JL (2002) RIN4 interacts with Pseudomonas syringae type III effector molecules and is required for RPM1-mediated resistance in Arabidopsis. Cell 108:743–754
Madhukar HM (2001) Ecology and management of fusarium wilt of Chilli. M.Sc (Agri) thesis, Univ. Agric. Sci. Dharwad, Karnataka (India)
Mahdavi F, Sariah M, Mazih M (2012) Expression of rice thaumatin-like protein gene in transgenic banana plants enhances resistance to Fusarium wilt. Appl Biochem Biotechnol 166 (4):1008–1019
Majeed M, Mir GH, Hassan M, Mohuiddin FA, Paswal S, Farroq S (2018) Damping off in Chilli and its biological management-a review. Int J Curr Microbiol App Sci 7(4):2175–2185. https://doi.org/10.20546/ijcmas.2018.704.247
Mazzola M (2004) Assessment and management of soil microbial community structure for disease suppression. Annu Rev Phytopathol 42:35–59
Meyer JB, Lutz MP, Frapolli M, Péchy-Tarr M, Rochat L, Keel C, Défago G, Maurhofer M (2010) Interplay between wheat cultivars, biocontrol pseudomonads, and soil. Appl Environ Microbiol 76:6196–6204
Molla KA, Karmakar S, Chanda PK, Ghosh S, Sarkar SN, Datta SK, Datta K (2013) Rice oxalate oxidase gene driven by green tissue-specific promoter increases tolerance to sheath blight pathogen (Rhizoctonia solani) in transgenic rice. Mol Plant Pathol 14:910–922

Molla KA, Debnath AB, Ganie SA et al (2015) Identification and analysis of novel salt responsive candidate gene based SSRs (cgSSRs) from rice (*Oryza sativa* L.). BMC Plant Biol 15:122. https://doi.org/10.1186/s12870-015-0498-1

Nikahm P, Jagtap GP, Bidwe KU (2008) Survey and Surveillance of chick pea wilt complex caused by *Fusarium oxysporum* f. sp. *ciceri*. J Pl Dis Sci 3(1):21–33

Nowicki M, Foolad MR, Nowakowska M, Kozik EU (2012) Potato and tomato late blight caused by Phytophthora infestans: an overview of pathology and resistance breeding. Plant Dis 96 (1):4–17. https://doi.org/10.1094/pdis-05-11-0458

Pandey A, Dubey S (2017) Evaluations of brinjal germplasm for resistance to fusarium wilt disease. Int J Sci Res Publ 7(7):435–438

Raaijmakers JM, Paulitz T, Steinberg C, Alabouvette C, Moënne-Loccoz Y (2009) The rhizosphere: a playground and battlefield for soil borne pathogens and beneficial microorganisms. Plant Soil 321:341–361

Raghu (2011) Studies on management of rhizome wilt of Ginger with special reference to Ralstonia solanacearum (E. F. Smith) Yabuuchi et al. MSc (Agri) thesis submitted to University of Agricultural Sciences, Dharwad, Karnataka

Raghu S (2014) Studies on chilli wilt complex disease. PhD (Agri) thesis, Univ. Agric. Sci. Dharwad, Karnataka (India)

Raghu S, Yadav MK, Prabhukarthikeyan SR, Baite MS, Lenka S, Jena M (2018) Occurrence, pathogenicity, characterization of *Fusarium fujikuroi* causing rice bakanae disease from Odisha and in vitro management. Oryza 55:214–223

Raghu S, Baite MS, Patil NB, Sanghamitra P, Yadav MK, Prabhukarthikeyan SR, Keerthana U, Gurupirasanna Pandi G, Aravindan S, Rath PC (2020) Grain discoloration in popular rice varieties (*Oryza sativa* L) in eastern India, associated mycoflora, quality losses and management using selected biocontrol agents. J Stored Prod Res 88:101682. https://doi.org/10.1016/j.jspr.2020.101682

Ranganathswamy M, Naik ST, Motagi BN, Sudini H (2014) Evaluation of groundnut genotypes for resistance to aflatoxin contamination. In: Advances in arachis through genomics & biotechnology (AAGB 2014). 7th International Conference of the Peanut Research Community, November 11–14, 2014. Savannah Marriott Riverfront, Savannah, Georgia, USA

Robleto EA, Borneman J, Triplett EW (1998) Effects of bacterial antibiotic production on rhizosphere microbial communities from a culture-independent perspective. Appl Environ Microbiol 64:5020–5022

Roy-Barman S, Sautter C, Chattoo BB (2006) Expression of the lipid transfer protein Ace-AMP1 in transgenic wheat enhances antifungal activity and defense responses. Transgenic Res 15:435–446

Sangeetha TV (2011) Studies on root rot/wilt of soybean. M.Sc (Agri) thesis, Univ. Agric Sci. Dharwad, Karnataka, (India)

Savary S, Ficke A, Aubertot JN, Hollier C (2012) Crop losses due to diseases and their implications for global food production losses and food security. Food Secur 4:519–537. https://doi.org/10.1007/s12571-012-0200-5

Savita SA (2019) Fungi as biological control agents. In: Giri B, Prasad R, Wu QS, Varma A (eds) Biofertilizers for sustainable agriculture and environment. Soil biology, vol vol 55. Springer, Cham. https://doi.org/10.1007/978-3-030-18933-4_18

Schisler DA, Slininger RJ, Behle RW, Jackson MA (2004) Formulation of Bacillus spp. for biological control of plant diseases. Phytopathology 94:1267–1271

Shadle GL, Wesley SV, Korth KL, Chen F, Lamb C, Dixon RA (2003) Phenylpropanoid compounds and disease resistance in transgenic tobacco with altered expression of l-phenylalanine ammonia-lyase. Phtyochemistry 64:156–161

Simmonds NW, Smartt J (1999) Principles of crop improvement. Blackwell Sci, Oxford

Simontacchi M, Galatro A, Ramos-Artuso F, Santa-Maria GE (2015) Plant survival in a changing environment: the role of nitric oxide in plant responses to abiotic stress. Front Plant Sci 6:977. https://doi.org/10.3389/fpls.2015.00977

Singh AK, Kamal S (2012) Chemical control of wilt in tomato (*Lycopersicon esculentum* L.). Int J Horticult 2(2):5–6

Singh D, Ambrosise A, Haicour R, Sihachakar D, VenkatRajam V (2014) Increased resistance to fungal wilts in transgenic eggplant expressing alfalfa glucanase gene. Physiol Mol Biol Plants 20(2):143–150. https://doi.org/10.1007/s12298-014-0225-7

Snapp S, Kirk W, Roman-Aviles B, Kelly J (2003) Root traits play a role in integrated management of fusarium root rot in snap beans. Hort Sci 38(2):187–191

Snehalatharani A, Meheswarapa HP, Devappa V, Malhotra SK (2016) Status of coconut basal stem rot disease in India – a review. Indian J Agric Sci 86(12):1519–1529

Sonyal S, Nargund VB, Puneeth ME, Benagi VI, Palanna KB, Giri MS, Shivalingappa H, Mahesha HS, Dev D, Pappachan A, Jagarkall Y (2016) Survey for pomegranate wilt complex caused by *Ceratocystis Fimbriata* and *Meloigogyne incognita* in Northern Karnataka. J Pure Appl Microbiol 10(1):253–257

Stark-Lorenzen P, Nelke B, Hanbler G, Mühlbach HP, Thomzik JE (1997) Transfer of a grapevine stilbene synthase gene to rice (Oryza sativa L.). Plant Cell Rep 16:668–673

Strange RN, Scott PR (2005) Plant disease: a threat to global food security. Annu Rev Phytopathol 43:83–116

Stuiver MH, Custers JH (2001) Engineering disease resistance in plants. Nature 411(6839):865

Umesha S, Kavitha R, Shetty HS (2005) Transmission of seed-borne infection of chilli by *Burkholderia solanacearum* and effect of biological seed treatment on disease incidence. Arch Phytopathol Plant Prot 38(4):281–293

Van Loon C, Van Strien EA (1999) The families of pathogenesis-related proteins, their activities, and comparative analysis of PR-1 type proteins. Physiol Mol Plant Pathol 55(2):85–97. https://doi.org/10.1006/pmpp.1999.0213

Van Wees SCM, Pieterse CMJ, Trijssenaar A, Van't Westende Y, Hartog F, van Loon LC (1997) Differential induction of systemic resistance in Arabidopsis by biocontrol bacteria. Mol Plant-Microbe Interact 10(6):716–724

Van Wees SCM, van der Ent S, Pieterse CMJ (2008) Plant immune responses triggered by beneficial microbes. Curr Opin Plant Biol 11:443–448

Vasantaraj DB, Ramamurthy VV (2016) Elements of economic entomology, 8th edn. Brillion Publishing, New Delhi

Waddington SR, Li X, Dixon J, Hyman G, de Vicente MC (2010) Getting the focus right: production constraints for six major food crops in Asian and African farming systems. Food Secur 2(1):27–48

Wheeler TA, Woodward JE, Mullinix BG Jr (2010) Effect of seeding rate on verticillium wilt incidence, yield, and value for three cotton cultivar. J Cotton Sci 14:173–180

Whipps JM (2001) Microbial interactions and biocontrol in the rhizosphere. J Exp Bot 52:487–511

Wu H, Shabala L, Barry K, Zhou M, Shabala S (2013) Ability of leaf mesophyll to retain potassium correlates with salinity tolerance in wheat and barley. Physiol Plant 149(4):515–527. https://doi.org/10.1111/ppl.12056. Epub 2013 Apr 24

Zhu YJ, Agbayani R, Moore PH (2007) Ectopic expression of Dahlia merckii defensin Dm-AMP1 improves papaya resistance to Phytophthora palmivora by reducing pathogen vigor. Planta 226: 87–97

Interactions Between Plant Genotypes and PGPR are a Challenge for Crop Breeding and Improvement Inoculation Responses

14

Inés E. García de Salamone and Luciana P. Di Salvo

Abstract

The world population for the year 2030 has been estimated in 8500 million people. It is a challenge to increase, with environmental awareness, the productivity of food crops reducing the use of pesticides and fertilizers to promote agricultural sustainability. This contrasts with the objectives of the green revolution of the 1960s. Cereal crops are the basis of human food but most of them have reached their peak of productive potential through traditional breeding. For this reason, a new green revolution is necessary, which achieves greater harvests on the basis for a better use of available natural resources with less loss from pests and diseases. Thus, it is necessary to consider the acquired knowledge of the roots and the thin layer of soil that surrounds them, which is the rhizosphere.

Keywords

Azospirillum brasilense · Sustainability · Plant Growth-Promoting Rhizobacteria · *Pseudomonas fluorescens* · Rhizosphere microbial communities

I. E. García de Salamone (✉)
Universidad de Buenos Aires, Facultad de Agronomía, Departamento de Biología Aplicada y Alimentos, Cátedra de Microbiología Agrícola, Buenos Aires, Argentina
e-mail: igarcia@agro.uba.ar

L. P. Di Salvo
Universidad de Buenos Aires, Facultad de Agronomía, Departamento de Biología Aplicada y Alimentos, Cátedra de Microbiología Agrícola, Buenos Aires, Argentina

CONICET, Buenos Aires, Argentina

R. Soni et al. (eds.), *Microbiological Activity for Soil and Plant Health Management*, https://doi.org/10.1007/978-981-16-2922-8_14

14.1 Introduction

The soil, particularly in the rhizosphere, contains a great diversity and quantity of microorganisms. The rhizosphere corresponds to the volume of the soil that surrounds plant roots, and which is under their direct influence (Philippot et al. 2013). It is one of the most dynamic habitats on the planet and its physicochemical and biological properties show great spatial and temporal heterogeneity (Shen et al. 2013). Numerous microorganisms coexist in it, many beneficial to plants (Den Herder et al. 2010; Gewin 2010; Aeron et al. 2011), and the quality and quantity of resources for human nutrition have defined this environment (Di Salvo et al. 2013a).

Plants and the rest of the organisms that develop in any ecosystem depend on the functions of edaphic and rhizosphere microorganisms to carry out their own biological processes. Plants make up the base of the trophic pyramid, as they represent the source of energy and nutrients available to heterotrophic organisms, which mainly decompose their residues. Thus, soil microorganisms control mineralization and regulate nutrient recirculation. The influence of plants on the structure and physiology of the rhizosphere microbial community has important ecological implications for soil functionality, including biogeochemical cycles (Morrissey et al. 2002; Minz et al. 2013).

In natural ecosystems, microbial communities are in harmony and guarantee the quality and health of the soil. On the other hand, in artificial systems such as agroecosystems, there is a decrease in soil organic matter that affects their physicochemical and biological quality altering microbial functionality (Cassman 1999). This can lead to the loss of beneficial microorganisms and the proliferation of pathogens (Suleiman et al. 2015). This can have a devastating effect on crop productivity and soil properties (Avis et al. 2008). Taking into account that the world population forecast for the year 2030 has been estimated at more than 8500 million people, the challenge of modern agriculture is to increase the productivity of crops that provide food with environmental awareness, reducing the use of pesticides and synthetic fertilizers (Avis et al. 2008; Leach and Mumford 2008; United Nations 2018) and favoring the sustainability of agroecosystems (Gamalero and Glick 2011). This contrasts with what was established in the green revolution of the last century, which sought to increase yields based on traditional plant breeding and the addition of massive use of agrochemicals. The main cereal crops, the basis of human nutrition, have already reached their productive potential peak through traditional plant breeding (Breseghello and Guedes Coelho 2013). The twenty-first century demands a new green revolution, which can achieve increasing harvests based on better use of available natural resources and lower losses due to pests and diseases. For this reason, it is necessary to achieve, among others, more detailed knowledge of the rhizosphere microbial ecology (Lynch 2007). Knowing and maintaining microbial functionality in the rhizosphere is essential in optimal conditions for plant growth and in situations of environmental stress, to improve food production worldwide (Sessitsch and Mitter 2015).

14.2 Plant Breeding and Rhizosphere Microorganisms

Traditional plant breeding is based on the selection of varieties with superior characteristics, mainly related to yield, health, and quality of the crop (Marschner and Rengel 2010). The concept of "ideotype" is frequently considered, which describes those morphophysiological characters that a genotype of a crop should have to achieve high yields under different production conditions (Cirilo et al. 2009; Breseghello and Guedes Coelho 2013). In the case of corn, plant breeding was successful in increasing yields, mainly in the development of hybrid genotypes. The most modern hybrids show high yields and great stability because they have the ability to sustain grain yield per plant when the available resources are reduced due to some type of stress (Tollenaar and Lee 2002; Valentinuz 2006). In the case of the wheat, the analysis of historical collections of varieties of this crop has shown that traditional plant breeding achieved increases in grain yield associated with increases in the harvest index (Slafer and Andrade 1989) and a systematic reduction in the competitive ability against environmental stress (Sadras and Lawson 2011). In relation to this, it has been shown that the rate of genetic gain is generally higher in favorable environments (Slafer et al. 2015).

Traditional plant breeding has largely ignored the role of rhizosphere microbial communities in aspects such as nutrition, stress tolerance, and maintenance of biodiversity (Morrissey et al. 2002; Rengel 2002). The first agricultural revolution introduced crop rotation in the eighteenth century to harness and manipulate microbial populations in the soil, although at that time it was not known why this have benefited plant health and growth. The second revolution began in the second part of the twentieth century and is usually described as the "green revolution." It was based on traditional plant breeding techniques and the development of hybrid genotypes without taking into account the microbial processes of the rhizosphere but with increasing applications of fertilizers. Currently, we are already in the era of genetic engineering of plants, and despite this, we continue with a high dependence on the use of agrochemicals, although recently the negative impact on the associated microbial communities has been shown (Newman et al. 2016; Escobar Ortega and García de Salamone 2017; Escobar Ortega et al. 2020). Thus, some authors have pointed out that the increase of crop yields based on traditional plant breeding represents an onerous practice that generates negative environmental impacts for the ecosystem (Tilman et al. 2002). Based on this, it is important to seek and improve agronomic practices aimed at increasing and maintaining high production levels in a more sustainable way (Altieri and Nicholls 2000; Tilman et al. 2002; Tester and Langridge 2010). Thus, the use of beneficial microorganisms in the rhizosphere to increase crop yield is considered the "new green revolution" (Tilman 1999; Rengel 2005; Den Herder et al. 2010; Gewin 2010; Aeron et al. 2011). Optimizing plant-associated microbial communities offers a challenging approach to increase productivity without environmental harm (Reid and Greene 2012). Basically, it is necessary to change the plant model to be improved, in such a way that associations with microorganisms both native and applied to the system by inoculation are included. This would improve the efficiency of the use of available resources and increase the potential of the soil-plant-environment system (García de Salamone 2012a; Pedraza et al. 2011).

14.3 Plant Growth-Promoting Rhizobacteria and Sustainable Agriculture

Bacteria that colonize the rhizosphere of plants and produce beneficial effects on their growth and nutrition are known as plant growth-promoting rhizobacteria (PGPR) (Bashan et al. 2004). In the literature, a huge amount of evidence indicates that the use of PGPR may have a significant role in the sustainability of agroecosystems (Antoun and Prevost 2005; Vejan et al. 2016). Among the most studied and already used as "biofertilizers" in the form of commercial inoculants are *Azospirillum brasilense* and *Pseudomonas fluorescens*. These and other PGPR were shown to be capable of improving the growth and nutrition of various crops, as well as their health and water status under a deficit situation (Lugtenberg et al. 2001; de Bashan et al. 2007; Pedraza et al. 2011; Glick 2012; García de Salamone 2013). Thus, they constitute an economic and ecological alternative to increase food production (Bashan et al. 2004; Lucy et al. 2004; Díaz-Zorita and Fernández Canigia 2009; Naiman et al. 2009) and a tool for more efficient use of available resources (Altieri and Nicholls 2000; Di Salvo et al. 2013a, b).

Inoculation with PGPR contributes to the implantation, development, and production of crops, such as rice, wheat, and corn (Lucy et al. 2004; Siddiqui 2005; García de Salamone 2012a, b). Biological nitrogen fixation acquires relevance because certain plant–PGPR associations can incorporate nitrogen through this process, which depends on specific combinations between plants and bacteria (García de Salamone et al. 1996, 2010; Urquiaga et al. 2004). In this regard, six cultivars of wheat showed different percentages of grain yield increases (García de Salamone 2012a) and nitrogen derived from biological fixation (García de Salamone 2012b) due to PGPR inoculation with an experimental inoculant of two strains of *A. brasilense* (Fig. 14.1). Wheat genotypes showed their differences in these two plant attributes associated with the economic and ecological pillars of agricultural sustainability.

Crop inoculation with PGPR, such as *A. brasilense*, should be associated with other recommended management practices to achieve high yields or to collaborate in a more efficient use of available resources (García de Salamone and Monzón de Asconegui 2008; Naiman et al. 2009). This PGPR (Cassán and García de Salamone 2008) and *P. fluorescens* (García de Salamone et al. 2001, 2012) can associate with species of agronomic interest and can produce direct beneficial effects on their growth and nitrogenous and phosphorous nutrition (Naiman et al. 2009; Pedraza et al. 2011; García de Salamone 2012a). There is some evidence that, under controlled conditions, certain strains of *P. fluorescens* can change cytokinin balance in wheat plants and their rhizospheres (García de Salamone 2000; García de Salamone et al. 2005) and in *Arabidopsis thaliana* plants (Drobkinsky et al. 2016). *Azospirillum* strains can also change the balance of auxins (Okon 1994) and

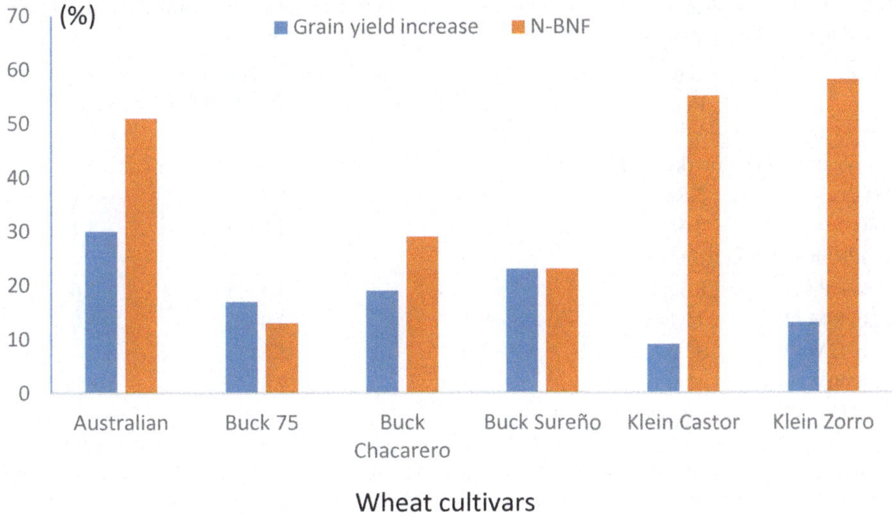

Fig. 14.1 Percentages of increase of grain yield responses and nitrogen derived from biological fixation (N-BNF) due to coinoculation with *Azospirillum brasilense* for six cultivars of wheat grown at field conditions. Compiled from García de Salamone (2012a, b)

gibberellins (Bottini et al. 2004) as well as phosphorus and iron availability (Pedraza et al. 2011).

PGPR action mechanisms are highly varied (Bashan and de Bashan 2010; Glick 2012) and remain a matter of debate. For more than four decades, numerous studies on *Azospirillum*–plant interactions have demonstrated several possible beneficial mechanisms which have led to the concept of the hypothesis of multiple mechanisms (Cassán et al. 2020). Regardless of the mechanisms involved in plant responses to PGPR inoculation, production increases are achieved, and they have allowed that this biotechnological alternative has emerged as a sustainable agriculture strategy (García de Salamone 2012a; Glick 2012).

Despite all the aforementioned advantages regarding the use of beneficial microorganisms, such as PGPR, the current model of agricultural production does not favor their use and, generally, the production increases that they generate under field conditions are scarce with respect to their potential shown under controlled conditions (Ferreira et al. 2013). Although the inoculation with PGPR of commercial varieties or hybrids of cereal crops often improves their yields in low percentages, these are considered "acceptable" by the farmers and by inoculant manufacturers. It has been proposed that the plant–bacteria interaction is less significant in wheat crops, compared to corn crops (Cassán and Díaz Zorita 2016). An explanation for this is based on wheat genotypes of European and Latin American origin presented similar levels of inoculation response (Díaz-Zorita and Fernández Canigia 2009) which is on average 5%. However, this should not be generalized, as shown in Fig. 14.1 for six wheat genotypes widely grown in Argentina where all the inoculation responses were higher than 9% reaching up to almost 30% in some genotypes.

Table 14.1 Percentages of change with respect to control of nitrogen fertilization and PGPR inoculation on maize plant attributes at different crop phenological stages

	Percentage of change with respect to control of each factor			
	Root Biomass at V5	Aerial Biomass at R3	Grain Yield at PM	MNFB at R3
Nitrogen fertilization (kg urea ha^{-1})				
90	5.8 (*)	4.3	25.8 (*)	−3.1
180	2.6	−1.8	38.5 (*)	−3.1
PGPR inoculation treatments				
A.b + P.f (1)	8.3	−23.7 (*)	5.3 (*)	−1.6
A.b 40M (2)	14.1	−6.1	4.9 (*)	−6.0
A.b 42M (2)	33.6	−7.0	5.5 (*)	3.2
A.b 40M, 42M (2)	10.0	−2.3	9.8 (*)	10.5 (*)

A.b*:* *Azospirillum brasilense*; P.f: *Pseudomonas fluorescens*. (1) Commercial inoculant. (2) Experimental inoculant. PM: Physiological maturity. MNFB: microaerophilic N_2 fixing bacteria (*) indicates statistical differences at $p \leq 0.05$, between absolute values with respect to factor control treatment. Compiled from Di Salvo et al. (2018a)

Thus, the levels of inoculation response observed in the field seem to be extremely far away from the potential of this biotechnology. Although it must consider that this biotechnology increases production at a significantly lower cost than the large-scale use of chemical fertilizers, it is known that the level of PGPR inoculation response depends on the interactions between the native and inoculated microorganisms (Bashan 1999), and between them and the genotype of the plants. Furthermore, these interactions are subject to prevailing physicochemical factors (Rani and Goel 2012a), such as the availability of nutrients (Dobbelaere et al. 2001) for both crop and rhizosphere microorganisms. In this sense, great variability in the response to inoculation was dependent on the associated plant genotypes (García de Salamone et al. 1996, 2009, 2012). As crops and their management systems represent changes in the quantity and quality of plant residues or radical exudates supplied to the soil, the magnitude of these changes could define the structure and function of rhizosphere microbial communities (Hartmann et al. 2009; Di Salvo et al. 2018a, b). Different genotypes of cereals will cause effects that will represent a combination of the amount of biomass supplied, the type of decaying substrates, and the distribution of the plants in the canopy (Escobar Ortega et al. 2020). There is evidence that indicates that *A. brasilense* and *P. fluorescens* inoculation can contribute to increasing the production of both aerial and radical biomass of cereal crops such as corn (Table 14.1) (García de Salamone 2012a; Di Salvo et al. 2013a, 2018a), rice (Table 14.2) (García de Salamone et al. 2010, 2012), wheat (Naiman et al. 2009; García de Salamone 2012a, b), and rye (Escobar Ortega et al. 2020).

Another factor that hinders or delays the use of PGPR on a large scale is the lack of adequate formulations that cause variability and inconsistency together with generally long and expensive registration processes (Malusá et al. 2012). The formulation of each bio input defines PGPR survival possibilities in the inoculant

Table 14.2 Percentages of change with respect to control PGPR inoculation and pair values of principal component analysis of rhizosphere microbial communities of three rice cultivars growing at field conditions

	Percentages of change with respect to control PGPR inoculation (1)						
	Panicles	Filled grains	Chaffy grains	Grain yield	1000 grain weight	HI (2)	Pair values PC1/PC2 (3) (4)
Camba INTA	−11.3	−6.6	−9.7	−7.3	0.4	1	2.1/0.7 Ba
Supremo	0.3	19.7	−14.0	20.2	1-0	5	−1.3/−0.2 Aa
Yerua	7.5	2.8	−18.3	11.0	4.5(*)	0	−0.8/−0.7 Aa

(1) Coinoculation with *Azospirillum brasilense* and *Pseudomonas fluorescens*. (2) Harvest index. (3) Pair values of principal component (PC) analysis of the microbial communities in the rhizosphere of three rice cultivars at physiological maturity. (4) Different uppercase letters indicate significant differences between means for 24 h absorbance values of each cultivar for PC1. Different lowercase letters indicate significant differences between means for absorbance values of each cultivar for PC2. Means were compared by Tukey's test at $P < 0.05$. (*) indicates statistical differences at $p \leq 0.05$, between absolute values with respect to control treatment without PGPR inoculation. Compiled from García de Salamone et al. (2012)

during the storage period and after inoculation on the seed. Bioencapsulation (Vejan et al. 2019) and the use of nanoparticles (Dikshit et al. 2013) have received enormous attention in creating new types of tools for biotechnology and life sciences to improve PGPR growth and survival with direct consequences on the quality of inoculants. In this sense, the physiological effects and possible cellular internalization of magnetite nanoparticles (nFe_3O_4) have been evaluated in plants grown in hydroponics (Iannone et al. 2013, 2016). The observation with electron microscopy of root sections showed the presence of Fe_3O_4 nanoparticles in the apoplast of the root tissues and a significant increase in the iron content only in the roots of wheat plants treated with these nanoparticles without modifying germination, growth, chlorophyll content, and other physiological and biochemical attributes of the seedlings (Iannone et al. 2016). However, chlorophyll content and aerial and root plant growth of soybean or alfalfa plants were increased when they were treated with these nanoparticles. Besides, a positive effect of magnetite nanoparticles was observed on the symbiotic systems soybean–*Bradyrhizobium japonicum* and *Medicago sativa*–*Sinorhizobium meliloti* because leghemoglobin content in the nodules and nitrogen content per plant were higher in the nanoparticle treated plants than in the control without nanoparticle treatment (Zawoznik et al. 2017a, b). For those reasons, these nanoparticles could be potentially useful for an innovative design to improve the formulations of PGPR inoculants.

It is important to bear in mind that traditional plant breeding of cereal crops has considered neither the associated rhizosphere nor endophytic microorganisms, nor the soil native microorganisms. As there is a close relationship between the plant genotype of the cultivated species and the microbial community that is established in their rhizospheres in a given environment (Cummings 2009), the traditional plant breeding should have taken into account the beneficial functions of rhizosphere microorganisms for efficient use of soil nutrients (Bakker et al. 2012; Rani and Goel 2012b). This historical mistake represents a complex problem since it is generally assumed that plants coevolved in association with soil microorganisms (Garcia and Kao-Kniffin 2018). An analysis of three rice genotypes that are widely cultivated in northeastern Argentina has demonstrated that genotypic differences between them were related to their inoculation responses to the PGPR A. *brasilense* and *P. fluorescens* and to significant differences among their rhizosphere microbial communities (Table 14.2).

The lack of significant crop responses to PGPR inoculation under field conditions is a worldwide recognized problem, which must be addressed in studies with an interdisciplinary perspective. It is necessary to analyze deeply each member of this association and their multiple interactions. Besides, it is necessary to change the model of plant genetic improvement by including the associations that occur between crops and soil microorganisms (Gopal and Gupta 2016; Pieterse et al. 2016), both the native ones and those introduced to the system through inoculation. Some authors have shown differences in the composition of the microbial communities associated with the roots of wild-type bean plants and modern genotypes (Pérez-Jaramillo et al. 2017), evidencing again an effect of crop breeding on established microbial associations. In this sense, other authors have shown changes in the composition of the exudates, and with them, in the structure of the associated microbial communities, due to the domestication of tetraploid wheat, or durum wheat (Iannucci et al. 2017). To include plant–microorganism associations as a sustainable agricultural practice, it is necessary to make additional efforts to describe in detail the microorganisms associated with the roots of different crops in different environments and the dynamics of microbial communities in crop rhizospheres (Rascovan et al. 2013). In this sense, the relationship between different corn genotypes and certain microbial OTUs present in their rhizospheres in different environments has been reported, highlighting the heritability of certain taxa or microbial functions between related genotypes (Walters et al. 2018).

14.4 PGPR and Rhizosphere Microbial Ecology

Regarding the need to improve the knowledge of the rhizosphere microbial ecology to increase the level of response to PGPR inoculation, numerous in vitro and in vivo studies have been carried out, in microcosm and under field conditions. The Az39, 40M, 42M strains of *A. brasilense* and the strain G20-18 of *P. fluorescens* have previously been characterized by their PGPR properties (García de Salamone et al. 2001, 2010; Di Salvo et al. 2014a) and by their capacities to promote plant growth (García de Salamone and Monzón de Asconegui 2008; Cassán and García de Salamone 2008; Naiman et al. 2009; García de Salamone et al. 1996, 2005, 2006, 2010, 2012; Drobkinsky et al. 2016; Di Salvo et al. 2018a, b). In many cases, colony morphology and the ability to fix N_2 showed significant variations among strains (Di Salvo et al. 2014a). These differences were also observed in vivo because the inoculation of certain maize genotypes with certain *Azospirillum* strains caused similar yields to those obtained with a fertilization with 100 kg of N ha^{-1} (García de Salamone 2012a). In addition, in numerous circumstances it has been possible to detect that *A. brasilense* can fix N_2 in the presence of carbon sources and in atmospheric oxygen concentrations not recommended in the literature (Di Salvo et al. 2010, 2014a) and in contrary to what it is established in the literature for the genus *Azospirillum* (Holguin and Glick 2001), some strains of *A. brasilense* (Di Salvo et al. 2014a) and *A. lipoferum* (Esquivel-Cote et al. 2010) have the ability to synthesize the ACC deaminase enzyme in vitro. This PGPR mechanism of growth promotion is beneficially related to plant nitrogen nutrition. It has also been shown that the profiles of fatty acids and the use of carbon sources, as well as the capacity to produce indole substances and siderophores, vary significantly between strains of this PGPR species (Pedraza et al. 2011; Baca et al. 2012; Di Salvo et al. 2014a). Likewise, differences have been demonstrated in the in vitro production of various cytokinins and in the carbon source utilization profiles among different strains of *P. fluorescens* (García de Salamone et al. 2001). Inoculation with the strain G20-18 of this PGPR produced direct effects on the growth of wheat plants associated with its ability to produce in vitro and in vivo significant amounts of cytokinins. Inoculated plants with strain G20-18 had high cytokinin concentrations in plant tissues, an aspect linked to the delay in foliar senescence (García de Salamone and Nelson 2006).

The evaluation of plant growth promotion capacity of *A. brasilense* and *P. fluorescens* under field conditions has shown differences in the inoculation response of different genotypes of maize (García de Salamone and Döbereiner 1996), rice (García de Salamone et al. 2012), and wheat (García de Salamone 2012a). Other authors have also found responses dependent on the interaction with maize genotypes and the *A. brasilense* strains in the generation of aerial biomass and the production of secondary metabolites (Walker et al. 2011). A technique for the evaluation of the functional diversity of bacterial communities has been adapted and standardized by obtaining the profiles of the use of carbon sources (Di Salvo and García de Salamone 2012). The use of this technique, and complementary methods of statistical analysis (Di Salvo et al. 2012, 2018a; Carlino et al. 2013), allowed significant advances in the functional and structural diversity characterization of the associated rhizosphere bacterial communities of crops, such as rice (García de Salamone et al. 2010, 2012), wheat (Naiman et al. 2009; Di Salvo et al. 2014b), corn (Di Salvo et al. 2013a; Carlino et al. 2013), and cover crops such as oats and rye (Escobar Ortega et al. 2020), which have been inoculated with these two PGPR. In certain cases, the differences in bacterial diversity were associated with differences between plant genotypes (García de Salamone et al. 2012) and in others, with the agronomic practices of PGPR inoculation and chemical fertilization, or the phenological stages of the crops (Carlino et al. 2013; Di Salvo et al. 2013c; Di Salvo and García de Salamone 2019). The presence of endophytic microorganisms in the seeds of different cultivars of barley (Zawoznik et al. 2014), wheat (Díaz Herrera et al. 2016), sunflower (Dominguez et al. 2019), and oats (Escobar Ortega et al. 2020) could compete with the inoculated PGPR strains and determine the lack of inoculation responses.

Although it is known that counting techniques for cultivable microorganisms quantify a very small proportion of the communities present, they are useful to determine impacts on soil health generated by anthropic activity (Ellis et al. 2003; García de Salamone et al. 2006, 2009). In this sense, it has been shown that different crop management conditions modify the number of various groups of microorganisms associated with rice, corn, wheat, oats, and rye plants (García de Salamone et al. 2004, 2010, 2012; Naiman et al. 2009; Di Salvo et al. 2012; Escobar Ortega et al. 2020). Furthermore, it has been detected that, under certain conditions, inoculation with certain PGPR modifies the activities of microorganisms linked to nitrogen mineralization and biological fixation in rice and wheat crops (D'Auria et al. 2012; García de Salamone et al. 2009; Ferrando 2013; Acosta et al. 2014). Other studies show that the combined inoculation of *A. brasilense* and *Bradyrhizobium japonicum* increased the number of nodules associated with the main root and the nitrogen content in soybean plants with respect to the plants inoculated only with *B. japonicum* (Groppa et al. 1998; Puente et al. 2010). Previous studies of the impact of cereals on microbial communities had focused on finding variables related to the activity of soil microorganisms that could indicate the influence of crop rotation in different soils of the Pampas Region managed under zero tillage. These studies indicated that the number of fungi and actinobacteria in the soil are biological indicators that allow the detection of seasonal variations in the annual wheat-soybean sequence in different locations of the Pampas region (Rorig et al. 2004; García de Salamone et al. 2006). These authors have shown that the genus *Pseudomonas* showed a significant number variation, which could indicate variations in the ability to carry out different functions in the agroecosystem, since this genus includes exponents that are environmental detoxifiers, PGPR, and biological controllers of pathogen microorganisms (Drobkinsky et al. 2016).

Carbon and nitrogen mineralization are carried out by cellulolytic and nitrifying microorganisms that vary with the conditions imposed by crop sequence in different edaphoclimatic situations. Thus, wheat crops favor the activity of these functional groups of microorganisms, while soybean crops reduce them (Rorig et al. 2004). In a comparison between oats and rye, in annual succession with soybeans, differences were observed in the most probable numbers of cellulolytic and nitrifying microorganisms of the rhizosphere (Escobar Ortega et al. 2020). Likewise, modifications were also observed in the number of these microbial groups and the microaerophilic nitrogen-fixing microorganisms in different phenological stages of corn when nitrogen fertilization and PGPR inoculation did not change their numbers (Table 14.1). On the contrary, the most probable numbers of the microaerophilic diazotrophs did not show change due to nitrogen fertilization or inoculation with *A. brasilense* in any of the phenological stages of wheat (Di Salvo et al. 2018b). This is an indication that there are important aspects to consider when selected agricultural practices associated with crop management are applied.

14.5 Biological Indicators of Soil Quality

Among the biological indicators of soil quality, some soil microbial enzymes are interestingly relevant because they are related to metabolic activity and bioavailability of essential nutrients, and they can also respond rapidly to changes caused by both natural and anthropogenic factors (Ferreras et al. 2009). Regarding the link of the enzymatic activity of the soil with the inoculation with PGPR, *Azospirillum* plant colonization plays a role in stress mitigation and in the activity of the enzyme nitrate reductase which is associated with the plant nitrogen assimilation, as well as the role of enzymes involved in the detoxification of active oxygen species, such as ascorbate peroxidase, catalase, glutathione reductase, and superoxide dismutase, among others, in wheat plants (Zawoznik et al. 2009) and rice (Ruíz-Sánchez et al. 2011) subjected to abiotic stresses (Iannone et al. 2012, 2013).

Among the bioindicators of soil quality is the symbiotic associations between certain soil fungi and plant roots which are called mycorrhizae. The arbuscular type of mycorrhiza is established in most plants (Koide and Mosse 2004). This symbiosis increases soil exploration, nutrient uptake, and enhances soil aggregation (Rillig et al. 2002; Richardson et al. 2009), for those reasons, it is important for the sustainability of ecosystems (Jeffries et al. 2003; García de Salamone et al. 2006). In this sense, it was determined that chemical fertilization affects the natural mycorrhization of wheat crops, while the inoculation with the 40M and 42M strains of *A. brasilense* did not affect the natural mycorrhization of the wheat or corn crops (Di Salvo et al. 2020). Likewise, it was determined that phosphorous fertilization of the wheat affected spore density and phenotypical diversity of the mycorrhiza forming fungi present in the rhizosphere at tillering and grain-filling phenological stages (Table 14.3). At the grain-filling phenological stage, the nitrogen fertilization changed the percentages of mycorrhiza fungal structures in the roots while the inoculation with *A. brasilense* did not change them (Table 14.4). This is an indication that PGPR inoculation favors ecosystem sustainability in contrast to chemical fertilization.

Table 14.3 Percentages of changes with respect to the control of phosphorous application on density of spores of arbuscular mycorrhiza and Shannon's diversity index of their morphotypes in wheat rhizospheres at tillering and grain-filling phenological stages

	Percentages of change with respect to control without fertilization			
	Spore density		Shannon diversity index	
	Tillering	Grain-filling	Tillering	Grain-filling
Triple superphosphate application (kg ha $^{-1}$)				
60	94.4 (*)	71.4 (*)	11.1	25.0 (*)
120	−44.4	−7.2	−11.9	−8.3

(*) indicates statistical differences at $p \leq 0.05$, between absolute values with respect to control treatment without fertilization. Compiled from Di Salvo et al. (2020)

Table 14.4 Percentages of change with respect to control of nitrogen fertilization and PGPR inoculation on structures of wheat mycorrhiza at grain-filling phenological stage

	Percentage of change with respect to control of each factor		
	Arbuscules	Vesicles	Spores
Nitrogen fertilization (kg urea ha^{-1})			
46	−9 (*)	−0.4	5.5 (*)
PGPR inoculation treatments (1)			
A.b 40 M	−2.2	−0.4	0.4
A.b 42 M	−1.0	−2.5	2.4
A.b 40 M + 42 M	−7.2	0.4	2.2

A.b: Azospirillum brasilense (1) Experimental inoculant. (*) indicates statistical differences at $p \leq 0.05$, between absolute values with respect to factor control treatment. Compiled from Di Salvo et al. (2020)

14.6 Conclusion

There are biotechnological tools to promote sustainable agriculture. Inoculation with good quality formulations of PGPR that can establish effective beneficial associations is one of them. Another tool is linked to plant breeding since it should focus on obtaining high-yield genotypes that favor the establishment of rhizosphere microorganisms with high potential to increase agricultural production in response to the growing demand for good quality grains. Furthermore, it is necessary to monitor microbial ecology to reduce the negative impacts of traditional agricultural practices with the goal of achieving sustainable development objectives.

References

Acosta C, Di Salvo LP, Escobar Ortega JS, Zubillaga MM, García de Salamone IE (2014) Nitrógeno potencialmente mineralizable y producción de trigo fertilizado e inoculado con rizobacterias. XXIV CACS. ISBN: 978-987-24771-6-5

Aeron A, Kumar S, Pandey P, Maheshwari DK (2011) Emerging role of plant growth promoting rhizobacteria in agrobiology. In: Maheshwari DK (ed) Bacteria in agrobiology: crop ecosystems. Springer, Berlin, pp 1–36

Altieri MA, Nicholls CI (2000) Applying agroecological concepts to development of ecologically based pest management strategies. In: National Research Council (ed) Professional societies and ecologically based pest management: proceedings of a workshop. National Academy Press, Washington DC, 60 pp

Antoun H, Prevost D (2005) Ecology of plant growth promoting rhizobacteria. In: Siddiqui ZA (ed) *PGPR:* biocontrol and biofertilization. Springer, Dordrecht, pp 1–38

Avis TJ, Gravel V, Antoun H, Tweddell RJ (2008) Multifaceted beneficial effects of rhizosphere microorganisms on plant health and productivity. Soil Biol Biochem 40:1733–1740

Baca BE, Jijón-Moreno S, Pedraza RO, Campero-Romero AN, Aura N, Soto-Urzúa L, García de Salamone IE, Bonilla-Buitrago R (2012) Genetic diversity of genes involved in indole-3-acetic acid production in the plant growth promoting bacterium *Azospirillum*. VIII Congreso Argentino de Microbiología General SAMIGE. Mar del Plata

Bakker MG, Manter DK, Sheflin AM, Weir TL, Vivanco JM (2012) Harnessing the rhizosphere microbiome through plant breeding and agricultural management. Plant Soil 360:1–13

Bashan Y (1999) Interactions of *Azospirillum* spp. in soils: a review. Biol Fertil Soils 29:246–256

Bashan Y, de Bashan LE (2010) How plant growth-promoting bacterium *Azospirillum* promotes plant growth—a critical assessment. In: Spark D (ed) Advances in agronomy, vol 108. Academic Press, Burlington, pp 77–136

Bashan Y, Holguin G, de-Bashan LE (2004) *Azospirillum*-plant relationships: physiological, molecular, agricultural, and environmental advances (1997–2003). Can J Microbiol 50:521–577

Bottini R, Cassán F, Piccoli P (2004) Gibberellin production by bacteria and its involvement in plant growth promotion and yield increase. Appl Microbiol Biotechnol 65:497–503

Breseghello F, Guedes Coelho AS (2013) Traditional and modern plant breeding methods with examples in Rice (*Oryza sativa* L.). J Agric Food Chem 61:8277–8286

Carlino ME, Di Salvo LP, Cellucci GC, García de Salamone IE (2013) Funcionalidad de las comunidades microbianas rizosféricas de maíz a campo en relación a la fertilización y la inoculación con *Azospirillum brasilense*. XIII Congreso Argentino de Microbiología y II Congreso Argentino de Microbiología Agrícola y Ambiental. http://aam.org.ar/src/img_up/18102014.0.pdf

Cassán F, Coniglio A, López G, Molina R, Nievas S, de Carlan CLN, Donadio F, Torres D, Rosas S, Olivera Pedrosa F, de Souza E, Díaz Zorita M, de-Bashan L, Mora V (2020) Everything you must know about *Azospirillum* and its impact on agriculture and beyond. Biol Fertil Soils. https://doi.org/10.1007/s00374-020-01463-y

Cassán FD, Díaz Zorita M (2016) *Azospirillum* sp. in current agriculture: From the laboratory to the field. Soil Biol Biochem 103:117–130

Cassán FD, García de Salamone IE (2008) *Azospirillum* sp.: Cell physiology, plant interactions and agronomic research in Argentina. Asociación Argentina de Microbiología, Buenos Aires, Argentina. http://www.aam.org.ar/vermas-otras_publicaciones.php?n=69

Cassman KG (1999) Ecological intensification of cereal production systems: yield potential, soil quality, and precision agriculture. Proc Nat Acad Sci 96:5952–5959

Cirilo AG, Dardanelli J, Balzarini M, Andrade FH, Cantarero M, Luque S, Pedrol HM (2009) Morpho-physiological traits associated with maize crop adaptations to environments differing in nitrogen availability. Field Crop Res 113:116–124

Cummings SP (2009) The application of plant growth promoting rhizobacteria (PGPR) in low input and organic cultivation of graminaceous crops; potential and problems. Environ Biotechnol 5: 43–50

D'Auria F, Di Salvo LP, García de Salamone IE (2012) Microbiología de la fitorremediación de suelos contaminados con hidrocarburos. XIX Congreso Latinoamericano de la Ciencia del Suelo. XXIII Congreso Argentino de la Ciencia del Suelo. ISBN 978-987-1829-11-8

de Bashan LE, Holguin G, Glick BR, Bashan Y (2007) Bacterias promotoras de crecimiento en plantas para propósitos agrícolas y ambientales. In: Ferrera-Cerrato R, Alarcón A (eds) Microbiología Agrícola: Hongos, Bacterias, micro y macrofauna, control biológico y planta-microorganismo. Editorial Trillas, Mexico, DF., pp 161–215

Den Herder G, Van Isterdael G, Beeckman T, De Smet I (2010) The roots of a new green revolution. Trends Plant Sci 15:600–607

Di Salvo LP, García de Salamone IE (2012) Evaluation of soil-microbial communities by their CLPP. Standardization of a laboratory technique to replace commercial available microplates. Ecol Austral 22:129–136

Di Salvo LP, García de Salamone IE (2019) PGPR inoculation and chemical fertilization of cereal crops, how do the plants and their rhizosphere microbial communities' response? (Chapter 4) In: Singh DP, Gupta VK, Prabha R (eds) Microbial Interventions in Agriculture and Environment. Vol. 2. Rhizosphere, Microbiome and Agro-ecology. Springer, Singapore, pp 123–148

Di Salvo LP, Silva E, Tenório JO, Teixeira KRS, Garcia de Salamone IE (2010) The *Azospirillun brasilense* Can Fix N in Presence of Non-Preferable Carbon Sources and Under Oxygen Non-Controlled Conditions. 12th International Symposium on Biological Nitrogen Fixation with Non-Legumes and 2nd International INCT Symposium on Biological Nitrogen Fixation, Armação de Buzios, RJ

Di Salvo LP, Carlino ME, Cellucci GC, García de Salamone IE (2012) *Azospirillum brasilense* y fertilización nitrogenada: incremento sustentable del rendimiento de maíz. XIX Congreso Latinoamericano de la Ciencia del Suelo. XXIII Congreso Argentino de la Ciencia del Suelo. ISBN 978-987-1829-11-8

Di Salvo LP, Cellucci GC, Carlino ME, García de Salamone IE (2013a) Inocular el cultivo de maíz con *Azospirillum brasilense* y fertilizarlo con urea incrementa el rendimiento y altera temporalmente los microorganismos rizosféricos asociados a los ciclos del carbono y el nitrógeno. Simposio Fertilidad 2013. Rosario

Di Salvo LP, Campillay MA, García de Salamone IE (2013b) La inoculación con *Azospirillum brasilense* como herramienta para mejorar la eficiencia de uso de los fertilizantes a campo. IX Reunión Nacional Científico Técnica de Biología de Suelos. I Congreso Nacional de Biología Molecular de Suelos. Santiago del Estero

Di Salvo LP, Carlino ME, Cellucci GC, Perdoménico P, García de Salamone IE (2013c) Análisis de perfiles fisiológicos en cultivos de grano bajo inoculación y fertilización. Tercera Jornadas del INBA "Aportes de la Microbiología a la Producción de Cultivos". ISBN: 978-987-29338-2-1

Di Salvo LP, Silva E, Teixeira KRS, Esquivel-Cote R, Pereyra MA, García de Salamone IE (2014a) Physiological and biochemical characterization of *Azospirillum brasilense* strains commonly used as plant growth-promoting rhizobacteria. J Basic Microbiol 54:1310–1321

Di Salvo LP, Campillay MA, García de Salamone IE (2014b) *Azospirillum brasilense* inoculation and fertilization increase wheat growth and grain yield but do not modify the number of some cultivable rhizosphere microorganisms associated with carbon and nitrogen cycles. II Workshop Latinoamericano de PGPR

Di Salvo LP, Cellucci GC, Carlino ME, García de Salamone IE (2018a) Lant growth-promoting rhizobacteria inoculation and nitrogen fertilization increase maize (*Zea mays* L.) grain yield and modified rhizosphere microbial communities. App Soil Ecol 126:113–120

Di Salvo LP, Ferrando L, Fernández Scavino A, García de Salamone IE (2018b) Microorganisms reveal what plants do not: wheat growth and rhizosphere microbial communities after *Azospirillum brasilense* inoculation and nitrogen fertilization under field conditions. Plant Soil 424:405–417

Di Salvo LP, Groppa MD, García de Salamone IE (2020) Natural arbuscular mycorrhizal colonization of wheat and maize crops under different agricultural practices (Chapter 4) In: Yadav AN, Mishra S, Kour D, Yadav N, Kumar A (eds) Agriculturally important fungi for sustainable agriculture. Vol. 1. Perspective for diversity and crop productivity. Springer, pp 89–108

Díaz Herrera SM, Grossi CEM, Zawoznik MS, Groppa MD (2016) Wheat seeds harbour bacterial endophytes with potential as plant growth promoters and biocontrol agents of *Fusarium graminearum*. Microbiol Res 186:37–43

Díaz-Zorita M, Fernández Canigia MV (2009) Field performance of a liquid formulation of *Azospirillum brasilense* on dryland wheat productivity. Eur J Soil Biol 45:3–11

Dikshit A, Shukla SK, Mishra RK (2013) Exploring nanomaterials with PGPR in current agricultural scenario. Lambert Academic Publishing Saarbrucken Germany. 51 pp

Dobbelaere S, Croonenborghs A, Thys A, Ptacek D, Vanderleyden J, Dutto P, Labandera-González C, Caballero-Mellado J, Aguirre J, Kapulnik Y, Brener S, Burdman S, Kadouri D, Sarig S, Okon Y (2001) Responses of agronomically important crops to inoculation with *Azospirillum*. Aust J Plant Physiol 28:871–887

Dominguez CP, Rodríguez MV, Batlla D, García de Salamone IE, Mantese AI, Andreani AL, Benech-Arnold RL (2019) Sensitivity to hypoxia and microbial activity are instrumental in pericarp-imposed dormancy expression in sunflower (*Helianthus annuus* L.). Seed Sci Res 29: 85–96

Drobkinsky DK, Tafner R, Moreno MV, Stenglein SA, García de Salamone IE, Nelson LM, Novák O, Strnad M, van der Graaff E, Roitsch T (2016) Cytokinin production by *Pseudomonas fluorescens* G20-18 determines biocontrol activity against *Pseudomonas syringae* in *Arabidopsis*. Sci Rep 6:23310. https://doi.org/10.1038/srep23310. 1–11

Ellis RJ, Morgan P, Weightman AJ, Fry JC (2003) Cultivation-dependent and -independent approaches for determining bacterial diversity in heavy-metal-contaminated soil. App Environ Microbiol 69:3223–3230

Escobar Ortega JS, García de Salamone IE (2017) Dynamics of rhizosphere microbial communities of cover crops dried with glyphosate. In: Dhananjaya Pratap Singh et al. (eds) Plant-microbe interactions in agro-ecological perspectives. Springer, pp 17–34

Escobar Ortega JS, Aguilar Vásquez NN, Ávila Alba T, García de Salamone IE (2020) Impact of management of cover crop-soybean agroecosystems on rhizosphere microbial communities. Eur J Soil Sci. https://doi.org/10.1111/ejss.13057

Esquivel-Cote R, Ramírez-Gama RM, Tsuzuki-Reyes G, Orozco-Segovia A, Huante P (2010) *Azospirillum lipoferum* strain AZm5 containing 1-aminocyclopropane-1-carboxylic acid deaminase improves early growth of tomato seedlings under nitrogen deficiency. Plant Soil 337:65–75

Ferrando L (2013) Factores que afectan la composición y diversidad de la comunidad de bacterias endófitas en arroz (*Oryza sativa*). Doctoral Thesis Universidad de la Republica, Montevideo, Uruguay

Ferreira AS, Pires RR, Rabelo PG, Oliveira RC, Luz JMQ, Brito CH (2013) Implications of *Azospirillum brasilense* inoculation and nutrient addition on maize in soils of the Brazilian Cerrado under greenhouse and field conditions. App Soil Ecol 72:103–108

Ferreras L, Toresani S, Bonel BA, Fernández E, Bacigaluppo S, Faggioli VS, Beltrán C (2009) Parámetros químicos y biológicos como indicadores de calidad del suelo en diferentes manejos. Rev Ciencia del Suelo 27:103–114

Gamalero E, Glick BR (2011) Mechanisms used by plant growth-promoting bacteria. In: Maheshwari DK (ed) Bacteria in agrobiology: plant nutrient management. Springer, Berlin

García de Salamone IE (2000) Direct Beneficial Effects of Cytokinin-Producing Rhizobacteria on Plant Growth. Ph.D. Thesis: University of Saskatchewan, Canada. https://harvest.usask.ca/handle/10388/etd-10212004-001737

García de Salamone IE (2012a) Use of soil microorganisms to improve plant growth and ecosystem sustainability. In: Caliskan M (ed) The molecular basis of plant genetic diversity. INTECH, Rijeka, Croatia, pp 233–258. ISBN 978-953-51-0157-4. open access: http://www.intechopen.com/articles/show/title/use-of-soil-microorganisms-to-improve-plant-growth-and-ecosystem-sustainability

García de Salamone IE (2012b) Microorganismos promotores del crecimiento vegetal. Informaciones Agronómicas de Hispanoamérica 5:12–16

García de Salamone IE (2013) Utilización de la rizobacteria *Azospirillum brasilense* para aumentar el crecimiento vegetal y la sostenibilidad agrícola. In: Albanesi A (ed) Microbiología Agrícola, Un aporte de la investigación en la Argentina. Ediciones Magna, S.M. de Tucuman, pp 367–381

García de Salamone IE, Döbereiner J (1996) Maize genotype effects on the response to *Azospirillum* inoculation. Biol Fertil Soils 21:193–196

García de Salamone IE, Monzón de Asconegui MA (2008) Ecofisiología de la respuesta a la inoculación con *Azospirillum* en cultivos de cereales. In: Cassán F, García de Salamone IE (eds) *Azospirillum* sp.: cell physiology, plant interactions and agronomic research in Argentina, pp 209–226. http://www.aam.org.ar/vermas-otras_publicaciones.php?n=69

García de Salamone IE, Nelson LM (2006) Inoculación de plantas de trigo con una cepa de *Pseudomonas fluorescens* con probada capacidad de producir citoquininas. XXVI Reunión Argentina de Fisiología Vegetal

García de Salamone IE, Dobereiner J, Urquiaga S, Boddey RM (1996) Biological nitrogen fixation in *Azospirillum* strain-maize genotype associations as evaluated by the ^{15}N isotope dilution technique. Biol Fertil Soils 23:249–256

García de Salamone IE, Hynes RK, Nelson LM (2001) Cytokinin production by plant growth promoting rhizobacteria and selected mutants. Can J Microbiol 47:404–411

García de Salamone IE, Rorig M, Bordato F, Michelena R (2004) Actividad microbiana luego de la aplicación de compactación sobre un suelo Haplustol éntico bajo siembra directa. XVI y XII Congresos Latinoamericano y Colombiano de la Ciencia del suelo

García de Salamone IE, Hynes RK, Nelson LM (2005) Role of cytokinins in plant growth promotion by rhizosphere bacteria. In: Siddiqui ZA (ed) PGPR: biocontrol and biofertilization. Springer, Dordrecht, pp 173–195

García de Salamone IE, Rorig M, Di Salvo L, Michelena R (2006) Comunidades microbianas en capas superficiales de un suelo Haplustol éntico bajo siembra directa. XX Congreso Argentino de la Ciencia del Suelo. Salta, Argentina. 19-22 de Setiembre

García de Salamone IE, Escobar Ortega JS, Gatica M, Di Salvo LP, Vilches MV, Zubillaga MM, Urquiaga S (2009) Effect of *Azospirillum* inoculation on N-cycling microorganisms associated with rice and wheat crops. 16th Nitrogen Workshop, http://www.nitrogenworkshop2009.org/

García de Salamone IE, Di Salvo LP, Escobar Ortega JS, Boa Sorte MP, Urquiaga S, Dos Santos Teixeira KR (2010) Field response of rice paddy crop to *Azospirillum* inoculation: physiology of rhizosphere bacterial communities and the genetic diversity of endophytic bacteria in different parts of the plants. Plant Soil 336:351–362

García de Salamone IE, Funes JM, Di Salvo LP, Escobar Ortega JS, D'Auria F, Ferrando L, Fernandez Scavino A (2012) Inoculation of paddy rice with *Azospirillum brasilense* and *Pseudomonas fluorescens*: impact of plant genotypes on rhizosphere microbial communities and field crop production. Appl Soil Ecol 61:198–204

Garcia J, Kao-Kniffin J (2018) Microbial group dynamics in plant rhizospheres and their implications on nutrient cycling. Front Microbiol 9:1516

Gewin V (2010) An underground revolution: plant breeders are turning their attention to roots to increase yields without causing environmental damage. Nature 466:552–553

Glick BR (2012) Plant growth-promoting bacteria: mechanisms and applications. Scientifica https://doi.org/10.6064/2012/963401

Gopal M, Gupta A (2016) Microbiome selection could spur next-generation plant breeding strategies. Front Microbiol 7:1971

Groppa MD, Zawoznik MS, Tomaro ML (1998) Effect of co-inoculation with *Bradyrhizobium japonicum* and *Azospirillum brasilense* on soybean plants. Eur J Soil Biol 34:75–80

Hartmann A, Schmid M, van Tuinen D, Berg G (2009) Plant-driven selection of microbes. Plant Soil 321:235–257

Holguin G, Glick BR (2001) Expression of the ACC Deaminase Gene from *Enterobacter cloacae* UW4 in *Azospirillum brasilense*. Microbiol Ecol 41:281–288

Iannone MF, Rosales EP, Groppa MD, Benavides MP (2012) Reactive oxygen species formation and cell death in catalase-deficient tobacco leaf discs exposed to paraquat. Biol Trace Elem Res 146:246–255

Iannone MF, Rosales EP, Groppa MD, Benavides MP (2013) H_2O_2 involvement in polyamine-induced cell death in tobacco leaf discs. J Plant Growth Regul 32:745–757

Iannone MF, Groppa MD, de Sousa ME, Fernández van Raap MB, Benavides MP (2016) Cadmium induces different biochemical responses in wild type and catalase-deficient tobacco plants. Environ Exp Bot. https://doi.org/10.1016/j.envexpbot.2016.07.004

Iannucci A, Fragasso M, Beleggia R, Nigro F, Papa R (2017) Evolution of the crop rhizosphere: impact of domestication on root exudates in tetraploid wheat (*Triticum turgidum* L.) Frontiers. Plant Sci 8:2124

Jeffries P, Gianinazzi S, Perotto S, Turnau K, Barea JM (2003) The contribution of arbuscular mycorrhizal fungi in sustainable maintenance of plant health and soil fertility. Biol Fertil Soils 37:1–16

Koide RT, Mosse B (2004) A history of research on arbuscular mycorrhiza. Mycorrhiza 14:145–163

Leach AW, Mumford JD (2008) Pesticide environmental accounting: a method for assessing the external costs of individual pesticide applications. Environ Pollut 151:139–147

Lucy M, Reed E, Glick BR (2004) Applications of free living plant growth-promoting rhizobacteria. Antonie van Leewenhoek 86:1–25
Lugtenberg BJJ, Dekkers L, Bloemberg GV (2001) Molecular determinants of rhizosphere colonization by *Pseudomonas*. Annu Rev Phytopathol 39:461–490
Lynch JP (2007) Roots of the second green revolution. Aust J Bot 55:493–512
Malusá E, Sas-Paszt L, Ciesielska J (2012) Technologies for beneficial microorganisms inocula used as biofertilizers. Sci World J Article ID 491206, https://doi.org/10.1100/2012/491206
Marschner P, Rengel Z (2010) The effects of plant breeding on soil microbes. In: Dixon GR, Tilston EL (eds) Soil microbiology and sustainable crop production. Springer Science & Business, New York. https://doi.org/10.1007/978-90-481-9479-7. 436 pp.
Minz D, Ofek M, Hadar Y (2013) Plant rhizosphere microbial communities. In: Rosenberg E, DeLong EF, Lory S, Stackebrandt E, Thompson F (eds) The prokaryotes. Springer, Berlin. https://doi.org/10.1007/978-3-642-30123-0_38
Morrissey JP, Dow JM, Mark GL, O'Gara F (2002) Are microbes at the root of a solution to world food production? EMBO 5:922–926
Naiman AD, Latrónico AE, García de Salamone IE (2009) Inoculation of wheat with *Azospirillum brasilense* and *Pseudomonas fluorescens*: impact on the production and culturable rhizosphere microflora. Eur J Soil Biol 45:44–51
Newman MM, Hoilett N, Lorenz N, Dick RP, Liles MR, Ramsier C, Kloepper JW (2016) Glyphosate effects on soil rhizosphere-associated bacterial communities. Sci Total Environ 543:155–160
Okon Y (1994) *Azospirillum*/plant associations. CRC Press. Florida, ISBN: 0-8493-4925-7
Pedraza RO, Teixeira KRS, Scavino AF, García de Salamone IE, Baca BE, Azcón R, Baldani VLD, Bonilla R (2011) Microorganismos que mejoran el crecimiento de las plantas y la calidad de los suelos. Revista Corpoica - Ciencia y Tecnología Agropecuaria 11:155–164
Pérez-Jaramillo JE, Carrión VJ, Bosse M, Ferrão LFV, de Hollander M, Garcia AAF, Ramírez CA, Mendes R, Raaijmakers JM (2017) Linking rhizosphere microbiome composition of wild and domesticated Phaseolus vulgaris to genotypic and root phenotypic traits. ISME J 11:2244–2257
Philippot L, Raaijmakers J, Lemanceau P, van der Putten WH (2013) Going back to the roots: the microbial ecology of the rhizosphere. Nat Rev Microbiol 11:789–799
Pieterse CMJ, de Jonge R, Berendsen RL (2016) The soil-borne supremacy. Trends Plant Sci 21: 171–173
Puente M, Zawoznik M, Perticari A, Carletti S (2010) Determinación de la concentración óptima de *Azospirillum brasilense* co-inoculado con *Bradyrhizobium japonicum* en soja (*Glycine max* (L.) Merr.) XII Congreso Argentino de Microbiología. I Congreso de Microbiología Agrícola y Ambiental
Rani A, Goel R (2012a) Role of PGPR under different agroclimatic conditions. In: Maheshwari DK (ed) Bacteria in agrobiology: plant probiotics. Springer, Berlin, pp 169–183
Rani A, Goel R (2012b) Role of PGPR under different agroclimatic conditions. In: Maheshwari DK (ed) Bacteria in agrobiology: plant probiotics. Springer, Berlin, pp 169–183
Rascovan N, Carbonetto B, Revale S, Reinert MD, Alvarez R, Godeas AM, Colombo R, Aguilar M, Novas MV, Iannone L, Zelada AM, Pardo A, Schrauf G, Mentaberry A, Vazquez MP (2013) The PAMPA datasets: a metagenomic survey of microbial communities in Argentinean pampean soils. Microbiome 1:21
Reid A, Greene S (2012) How microbes can help feed the world. Report on an American Academy of Microbiology Colloquium Washington, DC. https://www.ncbi.nlm.nih.gov/books/NBK559436/pdf/Bookshelf_NBK559436.pdf
Rengel Z (2002) Breeding for better symbiosis. Plant Soil 245:147–162
Rengel Z (2005) Breeding crops for adaptation to environments with low nutrient availability. In: Ashraf M, Harris PJC (eds) Abiotic stresses: plant resistance through breeding and molecular approaches. The Haworth Press, New York, pp 239–276

Richardson AE, Barea JM, McNeill AM, Prigent-Combaret C (2009) Acquisition of phosphorus and nitrogen in the rhizosphere and plant growth promotion by microorganisms. Plant Soil 321: 305–339

Rillig MC, Wright SF, Eviner VT (2002) The role of arbuscular mycorrhizal fungi and glomalin in soil aggregation: comparing effects of five plant species. Plant Soil 238:325–333

Rorig M, Alderuccio S, Malcolm V, Olsen D, Michelena R, García de Salamone IE 2004 Estimaciones del número de microorganismos, producción de nitratos y actividad de fosfatasa alcalina en un suelo Argiudol vértico de la localidad de Ramírez, Entre Ríos, bajo siembra directa y durante la rotación trigo-soja. In: Monzón de Asconegui M, García de Salamone IE, Miyazaki S (eds) Biología del Suelo. Editorial Facultad de Agronomía, Buenos Aires, Argentina

Ruíz-Sánchez M, Armada E, Muñoz Y, García de Salamone IE, Aroca R, Ruíz-Lozano JM, Azcón R (2011) *Azospirillum* and arbuscular mycorrhizal colonization enhance rice growth and physiological traits under well-watered and drought conditions. J Plant Physiol 168:1031–1037

Sadras V, Lawson C (2011) Genetic gain in yield and associated changes in phenotype, trait plasticity and competitive ability of South Australian wheat varieties released between 1958 and 2007. Crop Pasture Sci 62:533–549

Sessitsch A, Mitter B (2015) 21st century agriculture: integration of plant microbiomes for improved crop production and food security. Microb Biotechnol 8:32–33

Shen J, Li C, Mi G, Li L, Yuan L, Jiang R, Zhang F (2013) Maximizing root/rhizosphere efficiency to improve crop productivity and nutrient use efficiency in intensive agriculture of China. J Exp Bot 64:1181–1192

Siddiqui ZA (2005) PGPR: biocontrol and biofertilization. Springer, Dordrecht, p 318

Slafer GA, Andrade FH (1989) Genetic improvement in bread wheat (*Triticum aestivum*) yield in Argentina. Field Crops Res 21:289–296

Slafer GA, Kantolic AG, Appendino ML, Tranquilli G, Miralles DJ, Savin R (2015) Genetic and environmental effects on crop development determining adaptation and yield. Crop Physiol:285–319. https://doi.org/10.1016/b978-0-12-417104-6.00012-1

Suleiman B, Muhammad BL, Jakada BH, Vyas NL (2015) Rhizosphere microbiome and plant nutrition. IJETST 2:3208–3216

Tester M, Langridge P (2010) Breeding technologies to increase crop production in a changing world. Science 327:818–822

Tilman D (1999) Global environmental impacts of agricultural expansion: the need for sustainable and efficient practices. Proc Natl Acad Sci U S A 96:5995–6000

Tilman D, Cassman KG, Matson PA, Naylor R, Polasky S (2002) Agricultural sustainability and intensive production practices. Nature 418:671–677

Tollenaar M, Lee E (2002) Yield potential, yield stability and stress tolerance in maize. Field Crop Res 75:161–169

United Nations (2018) The Sustainable Development Goals Report. https://unstats.un.org/sdgs/files/report/2018/TheSustainableDevelopmentGoalsReport2018-ES.pdf

Urquiaga S, Jantalia CP, Alves BJR, Boddey RM (2004) Importancia de la FBN en el secuestro de carbono en el suelo y en la sustentabilidad agrícola. In: Monzón de Asconegui MA, García de Salamone IE, Miyazaki SS (eds) Biología del Suelo. Editorial Facultad de Agronomía, Buenos Aires, Argentina

Valentinuz O (2006) Maíz: uniformidad de plantas y tolerancia al estrés. In: Larrea DN, vol II, 1st edn. Ediciones INTA, Buenos Aires, pp 119–125

Vejan P, Abdullah R, Khadiran T, Ismail S, Nasrulhaq Boyce A (2016) Role of plant growth promoting Rhizobacteria in agricultural sustainability—a review. Molecules 21:573

Vejan P, Khadiran T, Abdullah R, Ismail S, Dadrasnia A (2019) Encapsulation of plant growth promoting Rhizobacteria—prospects and potential in agricultural sector: a review. J Plant Nutr 42:2600–2623

Walker V, Bertrand C, Bellvert F, Moënne-Loccoz Y, Bally R, Comte G (2011) Host plant secondary metabolite profiling shows a complex, strain-dependent response of maize to plant growth-promoting rhizobacteria of the genus *Azospirillum*. New Phytol 189:494–506

Walters WA, Jin Z, Youngblut N, Wallace JG, Sutter J, Zhang W, González-Peña A, Peiffer J, Koren O, Shi Q, Knight R, Galvina del Rio T, Tringe SG, Buckler ES, Dangl JL, Ley RE (2018) Large-scale replicated field study of maize rhizosphere identifies heritable microbes. PNAS 115: 7368–7373

Zawoznik MS, Rosales EP, Benavides MP, Groppa MD (2009) Colonización radical con *Azospirillum* como factor mitigador del estrés salino. Influencia de la cepa microbiana. II Congreso Internacional REDBIO-Argentina y VI Simposio Nacional de Biotecnología REDBIO

Zawoznik MS, Díaz Herrera S, Vázquez S, Groppa MD (2014) Search for endophytic diazotrophs in barley seeds. Brazilian J Microbiol 45:621–625

Zawoznik MS, Groppa MD, de Valois N, Alfonzo CE, Di Baggio Vega EG, Benavides MP, Iannone MF (2017a) ¿Podrán las nanopartículas de magnetita optimizar la simbiosis soja-*Bradyrhizobiumbium japonicum*? Jornada Interdisciplinaria sobre Cambio Climático. "PIUBACC Una década de desafíos". Ciudad Autónoma de Buenos Aires, Argentina. 21 de Noviembre de 2017. http://cyt.rec.uba.ar/piubacc/Paginas/Posters-jornadas.aspx

Zawoznik MS, Groppa MD, de Valois N, Alfonzo CE, Di Baggio Vega EG, Benavides MP, Iannone MF (2017b) Uso de nanopartículas de magnetita para optimizar la simbiosis soja-*Bradyrhizobiumbium japonicum.* Jornada Científica - Academia Nacional de Farmacia y Bioquímica 2017. Ciudad Autónoma de Buenos Aires, Argentina. 10 de Agosto de 2017. http://www.anfyb.com.ar/wp-content/uploads/2016/07/ANALES-2017.pdf

Significance of Microbial Enzyme Activities in Agriculture 15

Dixita Chettri, Bhaskar Sharma, Ashwani Kumar Verma, and Anil Kumar Verma

Abstract

Soil enzymatic processes that involve the decomposition of organic matter, geochemical nutrient cycling, and removal of contaminants are essential for the functioning of the soil environment. These activities help to maintain the fertility, health, and ecology of the soil necessary for sustainable agriculture. These enzymes promote the growth and production of agriculturally valuable crops by providing an interrelation between plant–soil environments for nutrient supply. From the production of various phytohormones and the acquisition of nutrients to biocontrol of pests and plant pathogens for crop protection, a diverse group of these enzymes is present in the soil environment playing a pivotal role in crop production. They can also act as an indicator of changes in soil quality as a measure of the level of pollution while some enzymes facilitate degradation of the hazardous aromatic compounds and pesticides. Microorganisms are the major source of soil enzymes. The enzymes are produced at either intracellular or extracellular locations. The soil characteristics such as the quantity and structure of organic matter along with the inhabitant organisms influence the enzyme levels and activities. Understanding of these microbial enzymes for their biotechnological application to increase production in the agriculture sector is necessary to provide sufficient food and fodder while giving a boost to the economy.

D. Chettri · A. K. Verma · A. K. Verma (✉)
Department of Microbiology, Sikkim University, Sikkim, India
e-mail: akverma@cus.ac.in

B. Sharma
School of Life and Environmental Sciences, Faculty of Science, Engineering, and Built Environment, Deakin University, Geelong, VIC, Australia

TERI School of Advanced Studies, New Delhi, India

R. Soni et al. (eds.), *Microbiological Activity for Soil and Plant Health Management*, https://doi.org/10.1007/978-981-16-2922-8_15

Keywords

Soil enzymes · Anthropogenic activities · Biogeochemical cycling · Pesticide · Bioindicator

15.1 Introduction

In the terrestrial ecosystems, the soil is one of the major constituents and the key matrix for agricultural activities. Since the soil supports the processes related to the biogeochemical cycling of the various macronutrients for the growth and survival of the living organisms, its proper functioning is essential (Bünemann et al. 2018). The soil is considered to be a dynamic system with the presence of enzymes in different forms, i.e. cell-associated, intracellularly associated with living cells, and as cell-free enzymes in the soil environment (Skujins 1978). In this context, the knowledge of the soil enzymes are essential as they play an active role in nutrient recycling, biotransformation processes as well as maintaining the structure and stability of soil along with the removal of pollutants (Piotrowska-Długosz 2014).

The enzymes present in the soil are mainly of microbial origin with a small amount coming from the vegetation, animals, and insects residing in the soil environment and can be both intracellular and extracellular. The intracellular enzymes are present within the living cells, while the extracellular enzymes can either be associated with the outer surfaces of the cell as ectoenzymes or as free enzymes in the soil. The free enzymes are then absorbed by the organic content and soil matrix making them resistant to denaturation. An enzyme may have multiple sources in the soil, making it hard to establish its exact origin.

These enzymes are analogous to the enzymes found in other ecosystems with different environmental factors influencing their activities such as the pH, temperature, soil texture, moisture content, and presence of different chemicals (Tabatabai 1994). The natural factors like geographical location, physicochemical properties, and organic matter content directly affect the enzyme activity by controlling the microbial count and the amount of enzyme produced along with its stability. The organic compounds though are important for the stability of the enzyme, they reduce its activity. The anthropogenic activities may have an impact on the soil physiochemical properties which indirectly influences the enzyme activity (Gianfreda and Bollag 1996).

The soil enzymes are important to maintain the balance of the ecosystem and are very sensitive to any change. Thus, they have also found their use in studies of soil health profile, especially to study the effect of various pollutants like fertilizers, heavy metals, pesticides, etc., and for the development of strategies for sustainable agriculture. These enzymes are critical for various fundamental soil activities; from cycling of nutrient flow and biotransformation of organic debris to the removal of hazardous chemicals. From an agricultural perspective, these activities are significant for increased yield of crops and thus the enzymes have long been studied for their role and implication in the soil environment (Sinsabaugh et al. 2008).

15.2 Principal Soil Enzymes in Sustainable Agriculture

There are a wide range of enzyme present in the soil such as phosphatases, sulfatase, chitinase, cellulase, urease, nitrogenase, amylase, invertase, amidase, protease, catalase, phenol oxidase, etc., with equal importance in soil health and fertility. Some of these enzymes have been discussed here.

15.2.1 Phosphatase Enzyme

The phosphatase enzyme is one of the most extensively studied soil enzymes. It is considered to be an important soil enzyme because of its ability to catalyze the hydrolysis of esters and anhydrides of phosphoric acid. Phosphatase enzyme along with other extracellular enzymes in soil environment bring about organic matter degradation, for example, organic phosphate mineralization. The reaction mediated by the soil phosphatase enzyme releases phosphate (P), which is taken up by the plants and soil microorganisms. These enzymes can be periplasmic (example in *Escherichia coli*), membrane bound, or extracellular. Many of the extracellular phosphatase enzymes are monomer but a few, for example, the enzyme from *Thermus aquaticus* is a trimer of 143-kD molecular weight (Obidi et al. 2018). A trimeric orthophosphoric monoester phosphohydrolase has been isolated from *Thermus* sp. strain Rt41A (Hartog and Daniel 1992). The name "phosphatase" includes a broad group of enzymes capable of carrying out similar functions. The Nomenclature Committee of the International Union of Biochemistry and Molecular Biology (IUBMB) classifies these phosphatase enzymes into five major groups; **(i)** phosphoric monoester hydrolases or phosphomonoesterases (EC 3.1.3), **(ii)** phosphoric diester hydrolases or phosphodiesterases (EC 3.1.4), **(iii)** triphosphoric monoester hydrolases (EC 3.1.5), **(iv)** enzymes acting on the phosphoryl-containing anhydrides (EC 3.6.1), and **(v)** enzyme acting on P-N bonds (EC 3.9), for example, phosphoamidase (EC 3.9.1).

These groups of phosphatase enzymes can be further subdivided according to their regulation, sensitivity to different phosphatase inhibitors, and the requirements of various metal ions such as Ca^{2+} and Mg^{2+} for the enzyme activity. Phosphoric monoester hydrolases include acid phosphomonoesterase (EC3.1.3.2) and alkaline phosphomonoesterase (EC 3.1.3.1), phosphoprotein phosphatase, phytases (EC 3.1.3.8 for 3-phytases and EC 3.1.3.26 for 4-phytases), and also nucleotidase. Phosphodiesterases hydrolysis involves one or two ester bonds of phosphodiester compounds. This group includes nuclease which causes hydrolysis of phosphodiester bond present in nucleic acid. Pyrophosphate phosphohydrolases (EC 3.6.1.1) are largely discussed because of its ability to hydrolyze pyrophosphate which can be used as a fertilizer (Dick and Tabatabai 1978).

Three components of rhizosphere are associated with active phosphatase enzymes; soil, plant roots, and the microorganisms. The phosphatases are quite adaptive enzymes and their activity often vary with the plant species. The intensity of exudation of such enzymes is dependent upon phosphorus (P) need of the plant species. Some plant roots also specifically stimulate phosphatase producing microorganisms (Greaves and Webley 1965). The soil surrounding the plant roots greatly supports the development of a large number of soil microorganisms including bacteria and fungi. Different phosphatase producing bacteria were isolated from Sangga Buana forest and their capability of organic phosphate hydrolysis were demonstrated. Three pre-eminent isolates, that included *Bacillus mycoides, Bacillus laterosporus, and Flavobacterium balustinum*, have shown strong ability to mineralize organic phosphate (Fitriatin et al. 2011). The common phosphatase producing fungi are from genus *Aspergillus*, e.g. *A. niger, A. terreus,* etc. The phosphatase activity is influenced by organic P substrate, for instance, the phosphatase activity of some soil organisms such as *Pseudomonas mallei, B. subtilis, A. niger,* and *Penicillium* sp. was found to be higher in medium containing phytic acid compared to medium with α-glycerophosphate, phenyl phosphate, and D-glucose-1-phosphate (Fitriatin et al. 2008).

Determination of phosphatase enzyme activity is necessary to broaden the understanding of various phosphatase enzymes and their further applications in agriculture and other fields. Two approaches commonly followed to determine the phosphatase enzyme activity include; (i) estimation of the released inorganic P after incubation of sample soil with organic P such as *β*-glycerophosphate and phytin, **(ii)** estimation of released organic moiety following the incubation of the soil sample with organic P such as *β*-naphthyl phosphate, *p*-nitrophenyl phosphate, and phenyl phosphate. Second approach is considered more reliable as the first approach is unsatisfactory and does not allow inorganic P fixation by the soil. One of the best methods based on the second approach was developed by (Tabatabai and Bremner 1969). The method involves colorimetric estimation of the released *p*-nitrophenol by incubating the soil sample with buffered solution of sodium *p*-nitrophenyl phosphate and toluene at 37 °C for a period of 1 h. The activity is dependent on soil organic matter content, organic P, inorganic P, Nitrogen (N) availability and is affected by several biotic and abiotic factors such as pH, soil water content and salinity, mycorrhizal association, etc. The correlation of acid and alkaline phosphatase with the soil pH was studied where it was concluded that with the increase in soil pH, alkaline phosphatase activity increases while acid phosphatase activity was found to be decreasing (Dick et al. 2000). Mycorrhizal association has also found to have a positive impact on the activity of acid phosphomonoesterase enzyme, thus supporting the enzyme in degrading soil bound P (van Aarle and Plassard 2010). Increase in soil moisture content also increases phosphatase activity, and this effect is more pronounced on alkaline phosphatase activity.

The phosphatase enzyme is considered to be a good indicator of soil condition reflecting the soil fertility and quality. In agricultural soil, phosphatase enzyme plays an important role being directly connected with phosphorus cycle. Several management practices on land affect the activity of phosphatases enzymes, hence it acts as an index for soil quality. Leguminous plants such as *Aspalathus, Cyclopia,* etc., require more P during the nitrogen fixation process. The need is fulfilled by the release of more phosphatase enzyme by the legume roots. The increase in enzyme activity allows assimilation of inorganic P to be available for the plant growth. This high activity of phosphatase enzyme in the soil is critical in increased yield from lands used for crop rotation system. Phosphatase enzyme activity increases the nutrient availability in the case of land used for crop rotation system. Such bi-culture crop system can be used to improve the P cycle and also can improve the physiochemical properties of soil. A combined application of vermicompost and mineral N fertilizer showed an increase in phosphatase activity when compared to the use of fertilizers alone. The activity increased after the addition of P fertilizer to low organic matter soil. Thus, phosphatase enzyme is critical in increasing the available P reserves, when this nutrient is limited. The phosphatase enzyme is important in soil conservation, increased P release, and agricultural sustainability.

15.2.2 Dehydrogenase Enzyme

Dehydrogenase enzyme is one of the important soil enzymes that is commonly used to assess the biological activity in the soil. They are categorized under the oxidoreductase enzyme class (E.C. 1.1.1). Such oxidoreductase enzyme brings about oxidation-reduction reaction, i.e. electron transfer reactions. This enzyme is found to be associated intracellularly with intact microbial cells and do not accumulate extracellularly. Dehydrogenase enzyme is an integral part of the microbial enzyme system, which includes the enzymes of respiratory metabolism, nitrogen metabolism, and citrate cycle. This allows them to reflect the overall microbial activities in the soil environment. The dehydrogenase enzyme carries out the oxidation of the soil organic matter by the transfer of hydrogen ions (proton) or electrons from organic substrates to the inorganic acceptors. Many dehydrogenases transfer hydrogen ion to nicotinamide adenine dinucleotide (NAD) or nicotinamide adenine dinucleotide phosphate (NADP). Active dehydrogenase enzyme has the capability to utilize O_2 and other compounds as the terminal electron acceptor, although most dehydrogenases are produced by anaerobic microorganisms. Therefore, the dehydrogenase activity (DHA) can reflect the overall metabolic ability of the soil. The DHA is often considered proportional to the biomass of soil microorganisms (Wolińska and Stępniewska 2012).

Soil dehydrogenase enzyme is produced by the soil microorganisms including bacteria, fungi, actinomycetes. The bacterium *Pseudomonas entomophila* is an abundant producer. DHA is dependent on the microbial host and does not carry out its function on its own. There are various methods for the determination of DHA in soil; the first method makes the use of 2,3,5-triphenyltetrazolium chloride (TTC). It is based on the assumption that, if acceptor O_2 is absent then TTC will act as a terminal acceptor of hydrogen ion in the dehydrogenase reaction system. TTC after accepting the proton forms a red colored compound triphenyltetrazolium formazan (TPF). Another method uses the reduction of Iodo-Nitro-Tetrazolium chloride (INT) substrate for determining DHA in soil. INT, similar to TTC is used as a terminal acceptor. In addition, INT is reduced to form formazan, and formazan calibration curve is used to measure the reduced INT. However, some limitations have been reported in both of these methods. TTC has low reactivity and requires a longer incubation time. It is toxic for microorganisms and is reduced by them at different rates. TTC reduction is often limited by O_2 availability and the transfer rate of electron is also recorded to be slow. The use of INT overcomes some of the limitations of TTC such as it is reduced rapidly and is less sensitive to oxygen. INT is also less toxic to the microorganisms but their water solubility is lower than TTC and their reduction is also dependent on soil composition. The third method of DHA estimation uses oxidation of NADH. Sample taken for this approach is 0.02 g and 0.04 g of freeze-dried soil or 0.2 g and 0.4 g of shifted biological culture soil. The soil sample is treated with NADH-Tris buffer and in later stages of the method, TRIS-carbonate buffer is added to arrest the oxidation of NADH. After several incubation stages the absorbance of supernatant of the solution is measured at 340 nm. The difference in absorbance between initial NADH concentration and concentration after 15–30 min incubation is converted in mg or millimole (mmol) of NADH with the help of a calibration curve. The DHA is then determined as mg or millimole (mmol) of oxidized NADH by 1 g of the soil sample during 1 h incubation.

Several soil factors stimulate and inhibit the DHA in soil. The factors stimulating the DHA include; the moisture content of soil, soil aeration rate (redox potential and O_2 diffusion rate), temperature, organic matter (OM) content, pH, and the changing seasons. A relation between soil oxygen status and dehydrogenase activity was established. Soil samples from Ap horizons of 11 Orthic Luvisols and 10 Haplic Phaeozems which were developed from loess were subjected to conditioning of 14 days at various temperatures (10, 20, and 30 °C) and soil water tension of 15.9 kPa. It was concluded that the DHA was increasing as the soil water content and conditioning temperature were increasing. The DHA was found to be increasing on an average of 4.6 fold for flooded treatment and 2.6 times for 15.9 kPa treatment, when the temperature was increased by 10 °C. The flooding treatment combined with an increase in temperature to 30 °C showed the increase of DHA on average of 129-fold as compared to 15.9 kPa treatment with 10 °C temperature (Brzezińska et al. 1998). A strong relationship between the DHA and the soil carbon content was established by Koper et al. (2008) in *Haplic podzol* soil sample (Koper et al. 2008). The correlation coefficient (r) was found out to be ranging between 0.56 and 0.98. Positive correlations between soil DHA and pH in the range of 4.1 (Ph_{KCl}) and 4.9 (Ph_{water}) were established. The result suggested that acidic conditions suppressed DHA (Fernández-Calviño et al. 2010). Soil factors inhibiting DHA include; depth of soil profile, use of pesticides in the soil, presence of heavy metals, etc. DHA of sludge amended soil was examined at different incubation periods. The inhibition of DHA was observed at all concentrations of the sludge. It was concluded that the lower DHA in sludge amended soil might be related to the heavy metal concentrations in the samplings (Reddy and Faza 1989).

Dehydrogenase enzyme is considered to be a sensitive bioindicator accessing the soil quality and fertility. Measurement of DHA of a particular soil sample allows researchers to examine and understand the agricultural practices conducted on the soil, such as use of pesticides, herbicides, organotins, sludges, deicing salts, industrial wastes, pentachlorophenol, etc., and other management practices such as crop rotation that affects the soil quality. Higher DHA indicates low doses of pesticides and vice-versa (Baruah and Mishra 1986). DHA has been used in the assessment of calcium peroxide and citric acid effects in petrol contaminated soil samples. The findings concluded that DHA is a good indicator of the H_2O_2 effects on the biochemical activity of soil polluted with petrol (Curyło and Telesiński 2020). DHA is also applied to evaluate the limiting nutrients for plant growth (Rossel et al. 1997) and also has been used as a criterion for determination of effects of toxic standards (such as Ag^+, Hg^{++}, Cr (VI), phenol, formaldehyde, etc.) in biological systems (Lenhard 1965). Dehydrogenase enzyme also shows the potential role in maintaining the soil health.

15.2.3 Sulfatase and Arylsulfatase

Sulfatase, a subclass of esterase enzymes converts organic sulfur form into an inorganic form in the soil through a chemical reaction as follows: $R.O.SO_3^- + H_2O = R.OH + H^+ + SO_4^{2-}$. As the majority of the sulfur in the soil exists as the sulfate esters, these enzymes are critical in sulfur cycling in the soil, i.e. their mineralization and bioavailability for the plants. These enzymes are widely dispersed throughout the soil and have been further classified according to the ester types, i.e. alkylsulfatases, arylsulfatases, glucosulfatases, myrosulfatases, steroidsulfatases, and chondrosulfatases. Among these different types of the sulfatases, the enzyme that beings about the hydrolysis of the aromatic esters (phenolic esters of sulfuric acid) of sulfur are the Arylsulfatase [23]. These enzymes hydrolyze the linkage between oxygen and sulfur (O-S) in the aromatic ester of sulfate ($R\text{-}O\text{-}SO_3^-$) to release phenol (R-OH) and sulfate (SO_4^{2-}) or sulfate sulfur (SO_4^-S) and were initially termed as phenolsulfatases. Since these enzymes were the first to be identified, greater studies have been done regarding their activity and importance in the soil (Tabatabai and Bremner 1970; Verma et al. n.d.).

On the basis of their catalytic domain, the sulfatases enzymes have been classified into three types. The type I sulfatases have 3-oxo-L-alanine as the key amino acid residue in their catalytic site, while the type II enzymes are dioxygenase dependent on non-heme iron and type III have a metallo-β-lactamase fold in their catalytic domain with the requirement for two zinc ions as the cofactor. The arylsulfatase (EC 3.1.6.1) belongs to the type I sulfatase with the 3-oxo-L-alanine bringing about nucleophilic attack on the sulfur atoms of their substrate (https://enzyme.expasy.org/EC/3.1.6.1) (Bairoch 2000). The different chemicals that include p-nitrophenyl sulfate, potassium phenolphthalein sulfate, potassium phenyl sulfate, and potassium nitrocatechol sulfate are also used as substrate by this enzyme (Verma et al. n.d.).

Sulfur is one of the macronutrients essential for growth of plants and all the organism in the soil. It is a fundamental part of the amino acid such as methionine and cysteine, vitamins (biotin, thiamine), plant alkaloids (alicin), defense compounds (mercapto), antioxidants (glutathionine), and sulfated carbohydrates that contains S in their structure. Most of the sulfur exists in organic form in the soil with esters of sulfate accounting for about 30 to 75% of this organic Sulfur (Chen et al. 2016). The microbes synthesize the sulfatase enzyme that releases inorganic sulfur from these organic sources and its production is induced by the low S conditions (Dotaniya et al. 2019).

The presence of sulfates in the soil is restricted by the process of leaching and atmospheric loss due to microbial activities like sulfate respiration. As the plants can only take up the sulfur only in its inorganic sulfate form, the role of arylsulfatase becomes very important in the plant nutrition as well as maintaining the geochemical cycle of sulfate. Inadequacy of S slows the synthesis of amino acids that reduces crop growth and ultimately has a negative impact on crop yields. Thus, it is extremely important to emphasize the role of arylsulfatases in S mobilization (Dotaniya et al. 2019).

The arylsulfatases are found to be present in both the prokaryotes and the eukaryotes; however, their presence in the higher plants is a subject of argument. In soil the microbes are the major source of these enzymes and are found to be present in the periplasmic space of the microbial cell (Knauff et al. 2003). However, knowledge regarding the specific microbial genera and species of microorganism involved in synthesis of arylsulfatase and mineralization of the organic sulfur is limited (Verma et al. n.d.). These enzymes can also indicate the presence of fungi, since only fungi contain ester sulfates, an arylsulfatase activity substrate (Kertesz and Mirleau 2004) and their activity is controlled by the C and N content of the soil (Dick 1997). Its presence in various soil systems is often directly linked with microbial biomass and immobilization rate of S (Dotaniya et al. 2019). It has also been noticed that activity decreases with increased soil depth (Baligar and Wright 1991). Along with the different environmental factors such as accumulation of heavy metals, pH, presence of various organic compounds in different quantity, soil texture, microbial load and composition, the activity of aryl sulfatase is also influenced by the concentration of the extra cellular arylsulfatases available in the soil (Chen et al. 2019; Verma et al.).

The past few decades have seen a constant decrease in sulfur concentration in the agriculture field. The suggested reasons have been extensive cultivation of crops demanding high sulfur, use of fertilizers with no sulfur, and development of technologies that reduces the rate of SO_2 emission in the environment. Considering the importance of this element, studies focused on arylsulfatase need to be conducted (Knauff et al. 2003).

Recently a lot of studies have been carried out on identification of the factors that affect the arylsulfatase activity. The results obtained from different experiments, however, were conflicting where laboratory-based experiments showed non-linear co-relation between the pH and the activity of arylsulfatase while field studies showed them to have a linear relationship. Similarly, different experiments showed contradictory results for the relationship between the soil properties (pH, organic content, texture, microbial load, carbon and nitrogen content) and arylsulfatase activity. Though the reasons for this inconsistency are not clear yet, it has been hypothesized that as these experiments were conducted either on the lab scales the results may not apply to the situation in the field. More of experiments on regional levels will provide a better idea of the factors controlling the activity of arylsulfatase.

Till date two research experiments have been reported to be performed on regional scale (Chen et al. 2019).

15.2.4 Urease Enzyme

Urease is N-related enzyme that is widely distributed in soil. These are partly extracellular and are released from the plant and microbial metabolism or from dead cells. They can also be intracellular, being associated with cytoplasm components and cell membrane. It is an important soil enzyme that is involved in the degradation of urea. Urease is among the first soil enzyme that was evaluated experimentally. The enzyme urease catalyzes the hydrolysis of urea (NH_2CONH_2) into ammonia (NH_3) and carbon dioxide (CO_2) in the following reaction: $NH_2CONH_2 + H_2O \rightarrow 2NH_3 + CO_2$. Urea is among the most commercially available nitrogen fertilizers and urease enzyme activity on urea imposes a significant problem of nitrogen losses in agricultural fields. In arable soils, urease activity often leads to ammonia and nitrite toxicity. Such toxicities are harmful for young seedlings. Ammonia generated by the activity of soil urease is often lost through volatilization which causes loss in applied nutrients and affects the environment negatively (such as soil acidification). Thus, new technologies are required for increasing the efficiency of urea (efficiency rarely exceeding 50% in normal conditions) as a fertilizer and also minimizing negative impacts of urea hydrolysis.

Sumner in 1926 obtained urease (urea amidohydrolase EC 3.5.1.5) in crystallized form from jack bean (*Canavalia ensiformis*) meal. Jack bean urease has the molecular weight of 480 kDa and has 47—sulfhydril (-SH) groups. 4 to 8 of such –SH groups are necessary for urease activity. Urease can also act on other substrates such as semicarbazide, hydroxyl urea, and dihydroxy urea, but the highest specificity is shown for urea. The hydrolysis involves non-peptide C-N bonds in the linear amides. Further studies on jack meal urease confirmed the presence of two essential atoms of bound metal ion (Ni^{2+}) per enzyme molecule and it is been considered as a metallo-enzyme (https://www.rcsb.org/3d-view/3LA4). The specific role of the metal ion is still to be confirmed, but it was believed that the metal ion (Ni^{2+}) is chelated into the amino acid residues of the enzyme and it is positioned at the active site. Thus, during the formation of enzyme–substrate complex, the metal ion might be involved in stabilization of polarized positive charge present on the carbonyl group and this may promote nucleophilic displacement reaction of N atom (Ladd and Jackson 1982).

Soil urease enzyme usually originates from soil microorganisms, plant residues and from animal wastes. 21 diverse families of plants Gramineae and Leguminaceae were reported to have urease and amidase enzyme activity (Frankenberger and Tabatabai 1982). Members of genus *Bacillus* are known for their high level of urease, particularly *B. pasteurii* (also known as *Sporosarcina pasteurii*). According to the response of the ureolytic bacteria to ammonium, they can be classified into two groups; **(i)** those bacteria whose urease activity is not inhibited in the presence of ammonia such as *Proteus vulgaris, S. pasteurii, Proteus mirabilis,* and *Helicobacter* *pylori*, **(ii)** those bacteria whose urease activity is affected by ammonia as an inhibitor like *P. aeruginosa, B. megaterium, Alcaligenes eutrophus,* and *Klebsiella aerogenes*. Soil fungi also produces a considerable amount of urease enzyme, 78–98% of soil fungi showed the ability to hydrolyze urea (Lloyd and Sheaffe 1973).

A better knowledge of urease enzyme activity can be implemented in understanding N element mineralization process and its response upon the use of inorganic fertilizers. It can also be applied to reflect the land use systems and several management practices. Different methods are used to determine urease activity. One important method still used today was developed by Kandeler and Gerber in 1988 (Kandeler and Gerber 1988). The method involved incubation of soil samples in 100 mL Erlenmeyer flasks with aqueous or buffered urea solution. Further stages of this approach include ammonium extraction with 1N KCl and 0.01N HCl, and colorimetric determination of NH_4^+ using a modified indophenol reaction. The method is characterized by high stability and sensitivity of the formed colored complex. However, this method is quite time consuming and hence quicker methods are also used. Recently this method has found its application as a common technique for evaluation of soil enzymes such as N-Acetyl-*β*-D-glucosaminidase and *β*-glucoside. Hoffman and Teicher also developed a method for assaying urease enzyme activity. The method involved modification of colorimetric technique. Following this method, the activity was expressed as mg NH_4^+-N (nitrogen content of ammonium ion) which was released per 100 g of oven-dried soil during the incubation of 4 h at 30 °C. A high throughput and reproducible colorimetric microplate method for determining urease activity was suggested by Corderó et al. (2019). The method followed incubation of soil slurries sample in 96 deep well blocks and urea solutions. Colorimetric technique was used for measurement of ammonium produced. The method provided several advantages over the original method by Kandeler and Gerber (1988) in being quicker and more accurate alternative (Cordero et al. 2019).

There are various factors affecting soil urease activities such as soil moisture content, pH, temperature, urea concentration in soil, organic matter, oxygen content, and various soil amendments (liming, herbicides, organic materials). The interrelationship between urease activity and variety of physical and chemical characteristics of soil was examined periodically over 1 year. Samples were taken from six field sites in Benton country, western Oregon: one site under native grassland, two sites under forest vegetation, and three sites were under the clover/grass pastures. The study indicated a positive correlation between soil organic matter and urease activity ($r = 0.59$). A set of regression analysis (statistical processes to estimate relationship between a dependent variable and one or multipe indipendent variable) indicated partial variability of urease enzyme by the monitored soil parameters ($r^2 = 0.509$). With the help of principal component analysis (PCA) 20 factors were generated originally, although only four of these factors were highly significant. These four parameters with high significant factor loadings included Ca^{++}, Mg^{++}, Na^+, and C: N ratio. PCA combined with these four factors represented 64.9% variability in urease activity. First factor represented 31.9% of variability, second factor expressing 15% variability and had high loadings for soil organic matter content, NH_4^+-N and organic-N, third factor with 9.7% variation, and the fourth factor expressing variation of 8.1%. Thus, availability of certain nutrients proved to be important parameters affecting levels of urease activity (Stott and Hagedorn 1980). As a primary goal, urease inhibitors have been extensively studied to increase the supply of N to soil in compliance with plant absorption as well as after reducing its volatile loss. Such inhibitors include quinine class of compounds, phenyl phosphorodiamidate (PPD), N-(n-butyl) thiophosphorictriamide (nBTPT), ammonium thiosulfate (ATS), etc. The effectiveness of such urease inhibitors depends upon their ability of improving the diffusion of NH_4^+-N away from high pH zones which are created upon urea hydrolysis. The application of nBTPT was studied on sandy soil (pH 5.2), where the inhibitor lowered both the pH and the ammonium ion concentration at the test site and an increased diffusion of urea away from the fertilizer microsite was also observed. The effect of this inhibitor is dependent on capacity of diffusion by the soil, as no pronounced effect of nBTPT was seen on clay soil (pH 8.2) (Christianson et al. 1993).

Environmental urease activity is often considered as a good indicator of the microbial communities in soil and soil pollution. Urease enzyme is also been studied to understand biomineralization process, i.e. $CaCO_3$ precipitation in soil, natural waters, and geological sediments. These enzymes can be studied for their role in nutrient cycling and yield sustainability. The proper application of urease inhibitors can lead to an increased efficiency of inorganic fertilizers on the agricultural land.

15.2.5 Cellulase

Cellulases are a family of enzymes that are known to catalyze cellulose degradation, a polysaccharide composed of glucose units connected via β-1,4 linkage. Structurally, cellulose is a linear polymer of D-glucose subunits liked by β-1,4-glycosidic bonds and has wide variation in their degree of polymerization. The cellobiose represents a disaccharide of β-1,4-linked glucose units (Ciolacu et al. 2011). Though conflicting reports concerning the composition of the cellulase system were provided in the early literature, in the present scenario, it has been acknowledged that there are three main types of enzymes in the cellulase system which are as follows: **(i)** endo-1,4- β-glucanase (EC 3.2.1.4), which randomly attacks and brings about the hydrolysis of the cellulose chains; **(ii)** exo-1,4-β-glucanase or the cellobiohydrolase (EC 3.2.1.91), which targets the non-reducing end of the cellulose chains and removes glucose monomers or cellobiose; and **(iii)** β-D-glucosidase (EC 3.2.1.21), which hydrolyzes the smaller chains of cellobiose and cellodextrins generated by the action of the two earlier enzymes into glucose monomer (Deng and Tabatabai 1994). The synergistic action of the endo- and the exo-glucanase generates cello-oligosaccharides and cellobiose which are then finally converted into simple sugar. Thus, the products generated by the degradation of cellulose are glucose along with cellobiose and other oligosaccharides of high molecular weight which ultimately gets converted to glucose (Sindhu et al. 2016). As these groups of enzymes are important in recycling of cellulose, they have an important role in nature.

The cellulose which encompasses almost 50% of the biomass generated by carbon fixation of CO_2 by the plant via photosynthesis is the major organic polymer present in the biosphere. Plant residues are the major source of soil cellulose as it is the primary component of the plant cell wall. A small amount of the cellulose is also derived from the fungi and other microorganisms residing in the soil (Deng and Tabatabai 1994). The celluloses are an important source of carbon for microbial growth in the soil environment, especially for agriculturally valuable microbes. However, these organisms cannot directly incorporate the cellulose itself and use its degradation products. Here the role of the cellulase enzyme becomes critical (Verma et al. n.d.).

The cellulases enzymes which are produced naturally by bacteria and fungi can degrade the cellulose from the plant, lichens, and β-D-glucans found in cereals. The saprophytes organisms are particularly the important producers of this enzyme. The bacterial species like *B. subtilis*, *Thermobifida fusca*, *B. amyloliquefaciens*, *Cellulomonas fimi*, *Ruminococcus albus*, *Acetivibrio cellulolyticus*, *Thermotoga maritima*, *P. fluorescens* are among the important cellulases producers (Turan et al. 2017). The fungal strains belonging to *Trichoderma* sp., *Gliocladium* sp., *Penicillium* sp., *Chaetomium* sp., that produce cellulase enzyme have also been found to promote plant development processes such as flowering, germination, strengthening the roots system, plant growth, and increased yield of the crops (Jayasekara and Ratnayake 2019).

The cellulase is widely used for plant growth promotion and as a control agent for various diseases of plants. A synergistic action of the enzyme cellulase along with hemicellulase and pectinase has been applied for the disintegration of the cell wall of various plant pathogen. Different bacterial and fungal species have been applied as plant growth-promoting rhizobacteria (PGPR) and biocontrol agents. The mechanism though not clearly understood, it was observed that these organisms produce the cellulase and other related enzymes. Moreover, synergy was observed between the production of antibiotics and cellulase by the bacterial species that were acting as biocontrol agents against plant pathogens of fungal origin.

In different countries, the farmers practicing traditional methods of farming have being using the leaves of *Gliricidia* sp. and straws which are a rich source of cellulose. Their addition has shown to increase nutrient availability and improve soil quality thus increasing crop yield. It clearly implies that the cellulolytic microorganisms contribute directly to this type of process. The research conducted on the cellulolytic organisms suggests that these microorganisms are involved in various processes in the soil. They help in the decomposition of the plant and other residues thereby increasing the amount of available nutrient, protect against pathogenic invasion and also help in colonization and penetration of the roots cereal crops for increasing their yield and nutritional value. However, lack of proper evidence and mechanism involved encourage further research in this area to characterize and improve the use of this enzyme in the agricultural field (Jayasekara and Ratnayake 2019).

Factors that affect cellulose degradation in soils should be understood, as the reactions involved provide readily accessible C for microorganism growth. The environmental factors such as pH, temperature, soil profile play an important role in determining the cellulase activity. One report suggests that the cellulase stimulates chitinase synthesis along with other cell wall degrading enzymes in the black soil compared to red soil. Different experiments also suggest that there is a negative correlation between the amount of different fungicides used and the cellulase activity in the soil. It was observed that the increased dosage of the fungicides cosan, thiram, sandolex, and zinels reduced the cellulase activity while calixin, captan, and benlate completely inhibited the enzyme. Thus, this enzyme can indicate regarding the soil properties and help in developing strategies for soil management. Considering their importance in plant growth and carbon cycling more information regarding the factors that influence cellulase activity needs to be studied (Verma et al. n.d.).

15.2.6 Chitinase Enzyme

Enzyme chitinase belongs to glycosyl hydrolases (GH) family and is present in a number of organisms such as bacteria, fungi, plants, actinomycetes, yeasts, arthropods, and humans. The enzyme was first observed by Bernard (Felse and Panda 2000). They have molecular weights ranging from 20 kDa to 90 kDa and perform diverse roles including physiological and bioconversion processes. They catalyze the hydrolysis of chitin $(C_8H_{13}O_5N)_n$ by breaking the β-1-4 glycosidic bonds and convert chitin to its monomer N-acetyl glucosamine. Chitin is one of the most widely distributed biopolymers. It is an inelastic linear polymer that is composed of the units of β-1,4-N-acetyl glucosamine (GlcNAc). Chitin is an important component of fungal cell wall, internal shells of cephalopods, cuticle of insects, arthropod exoskeleton, etc. The ability of chitinase to hydrolyze chitin has been receiving increased attention in the fields of agriculture, medicine, drug industries, food industries, and environment management.

Enzyme chitinases are classified into two major groups; Endo-chitinases and Exo-chitinases. The endo-chitinases (E.C 3.2.1.14) carry out random cleavage at the internal sites of chitin. This cleavage produces dimer di-acetylchitobiose and low molecular weight multimers, chitotriose and chitotetraose. The exo-chitinase is further classified into two subcategories; chitobiosidases (E.C 3.2.1.29) and β-1,4-N-acetyl glucosaminidases (E.C 3.2.1.30). Chitobiosidases catalyzes the progressive release of di-acetylchitobiose. This reaction begins at the non-reducing end of chitin microfibril. β-1,4-N-acetyl glucosaminidase catalyzes the cleavage of oligomeric products of chitobiosidases and endo-chitinases. The reaction results in generating GlcNAc monomers. Chitinolytic enzymes in the CAZY database (http://www.cazy.org/) (Lombard et al. 2014) are grouped in GH18, GH19, and GH20 on the basis of similarity in amino acid sequences. However, these enzymes differ markedly in their catalytic properties and amino acid sequences. GH18 and GH19 bring about chitin hydrolysis and are regarded as chitinases while GH20 includes enzymes such as chitobiase (breakdown of N-acetylglucosamine or chitobiose) and β-N-acetylhexosaminidase (breakdown of N-acetylgalactosamine or glucosamine). GH18 chitinases are categorized based on a catalytic region. This catalytic region contains TIM barrel $(\beta/\alpha)_8$ domain. The catalytic region of GH19 consist of lysozyme like domain rich in α-helix and the region has a characteristic deep cleft. The mechanisms of action are different in GH18 and GH19 family enzymes. GH18 uses substrate assisted mechanism, while GH19 uses acid-base mechanisms. GH18 includes chitinases produced from bacteria, fungi, and insects. GH19 includes the enzyme mainly produced by plants, bacteria, few viruses, and only one fungus (Han et al. 2016). Chitinases are further divided into five classes based on N-terminal sequences, signal peptides, inducer, and isoelectric pH. Class I is plant-based, class II found in bacteria, fungi, and some plants, Class III is dissimilar to class I and class II sequences, class IV is similar to class I, and Class V is associated with the plant-microbe interactions.

Microorganisms are important producers of hydrolytic enzyme chitinases. Actinobacteria, firmicutes, and certain members of proteobacteria are well known for their chitinolytic enzyme production and activity. Fungal chitinases are also produced at a very significant amount (Roopavathi et al. 2015). Numerous bacteria produce enzyme chitinases such as *Aeromonas, Alteromonas, Streptomyces, Klebsiella, Pseudomonas, Chromobacterium, Arthrobacter, Escherichia, Clostridium, Vibrio, Beneckea,* and *Serratia*. The molecular weight of the bacterial chitinases ranges from 20–60 kDa and they are active over a wide range of pH and temperature. Fungal chitinases contain five domains; N-terminal signal peptide region, catalytic domain, chitin-binding domain, serine/threonine rich-region, and C-terminal extension region. However, chitin-binding domain, C-terminal extension region, and serine/threonine rich-region are not present in most of the fungal chitinases, and these three domains seem to be unnecessary for the chitinase enzyme activity. This can be ascertained because naturally-occurring chitinases that lack these three regions are still found to be enzymatically active. Some important chitinases producing fungi include *Trichoderma harzianum* and *Aspergillus niger, Penicillium, Neurospora, Agaricus, Mucor, Conidiobolus, Beauveria, Myrothecium, Metarhizium, Stachybotrys, Lycoperdon, Lecanicillium,* etc. The production of plant chitinases is dependent upon the infection by phytopathogens and is also induced by growth regulator ethylene. Some plant chitinases are also produced in response to some environmental stress conditions such as drought, cold, high salt concentration, etc.

The chitinase activity can be determined to reflect further applications of this widely distributed enzyme. Rodriquez-Kabana et al. (1983) determined soil chitinase activity following the incubation of toluene treated soil sample with 1% (w/w) suspension of colloidal chitin for a period of 18 h at 37 °C temperature. After dilution, the amount of released N-acetyl glucosamine was assayed (Rodriguez-Kabana et al. 1983). In an experiment forest soil chitinase activity was determined by a highly sensitive method. This method involved the use of 4-methylumbelliferyl (4MU) derivatives of oligomers substrate N-acetyl-D-glucosamine. The soil sample and substrate were incubated with a buffer solution and 4MU released was subjected to fluorometric estimation (Ueno et al. 1991). Chitinase production is influenced by number of factors such as temperature, pH, concentration of chitin (substrate), carbon and nitrogen content of soil, soil depth, moisture content of soil, etc.

Microbial destruction of chitin in soil under different moisture conditions was studied. Using gas-chromatographic and luminescent-microscopic studies it was found that specific activity of respiration of chitinolytic microorganisms was higher at low redox potential with soil moisture content close to the total water capacity. It was also observed that the most intense chitin degradation was achieved in clay or clay loamy soil when compared to sandy soils (Yaroslavtsev et al. 2009). Concentration of chitin and incubation temperature also affects chitinase activity. Increase in concentration of chitin from 0–1.5% increased chitinase activity, above 1.5%, there was no further increase. Increasing the incubation temperature from 16.5 to 45.5 °C increased chitinase activity. The temperature and chitinase activity showed a positive correlation ($r = 0.980$) (Rodriguez-Kabana et al. 1983).

Chitinases can be used for biocontrol of phytopathogens. Number of chitin degrading bacteria have shown plant pathogen inhibiting abilities, for instance, *Streptomyces* sp. and *Paenibacillus* sp. against the effects of *Fusarium oxysporum* f. sp. *cucumerinum* on *Cucumis sativus* L. (fusarium wilt of cucumber). Chitinases can be used to decrease the chemical content of insecticides or pesticides (Yano et al. 2006). Different properties of enzyme chitinase can be used for removal of toxic compounds from the soil along with crop protection in agricultural practices that can be implemented with proper studies.

15.3 Factors Influencing Enzyme Activity

There are several environmental factors influencing the enzyme activity in soil which may maybe natural or anthropogenic in origin (Fig. 15.1).

The different natural factors include the geographical, physicogeological, and physico-chemical properties of soils, while the anthropogenic factors include environmental pollution, the use of fertilizer, biocontrol agents, and other agricultural practices. These variables or the factors can directly or indirectly alter the production and efficiency of catalytic activity of the soil enzymes. The reversible or irreversible

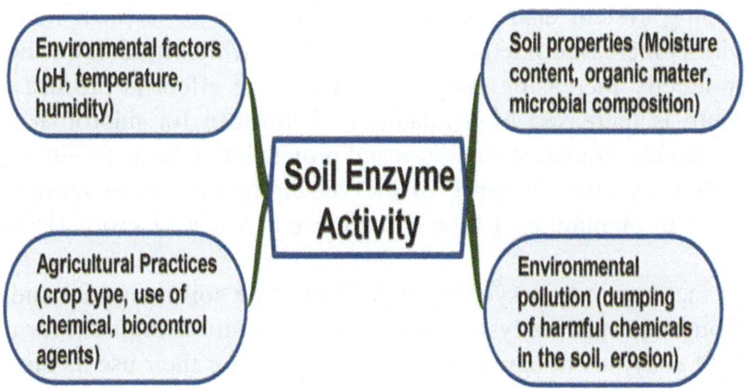

Fig. 15.1 The different factors influencing the soil enzymatic activity

modification in the structural conformation and enzyme catalysis is the direct impact of the environmental factors on enzyme activity. Similarly, the alteration of the size, composition, and function of the microbial communities in the soil directly influences the rate of enzyme production. Whereas the change in the climate and the different anthropogenic activities and agricultural practices like the use of pesticides and chemical fertilizers have an indirect impact on the enzyme activities. Since the function of the enzyme is easily influenced by these variables, the enzymes have been commonly used to recognize changes in soil ecology and hence used as a bioindicator of soil health.

The temperature, amount of precipitation, moisture content, soil texture, the content of the available forms of organic matter and nutrient, microbial load and composition, topological distribution, etc., can have both negative and positive influence on the enzyme. The features like a high content of colloidal particles in the soil, especially the organic compounds immobilize the enzymes helping in stabilizing and protecting them (Demkina et al. 2017). As most of these factors are interrelated among themselves and the seasonal variation, it is hard to clearly state their influence individually on the soil enzymes. A study revealed the influence of these factors on enzyme activity with an overall state of equilibrium of high activity for the enzymes under the natural condition (Paz-Ferreiro et al. 2011). Another study showed that during winter the soil temperature had a positive correlation with the enzyme activity in presence of high moisture content while during the summer the correlation was negative (Sardans et al. 2008).

The effect of different anthropogenic activities has also been analyzed. There are several studies where pesticides have shown a negative impact on the activity of different enzymes disturbing the local metabolism process. However, the reverse impact has also been reported. For instance, the use of quinalphos, an organothiophosphate insecticide, negatively influenced the dehydrogenase and alkaline phosphomonoesterase activity (Mayanglambam et al. 2005). Another study had a contradictory result with the insecticide acetamiprid showing a positive effect with a 22% increase in dehydrogenase activity after the first application (Singh and Kumar 2008). Similarly, a transient increase was seen in arylsulfatase phosphatase, dehydrogenase, and β-glucosidase activity by the fungicide metalaxyl (Sukul 2006).

The cropping system also influences enzyme activity as with increased plant biomass there is increased release of root exudates which modulates the growth of microbes indirectly increasing enzyme activity. The effect is profuse for legume plants as there is increased bioavailability of nitrogen for microbial growth and activity with reports of increased urease and protease activity with nitrogen fixation. Another evidence of the influence of the vegetation type is increased activity of arylsulfatase with plantation of high S requiring cruciferous crops (Dotaniya et al. 2019).

All these factors work in synergy to influence the soil microbial and enzymatic activities. Thus, a proper analysis of these factors in-situ is required for understanding the role of enzymes in agricultural practices and for their use as bioindicators.

15.4 Soil Enzymes as a Bioindicator of Soil Health

Soil is a lively natural medium for the growth of the plant and a functional source for terrestrial ecosystems. Soil quality is a measure of necessary elements for sustainable biological production to nourish humans, plants, and animals while balancing environmental integrity (Wienhold et al. 2005). The soil quality assessment is based on a unique balance of the chemical, physical, and biological components that serves as a dynamic and natural system for ecological activities. The soil's physical characteristics are density, texture, porosity, water holding capacity, particle size, and shape while chemical characteristics are electrical conductivity, pH, and nutrient levels. Soil biological components are the most sensitive parameters and largely include microbial activities in the soil. The soil is a sink of countless microorganisms that are necessary for maintaining soil health through numerous biochemical activities (Gundale et al. 2005). Microorganisms are playing a major role in ecological equilibrium and are the most suitable indicator for the detection of soil health and quality. The soil's biological activities are sensitive to numerous factors while soil physical and chemical parameters are mostly affected by extreme environmental conditions. The ever-increasing population and land degradation are continuously reducing the arable land area. The use of chemical fertilizers and pesticides has revolutionized agricultural production but the excess use of agrochemicals on the soil has severely damaged the soil health and quality. The inappropriate agricultural practices for high crop productivity sourced deterioration of the physical, chemical, and biological properties of the soil that posed serious environmental risks and threats to human, animal, and plant health. The extensive use of agrochemicals damages the soil health, causes soil, water, and air pollution, and presents a risk to the ecosystem (McLaughlin and Mineau 1995).

The soil enzymes play a vital role in the nutrient cycling in the soil and the source of soil enzyme can be a plant, microorganisms, and animals. The soil enzymes added by the plants and animal organic matter are degraded by proteases and are used to enrich humus. The biochemical reactions in the soil are catalyzed by the enzymes present in the soil and therefore can be used as an indicator for the soil health and quality assessment. Microorganisms are the major contributors to nitrogen, phosphorus, and sulfur nutrient cycling, removal of complex pollutant organic compounds, and absorption of heavy metals. Microbes are either beneficial to the crop grown or pathogenic for the plants. Soil enzymes are useful indicators for the estimation of pollutants, soil health, and remediation strategies. The soil enzyme activities rapidly change over time, respond to vital soil physical and chemical parameters (Badiane et al. 2001). The addition of organic matter increases the soil's biological activities and protects the soil enzyme activities. The soil enzymes indicate the soil fertility that is directly related to crop performance and various other environmental factors. Enzyme activities are dependent on the climatic conditions, chemical properties of compounds present in the cell, and soil depth. The commonly found soil enzymes are hydrolases, lyases, transferase, and oxidoreductases and which used for the assessment of soil quality. The hydrolytic enzymes are involved in the transformation of the nutrient molecules such as sulfur, nitrogen, carbon, and phosphorus through environmental cycles (Trasar-Cepeda et al. 2008). Soil dehydrogenase is the most assessed enzyme for the management of soil health and are major indicators of the microbial oxidative activities in living cells and are key factors for oxidation activities related to soil organic compounds (Zhang et al. 2010). The dehydrogenases transfer the hydrogen molecules from organic molecules to inorganic molecules and perform biological oxidation. The soil dehydrogenase provides information on microbial activity, pesticide applications, and soil management. The dehydrogenase is mostly produced by the *Pseudomonas sp.* in the soil (Wolińska and Stępniewska 2012). The urea is mostly used as a nitrogen fertilizer for crop production and is a major source of nitrogen. Urease enzyme catalyzes the degradation of the urea molecules into simpler form ammonia and carbonic acids (Tabatabai and Bremner 1969). The urease enzyme is used as an indicator of soil activity and can be influenced by extreme environmental factors and cropping cycles. The urease enzyme is produced by numerous microorganisms such as *Aerobacter aerogenes*, *Klebsiella sp.*, *Staphylococcus sp.*, *Aspergillus sp.*, plant species such as *Glycine max*, and *Gossypium hirsutum*, and aquatic alga such as *Nitellopsis obtusa* (Krajewska 2009). Cellulases are hydrolytic enzymes that degrade cellulose from soil organic content and the carbon released is available for the microorganisms for their survival. The most common microorganisms producing cellulase are *Thermomonospora*, *Aspergillus*, *Clostridium*, *Bacillus*, and *Trichoderma* (Kuhad et al. 2011).

The plants are a major source of abundant polymer lignin present in the soil. The peroxidases break lignin molecules and release nitrogen and carbon nutrient molecules that contribute to the survival of the microorganisms and balancing of the soil carbon and nitrogen pool (Sinsabaugh 2010). Moreover, the peroxidases can neutralize the reactive oxygen species and the toxic effect of phenolics and metal ions. The peroxidases are well known for their antioxidant properties in eukaryotes and are important indicators for estimation of the soil quality and health (Mangler and Tate 1982). Starch is another widely available polysaccharide and usually is present in organic matter. The amylases are responsible for the degradation of the starch molecules. The amylases are categorized into three major classes: alpha-amylase, beta-amylase, and glucoamylase. The alpha-amylase is extensively found in plants, animals, and soil environments (Thoma et al. 1971). The alpha-amylase degrades the starch molecules to produce glucose, maltose, and oligosaccharides. The beta-amylase breaks the starch into maltose and dextrin, while glucoamylase converts the maltose into glucose. In the soil, the amylases are produced by the *Bacillus sp.*, *Aspergillus sp.*, *Penicillium expansum*, plants, and animals (Singh and Kumari 2016). The phosphorus is directly related to the plant growth and soil ecosystem. The phosphorus is transformed into different forms during the phosphorus cycle and a class of enzymes called "phosphatase" is responsible for the production of the phosphorus ion from the substrate phosphorus complex (Nannipieri et al. 2011). Phosphorus is a critical nutrient for plant growth and is often found immobilized into organic and inorganic complexes. The phosphatase such as phosphomonoesterase is a versatile class that acts on low molecular weight compounds such as sugar phosphates, polyphosphates, and nucleotides (Pang and

Kolenko 1986). The phosphomonoesterase is stable and active in both acidic and alkaline environments. The phosphatase is mostly produced by *Aspergillus sp*, *Bacillus sp.*, *Penicillium sp.*, and *Pseudomonas sp.* in the soil (Tarafdar and Chhonkar 1979).

The extraction of the soil enzyme is a tedious and sensitive process. The traditional methods utilize spectrophotometric methods for the estimation of the enzyme activities. The enzymes are very sensitive to environmental stress and usually lose their activity at unfavorable temperature, pH, and substrate concentration. The latest developments in the field of genomics and proteomics have revolutionized the biological and biochemical activities of the soil through various sequencing methods such as metagenomics, metatranscriptomics, metaproteomics analyses. The sequencing methods can not only identify microbial species present in the soil but also the genes, enzymes, and metabolic compounds that are responsible for the balancing of the soil quality. There is a need to develop rapid and sophisticated methods to handle soil and soil enzyme activity for accurate estimation of soil quality.

15.5 Conclusion

The soil enzymes are an important component of the soil ecosystem and play a key role in soil health and management. They carry out all the biochemical reactions in the soil and control the recycling of all the important nutrients in the biosphere. These enzymes have a wide potential in the field of agriculture. They determine the fertility of the soil and also act as a protective barrier for the plants thus promoting faster growth with increased yield. Since these enzymes are naturally occurring biodegradable compounds, their use in the concept of sustainable agriculture is desirable. However, proper understandings of the different variables that have an influence on the activity of these enzymes need to be properly analyzed to enhance their efficiency. Since the different factors are affected by seasonal variations, a proper in-situ study on a regional scale can give a better understanding than the laboratory or the small field-based study. Also, the enzymes being extremely sensitive to the changes in the soil ecosystem are being explored for their application as indicators of soil health. Researches are still needed to be conducted for these factors and the enzyme distribution in the different soil profile.

Acknowledgments The authors would like to thank the Department of Microbiology, Sikkim University, TERI School of Advanced Studies, New Delhi, and School of Life and Environmental Sciences, Deakin University, Geelong, Australia for providing the computational infrastructure and central library facilities for procuring references and plagiarism analysis (Turnitin: Plagiarism Detection Software).

Conflict of Interest Statement The authors declare that the study was conducted in the absence of any commercial or financial relationships that could be construed as a potential conflict of interest.

References

Badiane NN, Chotte JL, Pate E, Masse D, Rouland C (2001) Use of soil enzyme activities to monitor soil quality in natural and improved fallows in semi-arid tropical regions. Appl Soil Ecol 18(3):229–238

Bairoch A (2000) The ENZYME database in 2000. Nucleic Acids Res 28(1):304–305

Baligar VC, Wright RJ (1991) Enzyme activities in appalachian soils: 1. Aryl-sulfatase. Commun Soil Sci Plant Anal 22(3–4):305–314

Baruah M, Mishra RR (1986) Effect of herbicides butachlor, 2, 4-D and oxyfluorfen on enzyme activities and CO 2 evolution in submerged paddy field soil. Plant Soil 96(2):287–291

Brzezińska M, Stępniewska Z, Stępniewski W (1998) Soil oxygen status and dehydrogenase activity. Soil Biol Biochem 30(13):1783–1790

Bünemann EK, Bongiorno G, Bai Z, Creamer RE, De Deyn G, de Goede R, Fleskens L, Geissen V, Kuyper TW, Mäder P, Pulleman M (2018) Soil quality–a critical review. Soil Biol Biochem 120:105–125

Chen H, Yang L, Wen L, Luo P, Liu L, Yang Y, Wang K, Li D (2016) Effects of nitrogen deposition on soil sulfur cycling. Glob Biogeochem Cycles 30(11):1568–1577

Chen H, Liu J, Li D, Xiao K, Wang K (2019) Controls on soil arylsulfatase activity at a regional scale. Eur J Soil Biol 90:9–14

Christianson CB, Baethgen WE, Carmona G, Howard RG (1993) Microsite reactions of urea-nBTPT fertilizer on the soil surface. Soil Biol Biochem 25(8):1107–1117

Ciolacu D, Ciolacu F, Popa VI (2011) Amorphous cellulose—structure and characterization. Cellul Chem Technol 45(1):13

Cordero I, Snell H, Bardgett RD (2019) High throughput method for measuring urease activity in soil. Soil Biol Biochem 134:72–77

Curyło K, Telesiński A (2020) Use of phosphatase and dehydrogenase activities in the assessment of calcium peroxide and citric acid effects in soil contaminated with petrol. Open Life Sci 15 (1):12–20

Demkina EV, Shanenko EF, Nikolaev YA (2017) Model of the regulation of activity of immobilized enzymes (amylases) in soil. Microbiology 86(2):231–240

Deng SP, Tabatabai MA (1994) Cellulase activity of soils. Soil Biol Biochem 26(10):1347–1354

Dick RP (1997) Soil enzyme activities as integrative indicators of soil health. In: Biological indicators of soil health, 1st edn. CAB International, New York

Dick WA, Tabatabai MA (1978) Hydrolysis of organic and inorganic phosphorus compounds added to soils. Geoderma 21(3):175–182

Dick WA, Cheng L, Wang P (2000) Soil acid and alkaline phosphatase activity as pH adjustment indicators. Soil Biol Biochem 32(13):1915–1919

Dotaniya ML, Aparna K, Dotaniya CK, Singh M, Regar KL (2019). Role of soil enzymes in sustainable crop production. In: Enzymes in food biotechnology. Academic Press pp 569–589

Felse PA, Panda T (2000) Production of microbial chitinases–a revisit. Bioprocess Eng 23 (2):127–134

Fernández-Calviño D, Soler-Rovira P, Polo A, Díaz-Raviña M, Arias-Estévez M, Plaza C (2010) Enzyme activities in vineyard soils long-term treated with copper-based fungicides. Soil Biol Biochem 42(12):2119–2127

Fitriatin BN, Joy B, Subroto T (2008) The Influence of organic phosphorous substrate on phosphatase activity of soil microbes. In: Proceeding of International Seminar of Chemistry, pp 30–31

Fitriatin BN, Arief DH, Simarmata T, Santosa DA, Joy B (2011) Phosphatase-producing bacteria isolated from Sanggabuana forest and their capability to hydrolyze organic phosphate. J Soil Sci Environ Manage 2(10):299–303

Frankenberger WT, Tabatabai MA (1982) Amidase and urease activities in plants. Plant Soil 64 (2):153–166

Gianfreda L, Bollag JM (1996) Influence of natural and anthropogenic factors on enzyme activity in soil. Soil Biochem 9:123–194

Greaves MP, Webley DM (1965) A study of the breakdown of organic phosphates by micro-organisms from the root region of certain pasture grasses. J Appl Bacteriol 28(3):454–465

Gundale MJ, DeLuca TH, Fiedler CE, Ramsey PW, Harrington MG, Gannon JE (2005) Restoration treatments in a Montana ponderosa pine forest: effects on soil physical, chemical and biological properties. For Ecol Manag 213(1–3):25–38

Han B, Zhou K, Li Z, Sun B, Ni Q, Meng X, Pan G, Li C, Long M, Li T, Zhou C (2016) Characterization of the first fungal glycosyl hydrolase family 19 chitinase (NbchiA) from *Nosema bombycis* (Nb). J Eukaryot Microbiol 63(1):37–45

Hartog AT, Daniel RM (1992) An alkaline phosphatase from *Thermus* sp. strain Rt41A. Int J Biochem 24(10):1657–1660

Jayasekara S, Ratnayake R (2019) Microbial cellulases: an overview and applications. In: Cellulose. Intechopen

Kandeler E, Gerber H (1988) Short-term assay of soil urease activity using colorimetric determination of ammonium. Biol Fertil Soils 6(1):68–72

Kertesz MA, Mirleau P (2004) The role of soil microbes in plant Sulphur nutrition. J Exp Bot 55:1939–1945

Knauff U, Schulz M, Scherer HW (2003) Arylsufatase activity in the rhizosphere and roots of different crop species. Eur J Agron 19(2):215–223

Koper J, Piotrowska A, Siwik-Ziomek A (2008) Dehydrogenase and invertase activities in a rusty soil in the neighbourhood of the Włocławek nitrogen plant "Anwil". Proc ECOpole 2 (1):197–202

Krajewska B (2009) Ureases I. functional, catalytic and kinetic properties: a review. J Mol Catal B Enzym 59(1–3):9–21

Kuhad RC, Gupta R, Singh A (2011) Microbial cellulases and their industrial applications. Enzyme Res 2011

Ladd JN, Jackson RB (1982) Biochemistry of ammonification. In: Stevenson FJ (ed) Nitrogen in agricultural soils. American Society Agron, Madison Publication, pp 173–228

Lenhard G (1965) Dehydrogenase activity as criterion for the determination of toxic effects on biological purification systems. Hydrobiologia 25(1–2):1–8

Lloyd AB, Sheaffe MJ (1973) Urease activity in soils. Plant Soil 39(1):71–80

Lombard V, GolacondaRamulu H, Drula E, Coutinho PM, Henrissat B (2014) The carbohydrate-active enzymes database (CAZy) in 2013. Nucl Acids Res 42(D1):D490–D495

Mangler JE, Tate RL III (1982) Source and role of peroxidase in soil organic matter oxidation in Pahokee muck. Soil Sci 134(4):226–232

Mayanglambam T, Vig K, Singh DK (2005) Quinalphos persistence and leaching under field conditions and effects of residues on dehydrogenase and alkaline phosphomonoesterases activities in soil. Bull Environ Contam Toxicol 75(6):1067–1076

McLaughlin A, Mineau P (1995) The impact of agricultural practices on biodiversity. Agric Ecosyst Environ 55(3):201–212

Nannipieri P, Giagnoni L, Landi L, Renella G (2011) Role of phosphatase enzymes in soil. In: Phosphorus in action. Springer, Berlin, pp 215–243

Obidi OF, Awe OO, Igwo-Ezikpe MN, Okekunjo FO (2018) Production of phosphatase by microorganisms isolated from discolored painted walls in a typical tropical environment: a non-parametric analysis. Arab J Basic Appl Sci 25(3):111–121

Pang PC, Kolenko H (1986) Phosphomonoesterase activity in forest soils. Soil Biol Biochem 18 (1):35–39

Paz-Ferreiro J, Trasar-Cepeda C, del Carmen Leirós M, Seoane S, Gil-Sotres F (2011) Intra-annual variation in biochemical properties and the biochemical equilibrium of different grassland soils under contrasting management and climate. Biol Fertil Soils 47(6):633–645

Piotrowska-Długosz A (2014) Enzymes and soil fertility. In: Gianfreda L, Rao MA (eds) Enzymes in agricultural sciences. OMICS Group E-Books, pp 44–79

Reddy GB, Faza A (1989) Dehydrogenase activity in sludge amended soil. Soil Biol Biochem 21(2)

Rodriguez-Kabana R, Godoy G, Morgan-Jones G, Shelby RA (1983) The determination of soil chitinase activity: conditions for assay and ecological studies. Plant Soil 75(1):95–106

Roopavathi AS, Vigneshwari R, Jayapradha R (2015) Chitinase: production and applications. J Chem Pharm Res 7(5):924–931

Rossel D, Tarradellas J, Bitton G, Morel JL (1997) Use of enzymes in soil ecotoxicology: a case for dehydrogenase and hydrolytic enzymes. Soil Ecotoxicol:179–206

Sardans J, Peñuelas J, Estiarte M (2008) Changes in soil enzymes related to C and N cycle and in soil C and N content under prolonged warming and drought in a Mediterranean shrubland. Appl Soil Ecol 39(2):223–235

Sindhu R, Binod P, Pandey A (2016) Biological pretreatment of lignocellulosic biomass–an overview. Bioresour Technol 199:76–82

Singh DK, Kumar S (2008) Nitrate reductase, arginine deaminase, urease and dehydrogenase activities in natural soil (ridges with forest) and in cotton soil after acetamiprid treatments. Chemosphere 71(3):412–418

Singh P, Kumari P (2016) Isolation and characterization of amylase producing bacillus spp. from selected soil sample. Int J Res Biosci 5(2):24–29

Sinsabaugh RL (2010) Phenol oxidase, peroxidase and organic matter dynamics of soil. Soil Biol Biochem 42(3):391–404

Sinsabaugh RL, Lauber CL, Weintraub MN, Ahmed B, Allison SD, Crenshaw C, Contosta AR, Cusack D, Frey S, Gallo ME, Gartner TB (2008) Stoichiometry of soil enzyme activity at global scale. Ecol Lett 11(11):1252–1264

Skujins J (1978) History of abiontic soil enzyme research. Soil enzymes. Academic Press

Stott DE, Hagedorn C (1980) Interrelations between selected soil characteristics and arylsulfatase and urease activities. Commun Soil Sci Plant Anal 11(10):935–955

Sukul P (2006) Enzymatic activities and microbial biomass in soil as influenced by metalaxyl residues. Soil Biol Biochem 38(2):320–326

Tabatabai MA (1994) Soil enzymes. In: Methods of soil analysis: Part 2 microbiological and biochemical properties, vol 5. Soil Science Society America, pp 775–833

Tabatabai MA, Bremner JM (1969) Use of p-nitrophenyl phosphate for assay of soil phosphatase activity. Soil Biol Biochem 1(4):301–307

Tabatabai MA, Bremner JM (1970) Arylsulfatase activity of soils. Soil Sci Soc Am J 34(2):225. https://doi.org/10.2136/sssaj1970.03615995003400020016

Tarafdar JC, Chhonkar PK (1979) Phosphatase production by microorganisms isolated from diverse types of soils. Zen Bac Natur 134(2):119–124

Thoma JA, Spradlin JE, Dygert S (1971) Plant and animal amylases. In: The enzymes, vol 5. Academic Press, pp 115–189

Trasar-Cepeda C, Leirós MC, Gil-Sotres F (2008) Hydrolytic enzyme activities in agricultural and forest soils. Some implications for their use as indicators of soil quality. Soil Biol Biochem 40 (9):2146–2155

Turan M, Nikerel E, Kaya K, Kitir N, Gunes A, Mokhtari NE, Tüfenkçi Ş, Karaman MR, Çimrin KM (2017) Enzyme dynamic in plant nutrition uptake and plant nutrition. In: Enzyme Inhibitors and Activators. Inctechopen, pp 225–242

Ueno H, Miyashita K, Sawada Y, Oba Y (1991) Assay of chitinase and N-acetylglucosaminidase activity in forest soils with 4-methylumbelliferyl derivatives. J Plant Nutr Soil Sci 154 (3):171–175

van Aarle IM, Plassard C (2010) Spatial distribution of phosphatase activity associated with ectomycorrhizal plants is related to soil type. Soil Biol Biochem 42(2):324–330

Verma SS, Kumar D, Naresh RK, Shukla AK (n.d.). Chapter-1 Conservation Agriculture and Role of Enzymes in Maintaining Soil Health

Wienhold BJ, Varvel GE, Doran JW (2005) Quality of soil. In: Hillel D (ed) Encyclopedia of soils in the environment 3. Academic Press, pp 349–353

Wolińska A, Stępniewska Z (2012). Dehydrogenase activity in the soil environment. In: Dehydrogenases. InTech Rijeka, pp 183–210

Yano S, Wakayama M, Tachiki T (2006) Cloning and expression of an α-1, 3-glucanase gene from Bacillus circulans KA-304: the enzyme participates in protoplast formation of *Schizophyllum commune*. Biosci Biotechnol Biochem 70(7):1754–1763

Yaroslavtsev AM, Manucharova NA, Stepanov AL, Zvyagintsev DG, Sudnitsyn II (2009) Microbial destruction of chitin in soils under different moisture conditions. Eurasian Soil Sci 42 (7):797–806

Zhang N, Xing-Dong HE, Yu-Bao GA, Yong-Hong LI, Hai-Tao WA, Di MA, Zhang R, Yang S (2010) Pedogenic carbonate and soil dehydrogenase activity in response to soil organic matter in Artemisia ordosica community. Pedosphere 20(2):229–235

Omics Technology for Plant Stress Management

16

U. M. Aruna Kumara and Nagarathnam Thiruchchelvan

Abstract

Plants are biosystems, they made-up with cells, which are responsible for all biological actions as well as responding to their micro- and macro-environment. Therefore, environment has direct and indirect influence on plant growth and development. The global climatic changes have created many harmful effects on crop production systems. However, global population will increase by nine billions in 2050. As a result of that food production has to be increased by 70% than today because human and animals largely depend upon the plants derived foods. These goals can only be achieved through sustainable technological innovations to develop higher yielding, nutritionally rich crop cultivars with resistance to biotic and abiotic stress factors. In order to develop such resistance crop cultivars, proper understanding of systems biological approaches to find out genomics, transcriptomics, proteomics, and metabolomics regulators, signal molecules, and their functional attributes within the cells is essential. Present findings of biological researches highlighted that introduction of novel omics technologies has contributed immensely to overcome the many bottle neck drawbacks in the field of crop improvement. This chapter will discuss the potential roles of genomics, transcriptomics, proteomics, metabolomics, lipidomics, proteogenomics, ionomics, bioinformatics, prime-omics, miRNA omics, and phenomics in plant stress management.

U. M. Aruna Kumara (✉)
Department of Agricultural Technology, Faculty of Technology, University of Colombo, Homagama, Sri Lanka
e-mail: umarunakumara@at.cmb.ac.lk

N. Thiruchchelvan
Department of Agricultural Biology, Faculty of Agriculture, University of Jaffna, Kilinochchi, Sri Lanka

R. Soni et al. (eds.), *Microbiological Activity for Soil and Plant Health Management*,
https://doi.org/10.1007/978-981-16-2922-8_16

Keywords

Biotic and abiotic stress · Data · Environment · Gene · Molecular mechanism

16.1 Introduction

Plants are biosystems, they made-up with cells, which are responsible for all biological actions as well as responding to their micro- and macro-environmental conditions of their complete life cycle (Mosa et al. 2017; Parida et al. 2018; Mehta et al. 2019). Such environmental factors have harmful effect on the growth, development, and productivity of the crop. The imbalance created by environmental stress at cellular, molecular, physiological, and developmental levels of plants has direct and indirect effect on their productivity (Singh et al. 2018; Mehta et al. 2019). Based on the environmental factor that generates stress, stress can be categorized into two groups, such as abiotic and biotic stress. The extreme temperatures, drought, salinity, freezing, toxic metals, high irradiance and ultraviolet (UV) light, and oxygen deficient conditions are categorized under the abiotic stress factors (Singh et al. 2018; Mehta et al. 2019). The pests such as microbes, insects, nematodes, rodents, etc. are termed as main biotic stress factors. At the moment, both biotic and abiotic stresses are poised to be most detrimental as they harshly reduce crop yield and productivity of agricultural commodities. Global warming, reduction of water resources, deforestation, and anthropogenic activities like abiotic stress factors will delimit the productivity of standing crops more adversely than today (Singh et al. 2018).

In recent years, the omics technologies have considerably contributed to studying biotic and abiotic stresses responses in plants (Zhang et al. 2016; Ibraheem et al. 2018). Omics tools have played significant role in crop quality improvement and protection which has brought about increase in agricultural food production by enhancing the quality, taste, and nutritional composition of food crops. The consistency and predictability in modern plant breeding have been improved by reducing the time and expense of producing better quality food crops having resistant to stress factors and showing a high nutritional value with the use of omics technologies such as genomics, transcriptomics, proteomics, metabolomics, and bioinformatics (Van Emon 2016; Ibraheem et al. 2018).

Several biochemical, physiological, and metabolic strategies have been developed by the plants in order to manage biotic and abiotic stresses. Through evolutionary adaptation, plants possess complex signal transduction pathways for different stress conditions but often it is difficult to predict the activation or deactivation of such pathways during the stress responses (Chawla et al. 2011). The development of system biology model is essential by attempting to integrate multi-dimensional biological information in a network. Genomics, proteomics, transcriptomics, and metabolomics are the four main axis of the plant system biology approaches, which provide us with a detailed knowledge about the topology and dynamic function of a molecular system (Yuan et al. 2008).

Understanding of the key molecular targets, regulators, their signaling involved in plant interactions with their environment are more important factors to be considered in novel research strategies in order to develop stress tolerant abilities in agricultural crops to enhance their productivity to meet the need of future generation (Mosa et al. 2017; Singh et al. 2018; Parida et al. 2018; Mehta et al. 2019). A new integrative "omics" approach has gained momentum in the past two decades, in the plant biology research field, fueled by advancements in nucleic-acid sequencing platforms, peptide-sequencing platforms, mass spectrometry (MS) technology, advanced computational capabilities, and statistical methodologies. As described by Mehta et al. (2019), integrative "omics" method gives a snapshot of the development, functioning, and interactions of a cell, tissue, or organism by characterizing and quantifying all its biomolecules in a high throughput approach (Mosa et al. 2017; Parida et al. 2018; Mehta et al. 2019).

At the moment, a huge amount of genetic information to help identification of mechanisms of biotic and abiotic stress responses in plants can be attained by the use of genomics knowledge such as next generation sequencing (NGS), gene editing, gene silencing, and overexpression methods (Luan et al. 2015; Ibraheem et al. 2018). At the transcriptome level, technological innovations have made it possible to overview the changes that occur at the transcriptomic level under different environmental stress conditions (Ibraheem et al. 2018).

16.2 Insights into Omics Technologies Used in Plant Stress Management

In the past two decades, scientists have performed numerous research to identify molecular and genetic basis of biotic and abiotic stress tolerant in plants, it revealed upregulation and downregulation of genes dynamically during plant stress responses. To see the integrated view of the response of plants to various biotic and abiotic stresses factors, scientists have used various omics approaches (Gupta et al. 2018; Mehta et al. 2019). Development of omics technology provides scientists to gain understanding of systems biological approach in plant rather than studding as a single event under different environmental regulations. Through these approaches, molecular biological systems, signal transduction, and other cellular functions become much clearer than ever (Parida et al. 2018; Shen et al. 2018; Mehta et al. 2019).

Identification of molecular mechanisms in plants responding to their environmental regulators now can easily be found with the help of innovative novel technologies derived from the omics technological hub (Ibraheem et al. 2018). Introduction of computational biological approaches such as bioinformatics provides good platform to develop systems biological process that happens inside the cell to gain understanding through computational modeling of biological systems (Bajwa et al. 2018). The knowledge extracted from the different omics technologies can be integrated to see real effect of different environmental factors on plant growth and development in order to engineering stress resistance in crops (Mehta et al. 2019).

16.3 Genomics

Scientific study of genome of given organism is referred as genomics (Gilliham et al. 2017). In 1970s first generation sequencing was started and next generation sequencing technology was stared in 1990. In twenty-first century, third generation sequencing has evolved and drastically increased the number of genomic studies to find structural and functional attributes of genes and genomes of many flora and fauna (Duque et al. 2013; El-Metwally et al. 2013, 2014). With the development of sequencing technology, genomics become most important omics technology among the researches. Genomics has close relationship with transcriptomics and proteomics as integrative approaches to see the functional effect of genome.

The recent advances in application of DNA marker technologies for detection of single nucleotide polymorphism (SNP) have been used for detection of desirable traits. The next generation sequence (NGS) technologies using Illumina/Solexa, Ion Torrent Personal Genome Machine (PGM), and Pacific Biosciences (PacBio) sequencing methods have also revolutionized genomic and genetic research through application of a precise and fast phenotyping which is often more expensive and durable than getting SNP data for thousands of markers (Heather and Chain 2016).

Application of biparental quantitative trait loci (QTL) mapping or association mapping techniques to understand the inheritance and genetic variance of complex quantitative traits is also currently in use (Bekele et al. 2014). Quantitative trait loci mapping is fast and accurate and can possibly discover rare allele and give good estimates of the allelic effects in a given genetic background.

Single nucleotides polymorphism genotyping methods are gaining wide popularity due to its arrival of cost-efficient and high-throughput genotyping ability. The genotyping by sequencing (GBS) approach of SNP-based genotyping methods is a highly multiplexed system for constructing reduced representation libraries (RRL), molecular marker discovery, and genotyping for crop improvement (Elbasyoni et al. 2018; Eltaher et al. 2018). The GBS has been applied to several crop species due to its low cost and advancing technologies (Kim et al. 2010). As an example, a tomato GBS study led to the discovery of 8784 SNPs based on an NGS approach and out of these SNPs, 88% are frequently observed in the tomato germplasm (Sim et al. 2012; Gupta et al. 2013).

Plant microbial interaction is one of the important relationships in the agro ecosystem (Vorholt 2012; Bulgarelli et al. 2013). Some interactions may be beneficial and some are detrimental. Some interaction can be endophytic or epiphytic (Newton et al. 2010). Therefore, identification of gene expression and regulatory mechanisms involves in plant microbial interactions are crucial for the development of biotic stress tolerant crop through genetic engineering technology (Sarowar et al. 2011; Imam et al. 2016; Ibraheem et al. 2018).

Plant pathogenic microorganisms can change their genetic material time to time. It may lead to develop adaptations in plant pathogens to host environment to facilitate pathogenesis (Benson et al. 2012; Thynne et al. 2015). To study this phenomenon (i.e. accelerated genome adaptation) the horizontal gene transfer (HGT) and inter-specific hybridization can be used (Raffaele and Kamoun 2012;

Imam et al. 2016). According to the approaches in systems biology, we have to find the cellular level mechanism for specific situation like stress then we have to study the genomic information through sequencing like technologies, finally, we have to develop computational or mathematical modeling to understand the integrative effect of such stress conditions. Mathematically modeling such as genome-scale models ("GEMing") is frequently used to study the phenomic or phenotypic changes during host–pathogen interactions (Collakova et al. 2012; Imam et al. 2016).

16.4 Transcriptomics

The scientific study of express sequence tag (ETS) or RNA express under certain condition is known as "transcriptomics" (Shen et al. 2018). Transcript abundance basically depends upon the environmental factors, plant growth and development, age and maturation level, and nutritional status of the plant (El-Metwally et al. 2014). The RNA sequencing, microarray platforms, digital gene expression profiling, and serial analysis of gene expression (SAGE) are the main technologies used in transcriptomics analysis (Kreszies et al. 2018). Transcriptomics provide good opportunity to find putative gene or genes express under certain stress condition (Zhang et al. 2013). These candidate genes can be incooperated in to crop improvement program to develop stress tolerant crop cultivars for future demand (Jogaiah et al. 2013; Agarwal et al. 2014).

Gene expression studies identify functional gene products that give rise to the phenotype, an information that can be used to create hybrid plant. The task of any breeder is to find reliable traits that can be used to select genotypes that best fit a target environment (Falconer 1989). Introducing a specific gene or genes to a plant or a gene knocking down with RNAi, the desirable phenotype can be produced more quickly than through traditional plant breeding (Dhondt et al. 2013). These techniques are also applied in biotechnology sector for biopharmaceuticals and industrial compounds.

Elucidation of sequence information of expressed sequence tags (EST) of plants under biotic stress condition is important to understand reasons for resistance or tolerant and susceptibility to plant pathogenic microorganisms (Kumara and De Costa 2015). The comparative analysis of differentially expressed gene profiles of plant species or their genotypes is useful to identify upregulated and downregulated genes under biotic or abiotic stress conditions (Freitas-Astúa et al. 2007). Such analyses of differential gene expression have been reported for tomato-*Alternaria solani* (Upadhyay et al. 2014), citrus leprosis virus (Freitas-Astúa et al. 2007), rice-*Magnaporthe grisea* (Xiong et al. 2001), and strawberry-*Colletotrichum* spp. (Casado-Díaz et al. 2006) pathosystems.

The transcriptomic analysis provides useful information about biochemical pathways of plant under biotic and abiotic stress conditions. These information can be manipulated to develop molecular markers in order to use for marker assisted selection in molecular plant breeding and screening and identification of potential genes through transcriptomics (Passos et al. 2012). Furthermore, Passos et al. (2012) have studied the candidate genes expressed under infection with *Mycosphaerella musicola*, the causal organism of Sigatoka leaf spot disease among banana genotypes of *M. acuminata* ssp. burmannicoides "Calcutta 4" and *M. acuminata* subgroup Cavendish cv. Grande Naine. The results revealed contrasting nature in resistance among banana genotypes.

The scientific investigation carried out by Kumara and De Costa (2015) was able to identify differentially express putative genes and proteins under the infection of *C. musae* in relatively anthracnose resistance banana cultivar Seenikesel (*Musa* spp. ABB). Differential hybridization of cDNA library derived under infected condition with *C. musae* was able to identify 88% upregulated and 11% downregulated gene transcripts after hybridization with treatment and control probes derived from inoculated and non-inoculated conditions, respectively. In addition to that, large majority of upregulated gene transcripts belongs to plant defense responsive genes such germin-like proteins (GLPs), beta-glucanase, chitinase, etc. (Kumara and De Costa 2015). Germin-like proteins are known to express differentially in plant tissues due to infections of biotic factors (e.g. viruses, bacteria, mycorrhizae, fungi, insects, nematodes, and parasitic plants) and abiotic factors such as salt, heat/cold, drought, nutrient, and metal stress (Lane 2002; Dunwell et al. 2008). Hence, H_2O_2, which produced by the enzymatic activity of GLPs, is a known signaling molecule that is involved in multiple pathways of plant innate immunity (Davidson et al. 2009).

Plants are exposed to various environmental and biological factors during their life time. These biotic and abiotic factors have detrimental effect on plant growth, development, and reproduction. Therefore, screening and identification of potential gene transcript expressed under stress conditions are essential for crop improvement through molecular plant breeding. In contrast, necessary transcriptomics data required for molecular breeding of *Musa* spp. are limited (Passos et al. 2013). Therefore, novel transcriptomics approach will provide the necessary technology to fill this gap in order to engineer plants against biotic and abiotic stress tolerant.

Some of transcriptomes identified from relatively resistant banana cultivar Seenikesel (*Musa* spp. ABB) in response to *C. musae* infection through research conducted by Kumara and De Costa (2015) is tabulated below in Table 16.1 (i.e. analysis of differentially expressed gene profiles in a resistant banana cultivar after infection by *C. musae*).

16.5 Proteomics

During the translational process, nucleotide sequence is converted into an amino acid sequence. Further dimerization and polymerization process through posttranslational modification gives birth to protein. The quantitative and qualitative scientific study of protein is referred as "proteomics" (Tyers and Mann 2003; Luan et al. 2018). Some of technologies can be used to analysis of proteomics with the help of bioinformatics tool to recognize the protein profiles under different biotic and abiotic stress conditions. Such technologies can be named as two-dimensional gel

Table 16.1 Putative genes/proteins and probable functions of the upregulated cDNA transcripts in a relatively resistant banana cultivar Seenikesel (*Musa spp.* ABB) in response to *C. musae* infection

Clone	Length (bp)	EMBL accession number	Homology/Host genome	Function	Identity %	Score (bits)	E-Value
Ma SINI 429	960	HF567821	Germin-like protein [*Musa acuminata*]	Defense response	94	363	7.00E-101
Ma SINI 433	956	HF567822	Germin-like protein [*Musa acuminata*]	Defense response	94	363	7.00E-101
Ma SINI 467	485	HF567823	Germin-like protein [*Musa acuminata*]	Defense response	94	348	8.00E-97
Ma SINI 507	423	HF567824	Mitochondrial ATP synthase g subunit family protein	Stress responses	92	237	5.00E-61
Ma SINI 586	586	HF567775	26S proteasome regulatory particle triple-A ATPase subunit4b [*Oryza sativa* (japonica cultivar-group)]	Defense response	96	227	3.00E-58
Ma SINI 594	660	HF567776	*Musa acuminata* 1-aminocyclopropane-1 carboxylate oxidase (MAO1B) gene, complete cds	Energy and metabolism	98	94	4.00E-47
Ma SINI 610	774	HF567779	Ubiquitin conjugating enzyme E2 [*Adiantum capillus-veneris*]	Protein synthesis, folding, and stabilization	96	273	4.00E-72
Ma SINI 617	649	HF567780	Ribulose 1,5-bisphosphate carboxylase/ oxygenase large subunit [*Musa acuminata*]	Energy and metabolism	94	476	1.00E-132
Ma SINI 621	704	HF567781	fra1 mRNA, complete sequence [*Musa acuminata*]	Growth and development	99	288	9.00E-75
Ma SINI 640	201	HF567783	Putative chitinase [*Musa acuminata*]	Defense response	98	273	6.00E-72
Ma SINI 643	205	HF567784	Hypothetical protein OsJ_017968 [*Oryza sativa* (japonica cultivar-group)]	Unknown function	95	307	3.00E-82
Ma SINI 676	812	HF567832	Glutamine synthetase [*Hevea brasiliensis*]	Growth and development	94	186	6.00E-46
MaSINI 693	1125	HF567835	Endochitinase [*Musa acuminata*]	Defense response	96	474	7.00E-132
Ma SINI 715	613	HF953990	Translation initiation factor; GOS2 [*Zea mays*]	Defense response	91	211	5.00E-53

(continued)

Table 16.1 (continued)

Clone	Length (bp)	EMBL accession number	Homology/Host genome	Function	Identity %	Score (bits)	E-Value
Ma SINI 752	516	HF567839	Germin-like protein [*Musa acuminata*]	Defense response	94	348	8.00E-97
Ma SINI 763	852	HF567841	Germin-like protein [*Musa acuminata*]	Defense response	95	365	6.00E-103
Ma SINI 891	722	HF95395	Ribosomal protein S11 [*Musa acuminata*]	Cellular metabolism	100	227	9.00E-58
Ma SINI 1027	793	HF953999	Germin-like protein [*Musa acuminata*]	Defense response	98	347	2.00E-94
Ma SINI 1029	475	HF954001	Actin [*Musa acuminata*]	Defense signaling	99	253	7.00E-66
Ma SINI 1093	*808*	HF567786	*Musa* cold stress induced protein	Stress responses	95.7	1702	1.80E-46
Ma SINI 1101	856	HF567787	*Musa* ABB stress induced (mRNA sequence) protein	Stress responses	99.8	2256	3.70E-38
Ma SINI 1247	475	HF567789	Pectate lyase 1 [*Musa acuminata*]	Growth and development	94.2	2035	7.60E-72
Ma SINI 1259	682	HF567790	Eukaryotic elongation factor 1A [*Bruguiera sexangula*]	Protein synthesis, folding, and stabilization	97	861	0
Ma SINI 1272	423	HF567791	*Musa* stress induced (mRNA sequence) protein	Stress responses	96.	1803	7.30E-61
Ma SINI 1336	948	HF567793	Germin-like protein mRNA complete cds [*Musa acuminata*]	Defense response	94 0.1	3294	3.0E-124
Ma SINI 1341	819	HF567795	*Musa*.ABB stress induced protein, cDNA, mRNA sequence	Stress responses	98 0.4	2070	1.30E-75
Ma SINI 1345	960	HF567796	*Musa*.ABB stress induced protein (mRNA sequence)	Stress responses	92.7	2252	3.30E-64
Ma SINI 1354	545	HF567797	β,1,3-Glucanase mRNA complete (ds) [*Musa spp*]	Defense response	95.4	2134	2.8E-78
Ma SINI 1389	612	HF567798	Methionine synthase protein [*Sorghum bicolor*]	Defense response	94	260	4.00E-68
Ma SINI 1396	1365	HF567800	Musa endochitinase (ds)	Defense response	92.7	2573	2.90E-97
Ma SINI 1407	850	HF567801	Germin-like protein [*Musa acuminata*]	Defense response	94	365	6.00E-103

electrophoresis (2D), mass spectrometry (MS), western blot, matrix assisted laser desorption ionization time of flight (MALDITOF), and enzyme-linked immunosorbent assay (ELISA). Proteomics analysis of fruit and root tissues has limitations due to effectiveness and efficiency of the available technologies. Mass spectrometry technologies have wider adaptability in protein and peptide analysis with computational aids under label free quantification procedures (Shao et al. 2014; Luan et al. 2018). These methods are usually implemented in protein expression profiling of tomato cultivars under different environmental conditions such as heat, drought, salinity, and chilling effect (Neilson et al. 2011; Mora et al. 2013).

The proteome profile basically depends upon the environmental factors, plant growth and development, age and maturation level, and nutritional status of the plant. Proteins are biomolecules, they act as an activator, receptor, elicitor, signaling molecules, enzymes, and hormones to facilitate biological reactions in the biosystems of micro- and macro-organisms. Therefore, elucidation of expressed protein molecules and their functional attributes under stress responses through high throughput technologies are important event in proteomics analysis (Liu et al. 2015; Kosová et al. 2018). Crop improvement based on the nutritional profiling through proteomics analysis is now been popular among the plant breeders due to its importance as a food safety measure rather than the increase in total biomass production to overcome nutritional deficiencies in the developing countries (Cilindre et al. 2008; Agrawal et al. 2012). Therefore, comparative analysis of proteome profiles in resistance and susceptible crop cultivar has gained attention among molecular plant breeders to develop high yielding, nutritionally rich crop cultivars with biotic and abiotic stress tolerant abilities (Agrawal et al. 2012). All plant cellular mechanisms are interconnected as a system, that is simply say systems biology of plant. It means, there is an interaction among several metabolic activities to perform single function such as photosynthesis. Therefore, external or internal signaling molecules such as jasmonate and cyclopentanone have tremendous influence on plant responses to its environment such as biotic and abiotic factors. The proteome profiles revealed expression of different types of proteins responsible for plant physiological functions under artificial treatment with methyl jasmonate (Chen et al. 2011; Evers et al. 2012). Luan et al. (2018) explain his experimentation on proteome profiling, he provided waterlogging conditions to different barley cultivars and analyzed the expressed protein profile of different plant parts with the help of two-dimensional gel electrophoresis and mass spectroscopy. The results revealed a reduction of total biomass, photosynthetic performance in the susceptible genotype. Their results highlighted our knowledge about the major regulatory genes involving in waterlogging tolerance in barley. These results can be used to develop new crop varieties with tolerant to waterlogging conditions (Luan et al. 2018).

Several research studies have been conducted by different scientific communities to elucidate the proteomics profile of tomato under different biotic and abiotic stress conditions but still there is a problem in data analysis and data interpretation due to technological gap in the field of proteomics. The development of public databases with high throughput protein annotations will help to fill this technological gap in the field of proteomics. Recent advancement in proteomics profiling will help to develop sustainable crop cultivars with resistance to microbial pathogens, tolerant to environmental stresses, and elevated properties of nutritional values (Chaudhary et al. 2019a, b).

Induction of plant immune responses under external elicitation by different key regulatory factors and understanding internal signal transduction mechanism to mitigate the external environmental factors such as biotic and abiotic stress is a current challenge faced by the proteomics research to understand proper cellular functions. To overcome these challenges mass spectrometry based proteomics tool has given tremendous support over the available technologies. Recent findings of proteomics researches have identified the importance of posttranslational modifications (PTMs) and spatial proteome in plant immune response under microbial pathogenic invention and global climatic changes. Hence development of systems biology approaches to study the plant responses towards biotic and abiotic stress factors mainly depends upon the understanding of posttranslational modification and subcellular localization of the proteins (Liu et al. 2019). Different proteomics technologies such as activity based protein profiling, engineered ascorbate peroxidase, organelle locatable reactive molecules, and proximity dependent biotin identification have been developed by the various research groups to elucidate the information about subcellular location and specialized proteomic behavior of the plant organs (Wiedner et al. 2014; Rhee et al. 2013; Yasueda et al. 2016; Le Sage et al. 2016).

Microbes are tiny creatures and composed with fundamentally less complexity with relatively small genome. Therefore, studying and understanding of their gene regulation mechanisms such as transcription, translation, and posttranslational modifications are much more methodical to gain understanding of these processes in multicellular organisms such as animal and plants. At present proteomics tool is employed to unveil the protein expression pattern of microorganisms under different conditions such as stress responses, microbial pathogenicity, extreme environmental adaptation, and metabolic engineering (He et al. 2016).

Almost all of the biological reactions are intermediate by proteins as enzymes or hormones in the biological systems. Therefore, microbial specificity as a mesophilic (+11° to +45 °C), thermophilic (+46 ° to +75 °C), hyperthermophilic (+76 °C) or psychrotrophs, also known as psychrotolerant organisms mainly depend upon the proteins profile of these creatures. Elucidation of these specialty proteins would be a great importance in metabolic engineering to develop sustainable methods for industrial uses such as enzymes with higher temperature tolerant abilities. Uses of fossil fuels in many industries lead to an environmental degradation and creating unfavorable condition for human civilization. Therefore, introduction of microbial proteins as an alternative energy source for fossil fuel in industrial establishment will lead to a biological economy in the world while creating earth as a fashionable place for its inhabitants. In addtion to that, invention of PCR technology was not realized until discover the Taq-Polymerase enzyme from *Thermos aquaticus* bacterium live in hot springs (Wilkins et al. 1996; Han et al. 2011; He et al. 2016).

16.6 Metabolomics

Metabolites are the intermediate molecules which facilitate the cellular functions to attain overall functions of living organisms. Metabolomics is the scientific investigation about screening and identification, quantification, and characterization of metabolites derived in different plant tissues under specific environmental condition (Freund and Hegeman 2017; Parida et al. 2018). Metabolomics can generate more information than transcriptomics and proteomics (Dos Santos et al. 2017). Therefore, metabolomics try to understand interconnection between genotypes and phenotypes. Genotypes or genes contain chemical information necessary for phenotypic expression under different environmental conditions in an organism; meanwhile, metabolomics studies try to identify biochemical processes of metabolites involved in these reactions in the cytoplasmic level (Aliferis and Chrysayi-Tokousbalides 2011; Dixon et al. 2006).

Plants generate different types of chemically active compound such as reactive oxygen species (ROS), suberins, beta-glucanases, and chitinases that help plants to mitigate the stress generated by microbial plant pathogens such as bacteria, fungi, viruses, and environmental factors including drought, chilling, and salinity, (Gupta et al. 2013). These chemically active metabolites can be identified by various technologies such as mass spectrometry, chromatography (i.e. HPLC, GC), direct injection mass spectrometry (DIMS), and nuclear magnetic resonance spectroscopy (NMR) (Wolfender et al. 2013; Parida et al. 2018).

Various scientific community in the world investigated the changes of metabolomics profile of newly developed varieties using cultivated types and their wild relatives under different biotic and abiotic stress conditions. Such as, water stress in cherry tomato had decrease the shikimate and phenolic compound, and low oxygen level was induced the accumulation of glycolysis intermediates and increased the level of lactate and sugar alcohol. (SánchezRodríguez et al. 2011; Ampofo Asiama et al. 2014).

To analyze the difference in metabolomics profile of maize, plants were exposed to different environmental stress factors such as heat stress, salinity, and drought, results revealed that individual stress is different from the combination of stress based on the metabolomics profile (Sun et al. 2016a, b). As same as the above experimentation, Khan et al. (2019) employed a research to apply untargeted metabolic profiling technology to reveal the effect of drought on metabolomics profile of chickpea varieties and results indicated significant reduction in growth, dry matter, relative water, and chlorophyll content of the all chickpea varieties (Mehta et al. 2019).

Knowledge of both extracellular and intracellular metabolites is required for quantitative understanding of microbial metabolism and its in vivo regulation. Conventionally, fast sampling, instant arrest of metabolic activity and deactivation of endogenous enzymatic activity, metabolite extraction, and subsequent quantification of intracellular reactants were performed to gain understanding of microbial metabolism and its in vivo regulation. Filtration or centrifugation at low temperatures can be used to quantify the extracellular metabolites in cell free supernatant. Quantification of the total complement of endometabolome and exometabolome in a given cell under different growth conditions is the future challenge in recent rapid development in the metabolomics studies. It will further clear the understanding of in vivo regulation of microbial networks (Mashego et al. 2007).

16.7 Lipidomics

The scientific study about the global characterization and quantification of lipids in biological matrices such as biofluid, cells, whole organs, and tissues is referred as lipidomics. During the stress responses, plants produce lipid molecular species that are showing changes from normal condition, it indicated effect of genes in lipid metabolism and lipid signaling in plant under constitutively express and stress induced conditions (Yu et al. 2020). The comparative analysis of lipid profiles and remodeling of lipids under various biotic and abiotic stress conditions are needed to study extensively to improve the area of plant lipidomics (Moradi et al. 2017; Zhang et al. 2018). Burgos et al. (2011) reported the glycerolipid remodeling and saturation profile of fatty acid. In his study *Arabidopsis* plants were exposed to eight different stress factors (Burgos et al. 2011). Tarazona et al. (2015) has developed a multiplexed LC-MS lipidomics platform for the better coverage of plant lipidomes. During their study, plant was exposed to cold and drought conditions to study the leaf lipidome by using their own platform, results revealed 23 different classes of lipids (Tarazona et al. 2015).

The plant molecular biologists have put their attention to plant lipidomics study to understand the metabolic changes in cytoplasmic lipids even including plasma membrane during different stress factors. In addition to that, they have continued their study on genetically modified organisms and changes of lipid profiles due to introduction of new genes into the plant genome.

Scientists have observed the increase level of unsaturated phospholipid when plants undergo low non-freezing temperature leading to changes in the membrane integrity (Welti et al. 2007; Thomashow 1999). Plant has ability to regulate the stability of plasma membrane with the help of membrane lipids during stress condition to maintain the cellular integrity (Barkan et al. 2006). Forward genetic approaches were used to understand the diverse mechanisms available in plant to regulate temperature fluctuations during their lifespan. *Arabidopsis thaliana* thermosensitive mutant (atts) was used by the researchers to investigate thermotolerant ability under different temperature levels (Burke et al. 2000). Oxylipins are chemical agents of plastid localized polar complex lipids in *A. thaliana* with special thermotolerant ability. T-DNA tagged knock out mutant and genetic alteration of lipid involve in biochemical processes are the novel approaches to identify the gene involved in lipid metabolism under in vivo condition (Hisamatsu et al. 2005; Welti et al. 2007).

Salinity condition develops in the soil due to low water availability or drought condition. Plants have developed many mechanisms to resist for salinity condition by accumulation of biomolecules in cytoplasm to decrease the water potential leading to stabilize the membrane fluidity and integrity by lipid rearrangement (Sarabia et al. 2018). Regulation of influx of sodium ions and osmolytes by protein molecule in plasma membranes is facilitated by its semipermeable nature and gatekeeping ability (Upchurch 2008). Alterations of the unsaturation degree of polar glycerol-based membrane lipids have the ability to regulate membrane fluidity (Sui and Han 2014). Salinity or salt stress can induce oxidation of vital molecules including glycerol-based membrane lipids through the enhanced production of reactive oxygen species (Mosblech et al. 2009). Oxylipins, which are considered as biochemical marker of oxidative stress in plant are generated by lipid peroxidation process during the reaction in between excessive ROS molecules with cellular lipids (Yu et al. 2020).

It is identified that, accumulation of cardiolipin (CL) in the mitochondrial membrane, when the plants are exposed to osmotic stress (Pan et al. 2014). Cardiolipin is a class of dimeric glycerophospholipids (GPs) and primarily localized in the inner membrane of mitochondria where they account for around 10% of the total lipid content (Schwarzländer and Fuchs 2017). Cardiolipin has ability to produce soluble proteins need to stabilize the respiratory chain super complexes in the mitochondrial membrane other than its support to improve permeability of the mitochondrial membrane (Mårtensson et al. 2017).

Comprehensive characterization of membrane lipids under different stress responses can be archived by high throughput and high sensitivity mass spectrometry based lipidomics platforms. Oxidized lipids in wheat seeds have been studied under different environmental parameters by using MS/MS spectra. It is necessary to have extensive computation to process the data (Riewe et al. 2017). Scheduled multiple reaction monitoring (sMRM) system combined with accurate mass of the precursor allows to identify the structural identification of the lipid. Yu et al. (2020) explained the use of sMRM system to profile oxidized lipids in roots of Australian barley cultivars in response to salt stress. Main lipid storage of bacteria is the plasma membrane. The accurate lipid composition in the plasma membrane has given ability to bacterial cell to bear or regulate external stress factors like temperature and pH while regulating the membrane permeability for material transport with the help of properly distributed membrane bound proteins (Zhang and Rock 2008; Chang and Cronan 1999). Fatty acyl composition has direct effect of the membrane transport than head group of the lipids. Branching, double bonds, and cyclopropyl modifications can stimulate membrane fluidity by increasing the space in lipid copies and fully saturated straight chain fatty acids can give the tightly packed rigid nature to bacterial cell membrane (Legendre et al. 1980; Poger and Mark 2015). Lipid II and lipoteichoic acid in Gram positive bacteria contain a lipid moiety that is attached to plasma membrane and other portion is connected to the cell wall for structural integrity (Percy and Grundling 2014; De Kruijff et al. 2008). Gram negative bacteria utilize lipopolysaccharides for structural integrity of their cell membrane and have a large antigenic glycan that extends beyond the cell surface (Erridge et al. 2002).

16.8 Proteogenomics

Information gather from the proteomics and genomic study will facilitate the unidirectional view about plant response to their external environmental factors. However, integration of genomics and transcriptomics data to elucidate novel peptides involved in plant response to biotic and abiotic stress will generate much clear picture about proteome profiling of plants. This integrative technology is referred as proteogenomics (Nesvizhskii 2014; Helmy et al. 2012; Mosa et al. 2017). In proteogenomic studies, high throughput technologies used to identify the structural and functional properties of protein and reverse process is used to map the transcriptomic and genomic information. *Arabidopsis thaliana* is used as an ideal model for flowering plant to investigate proteogenomics properties. Present findings identified 57 new genes and annotated hundreds of genes through intensive sampling from *Arabidopsis* (Baerenfaller et al. 2008).

The alternative splicing pattern of gene under the different concentrations of abscisic acid was studied by Zhu et al. (2017) to identify the different isoform of proteins. Results revealed that 83.4% of total intron containing genes undergo alternatively splicing. Proper understanding of proteogenomic profile expressed under different stress conditions will provide sufficient information to plant breeders for development of crop cultivars with improved qualities of nutrient values in edible portion to eradicate malnutrition and mineral deficiencies in developing nations.

Next generation sequencing technology (NGS) and transcriptome sequencing (RNA-Seq) have generated large number of data and those data are deposited in public domains in order to enhance the accessibility of researchers those who work on proteogenomics studies (Wang et al. 2009). RNA-Seq technology has now been popular among the researchers working on translational products such as ribosome profiling (Ingolia 2014). RefSeq or GENCODE like genome annotation methods provide platform to identify proteome data with the help of thousands of reference transcripts (Desiere et al. 2005). Transcriptomics and proteomic data are popular among the molecular biologists due to its applicability in identification of peptides and proteins with the help of sample specific protein sequence databases (Nesvizhskii 2014).

Multi-stage data analysis strategy provides high throughput sensitivity in peptide identification in proteogenomics studies. By using this strategy, mass spectrometric spectra can be compared with reference protein sequence databases and then unidentified mass spectrometric spectra will be identified from the large proteome databases (Helmy et al. 2012). At the initial stage majority of generally regulated peptide can be identified. Furthermore alternative splice junction can be considered only if both corresponding exons are supported by high scoring exon mapping peptides (Blakeley et al. 2012).

Accuracy of available proteogenomics tools has been studied by using many model organisms including *A. thaliana, Drosophila melanogaster, Caenorhabditis elegans,* etc. by various proteogenomics research teams under the well-established laboratory conditions (Nesvizhskii 2014). Before introduction of these novel proteogenomic tool to analyze the genotypic and phenotypic relationship at once, it was difficult to achieve the expected result in proteogenomic studies due to lack of sufficient proteomic data leading to poor understanding about proteomic data by molecular biologist or genomic community. These novel proteogenomics tools provide good background to identify protein level expression of putative genes derived from transcriptomics analysis under different biotic and abiotic stress condition even including microbial pathosystems in economically important agricultural crops, namely rice, maize, wheat, barley, and oats (Wilhelm et al. 2014). Rice is a staple food of many Asian countries but production is not at the expected levels due to many pest and diseases and effects of unexpected weather conditions. Therefore, development of high yielding, disease resistance cultivars with ability to stand upon under erratic weather pattern such as submergence tolerance is the challenge faced by Asian plant breeders. As a result of that, molecular breeding of rice gained popularity among the rice breeders but their effort was not succeeded due to inaccurate genomic annotations in rice. Single molecule long read RNA sequencing (lrRNA seq) technology was recently introduced to analyze the transcriptomics complexity in rice. According to the result obtained by Chen et al. (2020), lrRNA seq was able to identify 60% of loci associated with natural antisense transcripts (NATs) with multifunctional ability in gene regulation under different environmental factors. This technology was able to support progressive development in the field of proteogenomics with special attention to rice.

Comparative to other developed organisms, genome of bacteria is small and easy to sequence to obtain genomic data. With the help of genomic data of bacterium, construction of protein databases can be done. Proteins derived from direct translation is referred as pseudo proteins and annotated proteins are referred as known proteins. This pseudo and known proteins can be used to annotate the spectra derived from mass spectrometry of bacterial protein profiles (Renuse et al. 2011; Küster et al. 2001). This methodology will help to improve the available proteome databases (Uszkoreit et al. 2014).

16.9 miRNA Omics

Referencing to the relative cellular RNA abundance, 3% of RNA belongs to signal recognition particle (SRP) RNA, small nuclear RNAs (snRNAs), small nucleolar RNAs (snoRNAs), and micro-RNAs (miRNAs). Micro-RNAs involves in the structural modification of pre-mRNA after the transcriptional process (Sharma et al. 2017; Hernandez and Sanan-Mishra 2017). Differentially expressed genes of *A. thaliana* were identified during cold stress through scientific investigation. Results revealed that some of promoter sequences in upregulated genes show similarity with cis-element of stress regulators (Zhou et al. 2008). Based on the scientific investigation, scientists have identified the involvement of the cold responsive micro-RNA genes in signal transduction pathways during response to abiotic and biotic stress. As an example, when *A. thaliana* was exposed to stress condition sharp decreased rate of photosynthesis was observed due to activation of other energy promoting networks via reprograming of the energy associated transcriptome (Avin-Wittenberg et al. 2012). Up to date more than 400 micro-RNA molecules have been identified by different scientific investigations against stresses in plants species from different families, namely Euphorbiaceae, Apocynaceae, Brassicaceae, Amaranthaceae, Papaveraceae, Solanaceae, Poaceae, and Rosaceae (Zhang and Wang 2015).

Recent scientific research revealed that plant derived miRNA has critical role in structural modification of pre-mRNA after the transcriptional process in order to reduce plant growth and development under biotic and abiotic stress conditions. But molecular mechanism of these functions has not been discovered yet. The miRNA is composed of different types of small endogenous RNA molecules in cytoplasm, which controls the different activities carried out during posttranslational modifications such as pre-mRNA cleavage, transcriptional repression, DNA methylation, and chromatin remodeling (Zhang and Wang 2015). Plants accelerate to produce large copy numbers of miRNAs belongs to different families in plants when they are exposed to extreme environmental conditions such as drought, salinity and chilling. Information derived from these research findings will be an added advantage for molecular plant breeder to engineering of stress tolerant ability in agriculturally importance crops such as field crops and horticultural crops (Shriram et al. 2016).

Several miRNAs have been identified as abiotic stress regulators in important crops and model plants under different abiotic stress factors such as salinity, nutrient deficiency, UV-B radiation, heat, and metal stress by several research studies, respectively (Gao et al. 2011; Liang et al. 2015; Casadevall et al. 2013; Goswami et al. 2014; Gupta et al. 2014). Zhang et al. (2013) observed deferential expression of 1062 miRNA in 41 plant species under 35 different types of abiotic stress factors. Recent study on sugarcane abiotic stress responses revealed that miRNA expression pattern depends on the plant species, type of stress, plant parts such as seedlings, leaves, spikelets and roots, growth conditions such as field, greenhouse, hydroponic culture system, etc. In addition to that, they have recognized that miR396 and miR171 were differentially expressed in the most of the cases (Gentile et al. 2015). Differentially expressed gene profiles of radish under salt stress condition were able to identify new miRNA molecules showing putative functions as regulatory factors of iron homeostasis and signal transduction mechanisms in salt stress condition (Gao et al. 2011; Sun et al. 2015).

Transgenic strategies are used by researchers to improve the yield and quality attribute of agricultural crops through integration of resistance genes. Scientific investigations identified that micro-RNA has major regulatory role in transcriptional modifications and posttranslational modifications. Therefore, utilization of miRNA based technologies for genetic engineering of crops has greater success than other technological applications in crop improvements (Privalle et al. 2012; Buiatti et al.

2013). Involvement of miRNAs in growth and development and signal transduction pathways of plants have been identified by various scientific investigations (Zhang and Wang 2015). The miRNA156 is the first documented micro-RNA in plants, it has specific importance in plant development and signaling for floral primordial initiation. In addition to that overexpression of miRNA156 in plant showed less tolerant to low temperatures (Cui et al. 2015; Zhao et al. 2013). The miRNA319 was upregulated during many abiotic stress conditions applied through various scientific investigations showing its association with multiple stress responses (Zhou et al. 2010; Li et al. 2016). Transgenic plants of creeping bentgrass (*Agrostis stolonifera*) with osa-miRNA319a overexpression ability showed significant improvement in tolerant to salinity and drought (Zhou et al. 2013).

Developed novel high throughput genomics technology such as sequencing has given opportunities to scientists to study micro-RNA expression patterns across large number of organisms under valuable biotic and abiotic stress conditions (Shriram et al. 2016). Bioinformatics tools have been developed through various scientific study to support molecular biological analysis of micro-RNA expressed under different stress factors. Degradome sequencing like novel technological innovation has support to improve the time use efficiency in elucidation of different micro-RNA molecules. In addition to that, there are some limitations in identification of target miRNAs, particularly those target gene expression through translational repression. The miRNAs can act as a single or multiple stress regulators when plants are exposed to different environmental conditions including biotic and abiotic stress (Zhang and Wang 2015).

Virological investigation revealed accumulation of large copy number of virally encoded micro-RNAs in virus infected cells that undergo replication of viral genome (Cullen 2013; Grundhoff and Sullivan 2011). Recent findings highlighted that plant pathogenic fungus has ability to utilized plant cell owned RNAi to regulate cellular gene expression to facilitate pathogenicity in plant tissues (Weiberg et al. 2013). Usually prokaryotes do not produce miRNA but instead of that they can produce different types of small, non-coding RNAs (sRNAs) to regulate biochemical process in bacterial cell (Hoe et al. 2013; Lalaouna et al. 2013). As an example for above situation, recognition of foreign nucleic acid as a defense mechanism and secretion of protein within the bacterial cell can be done by the ribonucleoproteins derived from the bacterial sRNAs (Driessen and Nouwen 2008; Marraffini and Sontheimer 2010). In addition to that, sRNAs has ability to form sRNA/mRNA duplexes. These duplexes can alter or change the mRNA stability and translational efficiency of particular gene in their genome and utilize that situation to express other useful genes more rapidly under the environmental stress factors, namely extreme temperature, extreme pH, and lack of nutrients. This special ability develops antibiotic resistivity and fighting against plant defense mechanism to continue their pathogenicity in plants and animals as a virulence pathogens (Lalaouna et al. 2013; Gripenland et al. 2010; Sampson et al. 2013). There is a necessity to explore the ability of intracellular bacteria to secrete miRNAs or miRNA precursor molecule to host cytoplasm to inactivate the activity of RNAi mechanism and accelerate the pathogenicity through development of favorable conditions for bacterial replication in host cell (Furuse et al. 2014).

16.10 Prime-Omics

As described by Walters et al. (2005) priming agents or plant activators are class of agrochemicals that act as a resistance elicitors to enhance plant defense against different biotic and abiotic factors. This technology was originated like a novel technology during last 10 years. With the application of these external elicitors, plant reacts more powerfully to the stress generated by its environmental and other factors which affect plant growth (Balmer et al. 2015; Hussain et al. 2016; Lal et al. 2018). Priming or induction of plant defense against different types of stress agents would be able to do with the help of natural or synthetic compounds such as jasmonic acid, salicylic acid, 2-6-dichloroisonicotinic acid, benzothiadiazole, and beta-aminobutyric acid (Conrath 2009; Kauss et al. 1992; Katz et al. 1998; Oostendorp et al. 2001). Conrath (2011) has performed a scientific investigation to see the effect of same plant population under primed and unprimed conditions, results revealed higher induced level of plant defense responses to the given stress factors in primed plants when compared with the unprimed control.

The molecular basis of priming was investigated through various advance molecular biological technologies such as transcriptomics and genomics to elucidate the novel candidate genes or gene regulators under different stress conditions. Beckers et al. (2009) described the accumulation of inactive mitogen-activated protein kinases in *Arabidopsis* under chemically induced priming. In addition to that, priming has also been linked to di- or tri-methylation at lysine 4 of histone H3 and lysine acetylation of histone H3 at lysine 9, 5, 8, or 12 of the histone H4 in the promoter regions of defense related genes (Jaskiewicz et al. 2011).

Plants need to make changes in their genome with more resistance and potential genes to mitigate environmental stress generated by global climatic changes for their well been through evolutionary adaptation. As a result of that, genome complexity will be generated in plants to develop resistance (R) genes. Introduction of these resistance R genes to commercial crop cultivars with higher yield is the way to improve disease resistance ability in commercial cultivars usually having poor disease resistance ability (Rhee et al. 2016). Scientists, still they are unable to find out a gene or gene cascade which directly regulates crop yield component. This is a big limitation in crop improvement through molecular breeding. In contrast, selecting inducible epialleles that contribute to desired traits is highly desirable for crop production and can be effective against pathogenic attack through activation of plant defense mechanism. Therefore, screening and identification of compatible priming activators has ability to induce such epialleles in plant genome through natural or artificial substances. The integrated disease management approach is one of the great challenge faced by prime-omics technology with recent advancement in omics technology (Bruce 2010).

Through evolution, plant has developed various mechanisms of resistance that prime their innate immunity for more robust and active induction of defense responses against different kinds of stress. This priming ability can be transferred from one generation to another through sexual reproduction of plants. This phenomenon is called as transgenerational priming. There is an importance in selecting plants containing "marker" genes such as beta-glucanase, chitinase, PR-1, etc. (Martínez-Aguilar et al. 2016).

Plant scientists have done many scientific investigations to evidence transgenerational priming, in general they have subjected mother plant with priming agent and challenged with the particular pathogen. The seeds and seedlings resulted from that experimentation were challenged with same pathogen without priming and observed the transcript accumulation pattern through transcriptomics analysis. Rapid and more efficient level of plant defense related transcript accumulation was resulted in unprimed F1 generation (Ramírez-Carrasco et al. 2017; Fang et al. 2018).

16.11 Bioinformatics

Bioinformatics is a computer based technology that support the scientists to manipulate big data derived from various omics approaches such as genomics, transcriptomics and proteomics etc. Bioinformatics enables to study and identify the genetic elements behind the system complexity (Pérez-Clemente et al. 2013; Singh et al. 2015). It is highlighted that, global population will increase by nine billion in the year 2050 resulting production of 70% more than today (Joshi et al. 2016). There is a necessity to develop sophisticated bioinformatics platform for analysis of big data to improve the time used efficiency in biological researches (Ambrosino et al. 2020).

Computational modeling or computational application in biological research facilitates to study biological activities as a system to get holistic view. These holistic views can be incooperated with different crop improvement programs (Esposito et al. 2016). Bioinformatics is a blended system of computer science, statistics, and engineering, which can manipulate huge amount of biological information to prognosticate the future of some biological phenomenon (Upadhyay et al. 2017). Development of in silico-omics tools in recent years may lead to understand useful stress related genes and related molecules under global climatic changes. This novel omics technology can support plant breeders to produce improved crop varieties having high quality and productivity showing elevated resistance to microbial plant pathogens, insect pests, extreme heat, drought, salinity, and chilling effects (Singh et al. 2011). Bioinformatics deals with exchange, comparison, confirmation, storage, and analysis of biological large data globally but this technology is young enough to grow further in the field of information technology (Gibson and Muse 2002).

Bioinformatics serves as an essential tool in many of omics technology such as genomics, proteomics, and transcriptomics. If once genome or cDNA library is sequenced we need the help of bioinformatics tools to annotate the putative genes and putative function of the sequence data (Eyras et al. 2005). The above mentioned requirements can be achieved by bioinformatics tool such as Basic Local Alignment Searching Tools (BLAST). Construction of dendrogrames or phylogenetic trees to study the evolutionary relationship of organisms can be performed by Molecular Evolutionary Genetics Analysis (MEGA) software (Dubey et al. 2010; Tamura et al. 2007). Development of new bioinformatics software tools such as PDQuest and PEDRo for two-dimensional (2D) electrophoresis analysis helps to cut down the limitation in qualitative and quantitative analysis of protein profile under different environmental conditions (Rhee et al. 2006; Pomastowski and Buszewski 2014). Differential proteome profiles of plasma membranes of *Arabidopsis*, rice, and algae were analyzed under different stress responses like low temperature, salinity, elicitors of plant pathogens such as bacteria (Cheng et al. 2009; Minami et al. 2009). There are no special databases dealing with plant pathways with special reference to data collected under different stress conditions in order to identify tolerant or resistant genes from broad germplasm set including landraces and crop wild relatives (Jaiswal and Usadel 2016). The tremendous development in molecular microbiology provides useful venue to generate new research approaches leading to screening and identification of potential microorganism through metagenomics. Those sequence data provide excellent development in the field of microbial bioinformatics in twenty-first century (Pallen 2020).

16.12 Ionomics

Consequences of nutritionally poor staple food crops like wheat, rice, and maize have created a massive impact on human health. Nearly two billion people in the developing and developed nations suffer from malnutrition and micronutrient deficiencies (Tulchinsky 2010). Development of nutritionally enrich agricultural crop cultivars through novel agricultural technologies such as genetic enhancement or biofortification is a present challenge faced by the plant breeders in developing and developed countries. Elucidation of the genetic mechanisms involve in accumulation of mineral compound to the edible part of the plant such as grain or seed and fruit is the requirement in the process of production of biofortified agriculturally important crops (Singh et al. 2013).

Ionomic is the scientific study of all the mineral nutrients and trace elements found in the biological organisms including prokaryotes and eukaryotes (Salt et al. 2008). Ionome is the dynamic network in cell which facilitates the physiological and biochemical functions in cytoplasm with the instructions received through genetic and environmental regulators (Baxter 2010). Ionomics has interconnection between transcriptome, proteome, and metabolome (Salt 2004). Plant metabolomics provided the precursor to originate novel ionomics technology in the late 1960s and early 1970s as a result of scientific investigations carried out by Robinson and Pauling (1974). By the time of nineteenth century it is became a popular omics technology among the scientific community those who work on nutritional aspect of crops (Marschner 2011).

High throughput element analysis technology coupled with bioinformatics platform and genomics tool provides necessary information to discover the transporter genes involved in mineral transport mechanism of plant cell (Baxter 2009). External and internal factors such as environmental, biochemical, physiological, and genetics can alter the ionomic profile of any organism (White et al. 2012a, b). Understanding the gene regulation under these environmental factors can be done through the modern high throughput ionome analysis tool with the intension of identifying the interaction among ionome and genome (Singh et al. 2013).

Genes involved in ionomic activities can be elucidated by the help of forward genetics. In addition to that screening for mutant with altered ionomic functions can be utilized to identify candidate gene involved in ionomics (Chen et al. 2009). The enhanced suberin 1-1(esb 1-1) mutant is the first ionomic mutant developed through the forward genetic approach (Lahner et al. 2003; Baxter 2009; Hosmani et al. 2013). Transcriptional factor MYB36 regulates the gene involved in Casparian strip biosynthesis pathway. Mutation in MYB36 creates absence of Casparian strip formation and these changes will be leading to increased level of Na, Mg, and Zn and reduce the abundance of Ca, Mn, and Fe in plant tissues (Kamiya et al. 2015; Huang and Salt 2016). In addition to that, soluble metal binding protein encoded by NaKR1 has regular function to upload Na and K ions into phloem of plants to facilitate movement of material generated through many biochemical reactions such as photosynthesis but mutant NaKR1-1 accelerates the uploading of Na and K into the phloem than normal circumstances (Tian et al. 2010). Pathogenesis related gene 5 (CPR5) has ability to accumulate low level of K in plant leaves. Relationship between CPR5 and low K phenotypes is not yet been fully understood by the plant molecular biologists. However, there is no necessity to use elevated level of salicylic acid and jasmonic acid to induce resistance against plant pathogens under the mutated condition of CPR5 in plants. Borghi et al. (2011) and Huang and Salt (2016) highlighted that in their research findings, cyclic nucleotide gated channels and higher affinity potassium transporter HAK5 are the some of molecules involve in CPR5 regulation mechanism in plant.

Elucidation of genetic basis for variation in ionomic traits can be successfully studied by the linkage or QTL mapping with biparental cross between phenotypic extremes. The research conducted based on the few accessions of *A. thaliana* revealed that HKT1gene act as causal gene which control the Na ions accumulation in leaf tissue (Rus et al. 2006). The regulatory mechanism of HKT1gene in Na ions accumulation was confirmed by Baxter (2010), results highlighted that, deletion of the promoter region of HKT1 gene has shown the reduced expression of HKT1 resulting over accumulation of Na ions in leaf of the Arabidopsis accessions. The linkage or QTL mapping is limited to explore allelic variation among large number of genotypes at once but development of genome wide association analysis technology has given ability to expire extensive allelic variation in the large number of plants in natural population to discover linkage between genotype to phenotype in ionomic studies.

Evolutionary adaptation of biological organism to their environmental condition for survival is a continuous process. Therefore, plant has developed special structural and functional attributes to sustain in their environment through natural genetic engineering. Comparative ionomics can be used to analyze ionomic profile of plant under common environmental conditions and native local environmental conditions to reveal natural adaptations (Baxter 2010). Molecular biological analysis to discover the natural alleles involved in ionomic traits is an important step in ionome research. The involvement of the alleles in local adaptation of *A. thaliana* has been extensively studied by Busoms et al. (2015). They have cultivated salt tolerant costal population and salt sensitive inland population of *A. thaliana* in the Catalonian costal belt with elevated soil salinity with 30 km distance among two population to study the ionomic loci which facilitate the local adaptation in plants (Busoms et al. 2015).

As a branch of omics technology, ionomic has gained popularity among plant scientists during last decade to understand ionomic variations in plant under biotic and abiotic stress generated by internal and external factors of plants and its environment. The loci involve in ionomic variation of plants have been studied extensively by using a model plant A. thaliana and its accessions such as 1307. The broad spectrum of genomic data derived from the study has been provided a useful venue to the molecular biologists to work in plant ionome (Horton et al. 2012; Huang and Salt 2016). Inductively coupled plasma optical emission spectroscopy (ICP-OES) technology was introduced to analyze the ionome of ten lettuce cultivars with three level of resistance to identify the resistance against bacterial leaf spot caused by most destructive plant pathogen *Xanthomonas campestris pv. Vitians* (*Xcv*). This technology has ability to understand which nutrient/s balances are linked to resistance against *X. campestris pv. Vitians* (*Xcv*). Results revealed that in a scientific investigation, nitrogen, sulfur, and phosphorus balance in lettuce ionome has significant effect on plant resistance to bacterial leaf spot pathogen (Nicolas et al. 2019).

16.13 Phenomic

The global food production systems are threaten by many environmental and ecological factors, which create uncertainty in achieving future food demand (Ray et al. 2013). The replacement of conventional agricultural system with modern high-tech agricultural technologies such as molecular plant breeding will provide effective and efficient solutions to meet the need of future food demand. Selection of parental population for the molecular breeding is mainly depend on genotypic traits with special reference to their phenotype (Zhao et al. 2019; Araus and Cairns 2014; Feng et al. 2017). The development of high throughput phenotypic technologies would be useful to overcome limitations exist in conventional plant breeding. The data collected from different levels of organization of biological organisms' namely cellular level, organismal level, population level etc. will determine the phenotypic traits of a given organism. (Lobos et al. 2017). Genotype together with environmental factors determines the phenotype of an organism. Therefore, use of high throughput technology to analyze phenotypic variation under these conditions is referred as phonemics (Ichihashi and Sinha 2014).

Characterization of quantitative phenotypic traits such as tolerant to biotic and abiotic factors was difficult in past decades due to low technological innovations but introduction of novel omics technologies such as genotyping has given birth to advanced phonemics (White et al. 2012a, b). In contrast biosensors, automations, advance image processing systems, and computational bioinformatics platforms provide more comprehensive strategies to analysis of big data coming from genomic research for comparative analysis of genotypes to evaluate large number of phenotypic traits at once (Tardieu et al. 2017).

Phenotyping systems and tools which are developed at the moment mainly depend upon the environmental conditions such as control or field level (Dhondt et al. 2013). To expose crop micro-phenotypes, the pretreatment of plant is typically destructive when sampled in a cumbersome and includes multi-step procedure and high resolution imaging of samples is also inefficient in view of the output in micro-level. Most plant phenotypic platforms concentrated the high throughput of individual plants (Dhondt et al. 2013; Cabrera Bosquet et al. 2016). The morphological data, namely leaf shape, leaf dimension, leaf arrangement, and fruit shape and size are basic phenotyping indexes in plants (Zhang et al. 2017). Drone based technology and image analysis technology are heavily used in the field of phenotyping (Fahlgren et al. 2015; Confalonieri et al. 2017). The phonemics platforms such as scan analyzer 3D, used to investigate the physiological parameters in tomatoes under drought conditions (Laxman et al. 2018).

Different phenotypic traits such as structure, morphology, color, and physiological information are measured by using multi-model sensors in phenotypic investigation; therefore, phenomic experiments are not reproducible (Zhao et al. 2019). Further, discovery of the plant mechanisms involved in stress tolerant abilities will be depending upon the proper understanding about phenotypic traits which are associated with the stress tolerant in crops under different stress generated by controlled and field conditions. In the image processing technology with the help of visible to near infrared spectrum, hyperspectral imaging, X-ray computed tomography and fluorescence imaging help to analyze phenotypic information in non-destructive manner by using advance and sophisticated software systems (Sozzani et al. 2014; Rahaman et al. 2015).

When the plant are exposed to various biotic and abiotic stress conditions, plant processes several metabolic pathways to regulate environmental stress to control internal balance at a same time (Saito and Matsuda 2010). The bioaccumulation of proline or soluble carbohydrates like glycine betaine and γ-aminobutyric acid in cytoplasm to persist against dehydration responses generated by the drought and salinity in plant through changes in their physiological, biochemical and phenotypic status (Joshi et al. 2017). The extreme environmental and soil conditions such as drought and salinity respectively have major threat to agriculture. In cooperation of strategies to improve resistance or tolerant crop cultivars to these factors have added advantage in crop improvement to neutralize the abiotic stress (Acosta-Motos et al. 2017).

There is a limitation in high throughput phenotyping methods for molecular plant breeding. Therefore, progressive development in the field of plant breeding to develop high yielding crop cultivars with resistance to biotic and abiotic environmental factors will be restricted by manual plant breeding and insufficient genomic information for crop improvement (Cobb et al. 2013). The new next generation phenotyping platforms with efficient computer software in recent past has given considerable opportunity to evaluate drought tolerant abilities in several crops like Phinopsis and Wiwam in Arabidopsis and LemnaTech in barley, maize, tomatoes and wheat (Granier et al. 2006; Honsdorf et al. 2014; Petrozza et al. 2014; Fehér-Juhász et al. 2014; Ge et al. 2016; Singh et al. 2018).

There is a correlation in between metabolic traits and the phenotype of microbial consortia because phonemics has ability to uprising functional genomics (Acin-Albiac et al. 2020). Metagenomics has ability to identify taxonomy and functionality of food microbial consortia without culture dependent approaches. In addition to that, meta-transcriptomics can identify the constitutively and differentially expressed genes under environmental stress at population level. To see the overall effect of environmental stress regulators on plant phenome can be achieved by integrating meta-metabolomics, meta-transcriptomics, and meta-proteomics of microbial consortia. Integrated omics technologies try to understand the functional expression of meta-genome of microbiota under different regulatory conditions to see linear movement of biochemical mechanism. Meta-phenomics try to understand the physiological condition of microbial meta-communities. The meta genomics and meta phonemics are intermediate by several other omics approaches (Acin-Albiac et al. 2020).

16.14 Integration of Omics Technologies

The novel genetic and chemical elements that regulate the most important physiological processes of plant can be identified by using various omics technologies such as genomics, transcriptomes, proteomics, metabolomics, ionomics, and phenomics under different stress factors (Cohen et al. 2017). Complexity of stress responses in plants can be identified more precisely by integrating the relevant omics approaches. Development of suitable software tools for better big data analysis is one of the great challenges faced by the scientific community. Therefore, collaborative approaches can be implemented to study complex physiological and biochemical responses of organisms under biotic and abiotic stress conditions (Chaudhary et al. 2019a, b; Shivaraj et al. 2019).

Development of omics technology, integration of data in a usable format, and analysis of data with bioinformatics tool are the tree main domains that must be addressed to get full advantage of plant system biology (Urano et al. 2010; Gupta et al. 2013). These omics integration together with statistical and computational software generated higher success rates in the prediction in developing a more robust hybrid. Therefore, farmers can cultivate improved varieties of crops if plant molecular biologists, plant physiologists, and plant breeders are get together with one

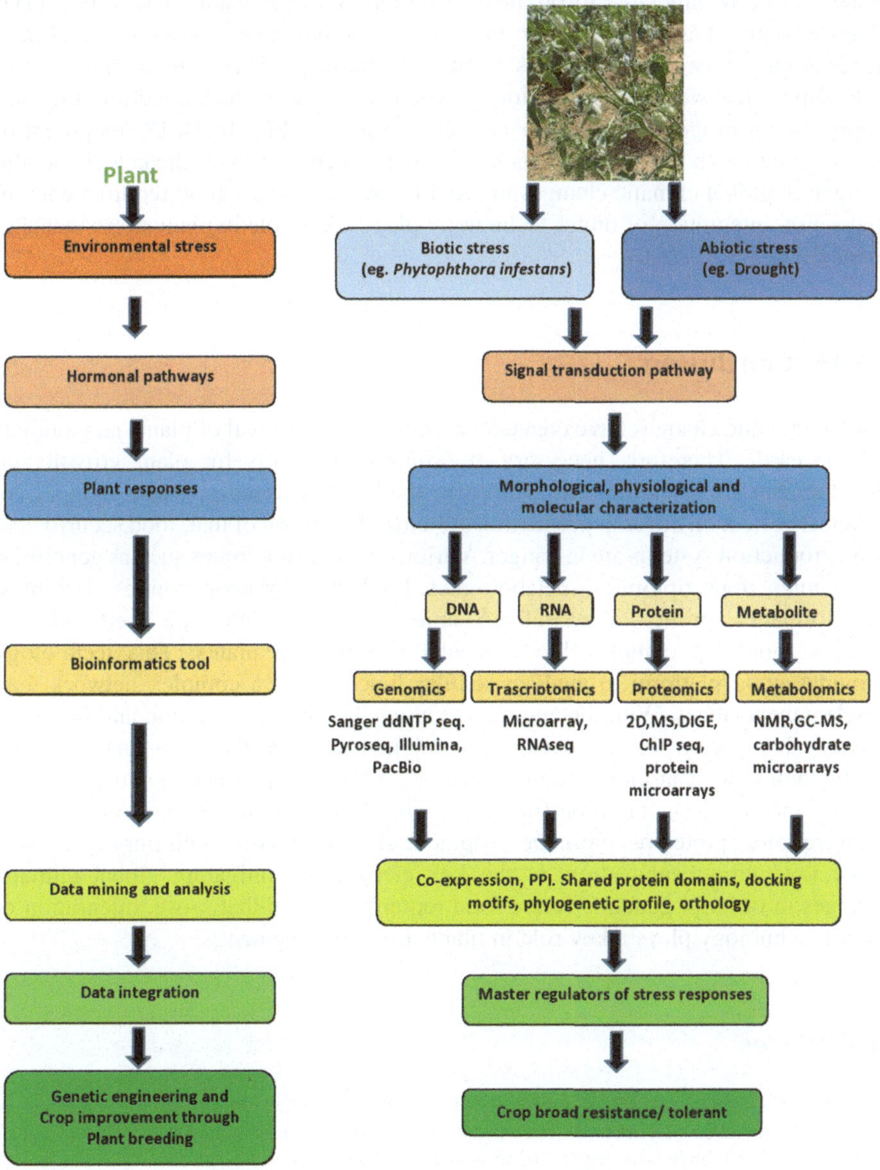

Fig. 16.1 Omics technologies for biotic and abiotic stress tolerance in plant

objective. If possible, farmers should be given the seeds at no cost and full implementations should be closely monitored by government. Through a strong dialog and cooperatives, it will be possible to deliver better agricultural products that utilize less input, have lower environmental costs, and provide higher levels of social well-being.

Elucidation of functional genes is a main purpose of omics data. Therefore, these genes can be manipulated to produce new crops for sustainable agriculture. QTL cloning, omics based association analysis, and rapid identification of important genes are the three major branches of omics technology. These omics technologies will help to discover novel regulatory genes that can be in cooperated into the crop improvement program to engineer new crop cultivars (Fig. 16.1). Development of new cultivars with improved resistance to environmental stress is the key element in mitigating global climatic changes in order to generate future food requirements. In that sense, integration of omics technology plays a key role in plant stress management (Li and Yan 2020).

16.15 Conclusion

Global climatic changes have created a big impact on survival of plants and animals on the earth. Therefore, necessary environmental factors for plant growth and development are changing drastically over the last few decades. These changes are adversely effect on the crop production systems. As a result of that, food security and food production systems are in danger. Various omic technologies such as genomics, proteomics, transcriptomics, metabolomics, lipidomics, proteogenomics, ionomics, bioinformatics, prime-omics, miRNA omics, and phenomics are used today to discuss about the complex bio-molecular activities in plants. This technology provides good platform to analyze cellular functions as a complex network than single gene analysis. With the enormous knowledge of domestication and improvement gained from omics data, in combination with the new gene editing technologies, we can create future crops via a three steps roadmap such as scale (micro, macro, etc.), condition (biotic and abiotic stress), and dimension (genome, transcriptome, proteome, etc.). Development of new cultivars with improved resistance to environmental stress is the key element in mitigating global climatic changes in order to generate future food requirements. In that sense, integration of omics technology plays a key role in plant stress management.

References

Acin-Albiac M, Filannino P, Gobbetti M, Di Cagno R (2020) Microbial high throughput phenomics: the potential of an irreplaceable omics. Comput Struct Biotechnol J 18:2290–2299. https://doi.org/10.1016/j.csbj.2020.08.010

Acosta-Motos JR, Ortuño MF, Bernal-Vicente A, Diaz-Vivancos P, Sanchez-Blanco MJ, Hernandez JA (2017) Plant responses to salt stress: adaptive mechanisms. Agronomy 7:18

Agarwal P, Parida SK, Mahto A, Das S, Mathew IE, Malik N, Tyagi AK (2014) Expanding frontiers in plant transcriptomics in aid of functional genomics and molecular breeding. Biotechnol J 9(12):1480–1492. https://doi.org/10.1002/biot.201400063

Agrawal GK, Pedreschi R, Barkla BJ, Bindschedler LV, Cramer R, Sarkar A, Renaut J, Job D, Rakwal R (2012) Translational plant proteomics: a perspective. J Proteome 75(15):4588–4601. https://doi.org/10.1016/j.jprot.2012.03.055

Aliferis K, Chrysayi-Tokousbalides M (2011) Metabolomics in pesticide research and development: review and future perspectives. Metabolomics 7:35–53

Ambrosino L, Colantuono C, Diretto G, Fiore A, Chiusano ML (2020) Bioinformatics resources for plant abiotic stress responses: state of the art and opportunities in the fast evolving -omics era. Plan Theory 9(591):1–36. https://doi.org/10.3390/plants9050591

Ampofo Asiama J, Baiye V, Hertog M, Waelkens E, Geeraerd A, Nicolai BJPB (2014) The metabolic response of cultured tomato cells to low oxygen stress. Plant Biol 16:594–606

Araus JL, Cairns JE (2014) Field high-throughput phenotyping: the new crop breeding frontier. Trends Plant Sci 19:52–61. https://doi.org/10.1016/j.tplants.2013.09.008

Avin-Wittenberg T, Tzin V, Angelovici R, Less H, Galili G (2012) Deciphering energy-associated gene networks operating in the response of Arabidopsis plants to stress and nutritional cues. Plant J 70:954–966

Baerenfaller K, Grossmann J, Grobei MA, Hull R, Hirsch-Hoffmann M, Yalovsky S, Zimmermann P, Grossniklaus U, Gruissem W, Baginsky S (2008) Genome-scale proteomics reveals Arabidopsis thaliana gene models and proteome dynamics. Science 320(5878):938–941. https://doi.org/10.1126/science.1157956

Bajwa AA, Farooq M, Nawaz A (2018) Seed priming with sorghum extracts and benzyl aminopurine improves the tolerance against salt stress in wheat (*Triticum aestivum* L.). Physiol Mol Biol Plants 24(2):239–249

Balmer A, Pastor V, Gamir J, Flors V, Mauch-Mani B (2015) The 'prime-ome': towards a holistic approach to priming. Trends Plant Sci 20(7):443–452

Barkan L, Vijayan P, Carlsson AS, Mekhedov S, Browse J (2006) A suppressor of fab1 challenges hypotheses on the role of thylakoid unsaturation in photosynthetic function. Plant Physiol 141:1012–1020

Baxter I (2009) Ionomics: studying the social network of mineral nutrients. Curr Opin Plant Biol 12:381–386

Baxter I (2010) Ionomics: the functional genomics of elements. Brief Funct Genomics 9:149–156

Beckers GJ, Jaskiewicz M, Liu Y, Underwood WR, He SY, Zhang S, Conrath U (2009) Mitogen-activated protein kinases 3 and 6 are required for full priming of stress responses in Arabidopsis thaliana. Plant Cell 21(3):944–953. https://doi.org/10.1105/tpc.108.062158

Bekele WA, Wieckhorst S, Friedt W, Snowdown RJ (2014) High-throughput genomics in sorghum: from whole- genome resequencing to a SNP screening array. Plant Biotechnol J 11:1112–1125

Benson DA, Karsch-Mizrachi I, Clark K, Lipman DJ, Ostell J, Sayers EW (2012) GenBank. Nucleic Acids Res 40:48–53. https://doi.org/10.1093/nar/gkr1202

Blakeley P, Overton IM, Hubbard SJ (2012) Addressing statistical biases in nucleotide-derived protein databases for proteogenomic search strategies. J Proteome Res 11:5221–5234

Borghi M, Rus A, Salt DE (2011) Loss-of-function of constitutive expresser of pathogenesis related Genes5 affects potassium homeostasis in Arabidopsis thaliana. PLoS One 6(10):e26360. https://doi.org/10.1371/journal.pone.0026360

Bruce TJA (2010) Tackling the threat to food security caused by crop pests in the new millennium. Food Secur 2:133–141. https://doi.org/10.1007/s12571-010-0061-8

Buiatti M, Christou P, Pastore G (2013) The application of GMOs in agriculture and in food production for a better nutrition: two different scientific points of view. Genes Nutr 8 (3):255–270. https://doi.org/10.1007/s12263-012-0316-4

Bulgarelli D, Schlaeppi K, Spaepen S, Ver Loren Van Themaat E, Schulze-Lefert P (2013) Structure and functions of the bacterial microbiota of plants. Annu Rev Plant Biol 64:807–838. https://doi.org/10.1146/annurev-arplant-050312-120106

Burgos A, Szymanski J, Seiwert B, Degenkolbe T, Hannah MA, Giavalisco P (2011) Analysis of short-term changes in the *Arabidopsis thaliana* glycerolipidome in response to temperature and light. Plant J 66:656–668

Burke JJ, O'Mahony PJ, Oliver MJ (2000) Isolation of Arabidopsis mutants lacking components of acquired thermotolerance. Plant Physiol 123:575–588

Busoms S, Teres J, Huang XY, Bomblies K, Danku J, Douglas A, Weigel D, Poschenrieder C, Salt DE (2015) Salinity is an agent of divergent selection driving local adaptation of *Arabidopsis* to coastal habitats. Plant Physiol 168:915–929

Cabrera Bosquet L, Fournier C, Brichet N, Welcker C, Suard B, Tardieu F (2016) High-throughput estimation of incident light, light interception and radiation-use efficiency of thousands of plants in a phenotyping platform. New Phytol 212:269–281. https://doi.org/10.1111/nph.14027

Casadevall R, Rodriguez RE, Debernardi JM, Palatnik JF, Casati P (2013) Repression of growth regulating factors by the microRNA396 inhibits cell proliferation by UV-B radiation in Arabidopsis leaves. Plant Cell 25:3570–3583. https://doi.org/10.1105/tpc.113.117473

Casado-Díaz A, Encinas-Villarejo S, de los Santos B, Schilirò E, Yubero-Serrano EM, Amil-Ruíz F, Pocovi MI, Pliego-Alfaro F, Dorado G, Rey M, Romero F, Muñoz-Blanco J, Caballero JL (2006) Analysis of strawberry genes differentially expressed in response to *Colletotrichum* infection. Physiol Plant 128:633–650. https://doi.org/10.1111/j.1399-3054.2006.00798.x

Chang YY, Cronan JE (1999) Membrane cyclopropane fatty acid content is a major factor in acid resistance of Escherichia coli. Mol Microbiol 33(2):249–259

Chaudhary J, Khatri P, Singla P, Kumawat S, Kumari A, Vinaykumar R, Vikram A, Jindal SK, Kardile H, Kumar R, Sonah H, Deshmukh R (2019a) Advances in omics approaches for abiotic stress tolerance in tomato. Biology 8:90. https://doi.org/10.3390/biology8040090

Chaudhary J, Shivaraj S, Khatri P, Ye H, Zhou L, Klepadlo M, Dhakate P, Kumawat G, Patil G, Sonah H (2019b) Approaches, applicability, and challenges for development of climate-smart soybean. In: Genomic designing of climate-smart oilseed crops. Springer, Berlin, pp 1–74

Chawla K, Barah P, Kuiper M, Bones AM (2011) Systems biology: a promising tool to study abiotic stress responses. In: Omics and plant abiotic stress tolerance. Bentham Science, Sharjah, pp 163–172

Chen X, Wang Y, Li J, Jiang A, Cheng Y, Zhang W (2009) Mitochondrial proteome during salt stress-induced programmed cell death in rice. Plant Physiol Biochem 47(5):407–415. https://doi.org/10.1016/j.plaphy.2008.12.021

Chen Y, Pang Q, Dai S, Wang Y, Chen S, Yan X (2011) Proteomic identification of differentially expressed proteins in Arabidopsis in response to methyl jasmonate. J Plant Physiol 168 (10):995–1008. https://doi.org/10.1016/j.jplph.2011.01.018

Chen MX, Zhu FY, Gao B, Ma KL, Zhang Y, Fernie AR, Chen X, Dai L, Ye NH, Zhang X, Tian Y, Zhang D, Xiao S, Zhang J, Liu YG (2020) Full-length transcript-based proteogenomics of rice improves its genome and proteome annotation. Plant Physiol 182(3):1510–1526. https://doi.org/10.1104/pp.19.00430

Cheng Y, Qi Y, Zhu Q, Chen X, Wang N, Zhao X, Chen H, Cui X, Xu L, Zhang W (2009) New changes in the plasma-membrane-associated proteome of rice roots under salt stress. Proteomics 9(11):3100–3114

Cilindre C, Jegou S, Hovasse A, Schaeffer C, Castro AJ, Clément C, Dorsselaer AV, Jeandet P, Marchal R (2008) Proteomic approach to identify champagne wine proteins as modified by Botrytis cinerea infection. J Proteome Res 7(3):1199–1208. https://doi.org/10.1021/pr070419p

Cobb JN, DeClerck G, Greenberg A, Clark R, McCouch S (2013) Next-generation phenotyping: requirements and strategies for enhancing our understanding of genotype–phenotype relationships and its relevance to crop improvement. Theor Appl Genet 126:867–887

Cohen H, Aharoni A, Szymanski J, Dominguez E (2017) Assimilation of 'omics' strategies to study the cuticle layer and suberin lamellae in plants. J Exp Bot 68:5389–5400

Collakova E, Yen JY, Senger RS (2012) Are we ready for genome-scale modeling in plants. Plant Sci 19:53–70. https://doi.org/10.1016/j.plantsci.2012.04.010

Confalonieri R, Paleari L, Foi M, Movedi E, Vesely FM, Thoelke W, Agape C, Borlini G, Ferri I, Massara F, Motta R, Ravasi RA, Tartarini S, Rossini L (2017) Pocketplant3d: analyzing canopy structure using a smartphone. Biosyst Eng 164:1–12. https://doi.org/10.1016/j.biosystemseng.2017.09.014

Conrath U (2009) Priming of induced plant defense responses. In: Van Loon LC (ed) Advances in botanical research, vol 51. Elsevier, London, pp 361–395

Conrath U (2011) Molecular aspects of defense priming. Trends Plant Sci 16:524–531. https://doi.org/10.1016/j.tplants.2011.06.004

Cui N, Sun X, Sun M, Jia B, Duanmu H, Dekang LV, Xu D, Zhu Y (2015) Overexpression of OsmiR156k leads to reduced tolerance to cold stress in rice (*Oryza sativa*). Mol Breed 35:214. https://doi.org/10.1007/s11032-015-0402-6

Cullen BR (2013) MicroRNAs as mediators of viral evasion of the immune system. Nat Immunol 14:205–210

Davidson RM, Reeves PA, Manosalva PM, Leach JE (2009) Germins: a diverse protein family important for crop improvement. Plant Sci 177(6):499–510. https://doi.org/10.1016/j.plantsci.2009.08.012

De Kruijff B, van Dam V, Breukink E (2008) Lipid II: a central component in bacterial cell wall synthesis and a target for antibiotics. Prostaglandins Leukot Essent Fat Acids 79(3–5):117–121

Desiere F, Deutsch EW, Nesvizhskii AI, Mallick P, King NL, Eng JK, Aderem A, Boyle R, Brunner E, Donohoe S, Fausto N, Hafen E, Hood L, Katze MG, Kennedy KA, Kregenow F, Lee H, Lin B, Martin D, Ranish JA, Rawlings DJ, Samelson LE, Shiio Y, Watts JD, Wollscheid B, Wright ME, Yan W, Yang L, Yi EC, Zhang H, Aebersold R (2005) Integration with the human genome of peptide sequences obtained by high-throughput mass spectrometry. Genome Biol 6(1):R9. https://doi.org/10.1186/gb-2004-6-1-r9

Dhondt S, Wuyts N, Inzé D (2013) Cell to whole-plant phenotyping: the best is yet to come. Trends Plant Sci 18:428–439. https://doi.org/10.1016/j.tplants.2013.04.008

Dixon RA, Gang DR, Charlton AJ, Fiehn O, Kuiper HA, Reynolds TL, Tjeerdema RS, Jeffery EH, German JB, Ridley WP, Seiber JN (2006) Applications of metabolomics in agriculture. J Agric Food Chem 54(24):8984–8994. https://doi.org/10.1021/jf061218t

Dos Santos VS, Macedo FA, do Vale JS, Silva DB, Carollo CA (2017) Metabolomics as a tool for understanding the evolution of *Tabebuia sensu lato*. Metabolomics 13:72

Driessen AJ, Nouwen N (2008) Protein translocation across the bacterial cytoplasmic membrane. Annu Rev Biochem 77:643–667

Dubey AK, Yadav S, Kumar M, Singh VK, Sarangi BK, Yadav D (2010) In silico characterization of pectate lyase protein sequences from different source organisms. Enzym Res 2010:950230

Dunwell JM, Gibbings JG, Mahmood T, Naqvi SMS (2008) Germin and germin- like proteins: evolution, structure and function. Crit Rev Plant Sci 27:342–375. https://doi.org/10.1080/07352680802333938

Duque AS, de Almeida AM, da Silva AB, da Silva JM et al (2013) Abiotic stress—plant responses and applications in agriculture. Intech, Rijeka. https://doi.org/10.5772/45842

Elbasyoni IS, Lorenz A, Guttieri M, Frels K, Baenziger P, Poland J, Akhunov E (2018) A comparison between genotyping-by-sequencing and array-based scoring of SNPs for genomic prediction accuracy in winter wheat. Plant Sci 270:123–130

El-Metwally S, Hamza T, Zakaria M, Helmy M (2013) Next-generation sequence assembly: four stages of data processing and computational challenges. PLoS Comput Biol 9(12):e1003345. https://doi.org/10.1371/journal.pcbi.1003345

El-Metwally S, Ouda OM, Helmy M (2014) First- and next-generations sequencing methods. In: Next generation sequencing technologies and challenges in sequence assembly, SpringerBriefs in systems biology, vol vol 7. Springer, New York. https://doi.org/10.1007/978-1-4939-0715-1_3

Eltaher S, Sallam A, Belamkar V, Emara HA, Nower AA, Salem KFM, Poland J, Baenziger PS (2018) Genetic diversity and population structure of F3:6 Nebraska winter wheat genotypes using genotyping-by-sequencing. Front Genet 9:76

Erridge C, Bennett-Guerrero E, Poxton IR (2002) Structure and function of lipopolysaccharides. Microbes Infect 4(8):837–851

Esposito A, Colantuono C, Ruggieri V, Chiusano ML (2016) Bioinformatics for agriculture in the next-generation sequencing era. Chem Biol Technol Agric 3:9. https://doi.org/10.1186/s40538-016-0054-8

Evers D, Legay S, Lamoureux D, Hausman JF, Hoffmann L et al (2012) Towards a synthetic view of potato cold and salt stress response by transcriptomic and proteomic analyses. Plant Mol Biol 78:503–514. https://doi.org/10.1007/s11103-012-9879-0

Eyras E, Reymond A, Castelo R, Bye JM, Camara F, Flicek P, Huckle EJ, Parra G, Shteynberg DD, Wyss C, Rogers J, Antonarakis SE, Birney E, Guigo R, Brent MR (2005) Gene finding in the chicken genome. BMC Bioinformatics 6:131. https://doi.org/10.1186/1471-2105-6-131

Fahlgren N, Gehan MA, Baxter I (2015) Lights, camera, action: high throughput plant phenotyping is ready for a close-up. Curr Opin Plant Biol 24:93–99. https://doi.org/10.1016/j.pbi.2015.02.006

Falconer DS (1989) Introduction to quantitative genetics. Longman, Scientific & Technical, Burnt Mill

Fang Y, Nazaries L, Singh BK, Singh BP (2018) Microbial mechanisms of carbon priming effects revealed during the interaction of crop residue and nutrient inputs in contrasting soils. Glob Chang Biol 24(7):2775–2790. https://doi.org/10.1111/gcb.14154

Fehér-Juhász E, Majer P, Sass L, Lantos C, Csiszár J, Turóczy Z, Mihály R, Mai A, Horváth GV, Vass I, Dudits D (2014) Phenotyping shows improved physiological traits and seed yield of transgenic wheat plants expressing the alfalfa aldose reductase under permanent drought stress. Acta Physiol Plant 36:663–673

Feng H, Guo Z, Yang W, Huang C, Chen G, Fang W, Xiong X, Zhang H, Wang G, Xiong L, Liu Q (2017) An integrated hyperspectral imaging and genome-wide association analysis platform provides spectral and genetic insights into the natural variation in rice. Sci Rep 7:4401. https://doi.org/10.1038/s41598-017-04668-8

Freitas-Astúa J, Bastianel M, Locali-Fabris EC, Novelli VM, Silva-Pinhati AC, Basflio-Palmieri AC, Targon MLPN, Machado MA (2007) Differentially expressed stress-related genes in the compatible citrus-citrus leprosis virus interaction. Genet Mol Biol 30(3):980–990. https://doi.org/10.1590/S1415-47572007000500026

Freund DM, Hegeman AD (2017) Recent advances in stable isotope-enabled mass spectrometry-based plant metabolomics. Curr Opin Biotechnol 43:41–48. https://doi.org/10.1016/j.copbio.2016.08.002

Furuse Y, Finethy R, Saka HA, Xet-Mull AM, Sisk DM, Smith KLJ, Lee S, Coers J, Valdivia RH, Tobin DM, Cullen BR (2014) Search for microRNAs expressed by intracellular bacterial pathogens in infected mammalian cells. PLoS One 9(9):e106434. https://doi.org/10.1371/journal.pone.0106434

Gao P, Bai X, Yang L, Lv D, Pan X, Li Y, Cai H, Ji W, Chen Q, Zhu Y (2011) osa-MIR393: a salinity and alkaline stress-related microRNA gene. Mol Biol Rep 38:237–242. https://doi.org/10.1007/s11033-010-0100-8

Ge Y, Bai G, Stoerger V, Schnable JC (2016) Temporal dynamics of maize plant growth, water use, and leaf water content using automated high throughput RGB and hyperspectral imaging. Comput Electron Agric 127:625–632

Gentile A, Dias LI, Mattos RS, Ferreira TH, Menossi M (2015) MicroRNAs and drought responses in sugarcane. Front Plant Sci 6:58. https://doi.org/10.3389/fpls.2015.00058

Gibson G, Muse S (2002) A primer in genome science. Sinauer Ass, Sunderland, p 347

Gilliham M, Able JA, Roy SJ (2017) Translating knowledge about abiotic stress tolerance to breeding programmes. Plant J 90(5):898–917

Goswami S, Kumar RR, Rai RD (2014) Heat-responsive microRNAs regulate the transcription factors and heat shock proteins in modulating thermo-stability of starch biosynthesis enzymes in wheat (*Triticum aestivum* L.) under the heat stress. Aust J Crop Sci 8:697–705

Granier C, Aguirrezabal L, Chenu K, Cookson SJ, Dauzat M, Hamard P, Thioux JJ, Rolland G, Bouchier-Combaud S, Lebaudy A, Muller B (2006) PHENOPSIS, an automated platform for reproducible phenotyping of plant responses to soil water deficit in *Arabidopsis thaliana* permitted the identification of an accession with low sensitivity to soil water deficit. New Phytol 169:623–635

Gripenland J, Netterling S, Loh E, Tiensuu T, Toledo-Arana A, Johansson J (2010) RNAs: regulators of bacterial virulence. Nat Rev Microbiol 8(12):857–866. https://doi.org/10.1038/nrmicro2457

Grundhoff A, Sullivan CS (2011) Virus-encoded microRNAs. Virology 411:325–343

Gupta B, Saha J, Sengupta A, Gupta K (2013) Plant abiotic stress: 'omics' approach. J Plant Biochem Physiol 1:e108. https://doi.org/10.4172/23299029.1000e108

Gupta OP, Sharma P, Gupta RK, Sharma I (2014) MicroRNA mediated regulation of metal toxicity in plants: present status and future perspectives. Plant Mol Biol 84:1–18. https://doi.org/10.1007/s11103-013-0120-6

Gupta N, Singh A, Zahra S, Kumar S (2018) PtRFdb: a database for plant transfer RNA-derived fragments. Database (Oxford). https://doi.org/10.1093/database/bay063

Han MJ, Lee JW, Lee SY (2011) Understanding and engineering of microbial cells based on proteomics and its conjunction with other omics studies. Proteomics 11(4):721–743

He F, Wu F, Zhong F (2016) Microbial proteomics: approaches, advances, and applications. Bioinfo Proteom Img Anal 2(1):85–91. https://doi.org/10.15436/2381-0793.16.004

Heather JM, Chain B (2016) The sequence of sequencers: the history of sequencing DNA. Genomics 107:1–8

Helmy M, Sugiyama N, Tomita M, Ishihama Y (2012) Mass spectrum sequential subtraction speeds up searching large peptide MS/MS spectra datasets against large nucleotide databases for proteogenomics. Genes Cells 17:633–644. https://doi.org/10.1111/j.1365-2443.2012.01615.x

Hernandez Y, Sanan-Mishra N (2017) miRNA mediated regulation of NAC transcription factors in plant development and environment stress response. Plant Gene 11:190–198

Hisamatsu Y, Goto N, Sekiguchi M, Hasegawa K, Shigemori H (2005) Oxylipins arabidopsides C and D from *Arabidopsis thaliana*. J Nat Prod 68:600–603. https://doi.org/10.1021/np0495938

Hoe CH, Raabe CA, Rozhdestvensky TS, Tang TH (2013) Bacterial sRNAs: regulation in stress. Int J Med Microbiol 303:217–229

Honsdorf N, March TJ, Berger B, Tester M, Pillen K (2014) High-throughput phenotyping to detect drought tolerance QTL in wild barley introgression lines. PLoS One 9(5):e97047. https://doi.org/10.1371/journal.pone.0097047

Horton MW, Hancock AM, Huang YS, Toomajian C, Atwell S, Auton A, Muliyati NW, Platt A, Sperone FG, Vilhjálmsson BJ, Nordborg M, Borevitz JO, Bergelson J (2012) Genome-wide patterns of genetic variation in worldwide Arabidopsis thaliana accessions from the RegMap panel. Nat Genet 44(2):212–216. https://doi.org/10.1038/ng.1042

Hosmani PS, Kamiya T, Danku J, Naseer S, Geldner N, Guerinot ML, Salt DE (2013) Dirigent domain-containing protein is part of the machinery required for formation of the lignin-based Casparian strip in the root. Proc Natl Acad Sci U S A 110(35):14498–14503. https://doi.org/10.1073/pnas.1308412110

Huang XY, Salt DE (2016) Plant Ionomics: from elemental profiling to environmental adaptation. Mol Plant 9:787–797

Hussain S, Yin H, Peng S, Faheem A, Khan FA, Khan F, Sameeullah M, Hussain HA, Huang J, Kehui Cui K, Nie L (2016) Comparative transcriptional profiling of primed and non-primed rice seedlings under submergence stress. Front Plant Sci 7:1125. https://doi.org/10.3389/fpls.2016.01125

Ibraheem O, Adigun RO, Olatunji IT (2018) Omics technologies in unraveling plant stress responses; using sorghum as a model crop, how far have we gone? Vegetos 31:1–18

Ichihashi Y, Sinha NR (2014) From genome to phenome and back in tomato. Curr Opin Plant Biol 18:9–15

Imam J, Singh PK, Shukla P (2016) Plant microbe interactions in post genomic era: perspectives and applications. Front Microbiol 7:1488. https://doi.org/10.3389/fmicb.2016.01488

Ingolia NT (2014) Ribosome profiling: new views of translation, from single codons to genome scale. Nat Rev Genet 15:205–213

Jaiswal P, Usadel B (2016) Plant pathway databases. In: Edwards D (ed) Plant bioinformatics. Methods in molecular biology, vol vol 1374. Humana Press, New York, NY. https://doi.org/10.1007/978-1-4939-3167-5_4

Jaskiewicz M, Conrath U, Peterhänsel C (2011) Chromatin modification acts as a memory for systemic acquired resistance in the plant stress response. EMBO Rep 12:50–55. https://doi.org/10.1038/embor.2010.186

Jogaiah S, Govind SR, Tran L-SP (2013) Systems biology-based approaches toward understanding drought tolerance in food crops. Crit Rev Biotechnol 33:23–39

Joshi R, Karan R, Singla-Pareek SL, Pareek A (2016) Ectopic expression of Pokkali phosphoglycerate kinase-2 (OsPGK2-P) improves yield in tobacco plants under salinity stress. Plant Cell Rep 35(1):27–41

Joshi R, Sahoo KK, Tripathi AK, Kumar R, Gupta BK, Pareek A, Singla-Pareek SL (2017) Knockdown of an inflorescence meristem-specific cytokinin oxidase—OsCKX2 in rice reduces yield penalty under salinity stress condition. Plant Cell Environ 41:936–946. https://doi.org/10.1111/pce.12947

Kamiya T, Borghi M, Wang P, Danku JM, Kalmbach L, Hosmani PS, Naseer S, Fujiwara T, Geldner N, Salt DE (2015) The MYB36 transcription factor orchestrates Casparian strip formation. Proc Natl Acad Sci U S A 112(33):10533–10538. https://doi.org/10.1073/pnas.1507691112

Katz VA, Thulke OU, Conrath U (1998) A benzothiadiazole primes parsley cells for augmented elicitation of defense responses. Plant Physiol 117:1333–1339. https://doi.org/10.1104/pp.117.4.1333

Kauss H, Theisinger-Hinkel E, Mindermann R, Conrath U (1992) Dichloroisonicotinic and salicylic acid, inducers of systemic acquired resistance, enhance fungal elicitor responses in parsley cells. Plant J 2:655–660. https://doi.org/10.1111/j.1365-313X.1992.tb00134.x

Khan N, Bano A, Rahman MA, Guo J, Kang Z, Babar MA (2019) Comparative physiological and metabolic analysis reveals a complex mechanism involved in drought tolerance in chickpea (*Cicer arietinum* L.) induced by PGPR and PGRs. Sci Rep 9:1–19. https://doi.org/10.1038/s41598-019-38702-8

Kim HK, Choi YH, Verpoorte R (2010) NMR-based metabolomic analysis of plants. Nat Protoc 5:536–549. https://doi.org/10.1038/nprot.2009.237

Kosová K, Vítámvás P, Urban MO, Prášil IT, Renaut J (2018) Plant abiotic stress proteomics: the major factors determining alterations in cellular proteome. Front Plant Sci 9:122. https://doi.org/10.3389/fpls.2018.00122

Kreszies T, Shellakkutti N, Osthoff A, Yu P, Baldauf JA, Zeisler-Diehl VV, Ranathunge K, Hochholdinger F, Schreiber L (2018) Osmotic stress enhances suberization of apoplastic barriers in barley seminal roots: analysis of chemical, transcriptomic and physiological responses. New Phytol 221(1):180–194

Kumara UMA, De Costa DM (2015) Analysis of differentially expressed gene profiles in a resistant banana cultivar after infection by *Colletotrichum musae*. Physiol Mol Plant Pathol 92:88–100. https://doi.org/10.1016/j.pmpp.2015.08.011

Küster B, Mortensen P, Andersen JS, Mann M (2001) Mass spectrometry allows direct identification of proteins in large genomes. Proteomics 1(5):641–650. https://doi.org/10.1002/1615-9861(200104)

Lahner B, Gong J, Mahmoudian M, Smith EL, Abid KB, Rogers EE, Guerinot ML, Harper JF, Ward JM, McIntyre L, Schroeder JI, Salt DE (2003) Genomic scale profiling of nutrient and trace elements in Arabidopsis thaliana. Nat Biotechnol 21(10):1215–1221. https://doi.org/10.1038/nbt865

Lal SK, Kumar S, Sheri V, Mehta S, Varakumar P, Ram B, Borphukan B, James D, Fartyal D, Reddy MK (2018) Seed priming: an emerging technology to impart abiotic stress tolerance in crop plants. In: Advances in seed priming. Springer, Singapore, pp 41–50

Lalaouna D, Simoneau-Roy M, Lafontaine D, Masse E (2013) Regulatory RNAs and target mRNA decay in prokaryotes. Biochim Biophys Acta 1829:742–747

Lane BG (2002) Oxalate, germins, and higher-plant pathogens. IUBMB Life 53(2):67–75. https://doi.org/10.1080/15216540211474

Laxman RH, Hemamalini P, Bhatt RM, Sadashiva AT (2018) Non-invasive quantification of tomato (*Solanum lycopersicum* L.) plant biomass through digital imaging using phenomics platform. Indian J Plant Physiol 23:369–375

Legendre S, Letellier L, Shechter E (1980) Influence of lipids with branched-chain fatty acids on the physical, morphological and functional properties of Escherichia coli cytoplasmic membrane. Biochim Biophys Acta 602(3):491–505

Le Sage V, Cinti A, Mouland AJ (2016) Proximity-dependent biotinylation for identification of interacting proteins. Curr Protocols Cell Biol 73:17191–171912. https://doi.org/10.1002/cpcb.11

Li Q, Yan J (2020) Sustainable agriculture in the era of omics: knowledge-driven crop breeding. Genome Biol 21:154. https://doi.org/10.1186/s13059-020-02073-5

Li W, Wang T, Zhang Y, Li Y (2016) Overexpression of soybean miR172c confers tolerance to water deficit and salt stress, but increases ABA sensitivity in transgenic *Arabidopsis thaliana*. J Exp Bot 67:175–194. https://doi.org/10.1093/jxb/erv450

Liang G, Ai Q, Yu D (2015) Uncovering miRNAs involved in crosstalk between nutrient deficiencies in *Arabidopsis*. Sci Rep 5:11813. https://doi.org/10.1038/srep11813

Liu B, Zhang N, Zhao S, Chang J, Wang Z, Zhang G, Si H, Wang D (2015) Proteomic changes during tuber dormancy release process revealed by iTRAQ quantitative proteomics in potato. Plant Physiol Biochem 86:181–190. https://doi.org/10.1016/j.plaphy.2014.12.003

Liu Y, Lu S, Liu K, Wang S, Huang L, Guo L (2019) Proteomics: a powerful tool to study plant responses to biotic stress. Plant Methods 15:135. https://doi.org/10.1186/s13007-019-0515-8

Lobos GA, Camargo AV, DelPozo A, Araus JL, Ortiz R, Doonan JH (2017) Editorial: plant phenotyping and phenomics for plant breeding. Front Plant Sci 8:2181. https://doi.org/10.3389/fpls.2017.02181

Luan Y, Cui J, Zhai J, Li J, Han L, Meng J (2015) High-throughput sequencing reveals differential expression of miRNAs in tomato inoculated with *Phytophthora infestans*. Planta 241(6):1405–1416. http://www.jstor.org/stable/43564234

Luan H, Shen H, Pan Y, Guo B, Lv C, Xu R (2018) Elucidating the hypoxic stress response in barley (*Hordeum vulgare* L.) during waterlogging: a proteomics approach. Sci Rep 8(1):9655

Marraffini LA, Sontheimer EJ (2010) CRISPR interference: RNA-directed adaptive immunity in bacteria and archaea. Nat Rev Genet 11:181–190

Marschner P (2011) Marschner's mineral nutrition of higher plants, 3rd edn. Academic Press, London

Mårtensson CU, Doan KN, Becker T (2017) Effects of lipids on mitochondrial functions. Biochim Biophys Acta Mol Cell Biol Lipids 1862(1):102–113. https://doi.org/10.1016/j.bbalip.2016.06.015

Martínez-Aguilar K, Ramírez-Carrasco G, Hernández-Chávez JL, Barraza A, Alvarez-Venegas R (2016) Use of BABA and INA as activators of a primed state in the common bean (*Phaseolus vulgaris* L.). Front Plant Sci 7:653. https://doi.org/10.3389/fpls.2016.00653

Mashego MR, Rumbold K, De Mey M, Vandamme E, Soetaert W, Heijnen JJ (2007) Microbial metabolomics: past, present and future methodologies. Biotechnol Lett 29(1):1–16. https://doi.org/10.1007/s10529-006-9218-0

Mehta S, James D, Reddy MK (2019) Omics technologies for abiotic stress tolerance in plants: current status and prospects. In: Wani S (ed) Recent approaches in omics for plant resilience to climate change. Springer, Cham, pp 1–34

Minami A, Fujiwara M, Furuto A, Fukao Y, Yamashita T, Kamo M, Kawamura Y, Uemura M (2009) Alterations in detergent-resistant plasma membrane microdomains in Arabidopsis Thaliana during cold acclimation. Plant Cell Physiol 50(2):341–359

Mora L, Bramley PM, Fraser PD, Soler LM (2013) Development and optimisation of a label-free quantitative proteomic procedure and its application in the assessment of genetically modified tomato fruit. Proteomics 13:2016–2030

Moradi P, Mahdavi A, Khoshkam M, Iriti M (2017) Lipidomics unravels the role of leaf lipids in thyme plant response to drought stress. Int J Mol Sci 18(10):2067. https://doi.org/10.3390/ijms18102067

Mosa KA, Ismail A, Helmy M (2017) Omics and system biology approaches in plant stress research. In: Plant stress tolerance. Springer, Cham, pp 21–34

Mosblech A, Feussner I, Heilmann I (2009) Oxylipins: structurally diverse metabolites from fatty acid oxidation. Plant Physiol Biochem 47:511–517. https://doi.org/10.1016/j.plaphy.2008.12.011

Neilson KA, Ali NA, Muralidharan S, Mirzaei M, Mariani M, Assadourian G, Lee A, VanSluyter SC, Haynes PA (2011) Less label, more free: approaches in label-free quantitative mass spectrometry. Proteomics 11:535–553

Nesvizhskii AI (2014) Proteogenomics: concepts, applications and computational strategies. Nat Methods 11:1114–1125. https://doi.org/10.1038/nmeth.3144

Newton AC, Fitt BDL, Atkins SD, Walters DR, Daniell TJ (2010) Pathogenesis, parasitism and mutualism in the trophic space of microbe-plant interactions. Trends Microbiol 18:365–373. https://doi.org/10.1016/j.tim.2010.06.002

Nicolas O, Charles MT, Jenni S, Toussaint V, Parent SÉ, Beaulieu C (2019) The ionomics of lettuce infected by *Xanthomonas campestris* pv. *vitians*. Front Plant Sci 10:351. https://doi.org/10.3389/fpls.2019.00351

Oostendorp M, Kunz W, Dietrich B, Staub T (2001) Induced disease resistance in plants by chemicals. Eur J Plant Pathol 107:19–28. https://doi.org/10.1023/A:1008760518772

Pallen MJ (2020) Microbial bioinformatics 2020. Microb Biotechnol 9(5):681–686. https://doi.org/10.1111/1751-7915.12389

Pan R, Jones AD, Hu J (2014) Cardiolipin-mediated mitochondrial dynamics and stress response in Arabidopsis. Plant Cell 26(1):391–409. https://doi.org/10.1105/tpc.113.121095

Parida AK, Panda A, Rangani J (2018) Metabolomics-guided elucidation of abiotic stress tolerance mechanisms in plants. In: Plant metabolites and regulation under environmental stress. Academic, San Diego, CA, pp 89–131

Passos MA, de Oliveira CV, Emediato FL, de Camargo TC, Souza MT, Jr Matsumoto T, Rennó Azevedo VC, Ferreira CF, Amorim EP, de Alencar FLF, Martins NF, de Jesus Barbosa Cavalcante M, Baurens FC, da Silva OB, Jr Pappas GJ, Jr Pignolet L, Abadie C, Ciampi AY, Piffanelli P, Miller RN (2012) Development of expressed sequence tag and expressed sequence tag-simple sequence repeat marker resources for *Musa acuminata*. AoB PLANTS:pls030. https://doi.org/10.1093/aobpla/pls030

Passos MAN, Cruz VDO, Emediato FL, Camargo de Teixeira C, Rennó Azevedo VC, Brasileiro ACM, Amorim EP, Ferreira CF, Martins NF, Togawa RC, Pappas Júnior GJ, da Silva Jr OB, Miller RNG (2013) Analysis of the leaf Transcriptome of *Musa acuminata* during interaction with *Mycosphaerella musicola*: gene assembly, annotation and marker development. BMC Genomics 14(1):78

Percy MG, Grundling A (2014) Lipoteichoic acid synthesis and function in gram-positive bacteria. Annu Rev Microbiol 68:81–100

Pérez-Clemente RM, Vives V, Zandalinas SI, López-Climent MF, Muñoz V, Gómez-Cadenas A (2013) Biotechnological approaches to study plant responses to stress. Biomed Res Int 2013:654120

Petrozza A, Santaniello A, Summerer S, Di Tommaso G, Di Tommaso D, Paparelli E, Piaggesi A, Perata P, Cellini F (2014) Physiological responses to Megafol treatments in tomato plants under drought stress: a phenomic and molecular approach. Sci Hortic 174:185–192

Poger D, Mark AE (2015) A ring to rule them all: the effect of cyclopropane fatty acids on the fluidity of lipid bilayers. J Phys Chem B 119(17):5487–5495

Pomastowski P, Buszewski B (2014) Two-dimensional gel electrophoresis in the light of new developments. TrAC Trends Anal Chem 53:167–177

Privalle LS, Chen J, Clapper G, Hunst P, Spiegelhalter F, Zhong C (2012) Development of an agricultural biotechnology crop product: testing from discovery to commercialization. J Agric Food Chem 60:10179–10187. https://doi.org/10.1021/jf302706e

Raffaele S, Kamoun S (2012) Genome evolution in filamentous plant pathogens: why bigger can be better. Nat Rev Microbiol 10:417–430. https://doi.org/10.1038/nrmicro2790

Rahaman MM, Chen D, Gillani Z, Klukas C, Chen M (2015) Advanced phenotyping and phenotype data analysis for the study of plant growth and development. Front Plant Sci 6:619. https://doi.org/10.3389/fpls.2015.00619

Ramírez-Carrasco G, Martínez-Aguilar K, Alvarez-Venegas R (2017) Transgenerational defense priming for crop protection against plant pathogens: a hypothesis. Front Plant Sci 8:696. https://doi.org/10.3389/fpls.2017.00696

Ray DK, Mueller ND, West PC, Foley JA (2013) Yield trends are insufficient to double global crop production by 2050. PLoS One 8(6):e66428. https://doi.org/10.1371/journal.pone.0066428

Renuse S, Chaerkady R, Pandey A (2011) Proteogenomics. Proteomics 11(4):620–630. https://doi.org/10.1002/pmic.201000615

Rhee SY, Dickerson J, Xu D (2006) Bioinformatics and its applications in plant biology. Annu Rev Plant Biol 57:335–360

Rhee HW, Zou P, Udeshi ND, Martell JD, Mootha VK, Carr SA, Ting AY (2013) Proteomic mapping of mitochondria in living cells via spatially restricted enzymatic tagging. Science 339 (6125):1328–1331

Rhee SY, Parker JE, Mockler TC (2016) A glimpse into the future of genome-enabled plant biology from the shores of Cold Spring Harbor. Genome Biol 17:3. https://doi.org/10.1186/s13059-016-0870-y

Riewe D, Wiebach J, Altmann T (2017) Structure annotation and quantification of wheat seed oxidized lipids by high resolution LC-MS/MS. Plant Physiol 175:600–618. https://doi.org/10.1104/pp.17.00470

Robinson AB, Pauling L (1974) Techniques of orthomolecular diagnosis. Clin Chem 20(8):961–965

Rus A, Baxter I, Muthukumar B, Gustin J, Lahner B, Yakubova E, Salt DE (2006) Natural variants of AtHKT1 enhance Na+ accumulation in two wild populations of Arabidopsis. PLoS Genet 2 (12):e210. https://doi.org/10.1371/journal.pgen.0020210

Saito K, Matsuda F (2010) Metabolomics for functional genomics, systems biology, and biotechnology. Annu Rev Plant Biol 61:463–489

Salt DE (2004) Update on ionomics. Plant Physiol 136:2451–2456

Salt DE, Baxter I, Lahner B (2008) Ionomics and the study of the plant ionome. Annu Rev Plant Biol 59:709–733

Sampson TR, Saroj SD, Llewellyn AC, Tzeng YL, Weiss DS (2013) A CRISPR/Cas system mediates bacterial innate immune evasion and virulence. Nature 497:254–257

SánchezRodríguez E, Moreno DA, Ferreres F, RubioWilhelmi MDM, Ruiz JM (2011) Differential responses of five cherry tomato varieties to water stress: changes on phenolic metabolites and related enzymes. Phytochemistry 72:723–729

Sarabia LD, Boughton B, Rupasinghe T, Allison ML, Van de Meene AML, Callahan DL, Hill CB, Roessner U (2018) High-mass-resolution MALDI mass spectrometry imaging reveals detailed spatial distribution of metabolites and lipids in roots of barley seedlings in response to salinity stress. Metabolomics 14:63. https://doi.org/10.1007/s11306-018-1359-3

Sarowar S, Zhao Y, Elena E, Soria-Guerra Ali S, Zheng D, Wang D, Korban SS (2011) Expression profiles of differentially regulated genes during the early stages of apple flower infection with *Erwinia amylovora*. J Exp Bot 62:4851–4861

Schwarzländer M, Fuchs P (2017) Plant mitochondrial membranes: adding structure and new functions to respiratory physiology. Curr Opin Plant Biol 40:147–157. https://doi.org/10.1016/j.pbi.2017.09.002

Shao S, Guo T, Aebersold R (2014) Mass spectrometry-based proteomic quest for diabetes biomarkers. Biochim Biophys Acta 1854(6):519–527. https://doi.org/10.1016/j.bbapap.2014.12.012

Sharma N, Mittal D, Mishra NS (2017) Micro-regulators of hormones and stress. In: Mechanism of plant hormone signaling under stress. Wiley, New York, pp 319–351

Shen W, Li H, Teng R, Wang Y, Wang W, Zhuang J (2018) Genomic and transcriptomic analyses of HD-zip family transcription factors and their responses to abiotic stress in tea plant (*Camellia sinensis*). Genomics 11(5):1142–1151. https://doi.org/10.1016/j.ygeno.2018.07.009

Shivaraj SM, Dhakate P, Sonah H, Vuong T, Nguyen HT, Deshmukh R (2019) Progress toward development of climate-smart flax: a perspective on omics-assisted breeding. In: Genomic designing of climate-smart oilseed crops. Springer, Berlin, pp 239–274

Shriram V, Kumar V, Devarumath RM, Khare TS, Wani SH (2016) MicroRNAs as potential targets for abiotic stress tolerance in plants. Front Plant Sci 7:817. https://doi.org/10.3389/fpls.2016.00817

Sim SC, Durstewitz G, Plieske J, Wieseke R, Ganal MW, Van Deynze A, Hamilton JP, Buell CR, Causse M, Wijeratne S, Francis DM (2012) Development of a large SNP genotyping array and generation of high-density genetic maps in tomato. PLoS One 7(7):e40563

Singh VK, Singh AK, Chand R, Kushwaha C (2011) Role of bioinformatics in agriculture and sustainable development. Int J Bioinforma Res 3(2):221–226

Singh UM, Sareen P, Sengar RS, Kumar A (2013) Plant ionomics: a newer approach to study mineral transport and its regulation. Acta Physiol Plant 35:2641–2653. https://doi.org/10.1007/s11738-013-1316-8

Singh BP, Jayaswal PK, Singh B, Singh PK, Kumar V, Mishra S, Singh N, Panda K, Singh NK (2015) Natural allelic diversity in OsDREB1F gene in the Indian wild rice germplasm led to ascertain its association with drought tolerance. Plant Cell Rep 34(6):993–1004

Singh B, Mishra S, Bohra A, Joshi R, Siddique KHM (2018) Crop phenomics for abiotic stress tolerance in crop plants. In: Wani SH (ed) Biochemical, physiological and molecular avenues for combatting abiotic stress in plants, 3rd edn. Academic Press, Elsevier, London, pp 277–296

Sozzani R, Busch W, Spalding EP, Benfey PN (2014) Advanced imaging techniques for the study of plant growth and development. Trends Plant Sci 19:304–310

Sui N, Han G (2014) Salt-induced photoinhibition of PSII is alleviated in halophyte *Thellungiella halophila* by increases of unsaturated fatty acids in membrane lipids. Acta Physiol Plant 36:983–992. https://doi.org/10.1007/s11738-013-1477-5

Sun X, Xu L, Wang Y, Yu R, Zhu X, Luo X, Gong Y, Wang R, Limera C, Zhang K, Liu L (2015) Identification of novel and salt-responsive miRNAs to explore miRNA-mediated regulatory network of salt stress response in radish (*Raphanus sativus* L.). BMC Genomics 16:197. https://doi.org/10.1186/s12864-015-1416-5

Sun CX, Li MQ, Gao XX, Liu LN, Wu XF, Zhou JH (2016a) Metabolic response of maize plants to multi-factorial abiotic stresses. Plant Biol 18(1):120–129

Sun C, Gao X, Chen X, Fu J, Zhang Y (2016b) Metabolic and growth responses of maize to successive drought and re-watering cycles. Agric Water Manag 172:62–73. https://doi.org/10.1016/j.agwat.2016.04.016

Tamura K, Dudley J, Nei M, Kumar S (2007) MEGA4: molecular evolutionary genetics analysis (MEGA) software version 4.0. Mol Biol Evol 24(8):1596–1599

Tarazona P, Feussner K, Feussner I (2015) An enhanced plant lipidomics method based on multiplexed liquid chromatography–mass spectrometry reveals additional insights into cold- and drought-induced membrane remodeling. Plant J 84(3):621–633

Tardieu F, Cabrera-Bosquet L, Pridmore T, Bennett M (2017) Plant phenomics, from sensors to knowledge. Curr Biol 27:R770–R783. https://doi.org/10.1016/j.cub.2017.05.055

Thomashow MF (1999) Plant cold acclimation: freezing tolerance genes and regulatory mechanisms. Annu Rev Plant Physiol Plant Mol Biol 50:571–599

Thynne E, McDonald MC, Solomon PS (2015) Phytopathogen emergence in the genomics era. Trends Plant Sci 20:246–255. https://doi.org/10.1016/j.tplants.2015.01.009

Tian H, Baxter IR, Lahner B, Reinders A, Salt DE, Ward JM (2010) Arabidopsis NPCC6/NaKR1 is a phloem mobile metal binding protein necessary for phloem function and root meristem maintenance. Plant Cell 22(12):3963–3979. https://doi.org/10.1105/tpc.110.080010

Tulchinsky TH (2010) Micronutrient deficiency conditions: global health issues. Public Health Rev 32(1):243–255

Tyers M, Mann M (2003) From genomics to proteomics. Nature 422:193–197

Upadhyay P, Rai A, Kumar R, Singh M, Sinha B (2014) Differential expression of pathogenesis related protein genes in tomato during inoculation with A. Solani. J Plant Pathol Microb 5:217. https://doi.org/10.4172/2157-7471.1000217

Upadhyay J, Joshi R, Singh B, Bohra A, Vijayan R, Bhatt M, Bisht SPS, Wani SH (2017) Application of bioinformatics in understanding of plant stress tolerance. In: Hakeem K, Malik A, Vardar-Sukan F, Ozturk M (eds) Plant bioinformatics. Springer, Cham, pp 347–374. https://doi.org/10.1007/978-3-319-67156-7_14

Upchurch RG (2008) Fatty acid unsaturation, mobilization, and regulation in the response of plants to stress. Biotechnol Lett 30(6):967–977. https://doi.org/10.1007/s10529-008-9639-z

Urano K, Kurihara Y, Seki M, Shinozaki K (2010) 'Omics' analyses of regulatory networks in plant abiotic stress responses. Curr Opin Plant Biol 13:132–138

Uszkoreit J, Plohnke N, Rexroth S, Marcus K, Eisenacher M (2014) The bacterial proteogenomic pipeline. BMC Genomics 15(Suppl 9):S19. https://doi.org/10.1186/1471-2164-15-S9-S19

Van Emon JM (2016) The omics revolution in agricultural research. J Agric Food Chem 64 (1):36–44. https://doi.org/10.1021/acs.jafc.5b04515

Vorholt JA (2012) Microbial life in the phyllosphere. Nat Rev Microbiol 10:828–840. https://doi.org/10.1038/nrmicro2910

Walters D, Walsh D, Newton A, Lyon G (2005) Induced resistance for plant disease control: maximizing the efficacy of resistance elicitors. Phytopathology 95:1368–1373. https://doi.org/10.1094/PHYTO-95-1368

Wang Z, Gerstein M, Snyder M (2009) RNA-Seq: a revolutionary tool for transcriptomics. Nat Rev Genet 10:57–63

Weiberg A, Wang M, Lin FM, Zhao H, Zhang Z, Kaloshian I, Huang HD, Jin H (2013) Fungal small RNAs suppress plant immunity by hijacking host RNA interference pathways. Science 342(6154):118–123. https://doi.org/10.1126/science.1239705

Welti R, Shah J, Li W, Li M, Chen J, Burke JJ, Fauconnier ML, Chapman K, Chye ML, Wang X (2007) Plant lipidomics: discerning biological function by profiling plant complex lipids using mass spectrometry. Front Biosci 12:2494–2506

White JW, Andrade-Sanchez P, Gore MA, Bronson KF, Coffelt TA, Conley MM, Feldmann KA, French AN, Heun JT, Hunsaker DJ, Jenks MA, Kimball BA, Roth RL, Strand RJ, Thorp KR, Wall GW, Wang G (2012a) Field-based phenomics for plant genetics research. Field Crops Res 133:101–112

White PJ, Broadley MR, Thompson JA, McNicol JW, Crawley MJ, Poulton PR, Johnston AE (2012b) Testing the distinctness of shoot ionomes of angiosperm families using the Rothamsted Park Grass Continuous Hay experiment. New Phytol 196(1):101–109

Wiedner SD, Anderson LN, Sadler NC, Chrisler WB, Kodali VK, Smith RD, Wright AT (2014) Organelle-specific activity-based protein profiling in living cells. Angew Chem 53 (11):2919–2922

Wilhelm M, Schlegl J, Hahne H, Gholami AM, Lieberenz M, Savitski MM, Ziegler E, Butzmann L, Gessulat S, Marx H, Mathieson T, Lemeer S, Schnatbaum K, Reimer U, Wenschuh H, Mollenhauer M, Slotta-Huspenina J, Boese JH, Bantscheff M, Gerstmair A, Faerber F, Kuster B (2014) Mass-spectrometry-based draft of the human proteome. Nature 509(7502):582–587. https://doi.org/10.1038/nature13319

Wilkins MR, Sanchez JC, Williams KL, Hochstrasser DF (1996) Current challenges and future applications for protein maps and post-translational vector maps in proteome projects. Electrophoresis 17(5):830–838. https://doi.org/10.1002/elps.1150170504

Wolfender J-L, Rudaz S, Choi YH, Kim HK (2013) Plant metabolomics: from holistic data to relevant biomarkers. Curr Med Chem 20:1056–1090

Xiong L, Lee MW, Qi M, Yang Y (2001) Identification of defense-related rice genes by suppression subtractive hybridization and differential screening. Mol Plant-Microbe Interact 14:685–692

Yasueda Y, Tamura T, Fujisawa A, Kuwata K, Tsukiji S, Kiyonaka S, Hamachi I (2016) A set of organelle-localizable reactive molecules for mitochondrial chemical proteomics in living cells and brain tissues. J Am Chem Soc 138(24):7592–7602

Yu D, Boughton BA, Hill CB, Feussner I, Roessner U, Rupasinghe TWT (2020) Insights into oxidized lipid modification in barley roots as an adaptation mechanism to salinity stress. Front Plant Sci 11:1. https://doi.org/10.3389/fpls.2020.00001

Yuan JS, Galbraith DW, Dai SY, Griffin P, Stewart CN Jr (2008) Plant systems biology comes of age. Trends Plant Sci 13:165–171

Zhang YM, Rock CO (2008) Membrane lipid homeostasis in bacteria. Nat Rev Microbiol 6 (3):222–233

Zhang B, Wang Q (2015) MicroRNA-based biotechnology for plant improvement. J Cell Physiol 230:1–15. https://doi.org/10.1002/jcp.24685

Zhang S, Yue Y, Sheng L, Wu Y, Fan G, Li A et al (2013) PASmiR: a literature-curated database form miRNA molecular regulation in plant response to abiotic stress. BMC Plant Biol 13:33. https://doi.org/10.1186/1471-2229-13-33

Zhang Y, Nan J, Yu B (2016) OMICS technologies and applications in sugar beet. Front Plant Sci 7:900. https://doi.org/10.3389/fpls.2016.00900

Zhang L, Li X, Ma B, Gao Q, Du H, Han Y, Li Y, Cao Y, Qi M, Zhu Y, Lu H (2017) The tartary buckwheat genome provides insights into rutin biosynthesis and abiotic stress tolerance. Mol Plant 10(9):1224–1237

Zhang X, Yao C, Fu S, Xuan H, Wen S, Liu C, Li F, Liu A, Bi S, Zhang S, Li S (2018) Stress2TF: a manually curated database of TF regulation in plant response to stress. Gene 638:36–40. https://doi.org/10.1016/j.gene.2017.09.067

Zhao Q, Tian M, Li Q, Cui F, Liu L, Yin B, Xie Q (2013) A plant-specific in vitro ubiquitination analysis system. Plant J 74(3):524–533. https://doi.org/10.1111/tpj.12127

Zhao C, Zhang Y, Du J, Guo X, Wen W, Gu S, Wang J, Fan J (2019) Crop phenomics: current status and perspectives. Front Plant Sci 10:714. https://doi.org/10.3389/fpls.2019.00714

Zhou X, Wang G, Sutoh K, Zhu JK, Zhang W (2008) Identification of cold-inducible microRNAs in plants by transcriptome analysis. Biochim Biophys Acta 1779(11):780–788. https://doi.org/10.1016/j.bbagrm

Zhou L, Liu Y, Liu Z, Kong D, Duan M, Luo L (2010) Genome-wide identification and analysis of drought-responsive microRNAs in Oryza sativa. J Exp Bot 61:4157–4168. https://doi.org/10.1093/jxb/erq237

Zhou M, Li D, Li Z, Hu Q, Yang C, Zhu L, Luo H (2013) Constitutive expression of a miR319 gene alters plant development and enhances salt and drought tolerance in transgenic creeping bentgrass. Plant Physiol 161(3):1375–1391. https://doi.org/10.1104/pp.112.208702

Zhu FY, Chen MX, Ye NH, Shi L, Ma KL, Yang JF, Cao YY, Zhang Y, Yoshida T, Fernie AR, Fan GY, Wen B, Zhou R, Liu TY, Fan T, Gao B, Zhang D, Hao GF, Xiao S, Liu YG, Zhang J (2017) Proteogenomic analysis reveals alternative splicing and translation as part of the abscisic acid response in Arabidopsis seedlings. Plant J 91(3):518–533. https://doi.org/10.1111/tpj.13571

Rhizosphere Modeling and Engineering for Agricultural and Environmental Sustainability

17

Charles Oluwaseun Adetunji, Olugbemi T. Olaniyan, Osikemekha Anthony Anani, and Abel Inobeme

Abstract

The rhizosphere has been acknowledged as one of the most complex microbial habitats on earth that entail varieties of networks such as a consortium of beneficial microorganisms, plant roots, and soil. The adequate knowledge, prediction, and regulation of the function of a structure could enhance adequate microbe–plant interaction as well as other activities in the rhizosphere which could serve as a key that could be applied for regulation of plant ecosystem productivity, increase in plant biomass maintenance of healthy environment prevention of several scenarios around climate changes. The adequate understanding of the mechanism involves in rhizosphere engineering will go towards the production of an increase in food production and sustenance of a healthy environment. Therefore, this chapter intends to provide detailed information on the engineering plants through plant genetic engineering, manipulating rhizosphere pH, vacuole nitrate transporters, malate channels and metabolism, multidrug and toxic compound extrusion (MATE), enhancing organic anion efflux from roots, acquisition

C. O. Adetunji (✉)
Applied Microbiology, Biotechnology and Nanotechnology Laboratory, Department of Microbiology, Edo State University Uzairue, Iyamho, Edo, Nigeria
e-mail: adetunji.charles@edouniversity.edu.ng

O. T. Olaniyan
Laboratory for Reproductive Biology and Developmental Programming, Department of Physiology, Edo State University Uzairue, Iyamho, Edo, Nigeria

O. A. Anani
Laboratory of Ecotoxicology and Forensic Biology, Department of Biological Science, Faculty of Science, Edo State University Uzairue, Iyamho, Edo, Nigeria

A. Inobeme
Department of Chemistry, Edo State University Uzairue, Iyamho, Edo, Nigeria

R. Soni et al. (eds.), *Microbiological Activity for Soil and Plant Health Management*,
https://doi.org/10.1007/978-981-16-2922-8_17

of phosphorus, aluminum-induced secretion of organic acid anions from roots, acquisition of other nutrients and heavy metals, interactions with plant growth-promoting microorganisms.

Keywords

Genetic engineering · Rhizosphere modeling · Multidrug and toxic compound extrusion · Agricultural sustainability

17.1 Introduction

The natural environment which entails plants and animals gives several ecological niches for microorganisms that have to enable different stains to coexist which could lead to the development of complex microbial communities. Moreover, adequate knowledge on the influence of the environment where these microorganisms dwell majorly on their hosts and the relationship between the microbiota and the multicellular host could be referred to as holobiont and supra-/metaorganism (Bosch and McFall-Ngai 2011).

The rhizosphere is a region that exists between the roots and the soil particles where the soil microorganism and the plant influence habitat are interconnected together. Moreover, the rhizosphere involves a plant–soil relationship most especially in a dynamic populated soil area entailing, food web relationship, inter-and interspecies communication that plays a crucial function in the flow of carbon and their eventual transformation (Wang et al. 2018; Dessaux et al. 2016).

The rhizosphere entails three basic different regions such as the endorhizosphere, the rhizoplane, ectorhizosphere. Moreover, the rhizosphere has been identified as a region where there are several biological, chemical, and physical features around the root region (McNear Jr. 2013). It has been observed that the root is capable of producing root exudates which entail phytohormones which could enhance rhizosphere associations when they are utilized as a source of energy for the sustainability of microorganisms and they could also serve as repellents and chemical attractants (Bais et al. 2001).

Moreover, these molecules could serve as means of communication molecules to stimulate physiological and biological relationships that exist between the root of a plant and different varieties of coil microorganisms which palsy several beneficial roles or synergetic function in the soil. Typical examples of such microorganisms entail mycorrhizal fungi, nitrogen-fixing bacteria, epiphytes, and many other important beneficial soil microorganisms (Nardi et al. 2000). Interestingly, there are several factors that could support the maintenance and sustainability of the ecosystem services which entails carbon uptake and storage, carbon and water cycling, crop production, nutrient trapping (Adl 2016).

The global population has been stipulated to increase tremendously; therefore, there is a need to place proper strategy in place that could lead to the sustenance of the human population through the identification of sustainable techniques (Laranjo et al. 2014). The identification of beneficial microorganisms and the reengineering of the beneficial microorganism have been shown to play a vital role in the increase in food production and alleviation of food insecurity. Some typical examples of such beneficial microorganisms include plant growth-promoting bacteria and soil endophytic microorganisms (Morel et al. 2012; Jackson et al. 2015). These organisms could enhance easy withdrawal of resources from the environment such as iron, phosphorus, and nitrogen as well as regulates numerous hormones for adequate plant development. Such hormones include ethylene, auxin, and cytokines (Santoyo et al. 2016).

Furthermore, they could help towards fighting against several plant pathogens and ensure enhanced plant immunity (Laranjo et al. 2014). Also, it must be noted that these beneficial microorganisms could also serve as important tools that could be applied in the manipulation of the gene expression and their other association with the other microorganism they have an encounter with for their own benefits (Pii et al. 2015).

Therefore, this chapter intends to provide detailed information on the engineering plants through plant genetic engineering, manipulating rhizosphere pH, vacuole nitrate transporters, malate channels and metabolism, multidrug and toxic compound extrusion (MATE), enhancing organic anion efflux from roots, acquisition of phosphorus, aluminum-induced secretion of organic acid anions from roots, acquisition of other nutrients and heavy metals, interactions with plant growth-promoting microorganisms. Moreover, detailed information on the encoding genes that enhance organic anion efflux from roots rhizosphere was also highlighted.

17.2 Engineering Plants Through Plant Genetic Engineering

Vital-López et al. (2016) investigated the inhibitory role of transgenic cultivar on the bacterial community in two genetically modified maize rhizosphere utilizing pat gene as resistance against glufosinate herbicide. The authors utilized polymerase chain reaction and single-strand conformation polymorphism for the amplification and characterization of the subunit rRNA gene genetic profile of the bacterial communities of the maize cultivars rhizosphere. Also, binary matrix analysis on similarity analysis of patterns revealed that both the genetically modified and the conventional maize have homogeneous bacterial communities profile. Thus concluded that the bacterial structural profile was not affected by the rhizosphere of the transgenic maize cultivar. Singhala et al. (2016) reported that the development of a plant is a complex system of the chemical process between the interface of soil and plant roots rhizosphere in several microbial environments. Studies have revealed that many environmental, physiochemical, biological, and chemical parameters influence the composition of the rhizosphere microbiome in soil and plant root systems. Various molecular biology techniques have been utilized to elucidate the microbe–microbe collaboration of rhizosphere and plant-microbe regulating the diversity plus shaping the nature of root exudations. The functional groups of several microorganisms can colonize, enhance, and develop the rhizosphere through indirect and direct mechanisms for sustainable agriculture and nature restoration.

Palmer et al. (2016) revealed that different array of strategies for improving disease resistance in plants and potential for reduced pesticide use by genetic engineering have been characterized. The application of transgenesis, mutagenesis, intragenesis, and cisgenesis for metabolic pathway analyses in genetic engineering varies significantly, thus the socioeconomic plus cultural perspectives must be considered. Singhala et al. (2016) reported that geranylgeranyl diphosphate synthase and 1-deoxy-D-xylulose-5-phosphate synthase were introduced into the hairy roots of *S. miltiorrhiza* for health-promoting effects. The study showed that these agents caused significant production of diterpenoids, tanshinones through genetic engineering with a high level of antitumor and antioxidant properties when associated with the control group. The further evaluation also showed that carotenoids, chlorophylls, gibberellins, and indoleacetic acid contents were enhanced in the transgenic group thus demonstrating the possibility of increasing the generation of diterpenoids and other natural plastid-derived isoprenoids in plants by utilizing genetic engineering through 2-C-methyl-D-erythritol-4-phosphate (MEP) pathway manipulations.

Su et al. (2019) reported that different breeding strategies can be adopted to enhance the physiological characteristics like flower colors, shapes, flowering times, plant architectures, postharvest quality, tolerance to biotic and abiotic stress of chrysanthemum such as conventional techniques, molecular breeding methods, and advanced high-throughput technologies like crossbreeding, mutation breeding, transgenic technology, genome editing, marker-assisted selection, genomics, proteomics, transcriptomics, microbiomics, and metabolomics. These approaches have resulted in the generation of a considerable amount of data thereby integrating these with other phenotypic information to discover pathways and genes of great significance. The authors reported that high polyploidy and heterozygosity of species have generated considerable difficulty in breeding chrysanthemum. Studies have shown that in the horticultural industry, serious attention has been given to enhance the safety of environmental and product protection. Application of advanced molecular techniques like CRISPR/Cas9 system, transcription activator-like effector nucleases plus zinc finger nucleases is urgently needed as potential tools for enhancing the ornamental horticulture industry. A more recent study on *C. seticuspe* and *C. nankingense* genomes have illuminated more findings on the mapping ratio in the hexaploid nature of cultivated chrysanthemum. Screening of the allele composition can equally be achieved by developing markers that can be incorporated into breeding programs to reduce time and cost. The reliable and efficient genomic sequence can be obtained utilizing algorithms, robotics, cameras, advanced sensors, image analysis tools, machine vision, and bioinformatics tools. Protein, metabolite, and transcript are phenotypic information from natural populations and linkage-mapping populations for the identification of loci responsible for gene expression in chrysanthemum cultivar.

Xu et al. (2020) reported that abiotic stresses like drought, salinity, temperature are contributory factors to the regulation of multifaceted transcriptional regulatory networks in cotton. Thus the identification of the molecular mechanisms involved in

Gossypium hirsutum biological process such as oxidation-reduction, carbohydrate metabolic of the alkali-salt stress response, will require molecular biology tools to elucidate the underlying physiological function. Previous studies have reported the role of genetically modified *Pinus radiata* in biolistic insertion *npt* II genes and LEAFY rhizosphere soil microbial communities. Culture-independent PCR-based approach was utilized to determine the microbial communities, thus they revealed that there is no significant difference between the control trees and the genetically modified ectomycorrhizal and fungal communities. Utilizing amplified rDNA restriction analysis, they showed that seasonal variations in ectomycorrhizal and fungal were seen in the two groups but no impact of genetic on rhizosphere microbial populations.

Rengel and Marschner (2005) showed that soil nutrients are improved by the application of fertilizers particularly to increase crop nutrition and yield. When a plant is suffering from nutritional deficiency, several mechanisms are activated to improve the nutrient available in the rhizosphere. This is done by plants increasing their plasma transport of nutrients, alteration in root morphology, generates organic molecules like carboxylates, carbohydrates, phenolics, enzymes, and protons. The authors showed that nutrients are enhanced and made available to crops through the addition of beneficial microorganisms to rhizosphere plant-microbe–soil interactions for sustainable crop production. Furthermore, Peter et al. (2014) reported that there are strategies like quorum quenching strategies available to manipulate plant root linked microbial interactions for improved productivity and health benefit targeting growth, rhizosphere components. The authors demonstrated that genetic engineering can be utilized to alter rhizosphere pH to release vital molecules to increase nutrients availability, defense against pathogens, stress, and microbial proliferation. Various stress-induced hormones like ethylene can be altered through engineered rhizobacteria to promote plant growth, root growth, antibiotic secretion, the release of lytic enzymes against soil-borne pathogens. Palmer et al. (2016) showed that genetic manipulation of gene expression in plants can be utilized to promote crop yield and productivity. This can be done at different levels of gene expression like transcription or translation utilizing endogenous plant promoters particularly those of ubiquitin and actin gene. Sustainable intensification has been described for the processes involved in the utilization of various resources to produce food, energy, and raw materials to meet the growing demand of the increased population. Increasing yield by the deployment of existing technology such as transgenic methods has been described by the authors as a powerful tool for manipulating plant's stress response.

John and Bonnie (2003) reported that over the last few decades, the R genes have been utilized to breed resistant programs for different pathogens focusing on downstream transduction pathways and signaling proteins to provide information on disease control mechanisms. Through the use of molecular biology methods such as gene editing technologies, next-generation sequencing, bioinformatics, and metagenomics, many entangled webs of microbiomes interactions with crop nutrients acquisition, stress adaptation, and response have been unraveled. Rhizosphere microbial engineering can alter the structural components of the plant-microbial community to improve the resistance to pathogens and upsurge the uptake of nutrients through the ATPase proton pump channels. Das et al. (2019) reported that different transcription factors like NAC, bHLH, WRKY, bZIP, AP2/ERF plus MYB are involved in plant stress response control. Zhao et al. (2018) revealed that bio-organic fertilizers from chicken and cow manure were utilized for controlling soil-borne diseases and stimulate plant growth in watermelon plant rhizosphere soil. The mechanisms behind the suppression of pathogens and plant enhancement were also investigated, thus new microbial communities were observed to form in this bio-organic fertilizer application with the increasing diversity of fungal and bacterial.

Besides, Lu et al. (2018) reported that several endosphere and rhizosphere bacteria perform a crucial role in promoting plant growth and health benefits in transgenic plants. Similarly, Manoj et al. (2018) revealed that rhizosphere microbial interactions in plants are responsible for fitness, nutrient delivery, pathogen or pest control, stress tolerance, for improved crop yield. Metagenomics and bioinformatics approaches are the latest development for elucidating microbial functions in the host rhizosphere community. Odelade and Babalola (2019) revealed that through genetic engineering on diverse beneficial microbes, various microbial metabolites have been generated to improve crop growth, productivity, and resistance to pests or pathogens. The authors demonstrated that microbes close to plant's roots such as bacteria, fungi, archaea interact with the host plants for increased nutrients acquisition in the root exudates. It was also revealed that plant rhizosphere microbiome interactions are impressively influenced by some environmental factors like soil type, pH, temperature climate change, salts, anthropogenic activities, plant cultivar. However, this scientific group also discovered that Azospirillum, *Pseudomonas*, *Bacillus*, and *Rhizobium* spp are capable of crop growth and biomass generation in fulfilling sustainable agriculture productions. The plant growth-promoting traits during the rhizosphere microbiome can be identified and characterized utilizing the next-generation sequencing. Through advanced molecular biology techniques, genetically engineered strains of plant growth-promoting rhizosphere using bio-inoculants can be effectively studied for the identification of genes responsible for drought tolerance.

Studies have revealed that rhizosphere bacteria influence plant growth through numerous mechanisms such as fixation of nitrogen, solubilization of minerals, generation of hormones, suppression of pathogens, uptake of nutrients through ATPase pump. Yonekura-Sakakibara et al. (2019) reported that analysis of plant metabolites utilizing metabolomics like chemo-informatics and genomics can elucidate many unique metabolic pathways and genes involved in various physiological mechanisms. They noted that understanding the chemodiversity will help during genetic engineering of the metabolic pathways to generate appropriate metabolites. Adesemoye and Egamberdieva (2013) and Quiza et al. (2015) showed that microbiome engineering like metatranscriptomics and metaproteomics facilitates microbial community to boost plant development through increased nutrient availability, reduced disease susceptibility, and tolerance to stress. The authors revealed that the plant rhizosphere secretes a natural biomolecule called semiochemicals which assist in the interaction between plant and microorganisms. Amir et al. (2017) reported that bioenergy and food production in an ever-changing climate can be sustained through the utilization of rhizosphere genetic engineering using re-designed plant metabolism. The authors revealed that root exudation in engineered rhizosphere regulates the efflux of organics anions and H^+ from the roots. Many genes are responsible for these physiological processes, for instance, in tomato and rice, Arabidopsis vacuolar H^+-pyrophosphatase gene AVP1 displays a greater amount of malate and citrate efflux.

In a recently published review article, it was reported that garlic crop improvement was carried out through the process of biotechnology like meristems culture, micropropagation, genetic transformation, genetic engineering, and somaclonal variation (Bikis 2018). Different genetic variations are produced to achieve resistance against pests, diseases, increase absorption of nutrients, and enhance the fabrication of hormones. Shenoy and Kalagudi (2005) suggested various processes involved in developing genetically enhanced plants for efficient supply of phosphorus to plants, phosphorylase stimulation in a deprived soil. Barret et al. (2011) described the essential role of the rhizosphere microbiome and their multitrophic interactions by genetically engineering and microarray-based techniques for probiotics microbial strains. Monica and Alison (1999) reported that recombinant proteins are utilized to provide valuable information on transcriptional and post-transcriptional control mechanisms in gene expression of transgenic plants. Venant et al. (2011) and Allah and Tahira (2015) reported that modern technology of genetic engineering has provided an opportunity to utilize different genetic pathways to improve crop production. Maziar and Afsaneh (2015) reported that transgenic plant tissue can be manipulated to express plant-based recombinant vaccines.

Chang et al. (2009) revealed that in plant biology, H^+-ATPases are the most important components involved in the maintenance of nutrients acquisition. Till today, only a few members of the proton pump family have been identified and characterized. Thus, the authors investigated the H^+-ATPase associated gene involved in rice nutrient uptake. They discovered that OsA8, an H^+-ATPases gene that is highly expressed in the root of rice was down-regulated in deprived soil, and in knockout showing a higher amount in the root than shoot. Thus it was concluded that these genes are involved in nutrient uptake in rice and also translocation of the nutrients from the root to shoot. Plant growth and development can be limited by heavy metal soil toxicity resulting in the production of acidic soil. Plants plasma membrane H^+-ATPase performs an important role in much physiological activity including plant Al-detoxification through phosphorylation and enzyme expression. H^+-ATPase mediated H^+ influx maintains the intracellular pH and Al-induced citrate anion efflux multi-drug and toxin extrusion or Al-activated malate transporter-co-transport system modulated by the application of an activator or by genetically engineered organisms or plants (Zhang et al. 2017).

17.3 Manipulating Rhizosphere pH

Bravin et al. (2009) demonstrated that root exudates display a significant function in changing rhizosphere bioavailability of many important biomolecules to plants. The authors investigated how root induced alteration in pH through redox-coupled reactions and other biomolecules changes copper levels in the rhizosphere of the wheat plant. They observed that due to soil alkalization, there was a significant increase in pH resulting in elevated labile-Cu concentration. Studies have revealed that rhizosphere acidification through the release of proton causes the dissolution of available phosphorus. It is known that modernized agricultural practices utilize fertilizers for the supply of necessary nutrients, though this has been challenged by the increasing cost of fertilizers, government policies, and poor quality land usage. Therefore, there is a need to acquire more knowledge on the role of root exudates in the rhizosphere region that could enhance the level of the nutritional value of the soil using genetic engineering in order to boost the health condition of the soil health and pH values. Muhammad et al. (2019) reported that different molecular markers can be utilized to enhance crop production. Different techniques are available for the inhibition of gene expression in the plant such as antisense RNA, mutagenesis plus RNAi. Plant genes are involved in the interaction between plant-microbial interplay. The authors suggested that this interaction could influence the transport and uptake of macro and micronutrients from the soil. It is known that calcareous and high pH soil levels would inhibit the bioavailability of many metabolites, nutrients, and biomolecules.

Studies have shown that the H^+-ATPase gene *ChvPMA* isolated from *Chloris virgata* roots and leaves shows tolerance to salt-induced stress. This gene was revealed to fuse with the N-terminus of the green fluorescence protein gene in transgenic yeast in high pH and salt conditions. The transgenic yeast shows overexpression of this gene, enhances root growth and high resistance to pH and salt conditions (Xinxin 2014). One of the vital modes of action through which plant shows resistance to soil toxicity and stress is by of efflux of organic anions like citrate and malate from roots.

17.4 Response of Plants to Aluminum Toxicity in Soil

The element aluminum is one of the most abundant elements and the metal with the highest abundance in the earth's crust. Most of the aluminum readily goes into solution under acidic conditions of soil usually at pH less than or equal to 5.5. The dissolution of this element poses challenges to the plants due to the toxicity associated with it. However, some plants have developed adaptation mechanisms towards surviving in such soil conditions (Kar et al. 2020).

One of the well-reported mechanisms in literature is the discharge of anions of organic acids such as oxalate, malate, and, citrate which occurs around the roots and induced by a high content of aluminum in the soil. The production of these anions brings about the formation of stable chelates with the aluminum ions around the range, thereby protecting the plant from aluminum toxicity. Other investigation also affirms that genes are responsible for the encoding of the organic anions transporters and there is also documented evidence showing that this is accountable for aluminum tolerance in plants (Yang et al. 2019a, b).

The high concentrations of aluminum ions around the plant roots under acidic soil conditions bring about the prevention of the elongation of primary root growth and productivity. The membrane potential is affected and critical ions are displaced within the apoplast resulting in the disruption of the intracellular ions content through targeting of solute transporter that is localized within the membrane. Aluminum ions affect the functioning of the various route channels most especially around the root. The functioning of various ions such as ammonium, nitrate, calcium, and magnesium is affected by high aluminum toxicity (Kar et al. 2020). Findings from their studies have revealed that the exposure of plants to aluminum toxicity in soil results in distortion of the elongation of primary roots (Kong et al. 2018). Aluminum ions in soil affect the structure and functionality of the primary root development in soil and have the potential in inhibiting its overall elongation. This usually brings about the rupturing around the cortical walls both internally and externally resulting in a decline in the elasticity (Yang and Horst 2015).

The primary sites of aluminum binding in the apoplast of the roots are the carboxylic acids groups that are negatively charged and are responsible for the moderation of the sensitivity of the plants to aluminum. There is also the interaction of aluminum ions with xyloglucans and hemicelluloses inhibiting the loosening of the cell wall at the elongation region of the root. The efficiency of photosynthesis and peroxidation of lipids is also affected by aluminum (Bojorquez-Quintal et al. 2017).

Rengel and Zhang (2003) in their work reported that the occurrence of high concentrations of aluminum in the soil brings about the disruption of ion fluxes that are transporter-mediated. Mostly around regions surrounding the root, the higher availability of aluminum ions brings about ions flux that is highly disproportional. The toxicity due to these ions also results in damage to the cell membrane of the cell. Usually one of the major responses to this is the passing of anions across the cell membrane. Most plants with low cation exchange capacity (CEC) are tolerant to aluminum toxicity. Aluminum toxicity also brings about the moderation of various ions such as nitrate, calcium ions, potassium, and other ions (Zheng et al. 2013).

17.5 Vacuole Nitrate Transporters, Malate Channels, and Metabolism

The two most common anions in plants are malate and nitrate in terms of quantitative presence. Both anions play a vital function in plant metabolism and are usually mainly accumulated around the lumen of the vacuole. Malate performs an indispensable function in plants aiding the balance of solute within the vacuole of the matured cells of glyoxylate and carboxylate. In Crassulacean Acid metabolism (CAM) there is the production of malate at night which is transferred towards the vacuole as a carbon storage molecule (de Alexis et al. 2007). The aluminum activated malate transporter class makes up a functionally different but similar group in terms of the structure of the ion channels. They are found in different plant species that are expressed in various tissues. Aluminum activated malate transporter (ALMT) is the first member of the family found in the tips of the wheat root and reported to be associated with aluminum ions adaptation through the exudation of malate into the soil. Much later many other members associated with this phenomenon have been documented (Palmer et al. 2016).

The multidrug and toxic compound extrusion (MATE), as well as the aluminum, activated malate transporter (ALMT) is a group of transporters that are vital for the exudation of citrate and malate, respectively, to the rhizosphere region so as to alleviate toxicity due to aluminum by bringing about the exclusion of the aluminum ions. Aquaporin and hydrogen ions aid the vascular sequestration of the aluminum ions resulting in tolerance of aluminum by the plants. Focusing on this solute transporter would play a potential role in the production of crops that are tolerant to aluminum toxicity in the nearest future (Kar et al. 2020).

Ions channels are vital factors in signaling the routes of adaptation of plant cells to environmental stresses in the controlling of metabolism and maintenance of electrochemical gradients. Anions channels are the major transport systems that are involved in the several physiological functioning within the membranes of the cells toward the osmoregulation, tolerance to metals like aluminum, and cell signaling. Anions channels are found in all plant's plasma membrane including tonoplast, mitochondria, endoplasmic reticulum, and chloroplasts and these are the most characterized when compared to those around the other membrane.

The homeostasis of auxin in Arabidopsis root is regulated by AtDTX30 which helps in the modulation of the development of the root and tolerance to aluminum. The DTX30 is expressed mainly around the root and localized around the root hair and primary root. Dtx30 mutants demonstrated a decrease in the primary root elongation as well as the lateral roots. More auxins are accumulated in the mutant seedlings usually around the tips of their root (Upadhyay et al. 2019).

17.6 Multidrug and Toxic Compound Extrusion (MATE)

Multidrug and toxic compound extrusion (MATE) channels are a group of cation antiporters that occur in most living organisms including eukaryotes and prokaryotes. This makes up the largest transporter of the family in plants. MATE transporters are vital since they are fundamental in the physiological functioning during the growth and development of the plants. They aid the transportation of a wide group of substrate materials including organic acids, secondary metabolites, and plant hormones (Takanashi et al. 2014). Multidrug and toxic compound extrusion (MATE) transporters are found in plants and function in the regulation of plants resistance to toxicity due to aluminum through the induction of citrate efflux (Ma et al. 2018).

One of the most categorized multidrug family transporters is the multidrug and toxic compounds extrusion (MATE). The proteins are coupled with the translocation of the substrate around the cell membrane with cations around electrochemical gradients (Ma et al. 2018).

In some studies, the transport activities of two anions transporters that are dependent on ATP around the tonoplast were isolated in barley while the vacuole characterization was made possible by N-ethylmaleimide glutathione as substrate. Different extent of sensitivity to organic anions was observed in the transporters. In the Arabidopsis, the extrusion of aluminum from the root is aided by aluminum activated root citrate and malate exudation through the malate transporter. The mechanism of tolerance internally is not well explored (Wang et al. 2018).

Various components that are linked to the exclusion of aluminum have been studied in various species of plants in recent times. One of the fundamental features for the produced gene that is resistant to aluminum is that they encode plasma membrane mainly, in some other times the citrate transporter and localized malate.

Ding et al. (2013) in their study identified some suppressor mutants which will be useful in revealing the signaling pathways that are involved in acidic conditions and aluminum resistant Arabidopsis (Wang et al. 2018). The family of the ALMT is made up of a functionally complex and structurally close class of ion channels. They are broadly found in different species of plants and found around different tissues with higher content in tonoplast or plasma membrane (Palmer et al. 2016). The homolog in Arabidopsis thaliana, AtALMT1 that was first identified and characterized was also reported to be connected with resistance to aluminum. In addition to the ALMT1, it was found to have 13 new members showing that they are connected to more than just resistance to Al. Some of them are now documented to have other functions (Palmer et al. 2016).

They documented the cloning of gene from wheat, coded as ALMT1 (aluminum activated malate transporter) which is associated with tolerance to aluminum obtained from the crossing between the different wheat lines with varying extent of tolerance (Sasaki et al. 2004).

17.7 Enhancing Organic Anion Efflux from Roots

One major challenge associated with acid soils is the limiting of agricultural yield because it leads to deficiencies of soil nutrients and mineral toxicities. For most plants found around areas with such soil, they are usually poorly adapted to such conditions resulting in weakly developed root systems which are due to the higher contents of the soluble aluminum ions which affect the elongation of the roots. Discharge of organic anions from roots is a vital strategy use for adaptation by different species of plants found in acidic soils. This trait is usually regulated by the genes that encode the transportation of proteins (Zhou et al. 2011).

Organic acids (OAs) excretion from the roots of plants has now been established to be highly beneficial and reported to be connected to the mechanism of the utilization of phosphorus in plant roots. Such an effect is presently known as the pleiotropic effect and is linked to the tolerance of stress in plants. When organic acids are excreted, they help in the recruitment of useful bacterium which would help in the induction of systemic resistance and also aid the modification or root structure for enhancing phosphorus starvation (Wu et al. 2018).

Oxidative stress in the root cells is also triggered by Al^{3+} through the formation of oxygen radicals. The mechanism of induction of this response whether by symplastic or apoplastic Al^{3+} is not very clear however it is known that these reactive species can result in the damaging of proteins, nucleic acids, and membrane. Oxidative stress can also bring about induction of callose production that can lead to an increase in the cell rigidity and reduces the symplastic flow of solute through the plasmodesmata.

Organic anions efflux from the roots of plants plays an indispensable role in plant nutrition. The discharge of anions such as malate, oxalate, and citrate have been identified as a mechanism for tolerating aluminum ions from soil and improvement of the intake of phosphorus. The metabolites cross the plasma membrane as multivalent ions and current findings have shown that channels that are permeable to anions help in the flow within the aluminum-dependent efflux of citrate and malates from maize and wheat. From current reports, the genes that are responsible for encoding the channels for the anions, as well as related proteins that promote the discharge of oxalate, citrate, and malate have not been isolated. This is one of the challenges with regard to the use of biotechnology for managing and combating toxicity due to aluminum ions as well as improvement of intake of phosphorus in plants (Ryan et al. 2003).

Some available studies have attempted to examine the association between effluxes of organic anions with variation in enzymatic activities. Li et al. (2000) carried out a study on the efflux of citrate from rye and associated this with a rise in synthase activity in citrate. Findings from the study have been used in substantiating that changes in metabolism are vital in the regulation of the efflux. Related studies by Neuman and Römheld (2000) reported weak correlations between enzymatic biosynthetic activities and efflux. The transport of organic anions through the cell membrane is a vital step during the process of exudation. The pH of the cytosol of the cell is close to neutral condition hence the organic acid is likely to move out of the cells as anions. In some studies carried out through the use of patch-clamp approach, it was observed that activation of anion permeable channels around the root cells of maize and wheat can be brought about by aluminum (Kollmeier et al. 2001).

The portion of soil that lies immediately after the root of the plant is the rhizosphere and it is modified by the activities within the root. This is the region where plants sense and respond to the environment. There is an exchange of materials between the soil and the root of plants which affects the physical and biochemical properties of the rhizosphere. In order to respond to stresses and signals from the environment, most plants tend to modify their rhizosphere. The detection of organic anions is usually at these locations and their exudation from the root regions is now connected with ion stresses and mineral deficiencies. The deficiency of phosphorus in most plants brings about the discharge of some organic anions like malate and citrate from the root of the plants. The anions are released to aid the bioavailability of phosphorus compounds for plant uptake through the mobilization of partially soluble forms. Findings from computational modeling reveal that the efficiency of citrate efflux in the mobilization of phosphorus in soil is higher with the rise in the volume of soil where it increases (Ryan et al. 2011).

17.8 Acquisition of Phosphorus

One of the major limiting factors to the growth of plants is the availability of phosphorus. Most plants tend to show a deficiency in phosphorus content even when there is an abundance of the element in the soil. The observed disparity is because the content of soluble phosphorus hence bioavailable is very small. Two major benefits are associated with having organic anions around the rhizosphere of the plants. Firstly is the competition with binding sites with phosphate groups thereby forming strong complexes with metallic ions such as calcium, iron, and aluminum.

Most especially, minerals containing Ca-P are capable of releasing phosphorus as organic anions complexes with calcium or hinder the absorption of phosphorus to the other sites (Lunstrom et al. 1995). Toxicity due to aluminum is a primary factor that is accountable for the reduction in the productivity of crops in most acidic soils around the tropic regions. One of the basic mechanisms through which plants respond and withstand toxicity due to this metal is the secretion of organic anions such as malate, oxalate, and citrate from the roots.

Further, Yang et al. (2013) identified numerous factors that are responsible for the release of organic acids through Al-induction, some of which include: the presence of anions transporters, the concentration of organic acids anions in the tissue of the plants, effect of temperature of the system, presence of mineral nutrients such as magnesium, phosphorus, and the nature of the root plasma membrane.

17.9 Aluminum-Induced Secretion of Organic Acid Anions from Roots

It has been reported that the production of various anions such as malate, oxalate, and citrate from the roots of most plants is one of the measures for aluminum ion tolerances. The anions secreted by the plants differ as well as the pattern of secretion, sensitivity to heat, dose-response, and response to inhibitors (Ma 2007). Aside from the first formal documentation of secretion of organic anions such as malates by wheat through aluminum induction, more studies have come up showing the presence of many aluminum tolerant species that produce a high concentration of malate and oxalate from the root as a result of exposure to the high content of aluminum (Inostroza-Blancheteau et al. 2012). The formation of this oxo anions from the roots can result in non-toxic complexes with aluminum around the rhizosphere which will prevent the binding of the aluminum to cellular components thereby protecting the plants from aluminum toxicity. Among the three major anions secreted, the highest chelating potential for aluminum is found in citrate followed by oxalate and then malate. This secretion induced by aluminum toxicity is mainly around the root tips. This is highly connected to aluminum (Lambers et al. 2018).

17.10 Acquisition of Other Nutrients and Heavy Metals

The content of various soil micronutrients could readily be influenced by organic anions in the soil which in turn will affect their bioavailability for plant uptake. Some available studies have linked a deficiency of Fe^{3+} with increasing efflux of organic anions.

17.11 Communications with Plant Growth-Promoting Microorganisms

Organic anions are a major source of carbon utilized by microorganisms in the soil (Oburger et al. 2011) this may in turn bring about stimulation of the proliferation of fungi hyphae and bacteria in soil (Yuan et al. 2015). Some bacteria in soil are capable of releasing organic anions and phosphatases for the mobilization of phosphorus such bacteria are capable of promoting the growth of plants (Miransari 2014).

17.11.1 Root-Released Organic Anions in Intercropping

Lambers et al. (2018) reported that there is accumulating evidence showing that biomass, the yield of grain, and acquisition of nutrient are enhanced during intercropping of legumes and cereal under soil condition of high nitrogen and low phosphorus and this is attributable to the exudation of anions by the legumes (Lambers et al. 2018). In a study carried out by Li et al. (2016), they observed that the content of citrate and malate of intercropped wheat was much higher significantly. The release of organic anions in some plants brings about a rise (Li et al. 2016).

17.11.2 The Encoding Genes that Enhance Organic Anion Efflux from Roots Rhizosphere

Ryan et al. (2003) established and evaluated the strategies involved in the transportation of organic anions (−) in plant roots via the soil. The authors reported that the transportation of organic anions in plant roots performs a vital function in the supply of functional nutrients chemicals like oxalate, malate, and citrate that are fingered to be influenced by the mode of action by Al (aluminum) and P (phosphorus) in soil.

These soil metabolites have polyvalent anions that can pass through cell membranes which are permeable to facilitate the transportation or flow to aluminum-dependent plants like maize and wheat. Besides, certain isolated genes (cRNAs and cDNAs) in the rhizosphere of the plant's roots are accountable for the programming of the anion networks based on the protein ladder which facilitates the release of oxalate, malate, and citrate in the soil. This process can enable the facilitation of phosphorus acquisition and the mitigation of aluminum toxicity when utilized in biotechnology and profiling of soil properties.

The functions of permease organic negative ion in the control of minerals and resistance of aluminum in the soil were evaluated by Delhaize et al. (2007). The solubility of Al^{3+} in the soil is one of the major limitations to the adequate development of plants. The Al^{3+} actions on plants have elicited the plants to develop a counter mechanism by using the efflux process via Al^{3+} chelation of organic anions to protect its roots. More so, the resistance of the plant towards the aluminum is spurred by some genes which belong to the ALMT and MATE (Al^{3+}-activated malate transporter and multidrug and toxin extrusion) protein-encoding groups. The authors suggested further work of the sequencing of the gene which will provide pieces of evidence on the link between various species, the evolvement of the genes, the molecular differences, the expression of the genotypes, and the induction of the genes. The functions of the ABC transporting genes in the resistant mechanisms need to be evaluated with a specific focus on the protein substrates and validating if the genes have underlying natural differences in the tolerance of aluminum ions. In conclusion, the authors opined that cloned tolerance and resistance Al^{3+} genes serve as a tool to transport flora-based acid to the soil via rhizospheric modeling and engineering breeding and markers.

Ryan et al. (2011) evaluated and identified the resistance genes of aluminum that give the opportunities for eliciting plant production on corrosive soils. The authors reported that acidic soils impede crop production because of the incidence of soluble Al^{3+} which tends to limit the growth of plant roots in micro-molar concentrations. However, some plant species possess a normal resistance to the toxicity of Al^{3+} and do better on corrosive soils. This ability is controlled by 14 genes like *TaALMT1*, *AtALMT1*, *SbMATE1*, *HvAACT1*, *ScALMT gene clu*, *ZmMATE1*, *AtMATE*, *AtSTOP1*, *OsSTAR1*, *Os*, *ART1*, *ALS3*, *ALS1*, and *AlSTAR1*, which account for the variations of the genotypic traits. These genes also depend on the influence of the organic anion's efflux like citrate and malate from the roots of crops such as wheat, *Arabidopsis*, sorghum, barley, maize, and rice. The controlling factor of the genes is encoded in protein membranes to facilitate the efflux of anions in the cell membrane of the crops. The proof of identity of these resistant genes gives a better platform in the enhancement of aluminum anions resistance in the crops via biotechnology and marker-supported breeding techniques. This method gives the crop root system transgenic transformation that is rapid in each root cells which correlates with the time of cell generations and the phenotype lines of the transgenic plants.

The level of aluminum anion tolerance by higher floras using organic acid as a buffer was investigated by Yang et al. (2013). The authors reported that about 30% of the globe's sum area of landmass and more than 50% of the globe's likely arable soil are corrosive (acidic). The concentration of the acids in the soils is increasing gradually based on the influence of human activities such as farming which have led to the increase of the environmental condition like acid rain. Al (aluminum) occurs basically as immiscible components in neutral and slightly acidic soils and it is significantly biologically inert. Besides, in subtropical and tropical corrosive soils, the toxicity of Al is a limiting factor in the productivity of crops. The induced exudation of OA ions (organic acid) like malate, oxalate, and citrate in the roots of plants by Al is one of the best modes of action of higher plants tolerance to environmental stress. Yang et al. (2013) opined that the induced exudation of OA ions can be correlated with the following factors genes like C2H2, STOP1, ART1, ALMT1, OsFRDL4, and MATE, AtMATE (transcription protein factors), phosphorus, magnesium, H^+-ATPase, root cell membrane permeability, temperature, and the concentration of the organic acid anions in the tissues of the plants. However, genetically engineered cells and plants with high aluminum tolerance with genes that are overexpressing have the potentials to biosynthesize and secrete organic acid anions for effective plant resistance towards environmental stressors. So, the application of this method will serve as a synergistic influence as a next-generation key in the future management of soil toxicity and stress against rape, *A. thaliana*, rye, and wheat plant production. In conclusion, the collaboration between plant physiologist and breeders is inevitable in the understanding of several transgenic crops and their abilities to tolerate Al influences in the soil and the rhizospheric roots regions.

Zhou et al. (2013) tested and evaluated HvAACT1 the MATE genes of *Hordeum vulgare* (barley) and *Triticum aestivum* (wheat) that are responsible for the tolerance of aluminum anion and also elicit the efflux of citrate in soil. The authors reported that the toxicity of Al^{3+} in soil has been known to reduce the growth of the plant's roots. The mode of action of the aluminum anions tolerance by crops comprises the release of OAs from the tips of roots. The activated aluminum anions discharge citrate from the tips of the roots in conjunction with the HvAACT1 the MATE genes, which encodes the transported citrate protein found at the root plasma membrane in the *Triticum aestivum* and *Hordeum vulgare* crops. The results of the biological experiment showed that the MATE genes elicited the expression of both *Triticum aestivum* and *Hordeum vulgare* crops which were linked with the elevation of efflux of citrate in the root tips as well, enhanced the crops tolerance to aluminum anion in the soil. The findings from the study therefore demonstrated that the native breeding method in conjunction with the modern biotechnology technique can be employed to improve the tolerance of different arable crops to soil toxicity caused by Al^{3+}.

The mode of actions of OAs with Al^{3+} at the apices of crop roots in soil were evaluated by Yang et al. (2013). The authors stated that at the average, acidic soil contains a pH of 5.5 at this level, it becomes solubilized as a solution into the soil. Aluminum which is an abundant element in the soil combines with the soil pH thus causing toxic influence and retarding the development and growth of crops. Nonetheless, some crops have developed a level of aluminum tolerance mode of action that permits the crops to counter the noxiousness produced by Al^{3+}. The crops use the secretion OAs like oxalate, malate, and citrate as well as the induction of Al^{3+} in the roots of the plants. The mechanisms involve the discharge of anions that can chelate the Al^{3+}, therefore defending the secreting crops from the noxiousness of aluminum. The authors opined that the transporting genes encoding the malate and citrate which are responsible for the discharge of the anions have been shown with shreds of evidence to regulate the plant tolerance to aluminum toxicity.

Ryan et al. (2009) evaluated the management and engineering modeling of rhizosphere for sustainable farming. The authors reported that the manipulation of the root of plants with engineered microorganisms to improve the productivity and health of plants is a novel strategy in modern agriculture. This new strategy is directly fingered on crop development and growth and the interaction with the soil, microorganisms, and the roots of the plants (rhizosphere). For example, some plants can be engineered to change pH or discharge compounds that can elevate or alleviate the abiotic and biotic stresses, the availability of nutrients, and the spread of beneficial microbes. The authors stated that rhizobacteria that improve the development of plants have also been modified to delay the production of hormones like ethylene the plant stress inducer. This hormone can also reduce the growth of plant roots and manufacture lytic and antibiotic enzymes that are active more against root and soil pathogens. The engineering modeling of the rhizosphere includes the selection of floras of important microbial consortia. For instance, plant cultivars or species that support the group of antibiotic-manufacturing strains play a significant function in the suppression of soil fungal diseases. Crops can also be influenced by the association of the community of the microbes found in their roots which also aid in the suppression of viral activities in the general Pectobacterium. The authors propose the utilization of novel molecular and biotechnological tools that will provide biological and chemical interactions thus ensuring the engineering of microbes in the roots of higher plants that will ensure the beneficial improvement and productivity of agricultural activities and soil functioning.

The significance of rhizospheric activities in the control and management of pathogenic microbe, human influence, and plant diseases was evaluated by Mendes et al. (2013). The authors recounted that microbial consortia play a basic role in the control and management of the effects of plants' growth and physiology. Some associates of the microbiome-rhizosphere are colonized pathogenic and beneficial forms determined to break through the cell membrane to bring about disease infections or defend the plant using some mechanized defense structures to protect the internal cells. The authors stated that to improve the health and growth of the plant, it is important to be abreast of the types of microorganisms present in the microbiome-rhizosphere and their activities therein. This will also enable humans to be protected from potential pathogenic strains dangerous to human health.

Ligaba et al. (2012) examined the transporter anionic gene ZmALMT2 of maize that intercedes between the root and the soil organic malate. The efflux of OAs in the roots of the plant triggers the main mode of action of Al^{3+} resistance on corrosive soils which is mediated by ALMT (Al-activated malate transporter) or MATE (multidrug and toxin extrusion) families. Ligaba et al. (2012) reported that the transporter anionic gene ZmALMT2 is involved in Al^{3+} resistance and mapping association for maize. The results from the biological experiment after the characterization of ZmALMT2 were shown to have been articulated in *Xenopus laevis* oocytes and a transgenic *Arabidopsis*. The ZmALMT2 oocytes facilitated an aluminum non-dependent electrogenic passage material of inorganic and OAs efflux. Overexpression of the presentation of ZmALMT2 transporter gene in hypersensitive-aluminum *Arabidopsis* KD/KO lines was considered deficient in the resistant genes AtMATE and AtALMT1, which caused an independent-aluminum malate constitutive root efflux and partially reinstated the level of the phenotypic aluminum tolerance. The authors reported that the non-relationship between the aluminum resistance and the gene expression of ZmALMT2 might lead to its questioning and role in the tolerance of aluminum. However, the functional chattels of the transporter gene in this study suggested that it is not tangled with aluminum tolerance in maize plants but may rather play a significant role in the transportation and acquisition of nutrients and mineral ions in plants.

The BnALMT2 and BnALMT1 transporter genes sourced from REA-AM (Rape Encode Aluminum Activated Malate) improve that the resistance to Al^{3+} toxicity in the cells of plant and soil was investigated by Ligaba et al. (2006). The authors recounted that the discharge of OAs from the plant, roots could serve as a defense against Al noxiousness and also aid to combat the deficiency of P in the soil. Previous studies of their work indicated that aluminum treatment provoked citrate and malate efflux in the roots of *Brassica napus* (rape plant) and the deficiency of P in soil did not provoke the anion efflux which is also similar to *Triticum aestivum* (wheat plant) regulated by the transporter gene TaALMT1. The result from the biological experiment from the isolated homologs transporter rape genes (BnALMT2 and BnALMT1), showed Al-induction in the roots and no deficiency of P was shown. However, these effects were not reflected in the shoot system. Numerous cations like erbium, ytterbium, and lanthanum also elevated the BnALMT2 and BnALMT1 genes in the root system. A further investigation of the BnALMT2 and BnALMT1 genes on *Nicotiana tabacum* (tobacco plant) cultured cells and in the oocytes of *Xenopus laevis* showed heterologic expression in both plants. The transfection effects on both systems indicated an improved capacity for efflux in malate but not in citrate when Al is exposed to them. Erbium and ytterbium treatment on the system also activated small fluxes amount of malate. It was observed that the cells of the transgenic *Nicotiana tabacum* significantly grew much better when compared to the control cells after 18 h of Al treatment. The findings indicated that BnALMT2 and BnALMT1 genes elicited the resistance of the crop cells to the stress caused by AL. This study demonstrated that the homologs transporter rape genes TaALMT1 sourced from *Triticum aestivum* performed a likely role as compared to other kinds.

Mariano et al. (2005) evaluated the exudation and root metabolism of OAs under the influence of Al stress. The authors recounted that many flora species can discharge OAs from their root system in reply to toxic Al ions found in the soil-root media. Theoretically, the complexes of the OAs and Al in the rhizosphere and the root apoplast avoid interactions in the plasma components into the symplast of the root. The authors stated that there are two patterns of the OAs exudation noted in the root system, the pattern I and II. In the first case (a pattern I), the OAs discharge is swiftly triggered after the interaction of the Al^{3+} with the roots. While in the latter, pattern II, a lag phase is noticed between the discharge of the OAs at the beginning at the adding of the Al. The authors stated that floras species like the tea and buckwheat have shown high Al mechanism and tolerance which prove them capable to store inactivate aluminum internally in the plant's leaves. The instability metabolic routes induced by aluminum are still not clear and their link to the change of the organic acid concentration noted in the plant roots under aluminum stress is not well defined. Mariano et al. (2005) stated that an upsurge in the concentration of the organic acid in the plant root does not often lead to the discharge of the organic acid even when the root axis and the spatial dispersal are taken into justification. Nonetheless, aluminum elicits high penetration ability to organic acid contents in the cell of young root and the channels of the anion located at the plasma membrane via mediated transportation were proposed by the authors.

The exclusion and tolerance of aluminum in woody floral were evaluated by Brunner and Sperisen (2013). The cation of aluminum was reported by the authors to be extremely rhizotoxic and portend chief stressor to the flora in corrosive soils. Globally, most acidic or corrosive soils are found in the boreal and tropical areas or regions. Numerous woody flora species are innate to corrosive soils and have been fingered to be well accustomed to extreme Al^{3+} settings. In the tropics, both aluminum and non-aluminum woody accumulator flora are found in this region as contrary to the arboreal where only non-aluminum woody accumulators are widely distributed. Their mode of action is divided into Al^{3+} prohibiting mechanisms; those that enable aluminum anions from the roots and the Al^{3+} internal prohibiting mechanisms; those that tolerate aluminum in their shoots and roots symplast. However, a clear examination of the arboreal and tropical woody plants showed similar mechanisms of action as aluminum and non-aluminum woody accumulator and clear-cut adaptors to abiotic and biotic stress factors. These plants also develop ectomycorrhizal assemblies which have the ability to absorb aluminum in the hyphal cell walls thus decreasing the level of interaction of the plant cells to the toxicity of the metal. The internal methods depend on the sequestration and transport of aluminum in the aerial regions of the flora. The aluminum may be accumulated in the tissues of the leaves, cell wall, and at the subcellular units or inclusions like the vacuoles and chloroplasts. Besides, some of the woody crops can accumulate an extreme amount of aluminum in their root cells and cell walls. The authors stated that the molecular and biochemical basis of these modes of actions has been extensively studied in many crops with *Arabidopsis* as a clear model. However, few studies done on woody plants have shown exceptions and proven to be good engineering testing candidates for the recognition of specific genes and traits that might perform an immense function in the association and adaptation of plants to the wide variation of Al^{3+} influenced soils.

17.12 Conclusion and Future Direction

This chapter has provided detailed information on detailed information on the engineering plants through plant genetic engineering. There is a need to still search for some other essential biotechnological tools that could lead to an increase in food production, management of instability in fluctuating climate changes, maintenance of a cleaner environment. Moreover, there is a need to understand the modes of action involving in plants could facilitate prognostic potentials to an effective prediction on how plants and rhizosphere ecosystem could help in the recycling of nutrients, maintenance of beneficial microorganism in the soils. The application of metabolomics, bioinformatics, proteomics, and genomics could help in the strain improvement that could lead to increase in the secretion of biocompounds that could act as a biostimulant that could help in the development of agricultural food production as well as maintenance of the cleaner environment. Moreover, there is a need to increase greater knowledge of the mycorrhizal-rhizospheric microbiome, plant-microbe. Identification of new techniques that could lead to adequate predation of novel strain that could help in feeding the ever-increasing populations as well as serves in establishing the amount of carbon cycle in the soil most especially under recent and future climates. The utilization of molecular and synthetic biology will play a robust function most especially a better understanding of root-microbe association for effective bioenergy and food production without allowing the interference of biotic and abiotic factors.

References

Adesemoye AO, Egamberdieva D (2013) Beneficial effects of plant growth-promoting rhizobacteria on improved crop production: prospects for developing economies. In: Maheshwari DK, Saraf M, Aeron A (eds) Bacteria in agrobiology: crop productivity. Springer, Berlin, pp 45–63

Adl S (2016) Rhizosphere, food security, and climate change: a critical role for plant-soil. Rhizosphere 1:1–3

Amir H, Ahkamia A, Richard W, Handakumburaa PP, Janssona C (2017) Rhizosphere engineering: enhancing sustainable plant ecosystem productivity. Rhizosphere 3:233–243

Bais HP, Loyola-Vargas VM, Flores HE, Vivanco JM (2001) Invited review: rootspecific metabolism: the biology and biochemistry of underground organs. In Vitro Cell Dev Biol Plant 37:730–741

Barret M, Morrissey JP, O'Gara F (2011) Functional genomics analysis of plant growth-promoting rhizobacterial traits involved in rhizosphere competence. Biol Fertil Soils 47:729–743

Bikis D (2018) Review on the application of biotechnology in garlic (Allium Sativum) improvement. Int J Res Stud Agric Sci (IJRSAS) 4(11):23–33. https://doi.org/10.20431/2454-6224.04011004

Bojórquez-Quintal E, Escalante-Magaña C, Echevarría-Machado I, Martínez-Estévez M (2017) Aluminum, a friend or foe of higher plants in acid soils. Front Plant Sci 8:1767. https://doi.org/10.3389/fpls.2017.01767

Bosch TC, McFall-Ngai MJ (2011) Metaorganisms as the new frontier. Zoology 114:185–190

Bravin MN, Tentscher P, Rose J, Hinsinger P (2009) Rhizosphere pH gradient controls copper availability in a strongly acidic soil. Environ Sci Technol 43(15):5686–5691. https://doi.org/10.1021/es900055k

Bravin M, Gamier C, Lenobe V, Gerard F, Hinsinger P (2012) Root-induced changes in pH and dissolved organic matter binding capacity affect copper dynamic speciation in the rhizosphere. Geochim Cosmochim Acta 84:256–268

Brunner I, Sperisen C (2013) Aluminum exclusion and aluminum tolerance in woody plants. Front Plant Sci 12:172

Chang C, Hu Y, Sun S, Zhu Y, Ma G, Xu G (2009) Proton pump OsA8 is linked to phosphorus uptake and translocation in rice. J Exp Bot 60(2):557–565. https://doi.org/10.1093/jxb/ern298

Das A, Pramanik K, Sharma R, Gantait S, Banerjee J (2019) In-silico study of biotic and abiotic stress-related transcription factor binding sites in the promoter regions of rice germin-like protein genes. PLoS ONE 14(2):e0211887. https://doi.org/10.1371/journal.pone.0211887

de Alexis A, Sébastien T, Jean-Marie F, Geneviève E, Franco G, Hélène B-B (2007) Anion channels and transporters in plant cell membranes. Plant Transp Channels 581 (12):2367–2374. https://doi.org/10.1016/j.febslet.2007.04.003

Delhaize E, Gruber BD, Ryan PR (2007) The roles of organic anion permeases in aluminium resistance and mineral nutrition. FEBS Lett 581:2255–2262

Dessaux Y, Grandclement C, Faure D (2016) Engineering the rhizosphere. Trends Plant 21 (3):266–278

Ding ZJ, Yan JY, Xu XY, Li GX, Zheng SJ (2013) WRKY46 functions as a transcriptional repressor of ALMT1, regulating aluminum-induced malate secretion in Arabidopsis. Plant J 76:825–835. https://doi.org/10.1111/tpj.12337

Inostroza-Blancheteau C, Rengel Z, Alberdi M, de la Luz Mora M, Aquea F, Arce-Johnson P et al (2012) Molecular and physiological strategies to increase aluminum resistance in plants. Mol Biol Rep 39:2069–2079. https://doi.org/10.1007/s11033-011-0954-4

Jackson C, Stone B, Tyler H (2015) Emerging perspectives on the natural microbiome of fresh produce vegetables. Agriculture 5:170–187

John MM, Bonnie JW (2003) Plant disease resistance genes: recent insights and potential applications. Trends Biotechnol 21:4

Kar D, Pradhan AN, Datta S (2020) The role of solute transporters in aluminum toxicity and tolerance. Physiol Plant. https://doi.org/10.1111/ppl.13214

Kollmeier M, Dietrich P, Bauer CS, Horst WJ, Hedrich R (2001) Aluminum activates a citrate-permeable anion channel in the aluminum-sensitive zone of the maize root apex. A comparison between an aluminum-sensitive and an aluminum-resistant cultivar. Plant Physiol 126:397–410

Kong X, Liu G, Liu J, Ding Z (2018) The root transition zone: a hot spot for signal crosstalk. Trends Plant Sci 23:403–409

Lambers H, Albornoz F, Kotula L, Laliberté E, Ranathunge K, Teste FP, Zemunik G (2018) How belowground interactions contribute to the coexistence of mycorrhizal and nonmycorrhizal species in severely phosphorus-impoverished hyperdiverse ecosystems. Plant Soil 424:11–34

Laranjo M, Alexandre A, Oliveira S (2014) Legume growth-promoting rhizobia: an overview on the Mesorhizobium genus. Microbiol Res 169:2–17

Li XF, Ma JF, Matsumoto H (2000) Pattern of Al-induced secretion of organic acids differs between rye and wheat. Plant Physiol 123:1537–1543

Li CJ, Dong Y, Li HG, Shen JB, Zhang FS (2016) Shift from complementarity to facilitation on P uptake by intercropped wheat neighboring with faba bean when available soil P is depleted. Sci Rep 6:18663

Ligaba A, Katsuhara M, Ryan PR, Shibasaka M, Matsumoto H (2006) The BnALMT1 and BnALMT2 genes from rape encode aluminum-activated malate transporters that enhance the aluminum resistance of plant cells. Environ Stress Adapt Stress. https://doi.org/10.1104/pp.106.085233

Ligaba A, Maron L, Shaff J, Kochian L, Pineros M (2012) Maize ZmALMT2 is a root anion transporter that mediates constitutive root malate efflux. Plant Cell Environ 35(7):1185–1200. https://doi.org/10.1111/j.1365-3040.2011.02479.x

Lu G-H, Tang C-Y, Hua X-M, Cheng J, Wang G-H, Zhu Y-L et al (2018) Effects of an EPSPS-transgenic soybean line ZUTS31 on root associated bacterial communities during field growth. PLoS ONE 13(2):e0192008. https://doi.org/10.1371/journal.pone.0192008

Lunstrom US, Van Breemen N, Jongmans AG (1995) Evidence for microbial decomposition of organic acids during podzolization. Eur J Soil Sci 46:489–496

Ma JF (2007) Syndrome of aluminum toxicity and diversity of aluminum resistance in higher plants. Int Rev Cytol 264:225–252

Ma Q, Yi R, Li L et al (2018) GsMATE encoding a multidrug and toxic compound extrusion transporter enhances aluminum tolerance in Arabidopsis thaliana. BMC Plant Biol 18:212. https://doi.org/10.1186/s12870-018-1397-z

Manoj V, Parakhia B, Golakiya A (2018) Manipulation of phytobiome: a new concept to control the plant disease and improve the productivity. J Bacteriol Mycol 6:322–324. https://doi.org/10.15406/jbmoa.2018.06.00227

Mariano ED, Jorge R, Willem G, MenossiI M (2005) Metabolism and root exudation of organic acid anions under aluminium stress. Raz J Plant Physiol 17:1. https://doi.org/10.1590/S1677-04202005000100013

McNear DH Jr (2013) The rhizosphere - roots, soil and everything in between. Nat Educ Knowl 4:1

Mendes R, Garbeva P, Raaijmakers J (2013) The rhizosphere microbiome: significance of plant beneficial, plant pathogenic, and human pathogenic microorganisms. FEMS Microbiol Rev 37 (5):634–663. https://doi.org/10.1111/1574-6976.12028

Miransari M (2014) Plant growth promoting rhizobacteria. J Plant Nutr 37:2227–2235

Morel MA, Braña V, Castro-sowinski S (2012) Legume crops, importance and use of bacterial inoculation to increase production. In: Goyal A (ed) Crop plant. InTech, Rijeka

Muhammad T, Zhang F, Zhang Y, Liang Y (2019) RNA interference: a natural immune system of plants to counteract biotic stressors. Cell 8(1):38. https://doi.org/10.3390/cells8010038

Nardi S, Concheri G, Pizzeghello D, Sturaro A, Rella R, Parvoli G (2000) Soil organic matter mobilization by root exudates. Chemosphere 41:653–658

Neumann G, Römheld V (2000) The release of root exudates as affected by the plant's physiological status. In: Pinton R, Varanini Z, Nannipieri Z (eds) The rhizosphere: biochemistry and organic substances in the soil-plant interface. Marcel Dekker, New York, pp 41–93

Oburger E, Leitner D, Jones DL, Zygalakis KC, Schnepf A, Roose T (2011) Adsorption and desorption dynamics of citric acid anions in soil. Eur J Soil Sci 62:733–742

Odelade KA, Babalola OO (2019) Bacteria, fungi and archaea domains in Rhizospheric soil and their effects in enhancing agricultural productivity. Int J Environ Res Public Health 16 (20):3873. https://doi.org/10.3390/ijerph16203873

Palmer A, Baker A, Muench P (2016) The varied functions of aluminium-activated malate transporters–much more than aluminium resistance. Biochem Soc Trans 44(3):856–862. https://doi.org/10.1042/BST20160027

Peter R, David C, Wang E (2014) Can citrate efflux from roots improve phosphorus uptake by plants. Testing the hypothesis with near isogenic lines of wheat. Physiol Plant 151(3):230–242. https://doi.org/10.1111/ppl.12150

Pii Y, Mimmo T, Tomasi N, Terzano R, Cesco S, Crecchio C (2015) Microbial interactions in the rhizosphere: Beneficial influences of plant growth-promoting rhizobacteria on nutrient acquisition process. A review. Biol Fertil Soils 2015(51):403–415

Quiza L, St-Arnaud M, Yergeau E (2015) Harnessing phytomicrobiome signaling for rhizosphere microbiome engineering. Front Plant Sci 6:507. https://doi.org/10.3389/fpls.2015.00507

Rengel Z, Zhang WH (2003) Role of dynamics of intracellular calcium in aluminium-toxicity syndrome. New Phytol 159:295–314. https://doi.org/10.1046/j.1469-8137.2003.00821.x

Rengel Z, Marschner P (2005) Nutrient availability and management in the rhizosphere: exploiting genotypic differences. New Phytol 168:305–312. https://doi.org/10.1111/j.1469-8137.2005.01558.x

Ryan PR, Dong B, Watt M et al (2003) Strategies to isolate transporters that facilitate organic anion efflux from plant roots. Plant Soil 248:61–69. https://doi.org/10.1023/A:1022314010798

Ryan P, Dessaux Y, Thomashow LS, Weller DM (2009) Rhizosphere engineering and management for sustainable agriculture. Plant Soil 321:363–383. https://doi.org/10.1007/s11104-009-0001-6

Ryan P, Tyerman R, Sasaki T, Furuichi T, Yamamoto Y, Zhang WH, Delhaize E (2011) The identification of aluminium-resistance genes provides opportunities for enhancing crop production on acid soils. J Exp Bot 62(1):9–20. https://doi.org/10.1093/jxb/erq272

Santoyo G, Moreno-Hagelsieb G, del Carmen Orozco-Mosqueda M, Glick BR (2016) Plant growth-promoting bacterial endophytes. Microbiol Res 183:92–99

Sasaki T, Yamamoto Y, Ezaki B (2004) A wheat gene encoding an aluminum-activated malate transporter. Plant J 37(5):645–653. https://doi.org/10.1111/j.1365-313X.2003.01991.x

Shenoy V, Kalagudi GM (2005) Enhancing plant phosphorus use efficiency for sustainable cropping V. Biotechnol Adv 23:501–513

Singhala P, Janc AT, Azmb M, Haqb Q (2016) Plant abiotic stress: a prospective strategy of exploiting promoters as alternative to overcome the escalating burden. Front Life Sci 9 (1):52–63. https://doi.org/10.1080/21553769.2015.1077478

Su J, Jiang J, Zhang F, Liu Y, Ding L, Chen S, Chen F (2019) Current achievements and future prospects in the genetic breeding of chrysanthemum: a review. Hortic Res 6:109. https://doi.org/10.1038/s41438-019-0193-8

Takanashi K, Shitan N, Yazaki K (2014) The multidrug and toxic compound extrusion (MATE) family in plants. Plant Biotechnol 31(5):417–430. https://doi.org/10.5511/plantbiotechnology.14.0904a

Upadhyay N, Kar D, Datta S (2019) A multidrug and toxic compound extrusion (MATE) transporter modulates auxin levels in root to regulate root development and promotes aluminium tolerance. Plant Cell Environ 43:3. https://doi.org/10.1111/pce.13658

Vital-López L, Cruz-Hernández MA, Fernández-Dávila S, Mendoza-Herrera A (2016) Bacterial diversity in the rhizosphere of a transgenic versus a conventional maize (Zea mays). Fyton 85: 210–217

Wang Y, Cai Y, Yu CY, Jiping LJ (2018) Aluminum-activated root malate and citrate exudation is independent of NIP1; 2-facilitated root-cell-wall aluminum removal in Arabidopsis. https://doi.org/10.1080/15592324.2017.1422469

Wu L, Yuriko K, Jun W, Hiroyuk K (2018) Organic acid excretion from roots: a plant mechanism for enhancing phosphorus acquisition, enhancing aluminum tolerance, and recruiting beneficial rhizobacteria. Soil Sci Plant Nutr 64(6):697–704. https://doi.org/10.1080/00380768.2018.1537093

Xinxin Z (2014) Cloning and characterization of a plasma membrane H^{+}-ATPase (PMA) gene from a salt-tolerant plant Chloris virgate. Mol Soil Biol. https://doi.org/10.5376/msb.2014.05.0003

Xu Y, Magwanga RO, Yang X et al (2020) Genetic regulatory networks for salt-alkali stress in Gossypium hirsutum with differing morphological characteristics. BMC Genomics 21:15. https://doi.org/10.1186/s12864-019-6375-9

Yang Z, Horst WJ (2015) Aluminum-induced inhibition of root growth: Roles of cell wall assembly, structure, and function. Alum Stress Adapt Plants Signal Commun Plants 24:253–274

Yang LT, Qi YP, Jiang HX, Chen LS (2013) Roles of organic acid anion secretion in aluminium tolerance of higher plants. Biomed Res Int 2013:173682. https://doi.org/10.1155/2013/173682

Yang J, Fan W, Zheng S (2019a) Mechanisms and regulation of aluminum-induced secretion of organic acid anions from plant roots. J Zhejiang Univ Sci B 20(6):513–527. https://doi.org/10.1631/jzus.B1900188

Yang JL, Fan W, Zheng SJ (2019b) Mechanisms and regulation of aluminum-induced secretion of organic acid anions from plant roots. Science 20(6):513–527. https://doi.org/10.1631/jzus.b1900188

Yonekura-Sakakibara K, Higashi Y, Nakabayashi R (2019) The origin and evolution of plant flavonoid metabolism. Front Plant Sci 10:943. https://doi.org/10.3389/fpls.2019.00943

Yuan J, Zhang N, Huang Q, Raza W, Li R, Vivanco JM, Shen Q (2015) Organic acids from root exudates of banana help root colonization of PGPR strain Bacillus amyloliquefaciens NJN-6. Sci Rep 5:13468

Zhang J, Wei J, Li D, Kong X, Rengel Z, Chen L, Yang Y, Cui X, Chen Q (2017) The role of the plasma membrane H^+-ATPase in plant responses to aluminum toxicity. Front Plant Sci 8:1757. https://doi.org/10.3389/fpls.2017.01757

Zhao J, Liu J, Liang H, Huang J, Chen Z, Nie Y et al (2018) Manipulation of the rhizosphere microbial community through application of a new bio-organic fertilizer improves watermelon quality and health. PLoS ONE 13(2):e0192967. https://doi.org/10.1371/journal.pone.0192967

Zhou G, Delhaize E, Zhou M, Ryan P (2011) Biotechnological solutions for enhancing the aluminium resistance of crop plants. Open Rev. https://doi.org/10.5772/25187

Zhou G, Delhaize E, Zhou M, Ryan P (2013) The barley MATE gene, HvAACT1, increases citrate efflux and Al3+ tolerance when expressed in wheat and barley. Ann Bot 112(3):603–612. https://doi.org/10.1093/aob/mct135

Factors Affecting Soil Ecosystem and Productivity

18

Jyoti Rajwar, Divya Joshi, Deep Chandra Suyal, and Ravindra Soni

Abstract

Soil is one of the most dynamic systems of earth. In the current scenario, exploding population, shrinking forest cover, undesirable climatic changes and shrinking agricultural landmass has imposed a lot of pressure on sustainable agricultural productivity. Therefore, it has become increasingly important to study the factors governing soil health and productivity. The study of soil health refers to a soil's capacity to sustain agricultural productivity by fostering plant growth while also preserving or enhancing the quality of the environment. Further, soil's health and productivity are an important function of soil's physio-chemical structure, nutrient cycles, and abiotic and biotic interactions. This chapter addresses the aspects of soil fertility and numerous abiotic and biotic factors that affect soil health and degradation.

Keywords

Soil health · Ecosystem · Microbial diversity · Abiotic factors

J. Rajwar (✉)
Department of Microbiology, Skyline University, Kano, Nigeria
e-mail: jyoti@sun.edu.ng

D. Joshi
Uttarakhand Pollution Control Board, Regional Office, Kashipur, Uttarakhand, India

D. C. Suyal
Department of Microbiology, Akal College of Basic Sciences, Eternal University, Sirmaur, Himachal Pradesh, India

R. Soni
Department of Agricultural Microbiology, College of Agriculture, Indira Gandhi Krishi Vishwa Vidyalaya, Raipur, Chhattisgarh, India

R. Soni et al. (eds.), *Microbiological Activity for Soil and Plant Health Management*, https://doi.org/10.1007/978-981-16-2922-8_18

18.1 Introduction

Soil ecosystem and productivity are vital for the continuity of life on Earth. Recent anthropogenic exploitation has exposed the land to several biotic and abiotic stressors. These have drastically degraded the soil ecosystems and physio-chemical processes. Soil is a dynamic media that facilitates numerous activities like soil organic carbon transformations, various nutrients cycling, maintaining self-structure, and controlling biotic diaspora. Activities like rigorous agriculture and mining operations, overgrazing, and indiscriminate waste creation and disposal often create an imbalance in the structure and functions of soil, especially at the expense of biodiversity (Morgado et al. 2018). Soil productivity is another aspect of soil health which includes soil fertilization and management factors associated with plant growth and development. The key factors that affect the productivity of the soil are its ability to hold water and air, along with ensuring the availability of essential micro and macronutrients. Mankind also plays a key part in soil productivity. Man-made activities such as fertilizer use, crop rotation, irrigation equipment, and drainage are some of the processes that affect land productivity (Karlen 2005).

Soil health is the study of various soil functions, which in turn depends on ecological boundaries, plant productivity, water-holding capacities, and quality of soil and atmospheric air. These activities ensure our well-being, and also provide a site for organisms to flourish (Trivedi et al. 2016). In the last few years, noteworthy efforts are made to upscale agricultural productivity via excessive use of fertilizers and pesticides, enhanced irrigation, soil management protocols, and gigantic land alterations (Tilman et al. 2002). This agricultural growth is creating pressure on soil's capability to sustain its additional functions and thus causing major ecosystem deprivation and long-term degradation of soil productivity (Vitousek et al. 2009). Current farming practices, hence, face difficulties not just as far as guaranteeing worldwide food security by expanding yields yet additionally moderating the ecological expenses, especially with regard to a variable environment and creating biological rivalry for energy, soil, and water (Chen et al. 2014). Furthermore, there is a pressing need to discover precursive markers to monitor the health of soil before it's too late so that the damage already done can be reversed. Different biomes respond differently to land exploitation like arid lands that contain lower levels of nutrients and have low productivity are therefore extremely risky to global environmental variations and desertification. They may additionally suffer a high level of reductions in nutrient accessibility due to rampant agricultural exploitation (Trivedi et al. 2016). Whereas extremely productive lands like those from tropics are extremely resistant/resilient to agriculture uses because of swift organic cycles and high moisture and water accessibility (Schlesinger and Bernhardt 2013). The important factors affecting the soil productivity are being discussed below:

18.2 Abiotic Factors

Abiotic aspects are important variables to assess the sustainability of the soil environment and productivity. Some of the mentioned factors are high temperatures, irradiation, waterlogging, dry matter, macro and micronutrient deficiency, precipitation, etc. This tension greatly changes the atmosphere-soil-plant cycle and results in deteriorating outcome of major crops. The important facet of climate change is a convoluted process with a drastically variable impacts on land, water, and air (Fig. 18.1).

18.2.1 Physical Factors

18.2.1.1 Soil Texture

Soil texture within the fine earth fraction is recognized as the relative composition of clay, silt, and sand. The content of texture and rough fragments is very significant in soils for several reasons (Li et al. 2020). Soil texture defines the pore spaces in soil which plays an important role in the flow of water and its preservation in the soil (Grant et al. 2019). More pore spaces in the sand and sandy-loam soils allow improved water flow and lower permeability of the clays and fine-textured soils due to lower pore surfaces limiting the movement of water suggesting better preservation of water (Hillel 2012). Surface area of soil is dependent upon the consistency of the soil and its coarse fragment material. The larger the soil surface area, the better it will be to retain soil-colored water, nutrients, cations, and anions

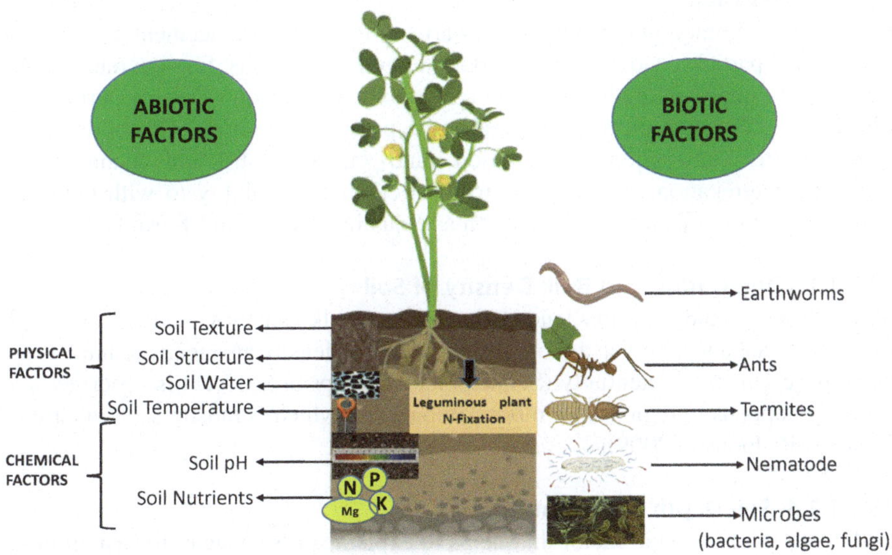

Fig. 18.1 Factors affecting soil health and productivity

(Shaetzl and Anderson 2005). Shape of soil particle plays an important role to understand the comparison between soil composition and quality. Sand particles are usually spherical, whereas clay and silt are generally smaller and flatter. Sandy textures are dominated by spherical/near-spherical particles which give better storage capacity and better movement of water and air thus ensuring more aeration and water-holding capacity but are vulnerable to drought. The clay textured soils similarly have smaller and flatter particle distribution facilitating smaller pores with lower fluid movements thus imparting prolonged retention of air and water. The soil with clayey textures also tends to hold more nutrients and facilitate greater ion exchange potential. They also contain more organic matter (Cornell University Cooperative Extension (CUCE) 2007). The particle size distribution is also important factor for development and stability of soil aggregates (Santamarina and Cho 2004). The study of soil texture is also important to understand the dynamics of soil erosion and compaction thus enriching our understanding for successful plant depth and rooting.

18.2.1.1.1 Soil Compaction

The compaction of soil is primarily depends upon the use of heavy agriculural equiments. Compaction induces systemic degradation and restricts root growth, capacity to store water, fertility, biological activity, and stability. It can also impact crop growth if the texture is too open, as roots will not stay properly on the ground. These soils are quickly whipped up by the wind, exposing air, and drought to the roots. When the soil is loose, it is demarcated as good soil to help in unregulated root penetration (Shah et al. 2017).

18.2.1.1.2 Erosion

Erosion is a common geological process arising from the displacement by water or wind of soil particles and their transportation to other locations. Erosion rates can be significantly improved by human activity and extreme erosion is largely permanent. The larger the coarse pieces, the greater the potential gap area. This adds to the resistance to compaction, degradation, and effective structure management. Enhanced aggregation and higher detritus strengthen the ability to withstand soil erosion, which has a beneficial effect on soil quality (Kumar and Karthika 2020)

18.2.1.1.3 Infiltration and Bulk Density of Soil

Soil with more sandy textures tends to have lower bulk densities and thus have high infiltration rate and vice versa. Soil with lower bulk density also tends to have more pore space which is potentially (Schaetzl and Thompson 2015). It is observed that flooding happens in region with more compact soil which is mostly seen in case of clayey soil (Jhonson 2009).

18.2.1.1.4 Soil Depth and Plant Rooting

Soil depth is an essential factor that helps to explain soil's capacity to sustain flora. The most effective depth of soil is recognized by the depth of soil which has maximum availability of water and nutrients to plant roots. Rooting depth differs from plant to plant, for example, plantation crops are deeply rooted while cereals are shallow rooted. The soil type is considered as healthy for a crop if it facilitates extensive root growth at the adequate depth (Jhonson 2009). As a consequence of the underlying chemical properties, soils typically have detrimental physical properties. Better root depth and improved soil texture are necessary for enhancing soil productivity (Bot and Benites 2005).

18.2.1.2 Soil Structure

Soil composition along with texture defines the structure of soil. The texture is the physical properties of soil such as water permeability, retention etc., while, composition defines the chemical structure of soil. The presence of organic matter is one of the most important characteristics of soil structure. Soil organic matter also defines some of the physical characteristics of soil like water absorption and surface crusting (Prescott et al. 2020). The structure of soil helps us to understand various physical cycles like wet and dry cycle, etc., root activity and soil flora and fauna (Xiao et al. 2020). Soil biota is also influenced by expansion of soil granular and subangular block structures (Shaetzl and Anderson 2005). A soil structure is considered stable if it has strong crumb structure and it breaks apart quickly. It should also be without clods. Soil water cycle which is crucial for enabling plant to have a favorable growth of rooting media is critically controlled by the soil structure (Kibblewhite et al. 2008).

18.2.1.2.1 Aggregation

The larger the accumulation, the greater the porosity of the soil would be. Among soil aggregates or peds, larger pores typically occur (Schimel et al. 2005). There would be a decrease in organic matter as the soil gets disrupted and aggregates become unstable thus making soil vulnerable to erosion (Nyawade et al. 2018). Stable soil aggregates ensure resistance to erosive action of water during heavy irrigation or precipitation (Bhat et al. 2019). Several microbial activities also prevent physical deterioration of soil by establishing soil aggregates and preserving soil structure (Nichols and Halvorson 2013). As evident from beach sand, the soil that is poor in organic matter and clay, will have weak accumulation and aggregation capacity (Usharani 2019).

18.2.1.3 Soil Water

Soil water quality is a result of soil composition, consistency of aggregates, bulk density, and numbers of earthworms. All of these are critical in understanding the soil tolerant capacity during rainfall and irrigation events. In the conservation of soil water quality, soil organization plays a significant function. Bigger grains make soil additional porous, suggesting greater permeability and penetration. Platy soil structure checks infiltration. Similarly, the sluggish permeability of water stems from prismatic structures. Increased compaction leads to declined percolation, contributing eventually to water-logged environments. The structure of platy results in water-logged soils, especially when there are significant downpours and severe flooding. Bulk density and biological behavior are also affected by surface water retention. Low bulk density is an indicator of bigger pore space, which implies improved soil water retention (Karlen et al. 1994). Burrowing activity of earthworms create more macropores in soil resulting in faster surface runoff (Hallam and Hodson 2020). Soil water stress incorporates salt stress and water scarcity stress, as soil chemical influences struggle with salt stress. Soil water governs the biology and chemistry of the soil, thereby suggesting its unavoidable role in deciding the soil's health and consistency.

18.2.1.4 Soil Temperature

Temperature affects the quality and biological function of soil moisture. Soil temperature acts as an insulator and shield soil fauna and flora from extreme atmospheric temperature (Yadav et al. 2020). The flow of water through the soil matrix is influenced by soil temperature. The content of soil water has an impact on the temperature of the soil as higher water content contributes to increased solar energy absorption (Shaetzl and Anderson 2005).

18.2.2 Chemical Factors

18.2.2.1 Soil pH

Soil pH demonstrates soil response. Based on it, soil acidity and alkalinity are described. In terms of nutrient supply, frequency, and behavior of soil microbes and plant nutrition, soil pH plays a very important role. Soil functions are significantly affected by acidification and salinization. Soil pH governs soil biogeochemical processes and has surging impacts on the composition and functions of the terrestrial environment (Slessarev et al. 2016). Soil pH influences nutrients retention and thereby provides a major contribution to regulating the productivity of terrestrial environments (Hong et al. 2018).

18.2.2.1.1 Soil Acidification

Soil acidity is an important aspect of soil quality as it influences various properties of soil that gives it a desired function (Schjonning et al. 2004). The level of soil acidification is influenced by the parent rock and the conditions under which the weathering process has undergone (Wilson 2019). Acidity also impacts the soil nutrients availability, the best pH range for maximum nutrient availability is 5–6, i.e., slightly acidic. However, the availability of most nutrients decreases drastically when pH falls below 5. Increased soil acidity also affects the ionic levels of aluminum and manganese to toxic levels at times. The soil of the tropical tropics is naturally acidic. Agricultural practices such as organic matter addition and chemical fertilizer acid production attenuate soil acidity. Soil micro and macroflora are stressed in acidic conditions. It is also observed that fine-textured soils coupled with higher quantities of soil organic matter have more hydrogen ions than coarse textures ones with low organic content (Kumar and Karthika 2020).

18.2.2.1.2 Soil Salinity and Sodicity

Salinity degrades soil quality and is one of the many ways of soil depletion in the semi-arid and arid regions of the world, which is primarily a concern. Agricultural intensification has accelerated soil salinity, coupled with adverse natural factors. As a consequence of normal and anthropogenic processes, salinity grows. The consequence of weathering or oceanic salt accumulation may be natural salinity. Random irrigation creates an imbalance between soil water and transpiration rates (Kumar 2013). Soil salinity is increased by deposition of soluble salts of calcium, sodium, and magnesium to harmful levels. Soil salinity is measured with Exchangeable Sodium Percentage (ESP) calculated in Electrical Conductivity (ECE) terminology. The system is also linked with water drainage and is rampant in arid, sub-humid, and semi-arid climates. Soil salinity contributes to a reduction in water available to plants and thus water stress. Salinity influences the physical characteristics of the soil through flocculation, caused by the joining of fine particles to form aggregates that have a beneficial effect on soil aeration, root infiltration, and root formation. Soil desalinization by aggregation and flocculation can increase soil structure; however, it deteriorates plant growth and yield. Moreover, it posed a serious threat of soil erosion and negatively affects impacts agricultural production. High salinity increases osmotic stress which results in poor plant growth, dietary disorders and low yield. Sodicity is another important chemical decay that occurs in soils. Further, decreased microbiological activities and high soil pH makes land unsuitable for agriculture (Kanwar and Bhumbla 1969). Sodic soil promotes surface clay dispersion, obstructs soil pores, and creates calcareous hardpan in sub-surface resulting in degradation of soil structure (Dougherty and Anderson 2001). It also decreases the rate and permeability of water penetration in the soil profile.

18.2.2.2 Nutrients

The state of soil nutrients and their changes is very important to the quality of the soil. Nutrient shortages adversely impact soil quality and decrease the production of crops. Soil pH influences the supply of nutrients and changes in the absorption of soil and nutrients by plants. The supply of critical nutrients for plant growth is impaired by high soil pH (Singh 2009) and nutrient toxicities are correlated with sodium production. The supply of phosphorous, boron, and manganese reduce when the soil pH is above 6. The supply of potassium, magnesium, copper, and zinc is expected to decline when the pH increases above 7. However, under alkaline conditions, the supply of molybdenum increases. Cations such as calcium, magnesium, potassium, and sodium are closely related to pH. Iron, manganese, and aluminum are strongly soluble at lower pH and reach toxic levels, whereas these ions become deficient at higher pH. In general cation availability (micronutrients), however, increase with soil acidity and decrease with the addition of anionic micronutrients, namely molybdenum and boron. Bacteria and actinomycetes thrive well in acidic and neutral soil environments while fungi can thrive in a wide range of pH conditions (McCauley et al. 2009).

18.2.2.2.1 Plant Nutrient Depletion

Production of crops also takes away soil nutrients. High-input farming of large inputs alone also contributes to secondary and micronutrient shortages. To identify such failures or excesses, routine monitoring is important. The two main nutrients that are particularly influential in the soil environment are nitrogen and phosphorus when it comes to soil conservation and environmental services, such as agricultural development. Soil microbes are also shown to be limited to nitrogen, but the main limiting factor for their operation is carbon availability (Schimel et al. 2005). Functional potential of soil is impaired by the availability of nitrogen. When, with minor leaching and pollution losses, the nitrogen inputs are smaller, demand of nitrogen rises in an undisturbed natural soil environment. Depletion of soil nitrogen is caused by decomposition of organic matter, mineralization of inorganic nitrogen, and consumption by land flora. One of the early markers of degrading soil health is the disparity between carbon and nitrogen biomass needs to sustain soil functions as well as plant growth. Therefore, in agricultural lands it is important to replenish nitrogen artificially in nitrogen deficient soils by the use of fertilizers. Similar criterion is applied for phosphorous and potassium. External inclusion of organic manures or mineral fertilizers is important for restoring and sustaining soil quality to fix nutrient losses. Soil health could be sustained, and production could be attained in effectively managed agricultural systems if adequate management methods for this are introduced and applied successfully. Further, under excess nutrients condition the soil environment appears unhealthy and contaminated (Kibblewhite et al. 2008).

18.3 Biotic Factors

18.3.1 Soil Organic Matter (SOM)

SOM consists of living organisms both macro and micro, secretions, and output of living flora and fauna, dead and decaying remains as a result of microbial decomposition from detritus to humus. SOM is important for soil functions and fertility. It provides the potential for soil binding and buffering, thereby limiting water contamination. SOM is the primary food source for soil microbes and therefore, is vital to soil health conservation. It supports the soil as a nutrient dump. Nitrogen, phosphorous, sulfur, and SOM strengthen the biophysical and biochemical properties of the soil. It also enhances porosity and water-holding capacity. Carbon present in SOM links the chain of carbon cycle by transforming atmospheric CO_2 and storing as SOC pool. It has been observed that soil has smaller bulk densities and a higher organic matter volume. Fungi and bacteria, the main decomposition agents, are a source of food for many microbivorous eaters. Studies have shown that the microorganisms contribute to the release of nutrients and promote the population of microbes, thus regulating the rate of SOM decomposition (Coleman and Hendrix 2000). In the decomposition process, earthworms, termites, and so on are also involved. The mechanism of decomposition is determined by many factors, such as temperature and soil conditions. SOM is found in its active and passive states, where the active state participates in nutrient cycling and the passive state contributes to soil structural characteristics. Many microorganisms that conduct decomposition processes often lead to soil structure modifications and nutrient cycles, for example, fungi spp., bacterial spp., earthworms, termites, etc. Deforestation for agricultural lands is leading to a drastic decline in SOM (Kibblewhite et al. 2008).

18.3.2 Soil Biota

Soils are inhabited by large varieties of living organisms. Soil bacteria, fungi, protozoa, and small organisms play an indispensable role in maintaining soil fertility. SOM degraded by microbes are broken up by higher organisms. Soil organisms are an important part of the biome food chain. Soil biota, especially multicellular soil fauna like worms and termites contributes to maintenance of soil porosity (Dexter 1978). Early indication of a healthy soil is fresh earthy smell coupled with residues of SOM at various stages of decomposition especially with the presence of earthworms. In the soil macropores, soil biota and plant roots thrive gracefully. Nutrient transformers such as decomposers by decomposers such as fungi, microbes, microbivores, detritivores, and nutrient cycling conduct biological processes that lead to carbon transition, functional assemblages of interacting species that are often referred to as "primary functional classes" (Lavelle 1997; Swift et al. 2004). Bio-controllers such as hyperparasites, rodents, and others regulate the biological population and soil organic cycle. Soil biological processes are determined by ecological services, which are inevitable in the concept of soil health together with biotic interactions (Young and Ritz 2005). The primary role of variable soil dynamics is providing opportunities for colonization and enhancing pore networks. Well-developed pore networks facilitate movement of soil biota. The process of pore formation consists of particle adhesion, covering, enmeshment, and orientation and locomotion (Lavelle et al. 1997; Ritz and Young 2004). The structural stability of these operations along with decomposition of SOM by soil biota reflects a two-dimensional collaborative interaction between the environment and biotic population and this can be called as "soil as a self-organizing mechanism" (Young and Crawford 2004). Therefore, potential for self-organization can be one of the markers for soil well-being.

18.3.2.1 Soil Fauna

Soil is the biggest home for terrestrial fauna which contains enchytraeids, nematodes, earthworms, chilopoda, symphyla, and pauropoda. The arthropods species present in soil are primarily spiders, ants, diplopods, and pseudo-scorpion. Coleman and Hendrix (2000) categorized the soil fauna as megafauna (above 20 mm), macrofauna (2 and 20 mm), mesofauna (200 μm and 2 mm), and microfauna (20 and 200 μm) based on their body size.

18.3.2.1.1 Earthworms

Earthworms are regarded as ecosystem engineers (Brussaard et al. 2012); because of their digging behaviors, soil digestion, and cast formation, they have significant effects on soil structure. Earthworms eat soil/SOM and the digestive enzymes in its gut form casts which are stored in the burrows as slurry. The cast formation is further stabilized by particle organic bonding of polymers by bacterial secretions and mechanically by plant fibers, fungi, wet and dry process, and age hardening (Coleman et al. 2012). These burrow linings and earthworm cast provide site for SOM mineralization and are referred as drilosphere (Brussaard et al. 2012). The biota of soils enhances soil health and productivity. In most habitats, earthworms are considered as a chief constituent of soil macrofauna and form a significant share of biomass. Their involvement is advantageous because they accelerate absorption of detritus into mineral soils, it can increase soil nutrient cycling. The formation of mucus accompanying water secretion in earthworm guts increases the activities of other useful soil microorganisms. This is followed by organic matter production. In the short term, the accumulation of large amounts of nutrients that are readily assimilated by plants in fresh cast deposits have a more important impact. Moreover, earthworms tend to enhance mineralization together with the increase in SOM (Bhadauria and Saxena 2010). It is observed that the cast produced by earthworms contains many plant growth-promoting chemicals. Earthworm castings promote soil alteration, which boost plant growth (Gilbert 2010). Earthworms burrowing also help to increase soil micropores numbers (Calderon-Cortes et al. 2012).

18.3.2.1.2 Formicidae (Ants)

Ants are significant arthropods that influence soil structure. They are also an active predator which helps to sustain the food chain and soil carbon cycle. Predatory competition is observed between ants and other similar predators like spiders and carabid beetles (Moore and DeRuiter 2012). The colony forming habit of ants make them ecosystem builders like earthworms due to their excessive digging and moving vast quantities of soil (Garcıa-Palacios et al. 2013). The soil structure is also affected by ants; this mechanism is especially significant in deserts because earthworm densities in deserts are poor (Scharroba et al. 2012). Ants play an important role in shaping the physical characteristics of the soil, such as soil structure and porosity, through nest-building and maintenance, organic matter accumulation, and interactions with soil fauna. These activities have localized influences on the hydrology of an area and can have larger influences on ecosystem hydrology depending on nest density. Increased porosity also increases water retention, healthier growth of plant roots, and improves primary productivity. They alter microclimates inside and around nests, altering other organisms' environment, including myrmecophiles living inside the nests. Activities of ants and ants could thus lead to more productive soil (Del toro et al. 2012).

18.3.2.1.3 Termitidae (Termites)

Besides ants, termites are also known as social insects and have an organized caste system (Bignell and Eggleton 2000). Main food source of termites is damp wood which mixes with the soil. Like ants' termites are also mount builders and natural earth movers. Their feeding habits help to convert cellulose into simpler forms (Bignell 2000). Many cellulose breaking bacteria's also resides in secretions of termites which further break down the degraded cellulose in adsorbable form thus making it available for plants. Termites are considered to have a substantial impact on the agroecosystems. As they can sustain, transform, and promote soil fertility, they are referred to as ecosystem engineers. By excavating and breaking down organic materials while building their mounds, termites make a substantial contribution to preserving the chemical and physical conditions of the soil. Soil fertility relies on the accessibility of stable nutrients in a manner that can be utilized by plants. In low-input cropping systems, the use of termite mound soil has been proposed as biofertilizers and inoculants because it is rich in nutrients and bacteria that encourage plant growth. It is important to notice that, compared to the adjacent soils, termite mound soils contain higher phosphorus levels. This is because of the phosphate-solubilizing bacteria found in termite mound soils that are highly productive. Termite mound soil provides beneficial bacteria proficient in decomposing lignin and cellulose, fixing nitrogen, phosphate solubilization, and suppressing pathogens from plant-soil (Enagbonma and Babalola 2019). These placed them in a position to act as biofertilizers and biocontrol agents.

18.3.2.1.4 Nematodes

Nematoda is a phylum of invertebrate roundworm species that appeared over 500 million years ago during the Precambrian or Cambrian explosion (Sudhaus 2008). Nematodes play important role in the soil system. They play significant roles in soil ecology and engage in nutrient cycling affecting crop plants. They also engage in diverse food webs with other soil species to provide critical functions and ecosystem services that include soil structure management, soil detoxification, and pest and disease biocontrol. Thanks to their central position in the soil food web and associated to ecological and soil processes, nematodes can be a method for evaluating ecological theories and understanding biological pathways in soil. In controlled systems, certain beneficial nematodes act as biological pest control agents, and others control the natural environment and the cycling of soil nutrients (Mekonen et al. 2017).

18.3.2.2 Soil Flora

Soil flora is also an important part of soil ecosystems and agricultural functions. It majorly helps in nutrients recycling by absorbing all the important nutrients and returning complex organic derivatives after decomposition. Because of the high N2 fixation capacity, nutrient cycling, and contribution to soil cover, leguminous plants used as green manure are a source of organic material with significant advantages for soil and crops (Cobo et al. 2002). The annual carbon fixation rate could be improved by 20–75 g C m^{-2} by sustainably modifying the land-use pattern by adding legumes and tree. Legumes are vital to soil ecology, sustainability, and nutritional security as it is also a valuable source of human food protein (Kumar et al. 2018a, b). From its emergence as a scientifically recognized science and practice, agroforestry's ability to optimize soil quality has been generally recognized as a significant advantage. In the tropics and temperate regions of the world, agroforestry activities have been advocated for decades for their hypothesized advantages by enhancing soil quality and delivering ecological services. Agroforestry raises the amount of organic carbon in the soil, soil biota, and soil nutrients (Dollinger and Jose 2018).

18.3.2.2.1 Soil Microbes

The microbes help to maintain/improve soil health by altering: (1) intercellular plant–microbe interactions, (2) intracellular metabolism, and (3) microbial species composition by crop rotation thus improving soil microbiome (Brussaard et al. 2007). Effective use of microorganisms helps to preserve soil quality, boost water retention, carbon storage, and root growth (Dubey et al. 2019). For our agricultural productivity, the conservation and growth of soil health and fertility are significant (Joshi et al. 2019; Rawat et al. 2019). In most of the world's important agricultural plants, soil-borne beneficial microbes have been extensively researched and introduced, but inadequate incorporation of these beneficial soil microbes into agriculture has limited the production of successful practices for disease control. Beneficial microbes provide new possibilities in regulating pathogen population (Bonanomi et al. 2018). Soil microorganisms are main component of SOM, plants are provided with nutrients and a multitude of soil microbes offer protection against outbreaks of pests and diseases. Popular agricultural practices can, however, deplete microbes in soil and general soil health, but this can be avoided by different methods of enhancing soil quality. Certain soil microbes are essential for plant nutrient uptake by providing two major macronutrients nitrogen and phosphorus, respectively (Suyal et al. 2014; Tomer et al. 2017; Rajwar et al. 2018). The microbiota also leads to the degradation of recalcitrant organic matter and the weathering of minerals that are supplied with carbon in root exudates and other rhizosphere deposits in exchange (Van Der Heijden et al. 2008; Kumar et al. 2018a, b). In a model grassland, several researchers researched AM fungi and PGPRs and found that both symbionts complement each other and resulted in increased plant diversity, enhanced seedling recruitment, and improved nutrient acquisition. Root microbiome with diversified symbionts complement each other in obtaining mutually important nutrients and driving ecosystem functions (Dubey et al. 2019).

18.3.2.2.2 Bacteria

Rhizosphere soil has millions of microorganisms that can impact plant growth in a variety of ways. Soil bacteria are one of the most essential component of the biome which dwells as free-living/symbiotic or as pathogens. Beneficial endophytes or free-living bacteria are associated with plant growth stimulation through direct or indirect functions are termed as plant growth promoting bacteria (PGPB) (Shameer and Prasad 2018). PGPB and plants share a complex relationship but their interaction with host plants is generally considered to be the product of root exudates constituting of organic compounds that serve as nutrients to the microbes (Kumar et al. 2019; Suyal et al. 2019). Colonized microbes can influence the status of plants in terms of biology, growth, nutrition, and health (Mantelin and Touraine 2004).

PGPB promotes plant growth through numerous mechanisms ranging from reclamation of soil, producing substances that encourage growth, suppressing harmful pathogens, solubilization of phosphorus and nitrogen, and mobilization of nutrients (Lugtenberg and Kamilova 2009). It has been widely documented that these beneficial microorganisms promote the fixation of nitrogen and enhance the process of nitrate uptake by roots, enable solubilization of phosphorus and its bioavailability in the rhizosphere along with mobilization and uptake of nutrients. They also contribute to the remediation of heavy metal in polluted soils, multi-activity soil improvement, the release of hormones, vitamins, volatile compounds, and stress management in case of soil salinity and droughts. They act as biocontrol agents against various pathogenic diseases (Majeed et al. 2018). As a result, PGPB-mediated soil bioavailability of nutrients contributes to improved plant growth, yield.

18.3.2.2.3 Fungi

Fungi have high plasticity and greater adaptability to adverse living conditions making them an amazingly effective inhabitants of soil biome (Sun et al. 2005). Due to their capacity to generate a wide range of enzymes extracellularly, they can break down all sorts of organic matter and thus maintaining the balance of plant nutrients (Zifcakova et al. 2016). Fungi transform dead organic content into simpler organic compounds, CO_2, and easily digestible biomass. Many species of fungi are also important bio-sorbents for metals considered toxic to soil flora and fauna thus helping in bioremediation (Baldrian 2003). Various biotic factors (plants and other organisms) and abiotic factors (soil pH, humidity, salinity, structure, and temperature) regulate the diversity and activity of fungi (López-Bucio et al. 2015; Rouphael et al. 2015). Fungi are cosmopolitan and can survive in a broad pH and temperature range (Frąc et al. 2015).

Soil fungi based on their activities can be classified as (a) biological controllers, (b) ecosystem regulators, and (c) organic matter decomposers and compound transformers (Swift 2005). Through the regulation of physiological processes in the soil environment, ecosystem regulators are responsible for the formation of stable soil structures soil structure formation and habitat alteration for other species. Diseases and the growth of other species can be regulated by biological controllers' Mycorrhizal fungi, for instance, promote plant growth by increasing nutrient absorption and protecting them against pathogens (Bagyaraj and Ashwin 2017). They also play a vital role in soil organic matter stabilization and residual decomposition (Treseder and Lennon 2015).

18.3.2.2.4 Algae

Algae are a very large and complex group of simple species usually autotrophic, capable of carrying out photosynthesis and capturing energy from sunlight. In sustainable and organic agriculture, where algae are used as biofertilizers and soil stabilizers, they play an important role. Algae, particularly seaweeds, are used as fertilizers, resulting in less drainage of nitrogen and phosphorous as compared to the livestock manure. In essence, this enhances the water quality that flows through rivers and oceans (Sharma et al. 2012). Sustainable agriculture is favorable over traditional agriculture because of its potential to satisfy the need for food by utilizing natural resources without adversely impacting them. The beneficial function of blue-green algae (BGA) is well known for promoting the nitrogen supply of paddy fields and improving the production of rice. The modest presence of BGA in the soil helps in the construction of soil aggregates, minimizing soil loss during the rainy season, and controlling the aeration and temperature of the soil, thus improving the physical and chemical properties of the soil concerning the physical environment of the yield. By secreting exopolysaccharides and bioactive compounds, BGA assists in recovering soil nutrients. They can mobilize the forms of inorganic phosphates that are insoluble. Algalization has been utilized for reducing the volume of exchangeable sodium, which results in altered soil pH and contributes to the reclamation of sodic soils. It has been found that some red algae are used as biofertilizers (Chatterjee et al. 2017).

18.3.3 Plant–Microbe Interactions: Key Player of Agricultural Sustainability

A balanced soil functions as a complex living environment that provides different services to the environment such as preserving water quality and plant fertility, regulating decomposition to regulate various soil cycles, and reducing the release of greenhouse gases in the atmosphere. Organic agriculture plays a vital role in determining soil quality as it provides a site for soil microbes to thrive and diversify which in turns affect the land productivity. Agricultural sustainability is broadly classified as the continuous ability of the soil to produce crops without losing its fertility. Arbuscular mycorrhizal fungi (AMF), nematodes, cyanobacteria, and other beneficial microbes increase the productivity of water usage and the supply of nutrients to plants, development of phytohormones, soil nutrients cycling, and plant tolerance to ecological stress. It has being observed that traditional framing practices, tillage, organic and terrace farming improve the health of the soil by promoting microorganisms richness, variety, and actions (Tahat et al. 2020). In sustainable agriculture, the importance of plant–microbe interactions is immense. Some pathogenic interactions cause plant diseases while the beneficial interaction between plants soil microbiota encourage plant growth, impart resistance to various kinds of environmental stress and assist plants to revitalize polluted and depleted loams (Abhilash et al. 2012). Moreover, the helpful microbes also play a vital role in regulating the energy distribution amid plant root and shoot, ecology, and intermediate the relationships and interaction between fauna living above and below the soil. Henceforth, the biological, physical, and biochemical properties of the soil can be changed by such dialogues between plants and microbes. Plant and microbial secretion comprise of simple carbohydrates, organic acids, polymerized sugars (e.g. mucilage), root boundary cells, amino acids and root exudates. The plant secretes phytosiderophores that assist in the removal from the soil of metallic micronutrients. Various secondary metabolites are secreted by plant roots and they play a vital role by attracting soil microbes and establishing a healthy plant-microbial interaction in the rhizospheric area. The rhizodeposits allow the plant-microbial interaction in various regions like rhizosphere, phyllosphere, adjacent soils, and endosphere and this help the plant to host a wide range of microbial communities and these microbes play a fundamental role in plant growth and development. To unravel the commitment of every player to the well-being of the ecosystem, in-depth knowledge of various interlinked activities is important. Therefore, in evaluating the importance of plant-related microbes to sustainable food production, habitat regeneration, biomass development, and curbing the hostile effects of climate change, it is crucial to consider the main mechanisms of plant–microbe interactions (Saleem and Moe 2014). Effective knowledge of rhizospheric interactions is helpful to increase soil productivity and in maintaining soil carbon along with checking trace gases release from agro-systems. Most notably, such skills can help to strengthen the relationships between plants and microbes, conferring resistance to pathogens, bioremediation, and regeneration of degraded lands. Recent experiments, however, have shown that evolving climatic factors will alter the biology of the rhizosphere by altering the rate of root exudation, the availability of resources, and biogeochemical cycling (Dubey et al. 2016). In brief, the main measure of sustainable land management is the measurement of soil quality or health and the course of progress over time.

18.4 Recent Trends and Future Perspectives

Many people consider soil quality and health as same but usually all the scientific studies are being done considering the soil quality, whereas the farmers are preferring the soil health. But in recent years, the term soil health has been increasingly adopted by scientists due to the inferred presence of biological activities. Using several techniques, including several mathematical approaches, fuzzy methods, biodiversity analysis, and spatial analysis, attempts are made to find standard measures defining soil quality/health. However, there is still a significant need to obtain additional soil details. Precise data related to soil and properties is required as a supplement to current soil mapping, classification, and analysis datasets. In the last decade, a variety of analytical methods have been developed, e.g., Soilflex and SOCOMO and set models such as Prager and Cap model, two- and three-dimensional finite models are developed with the creation of numerical computers in order to check soil compaction and pneumatic *tire–soil interaction* (Muñoz-Rojas et al. 2017). By raising the awareness of the role of soil in contemporary society, the need of implementation and importance of soil health is being appreciated by stakeholders, small and large scale farmers, land supervisors, municipal cooperation, and policymakers. The notion's simplicity helps multiple investors to support soil health and make its utility within their context. Rather than a property to measure, researchers must accept soil quality as a predominant standard for better understanding and awareness (Lehmann et al. 2020). By doing so, soil heath will gain more scientific importance and this will also encourage involvement of other disciplines.

A better understanding between scientist and stakeholders is needed in order to unify all concepts related to soil health to ensure better productivity along with sustainability.

18.5 Conclusion

One of the major challenges that is being faced is to satisfy the food needs of the ever increasing population in a sustainable way, i.e. to preserve soil quality levels and avoid soil depletion. Extensive information on soil erosion is already present in the agricultural sphere. As a result of a kind of agriculture whose priority was just high crop productivity, there are several thousands of km^2 of polluted soils all over the world. The intensive usage of synthetic fertilizers, systematic degradation, soil erosion and the impact of rainfall or storms, the depletion of organic matter, and many other factors have resulted in growing desertification, the loss of millions of tons of productive surface soil and, consequently, the silting up of rivers and dams, the salinization of soils, climate change, and the loss of biodiversity. Decline in soil fertility due to the loss of organic matter, sudden increased erosion, and increase in soil salinity, soil compaction and reduced biodiversity are the major threats affecting soil health directly. Delayed implementation of sustainable approaches and strategies may make the situation worse by increasing the rate of land degradation due to climate change. Current climate change patterns would influence the atmosphere and infrastructure. New transformation and moderation strategies for these effects should be immediately executed. This would incorporate suitable control, assessment, and understanding of expected situations at adequate fleeting and spatial scales. Further, cycle based and logical models give separate, however, integral strategies, every one of which would be utilized to help the assortment of field information and produce gauges under conceivable worldwide change situations.

References

Abhilash PC, Powell JR, Singh HB, Singh BK (2012) Plant–microbe interactions: novel applications for exploitation in multipurpose remediation technologies. Trends Biotechnol 30:416–420. https://doi.org/10.1016/j.tibtech.2012.04.004

Bagyaraj DJ, Ashwin R (2017) Soil biodiversity: role in sustainable horticulture. Biodiv Hortic Crops 5:1–18

Baldrian P (2003) Interactions of heavy metals with white-rot fungi. Enzym Microb Technol 32 (1):78–91. https://doi.org/10.1016/S0141-0229(02)00245-4

Bhadauria T, Saxena KG (2010) Role of earthworms in soil fertility maintenance through the production of biogenic structures. Appl Environ Soil Sci. https://doi.org/10.1155/2010/816073

Bhat SA, Dar MUD, Meena RS (2019) Soil erosion and management strategies. In: Meena R, Kumar S, Bohra J, Jat M (eds) Sustainable management of soil and environment. Springer, Singapore, pp 73–122

Bignell DE (2000) Introduction to symbiosis. In: Abe T, Bignell DE, Higashi M (eds) Termites: evolution, sociality, symbioses, ecology. Kluwer Academic, Dordrecht, pp 189–208

Bignell DE, Eggleton P (2000) Termites in ecosystems. In: Abe T, Bignell DE, Higashi M (eds) Termites: evolution, sociality, symbioses, ecology. Kluwer Academic, Dordrecht, pp 363–387

Bonanomi G, Lorito M, Vinale F, Woo SL (2018) Organic amendments, beneficial microbes, and soil microbiota: toward a unified framework for disease suppression. Annu Rev Phytopathol 56:1–20. https://doi.org/10.1146/annurev-phyto-080615-100046

Bot A, Benites J (2005) The importance of soil organic matter: key to drought-resistant soil and sustained food production (No. 80). Food & Agriculture Org

Brussaard L, de Ruiter PC, Brown GG (2007) Soil biodiversity for agricultural sustainability. Agric Ecosyst Environ. https://doi.org/10.1016/j.agee.2006.12.013

Brussaard L, Aanen DK, Briones MJ, Decaens T, De Deyn GB, Fayle TM, James SW, Nobre T (2012) Biogeography and phylogenetic community structure of soil invertebrate ecosystem engineers: global to local patterns, implications for ecosystem functioning and services and global environmental change impacts. In: Wall DH, Bardgett RD, Behan-Pelletier V, Herrick JE, Jones H, Ritz K, Six J, Strong DR, van der Putten WH (eds) Soil ecology and ecosystem services. Oxford University Press, Oxford, pp 201–232

Calderon-Cortes N, Quesada M, Watanabe H, Cano-Camacho H, Oyama K (2012) Endogenous plant cell wall digestion: a key mechanism in insect evolution. Annu Rev Ecol Evol Syst 43:45–71. https://doi.org/10.1146/annurev-ecolsys-110411-160312

Chatterjee A, Singh S, Agrawal C, Yadav S, Rai R, Rai LC (2017) Role of algae as a biofertilizer. In: Algal green chemistry. Elsevier, Amsterdam, pp 189–200

Chen X, Cui Z, Fan M, Vitousek P, Zhao M, Ma W (2014) Producing more grain with lower environmental costs. Nature 514:486–489. https://doi.org/10.1038/nature13609

Cobo JG, Barrios E, Kass DCL, Thomas RJ (2002) Decomposition and nutrient release by green manure in a tropical hillside agro-ecosystem. Plant Soil 240:331–342. https://doi.org/10.1023/A:1015720324392

Coleman DC, Hendrix PF (eds) (2000) Invertebrates as webmasters in ecosystems. CABI, Wallingford

Coleman DC, Gupta VV, Moore JC (2012) Soil ecology and agroecosystem studies. In: Microbial ecology in sustainable agroecosystems. Springer, New York, pp 1–21

Cornell University Cooperative Extension (CUCE) (2007) Cation Exchange Capacity (CEC). Agronomy Fact Sheet Series # 22. Department of Crop and Soil Sciences, College of Agriculture and Life Sciences, Cornell University

Del Toro I, Ribbons RR, Pelini SL (2012) The little things that run the world revisited: a review of ant-mediated ecosystem services and disservices (Hymenoptera: Formicidae). Myrmecol News 17:133–146

Dexter AR (1978) Tunnelling of soil by earthworms. Soil Biol Biochem 10:447–449. https://doi.org/10.1016/0038-0717(78)90075-5

Dollinger J, Jose S (2018) Agroforestry for soil health. Agrofor Syst 92(2):213–219. https://doi.org/10.1007/s10457-018-0223-9

Dougherty W, Anderson A (2001) Sodic soils: their properties and management. In: Cattle SR, George BH (eds) Describing, analysing and managing our soil. The University of Sydney and the Australian Soil Science Society, Sydney, pp 105–121

Dubey RK, Tripathi V, Dubey PK, Singh HB, Abhilash PC (2016) Exploring rhizospheric interactions for agricultural sustainability: the need of integrative research on multi-trophic interactions. J Clean Prod 115:362–365. https://doi.org/10.1016/j.jclepro.2015.12.077

Dubey A, Malla MA, Khan F, Chowdhary K, Yadav S, Kumar A, Khan ML (2019) Soil microbiome: a key player for conservation of soil health under changing climate. Biodivers Conserv 28(8-9):2405–2429. https://doi.org/10.1007/s10531-019-01760-5

Enagbonma BJ, Babalola OO (2019) Potentials of termite mound soil bacteria in ecosystem engineering for sustainable agriculture. Ann Microbiol 69(3):211–219. https://doi.org/10.1007/s13213-019-1439-2

Frąc M, Jezierska-Tys S, Yaguchi T (2015) Occurrence, detection, and molecular and metabolic characterization of heat-resistant fungi in soils and plants and their risk to human health. In: Sparks DL (ed) Advances in agronomy, vol 132. Academic, New York, pp 161–204

Garcıa-Palacios P, Maestre FT, Kattge J, Wall DH (2013) Climate and litter quality differently modulate the effects of soil fauna on litter decomposition across biomes. Ecol Lett 16:1045–1053. https://doi.org/10.1111/ele.12137

Gilbert HJ (2010) The biochemistry and structural biology of plant cell wall decomposition. Plant Physiol 153:444–455. https://doi.org/10.1104/pp.110.156646

Grant KN, Macrae ML, Ali GA (2019) Differences in preferential flow with antecedent moisture conditions and soil texture: Implications for subsurface P transport. Hydrol Process 33 (15):2068–2079. https://doi.org/10.1002/hyp.13454

Hallam J, Hodson ME (2020) Impact of different earthworm ecotypes on water stable aggregates and soil water holding capacity. Biol Fertil Soils 56:607–617. https://doi.org/10.1007/s00374-020-01432-5

Hillel D (2012) Soil and water: physical principles and processes. Elsevier, Amsterdam

Hong S, Piao S, Chen A, Liu Y, Liu L, Peng S, Zeng H (2018) Afforestation neutralizes soil pH. Nat Commun 9(1):520. https://doi.org/10.1038/s41467-018-02970-1

Jhonson C (2009) Biology of soil science. Oxford Book Company, Jaipur, p 298

Joshi D, Chandra R, Suyal DC, Kumar S, Goel R (2019) Impact of bioinoculants Pseudomonas jesenii MP1 and Rhodococcus qingshengii S10107 on Cicer arietinum yield and soil nitrogen status. Pedosphere 29(3):388–399

Kanwar JS, Bhumbla DR (1969) Physico chemical characteristics of sodic soils of Punjab and Haryana and their amelioration by the use of gypsum. Agrokem Talajatan 18:315–332

Karlen (2005) Productivity. In: Encyclopedia of soils in the environment. Springer, New York, pp 330–336. https://doi.org/10.1016/B0-12-348530-4/00241-1

Karlen DL, Wollenhaupt NC, Erbach DC, Berry EC, Swan JB, Eash NS, Jordahl JL (1994) Crop residue effects on soil quality following 10-years of no-till corn. Soil Tillage Res 31:149–167. https://doi.org/10.1016/0167-1987(94)90077-9

Kibblewhite MG, Ritz K, Swift MJ (2008) Soil health in agricultural systems. Philos Trans R Soc B Biol Sci 363:685–701. https://doi.org/10.1098/rstb.2007.2178

Kumar M (2013) Crop plants and abiotic stresses. J Biomol Res Ther 3:1. https://doi.org/10.4172/2167-7956.1000e125

Kumar KA, Karthika K (2020) Abiotic and biotic factors influencing soil health and/or soil degradation. In: Soil health. Springer, Cham, pp 145–161

Kumar S, Meena RS, Lal R, Yadav GS, Mitran T, Meena BL, Ayman ES (2018a) Role of legumes in soil carbon sequestration. In: Meena R, Das A, Yadav G, Lal R (eds) Legumes for soil health and sustainable management. Springer, Singapore, pp 109–138

Kumar S, Suyal DC, Bhoriyal M, Goel R (2018b) Plant growth promoting potential of psychrotolerant Dyadobacter sp. for pulses and finger millet and impact of inoculation on soil chemical properties and diazotrophic abundance. J Plant Nutr 41(8):1035–1046

Kumar S, Suyal DC, Yadav A, Shouche Y, Goel R (2019) Microbial diversity and soil physiochemical characteristic of higher altitude. PLoS One 14(3):e0213844

Lavelle P (1997) Faunal activities and soil processes: adaptive strategies that determine ecosystem function. Adv Ecol Res 27:93–132. https://doi.org/10.1016/S0065-2504(08)60007-0

Lavelle P, Bignell D, Lepage M, Wolters V, Roger P, Ineson P, Heal OW, Dhillion S (1997) Soil function in a changing world: the role of invertebrate ecosystem engineers. Eur J Soil Biol 33:159–193

Lehmann J, Bossio DA, Kögel-Knabner I, Rillig MC (2020) The concept and future prospects of soil health. Nat Rev Earth Environ 1(10):544–553. https://doi.org/10.1038/s43017-020-0080-8

Li J, Han Z, Zhong S, Gao P, Wei C (2020) Pore size distribution and pore functional characteristics of soils as affected by rock fragments in the hilly regions of the Sichuan Basin, China. Can J Soil Sci 2020:1–10. https://doi.org/10.1139/cjss-2019-0167

López-Bucio J, Pelagio-Flores R, Herrera-Estrella A (2015) Trichoderma as biostimulant: exploiting the multilevel properties of a plant beneficial fungus. Sci Hortic 196:109–123. https://doi.org/10.1016/j.scienta.2015.08.043

Lugtenberg B, Kamilova F (2009) Plant-growth-promoting rhizobacteria. Annu Rev Microbiol 63:541–556. https://doi.org/10.1146/annurev.micro.62.081307.162918

Majeed A, Muhammad Z, Ahmad H (2018) Plant growth promoting bacteria: role in soil improvement, abiotic and biotic stress management of crops. Plant Cell Rep 37(12):1599–1609. https://doi.org/10.1007/s00299-018-2341-2

Mantelin S, Touraine B (2004) Plant growth-promoting bacteria and nitrate availability: impacts on root development and nitrate uptake. J Exp Bot 55(394):27–34. https://doi.org/10.1093/jxb/erh010

McCauley A, Jones C, Jacobsen J (2009) Soil pH and organic matter. Nutr Manag Mod 8(2):1–12

Mekonen S, Petros I, Hailemariam M (2017) The role of nematodes in the processes of soil ecology and their use as bioindicators. Agric Biol J N Am 8(4):132–140

Moore JC, DeRuiter PC (2012) Soil food webs in agricultural ecosystems. In: Cheeke TE, Coleman DC, Wall DH (eds) Microbial ecology of sustainable agroecosystems. CRC Press, Boca Raton, pp 63–88

Morgado RG, Loureiro S, González-Alcaraz MN (2018) Changes in soil ecosystem structure and functions due to soil contamination. In: Soil pollution. Academic, New York, pp 59–87

Muñoz-Rojas M, Pereira P, Brevik EC, Cerdà A, Jordán A (2017) Soil mapping and processes models for sustainable land management applied to modern challenges. In: Pereira et al (eds) Soil mapping and process modeling for sustainable land use management. Elsevier, Amsterdam, pp 151–190

Nichols KA, Halvorson JJ (2013) Roles of biology, chemistry, and physics in soil macroaggregate formation and stabilization. Open Agric J 7:107–117

Nyawade S, Karanja N, Gachene C, Parker M, Schulte-Geldermann E (2018) Susceptibility of soil organic matter fractions to soil erosion under potato-legume intercropping systems in central Kenya. J Soil Water Conserv 73(5):567–576. https://doi.org/10.2489/jswc.73.5.567

Prescott CE, Katzensteiner K, Weston C (2020) Soils and restoration of forested landscapes. In: Soils and landscape restoration. Academic, New York, pp 299–331

Rajwar J, Chandra R, Suyal DC, Tomer S, Kumar S, Goel R (2018) Comparative phosphate solubilizing efficiency of psychrotolerant Pseudomonas jesenii MP1 and Acinetobacter sp. ST02 against chickpea for sustainable hill agriculture. Biologia 73(8):793–802

Rawat N, Sharma M, Suyal DC, Singh DK, Joshi D, Singh P, Goel R (2019) Psyhcrotolerant bio-inoculants and their co-inoculation to improve Cicer arietinum growth and soil nutrient status for sustainable mountain agriculture. J Soil Sci Plant Nutr 19(3):639–647

Ritz K, Young IM (2004) Interactions between soil structure and fungi. Mycologist 18:52–59. https://doi.org/10.1017/S0269915X04002010

Rouphael Y, Franken P, Schneider C, Schwarz D, Giovannetti M, Agnolucci M, Colla G (2015) Arbuscular mycorrhizal fungi act as biostimulants in horticultural crops. Sci Hortic 196:91–108. https://doi.org/10.1016/j.scienta.2015.09.002

Saleem M, Moe LA (2014) Multitrophic microbial interactions for eco and agro-biotechnological processes: theory and practice. Trends Biotechnol 32:529–537. https://doi.org/10.1016/j.tibtech.2014.08.002

Santamarina JC, Cho GC (2004) Soil behaviour: The role of particle shape. In: Santamarina JC, Cho GC (eds) Advances in geotechnical engineering: The Skempton conference: proceedings of a three day conference on advances in geotechnical engineering, organised by the Institution of Civil Engineers and held at the Royal Geographical Society. Thomas Telford Publishing, London, pp 604–617

Schaetzl RJ, Thompson ML (2015) Soils. Cambridge University Press, Cambridge

Scharroba A, Dibbern D, Hunninghaus M, Kramer S, Moll J, Butenschoen O, Bonkowski M, Buscot F, Kandeler E, Koller R, Kruge D, Lueders T, Scheu S, Ruess L (2012) Effects of resource availability and quality on the structure of the micro-food web of an arable soil across depth. Soil Biol Biochem 50:1–11. https://doi.org/10.1016/j.soilbio.2012.03.002

Schimel JP, Bennett J, Fierer N (2005) Microbial community composition and soil nitrogen cycling: is there really a connection? In: Bardgett RD, Usher MB, Hopkins DW (eds) Biological diversity and function in soils. Cambridge University Press, Cambridge, pp 172–188

Schjonning P, Elmholt S, Christensen BT (2004) Soil quality management-concept and terms. In: Schjonning P, Elmholt S, Christensen BT (eds) Managing soil quality challenges in modern agriculture. CABI Publishing, Wallingford, p 344

Schlesinger WM, Bernhardt ES (2013) Biogeochemstry. An analysis of global change, 3rd edn. Academic Press, New York

Shaetzl JR, Anderson S (2005) Soils: genesis and geomorphology. Cambridge University Press, New York, p 817

Shah AN, Tanveer M, Shahzad B, Yang G, Fahad S, Ali S, Souliyanonh B (2017) Soil compaction effects on soil health and crop productivity: an overview. Environ Sci Pollut Res 24 (11):10056–10067. https://doi.org/10.1007/s11356-017-8421-y

Shameer S, Prasad T (2018) Plant growth promoting rhizobacteria for sustainable agricultural practices with special reference to biotic and abiotic stresses. Plant Growth Regul 2018:1–13. https://doi.org/10.1007/s10725-017-0365-1

Sharma R, Khokhar MK, Jat RL, Khandelwal SK (2012) Role of algae and cyanobacteria in sustainable agriculture system. Wudpecker J Agric Res 1(9):381–388

Singh G (2009) Salinity related desertification and management strategies: Indian experience. Land Degrad Dev 20:367–385. https://doi.org/10.1002/ldr.933

Slessarev EW, Lin Y, Bingham NL, Johnson JE, Dai Y, Schimel JP, Chadwick OA (2016) Water balance creates a threshold in soil pH at the global scale. Nature 540(7634):567–569. https://doi.org/10.1038/nature20139

Sudhaus W (2008) Evolution of insect parasitism in rhabditid and diplogastrid nematodes. Adv Arach Dev Biol 12:143–161

Sun JM, Irzykowski W, Jedryczka M, Han FX (2005) Analysis of the genetic structure of Sclerotiniasclerotiorum (Lib.) de Bary populations from different regions and host plants by random amplified polymorphic DNA markers. J Integr Plant Biol 47(4):385–395. https://doi.org/10.1111/j.1744-7909.2005.00077.x

Suyal DC, Shukla A, Goel R (2014) Growth promotory potential of the psychrophilic diazotroph Pseudmonas migulae S10724 against Native Vigna radiata (L.) Wilczek. 3 Biotech 4 (6):665–668

Suyal DC, Kumar S, Joshi D, Yadav A, Shouche Y, Goel R (2019) Comparative overview of red kidney bean (Phaseolus valgaris) rhizospheric bacterial diversity in perspective of altitudinal variations. Biologia 74(10):1405–1413

Swift MJ (2005) Human impacts on biodiversity and ecosystem services: an overview. Mycol Ser 23:627

Swift MJ, Izac AMN, Van Noorwidjk MN (2004) Biodiversity and ecosystem services in agricultural landscapes—are we asking the right questions? Agric Ecosyst Environ 104:113–134. https://doi.org/10.1016/j.agee.2004.01.013

Tahat M, Alananbeh K, Othman Y, Leskovar D (2020) Soil health and sustainable agriculture. Sustain For 12(12):4859. https://doi.org/10.3390/su12124859

Tilman D, Cassman KG, Matson PA, Naylor R, Polasky S (2002) Agricultural sustainability and intensive production practices. Nature 418:671–677. https://doi.org/10.1038/nature01014

Tomer S, Suyal DC, Rajwar J, Yadav A, Shouche Y, Goel R (2017) Isolation and characterization of Phosphate solubilizing bacteria from Western Indian Himalayan soils. 3Biotech 7(2):1–8

Treseder KK, Lennon JT (2015) Fungal traits that drive ecosystem dynamics on land. Microbiol Mol Biol Rev 79(2):243–262. https://doi.org/10.1128/MMBR.00001-15

Trivedi P, Delgado-Baquerizo M, Anderson IC, Singh BK (2016) Response of soil properties and microbial communities to agriculture: implications for primary productivity and soil health indicators. Front Plant Sci 7:990. https://doi.org/10.3389/fpls.2016.00990

Usharani KV (2019) Role of soil physical, chemical and biological properties for soil health improvement and sustainable agriculture. J Pharm Phytochem 8(5):1256–1267

Van Der Heijden MGA, Bardgett RD, Van Straalen NM (2008) The unseen majority: soil microbes as drivers of plant diversity and productivity in terrestrial ecosystems. Ecol Lett 11:296–310. https://doi.org/10.1111/j.1461-0248.2007.01139.x

Vitousek PM, Naylor R, Crews T, David MB, Drinkwater LE, Holland E (2009) Nutrient imbalances in agricultural development. Science 324:1519. https://doi.org/10.1126/science.1170261

Wilson MJ (2019) The importance of parent material in soil classification: a review in a historical context. Catena 182:104131. https://doi.org/10.1016/j.catena.2019.104131

Xiao L, Yao K, Li P, Liu Y, Zhang Y (2020) Effects of freeze-thaw cycles and initial soil moisture content on soil aggregate stability in natural grassland and Chinese pine forest on the Loess Plateau of China. J Soils Sediments 20(3):1222–1230. https://doi.org/10.1007/s11368-019-02526-w

Yadav B, Krishnan P, Shafeeq PM, Parihar CM, Aggarwal P (2020) Modelling soil thermal regime in wheat using HYDRUS-2D under diversified maize-wheat-mungbean cropping system. Catena 194:104765. https://doi.org/10.1016/j.catena.2020.104765

Young IM, Crawford JW (2004) Interactions and self-organization in the soil-microbe complex. Science 304:1634–1637. https://doi.org/10.1126/science.1097394

Young IM, Ritz K (2005) The habitat of soil microbes. In: Bardgett RD, Usher MB, Hopkins DW (eds) Biological diversity and function in soils. Cambridge University Press, Cambridge, pp 31–43

Zifcakova L, Vetrovsky T, Howe A, Baldrian P (2016) Microbial activity in forest soil reflects the changes in ecosystem properties between summer and winter. Environ Microbiol 18 (1):288–301. https://doi.org/10.1111/1462-2920.13026

Microbial Genes, Enzymes, and Metabolites: To Improve *Rhizosphere* and Plant Health Management

19

Bahman Fazeli-Nasab and Ahmad Farid Rahmani

Abstract

In today's world, there are other important problems, such as climate change and environmental problems and the loss of genetic resources, in addition to the issue of human societies supplying food and creating more food. One of the reasons being the improper use of chemical products in agriculture (such as pesticides and chemical fertilizers, etc.). In recent decades, the growing population of the world and the growing market for food have led to a serious and imminent change from conventional agriculture to advanced agriculture in the agricultural and food sciences and to the use of modern genetic technologies in the production of crops and livestock. The application of the techniques of genetic evolution and molecular genetics in the use of microorganisms and microbial genes to improve the amount and efficiency of goods, on the one hand, and, on the other hand, to minimize costs and processing time, has made the use of these techniques very useful in the different branches of agriculture. So far, microorganisms have been used in different sectors of agriculture as follows; reducing the toxicity of antibiotics and herbicides (beta-lactam gene), production of fungicides and biocides (chitinase gene), resistance to pathogenic bacteria (WRKY gene family), resistance to citrus bacterial canker (pthA gene), dissolution of soil phosphorus (gabY, Mps, pKKY, pKG3791 and OlpA-Cm genes), tolerance to abiotic stresses (Flavdex gene (Fld), PR5 gene family), coexistence with plants in water and mineral production (mycorrhizal fungus), and editing the plant genome (CRISPR/

B. Fazeli-Nasab (✉)
Research Department of Agronomy and Plant Breeding, Agricultural Research Institute, University of Zabol, Zabol, Iran
e-mail: bfazeli@uoz.ac.ir

A. F. Rahmani
Department of Horticulture, Faculty of Agriculture, Herat University, Herat, Afghanistan

R. Soni et al. (eds.), *Microbiological Activity for Soil and Plant Health Management*,
https://doi.org/10.1007/978-981-16-2922-8_19

Cas9 gene). Various genes have also been used in the removal of soils contaminated with heavy metals and herbicides (*atzA*, *atzB*, *atzC*, *atzD*, *atzE* and *atzF*, *BPDO*, *CotA*, and *merA* genes). Production of sugar biopolymers (Asr gene), production of biofilms, production of dietary supplements for oil enrichment (fatty acyl-ACP thioesterase gene), development of immunity against gas spoilage (*alfa-Toxin* gene), bioethanol synthesis (*Cel6B* gene), Baker's yeast engineered to promote the bakery industry, and engineered yeasts for the production of engineering and industrial alcohol that have also been controlled by other genes. We also attempted to review in this chapter the form and manner in which microbial genes are used directly and indirectly to improve the quantity and efficiency of agricultural products.

Keywords

Beta-lactam · Chitinase · CRISPER · Flavodoxin Flv · Laccase · WRKY gene family

Abbreviations

ABC	ATP-binding cassette
AMF	Arbuscular mycorrhizal fungi
ASR	Alternate sucrase
BPDO	Biphenyl dioxygenase
CRISPR	Clustered regularly interspaced short palindromic repeats
crRNA	CRISPR RNA
Cry1Ac protoxin	Is a crystal protein produced by the gram-positive bacterium, Bacillus thuringiensis (*Bt*) during sporulation
DDD	Domain-driven design
DDT	Dichlorodiphenyltrichloroethane
GDH	Glucose dehydrogenase holoenzyme
GMOs	Genetically modified organisms
HPLC	High-performance liquid chromatography
LMOs	Living modified organisms
PAL	Phenylalanine ammonialyase
PPP1	Phosphoprotein phosphatase 1
PQQ	Pyrroloquinoline quinine
PthA	*Pathogenicity gene*
T-DNA	Transfer DNA
TracrRNA	Small trans-encoded RNA
WRKY gene family	The length of the WRKY domain is approximately 60 amino acids long and also they have one or two DNA binding domains that contain the conserved heptapeptide WRKYGQK and also they are responsible for the recognition with W-box sequence "(C/T)TGAC(T/C)"

19.1 Introduction

Food availability will be one of the most significant obstacles ahead as the population rises. Although the green revolution could help people get the food they need, but with the increasing population and the need for more food, the need for another green revolution is felt, with greater focus on environmental values and resource management and preservation, so that food will rise by 50% over the next 20 years (Khan et al. 2009). The Green Revolution, which was produced with the advent and delivery of chemical fertilizers, posed threats to the atmosphere and human beings along with increasing production. Therefore, by creating organic agriculture with a greater focus on soil capacity and biological capacities, human beings have chosen to use approaches that are more consistent with nature and uphold the ecological equilibrium of soil and climate to maintain productivity and conserve their basic resources (Sarikhani et al. 2014).

In recent decades, the growing population of the world and the growing food market have led to a serious and imminent change from conventional agriculture to advanced agriculture in the agricultural and food sciences and the use of modern genetic technologies in the production of crops and livestock. As we know, plants are the world's primary and most essential renewable resources that also fulfill non-nutritional, chemical, and industrial requirements, such as photosynthesis, in addition to supplying food for humans and animals. For this reason, the application of genetic engineering and molecular genetics methods in the use of microorganisms and microbial genes to increase the quantity and quality of products, on the one hand, and reduce costs and production time, on the other hand, the use of these methods in various branches of agriculture is very valuable.

All food cycles are related by bacteria to higher nutritional levels, so it would be a significant factor in deciding the role of an ecosystem because of the range of reactions that bacteria cause, role and likely bacterial diversity (Khodashenas et al. 2010).

In soil composition, bacteria play a significant role. Bacteria-produced polysaccharides bind soil particles together and help form the foundation of the soil. Bacterial humus also forms complexes of clay-organic matter that assist in the granulation of the soil. Actinomycete community bacteria create hyphae that bind soil particles together, thereby playing a role in the granulation of soil. Soil granulation lowers soil erosion, increases the infiltration of water and adequate aeration of the soil (Khodashenas et al. 2010).

Endophytic bacteria, by maintaining their survival in the host plant, in addition to not harming their host, but with the help of various mechanisms, directly and indirectly increase plant growth. Indirect stimulation of plant growth occurs when bacteria counteract the harmful effects of one or more plant pathogens, which can be achieved in two ways. In one method, bacteria stop the activity of the pathogen by secreting siderophore, producing hydrogen cyanide, and secreting extracellular enzymes such as chitinase, beta-one, and three gluconases, protease, and lipase. In another method, the bacterium activates the induced systemic resistance mechanism in the plant (Etminani and Etminani 2018). These include siderophores, lipopolysaccharides, and salicylic acid. Siderophore can specifically help improve the growth of the host plant in addition to its indirect influence.

In fact, iron absorption by microorganisms and plants in iron deficiency conditions usually depends on chelating agents for the breakdown and transport of inorganic (mineral) iron. The most diverse biosynthetic chelates are microbial siderophores and to a lesser extent phytosiderophores produced by Geramineaes. Siderophores are low molecular weight compounds (less than 1000 Da) with a high affinity for trivalent iron that are secreted by various bacteria to dissolve trivalent iron in the extracellular medium. *Pseudomonas* was introduced as the predominant endophytic genus with the ability to produce growth-promoting compounds in this study. Bacteria of *Pseudomonas* have spread extensively in nature and are isolated from most environments (Alexander and Zuberer 1993; Fazeli-Nasab and Sayyed 2019). These bacteria are important in terms of a broad range of metabolites supporting plant growth, such as production of hydrogen cyanide, production of siderophores, solubilization of phosphates, and production of auxins (Costa-Gutierrez et al. 2020; Schippers et al. 1987). This genus has been introduced as a growth-promoting endophytic bacterium in crops (Maheswari et al. 2013) and it has been shown that *Pseudomonas fluorescens* and *Pseudomonas putida* bacteria in pine and *Pseudomonas aureofaciens* bacteria in Fir plant can increase plant height and biomass (Ahmadzadeh 2013). *Pseudomonas* has also been identified in *Arabidopsis* and soybeans (Chaudhry and Patil 2020; Panchal and Ingle 2011).

19.2 Importance of Different Microbial Populations Associated with the Plant

The rhizosphere is a microecological area near the plant root, where rapid and numerous chemical interactions take place and its environment is more competitive than the soil mass. This environment is divided into three regions: internal, middle, and external, based on proximity to the root and the extent of its impact. Compounds added to the soil by the roots are classified into four categories: exudates (passively removed from the roots), secretions (actively removed from the roots), dead cells, and gaseous compounds. The compounds in the substances left by the roots, by acidifying or changing the redox conditions in the rhizosphere or directly chelating the elements, help to provide nutrients such as nitrogen, phosphorus, iron, etc. As the soil dries, the hydraulic potential decreases, after which the root seepage begins to return water to the soil, increasing the degree of stability of the rhizosphere. Conventional nutrient management strategies are highly dependent on the use of chemical fertilizers, and the potential biological potential of soil and plants has in many cases received less attention. Of course, the growing demand of the world for food and, as a result, the need to achieve high yields of agricultural products, has been a pressure lever for this lack of attention. In many cases, the expediency of producing more and obtaining food has even violated other environmental considerations. The chemical and biological processes that take place in the rhizosphere not only determine the mobility and uptake of soil nutrients but also control the efficiency of nutrient consumption. Establishing an integrated nutrient management strategy in the root zone is an effective way to solve the problem between high crop yield, nutrient efficiency, and environmental protection (Dadivar 2015).

Plants are an important source of organic matter in the soil and organic matter is a major source of energy for microbial activity. In most cases, it defines the basis of the tendency of microorganisms to the roots and the formation of the interaction of microorganisms and plants in the form of cooperation and coexistence. The formation of root–microorganism interactions in the rhizosphere has caused many physical, chemical, and biological properties of soil in this environment to be different from soil mass. This difference is very favorable for both the plant and the rhizosphere microorganisms (Dadivar 2015).

The uniqueness of rhizosphere conditions in terms of plant nutrition and also the difference of rhizosphere between different plants can provide effective management strategies for farmers and producers (Dadivar 2015; Ryan et al. 2009). Provision of nutrients locally and in the environment adjacent to the roots, instead of consumption as a spread in the soil mass, uses the capabilities and benefits of efficient plants in improving the nutrition and growth of inefficient or inefficient plants in the form of mixed cultivation of different plant species or different cultivars of a species can increase production and reduce the consumption of agricultural inputs, including chemical fertilizers (Dadivar 2015; Gqozo et al. 2020; Rehman et al. 2020). Also, adopting appropriate crop rotations by considering the rhizosphere characteristics of each plant, so that the appropriate conditions created in the soil by one crop can be used in cultivation and subsequent crop production, including cultivation management strategies. Emphasis is placed on the rhizosphere, which can increase productivity and reduce chemical fertilizers (Ayub et al. 2020; Dadivar 2015).

The reaction of calcium carbonate in calcareous soils leads to an increase in soil pH, especially in areas with low rainfall. These reactions in the surface horizon of calcareous soils limit the solubility and uptake of many elements such as Fe, Cu, Mn, Zn, P. It also impairs plant and root growth and ultimately reduces yields unless large amounts of chemical fertilizers are applied. Therefore, the solubility and low uptake of nutrients in calcareous soils have attracted the attention of many nutritionists due to the high cost of chemical fertilizers, the environment, and public health. Many studies have shown that in calcareous soils, organic acids from plant root secretions can act as an effective factor in extracting a significant portion of plant nutrients and improve the efficiency of fertilizer and water consumption in these soils (Khademi et al. 2009).

Different bacterial species protected by a polymer substrate are known as biofilms. In adverse environmental conditions, this polysaccharide coating plays a part in preserving the cells in the biofilm and giving them certain capabilities to maintain and withstand bacteria in adverse environmental conditions. The biofilm produced by them may have positive or negative effects on human life due to the presence of bacteria in different ecological environments. The production of biofilms in plant pathogens causes many problems in killing bacteria because biofilms prevent the effects of disinfectants, antibiotics, and chemical toxins on harmful bacteria. On the other hand, the prominent effect of heavy metal decomposing bacteria in contaminated soils and waters, wastewater, and air purifying bacteria, as well as beneficial bacteria effective in improving plant growth cannot be ignored. Considering the issues raised, by increasing the level of knowledge and examining various aspects of bacterial life, biofilm-producing bacteria can be used to improve the quality of human life (Khezri 2019).

The use of antagonistic bacteria in the biological regulation of plant diseases and the quantitative and qualitative enhancement of crop growth has recently been found to be of considerable importance by researchers. Increasing use of chemical compounds in the form of fertilizers and pesticides to control pests and plant diseases has caused serious pollution in the environment, human health and other organisms. For this reason, scientists are looking for alternative methods to these harmful compounds in controlling plant pests and diseases and improving plant growth. The use of biologically inhibitory agents that have high capabilities in the production of secondary metabolites effective in reducing or inhibiting plant diseases is one of the strategies that has been welcomed by researchers (Khezri 2019).

Most of the bacteria that inhibit plant diseases are located in the soil around the roots, called the *rhizosphere*. In the form of compounds rich in sugars and organic acids, the root of the plant secretes a large part of the stabilized compounds into the soil atmosphere through the roots. For this cause, the rhizosphere is an ideal location for various types of microorganisms to expand and multiply (Khezri 2019).

The findings of studies conducted by different researchers suggest that biofilms formed by beneficial bacteria may be useful for disease control (Younessi et al. 2017).

Bacillus subtilis, *B. cereus*, and *Pseudomonas fluorescent* bacteria can be mentioned as effective bacteria in biologically controlling plant diseases (Younessi et al. 2017). Different strains of these bacteria have high potential in producing a variety of secondary metabolites such as biofilms, biosurfactants, extracellular fluid secretions, antifungal volatile compounds, antibiotics, and various enzymes and reduce disease in different ways (Younessi et al. 2017). In one study, the probability of managing tomato blight bacterial disease was assessed using *B. subtilis* 6051 strain on the model plant of Arabidopsis and the results showed that the reduction of the disease relied directly on the development of biofilm by the antagonist bacterium (Younessi et al. 2017).

19.3 The Role of Microorganisms in Soil Protection

Soil conservation as a living organism is considered to be the main goal in bio-agriculture, so that other factors and institutions should be such that in the long run, they cause more soil fertility and revitalization of its living ecosystem. Production of high quality food, sufficient quantity, preservation and increase of soil fertility in the long run, preservation of genetic diversity, and consideration of the wider social and ecological effects of the crop system are the goals of bio-agriculture. Soil microorganisms, especially bacteria, cause many reactions that are necessary for the production of agricultural products. These reactions include the following; Food cycle, soil structure preservation, organic matter decomposition and food release, decomposition of agricultural chemicals, decomposition of other contaminants, production of plant humus, control of plant and animal pests. Soil bacteria are dynamic reservoirs of nutrients in all ecosystems and are directly or indirectly important in improving plant health (Forouzandeh et al. 2019; Jahantigh-Haghighi et al. 2020; Mehrban and Fazeli-Nasab 2017; Naghavi et al. 2004).

Many reactions that are essential for the production of agricultural products are caused by soil microorganisms, especially bacteria. These include the food cycle, soil structure preservation, decomposition of organic matter and food release, decomposition of agricultural chemicals, decomposition of other contaminants, humus production, and control of pests and plants and animals. Soil bacteria are dynamic sources and reservoirs of nutrients in all ecosystems and are critical for improving plant health directly or indirectly (Khodashenas et al. 2008).

Microbial decomposition has been introduced as the most important natural mechanism for removing non-volatile hydrocarbon pollutants from the environment. Although biodegradation occurs slowly, the use of microbial species that decompose pollutants more effectively or through improving environmental conditions such as food addition and aeration can increase biodegradation. Creating different conditions related to native microbial communities such as energy sources, pH, electron receivers and donors, food, temperature, etc., in contaminated sites is also required in the bioremediation method (Gerhardt et al. 2009; Megharaj et al. 2011).

Owing to its wide distribution in the biosphere compared with other living species, bacteria play a prominent role in Atrazine degradation. Biodegradation of the herbicide atrazine in microorganisms is mostly carried out by bacteria due to the presence of *atzA*, *atzB*, *atzC*, *atzD*, *atzE*, and *atzF* genes, which encode the degrading enzymes of this organic pollutant. Since the mid-1990s, there have been reports of atrazine degradation by a large number of degrading bacteria, including the genera *Pseudomonas*, *Rhizobium*, *Acinetobacter*, *Agrobacterium*, and *Pseudoaminobacter*. Atrazine is often used as a source of nitrogen and carbon by bacteria (Fernandes et al. 2014; Ma et al. 2017; Qingyan et al. 2008).

Biological oxidation of organic compounds is the main advantage of microbial decomposition. Mineralization is said to take place as organic compounds are converted to H_2O and CO_2. It should be remembered that mineralization never happens entirely because the cell is converted to part of the organic matter and an essential part of the cell mass is in some way non-degradable. With hazardous substances combined with H_2O, CO_2 and new bacterial cells will solve a number of problems, but this itself needs to be corrected. In biodegradation, mineralization does not always occur. Changes in the molecular structure of a contaminant during bioremediation may result in the production of different materials from raw materials that are still toxic or hazardous (Abdollahi and Abdollahi 2008).

In general, the three metabolic pathways by which contaminants are modified and degraded by microorganisms are: aerobic, anaerobic, fermentation or fermentation methods (King et al. 1997). The aerobic process can cause many of these contaminants to degrade, but when reacted, heavy halogen compounds are not readily affected by aerobic microorganisms and become more toxic. It is important to note that the rate of reactions for the aerobic metabolism process is often faster and typically easier to control (Gibson 1988; Heitkamp et al. 1988; Tabak et al. 1964).

Almost all methods of bioremediation are aerobic, but since they can catalyze most reactions and degrade certain compounds that are immune to aerobic decomposition, anaerobic bacteria should not be ignored. A wide variety of compounds can be used by anaerobic bacteria to generate energy. In fuel products, these compounds include carbohydrates, amino acids, fats, pesticides, and aromatic compounds. Anaerobic microbes can reduce (reduce) chlorogenic molecules that are resistant and less attacked by aerobic bacteria. This toxin, like other chlorine toxins, is very resistant and its decomposition in the soil contributes to DDD, which is also resistant and both compounds may be contained in fats and are detrimental to the nutrition of animal products. DDT is one of the toxins that is highly regarded in terms of environmental contamination. Some researches have shown that the breakdown of DDT and ddd in anaerobic conditions is much faster than in aerobic ones. Normally, only one to two percent of it remains after a few months. About 26 microbial species have the ability to degrade ddt to ddd. These microbes include *Aerobacter*, *Aerogenase*, *Enterobacter*, and *E. coli.* (Abdollahi and Abdollahi 2008; Morrison et al. 2000; Tang et al. 1999).

In the climate, hydrocarbons are degraded primarily by *filamentous fungi*, yeast, actinomycetes, and bacteria. Biodegradation efficiencies have been recorded for soil fungi at 6–82%, soil bacteria at 13–50%, and marine bacteria at 0.003–10% (Das and Chandran 2011). While many microorganisms are capable of degrading crude oil in the soil, bacteria are increasingly recognized for their biodegradability (Sebiomo et al. 2011), and are commercially available as lyophilized hydrocarbon decomposition bacteria. The way bacteria do the biological process is well known; and the bacteria that can break down petroleum products are *Pseudomonas*, *Aeromonas*, *Moraxella*, *Bijrinkia*, *Flavobacter*, *Korobacter*, *Nocardia*, *Corynebacteria*, *Acinetobacter*, *Mycobactena*, *Monococcus*, *Streptococcus*, *Streptococcus aureus*. Microbial deformation and mineralization are the most important methods of decomposing pesticides in soil. The size and activity of soil microbial biomass affect the rate of decomposition (Burken and Schnoor 1996; Burken et al. 2011).

Microorganisms have been used in different agricultural industries so far. Examples include the role of microorganisms in biofilms and the development of dietary supplements to enrich oil (Franklin et al. 2019, 2020), the reduction of antibiotic toxicity (Allen et al. 2009), the promotion of engineered bread yeast for the bakery industry (Prieto et al. 2005), and the development and industrialization of engineered alcohol yeast.

19.4 The Role of Microorganisms (Biofertilizers) in Sustainable Agriculture

The use of bio-fertilizers to increase soil fertility in agricultural production has been proposed to achieve sustainable agriculture, expand the production of products without toxins and chemical fertilizers as an alternative to chemical fertilizers.

Types of biofertilizers include molecular nitrogen-fixing bacteria (diazotrophs), root fungi (mycorrhiza), microorganisms (soluble microorganisms), and insoluble phosphates, rhizosphere bacteria that stimulate plant growth, and plant-farming microorganisms, microorganisms, plant-producing microorganisms. It is used to expand the biological regulation of plant diseases (Sabbagh and Valizadeh 2019).

One of the forced coexistences of plants that are useful in water absorption and nutrients, as well as in soil formation and soil stabilization, is arbuscular mycorrhizal fungi. The coexistence of arbuscular mycorrhizal fungi also increases plant production capacity. In this regard, the mycorrhizal relationship can be mentioned as a living structure in which there is coexistence between the fungus and the root and increases the potency of both (Xu et al. 2019). Since most mycorrhizal fungi do not have a specific host, large populations of plants coexist with these fungi. Approximately 240 species of arbuscular mycorrhizal fungi (AMF) have been identified using the morphology of their spores, and in the absence of spores, the presence of fungal organs within roots such as arbuscules and vesicles as well as their structural features is the best means of identification. In addition to decreasing the adverse effects of nutrient shortages and drought and salinity stresses, coexistence with mycorrhizal fungi often improves reversibility after plant stress. And their vegetation has declined and the biosynthesis of secondary metabolites in plants has also increased (Hatami et al. 2020).

In order to stimulate rooting and root production, in rosemary (*Rosmarinus officinalis*) a two-factor factorial experiment was performed in three replications (first factor (phosphorus fertilizer (ammonium phosphate) equivalent to zero, 25, 50, 75 and 100 kg / ha). And the second factor (two levels of *mycorrhizal fungus* treatment including *Glomus intraradices* and *Glomus mosseae*)). Some traits such as plant height, root length, stem diameter, shoot fresh weight, shoot dry weight, root wet weight, stem dry weight and number of leaves per plant were also investigated. The results showed that in addition to both types of fungi were effective on all morphological indices of the plant (Table 19.1), even fungi in the presence of phosphorus fertilizer could affect the morphological indices more so that the highest plant height (1/1) 195 mm) was obtained from the treatment of *glomus muse* A and the use of 100 kg/ha of ammonium phosphate and the lowest amount (110.1 mm) was obtained from the treatment of *glomus muse* A and no application of ammonium phosphate (control) (Table 19.2) (Bagheri et al. 2018).

The use of nitroxin, vermicompost fertilizers in sesame plants has had a positive and important impact on most of the characteristics evaluated, so that in some characteristics, including the number of capsules per plant, the yield of grain and the yield of oil, the use of these fertilizers have an impact equal to half the effect (Sajadi et al. 2011). Nitroxin biofertilizer contains a set of the most effective strains of nitrogen-fixing bacteria of *Azospirillum*, *Azotobacter* and phosphate solvent of *Pseudomonas*. *Azotobacter* and *Azospirillum* are the most important growth-promoting bacteria in the plant, which in addition to bio-stabilizing nitrogen and helping the plant easily access soil nutrients, produce significant amounts of growth-promoting hormones, especially auxin, gibberellin, and cytokine (Sabbagh and Valizadeh 2019).

Table 19.1 Comparison of the average simple effects of rosemary under the influence of *arbuscular mycorrhiza*

Sources of changes	Bush height (mm)	Root length (mm)	Stem diameter (g)	Wet weight of aerial parts (g)	Dry weight of aerial parts (g)	Root wet weight (g)	Dry weight of the stem (g)	Number of leaves per plant
Glomus Musa A	130.1^{b}	144.1^{a}	1.83^{a}	7.83^{a}	6.90^{a}	0.73^{a}	1.02^{b}	29.80^{b}
Glomus intraradices	143.4^{a}	126^{b}	1.92^{a}	8.00^{a}	7.20^{a}	0.76^{a}	1.19^{a}	34.86^{a}
Control without fungi	119.85^{c}	88.95^{c}	1.78^{b}	7.20^{b}	6.27^{b}	0.34^{b}	0.71^{c}	19.16^{c}

Similar letters in each column indicate no significant difference at the 5% level based on Duncan's multiple range test

Table 19.2 Mean interactions on growth characteristics of rosemary under the influence of *mycorrhiza* and ammonium phosphate fertilizer

Arbuscular mycorrhizae	Ammonium phosphate	Number of leaves per plant	Dry weight of the stem (g)	Root wet weight (g)	Bush height (mm)
Glomus intraradices	100 kg/hectare	31.00^{b}	1.37^{b}	1.21^{b}	147.02^{bc}
	75 kg/hectare	36.00^{b}	0.92^{dc}	0.71^{c}	124.96^{dc}
	50 kg/hectare	25.33^{c}	1.29^{b}	0.59^{dc}	121.66^{d}
	25 kg/hectare	37.00^{b}	0.71^{e}	0.65^{c}	127.79^{dc}
	No fertilizer application (control)	13.66^{d}	0.83^{de}	0.48^{d}	129.54^{dc}
Glomus musa A.	100 kg/hectare	45.00^{a}	1.65^{a}	1.59^{a}	195.10^{a}
	75 kg/hectare	35.66^{b}	1.81^{a}	0.69^{c}	159.21^{b}
	50 kg/hectare	40.00^{b}	0.76^{de}	0.62^{dc}	130.08^{dc}
	25 kg/hectare	29.00^{c}	0.72^{e}	0.68^{c}	122.89^{dc}
	No fertilizer application (control)	24.66^{c}	1.01^{c}	0.21^{e}	110.17^{d}

Similar letters in each column indicate no significant difference at the 5% level based on Duncan's multiple range test

19.5 The Role of Genetic Engineering in the Use of Microbial Genes in Agriculture

By engineering the fatty acyl-ACP thioesterase gene, they have reduced the carbon chain, increased the degree of saturation, quality and efficiency of the oil. Asparagus converts to aspartic acid and consequently reduces the formation of acrylamide (a carcinogen during cooking). Genes involved in the biosynthesis pathway of isobutanol and ethanol in *S. Cerevisiae* yeast have been engineered to increase production (FDA 2013). Later, recombinant microorganisms (such as recombinant probiotics) will appear in dairy products such as yogurt and cheese (Aguilera et al. 2013).

Numerous studies have been performed to increase the efficiency of pest control agents such as *Bacillus thuringiensis* (*Bt*) through genetic engineering, during which genes encoding crystalline proteins or other toxins are transferred to Bt bacteria and their expression is increased or for the first time, a specific gene has been transferred to the target strain because each strain of this bacterium has a specific number of such cry genes (Driss et al. 2011; McDade 2019). In 2011, with the transfer of the chitinase gene to the Bt bacterium, the insecticide capacity of this bacterium increased by 50% (Driss et al. 2011). Insect-specific neurotoxin gene along with *cry1Ac* gene was transferred to a Bt strain and it was found that its insecticidal rate was increased (Li et al. 2012).

Another approach is to transfer to other microbial agents the genes of a biological agent, such as *Bt*. The *vip3A* and *cry1I* genes, for example, have been isolated from the *Bt* bacterium and transferred to the *Pseudomonas fluorescens* bacterium, which is a biological agent to control plant diseases, so that it may also have insecticidal properties (Hernández-Rodríguez et al. 2013). The *cry9Aa* gene has also been transferred to *E. coli*, leading to insecticidal activity against *Spodoptera exigua* (Naimov et al. 2014).

Concerning the genetic engineering of plant disease antagonists, several studies have been performed so far to improve the production and controllability of plant pathogens, including *Trichoderma* sp. (Kowsari et al. 2013; Malmierca et al. 2012) and *Streptomyces* (Clermont et al. 2011), *Bacillus subtilis* (Leclere et al. 2005) and *Psuedomonas* sp. (Hernández-Rodríguez et al. 2013), respectively, observed. Genetic engineering has increased the efficiency of microbial growth stimulants (biofertilizers) through the production of recombinant nitrogen-fixing bacteria (Rhizobium) with the ability to bioremediate soil heavy metals (Ike et al. 2007), *Anabaena* sp. With higher growth stimulant and nitrogen fixation (Chaurasia and Apte 2011) and *Azospirillum* with high auxin production and higher growth stimulus (Baudoin et al. 2010; Malhotra and Srivastava 2006).

The *BADH* gene was transfected into the walnut (Juglans regia L.) cultivar "Chandler" by *Agrobacterium* LBA4404, plasmid pBI121, CaMV 35S promoter, and *NPTII* gene as a selectable marker (in turn driven by a NOS promoter) to confirm that *BADH* transgenes were successfully incorporated into the plant genome using PCR and Southern blot analysis. Transgenic and wild plants grown from embryos exposed to four levels of osmotic stress (i.e., zero, 2, 4, and 8% PEG) and four levels of salinity (i.e., zero, 50, 100, and 200 mM NaCl) and after 21 days, they found that the transgenic plants grew under almost extreme salinity and drought stress, but the wild-type plants showed a lagging growth rate and did not survive the cradle stage (Fig. 19.1) (Rezaei Qusheh Bolagh et al. 2020).

Results of research on *Rhizobium leguminosarum* showed that the introduction of a high catalase activity *VKT* gene into this bacterium resulted in a 1.7 to 2.3-fold increase in the nitrogen fixation activity of its nodes relative to control bacteria (Orikasa et al. 2010). The gcd gene was also cloned from *E. coli* in *Azotobacter vinelandii*, which resulted in increased phosphorus solubility and growth of sorghum (Sashidhar and Podile 2009). However, it should be noted that despite extensive studies and research on the creation of recombinant microorganisms for the biological control of pests and diseases, as well as biofertilizers, a significant number of such products have not yet been commercialized. The engineered strain of *Sinorhizobium meliloti*, called RMBPC-2, in which the *nifA* gene was inserted and nitrogen fixation increased, was one of the first commercialized strains (Bosworth et al. 1994). However, engineered nitrogen stabilizing strains have become increasingly commercialized in both the United States and Australia (Kunjapur and Prather 2015; Mindt et al. 2020).

Fig 19.1 The stages of transformation of *BADH* gene for salt and drought tolerance to Persian walnut using *Agrobacterium*-mediated (section A: Difference in growth of transgenic walnut (left) and non-transgenic (right) somatic embryos on DKW (Driver and Kuniyuki Walnut) containing 100 mM kanamycin); Shoot formation directly from transgenic somatic embryos of walnut 5 weeks after culture on selective medium in dark, 5 weeks on germination medium, and weekly on regeneration medium ((b) Selective medium, (e) Germination medium, (f) Proliferation medium)

19.5.1 The Role of WRKY Gene Family in Bacterial Resistance

The *WRKY* gene family is the most significant group of regulatory transcriptional factors (Xue et al. 2019). For the activation of *PR* genes, binding of transcription factors to DNA is necessary (van Loon et al. 2006). In rice, one hundred forms of WRKY genes were identified (Ross et al. 2007). Xoo's expression of the *WRKY12* gene in rice increases *NPR1*, *PR1b*, phenylalanine ammonialyase (PAL), and peroxidase (pox) expression. Increased expression of *WRKY70* in Arabidopsis increases resistance to *Pseudomonas syringae* and *Pectobacterium carotovorum* (Li et al. 2006). It was suggested that *WRKY* is a transcriptional regulator in JA and SA-dependent signal cascades (Song and Goodman 2001). The result of *WRKY* gene expression in tobacco increases the level of programmed cell death response and *HR* (Oh et al. 2008). It has been suggested that the *WRKY12* transcription activation factor in tobacco causes overexpression of the *PR1* gene by binding to a protected sequence in the *PR1* gene promoter called *WK-boxes* (TTTTCCAC) (van Verk et al. 2011).

19.5.2 The Role of *pth*A Gene in Developing Resistance to Chancre

One of the most significant genes for the pathogenesis of chancre-causing bacteria is the *pthA* gene (Shiotani et al. 2007). The homologs of this gene, including pthB and pthC, are present in the citrus chancroid bacterial strains that cause forms B and C (Mokhtari et al. 2015). This gene is part of the *avrBs3/pthA* (Transcriptional Activator-like (*TAL*) effectors) (*TAL* family) family of genes. This family is widely found in Xanthomonas species (Shiotani et al. 2007). The *pthA* gene in all citrus chancre bacteria is about three to four kilograms long. The gene has 17.5 consecutive repeating regions, each of which is 102 bp and is located in the central part of the gene. This repetitive region is necessary to determine host specificity as well as infectivity (Al-Saadi et al. 2007). *PthA* protein is composed of 1163 amino acids. The middle portion of the *PthA* protein contains 17.5 identical replicates of 34 amino acids, with the exact number and arrangement of replicate units differing in different bacteria. This difference results in function and specificity when resistance or infectivity develops in the respective host species in the absence of resistance genes (Al-Saadi et al. 2007).

The most effective way to control citrus chancre is producing disease-resistant plants. The lack of access to resistant plants and the restriction of the presence of natural genes for disease resistance have resulted in the use of methods based on the genetic modification for other sources of resistance. The use of the resistance mechanism of plant antibody expression (Plantibody-mediated resistance) is one of the new solutions for creating resistant plants. The use of this mechanism has so far led to resistance against a large number of viral, bacterial, and fungal diseases (Mokhtari et al. 2015; Yajima et al. 2010). In the latter mechanism, disruption of the pathogenic process is achieved by targeting essential pathogen proteins by specific antibody–antigen binding. Therefore, *PthA* protein, with its essential functions in causing disease, can be considered as one of the best candidates in creating resistance.

19.5.3 The Role of Beta-Lactam Gene in Reducing the Toxicity of Antibiotics

An unpleasant occurrence and a health alert is the presence of antibiotic resistance genes in bacteria. Studies have shown that human activity contributes to an increase in bacterial genes for antibiotic resistance. Antibiotic resistance genes spread to other species beyond particular bacterial species, and these genes are also known as bio-pollutants. In environments where pollutants put a lot of stress on bacteria, antibiotic resistance is more widespread (Abou-Shanab et al. 2007).

Genes for antibiotic resistance are very complex. To date, 95 distinct antibiotic resistance genes have been isolated from humans, of which only 69.5% are similar to known resistance genes, and other sequences are unknown. Genealogy of *beta-lactamase* genes derived from Alaskan soils, many similarities with known *beta-lactamase* genes (Allen et al. 2009; Huff et al. 2020), and in the phenotypic analysis, bacteria with these genes also had different responses. Because there is no restriction on the transport of plasmids and other genetic elements between bacterial species, the increase in gene contamination leads to the spread and spread of resistant bacteria. Numerous reports indicate that antibiotic resistance genes are highly similar in bacteria isolated from natural habitats to genes of human pathogenic bacteria, and therefore, infected natural habitats can be the origin of resistance genes (Aminov 2009; Sabri et al. 2020).

In most cases, genes associated with bacterial resistance to heavy metals are associated with antibiotic resistance genes. These genes control mechanisms such as detoxification by increasing the flow of substances out of the cell. Because these genes are nonspecific, they reduce the harmful effects of both metals and antibiotics on the cell, and the presence of one of these two genes is needed to motivate these genes and make this group of bacteria more abundant, although the concentration of metals still increases. The abundance of bacteria with antibiotic resistance genes in the soil is not known (Knapp et al. 2011).

Increased resistance to antibiotics, especially against *beta-lactams* (*beta-lactamases*, which break down beta-lactams in bacteria) has become more common in the last two decades. (Boyd et al. 2020; Bush and Jacoby 2010). Antibiotic resistance is primarily due to the transport and maturation of large plasmids, such as several *beta-lactamase* genes, capable of acquiring different resistance genes. Other mechanisms of resistance to *beta-lactams* can be harmful to bacteria (Allen et al. 2009); For example, reducing the efficiency of purines and increasing the flow of substances are mechanisms to reduce the toxicity of antibiotics that reduce the essential nutrients in the bacterium and cause problems for it.

Beta-lactamases are inactivating enzymes of penicillin and related compounds (Hemmati et al. 2015); the genes encoding these enzymes are found chromosomally in bacteria. The study of the frequency of genes that make this enzyme is one of the special methods in estimating the frequency of genes resistant to antibiotics such as penicillin, amoxicillin, and ampicillin (Younessi et al. 2017).

As the only source of carbon and electricity, *beta-lactamase*-producing bacteria are said to often use antibiotics. A variety of species of *Pseudomonas* can use benzyl penicillin, a process involving the synthesis and release of large quantities of *beta-lactamase*, as their sole source of carbon. Since breaking the *beta-lactam* ring in antibiotics is the first step in bacterial detoxification and thus antibiotic resistance, bacteria capable of producing the *beta-lactamase* enzyme show resistance to a wide variety of antibiotics (Younessi et al. 2017).

19.5.4 The Role of PR5 Gene Family in Responding to Stressful Situations

Due to their sequence similarity with the plant protein thaumatin (TL) (Liu et al. 2020), proteins belonging to the *PR5* family are known as Thaumatin-like proteins. So far, the PR5 protein group has been isolated from tobacco, Arabidopsis, rice, wheat, and many other plants (Baek et al. 2019). In response to stressful conditions, such as high salt concentrations, wounds, or pathogen attacks, the accumulation of these proteins in plants has been observed. The permeability of the pathogenic cell membrane is altered by these proteins (Boccardo et al. 2019; Kitajima and Sato 1999). Osmotins are similar to PR5 protein types of play. Osmotin is an inducible protein that has been found in tobacco by salinity tension. Game types, therefore, are referred to as Osmotins. In healthy plants, neutral *PR5* proteins are not present but are caused by ethylene (Liu et al. 2020).

According to studies on the involvement of phenylpropanoid biochemical pathway derivatives in different biological and abiotic stresses (Campbell and Ellis 1992; Chen et al. 2019), changes in enzymes like phenylalanine ammonialyase (*PAL*) in this pathway may be studied. By interacting with the synthesis pathways of phenylpropanoids and isoflavones that have phytoalexin activity, the *PAL* gene plays a very significant role in plant resistance. This gene is involved in the biosynthesis of salicylic acid and other defense-related compounds and is a key signaling compound for the activation of defense-dependent genes, catalysts, and transcription factors (Stotz et al. 2009).

19.5.5 *Bt* gene and Concern Management in Transgenic Crops

Growth in resistance management strategies requires that the biochemical and genetic mechanisms of resistance production are properly understood (Tohidfar and Khosravi 2015). The *Cry1Ac* protein receptor (Fabrick and Tabashnik 2007) is referred to as Cadherin. The formation of resistance in some studies has been correlated with Cadherin Locus. Due to mutations in cadherin protein (Gahan et al. 2010), *Cry1Ac* resistance has been observed in some essential cotton pests. This resistance has been attributed to the degradation by retrotransposons of a gene belonging to the broad cadherin gene family, producing several proteins necessary for larval development. However, there was no association between *Cry1Ac* resistance and cadherin gene orthologs in the two *P. xylostella* breeds isolated from the field, suggesting a separate genetic basis for resistance in farm-bred breeds (Tohidfar and Khosravi 2015).

Loss of adhesion and irreversible binding to the pest's precursor membrane have been due to resistance to *Cry1Ac*. But reducing bond strength is not the only resistance creation mechanism. Mutations in the second 12 cadherin proteins also induced resistance to *Cry1Ac*, but in laboratory assays, they did not fully inhibit toxin binding (Gahan et al. 2010). Another way to build resistance is to mutate ABC protein vectors. ABC proteins are inner membrane proteins that are involved in many activities, including the transfer of toxic molecules from the cell. In one study, an inactive mutation was made in one of the ABC proteins called ABCC2, which reduced the binding of *Bt* toxin to membrane vesicles. Decreased binding of Bt toxin to cell membranes following mutations in this protein led to the introduction of ABCC2 as one of the proteins involved in the integration of *Bt* protein with membranes (Gahan et al. 2010). In general, the problems associated with the use of *Bt* pesticides have been reduced by producing transgenic *Bt* products containing the *cry1Ac* gene (Tohidfar et al. 2013).

19.5.6 The Role of *rol* Gene Family in Increasing the Sensitivity of Plants to Certain Hormones

The family of rol-genes located on the T-DNA of bacterium *A. Rhizogenesis* is the main cause of capillary root syndrome. They have *rolA*, *rolB*, *rolC*, and *rolD* in these genes. Increasing plant exposure to hormones such as auxin, which has been confirmed by studies on plants such as *L. corniculatus* and *N. tabacum* are the most significant impact of position genes in plants. The *rolC* gene has been studied more than other genes by researchers in terms of its importance in improving decorative and horticultural traits on the plant. In terms of pathogenicity, rolC gene expression is associated with capillary root syndrome and changes such as the production of new secondary compounds, changes in plant hormonal balance, and chlorosis. Increased lateral branches, the formation of needle-shaped leaves, early flowering, and reduction of flower size, and creation of male sterility by reducing pollen grain production are other morphological changes caused by expression of *rolC* gene in plants. One of the most important regulators of *rolC* gene expression is sucrose. According to studies, the sucrose response region in the promoter of this gene is in the range of −94 to −135. The presence of high sources of sucrose in phloem tissues has led to high expression of rolC gene in these tissues (Gardoonpar et al. 2016).

Regarding the effect of rolC gene on plant hormones, it should be noted that this gene affects the number of cytokinins, auxins, and gibberellins (GA). Changes such as decreased vertebral dominance and increased branching indicate changes in cytokinin hormone levels (Boutigny et al. 2020; Zuker et al. 2001). Examination of *rolC* tobacco protoplasts has shown an increase in membrane hyperpolarization of these cells in the presence of auxin, or in other words, increased membrane excitability of protoplasts in the presence of this hormone (Maurel et al. 1991).

A decrease in plant size with a higher number of buds was observed in potato transgenic *rolC* (Bettini et al. 2016; Fladung et al. 1993). It has also been found that less photosynthesis occurs in transgenic *rolC* plants due to the decrease in chlorophyll content and the leaves appear yellowish-green (chlorosis) compared to natural plant organisms.

The 35S-*rolC* (*roleC* gene under the influence of the strong 35S promoter of CaMV virus) potato transgene plants also have a high expression of this gene in the leaves and, as a consequence, these leaves display more chlorosis (Bahramnejad et al. 2019; Schmülling et al. 1988). In addition to the above, other role models of the rolC gene in plants are stimulation of the development of secondary metabolites and defense proteins in transgenic plants and its relationship with oxygen-free radicals and cyclin-dependent protein kinases (*CDPK*) (Mark et al. 2019; Shkryl et al. 2008).

19.5.7 *PA* Gene Expression of *Bacillus anthracis* in Plants

The use of injectable vaccines in the mucosal tissues receiving the vaccine typically does not induce a sufficient immune response. The principal inputs of pathogenic species are mucosal surfaces, including the mouth and genitals. As a result, oral vaccines were produced which were shown to enhance the immune response of these tissues. There are several production advantages of oral vaccines, such as the fact that there is no need for cleaning anymore, so production costs are significantly reduced and a person receives the appropriate amount of daily fruit or vegetables (Jalali javaran et al. 2011).

Anthrax is a deadly disease common between livestock and humans, and the Protective Antigen (*PA*) gene from *Bacillus anthracis* has the highest potential for a vaccine against anthrax, which is transferred to the lettuce genome and *PA* gene expression is confirmed by ELISA (Honari 2008).

19.5.8 The Role of Gene Encoding the Enzyme *Asr* in Production of Glucose Biopolymers

In addition to plants, bacteria also synthesize different high molecular weight polysaccharides as follows; Alginates, Gellan, Xanthan, Alternan and Mutan (Venkatachalam et al. 2020). Lactic acid bacteria can also generate useful biopolymers such as alternan, which are used in the extracellular environment in industry, agriculture, and medicine, using sucrose, and expressing enzymes of glycosyltransferase. In general, extracellular polysaccharides given by lactic acid bacteria are divided into two groups: hemopolysaccharides and heteropolysaccharides. These enzymes use the energy from the hydrolysis of sucrose chains to transfer fructose or glucose to an acceptor molecule. Glycosyltransferase enzymes, after breaking down sucrose molecules, may transfer glucose units to a growing glucan chain or other substrate, such as maltose or isomaltose, to form glycolic oligosaccharides. In addition, these enzymes may play a hydrolyzing role (transfer of glucose units to water). (Nazarian-Firouzabadi et al. 2019).

Glucansucrase is mainly produced by *Leuconostoc mesenteroides*, *Streptococcus* oral flora, and Lactobacillus species (Yan et al. 2018). So far, more than 60 enzymes of this family have been identified, all produced by bacteria of the four genera *Leuconostoc*, *Streptococcus*, *Lactobacillus*, and *Weissella*. However, some genes encoding these enzymes have also been identified in some other lactic acid bacteria such as *Oenococcus*, *Fructobacillus*, and *Enterococcus* (Gangoiti et al. 2018).

From a medical point of view, alpha-glucans and exopolysaccharides are potentially valuable because they are not digested by human gastrointestinal enzymes. Alpha-glucans pass unchanged deoxopolysaccharide from the upper gastrointestinal tract and are fermented in the large intestine by colon bacteria; therefore, glucansucrase products can be used as fiber in human nutrition, because such biopolymers do not cause the rapid release of blood sugar. Besides, exopolysaccharide alpha-glucans are potentially prebiotic; In other words, these polymers are selectively fermented and cause certain changes in the composition or microbial activity of the gastrointestinal tract that are beneficial to the well-being and health of the host (Roberfroid 2007).

Alternan is an important biopolymer that is mentioned as an alternative to gum arabic, especially in low-viscosity diets. Alternan may also be used as a low-calorie or non-calorie food additive as a filler and bulking agent for food products, in the manufacture of inks, adhesives, cosmetics, creams and ointments, and as a coating for drug release (Nazarian-Firouzabadi et al. 2019). The production of new compounds in plants, using genetic engineering methods, is one of the goals of gene transfer methods. Therefore, Alternan can be produced instead of using bacterial bioreactors, using genetic engineering methods in plants.

The gene encoding the enzyme Alternan sucrase (*Asr*) of *Leuconostoc mesenteroides* was transferred to the sugar beet plant and sugar analysis of fresh beet plants showed that the control plant with 19.6% bridge (sucrose) had more sucrose than the wet bridge plants was with medium bridge 14.4. Also, the amount of Brix in wet transgenic plants was lower than control plants and the rate of reduction of sugar (sucrose) in transgenic plants with *Asr* gene was about 36.1% compared to the control. Bacterial alternan sucrase enzyme can produce 36.6 mg/g FW of Alternan biopolymer in sugar beet-roots and convert significant amounts of root sucrose to Alternan biopolymer for industrial and pharmaceutical applications (Nazarian-Firouzabadi et al. 2019).

19.5.9 The Role of Food Coloring Phycocyanin

Phycocyanin has two subunits of protein, alpha (a) and beta (β), of which there is one site in the alpha subunit, and two sites in the beta subunit to bind phycocyanobilin to the apoprotein. Full phycobilin protein synthesis depends on the synthesis of alpha and beta chains simultaneously, and the proper placement of phycobilin in these two chains. Therefore, the recombinant production of this phycobili-protein is more difficult than other proteins (Eriksen 2008). Although the halo protein of the alpha phycocyanin subunit belongs to the cyanobacterium *Synechocystis* sp. PCC6803 has been reported in Escherichia coli (Tooley et al. 2001).

In fruit, medicine, and cosmetics, phycocyanin is a blue pigment that can be used as a natural dye to substitute carcinogenic synthetic dyes. Today, it is used to a lesser degree in immunoassay and cytometry, in addition to being used as a food coloring. Because of its antioxidant, anti-inflammatory, and anti-cancer roles, phycocyanin in Spirulina is also beneficial to human health, which is why it has gained more attention in recent years. Considering that phycocyanin accounts for 20% of the total cellular protein of Spirulina platensis, it has been selected as a suitable model for commercial production of phycocyanin in phototrophic cultures. Phytocyanin production in phytotrophy is associated with problems. One way to reduce the problems of phycocyanin production is to produce heterotrophy (the production of recombinant protein is one way to produce heterotrophy). (Shoja et al. 2015).

19.5.10 The Role of *Alpha-toxin* Gene in Creating Immunity Against Gas Gangrene

Using the toxic protein *alpha-toxin* with various phospholipase, sphingomyelinase, and biological pathogenic activities, the bacterium *Clostridium perfringens* contributes to a series of cellular reactions and ultimately induces cell death, lethality, and death of the skin. This bacterium has caused diseases such as gas gangrene with symptoms of pain, fever, and swelling, and in this way has caused significant damage to livestock parts, including high casualties of sheep. Injecting alpha-toxin into animals such as rabbits and sheep can cause symptoms of the disease and reveal signs of tissue damage. On the other hand, vaccination of animals such as mice with portions of the *alpha-toxin* protein of the bacterium *Clostridium perfringens* leads to immunity and makes the animal resistant to infection with the pathogenic bacterium *Clostridium perfringens*, so that no symptoms of Gas gangrene have been observed in mice (Rasani et al. 2020).

In a *Clostridium perfringens* immunization analysis of broilers with a recombinant *alpha-toxin* toxin, it was found that birds vaccinated with recombinant *alpha-toxin* were 35.1% damaged by necrotic enteritis. The rate of damage was measured at an average of 37.2% for unvaccinated birds. The concentration of IgG antibody in vaccinated birds was five times higher than in unvaccinated birds. These results showed that in addition to its pathogenic role, alpha-toxin can also be used as an immunogen (Sakurai et al. 2009). In one study, the amplification, expression, and immunization of the alpha and beta combination gene of *Clostridium perfringens* were studied and the result showed that the alpha and beta protein produced could be resistant to the attack of alpha and beta toxins. It has been suggested that an alternative method for using *alpha-toxin* domains as a vaccine can be the natural form of the toxin or engineered various forms of the toxin with reduced toxicity (Bai et al. 2006).

19.5.11 The Role of *BPDO* Genes in Reducing Environmental Pollution

Polychlorinated biphenyls (PCBs) are chlorinated cyclic compounds which, because of their properties such as heat resistance and stability, were commonly used in different industries in the 1930s and 1980s. In the 1980s, this function of stability and resistance to decay, along with the adverse effects on human health, led to their cessation of development. The contamination of water and soil around the world with PCBs is one of the major environmental problems. Due to the toxicity of these compounds and their accumulation in the adipose tissue of living organisms and due to their adverse effects on humans, such as cancer, genetic abnormalities in infants, and liver and thyroid tumors, it is necessary to remove and degrade *PCBs* from the environment. One way to reduce PCBs contamination is to transfer and express the bacterial genes of biphenyl dioxygenase (*BPDO*) (which have the ability to break down *PCBs*) to plants. This enzyme has three components of oxygenase with two subunits *bphA* and *bphE*, a *bphF* ferredoxin and a *bphG* reductase (Alizadeh Arimii et al. 2015).

Most gene transfer experiments in plants involve the transfer and expression of a single gene, and the simultaneous transfer of several genes to plants for a biochemical pathway remains a difficult task. There are various ways to transfer several foreign genes into plant cells. One method is to create a polyprotein structure in which the coding sequence of several proteins joins together to form a single copy. It is also possible to use sequential or simultaneous transmission of several vectors carrying different genes. Another way to express several genes in a plant is by crossing translocated plants containing different genes or cloning several genes into one vector by assembling the cassettes (Tzfira et al. 2007).

Simultaneous transferring of *bphA*, *bphE*, and *bphG* genes, which are encoding components of *BPDO* enzyme, to Arabidopsis was investigated and based on the obtained results, 3 *bphA*, *bphE*, and *bphG* genes cloned in pGreen vector into *E. coli* and *Agrobacterium* LBA4404 and C58C1 and finally transferred to Arabidopsis. In terms of gene transfer efficiency to plants, there was a difference between the two strains of *Agrobacterium* LBA4404 and C58C1 used. The highest number of transgenic plants (0.85%) was obtained with LBA4404 strain.

The transgenicity of Arabidopsis seedlings was confirmed by selecting completely green plants in an environment containing 50 mg/l kanamycin as well as a PCR test and finally, the transgenic plants were successfully transferred to the soil and continued to grow (Alizadeh Arimii et al. 2015).

19.5.12 The Role of Laccases (Such as *CotA*) in Environmental Detoxification

Laccases (EC: 1.10.3.2) are N-glycosylated multi-copper oxidases belonging to a group of proteins containing copper (Hesampour and Mohandesi 2018). The four copper ions in the laccase enzyme structure are divided into three types: copper type 1 (T1), copper type 2 (T2), and copper type 3 (T3). The fungal lactase molecule typically contains four copper atoms, while some types of the lactase enzyme are also found to contain three copper atoms in their structure. Laccases have a molecular weight of about 50–100 kDa and when ABTS is used as a substrate, the optimal pH of the enzyme is in the range of 3–5 (Parand et al. 2015).

Laccases are one of the major proteins capable of catalyzing the oxidation of phenolic compounds used in biotechnology as biocatalysts to detoxify the atmosphere and explain food industry fruit juices (Hesampour and Mohandesi 2018). The first prokaryotic lactases identified belong to the *Azospirillum lipoferum* bacteria. The most important bacterial lactase that has been well studied and its physical and biochemical properties have been determined is the *CotA* protein of Bacillus subtilis. *CotA kDa65* protein belongs to the outer covering of spores. This protein participates in the biosynthesis of spore brown pigment, a melanin-like product, and appears to be responsible for protecting against UV light and hydrogen peroxide. This protein shows similarities with multi-copper oxidases and has high temperature stability (Zamani et al. 2014).

Other laccases have been isolated from *Escherichia coli* (*E. coli*), *Bacillus halodurans*, and *B. licheniformis*. Most of the lactases that have been identified so far and have biotechnological applications have been isolated from fungi. However, the efficient expression of recombinant fungal lactases, which are essential for biotechnological applications, is more difficult than the expression of bacterial enzymes. Problems and barriers to the use of these enzymes include information on sequences that are not accessible, the presence of exon and intron structures in eukaryotic genes, post-translational changes, and bridge formation. Disulfide noted long fermentation time and low efficiency. Despite the industrial applications of bacteria, so far little attention has been paid to bacterial laccases. Studies in genome analysis have shown that these enzymes are widely distributed in bacteria. The development of bacterial laccases for biotechnological applications has advantages because they have high temperature stability and are produced in a short time in cheap environments (Zamani et al. 2014), which is hoped to be used more in the future.

19.5.13 The Role of Flavodexin (*Fld*) Gene in Tolerance to Abiotic Stresses

In agriculture, the most important factors in reducing yield are abiotic stresses (high temperature, cold, frost, and dehydration due to drought or salinity, intensity of sunlight, flooding, ultraviolet light, and heavy metals). Most of these environmental stresses directly or indirectly generate reactive oxygen-free radicals and eventually lead to oxidative stress. Such stresses are the main reason for the decline in crops worldwide and lead to a reduction in the yield of the most important crops by more than 50%. One of the genes that play a key role in responding to such environmental stresses is the cyanobacterial fld gene. Cyanobacteria induce the expression of electron transporters such as flavodoxin (*fld*) that act similarly to ferredoxin (*fd*) in the plant to prevent the adverse effects of ferredoxin depletion. Flavodoxin plays a role similar to ferredoxin (*fd*) in the plant and acts as an antioxidant and increases plant tolerance when non-biological stress occurs by preventing disturbances and irregularities in the electron transfer cycle and the formation of reactive oxygen forms (Ghoreyshi et al. 2016).

Although ferrodoxins (*fd*) are found in all organisms, from prokaryotes to animals, flaudoxins (*flds*) are found only in some bacteria and oceanic algae.

Flavodoxin in these organisms can induce the function of ferredoxin under conditions of iron deficiency and environmental stresses that lead to a decrease in ferredoxin; therefore, they play an adaptive key role in photosynthetic microorganisms that allow rescue and reproduction under adverse conditions (Abdolmaleki et al. 2013).

Fld expression in tobacco plant chloroplasts compensates for the reduction of ferredoxin levels caused by adverse environmental conditions and increases the tolerance of transgenic lines to oxidative stress and a wide range of environmental challenges. Expression of the *fld* gene in tobacco also caused transgenic tobacco lines to tolerate iron deficiency. Transfer of *fld* to tomatoes and potatoes also increased tolerance to oxidative stress and drought stress (Tognetti et al. 2006; Zurbriggen et al. 2007).

Considering that such a system in crops has disappeared along the evolutionary pathway of vascular plants from the plant genome and the benefits of its expression and function have been lost, due to the successful results in increasing tolerance to non-biological stresses in plants. By transferring the bacterial *fld* gene with useful biotechnology tools, the mentioned crop can be taken to increase the yield of other crops, especially wheat. However, due to the multiplicity and genetic and physiological complexity of tolerance to abiotic stresses, trying to improve these traits with conventional breeding programs, while accompanied by many limitations, also requires a lot of time (Abdolmaleki et al. 2013).

19.5.14 The Role of Bacterial *merA* Gene in Environmental Purification

Mercury is a toxic heavy metal that is widely distributed in ecosystems through industrial pollutants and their sediments. Fossil fuels are one of the major sources of mercury pollution in the environment and wastewater is a major source of two types of organic and inorganic mercury, such as elemental mercury, methyl mercuric chloride, and dimethyl mercury. Mercury is a major environmental pollutant and is one of the bio-accumulative toxins that stay in the environment for a long time (its estimated duration is between 0.5 and 2 years). Mercury changes its chemical forms in the environment, moving from one place to another, and being buried in soil and sediments. Most marine plants and animals absorb mercury, and organisms in the lower branches of the food cycle (such as plankton) trap mercury in their bodies. When vegetarians or carnivores eat the higher branches of the plankton food chain, mercury is transferred to the body of the fish and eventually consumed by humans. Mercury is a mutagenic, growth-inhibiting agent with toxic effects and the cause of most important human diseases and syndromes. The effects of mercury on ecosystem performance are economically and hygienically significant (Dash and Das 2012; Khoshniyat et al. 2018; Teng et al. 2020).

Mercury resistance has been observed in a variety of gram-positive and gram-negative bacteria. In bacteria, mercury resistance genes are mostly located in the operon on the plasmid or transposons. The narrow-spectrum mer resistance to inorganic mercury (merRTPADE) is only resistant to inorganic mercury, and the broad-spectrum mer operon is resistant to broad-spectrum mercury. Resistance to organomercurials (merRTPAGBDE) resists both organic and inorganic mercury. In this operon, *merA* encodes the mercury ion reductase enzyme, and this cytoplasmic protein plays a key role in mercury removal. This enzyme converts Hg^{2+} to less toxic Hg^0 by the following mechanism. Hg^0 is highly volatile and passes freely through biological membranes out of the cell and back into the atmosphere. Bacteria with the *merA* gene encoding the mineral mercury-lowering enzyme have the genetic ability to remove mercury by reducing the mineral mercury to a gaseous form, thus clearing the contaminated area. This is the last step in the path of non-toxicity of mercury in mercury-resistant bacteria, and thus, the bacteria remove mercury from their environment (Dash and Das 2012; Khoshniyat et al. 2018; Pietro-Souza et al. 2020) and prepare the environment for the cultivation and cultivation of agricultural products.

19.5.15 The Role of Chitinase Gene in Production of Biological Fungicides and Insecticides

Chitinases (EC: 3.2.1.14) are bonds of carbon 1 and 4 between two back-molecules of N-acetylglucosamine (GlcNAc) in chitin chains ranging in size from 20 kDa to approximately 90 kDa; they hydrolyze. Secretory chitinases can be present in chitin-containing (insects, crustaceans, and fungi) or chitin-deficient (plants and bacteria) species. In certain bacterial species, including *Aeromonas*, *Serratia marcescens*, *Myxobacter*, *Vibrio*, *Streptomyces*, and *Bacillus* species, this enzyme is present in abundance. The chitinase enzyme has attracted a great deal of interest due to its wide variety of commercial, agricultural, and medical applications, and the isolation of these enzymes from microscopic organisms has found wide applications in the biocontrol of fungi and nematodes of agricultural pests (Mortazavi et al. 2017).

In addition to their role in the growth and division of fungal cells, *Trichoderma* filamentous fungi, due to the secretion of different chitinase enzymes, are used as a powerful factor in the biological control of fungal diseases. These fungi, chitinase enzymes, have many advantages over other organisms' chitinases, including plant chitinases. For example, plant chitinases, unlike fungal chitinases, only affect the tip of the pathogenic fungal hyphae and are unable to break down the hard chitin structures. Also, these enzymes alone have weak antifungal effects and are effective only on a limited number of fungal species. Studies have shown, on the other hand, that all chitin-containing pathogens in the wall are susceptible to *trichoderma* fungal chitinases, while high concentrations of these enzymes do not have any toxic effects on plants (Berini et al. 2018; Chen et al. 2017).

The possible risk of crop epidemics still exists today. Thus, the introduction of new genes encoding antimicrobials and fungi is what is required in new genetic engineering methods. Therefore, what is needed in new methods of genetic engineering is the introduction of new genes encoding antimicrobials and fungi. Chitinases are one gene class of this type. In insects, nematodes, and some other species, these genes encode beta-1-4 hydrolyzing enzymes in cell wall chitin of fungi and exoskeletons (Ahmadian et al. 2012).

Chitinases can be used as a biological fungicide and insecticide agent, as well as in animal SCP processing, aquatic organism feeding, fungal protoplast isolation, bioactive cyto-oligosaccharides preparation, and plant pathogens inhibition. Chitin material and environmental purification and its conversion into raw materials, and with population growth and limited natural resources, enzyme technology can be useful for many industries to overcome economic problems soon (Babashpour et al. 2011). Chitinase is a recombinant and decomposing chitin that due to this antifungal property can be used as an effective substance in the treatment of human fungal infections and a safe substance in the elimination of pests and biological contaminants (Ahmadian et al. 2012).

The native Iranian strain of gram-positive *Paenibacillus* sp. bacterium A01 in southern Iran, shrimp ponds have been used for the development of recombinant protein chitinase. After replication of the gene by specific primers for heterologous expression of the recombinant enzyme, cloning was performed in the expression vector pET26b and transferred to Escherichia coli to produce the desired protein. Protein extraction was performed with a nickel-Sepharz affinity column. Its biological activity has also been studied. The results showed that the degradation of chitin by the enzyme chitinase in the bacterium *Paenibacillus* sp. A01 has been developed and it has been suggested that this bacterium be used for mass production of this enzyme in industrial and agricultural applications such as biological pesticides (mortazavi et al. 2017).

19.6 The Role of Microorganisms in Dissolving Phosphate

Phosphorus is one of the most significant plant nutrients and has a little abundance in the soil. Phosphorus, in both organic and inorganic forms, is present in the soil. The ability of some microstructures to convert insoluble phosphorus into a usable form such as orthophosphate is an important feature of PGPR that enhances plant yield. *Pseudomonas*, *Bacillus*, *Pantoea*, and *Rhizobium* are the most potent phosphate solvents. Although the genetic knowledge of phosphate dissolution is still limited, several genes encoding phosphatase have been identified and cloned, and a number of genes involved in mineral phosphate dissolution have been isolated. Transfer and expression of genes involved in phosphate dissolution (organic or inorganic phosphate) in bacteria or plants is a new way to improve the capacity of microorganisms as a microbial inoculum (Sarikhani et al. 2014).

19.6.1 Dissolution of Mineral Phosphate

The activity of mineral phosphate dissolution is linked to the ability of microstructures to release metabolites such as hydrogen ion or proton (H^+) secretion and organic acid (Liu 2019; Surange et al. 1997). Some believe that organic and inorganic acids with their carboxyl and hydroxyl groups chelate the cations with phosphate anion (Al^{3+}, Fe^{3+}, Ca^{2+}) and thus help the dissolution of phosphate (Purakayastha et al. 2019; Stevenson and Cole 1999); some also believe that the dissolution of phosphate Anion exchange of PO_4^{3-} with anion is an organic acid (Jiang et al. 2019; Omar 1997).

The study of the production of organic acids is often done in liquid media and has been done by methods such as paper chromatography or thin-layer chromatography or by high-performance liquid chromatography (HPLC) and some specific enzymatic methods (Gupta et al. 2020; Gyaneshwar et al. 1998). Among different organic acids, gluconic acid seems to be the predominant and most important acid produced in gram-negative bacteria (de Oliveira Mendes et al. 2020; Goldstein et al. 1993). Production of this organic acid by bacteria such as *Pseudomonas* sp., *Erwinia herbicola*, *P. cepacia*, *Azospirillum* spp., *R. leguminosarum*, *R. meliloti*, *B. firmus*, and *Burkholderia cepacia* has been reported (Fazeli-Nasab and Sayyed 2019; Rodríguez and Fraga 1999; Rodriguez et al. 2004; Saia et al. 2020). Strains of *B. licheniformis* and *B. amyloliquefaciens* have been found to produce a mixture of acetic, lactic, isovaleric, and isobutyric acids. Other organic acids such as succinic, malonic, oxalic, and glycolic acids have also been identified among phosphate solvents (Rodríguez and Fraga 1999; Saia et al. 2020).

In addition to organic acids, phosphatase and phytase enzymes are also involved in the dissolution of phosphates. In the case of enzyme-destroying agent therapy, if there is no improvement in the release of phosphorus, it is suggested that the mechanism involved in the dissolution of phosphate is non-enzymatic and is connected to the processing of organic or inorganic acids. If the dissolution of phosphate is stopped if it is treated with a base, it indicates that the dissolution of phosphate is organic acid (Fazeli-Nasab and Sayyed 2019; Rodríguez and Fraga 1999; Saia et al. 2020). Based on these findings, the cloning of genes related to mineral phosphate dissolution was followed. Oxidation of glucose to gluconic acid and ketogluconic acid is the metabolic basis of mineral phosphate-solubilizing phenotypes in some gram-negative bacteria (Goldstein 1995; Khoshmanzar et al. 2020).

19.6.1.1 The Most Important Genetic Factors Involved in the Dissolution of Mineral Phosphate

In 1987, for the first time, an *Erwinia herbicola* gene involved in mineral phosphate dissolution was cloned in an environment containing hydroxyapatite as the sole source of phosphorus by screening antibiotic-resistant recombinants from the Genomic library. Expression of this gene has led to the production of gluconic acid and the dissolution activity of the mineral phosphate in *E. coli* HB101. Sequencing of this gene indicates its possible involvement in the synthesis of *Pyrroloquinoline quinone* (PQQ), which is an essential factor in the formation of the glucose dehydrogenase holoenzyme (GDH) (Table 19.3). GDH-PQQ catalyzes the formation of gluconic acid by direct oxidation of glucose. *E. coli* is able to produce GDH but cannot make PQQ, so gluconic acid is not produced (Dai et al. 2020; Rodríguez and Fraga 1999; Rodríguez et al. 2006; Saia et al. 2020).

Table 19.3 Inorganic phosphate solubilizing genes from different bacteria

Microorganism	Gene or plasmid	Characteristics	References
Erwinia herbicola	*Mps*	Produces gluconic acid and dissolves mineral phosphorus in *E. coli* HB101, possibly involved in PQQ synthesis	Rodríguez et al. (2006)
Pseudomonas cepacia	*gab Y*	In *E. coli* JM109 produces gluconic acid and dissolves mineral phosphorus, has no similarity with PQQ genes	Rodríguez et al. (2006)
Microbe-derived enzymes	*olpA, phoD, appA, phnX, and phnJ*	Able to release free orthophosphate from organic P form	Liang et al. (2020)
Enterobacter agglomerans	*pKKY*	Dissolve phosphorus in *E. coli* JM109, without lowering the pH	Rodríguez et al. (2006)
Microbe-derived enzymes	*NSAPs*	Releasing inorganic phosphates from nucleotides and sugar phosphates	Sarikhani et al. (2014)
Serratia marcescens	*pKG3791*	Produces gluconic acid and dissolves mineral phosphorus	Rodríguez et al. (2006)
Synechococcus PCC7942	*Pcc gene*	Produces phosphoenol carboxylase	Rodríguez et al. (2006)
Pseudomonas fluorescens	*pqqA,B,C,D,E, and F*	Produces gluconic acid and dissolves mineral phosphorus, possibly involved in PQQ synthesis	Orikasa et al. (2010)
Burkholderia multivorans WS-FJ9	*AP-2, GspE, GspF, PhoR, HlyB, PhoA, AP-1* and *AP-3*	Involved in the decomposition of organic and inorganic phosphates	Liu et al. (2020)
Ralnella aquatillis	*KIM10*	In *E. coli* DH5α coli produces gluconic acid and dissolves mineral phosphorus, possibly involved in PQQ synthesis	Rodríguez et al. (2006)
Prokaryotic genes	*phoX, phoA, and phoD*	Conversion of organic phosphate to mineral and accessibility for plants	Wan et al. (2020)

By taking a similar approach, another gene associated with mineral phosphate dissolution was isolated from *P. cepacia* (Table 19.3). Expression of this *gabY* gene, which led to the formation of a mineral phosphate-solubilizing phenotype through the production of gluconic acid in *E. coli* JM109, bore no apparent resemblance to the cloned gene synthesizing *PQQ*. The *gabY* gene was able to play an alternative role in expressing or regulating the direct oxidation pathway in *P. cepacia* (Dai et al. 2020; Rodríguez and Fraga 1999; Saia et al. 2020). Other isolated genes associated with MPS phenotypes do not appear to be limited to *pqqDNA* and *gab-synthesizing* genes. Genomic DNA fragments from *Enterobacter agglomerans* showed MPS activity in *E. coli* JM109, although the ambient pH did not change. These results indicate that acid production is an important method but not the only mechanism involved in the dissolution of phosphate by bacteria. Isolation of the phosphoenolpyruvate carboxylase encoding gene pcc in *Synechococcus* PCC7942 indicates that it is involved in MPS (Dai et al. 2020; Rodríguez et al. 2006).

19.6.2 Mineralization of Organic Phosphorus

Organic compounds of phosphorus such as phytic acid can account for 20–80% of soil phosphorus (Chiu and Paszkowski 2019; Raghothama and Karthikeyan 2005), although changes have been reported in the range between 4 and 90%. Approximately half of the microorganisms with their phosphatase activity present in the soil and roots of plants mineralize organic phosphorus (Tarafdar et al. 1988; Zhang et al. 2020). In the form of a substrate, acidic and alkaline phosphatases convert organic phosphate into a mineral (Beech et al. 2001; Xu et al. 2020).

Phosphorus can be released from organic compounds in the soil by three enzyme groups. (1) Nonspecific phosphatases that follow the dephosphorylation of phosphoester or phosphoanhydride bonds in organic matter. (2) Specific phosphatases such as phytases that release phosphorus from phytate. Although this division is not correct in terms of gene and protein sequence, but in practice it can be said that the main activity of organic phosphorus mineralization is the responsibility of the first two groups (Rodríguez et al. 2006).

Phytases (Myoinositol hexaphosphate phosphohydrolase) belong to a special group of phosphomonoesterases that are able to release phosphorus from phytate (Davis 2020; Zhao et al. 2019). Phytic acid was first discovered in 1903, and its salts are known as phytates (Haefner et al. 2005; Nobile et al. 2019).

19.6.2.1 The Most Important Genetic Factors Involved in the Dissolution of Organic Phosphorus

19.6.2.1.1 Phosphatases

Different phosphatase activity patterns have been observed in bacteria, and complex regulatory mechanisms such as specific environmental conditions often control the production of these enzymes (Liang et al. 2020; Wan et al. 2020). Probably the main mechanism for regulating the expression of phosphatases is an induction by the amount of mineral phosphate (Pi) present in the medium. This mechanism has been studied for alkaline phosphatase (*pho A*) *E. coli*. When Pi concentration decreases to 0.16 mM, the expression of this gene is induced. This mechanism includes a Pi transport operon Pi as a regulating element, as well as a sensor-activator operon as a solver and activator. Genes controlled by the amount of Pi, their expression is activated by PhoB, which is the main part of the PHO regulon (Rodríguez and Fraga 1999).

The grouping of bacterial phosphatases was focused on the enzyme's biochemical and biophysical properties, such as optimum pH (acidic, neutral, or alkaline), substrate profile (specific or nonspecific), and molecular weight (high versus low molecular weight) (Dai et al. 2020; Gaiero et al. 2020; Rossolini et al. 1998).

Bacterial nonspecific acid phosphatase (NSAP) belongs to three families called molecular groups A, B, and C (Rossolini et al. 1998). Attention to group A of these enzymes for bioremediation of heavy metals has increased in the last decade. Attention is also paid to NSAPs for the transfer and expression of these genes in PGPR bacteria to achieve improved phosphate-solubilizing strains using recombinant DNA technology (Rodríguez et al. 2006).

The existence of preserved domains is seen in the comparison of amino acid sequences of six recognized group A enzymes, and the existence of the GSYPSGH [TA] motif is characteristic of this family (Felts 2007; Rossolini et al. 1998). The presence of highly conserved fragments in them is shown by a comparison of amino acid sequences in group B. In this category (Gaiero et al. 2018; Rossolini et al. 1998), the FDIDDTVLFSSP sequence is suggested as a signature sequence pattern. Although group C is distinct from the other two groups, it is similar in sequence level to group B phosphatases and some plant acid phosphatases. The first identified member of this group is the OlpA-Cm protein, whose gene encoding was isolated from the bacterium Chryseobacterium meningosepticum (Rossolini et al. 1998; Wang et al. 2019). A comparison of the amino acid sequences of this group with other proteins has allowed the identification of protected and common areas between these sequences. The results show that group B and group C acid phosphatases, together with some plant acid phosphatases, can be included in a protected subfamily called "DDDD phosphohydrolase subfamily" due to the presence of four aspartate (D) amino acids (Rossolini et al. 1998; Wang et al. 2019).

Several acid phosphatase genes have been isolated from Gram-negative bacteria and characterized. For example, the *acpA* gene isolated from Francisella tularensis encodes acid phosphatase with optimal activity at 6 pH: with a wide range of action on specific substrates. Class A (*PhoC*) and class B (*NapA*) acid phosphatase encoding genes were also isolated from Morganella morganii. In addition, these enzymes are rebellious or non-inducible and show high activity at pH: 6 and 30 °C and action on various substrates (Rodríguez et al. 2006; Sarikhani et al. 2014).

Among rhizobacteria, a gene that facilitates phosphatase activity has been isolated from *Burkholderia cepacia*. This gene encodes an outer membrane-bound protein that increases its expression in the absence of solution phosphorus and may be involved in transporting phosphorus (Liu et al. 2019; Rodríguez et al. 2000). In addition, cloning of two nonspecific phosphatase genes of periplasmic acid (nap E and nap D) from *Sinorhizobium meliloti* was carried out. Also, cloning and transfer of *napA* phosphatase gene from Morganella morganii to *Burkholderia cepacia* IS-16 was performed using broad-host-range vector *pRK293* and an increase in the extracellular phosphatase activity of the recombinant strain was reported (Ferroni et al. 2011; Rodríguez et al. 2006). Cloning and expression of the phosphotriester gene (*hocA*) of *Pseudomonas monteilii* C11 have been performed (Horne et al. 2002; Latip et al. 2019). It is named because of the hydrolysis of coroxon as a source of organic phosphate. The gene is 501 bp in length and encodes a 19 Kd protein, enabling the bacterium to use this source of organic phosphate as the only source of phosphorus in the environment.

19.6.2.1.2 Phytases

Phytate (Mayo Inositol Hexaxis Phosphate) is one of the major forms of phosphorus in oilseeds, legumes and oilseeds (Mayo Inositol Hexaxis Phosphate). In many grains and oilseeds, phytic acid makes up 1–3% of the weight, and usually 60–80% of the phosphorus in these plants. Phytase is the enzyme that hydrates phytate to lower myoinositols and, in some cases, free myoinositol and mineral phosphorus (Ariyan Nezhad et al. 2013).

Phytase (myoinositol hexaxis phosphate phosphohydrolase) genetic studies began in 1984, and in the mid-1990s, the first commercial phytase developed by engineered microstructures entered the market under the name of Natuphos (Bavaresco et al. 2020; Corrêa and de Araújo 2020). In order to improve the diet of monogastric animals, most genetic engineering studies have concentrated on phytase science. It is also used to dissolve soil phytate, as inoculants with high phytase production are among the favorites to improve plant nutrition and reduce soil phosphorus contamination. Phytase genes are also cloned from fungi, plants, and bacteria. Heat-stable genes (phy) from *Bacillus* sp. DS11 and cloned from *B. subtilis* VTT E-68013 (Corrêa and de Araújo 2020; Konietzny and Greiner 2004; Rodríguez et al. 2006). So far, four distinct groups have been reported based on the sequence of genes and their conserved regions, three-dimensional structure, reaction mechanisms, and enzymatic properties for phytases abbreviated to *HAP*, *PAP*, *CP*, and *BPP* (Hou et al. 2020; Naghshbandi and Moghimi 2020; Sarikhani 2012).

Phytases are also divided into two general groups based on optimal pH of activity: acidic and alkaline phytases. The first group includes fungal phytases and the group of gram-negative bacteria, and *Bacillus* bacteria belong to the group of alkaline phytases (Oh et al. 2004; Vasudevan et al. 2019). In another classification, phytases are named based on which group of phosphates is first removed from the phytate by the enzyme. For example, 3-phytase or 6-phytase, which indicates that phosphate 3 and phosphate 6 are the first phosphates removed from the phytate molecule, respectively. *E. coli* phytase is of 6-phytase type, while fungal phytases and bacilli are of 3-phytase type (Jatuwong et al. 2020; Oh et al. 2004; Vats and Banerjee 2004; Vohra and Satyanarayana 2003).

Natural phytases are distinguished into constitutive and inducible phytases in terms of expression pattern (Shieh and Ware 1968; Vohra and Satyanarayana 2003).

Fungal extracellular phytase is induced and produced at low concentrations of mineral phosphate in the growth medium (Vohra and Satyanarayana 2003). Unlike fungal phytase, *B. subtilis* phytase is induced in the presence of phytate (Kerovuo 2000).

Many fungal and bacterial phytases have been studied. The enzyme phytase produced by gram-positive bacteria and fungi is extracellular (Ariyan Nezhad et al. 2013). Phytase-producing bacteria and fungi have been extensively studied, including gram-negative bacteria such as *E. coli*, *Pseudomonas* sp., *Klebsiella* sp. and from gram-positive bacteria to *Bacillus* sp. (Haefner et al. 2005; Oh et al. 2004; Vats and Banerjee 2004). Cloning, sequencing, and expression of the acid phosphatase/phytase gene (*appA2/appA*) have also been identified in pigs (Liang et al. 2020). The dual function of these enzymes makes them attractive for dissolving soil organic phosphorus. However, isolation of phytase genes from *Aspergillus Niger*, *Emericella nidulans*, and several other species has been previously reported. Alkaline to neutral phytase genes also derived from *B. subtilis* and *B. licheniformis* have been cloned (Kerovuo 2000).

Due to the importance of phytase enzyme in the dissolution of organic phosphate and the need of monogastric animals to use these food sources, the production of phytase and its addition as a food additive to the diet [feed additive] is considered (Kour et al. 2020; Pandey et al. 2001). In a study, isolation and sequencing of the *phyA* gene from the Obesumbacterium proteus genomic library was reported, and then cloning and expression of this gene in *E. coli* were investigated, and the characteristics of the produced phytase enzyme were evaluated (Zinin et al. 2004). In another study, after establishing the genomic library of *P. putida* strain P13 in *E. coli* strain DH5α strain, they screened in at least Sperber medium in the presence of BCIP and then isolated two genes related to organic phosphate mineralization called PPP1 and PPP2. Examination of the enzymatic properties of the mentioned genes has shown that PPP1 has a prominent phytase property, while PPP2 has shown more sugar phosphatase properties (Malboobi et al. 2013; Sarikhani et al. 2014).

19.7 The Role of Different Microbial Genes in the Synthesis of Bioethanol

Bioethanol (bioethanol, bio-ethyl alcohol) is a clear, colorless, bicarbonate (C_2H_5OH), with low toxicity, biodegradability and causes less pollution to the atmosphere. Ethanol combustion absorbs carbon dioxide and water. The biological raw materials used in bioethanol production are primarily plant-based. Bioethanol (Champagne 2007; Yu et al. 2020) may also be produced in some raw materials of animal origin. A large part of agricultural products, wastes and wastes of agricultural products, by-products of agricultural conversion industries, products and wastes of by-products of forest and related industries, and of course urban and industrial wastes and biological wastes are used for bioethanol production. Increasing the yield of ethanol along with improving production economic processes and technological advances is key parameters in ethanol production. In order to achieve a high yield of ethanol, strains must be produced that produce fewer unnecessary products and are able to metabolize all major sugars. One of the main obstacles to the development of bioethanol production has been the lack of suitable industrial microorganisms to convert biomass into ethanol fuel. Quantitative conversion of glucose to alcohol is done by yeasts as well as a number of bacteria. Lignocellulosic biomass contains complex carbohydrates that necessitate the use of associated microorganisms; for the yeast to ferment non-fermentable sugars (Zeinali et al. 2016).

The pyruvate decarboxylase gene from the bacterium Zymomonas mobilis is the primary and main gene for the development of bioethanol and, among all related enzymes in other microorganisms, the enzymes it produces have the highest enzymatic activity. It was ethanol (Zeinali et al. 2016). Also, the enzyme *Cel6B* from the bacterium *Thermobifida fusca*, a CBHII belonging to the family of celluloses that is very resistant to heat, which is ultimately effective in the production of ethanol (Heidari-Gharehsoo et al. 2018).

The hexose transporter family in the yeast *Saccharomyces cerevisiae* includes the proteins *Hxt1p-Hxt17p*, *Gal2p*, *Snf3p*, and *Rgt2p*. As the activity of these transducers increases, the accumulation of ethanol or lactic acid in the cells increases. *Hxt1* to *Hxt17* transducers are involved in glucose transport, *Gal2* acts as a glucose sensor for galactose transport, *Snf3* and *Rgt2*. This gene family has different expression patterns and their regulation is strongly influenced by the kinetics of transmitters and glucose is the first-factor controlling expression. *HXT1-HXT7* are among the most important transporters and are metabolically similar and interrelated. The effect of maximal expression of these genes in yeasts has been studied and ethanol production in the wild yeast strain has been compared with the engineered strain. The obtained data showed that overexpression of hexose transporters leads to increased glucose uptake. The researchers showed that by regulating the first stage of the glucose biosynthesis pathway, lactic acid accumulation could be observed, which observed a 15% increase in ethanol production compared to the wild strain (Azizi et al. 2016).

19.8 Use of Microbial Genes in Plant Genome Editing

Ensuring society's food security and the need to increase agricultural production on an ongoing basis depends on arming and efficiently integrating conventional plant breeding with modern biotechnology and powerful tools for genetic engineering (Kirillova et al. 2020; Yadav et al. 2019). The most complex branch of biotechnology is genetic engineering, which involves methods to pick the target gene, locate, isolate, purify, replicate, and transfer genes and test their expression in living organisms (Kim et al. 2020; Sedeek et al. 2019).

At present, the great importance and ability of genetic engineering by making purposeful changes in the genomes of plants and animals to remove many limitations of traditional plant breeding, create new characteristics, and improve the quantity and quality of food products by producing crops and gardens. Transgenics are not hidden from anyone. However, modern biotechnology has been accepted as a way of producing products that have wide applications in medicine, agriculture and industry. However, the safety aspects of genetically modified or transgenic organisms (LMOs) and their products must also be carefully considered before use. (Gabriel 2019; Haribabu 2019).

Selective genes such as antibiotic resistance and herbicide genes are mentioned as the most important considerations influencing the utilization of transgenic products, penetration of non-plant DNA fragments into the parent plant genome, gene escape, and vertical transfer of the target gene (Mackelprang and Lemaux 2020; Mathur 2018; Parray et al. 2019). Therefore, the need to use advanced methods of producing transgenic plants such as short intervals of the regular clustered palindrome

(*CRISPR*/*Cas9*), which is a genome editing method, to improve the quality and effective yield of high-yield crops is felt more than ever. So that this system can accelerate plant modification without transferring external genes by making mutations in multiple gene sites and creating large deletions. This method can improve the function and activity of plant genes and create new traits. Over the years, the process of this advanced technology has been extensively studied by numerous examples of targeted mutations and regulation of the copying process in a variety of crops, and this has demonstrated the effective aspect of this new system. To date, the *Cas9* system has been widely used in gene silencing, gene replacement, multiple gene editing, gene function identification, and regulating the replication process in animals and plants (Artegiani et al. 2020; Cunningham et al. 2018; Mackelprang and Lemaux 2020; Molla and Yang 2020; Si et al. 2020).

19.8.1 Overview of the *CRISPR*/*Cas9* System Mechanism

In the Escherichia coli genome, the *CRISPR*/*Cas*9 system was first identified in 1987 as an acquired immune system against the invasion of bacteriophages and the entry into the bacterial cell of foreign DNA such as plasmids. In all prokaryotes, 2000 *CRISPR* gene families were described (Si et al. 2020; Zhang and Zhang 2020). According to the *CRISPRdb* (http://www.crispr.i2bc.paris-saclay.fr) database, the locus of the *CRISPR* gene is approximately 84% in the archaeal genome and 45% in the bacterial genome (Makarova et al. 2020).

The *CRISPR* system consists of two regions, including genes encoding Cas nuclease enzymes and the gene locus of *CRISPR* arrays containing repeat sequences and spacer sequences between them. The length of duplicate sequences is about 25–50 bp and more than 249 bits and the spacing area containing non-duplicate sequences is about 26–72 bp (Arslan et al. 2014; Kunin et al. 2007; Pourcel et al. 2020). The leader sequence is approximately 200–500 bp in length, consisting of AT-rich sequences that are necessary as a promoter sequence for copying *CRISPR* locus arrays. The 4 (*Cas*) associated *CRISPR* genes (Cas 1–4) are located near the CRISPR array region, encoding proteins essential for inducing an immune response by bacteria against virus attack (Fonfara et al. 2016; Richter et al. 2012; Zhang and Ye 2017).

Streptococcus pyogenes has been widely used to edit the genomes of various species and cell types such as human, bacterial, yeast, laboratory mice, vinegar flies, nematodes, crops, insects, and monkeys. In the crRNA processing mechanism type II, the Trans-activating crRNA sequence binds to duplicate sequences to form the crRNA/tracrRNA complex. It then produces mature crRNAs with the *Cas9* protein and RNaseIII enzyme activity. This processing system is well known and widely used in genetic engineering and genome editing. The generated *crRNAs* have 20 nucleotide sequences at their 3′ end. In this system, adult *crRNAs* are generated after transcription of pre-crRNA sequences from the *CRISPR* gene locus by the *Cas9* protein (Fonfara et al. 2016; Hynes et al. 2017; Marraffini 2016; Zhang and Ye 2017).

CRISPR/Cas9 achievements in plants: The *CRISPR/Cas9* system produces stable and inherited mutations that can be easily distinguished from the *Cas9*/gRNA structure for further modification by *CRISPR/Cas9*. This leads to the development of non-transgenic homozygous plants that have been produced in only one generation (Fauser et al. 2014; Xu et al. 2015). A non-transgenic rice cultivar with a mutation in the target gene has been successfully produced by transgenic cleavage by causing T1 generation self-healing (Xu et al. 2015). Also, a series of dual vectors based on *CRISPR/Cas9* system with stable expression capability in plant systems and a series of vectors containing gRNA module have been designed. Therefore, the transfer of only Cas9 nuclease protein and gRNA into the host cell by genetic transfer methods is the only necessity for plant genome editing (Xing et al. 2014).

It has been suggested that viral glycemic replicates (GVRs) can be used to transfer the *Cas9*/gRNA structure into the host cell when the virus replication initiation protein (REP) gene is transferred along with the Cas9/gRNA structure (Baltes et al. 2014). In addition, in order to use this system in upgrading and discovering genetic traits, high-performance transfer methods such as DNA-based replication viruses are used to transfer genome engineering materials without the need for genetic engineering transfer methods (Ali et al. 2015; Yin et al. 2015). Based on the researches, direct transmission using tobacco rattle virus (TRV), cabbage leaf complex virus (*CaLCuV*), *Cas9*/gRNA transferability by different viruses in editing the genomes of different plants have been clearly shown (Yin et al. 2015).

19.9 Conclusion

In today's world, there are other significant issues such as climate change and environmental problems, and the loss of genetic resources, in addition to the issue of human communities supplying food and producing more food, one of the reasons for the improper use of chemical products in agriculture (such as toxins and chemical fertilizers, etc.).

Although pesticides and herbicides play an important role in controlling plant diseases and weeds, they have negative side effects on living organisms and their environment. They also cause resistance to pests and weeds and have adverse effects on desirable non-target microorganisms. Also, the limitation of the use of physical and chemical cleaning methods of plant toxins, such as reducing the quality and nutritional properties of the product and the high cost of the required equipment, has led researchers to focus on biological methods. In order to minimize the use of these toxic chemicals, the use of biological agents such as fertilizers and biotoxins can play a very important role in the protection of the environment and of agricultural farms. Moreover, by using modern biological technology, such as genetic engineering, increasing the performance of these biological factors will also improve their efficacy. However, the transfer of new genes to these biological agents and their release into agricultural environments can be associated with certain potential environmental hazards. Therefore, due to the acquisition and development of new molecular technologies to track and study the effects of such recombinant microorganisms in nature, it is necessary to carefully study the possible effects to safely and efficiently use the definite benefits of these organisms in agriculture.

It is suggested that given the standards for assessing the potential hazards of recombinant microorganisms as well as microbial compounds (genes, proteins, or metabolites) used, possible risk assessment on a case-by-case basis for each microorganism or microbial compound according to the purpose and type of environment used and done in a completely scientific manner and if there was a possibility of specific hazards for that microorganism or the microbial compound used, scientific and managerial solutions should be provided to prevent or reduce those risks so that we can take full advantage of the environmental and economic benefits of such biological products.

Acknowledgment We thank Dr. Ali Izadi-Darbandi from University of Tehran for providing the information in Fig. 19.1 and Dr. Mohammad Forouzandeh from the University of Zabol for providing the information in Tables 19.1 and 19.2.

References

Abdollahi K, Abdollahi A (2008) Bioremediation of hydrocarbons in soil (*chlorine hydrocarbons*). Theol Z 193:11–18

Abdolmaleki S, Tohidfar M, Fotovat R, Salehi-Jozani G (2013) Genetic engineering of wheat, using *fld* gene by gene gun method. J Crop Breed 5(11):1–10

Abou-Shanab R, Van Berkum P, Angle J (2007) Heavy metal resistance and genotypic analysis of metal resistance genes in gram-positive and gram-negative bacteria present in Ni-rich serpentine soil and in the rhizosphere of *Alyssum murale*. Chemosphere 68(2):360–367. https://doi.org/10.1016/j.chemosphere.2006.12.051

Aguilera J, Gomes AR, Olaru I (2013) Principles for the risk assessment of genetically modified microorganisms and their food products in the European Union. Int J Food Microbiol 167 (1):2–7. https://doi.org/10.1016/j.ijfoodmicro.2013.03.013

Ahmadian G, Keshavarz M, Ahmadi Zeydabadi M (2012) Cloning and expression of Recombinant antifungal chitinase enzyme of *Bacillus pumilus* in *Bacillus subtilis* 168. Koomesh J 13 (2):151–158

Ahmadzadeh M (2013) Biological control of plant diseases, plant probiotic bacteria. Uni Tehran Publication, Tehran, p 474

Alexander D, Zuberer D (1993) Responses by iron-efficient and inefficient oat cultivars to inoculation with siderophore-producing bacteria in a calcareous soil. Biol Fertil Soils 16(2):118–124. https://doi.org/10.1007/BF00369412

Ali Z, Abul-faraj A, Piatek M, Mahfouz MM (2015) Activity and specificity of TRV-mediated gene editing in plants. Plant Signal Behav 10(10):e1044191. https://doi.org/10.1080/15592324.2015.1044191

Alizadeh Arimii F, Chalavi V, Dehestani A (2015) Simultaneous transformation of three biphenyl dioxygenase bacterial genes into *Arabidopsis* plants. Iran J Field Crop Sci 46(1):147–155. https://doi.org/10.22059/ijfcs.2015.54054

Allen HK, Cloud-Hansen KA, Wolinski JM, Guan C, Greene S, Lu S, Boeyink M, Broderick NA, Raffa KF, Handelsman J (2009) Resident microbiota of the gypsy moth midgut harbors antibiotic resistance determinants. DNA Cell Biol 28(3):109–117. https://doi.org/10.1089/dna.2008.0812

Al-Saadi A, Reddy JD, Duan YP, Brunings AM, Yuan Q, Gabriel DW (2007) All five host-range variants of *Xanthomonas citri* carry one *pthA* homolog with 17.5 repeats that determines pathogenicity on citrus, but none determine host-range variation. Mol Plant-Microbe Interact 20 (8):934–943. https://doi.org/10.1094/MPMI-20-8-0934

Aminov RI (2009) The role of antibiotics and antibiotic resistance in nature. Environ Microbiol 11 (12):2970–2988. https://doi.org/10.1111/j.1462-2920.2009.01972.x

Ariyan Nezhad H, Nasiri MR, Aslami Nezhad AA, Asoude A, Dehqani H (2013) Cloning, over Expression and Characterization of Alkalin Phytase Enzyme in *Escherichia coli*. Biocatal Agric Biotechnol 5(2):1–16. https://doi.org/10.22103/jab.2013.1197

Arslan Z, Hermanns V, Wurm R, Wagner R, Pul Ü (2014) Detection and characterization of spacer integration intermediates in type IE *CRISPR–Cas* system. Nucleic Acids Res 42 (12):7884–7893. https://doi.org/10.1093/nar/gku510

Artegiani B, Hendriks D, Beumer J, Kok R, Zheng X, Joore I, de Sousa LSC, van Zon J, Tans S, Clevers H (2020) Fast and efficient generation of knock-in human organoids using homology-independent *CRISPR–Cas9* precision genome editing. Nature 530:340–343

Ayub MA, Usman M, Faiz T, Umair M, Haq MA, Rizwan M, Ali S, Rehman MZ (2020) Restoration of degraded soil for sustainable agriculture soil health restoration and management. Springer, New York, pp 31–81

Azizi S, Tarinejad A, Pazhang M (2016) Isolation, cloning and analysis of the hexose transporter 6 gene (*HXT6*) in a native strain of *Saccharomyces cerevisiae* IBRC-M30069. Biol J Microorg 5 (18):29–40

Babashpour S, Aminzadeh S, Farrokhi N, Khosroshahli M, Keshavarz M (2011) Isolation and cloning of chitinase gene from *Serratia marcescens* B4A from shrimp farming ponds. Vet Microbiol 7(1):41–45

Baek D, Kim MC, Kumar D, Park B, Cheong MS, Choi W, Park HC, Chun HJ, Park HJ, Lee SY (2019) AtPR5K2, a PR5-like receptor kinase, modulates plant responses to drought stress by phosphorylating protein phosphatase 2Cs. Front Plant Sci 10:1146. https://doi.org/10.3389/fpls.2019.01146

Bagheri A, Sirousmehr AR, Asgharipour MR, Forouzandeh M (2018) Effect of mycorrhizal inoculation and phosphorus supply on morphological traits of rosemary under greenhouse conditions. J Appl Hortic 20(2):97–102

Bahramnejad B, Naji M, Bose R, Jha S (2019) A critical review on use of *Agrobacterium rhizogenes* and their associated binary vectors for plant transformation. Biotechnol Adv 37 (7):107405. https://doi.org/10.1016/j.biotechadv.2019.06.004

Bai J-N, Zhang Y, Zhao B-H (2006) Cloning of α-β fusion gene from *Clostridium perfringens* and its expression. WJG 12(8):1229–1234. https://doi.org/10.3748/wjg.v12.i8.1229

Baltes NJ, Gil-Humanes J, Cermak T, Atkins PA, Voytas DF (2014) DNA replicons for plant genome engineering. Plant Cell 26(1):151–163. https://doi.org/10.1105/tpc.113.119792

Baudoin E, Lerner A, Mirza MS, El Zemrany H, Prigent-Combaret C, Jurkevich E, Spaepen S, Vanderleyden J, Nazaret S, Okon Y (2010) Effects of *Azospirillum brasilense* with genetically modified auxin biosynthesis gene *ipdC* upon the diversity of the indigenous microbiota of the wheat *rhizosphere*. Res Discl 161(3):219–226. https://doi.org/10.1016/j.resmic.2010.01.005

Bavaresco C, Krabbe E, Gopinger E, Sandi A, Martinez F, Wernik B, Roll V (2020) Hybrid phytase and carbohydrases in corn and soybean meal-based diets for broiler chickens: performance and production costs. Rev Bras Cienc 22(1):1178. https://doi.org/10.1590/1806-9061-2019-1178

Beech I, Paiva M, Caus M, Coutinho C (2001) Enzymatic activity and within biofilms of sulphate-reducing bacteria. Biofilm Community Interact 2001:231–239

Berini F, Katz C, Gruzdev N, Casartelli M, Tettamanti G, Marinelli F (2018) Microbial and viral chitinases: attractive biopesticides for integrated pest management. Biotechnol Adv 36 (3):818–838. https://doi.org/10.1016/j.biotechadv.2018.01.002

Bettini PP, Marvasi M, Fani F, Lazzara L, Cosi E, Melani L, Mauro ML (2016) *Agrobacterium rhizogenes rolB* gene affects photosynthesis and chlorophyll content in transgenic tomato (*Solanum lycopersicum* L.) plants. Plant Physiol 204:27–35. https://doi.org/10.1016/j.jplph.2016.07.010

Boccardo NA, Segretin ME, Hernandez I, Mirkin FG, Chacón O, Lopez Y, Borrás-Hidalgo O, Bravo-Almonacid FF (2019) Expression of pathogenesis-related proteins in transplastomic tobacco plants confers resistance to filamentous pathogens under field trials. Sci Rep 9 (1):1–13. https://doi.org/10.1038/s41598-019-39568-6

Bosworth AH, Williams MK, Albrecht KA, Kwiatkowski R, Beynon J, Hankinson TR, Ronson CW, Cannon F, Wacek TJ, Triplett EW (1994) Alfalfa yield response to inoculation with recombinant strains of *Rhizobium meliloti* with an extra copy of *dctABD* and/or modified *nifA* expression. Appl Environ Microbiol 60(10):3815–3832

Boutigny A-L, Dohin N, Pornin D, Rolland M (2020) Overview and detectability of the genetic modifications in ornamental plants. Hortic Res 7(1):1–12. https://doi.org/10.1038/s41438-019-0232-5

Boyd SE, Livermore DM, Hooper DC, Hope WW (2020) Metallo-β-lactamases: structure, function, epidemiology, treatment options, and the development pipeline. Antimicrob Agents Chemother 64(10):20. https://doi.org/10.1128/AAC.00397-20

Burken JG, Schnoor JL (1996) Phytoremediation: plant uptake of atrazine and role of root exudates. J Environ Eng 122(11):958–963. https://doi.org/10.1061/(ASCE)0733-9372(1996)122:11(958)

Burken JG, Vroblesky DA, Balouet JC (2011) Phytoforensics, dendrochemistry, and phytoscreening: new green tools for delineating contaminants from past and present. Environ Sci Technol 45:6218–6226

Bush K, Jacoby GA (2010) Updated functional classification of β-lactamases. Antimicrob Agents Chemother 54(3):969–976. https://doi.org/10.1128/AAC.01009-09

Campbell MM, Ellis BE (1992) Fungal elicitor-mediated responses in pine cell cultures: III. Purification and characterization of phenylalanine ammonia-lyase. Plant Physiol 98(1):62–70. https://doi.org/10.1104/pp.98.1.62

Champagne P (2007) Feasibility of producing bio-ethanol from waste residues: a Canadian perspective: feasibility of producing bio-ethanol from waste residues in Canada. Resour Conserv Recycl 50(3):211–230. https://doi.org/10.1016/j.resconrec.2006.09.003

Chaudhry V, Patil PB (2020) Evolutionary insights into adaptation of *Staphylococcus haemolyticus* to human and non-human niches. Genomics 112(2):2052–2062. https://doi.org/10.1016/j.ygeno.2019.11.018

Chaurasia AK, Apte SK (2011) Improved eco-friendly recombinant *Anabaena* sp. strain PCC7120 with enhanced nitrogen biofertilizer potential. Appl Environ Microbiol 77(2):395–399

Chen J, An Y, Kumar A, Liu Z (2017) Improvement of chitinase Pachi with nematicidal activities by random mutagenesis. Int J Biol Macromol 96:171–176. https://doi.org/10.1016/j.ijbiomac.2016.11.093

Chen R, Li Y, Huang K, Li H, Chen H, Xu T, Hao D (2019) Time-course transcriptomic study of phenolic metabolism and *P450* enzymes in *Pinus massoniana* Lamb. after feeding by *Monochamus alternatus* hope. Scand J For Res 34(7):569–576. https://doi.org/10.1080/02827581.2019.1628295

Chiu CH, Paszkowski U (2019) Mechanisms and impact of symbiotic phosphate acquisition. Cold Spring Harb Perspect Biol 11(6):a034603. https://doi.org/10.1101/cshperspect.a034603

Clermont N, Lerat S, Beaulieu C (2011) Genome shuffling enhances biocontrol abilities of *Streptomyces* strains against two potato pathogens. J Appl Microbiol 111(3):671–682. https://doi.org/10.1111/j.1365-2672.2011.05078.x

Corrêa TLR, de Araújo EF (2020) Fungal phytases: from genes to applications. Braz J Microbiol May:1–12. https://doi.org/10.1007/s42770-020-00289-y

Costa-Gutierrez SB, Lami MJ, Santo MCC-D, Zenoff AM, Vincent PA, Molina-Henares MA, Espinosa-Urgel M, de Cristóbal RE (2020) Plant growth promotion by *Pseudomonas putida* KT2440 under saline stress: role of eptA. Appl Microbiol Biotechnol 2020:1–16. https://doi.org/10.1007/s00253-020-10516-z

Cunningham FJ, Goh NS, Demirer GS, Matos JL, Landry MP (2018) Nanoparticle-mediated delivery towards advancing plant genetic engineering. Trends Biotechnol 36(9):882–897. https://doi.org/10.1016/j.tibtech.2018.03.009

Dadivar M (2015) Rhizosphere, an approach to management the soil around the roots. J Land Manage 3(1):25–35. https://doi.org/10.22092/lmj.2015.103670

Dai Z, Liu G, Chen H, Chen C, Wang J, Ai S, Wei D, Li D, Ma B, Tang C (2020) Long-term nutrient inputs shift soil microbial functional profiles of phosphorus cycling in diverse agroecosystems. ISME J 14(3):757–770. https://doi.org/10.1038/s41396-019-0567-9

Das N, Chandran P (2011) Microbial degradation of petroleum hydrocarbon contaminants: an overview. Biotechnol Res Int 2011:941810. https://doi.org/10.4061/2011/941810

Dash HR, Das S (2012) Bioremediation of mercury and the importance of bacterial mer genes. Int Biodeterior Biodegradation 75:207–213. https://doi.org/10.1016/j.ibiod.2012.07.023

Davis RC (2020) The purification and properties of peanut phytase and the identification of the myo-inositol phosphates from partial dephosphorylation of myo-inositol hexaphosphate by the enzyme. Texas A&M University. Libraries. Retrieved from http://hdl.handle.net/1969.1/DISSERTATIONS-172058

de Oliveira Mendes G, Murta HM, Valadares RV, da Silveira WB, da Silva IR, Costa MD (2020) Oxalic acid is more efficient than sulfuric acid for rock phosphate solubilization. Miner Eng 155:106458. https://doi.org/10.1016/j.mineng.2020.106458

Driss F, Rouis S, Azzouz H, Tounsi S, Zouari N, Jaoua S (2011) Integration of a recombinant chitinase into *Bacillus thuringiensis* parasporal insecticidal crystal. Curr Microbiol 62 (1):281–288. https://doi.org/10.1007/s00284-010-9704-4

Eriksen NT (2008) Production of phycocyanin—a pigment with applications in biology, biotechnology, foods and medicine. Appl Microbiol Biotechnol 80(1):1–14. https://doi.org/10.1007/s00253-008-1542-y

Etminani F, Etminani A (2018) Phylogenetic characterization of endophytic bacterium *Pseudomonas Protegens* with antagonistic ability isolated from privet. J Ilam Univ Med Sci 26(4):74–84. https://doi.org/10.29252/sjimu.26.4.74

Fabrick J, Tabashnik BE (2007) Binding of *Bacillus thuringiensis* toxin *Cry1Ac* to multiple sites of cadherin in pink bollworm. Mol Biol 37(2):97–106. https://doi.org/10.1016/j.ibmb.2006.10.010

Fauser F, Schiml S, Puchta H (2014) Both *CRISPR*/C as-based nucleases and nickases can be used efficiently for genome engineering in *Arabidopsis thaliana*. Plant J 79(2):348–359. https://doi.org/10.1111/tpj.12554

Fazeli-Nasab B, Sayyed R (2019) Plant growth-promoting rhizobacteria and salinity stress: a journey into the soil plant growth promoting rhizobacteria for sustainable stress management. Springer, New York, pp 21–34. https://doi.org/10.1007/978-981-13-6536-2_2

FDA (2013) Food and drug administration center for food safety and applied nutrition (CFSAN) office of food additive safety agency response letter GRAS notice No. GRN000120

Felts RL (2007) Structural studies of acid phosphatases from pathogenic bacteria. University of Missouri, Columbia

Fernandes AFT, da Silva MBP, Martins VV, Miranda CES, Stehling EG (2014) Isolation and characterization of a *Pseudomonas aeruginosa* from a virgin Brazilian Amazon region with potential to degrade atrazine. Environ Sci Pollut Res 21(24):13974–13978. https://doi.org/10.1007/s11356-014-3316-7

Ferroni FM, Rivas MG, Rizzi AC, Lucca ME, Perotti NI, Brondino CD (2011) Nitrate reduction associated with respiration in *Sinorhizobium meliloti* 2011 is performed by a membrane-bound molybdoenzyme. Biometals 24(5):891–902. https://doi.org/10.1007/s10534-011-9442-5

Fladung M, Ballvora A, Schmülling T (1993) Constitutive or light-regulated expression of the *rolC* gene in transgenic potato plants has different effects on yield attributes and tuber carbohydrate composition. Plant Mol Biol 23(4):749–757. https://doi.org/10.1007/BF00021530

Fonfara I, Richter H, Bratovič M, Le Rhun A, Charpentier E (2016) The *CRISPR*-associated DNA-cleaving enzyme *Cpf1* also processes precursor *CRISPR* RNA. Nature 532 (7600):517–521. https://doi.org/10.1038/nature17945

Forouzandeh M, Mohkami Z, Fazeli-Nasab B (2019) Evaluation of biotic elicitors foliar application on functional changes, physiological and biochemical parameters of fennel (*Foeniculum vulgare*). Int J Plant Prod 25(4):49–65. https://doi.org/10.22069/jopp.2018.14077.2262

Franklin S, Rakitsky W, Rudenko G, Zhao X, Rodriguez F A, Lu W, Wee J (2019) Microbial oils with lowered pour points, dielectric fluids produced therefrom, and related methods: Google Patents, US10167489B2

Franklin S, Somanchi A, Rudenko G, Bhat R, Zhao X, Bond R, Rakitsky W, Marangoni A, Braksmayer D (2020) Structuring fats and methods of producing structuring fats: Google Patents, US20200032302A1

Gabriel A-B (2019) Genetically modified food, intellectual property, and food security in Africa (Master's thesis), University of Turku

Gahan LJ, Pauchet Y, Vogel H, Heckel DG (2010) An ABC transporter mutation is correlated with insect resistance to *Bacillus thuringiensis Cry1Ac* toxin. PLoS Genet 6(12):e1001248. https://doi.org/10.1371/journal.pgen.1001248

Gaiero JR, Bent E, Fraser TD, Condron LM, Dunfield KE (2018) Validating novel oligonucleotide primers targeting three classes of bacterial non-specific acid phosphatase genes in grassland soils. Plant Soil 427(1-2):39–51. https://doi.org/10.1007/s11104-017-3338-2

Gaiero JR, Bent E, Boitt G, Condron LM, Dunfield KE (2020) Effect of long-term plant biomass management on phosphatase-producing bacterial populations in soils under temperate grassland. Appl Soil Ecol 151:103583. https://doi.org/10.1016/j.apsoil.2020.103583

Gangoiti J, Pijning T, Dijkhuizen L (2018) Biotechnological potential of novel glycoside hydrolase family 70 enzymes synthesizing α-glucans from starch and sucrose. Biotechnol Adv 36 (1):196–207. https://doi.org/10.1016/j.biotechadv.2017.11.001

Gardoonpar H, Mohajal H, Hosseinpour-feizi MA (2016) Study of the effect of *rolC* and *trolC* genes on chlorosis in Tobacco plants. J Cell Mol Res 28(4):579–587

Gerhardt KE, Huang X-D, Glick BR, Greenberg BM (2009) Phytoremediation and rhizoremediation of organic soil contaminants: potential and challenges. Plant Sci 176 (1):20–30. https://doi.org/10.1016/j.plantsci.2008.09.014

Ghoreyshi E, Tohidfar M, Mohsenpour M (2016) *Agrobacterium*-mediated transformation of rice using cyanobacteria *fld* gene. J Crop Breed 8(18):7–15. https://doi.org/10.29252/jcb.8.18.7

Gibson D (1988) Microbial metabolism of aromatic hydrocarbons and the carbon cycle. Microbial metabolism and the carbon cycle. Harwood Academic Publishers, Chur, pp 43–52

Goldstein AH (1995) Recent progress in understanding the molecular genetics and biochemistry of calcium phosphate solubilization by gram negative bacteria. Biol Agric Hortic 12(2):185–193. https://doi.org/10.1080/01448765.1995.9754736

Goldstein AH, Rogers RD, Mead G (1993) Mining by microbe. Biotechnology 11(11):1250–1254. https://doi.org/10.1038/nbt1193-1250

Gqozo MP, Bill M, Siyoum N, Labuschagne N, Korsten L (2020) Fungal diversity and community composition of wheat rhizosphere and non-rhizosphere soils from three different agricultural production regions of South Africa. Appl Soil Ecol 151:103543. https://doi.org/10.1016/j.apsoil.2020.103543

Gupta S, Didwania N, Srinivasa N (2020) Role of biofertilizer in biological management of fungal diseases of pigeon pea [(*Cajanus cajan*)(L.) Millsp.] management of fungal pathogens in pulses. Springer, New York, pp 205–217. https://doi.org/10.1007/978-3-030-35947-8_12

Gyaneshwar P, Kumar GN, Parekh L (1998) Effect of buffering on the phosphate-solubilizing ability of microorganisms. World J Microbiol Biotechnol 14(5):669–673. https://doi.org/10.1023/A:1008852718733

Haefner S, Knietsch A, Scholten E, Braun J, Lohscheidt M, Zelder O (2005) Biotechnological production and applications of phytases. Appl Microbiol Biotechnol 68(5):588–597. https://doi.org/10.1007/s00253-005-0005-y

Haribabu E (2019) Public perceptions of risk about LMOs: a sociological perspective socio-economic impact assessment of genetically modified crops. Springer, New York, pp 61–77. https://doi.org/10.1007/978-981-32-9511-7_4

Hatami N, Bazgir E, Sedaghati E, Darvishnia M (2020) The symbiosis study of *Arbuscular Mycorrhizal* fungi with some annual herbaceous plants and the morphological identification of dominant species of these fungi in Kerman Province. BJM 9(33):41–55. https://doi.org/10.22108/bjm.2020.120148.1242

Heidari-Gharehsoo F, Moshtaghi N, Meshkani B, Malekzade-Shafarudi S (2018) Cloning and expression of *cel6B* cellobiohydrolase gene in *Pichia pastoris*. J Genet Eng Biotechnol 7 (2):139–151

Heitkamp MA, Franklin W, Cerniglia CE (1988) Microbial metabolism of polycyclic aromatic hydrocarbons: isolation and characterization of a pyrene-degrading bacterium. Appl Environ Microbiol 54(10):2549–2555

Hemmati TB, Mehdipour Moghaddam MJ, Salehi Z, Habibzadeh SM (2015) Prevalence of CTX-M-type β-lactamases in multi-drug resistant *Escherichia coli* isolates from north of Iran, Rasht. BJM 3(12):69–78

Hernández-Rodríguez CS, de Escudero IR, Asensio AC, Ferré J, Caballero P (2013) Encapsulation of the *Bacillus thuringiensis* secretable toxins *Vip3Aa* and *Cry1Ia* in *Pseudomonas fluorescens*. Biol Control 66(3):159–165. https://doi.org/10.1016/j.biocontrol.2013.05.002

Hesampour A, Mohandesi N (2018) Purification of fungal laccase from trametes and characterization of recombinant laccase physicochemical properties: a laboratory study. J Rafsanjan Univ Med Sci 17(6):511–522

Honari H (2008) Expression of *PA* gene from Bacillus antracis in Iranian lettuce *Lactuca sativa*. Ph. D. thesis, Department of Plant Breeding, College of Agriculture, University of Tehran

Horne I, Sutherland TD, Oakeshott JG, Russell RJ (2002) Cloning and expression of the phosphotriesterase gene *hocA* from *Pseudomonas monteilii C11bb* The GenBank accession number for the *hocA* gene is AF469117. Microbiology 148(9):2687–2695. https://doi.org/10.1099/00221287-148-9-2687

Hou X, Shen Z, Li N, Kong X, Sheng K, Wang J, Wang Y (2020) A novel fungal beta-propeller phytase from nematophagous *Arthrobotrys oligospora*: characterization and potential application in phosphorus and mineral release for feed processing. Microb Cell Factories 19(1):1–13. https://doi.org/10.1186/s12934-020-01346-9

Huff R, Pereira RI, Pissetti C, de Araújo AM, d'Azevedo PA, Frazzon J, GuedesFrazzon AP (2020) Antimicrobial resistance and genetic relationships of enterococci from siblings and non-siblings Heliconius erato phyllis caterpillars. Peer J 8:e8647. https://doi.org/10.7717/peerj.8647

Hynes AP, Rousseau GM, Lemay M-L, Horvath P, Romero DA, Fremaux C, Moineau S (2017) An anti-*CRISPR* from a virulent *streptococcal* phage inhibits *Streptococcus pyogenes Cas9*. Nat Microbiol 2(10):1374–1380. https://doi.org/10.1038/s41564-017-0004-7

Ike A, Sriprang R, Ono H, Murooka Y, Yamashita M (2007) Bioremediation of cadmium contaminated soil using symbiosis between leguminous plant and recombinant *rhizobia* with the *MTL4* and the *PCS* genes. Chemosphere 66(9):1670–1676. https://doi.org/10.1016/j.chemosphere.2006.07.058

Jahantigh-Haghighi Z, Fahmideh L, Fazeli-Nasab B (2020) Genetic diversity in plant medicinal of Tomato Genotypes using RAPD and ISSR markers. https://doi.org/10.22084/ab.2019.17227.1386

Jalali Javaran M et al (2011) The success of molecular farming in Iran. Biocatal Agric Biotechnol 1 (1):19–47. https://doi.org/10.22103/jab.2011.164

Jatuwong K, Suwannarach N, Kumla J, Penkhrue W, Kakumyan P, Lumyong S (2020) Bioprocess for production, characteristics, and biotechnological applications of fungal phytases. Front Microbiol 11:188. https://doi.org/10.3389/fmicb.2020.00188

Jiang Z, Zhang X, Wang Z, Cao B, Deng S, Bi M, Zhang Y (2019) Enhanced biodegradation of atrazine by *Arthrobacter* sp. DNS10 during co-culture with a *phosphorus solubilizing* bacteria: *Enterobacter* sp. P1. Ecotoxicol Environ Saf 172:159–166. https://doi.org/10.1016/j.ecoenv.2019.01.070

Kerovuo J (2000) A novel phytase from *Bacillus*: characterization and production of the enzyme. Academic Dissertation, University of Helsinki

Khademi Z, Jones D, Malakouti M, Asadi F, Ardebili M (2009) Organic acid mediated nutrient extraction efficiency in three calcareous soils. Soil Res 47(2):213–220

Khan MS, Zaidi A, Wani PA (2009) Role of *phosphate solubilizing* microorganisms in sustainable agriculture-a review. In: Sustainable agriculture. Springer, New York, pp 551–570. https://doi.org/10.1007/978-90-481-2666-8_34

Khezri M (2019) Effects of biofilm formation in bacteria from different perspectives. Nova Biol Rep 6(1):70–78. https://doi.org/10.29252/nbr.6.1.70

Khodashenas A, Koocheki A, Rezvani-Moghadam P, Lakzian A (2008) Effect of agricultural practices on soil nematodes and bacteria in the winter wheat fields of Khorasan Province. Iran Environ Sci 5(3):53–64

Khodashenas A, Koocheki A, Rezvani Moghadam P, Lakzian A, Nassiri Mahallati M (2010) Evaluation of agricultural practices effect on soil bacterial diversity and abundance. J Soil Water Conserv 14(52):99–114

Khoshmanzar E, Aliasgharzad N, Neyshabouri M, Khoshru B, Arzanlou M, Lajayer BA (2020) Effects of *Trichoderma* isolates on tomato growth and inducing its tolerance to water-deficit stress. Int J Environ Sci Technol 17(2):869–878. https://doi.org/10.1007/s13762-019-02405-4

Khoshniyat P, Tarinejad A, Pazhang M (2018) Isolation and cloning of mercuric reductase gene (*merA*) from mercury-resistant bacteria. Biol J Microorg 7(25):19–31. https://doi.org/10.22108/bjm.2018.21822

Kim HS, Wang W, Kang L, Kim S-E, Lee C-J, Park S-C, Park WS, Ahn M-J, Kwak S-S (2020) Metabolic engineering of low-molecular-weight antioxidants in sweetpotato. Plant Biotechnol Rep 14:193–205. https://doi.org/10.1007/s11816-020-00621-w

King RB, Sheldon JK, Long GM (1997) Practical environmental bioremediation: the field guide. CRC Press, Boca Raton, p 208

Kirillova OV, Amirova EF, Kuznetsov MG, Valeeva GA, Zakharova GP (2020) Innovative directions of agricultural development aimed at ensuring food security in Russia. Paper presented at the BIO Web of Conferences. https://doi.org/10.1051/bioconf/20201700068

Kitajima S, Sato F (1999) Plant pathogenesis-related proteins: molecular mechanisms of gene expression and protein function. J Biochem 125(1):1–8

Knapp CW, McCluskey SM, Singh BK, Campbell CD, Hudson G, Graham DW (2011) Antibiotic resistance gene abundances correlate with metal and geochemical conditions in archived Scottish soils. PLoS One 6(11):e27300. https://doi.org/10.1371/journal.pone.0027300

Konietzny U, Greiner R (2004) Bacterial phytase: potential application, in vivo function and regulation of its synthesis. Braz J Microbiol 35(1-2):12–18. https://doi.org/10.1590/S1517-83822004000100002

Kour D, Kaur T, Yadav N, Rastegari AA, Singh B, Kumar V, Yadav AN (2020) Phytases from microbes in phosphorus acquisition for plant growth promotion and soil health. In: Trends of microbial biotechnology for sustainable agriculture and biomedicine systems: diversity and functional perspectives. Elsevier, Amsterdam, pp 157–176. https://doi.org/10.1016/B978-0-12-820526-6.00011-7

Kowsari M, Zamani M, Motallebi M, Joorabchi E (2013) Transformation of *Trichoderma Harzianum* T8 with *Gfp* gene to facilitate its monitoring in soil and use of sem to study interaction of *T. Harzianum* with *Rhizoctonia Solani Sclerotia*. Iranian J Plant Pathol 49 (4):153–156

Kunin V, Sorek R, Hugenholtz P (2007) Evolutionary conservation of sequence and secondary structures in *CRISPR* repeats. Genome Biol 8(4):R61. https://doi.org/10.1186/gb-2007-8-4-r61

Kunjapur AM, Prather KL (2015) Microbial engineering for aldehyde synthesis. Appl Environ Microbiol 81(6):1892–1901. https://doi.org/10.1128/AEM.03319-14

Latip W, Knight VF, Abdul Halim N, Ong KK, Mohd Kassim NA, Wan Yunus WMZ, Mohd Noor SA, Mohamad Ali MS (2019) Microbial phosphotriesterase: structure, function, and biotechnological applications. Catalysts 9(8):671. https://doi.org/10.3390/catal9080671

Leclere V, Béchet M, Adam A, Guez J-S, Wathelet B, Ongena M, Thonart P, Gancel F, Chollet-Imbert M, Jacques P (2005) *Mycosubtilin* overproduction by *Bacillus subtilis* BBG100 enhances the organism's antagonistic and biocontrol activities. Appl Environ Microbiol 71 (8):4577–4584. https://doi.org/10.1128/AEM.71.8.4577-4584.2005

Li J, Brader G, Kariola T, Tapio Palva E (2006) *WRKY70* modulates the selection of signaling pathways in plant defense. Plant J 46(3):477–491. https://doi.org/10.1111/j.1365-313X.2006.02712.x

Li W, Xia L, Ding X, Lv Y, Luo Y, Hu S, Yin J, Yan F (2012) Expression and characterization of a recombinant *Cry1Ac* crystal protein fused with an insect-specific neurotoxin ω-ACTX-Hv1a in *Bacillus thuringiensis*. Gene 498(2):323–327. https://doi.org/10.1016/j.gene.2012.01.034

Liang W, Wu R, Yang T, Shen H, Hu Z (2020) Effect of pathogenic bacteria on a novel C-type lectin, hemocyte and superoxide dismutase/alkaline phosphatase activity in *Onchidium reevesii*. Fish Shellfish Immunol. https://doi.org/10.1016/j.fsi.2020.04.001

Liu S (2019) Identification and characterization of the *Phosphate-Solubilizing* bacterium *Pantoea sp*. S32 in reclamation soil in Shanxi, China. Front Microbiol 10:2171. https://doi.org/10.3389/fmicb.2019.02171

Liu C, Mou L, Yi J, Wang J, Liu A, Yu J (2019) The eno gene of *Burkholderia cenocepacia* strain 71-2 is involved in *Phosphate Solubilization*. Curr Microbiol 76(4):495–502. https://doi.org/10.1007/s00284-019-01642-7

Liu Y, Cui J, Zhou X, Luan Y, Luan F (2020) Genome-wide identification, characterization and expression analysis of the *TLP* gene family in melon (*Cucumis melo* L.). Genomics 112 (3):2449–2509. https://doi.org/10.1016/j.ygeno.2020.02.001

Ma L, Chen S, Yuan J, Yang P, Liu Y, Stewart K (2017) Rapid biodegradation of atrazine by *Ensifer sp*. strain and its degradation genes. Int Biodeterior Biodegradation 116:133–140. https://doi.org/10.1016/j.ibiod.2016.10.022

Mackelprang R, Lemaux PG (2020) Genetic engineering and editing of plants: an analysis of new and persisting questions. Annu Rev Plant Biol 71:659–687. https://doi.org/10.1146/annurev-arplant-081519-035916

Maheswari T, Anbukkarasi K, Hemalatha T, Chendrayan K (2013) Studies on phytohormone producing ability of indigenous endophytic bacteria isolated from tropical legume crops. Int J Curr Microbiol App Sci 2(6):127–136

Makarova KS, Wolf YI, Shmakov SA, Liu Y, Li M, Koonin EV (2020) Unprecedented diversity of unique *CRISPR*-Cas-related systems and *Cas1* homologs in *Asgard Archaea*. CRISPR J 3 (3):156–163. https://doi.org/10.1089/crispr.2020.0012

Malboobi MA, Sarikhani MR, Greiner R (2013) Recombinant APases nucleic acid sequences: Google Patents

Malhotra M, Srivastava S (2006) Targeted engineering of *Azospirillum brasilense* SM with indole acetamide pathway for indoleacetic acid over-expression. Can J Microbiol 52(11):1078–1084. https://doi.org/10.1139/w06-071

Malmierca MG, Cardoza RE, Alexander NJ, McCormick SP, Hermosa R, Monte E, Gutiérrez S (2012) Involvement of *Trichoderma trichothecenes* in the biocontrol activity and induction of plant defense-related genes. Appl Environ Microbiol 78(14):4856–4868. https://doi.org/10.1128/AEM.00385-12

Mark R, Lyu X, Lee JJ, Parra-Saldívar R, Chen WN (2019) Sustainable production of natural phenolics for functional food applications. J Funct Foods 57:233–254. https://doi.org/10.1016/j.jff.2019.04.008

Marraffini LA (2016) The *CRISPR*-Cas system of *Streptococcus pyogenes*: function and applications *Streptococcus pyogenes*: basic biology to clinical manifestations: University of Oklahoma Health Sciences Center

Mathur R (2018) Genetic engineering and biosafety in the use of genetically modified foods. IJASRM 2018(I):76–82

Maurel C, Barbier-Brygoo H, Spena A, Tempe J, Guern J (1991) Single rol genes from the *Agrobacterium rhizogenes* TL-DNA alter some of the cellular responses to auxin in *Nicotiana tabacum*. Plant Physiol 97(1):212–216. https://doi.org/10.1104/pp.97.1.212

McDade KH (2019) Developing an in cellulo carrier-driven crystallization system using *Cry1Ac* from *Bacillus thuringiensis*. (Master's Thesis), Science

Megharaj M, Ramakrishnan B, Venkateswarlu K, Sethunathan N, Naidu R (2011) Bioremediation approaches for organic pollutants: a critical perspective. Environ Int 37(8):1362–1375. https://doi.org/10.1016/j.envint.2011.06.003

Mehrban A, Fazeli-Nasab B (2017) The effect of different levels of potassium chloride on vegetative parameters of sorghum variety KGS-29 inoculated with *mycorrhizal fungi* under water stress. J Microbiol 10(32):275–288

Mindt M, Walter T, Kugler P, Wendisch VF (2020) Microbial engineering for production of N-functionalized amino acids and amines. Biotechnol J 2020:1900451. https://doi.org/10.1002/biot.201900451

Mokhtari M, Safarnejad MR, Alavi SM, Torkamanzehi A (2015) Isolation, gene expression and *PthA* effector protein production of *Xanthomonas citri* subsp citri causal agent of citrus bacterial canker. J Agric Biotechnol 7(2):155–170. https://doi.org/10.22103/jab.2015.1370

Molla KA, Yang Y (2020) Predicting *CRISPR/Cas9*-induced mutations for precise genome editing. Trends Biotechnol 38(2):136–141. https://doi.org/10.1016/j.tibtech.2019.08.002

Morrison DE, Robertson BK, Alexander M (2000) Bioavailability to earthworms of aged DDT, DDE, DDD, and dieldrin in soil. Environ Sci Technol 34(4):709–713. https://doi.org/10.1021/es9909879

Mortazavi M, Aminzadeh S, Ghanbari A, Farrokhi N, Karkhane AA, Javaheri Z (2017) Recombinant chitinase produced from a thermophilic *Paenibacillus ehimensis*. J Cell Mol Res 30 (2):188–197

Naghavi MR, Mardi M, Ramshini HA, Fazeli-Nasab B (2004) Comparative analyses of the genetic diversity among bread wheat genotypes based on RAPD and SSR markers. Iran J Biotechnol 2 (3):195–202

Naghshbandi MP, Moghimi H (2020) Stabilization of phytase on multi-walled carbon nanotubes via covalent immobilization. In: Methods in enzymology, vol 630. Elsevier, London, pp 431–451. https://doi.org/10.1016/bs.mie.2019.10.013

Naimov S, Nedyalkova R, Staykov N, Weemen-Hendriks M, Minkov I, de Maagd RA (2014) A novel *Cry9Aa* with increased toxicity for *Spodoptera exigua* (Hübner). J Invertebr Pathol 115:99–101. https://doi.org/10.1016/j.jip.2013.11.003

Nazarian-Firouzabadi F, Goodarzi D, Ismaili A (2019) A study of possibility of expression of an alternasucrase gene in sugar beet to produce alternan biopolymer. Biocatal Agric Biotechnol 11 (4):105–120

Nobile C, Houben D, Michel E, Firmin S, Lambers H, Kandeler E, Faucon M-P (2019) Phosphorus-acquisition strategies of canola, wheat and barley in soil amended with sewage sludges. Sci Rep 9(1):1–11. https://doi.org/10.1038/s41598-019-51204-x

Oh B-C, Choi W-C, Park S, Kim Y-O, Oh T-K (2004) Biochemical properties and substrate specificities of alkaline and histidine acid phytases. Appl Microbiol Biotechnol 63 (4):362–372. https://doi.org/10.1007/s00253-003-1345-0

Oh SK, Baek KH, Park JM, Yi SY, Yu SH, Kamoun S, Choi D (2008) *Capsicum annuum WRKY* protein *CaWRKY1* is a negative regulator of pathogen defense. New Phytol 177(4):977–989. https://doi.org/10.1111/j.1469-8137.2007.02310.x

Omar S (1997) The role of rock-phosphate-solubilizing fungi and *vesicular–arbusular-mycorrhiza* (VAM) in growth of wheat plants fertilized with rock phosphate. World J Microbiol Biotechnol 14(2):211–218. https://doi.org/10.1023/A:1008830129262

Orikasa Y, Nodasaka Y, Ohyama T, Okuyama H, Ichise N, Yumoto I, Morita N, Wei M, Ohwada T (2010) Enhancement of the nitrogen fixation efficiency of genetically-engineered *Rhizobium* with high catalase activity. J Biosci Bioeng 110(4):397–402. https://doi.org/10.1016/j.jbiosc.2010.04.007

Panchal H, Ingle S (2011) Isolation and characterization of endophytes from the root of medicinal plant *Chlorophytum borivilianum* (Safed musli). J Adv Dev Res 2(2):205–209

Pandey A, Szakacs G, Soccol CR, Rodriguez-Leon JA, Soccol VT (2001) Production, purification and properties of microbial phytases. Bioresour Technol 77(3):203–214. https://doi.org/10.1016/S0960-8524(00)00139-5

Parand M, Ranaei Siadat SO, Yamchi A (2015) Cloning and extracellular expression of laccase enzyme from *bacillus* of iranian hot spring into yeast cell *Pichia pastoris*. Mod Genet J 10 (1):1–10

Parray JA, Mir MY, Shameem N (2019) Plant genetic engineering and GM crops: merits and demerits. In: Sustainable agriculture: biotechniques in plant biology. Springer, New York, pp 155–229. https://doi.org/10.1007/978-981-13-8840-8_4

Pietro-Souza W, de Campos PF, Mello IS, Stachack FFF, Terezo AJ, da Cunha CN, White JF, Li H, Soares MA (2020) Mercury resistance and bioremediation mediated by endophytic fungi. Chemosphere 240:124874. https://doi.org/10.1016/j.chemosphere.2019.124874

Pourcel C, Touchon M, Villeriot N, Vernadet J-P, Couvin D, Toffano-Nioche C, Vergnaud G (2020) *CRISPRCasdb* a successor of *CRISPRdb* containing *CRISPR* arrays and cas genes from complete genome sequences, and tools to download and query lists of repeats and spacers. Nucleic Acids Res 48(D1):D535–D544. https://doi.org/10.1093/nar/gkz915

Prieto JA, Aguilera J, Randez-Gil F (2005) Genetic engineering of baker's yeast: challenges and outlook. CRC Press, Boca Raton, pp 272–307

Purakayastha T, Bera T, Bhaduri D, Sarkar B, Mandal S, Wade P, Kumari S, Biswas S, Menon M, Pathak H (2019) A review on biochar modulated soil condition improvements and nutrient dynamics concerning crop yields: pathways to climate change mitigation and global food security. Chemosphere 227:345–365. https://doi.org/10.1016/j.chemosphere.2019.03.170

Qingyan L, Ying L, Xikun Z, Baoli C (2008) Isolation and characterization of atrazine-degrading *Arthrobacter* sp. AD26 and use of this strain in bioremediation of contaminated soil. Int J Environ Sci 20(10):1226–1230. https://doi.org/10.1016/S1001-0742(08)62213-5

Raghothama K, Karthikeyan A (2005) Phosphate acquisition. Plant Soil 274(1-2):37. https://doi.org/10.1007/s11104-004-2005-6

Rasani M, Sasan H, Karamooz S, Ali Esmailizadeh Koshkoiyeh A, Asadi FM, Ayatollahi MA (2020) Amplification and Cloning of alfa-Toxin Gene from *Clostridium Perfringenes* Bacterium in *E. coli*. J Agric Biotechnol 11(4):76–86. https://doi.org/10.22103/jab.2019.14129.1135

Rehman A, Almas HI, Mazhar K, Akbar F, Ali Q, Azhar MT, Du X (2020) Breeding approaches to generate biofortified rice for nutritional enhancement rice research for quality improvement: genomics and genetic engineering. Springer, New York, pp 509–540. https://doi.org/10.1007/978-981-15-5337-0_22

Rezaei Qusheh Bolagh F, Solouki A, Tohidfar M, Zare Mehrjerdi M, Izadi-Darbandi A, Vahdati K (2020) *Agrobacterium*-mediated transformation of Persian walnut using *BADH* gene for salt and drought tolerance. J Hortic Sci Biotechnol 2020:1–10. https://doi.org/10.1080/14620316.2020.1812446

Richter C, Chang JT, Fineran PC (2012) Function and regulation of clustered regularly interspaced short palindromic repeats (*CRISPR*)/*CRISPR* associated (Cas) systems. Viruses 4 (10):2291–2311. https://doi.org/10.3390/v4102291

Roberfroid M (2007) Prebiotics: the concept revisited. J Nutr 137(3):830S–837S. https://doi.org/10.1093/jn/137.3.830S

Rodríguez H, Fraga R (1999) Phosphate solubilizing bacteria and their role in plant growth promotion. Biotechnol Adv 17(4-5):319–339. https://doi.org/10.1016/S0734-9750(99)00014-2

Rodríguez H, Rossolini GM, Gonzalez T, Li J, Glick BR (2000) Isolation of a gene from *Burkholderia cepacia* IS-16 encoding a protein that facilitates phosphatase activity. Curr Microbiol 40(6):362–366. https://doi.org/10.1007/s002840010071

Rodriguez H, Gonzalez T, Goire I, Bashan Y (2004) Gluconic acid production and phosphate solubilization by the plant growth-promoting bacterium *Azospirillum* spp. Sci Nat 91 (11):552–555. https://doi.org/10.1007/s00114-004-0566-0

Rodríguez H, Fraga R, Gonzalez T, Bashan Y (2006) Genetics of phosphate solubilization and its potential applications for improving plant growth-promoting bacteria. Plant Soil 287 (1-2):15–21. https://doi.org/10.1007/s11104-006-9056-9

Ross CA, Liu Y, Shen QJ (2007) The *WRKY* gene family in rice (*Oryza sativa*). J Integr Plant Biol 49(6):827–842. https://doi.org/10.1111/j.1744-7909.2007.00504.x

Rossolini G, Schippa S, Riccio M, Berlutti F, Macaskie L, Thaller M (1998) Bacterial nonspecific acid phosphohydrolases: physiology, evolution and use as tools in microbial biotechnology. Cell Mol Life Sci 54(8):833–850. https://doi.org/10.1007/s000180050212

Ryan PR, Dessaux Y, Thomashow LS, Weller DM (2009) Rhizosphere engineering and management for sustainable agriculture. Plant Soil 321(1-2):363–383. https://doi.org/10.1007/s11104-009-0001-6

Sabbagh SK, Valizadeh S (2019) Bio-fertilizer application: molecular and biochemical changes in infected cucumber with *Phytophthora melonis*. Biol Contr Pests Plant Dis 8(1):1–15. https://doi.org/10.22059/jbioc.2019.267141.24

Sabri N, Schmitt H, Van der Zaan B, Gerritsen H, Zuidema T, Rijnaarts H, Langenhoff A (2020) Prevalence of antibiotics and antibiotic resistance genes in a wastewater effluent-receiving river in the Netherlands. J Environ Chem Eng 8(1):102245. https://doi.org/10.1016/j.jece.2018.03.004

Saia S, Aissa E, Luziatelli F, Ruzzi M, Colla G, Ficca AG, Cardarelli M, Rouphael Y (2020) Growth-promoting bacteria and *arbuscular mycorrhizal* fungi differentially benefit tomato and corn depending upon the supplied form of phosphorus. Mycorrhiza 30(1):133–147. https://doi.org/10.1007/s00572-019-00927-w

Sajadi NR, Yadavi A, Balouchi H, Farajee H (2011) Effect of chemical (urea), organic (vermicompost) and biological (nitroxin) fertilizers on quantity and quality yield of sesame (*Sesamum indicum* L.). J Agric Sci 21(2):87–101

Sakurai J, Nagahama M, Oda M, Tsuge H, Kobayashi K (2009) *Clostridium perfringens* iota-toxin: structure and function. Toxins 1(2):208–228. https://doi.org/10.3390/toxins1020208

Sarikhani MR (2012) Phytases and its application in agriculture. Paper presented at the 4th International Congress of the European Confederation of Soil Science Societies (ECSSS-EUROSOIL), 2-6 July 2012, Bari, Italy

Sarikhani M, Malboobi M, Ebrahimi M (2014) Phosphate solubilizing bacteria: Isolation of bacteria and phosphate solubilizing genes, mechanism and genetics of phosphate solubilization. J Agric Biotechnol 6(1):77–110. https://doi.org/10.22103/jab.2014.1294

Sashidhar B, Podile AR (2009) Transgenic expression of glucose dehydrogenase in *Azotobacter vinelandii* enhances mineral phosphate solubilization and growth of sorghum seedlings. Microb Biotechnol 2(4):521–529. https://doi.org/10.1111/j.1751-7915.2009.00119.x

Schippers B, Bakker AW, Bakker PA (1987) Interactions of deleterious and beneficial rhizosphere microorganisms and the effect of cropping practices. Annu Rev Phytopathol 25(1):339–358. https://doi.org/10.1146/annurev.py.25.090187.002011

Schmülling T, Schell J, Spena A (1988) Single genes from *Agrobacterium rhizogenes* influence plant development. EMBO J 7(9):2621–2629. https://doi.org/10.1002/j.1460-2075.1988.tb03114.x

Sebiomo A, Ogundero V, Bankole S (2011) Effect of four herbicides on microbial population, soil organic matter and dehydrogenase activity. Afr J Biotechnol 10(5):770–778

Sedeek KE, Mahas A, Mahfouz M (2019) Plant genome engineering for targeted improvement of crop traits. Front Plant Sci 10:114. https://doi.org/10.3389/fpls.2019.00114

Shieh T, Ware J (1968) Survey of microorganisms for the production of extracellular phytase. Appl Microbiol 16(9):1348–1351

Shiotani H, Fujikawa T, Ishihara H, Tsuyumu S, Ozaki K (2007) A pthA homolog from *Xanthomonas axonopodis* pv. citri responsible for host-specific suppression of virulence. J Bacteriol 189(8):3271–3279. https://doi.org/10.1128/JB.01790-06

Shkryl YN, Veremeichik GN, Bulgakov VP, Tchernoded GK, Mischenko NP, Fedoreyev SA, Zhuravlev YN (2008) Individual and combined effects of the *rolA*, *B*, and *C* genes on anthraquinone production in *Rubia cordifolia* transformed calli. Biotechnol Bioeng 100(1):118–125. https://doi.org/10.1002/bit.21727

Shoja Z, Rajabi-Memari H, Roayaei Ardakani M (2015) Isolation and cloning of *Phycocyanin Alpha* subunit gene and its production in *E. coli* expression system. J Mol Cell Res 28(3):352–359

Si X, Zhang H, Wang Y, Chen K, Gao C (2020) Manipulating gene translation in plants by *CRISPR–Cas9*-mediated genome editing of upstream open reading frames. Nat Protoc 15 (2):338–363. https://doi.org/10.1038/s41596-019-0238-3

Song F, Goodman RM (2001) Molecular biology of disease resistance in rice. Physiol Mol Plant Pathol 59(1):1–11. https://doi.org/10.1006/pmpp.2001.0353

Stevenson FJ, Cole MA (1999) Cycles of soils: carbon, nitrogen, phosphorus, sulfur, micronutrients. Wiley, New York, p 448

Stotz HU, Thomson J, Wang Y (2009) Plant defensins: defense, development and application. Plant Signal Behav 4(11):1010–1012. https://doi.org/10.4161/psb.4.11.9755

Surange S, Wollum Ii A, Kumar N, Nautiyal CS (1997) Characterization of *Rhizobium* from root nodules of leguminous trees growing in alkaline soils. Can J Microbiol 43(9):891–894. https://doi.org/10.1139/m97-130

Tabak HH, Chambers CW, Kabler PW (1964) Microbial metabolism of aromatic compounds I: decomposition of phenolic compounds and aromatic hydrocarbons by phenol-adapted bacteria. J. Bacteriology 87(4):910–919

Tang J, Robertson BK, Alexander M (1999) Chemical-extraction methods to estimate bioavailability of DDT, DDE, and DDD in soil. Environ Sci Technol 33(23):4346–4351. https://doi.org/10.1021/es990581w

Tarafdar J, Rao A, Bala K (1988) Production of phosphatates by fungi isolated from desert soils. Folia Microbiol 33(6):453–457. https://doi.org/10.1007/BF02925770

Teng D, Mao K, Ali W, Xu G, Huang G, Niazi NK, Feng X, Zhang H (2020) Describing the toxicity and sources and the remediation technologies for mercury-contaminated soil. RSC Adv 10 (39):23221–23232. https://doi.org/10.1039/D0RA01507E

Tognetti VB, Palatnik JF, Fillat MF, Melzer M, Hajirezaei M-R, Valle EM, Carrillo N (2006) Functional replacement of ferredoxin by a cyanobacterial flavodoxin in tobacco confers broad-range stress tolerance. Plant Cell 18(8):2035–2050. https://doi.org/10.1105/tpc.106.042424

Tohidfar M, Khosravi S (2015) Challenges for releasing *Bt* transgenic plants. Biocatal Agric Biotechnol 7(3):33–54. https://doi.org/10.22103/jab.2015.1131

Tohidfar M, Zare N, Jouzani GS, Eftekhari SM (2013) *Agrobacterium*-mediated transformation of alfalfa (*Medicago sativa*) using a synthetic *cry3a* gene to enhance resistance against *alfalfa weevil*. Plant Cell Tissue Organ Cult 113(2):227–235. https://doi.org/10.1007/s11240-012-0262-2

Tooley AJ, Cai YA, Glazer AN (2001) Biosynthesis of a fluorescent cyanobacterial C-phycocyanin holo-α subunit in a heterologous host. PNAS 98(19):10560–10565

Tzfira T, Kozlovsky SV, Citovsky V (2007) Advanced expression vector systems: new weapons for plant research and biotechnology. Am Soc Plant Biol 145(4):1087–1089. https://doi.org/10.1104/pp.107.111724

van Loon LC, Rep M, Pieterse CM (2006) Significance of inducible defense-related proteins in infected plants. Annu Rev Phytopathol 44:135–162. https://doi.org/10.1146/annurev.phyto.44.070505.143425

van Verk M, Neeleman L, Bol J, Linthorst H (2011) Tobacco transcription factor *NtWRKY12* interacts with TGA2. 2 in vitro and in vivo. Front. Plant Sci 2:32. https://doi.org/10.3389/fpls.2011.00032

Vasudevan UM, Jaiswal AK, Krishna S, Pandey A (2019) Thermostable phytase in feed and fuel industries. Bioresour Technol 278:400–407. https://doi.org/10.1016/j.biortech.2019.01.065

Vats P, Banerjee UC (2004) Production studies and catalytic properties of phytases (*myo*-inositolhexakisphosphate phosphohydrolases): an overview. Enzym Microb Technol 35 (1):3–14. https://doi.org/10.1016/j.enzmictec.2004.03.010

Venkatachalam G, Arumugam S, Doble M (2020) Industrial production and applications of α/β linear and branched glucans. Indian Chem Eng 2020:1–15. https://doi.org/10.1080/00194506.2020.1798820

Vohra A, Satyanarayana T (2003) Phytases: microbial sources, production, purification, and potential biotechnological applications. Crit Rev Biotechnol 23(1):29–60. https://doi.org/10.1080/713609297

Wan W, Wang Y, Tan J, Qin Y, Zuo W, Wu H, He H, He D (2020) Alkaline phosphatase-harboring bacterial community and multiple enzyme activity contribute to phosphorus transformation during vegetable waste and chicken manure composting. Bioresour Technol 297:122406. https://doi.org/10.1016/j.biortech.2019.122406

Wang Y, Wang L, Zhang J, Duan X, Feng Y, Wang S, Shen L (2019) *PA0335*, a novel gene encoding histidinol phosphate phosphatase mediates histidine auxotrophy in *Pseudomonas aeruginosa*. Appl Environ Microbiol 86(5):e02593–e02519. https://doi.org/10.1128/AEM.02593-19

Xing H-L, Dong L, Wang Z-P, Zhang H-Y, Han C-Y, Liu B, Wang X-C, Chen Q-J (2014) A *CRISPR/Cas9* toolkit for multiplex genome editing in plants. BMC Plant Biol 14(1):327. https://doi.org/10.1186/s12870-014-0327-y

Xu R-F, Li H, Qin R-Y, Li J, Qiu C-H, Yang Y-C, Ma H, Li L, Wei P-C, Yang J-B (2015) Generation of inheritable and "transgene clean" targeted genome-modified rice in later generations using the *CRISPR/Cas9* system. Sci Rep 5:11491. https://doi.org/10.1038/srep11491

Xu H, Shao H, Lu Y (2019) *Arbuscular mycorrhiza* fungi and related soil microbial activity drive carbon mineralization in the maize rhizosphere. Ecotoxicol Environ Saf 182:109476. https://doi.org/10.1016/j.ecoenv.2019.109476

Xu R, Li T, Shen M, Yang ZL, Zhao Z-W (2020) Evidence for a dark septate endophyte (*Exophiala Pisciphila*, H93) enhancing phosphorus absorption by maize seedlings. Plant Soil. https://doi.org/10.1007/s11104-020-04538-9

Xue C, Li H, Liu Z, Wang L, Zhao Y, Wei X, Fang H, Liu M, Zhao J (2019) Genome-wide analysis of the *WRKY* gene family and their positive responses to phytoplasma invasion in *Chinese jujube*. BMC Genomics 20(1):464. https://doi.org/10.1186/s12864-019-5789-8

Yadav SS, Redden RJ, Hatfield JL, Ebert AW, Hunter D (2019) Food security and climate change. Wiley, London, p 568

Yajima W, Verma SS, Shah S, Rahman MH, Liang Y, Kav NN (2010) Expression of anti-sclerotinia scFv in transgenic *Brassica napus* enhances tolerance against stem rot. Nat Biotechnol 27(6):816–821. https://doi.org/10.1016/j.nbt.2010.09.010

Yan M, Wang B-h XX, Chang P, Hang F, Wu Z, You C, Liu Z (2018) Molecular and functional study of a branching sucrase-like glucansucrase reveals an evolutionary intermediate between two subfamilies of the *GH70* enzymes. Appl Environ Microbiol 84(9):e02810–e02817

Yin K, Han T, Liu G, Chen T, Wang Y, Yu AYL, Liu Y (2015) A geminivirus-based guide RNA delivery system for *CRISPR/Cas9* mediated plant genome editing. Sci Rep 5:14926. https://doi.org/10.1038/srep14926

Younessi N, Safari Sinegani AA, Khodakaramian G (2017) Detection of *Beta-lactamase* gene in the culturable bacteria isolated from agricultural, pasture and mining soils around mines in Hamedan, Iran. Biol J Microorg 6(23):35–48. https://doi.org/10.22108/bjm.2017.21660

Yu KL, Chen W-H, Sheen H-K, Chang J-S, Lin C-S, Ong HC, Show PL, Ling TC (2020) Bioethanol production from acid pretreated microalgal hydrolysate using microwave-assisted heating wet torrefaction. Fuel 279:118435. https://doi.org/10.1016/j.fuel.2020.118435

Zamani P, Amoozegar MA, Khajeh K (2014) Cloning and expression of laccase enzyme from *B. pumilus* strain GAZ23. Biol J Microorg 3(9):1–10

Zeinali M, Hosseini B, Jahanbakhsh S, Rezazadeh-bari M, Tabatabaee M (2016) Cloning of affecting *pyruvate decarboxylase* gene in the production bioethanol of agricultural waste in the *E.coli* bacteria. Biol J Microorg 5(18):11–28. https://doi.org/10.22108/bjm.2016.20380

Zhang Q, Ye Y (2017) Not all predicted *CRISPR–Cas* systems are equal: isolated cas genes and classes of *CRISPR* like elements. BMC Bioinf 18(1):92. https://doi.org/10.1186/s12859-017-1512-4

Zhang D, Zhang B (2020) SpRY: engineered *CRISPR/Cas9* harnesses new genome-editing power. Trends Genet 36(8):546–548. https://doi.org/10.1016/j.tig.2020.05.001

Zhang Y, Liang J, Zeng G, Tang W, Lu Y, Luo Y, Xing W, Tang N, Ye S, Li X (2020) How climate change and eutrophication interact with microplastic pollution and sediment resuspension in shallow lakes: a review. Sci Total Environ 705:135979. https://doi.org/10.1016/j.scitotenv.2019.135979

Zhao Y, Zhu L, Lin C, Shen Z, Xu C (2019) Transgenic soybean expressing a thermostable phytase as substitution for feed additive phytase. Sci Rep 9(1):1–7. https://doi.org/10.1038/s41598-019-51033-y

Zinin NV, Serkina AV, Gelfand MS, Shevelev AB, Sineoky SP (2004) Gene cloning, expression and characterization of novel phytase from *Obesumbacterium proteus*. FEMS Microbiol Lett 236(2):283–290. https://doi.org/10.1111/j.1574-6968.2004.tb09659.x

Zuker A, Tzfira T, Scovel G, Ovadis M, Shklarman E, Itzhaki H, Vainstein A (2001) RolC-transgenic carnation with improved horticultural traits: quantitative and qualitative analyses of greenhouse-grown plants. J Am Soc Hortic Sci 126(1):13–18. https://doi.org/10.21273/JASHS.126.1.13

Zurbriggen MD, Tognetti VB, Carrillo N (2007) Stress-inducible flavodoxin from photosynthetic microorganisms. The mystery of flavodoxin loss from the plant genome. IUBMB Life 59 (4-5):355–360. https://doi.org/10.1080/15216540701258744

Recent Trends in Organic Farming 20

Charles Oluwaseun Adetunji, Osikemekha Anthony Anani, Olugbemi T. Olaniyan, Abel Inobeme, Frances N. Olisaka, Eseosa Oluwadamilare Uwadiae, and Omoregbe Nosa Obayagbona

Abstract

It has been clarified that the whole globe will experience a rapid increase in population. Therefore, there is a need to identify a sustainable solution that could help to meet the demand of the ever-growing population. Some of the rising question from the majority of the globe is how to proffer solution of the challenges of foods insecurity, climates changes, high level of anthropogenic activities in the environment, and high level of unemployment among the youths, as well as a higher level of mortality rate as a result of the recent pandemic activity

C. O. Adetunji (✉)
Applied Microbiology, Biotechnology and Nanotechnology Laboratory, Department of Microbiology, Edo Stae University Uzairue, Iyamho, Edo State, Nigeria
e-mail: adetunji.charles@edouniversity.edu.ng

O. A. Anani
Laboratory of Ecotoxicology and Forensic Biology, Department of Biological Science, Faculty of Science, Edo Stae University Uzairue, Iyamho, Edo State, Nigeria

O. T. Olaniyan
Laboratory for Reproductive Biology and Developmental Programming, Department of Physiology, Edo Stae University Uzairue, Iyamho, Nigeria

A. Inobeme · E. O. Uwadiae
Department of Chemistry, Edo Stae University Uzairue, Iyamho, Nigeria

F. N. Olisaka
Environmental and Public Health Microbiology Laboratory, Department of Biological Sciences, Faculty of Science, Benson Idahosa University, Benin City, Edo State, Nigeria

O. N. Obayagbona
Department of Environmental and Animal Biology, Faculty of Life Sciences, University of Benin, Benin City, Edo State, Nigeria

Dept. of Environmental Management and Toxicology, Faculty of Life Sciences, University of Benin, Benin City, Edo State, Nigeria

R. Soni et al. (eds.), *Microbiological Activity for Soil and Plant Health Management*,
https://doi.org/10.1007/978-981-16-2922-8_20

due to COVID-19. Furthermore, some other challenges include malnutrition and several nutrition challenges. Therefore, provides comprehensive details on the numerous microorganism that could influence soil health in promoting plant growth, and serves as potential bioremediation of polluted soil as well as provide detailed information on the application of plant growth-promoting rhizobacteria (PGPR) in sustainable agriculture and environment as well as provide detailed information on other beneficial microorganisms that could boost Agricultural production.

Keywords

Plant growth-promoting rhizobacteria · Sustainable agriculture · Agricultural production · Environment · Biological control agent · Biological fertilizers · Beneficial microorganisms

20.1 Introduction

The global farming practices have gradually been transformed with the emergence of organic farming. The rationale behind this approach to farming is the utilization of various kinds of organic-based fertilizers such as bone meal, compost manure, and green manure and replacement of chemical pesticides that are not environmentally friendly. Organic farming emerged with a view to promoting agricultural sustainability as an innovative technology for meeting the agricultural needs of the fast growing global population (Tsvetkov et al. 2018). This agricultural technique is remarkable for environmental friendliness. There are an increasing concern and demand for safe food which has resulted in the high demand for food and related products that are cultivated organically without the use of pesticides. Currently, the sale of such products is not quite accessible to farmers who cultivate on small scale and the sale of the excess production is usually done in local markets without certification (Nielsen 2019).

The advancement of organic farming is a result of the concerns about the negative consequences associated with industrialized farming practices of the twentieth century. The industrialized agricultural practices of the twentieth century were related to the utilization of agricultural chemicals which have a deleterious influence on the environment and reduction in soil fertility, decline in immunity of plants and overall biodiversity, which in all would affect the quality of food and human health.

According to the United Nations (2015), the human population is due to hit over 9.1 billion in 2050. This growth can be forecasted to tell negatively on the demand for food, land resources, and security of other agricultural products (Yanakittkul and Aungvaravong 2020).

Accelerated agriculture using chemicals like pesticides, insecticides, and fertilizers to boost plant products is an essential and critical sector that has assisted in the management of food products to meet the increasing request for food by the ever-evolving human population. It has been estimated that using this method of food production, the rate of meeting the global population food demand like meat 0.47 billion tonnes and grains 3 billion tonnes was billed to hit 70% (FAO 2015). However, this accelerated agriculture comes with a lot of economic toll resulting in the use of five million tonnes of farm chemicals annually Chakrabarty et al. (2014) and Fernando (2017), the yearly pollution and buildup of pesticides and heavy metals in the ecosystem, and possible health risk (Anani and Olomukoro 2019; Anani et al. 2020a, b; Adetunji and Anani 2020; Yanakittkul and Aungvaravong 2020).

The influence of farm chemicals on the soil, water, and air has greatly impacted human health via the accumulation and consumption of toxins in food which have caused several health diseases in humans like cancer, Alzheimer, Parkinson, diabetes, reproductive ailments, learning incapacities, autism, and asthma diseases (Katherine and Hendrik 2010; Owens et al. 2010; Onder et al. 2011; Costa et al. 2014; Sharma and Singhvi 2017). These problems can be averted by the practice of sustainable farming which will positively control the negative ecological influences on agriculture, thus ensuring good and quality food free from toxins. Such sustainable practice is organic farming (FAO 2013; Muller et al. 2017).

Organic farming is one of the existing forms of agriculture that refrains from the utilization of chemicals like pesticides such as fertilizers to promote nutrients, the fertility of the soil, pests, and diseases (Foley et al. 2011; Muller et al. 2017; Yanakittkul and Aungvaravong 2020). The benefits derived from organic farming as measured to the range of ecological indicators caused by the use of synthetic chemicals for farming have been recounted by various authors (Mäder et al. 2002; Schader et al. 2012; Seufert et al. 2012; Tuomisto et al. 2012; Meier et al. 2015; Reganold and Wachter 2016). So, organic farming is an appropriate answer to mitigate the utilization of artificial chemicals, thus avoiding environmental and health problems and food contamination (Sharma and Singhvi 2017; Sangkumchaliang and Huang 2012).

In recent times, organic food has gained popularity among farmers, consumers, and concerned stakeholders. In 2014, the revenue derived from the organic food hub market was estimated to be US$ 80 billion because of the economically driven policies set by the European Unions and the USA (FiBL and IFOAM 2016). Organic farming has been a novel area of agriculture that produces new products that are of great appeal in the food sector. These products have been estimated to cover an arable hectare of 43.7 million about 0.99% of the world's land (Yanakittkul and Aungvaravong 2020). The amalgamation of sustainable economic and environmental safety opportunities in agriculture using the organic method of farming has aided in soil-organic matter improvement via waste recycling, in turn, this will benefits humans because it assists in the reduction of noxious pollutants in the soil and the food (Ulm et al. 2019; Yanakittkul and Aungvaravong 2019).

The role of PGPR as a sustainable organic agriculture tool has been at the forefront of modern farming. Recent application of PGPR in organic farming has shown a positive influence on the soil structure and functions by way of the reduction of the impacts of herbicides, pesticides, and fertilizer chemicals (Vejan et al. 2016; Backer et al. 2018; dos Santos et al. 2020). The roots of the plant are colonized by beneficial bacterial that excite their growth via different mechanisms. The impacts and performances of these growth-stimulating microorganisms (fungal and bacterial) have been earlier reported by different studies (Vessey 2003; Perez-Montano et al. 2014; Meena et al. 2017; Backer et al. 2018; Alooa et al. 2019; Adetunji and Anani 2020; dos Santos et al. 2020).

PGPR is a perfect substitute for the agriculturist to combat serious social and ecological problems that stem up such as food shortages caused by epidemics, nutrient deficiencies, abiotic and biotic factors. The microbial consortia of PGPR (rhizomicrobiome) are linked to all parts of the plants like roots, leaves, stems, fruits, and flowers Berg et al. (2016) which assist these parts of plants to overcome these challenges (Smith et al. 2015). Though, these conditions differ across the various plant structures.

The rhizomicrobiome use two types of mode of actions (direct and indirect) in which they use to promote plant productivity and growth. The direct mode of action consists of phytohormones such as auxins, siderophores, phosphorus solubilization, and nitrogen fixing that are used for plant production (Riggs et al. 2001; Khalid et al. 2004; Cassán et al. 2009; Krey et al. 2013; Yu et al. 2019). The indirect mode of action is linked to biological control through the means of antagonistic action against pathogenic plant microorganisms by inducing resistance and systemic responses that interfere with the bacterial minimal sensing systems (Mahmood et al. 2016; dos Santos et al. 2020).

This chapter anticipates evaluating the application of plant growth-promoting rhizobacteria (PGPR) as sustainable agriculture and environmental tools. The authors highlighted the significance of PGPR in sustainable agriculture and in the bioremediation of polluted environments. Moreover, specific examples of beneficial attributes of some beneficial microorganisms such as *Rhizobium*, *Azospirillum*, *Bacillus*, *Pseudomonas*, *Serratia*, and *Stenotrophomonas* when applied as PGPR for boosting agricultural products were highlighted.

20.2 Roles of Microorganism as the Potential Rejuvenator of Polluted Soil, Soil Health Promoter, and Plant Growth Stimulator

Tahat et al. (2020) evaluated the sustainable application of microorganisms for the improvement of soil health. The authors stated that a fit soil behaves as an active biotic system that transports various ecological services like removal of GHGs (greenhouse gases) from the atmosphere, decomposition, regulation, and recycling of nutrients, the sustenance of plant productivity, and water quality. The health of the soil is closely linked to sustainable farming and the activities of microorganisms. Soil health sustainability is explicitly based on the ability of the plants to manufacture food products without the inference or influence of ecological factors. Nematodes, cyanobacteria, and AMF (arbuscular mycorrhizal fungi) play an essential role in the regulation of plant response to ecological stress, the cycling of soil nutrients, production of plant hormones, availability of nutrient to plants, and the efficient utilization of soil water. Agricultural activities have revealed that tillage and organic farming have efficiently improved soil health by increasing the numbers of soil microbes in activities, diversity, and abundance.

Li et al. (2017) evaluated the function of soil microorganisms in promoting flora growth. The ever-evolving growing population has elicited a great demand for more food. This has necessitated the need to employ modern farming to meet this demand. The need to cultivate novel crop assortments for an increase in the resistance against environmental stressors and insects as well as in the improvement of their yields is very important in the sustainable maintenance of the health of the soil and the plant at large. Though, crops are still requiring artificial nutrients like fertilizers and manure to boost their efficiencies and quality. Recent studies showed that microbes found in the soil provide more positive influence when compared to synthetic fertilizers. These microbes based on their great gene pool serve as a probable resource for the recycling of nutrients and biochemical actions for plant development. In conclusion, the authors recommend the modification of the microbiota in the soil to elicit the growth of the plants and improve the soil health.

Hayat et al. (2010) assessed the role of soil bacteria in promoting the health of the soil and the growth of plants. Soil microbes like bacteria are very significant in the biological, geographical, and chemical cycling of nutrients to plants. Bacterial and plant association in the root rhizoids as the rhizosphere are the major determining factors of soil and plant health as well as fertility. PGPR is usually known as beneficial bacteria to plant, commonly also referred to as plant growth-promoting organisms that have the efficiency to elicit plant development during the root colonizing process. They are important in the ecological sustenance of soil and plant health via symbiosis. Specific microbes that are significant in this process are *Mesorhizobium*, *Sinorhizobium*, *Allorhizobium*, *Azorhizobium*, *Bradyrhizobium*, and *Rhizobium*—a group of cyanobacteria that are nitrogen fixers. Typical examples of microorganisms that are nitrogen fixers include *Pseudomonas, Klebsiella, Enterobacter*, and *Azospirillum.* They have the capability to colonize and attach themselves to the root superficial region of plants thus promoting and facilitating indirectly and directly nutrients cycling and uptake, the reduction of and prevention of phytopathogens. This can be attained by the production of some metal-binding and small molecules called siderophores. PGPR can also produce HCN (hydrogen cyanide) which they can use to prevent the cell wall of plant pathogens with the aid of biodegradable enzymes like ß-1,3-glucanase and chitinase. PGPR can also affect the direct development of plants by the production of plant hormones like indole-3-acetic acid, abscisic acid, ethylene, gibberellins, cytokinins, and auxins which have been recounted for another genus of bacteria.

Some PGPR function as a reservoir for ammonia and ACC (1-aminocyclopropane-1-carboxylate) in higher flora, by resolving them into ammonia and α-ketobutyrate. This will enhance the development of the plants' roots, therefore decreasing the levels of ethylene in the microrhizome community. Moreover, the PGPR also aid in the solubilization of nutrients and phosphates, thus improving the organic matter constituents and stress resistance of the soil and plant to external and internal factors. They also retain enough organic soil nitrogen in the soil and plants, thereby releasing sufficient nutrients to them.

Sathya et al. (2016) evaluated in a review the function of soil microorganisms in the sustenance of soil health. The authors recounted that soil health is characterized by the persistent ability to carry significant biotic systems. The chief driver of soil health is to maintain all the factors controlling sustainable farming to preserve the natural contents therein. Microbes perform a crucial function in the health of the soil, thus influencing the various chemicals, geomaterials, and biological cycling of nutrients like phosphorus, sulfur, nitrogen, and carbon as well as other micro- and macronutrients that play an imperative role in the preservation of the biological and health of the soil. These microbes also have the ability to overpower directly or indirectly the soil-borne diseases and enhance their agricultural outputs. Their role in nutrient cycling also contributes directly or indirectly to the production and promotion of enzymes and phytohormones that aid in combating insects and plant diseases. The massive genetic assortment and role of different microbial consortia like actinomycetes, fungi, and bacteria are indelible assets in the functioning of the soil health and all the other biological entities that promote or contribute to as functional soil health pointers.

Most soil fertilities are usually tied with the effective role and actions of soil microbes. de Souza et al. (2015) evaluated the role of PGPB as special engineers in the restructuring of the health of agricultural soils. The authors reported that microbial and plant associations in the root nodules are the major driver of soil lushness, productivity, and heath of the plants. PGPB are organisms that can improve the protection and growth of plants towards ecological stressors, diseases, and other factors that have close links with plants like endophytes that could influence the development of the plants. Some significant characteristics of bacterial like the fabrication of siderophores, plant hormones, deaminase action of ACC, solubilization of phosphate, and biological fixation of nitrogen are special traits of the PGPB and solitary roles to improve the soil and flora fertility. PGPB inoculants also improve the agronomic efficacy by decreasing the rate of environmental contamination and economic cost via production on the ground that the utilization of fertilizer chemicals be phase out or reduced. For PGPB inoculants to attain success in eliciting the productivity and growth of the plant, there have to be several steps that can impact their efficacy of inoculation like the health of the soil, colonization of microbes in the root nodules, and plant root exudation.

Ojo et al. (2015) evaluated the impact of fertilizers on the population and growth of microbes in the soil. The putrefaction of organic matter and the accessibility of nutrients in the soil depend on the activities of microorganisms therein. Inorganic and organic fertilizers are important in the improvement of the needs of the soil microbes for the development of the floras. Ojo et al. (2015) opined that the low population of microbes is caused by the inefficiency of organic nutrients or matter to be readily remedied by altering the soil with organic nutrients and fertilizers, thus permitting more time for the microbial community to thrive reproductively. The microorganisms increase the soil components via the digestion of the organic matter and humus to aid in the fixating of nitrogen in the rhizosphere.

Seneviratne et al. (2011) evaluated the implications in the utilization of biofertilizer and compost in organic farming. The authors reported that compost and biofertilizers have been fingered as a promising substitute to the traditional induced chemical fertilizers because of their positive influence on the development of the plant as well as the improvement of the soil quality, health, and its functions. Compost and biofertilizers also have the ability to decrease the ecological and human risk or damages from pollutants by biodegradation. Biofertilizers derived from microorganisms have been known to date for their beneficial influence on the biological activities of plants such as economic and environmental viability, the structural buildup of soil richness, phytohormones production, manufacture of antibiotics, flora pathogen suppression, plant stimulation, and nitrogen fixation. Compost, however, is significant in the improvement of the chemical, biological, physical structure of the soil. The limitation of this influence depends on the derived source of the compost as well as the process it undergoes. However, one of the demerits of the utilization of compost includes the probable presence of heavy metals, the possible link of pathogens, high generation of ammonia, excessive production of leachate in bulky volumes, and inadequate result delivery. Therefore, there is a need to combine the compost and biofertilizers to achieve maximum soil health.

Javaid (2011) tested and evaluated in a rice plant pot assay the impact of biofertilizers (effective microorganisms and biopower) on the growth and development of various amendment of soils. The amendments were fertilizer-NPK, farmyard manure (FYM), and green manure (GM). The results from the biological experiment showed that the usage of the biopower negatively influenced the yield and growth of the plant in the fertilizer-NPK amendment. Conversely, the same biofertilizer sharply improved the yield and growth of the amended GM soil, whereas their influence on the FYM amendment was not significant. In the GM amendment, the application of effective microorganisms improved the yield of the rice by 46%. An amalgam of both biofertilizers markedly enhanced the shoot and root growth in the soil amended with the FYM. The findings from this study showed that effective microorganisms and biopower biofertilizers sharply improved the biomass of the shoot and roots as well as the yield of the grain in the amended GM soils.

Microorganisms in the soil perform significant roles in the control of the soil ecosystem. Alexander and Chong (2014) in a biological experiment tested and evaluated the influence of biological control messengers on the microbial consortia of soils in oil palm farmstead. The authors stated that the microbial consortia may play the role of a check-balance to different plant diseases. Besides, soil microorganisms are subtle to the vicissitude of soil factors.

Alexander and Chong (2014) recounted that the effects of these messengers on the diversity and population of soil microbes are still vague. The results from the study revealed that feasible microbes were seen scattered in the cultured media after counting using the CFU (colony forming unit). They were later recognized by employing molecular and Biolog techniques. The novel predominant species identified were *Trichoderma spp., Streptomyces spp., Yarrowia spp., Burkholderia spp., Microbacterium spp.,* and *Enterobacter spp.* The colony forming unit for the yeast was 10^2 to $10\ ^6$CFU/g and that for bacteria was 10^3 to 10^6CFU/g. They remained unchanged after treatment of the soils. Though, the colony forming unit for fungi was later amplified to 10 4 cfu/g in the amended soil. The utilization of biological messengers to regulate the root and stem disease of oil palm has served as a sustainable promising tool for the management of these pathogens. The authors concluded that the usage of biofertilizers in the soil can potentially enhance the evenness and richness of the distribution of microorganisms in the soil as biomessengers.

Sachidanand et al. (2019) evaluated in a review of the influences of microbes on soil structure ex situ. The authors reported that soil microbes help to preserve the ecosystem via structural and functional engineering of the soil. The interaction of the microbes with the chemical, biological, and physical features of the soil brings about control of the soil negative impacts and possibly agricultural ecosystem management. The authors stated that in some cases, the soil is the main determinant factor that controls the complexity and diversity of soil microbes as well as their symbiotic relationships with the other abiotic and biotic entities. In conclusion, the authors proposed a theoretical framework founded on the relative forces and strengths exerted by the soil microbes on the soil.

Lehman et al. (2015) assessed the bio-health of the enhancement of soil via reverse soil breakdown of pollutants by microbes. The authors recounted that soil health relates to the biological processes, effects, and properties that manipulate high yield and qualities of crop production, improves the availability of nutrients, protects the plant against pathogens, and manages and regulates ecological stressors like high temperature and drought. The authors opined that microbes serve as an engineering tool in addressing these factors with a perception of sustainable management, repair, and regulation of all abiotic and biotic constituents in the soil environment. In conclusion, the authors recommended novel researches that will project the sustainable productivity and utilization of soil microorganisms for soil health rejuvenation.

Globally, the degradation of land by the activities of humans and natural occurrences has become a bane to soil animals and the ecosystem at large. The drive for ecosystem sustainability is geared towards the conservation, management, and improvement of the agricultural land for the present and future generations. Singh et al. (2016) assessed the role of microbes as soil engineers in the ecorestoration of polluted land. The authors reported that to ensure total restoration of degraded soil or land, a systematic approach towards the establishment of set goals should be considered. This will enable timely reverse degradation, structural and aggregate growth, balanced micro-ecosystem, nutrient formation and cycling, and the degradation of litters by the activities of microbes. The sustainable enhancement of the agricultural soil relies on the biodiversity and bioprocess that are buildup in the ecosystem, which allow crop productivity and soil fruitfulness. This will enable the restoration of land that has already be degraded.

Nunes et al. (2012) tested and evaluated the activities of soil microbes in the degradation of impacted soil. The authors stated that soil degradation results in severe biological changes. This process can reduce the biomass of soil microbes. The result from the biological experiment showed that the enzyme and biomass activities of the soil microbes were reduced by the impacted land to about 8–10 times more than the natural vegetation. Besides, after the restoration of the land, the soil microbial biomass and the natural biomass improved by two- and five-fold correspondingly when compared to the highly impacted land. The findings showed that impacted land produced a low microbial consortium but the restored land may elicit both short- and long-run increases in the biomass and the soil consortia of microbes.

Masciandaro et al. (2013) in a synergic method evaluated the bioremediation of soil impacted with organic matter using microorganisms. The authors recounted that bioremediation which is a natural process depends on plants, fungi, and bacteria to remove, transform, and breakdown pollutants, thus ensuring ecosystem conservation of the biological and physical properties. The application of sludge and compost (organic matter) on soil has been seen to activators or messengers of microorganisms to improve on the degradation potentials of pollutants. Their presence in the soil aids in swift degradation, organic matter, and nutrient cycling via the processes of bioaugmentation, bio-enhancement, and biotransformation which are considered as possible accelerators of pollutants breakdown. Besides, during these processes, the activities of the microorganisms provide the platform for soil health, water retention ability of the soil, the porosity of the soil, and the exchange capacity of a cation. Masciandaro et al. (2013) also opined that plant species also serve as tools for the reclamation of impacted land by using the strategy of bioabsorption and biotransformation of pollutants as well as the promotion of the breakdown of organic matter by the activities of the microorganisms at the rhizosphere. The plants also provide a good microenvironment that is palatable for eliciting the activity and proliferation of the microorganisms.

Ajao et al. (2011) tested and evaluated the bioremediation potentials of microbial consortia on soils sourced from automobile mechanic workshops. The authors reported that the activities of microorganisms in the significant restoration of impacted hydrocarbons soils defined their ecological role in the mineralization and biotransformation of crude oil products into less toxic forms. Eighty-six polluted soil samples from 15 stations were collected and assayed for the bioremediation study. The isolation techniques carried on the samples resulted in the identification of five bacterial species which are *Bacillus sp.*, *Serratia sp.*, *Pseudomonas sp.*, *Flavobacterium sp.*, and *Acinetobacter sp.* The bioremediation setup was set for 2 months by employing amalgam isolates culture of lipase, dehydrogenase, protein, and TVC as bioindicators. The outcome from the experiment showed increased activity of the dehydrogenase, protein, and TVC recorded in the first month were 5.53 mg/g, 6.3×107, and 163.15μgTPFg^{-1} soil correspondingly at 7.17 pH concentration. However, a gradual reduction (4.72 unit/g) was observed in the activities of the lipase with a percentage increase (65.41%) of the crude oil at the sixth week. The findings from this study showed a possible ecological implication for designing a bioremediation procedure for the decontamination of crude oil polluted or impacted soil sites.

20.3 Features of Organic Farming

There are areas of differences between traditional farming and organic farming practices and this lies in most cases in the approaches utilized during the crop production processes. The environmental impact of agricultural activities is vital when looking at the organic farming approach. This, therefore, implies that one of the major focuses of organic farming is the preservation, conservation, and management of the ecosystem and concern on minimization of various synthetic materials that are not of organic origin (Skoufogianni et al. 2015).

This approach to farming also aids the improvement of soil health as the combined agricultural practices utilized in organic farming are known to bring about an increase in the carbon reservoir, most especially during mixed farming and crop rotational practices. In organic farming, there is efficient carbon friendly monitoring in terms of reduction of overall emissions with respect to sequestration (Smith et al. 2019).

The practice of organic farming brings about a reduction in the emission of nitrogen oxide since elementary nitrogen is not utilized as fertilizer and the content of nitrate ions in the soil is lower with greater aeration of the soil. This, therefore, implies that the rate of emission of nitrogen gas is lower when related to the traditional farming system as a result of the lower availability of nitrogen. The nitrogen obtained from the green manure during organic farming does not contribute to the overall emission of N_2O. There is an enhancement in the structure of the soil and a reduction in the emission of N_2O. Replacement of urea with other materials of organic sources is a unique management plan for the reduction of N_2O in the soil (Jalota et al. 2018).

Another factor that catalyzed the advancement of organic farming lies in intensive animal rearing that brought about an increase in medicaments, the poor health state of animals, and a decline in lifespan. At first, organic farming was developed primarily by the farmers, this was then fully supported by scientific findings. At present, there are national laws and government-approved trademarks for the consumption of organic-based food materials. There are government programs at present that boost organic farming.

The major advancement in the area of organic farming is related to the improvement in the quality of soil, enhancing efficient management of pests, the introduction of agencies with the major concern of certifying, and provision of labels so as to ensure the safety of such foods for human consumption. The sector of organic farming basically has developed to a US$60 billion sector of the global food production as of 2012 (Francis 2013).

The rapid advancement in organic farming has a strong impact on the emission of N_2O. Though some countries are likely to seriously depend on synthetic fertilizer, some others would have the capacity of reducing the use of mineral fertilizers. In organic farming, food is grown with the environmental constraints of the release of N_2O. There are also many organic materials that are produced as by-products which can be utilized in the production of biofertilizer such as animal manures, plant residues, biosolids, among others. The use of organic amendments gives remarkable trade-offs that can result in the emission of greenhouse gases (Lal 2016).

There have also been reports that the emission of greenhouse gases, as well as the use of energy in organic farming practices, are more pronounced when compared to the traditional production of crops. This is as a result of the greater intensity of highly demanded cash crops, frequent cultivation of farms, and higher fertilizer usage in organic farming. Storage methods and manure applications have a serious effect on the emission of greenhouse gases. The use of anaerobic digestion could be useful in the reduction of emission and storage. The spreading out of manures during the coolest times of the day also helps in the reduction of emissions.

The method used in the application of manure, whether as solid or slurry, incorporation or broadcasting among others also affects the emission of greenhouse gases. The use of grass residues during organic rotations can also increase the emissions of N_2O (Lal 2016).

Badgley et al. emphasized that organic farming practice would be relevant in feeding the rapidly growing population using the currently available landmass while ensuring total conservation of the fertility and structure of the soil. Farming practices that ensure conservation are currently being encouraged for the recovery of soils that have degraded. There is also an overall increase in productivity and food security.

20.4 Benefits of Organic Farming

Several agricultural benefits on the ecosystem have been attributed to conservation agriculture some of which include reduction of soil erosion, enhancement of water use, efficient cycling of nutrients, and reduction of soil organic carbon loss. There is also an enhancement of the organic matter content of the soil during organic farming, brought about by the activities of microorganisms on the organic materials. A proper comprehension of the processes involves requires researchers to focus more on and promote researches on soil structure, humus content, and microorganisms' profile.

20.4.1 Environment

The use of organic farming has been documented to be generally more beneficial to the ecosystem when compared to traditional farming practices. This is so because there is no form of contamination of soil, water, and other immediate environments during this type of farming practice. This also implies that there is no fear of leaching agrochemicals from topsoil into underground water bodies and rivers which could be taken into the bodies of aquatic organisms and passed across the food chain. In addition, the organic farming practice also ensures the preservation of wildlife, retreat for natural wildlife, rather than destruction of their natural habitats, avoidance of toxic chemicals, and maintenance of field margin.

Agricultural diversity of organisms is a vital component of the ecosystem that is affected by the particular method of farming adopted. Organic farming is known to enhance the level of agro diversity. The relegation of chemical pesticides and the use of other green substituent make it possible for different groups of plants and animals to flourish in the farmlands. This also helps in ensuring natural balance in the ecosystem. There are varying methods that have been used for the comparison of the impact of farming practices on the environment. Most researchers have used the assessment of biodiversity of organisms, nutrient emission, land use, and soil properties for comparing organic and conventional farming practices.

In a study conducted by Mondelaers et al. (2009) they carried out a comparison of the traditional farming and the organic farming system through the use of a meta-analysis approach with a focus on the impact on the environment, efficiency of land use, soil carbon content, leaching of materials into surround environment and water bodies, production of greenhouse gas as well as the general biodiversity.

Hole et al. (2005) in their review made a comparison on biodiversity in traditional and organic farming. They observed that organic practice generally brought about an improvement in biodiversity. In a related study carried out Hole et al. (2005), there was no negative environmental and population impact associated with organic agriculture especially in terms of biodiversity. Rather it was confirmed that this technique led to an increase in biodiversity especially the varieties of herbaceous plants when compared to conventional farming. It was also reported when used alone, organic farming is not sufficient for the preservation of species of animals (Bengtsson et al. 2005).

20.5 Production Requirements in Organic Farming

Organic farming is unique in that it utilizes the functional integrity of the system, unlike other traditional farming practices that require an abundance of various materials that are synthetic in nature and other man-made substances (Boelling et al. 2003). Another major advantage of organic farming is its dependence on water and soil which are localized materials that are readily available in the farming environment with less concern for heavy tools and equipment. Though there is variation in the actual production method, there are some general principles that are basically the same such as the soil management by addition of organic substances, avoidance of the use of chemical pesticides and fertilizers that are synthetic in nature, and utilization of crop rotation system.

20.6 Crop Requirements

Prior to the harvesting of organic crops, no prohibited substance must be applied to the farmland 3 years to the period of harvest (Escoba and Hue 2007). Genetic engineering, sludges from sewage, and ionizing radiation must not be used for farming. Preservation of the soil nutrients will be achieved through the use of permissible practices such as cover crops, crop rotation alongside animals, and plant materials that are not allowed. There will be preferential usage of organic stock and seeds, while farmers could only use nonorganic under certain permissible conditions. Agricultural weeds, pests, and disease could be managed through biological, physical, and mechanical approaches.

20.7 Effects on Soil Quality

The role of soil quality in sustainable farming cannot be overemphasized hence more recently, various researchers have carried out studies in this regard. It has also been reported that organic farming helps in the improvement of the quality of soil (Otutumi et al. 2004).

20.8 Advantages of Organic Farming

20.8.1 Sustainability

One of the major concerns of organic farming practice is the future implications of any agricultural practice on the ecosystem and environment at large. Food production is associated with the setting up of ecological balance so as to prevent problems of pests and soil fertility (Tsvetkov et al. 2018).

20.8.2 Ecological Services

There are more favorable interactions and balances between the agroecosystem and the variables within the environment during organic farming. Some of the associated phenomena include waste recycling, soil conditioning and forming, nutrient cycling, and sequestration of carbon.

20.8.3 Biodiversity

Organic farming promotes agricultural diversity as well as the preservation of the environment. There is an enhancement of various species of macro and micro fauna and flora in the environment since chemical pesticides and other toxic materials are not used during the farming practice. Various studies have documented that there is an increase in the level of biodiversity during organic farming when compared to other known farming practices.

20.9 Challenges Faced in Organic Farming

1. It demands much labor.
2. Organic materials commonly required may not be available in the appropriate quantity.
3. Poor adherence to standard practices in organic farming (Garg and Balodi 2014).

20.10 Role of PGPR as Biotechnological Tool for the Achievement of Sustainable Agriculture and Environment

It has been well reported that the utilization of beneficial microorganisms to boost agricultural production has expanded in the last few years due to population explosion, food shortage, and increased pest pathogenic attack. Qessaoui et al. (2019) showed that many biomolecules are extracted from beneficial microbes that act as plant growth-promoting rhizobacteria. The authors utilized soil inoculated by *Pseudomonas sp.* bacterial isolates to enhance the production of *Solanum lycopersicum*, thus it was revealed that there was a substantial upsurge in seed germination, which enlarged the collar diameter and increased leaf number. It was concluded that this isolate facilitated the growth of the plant. Gupta et al. (2015) demonstrated that soil health is an integral part of agricultural resources that demands critical attention. Pathogenic microbes affect the physicochemical status of soil, plants, and threaten the entire sustainable agricultural sector if not quickly reviewed. Over the years, chemical fertilizers have rendered soil integrity poor, wrecking serious havoc on the ecosystem. Recently, renewed interest in the use of biological fertilizers has increased. Many biological agents are emanating for promoting plant growth in an eco-friendly sustainable agricultural process. Tan et al. (2015) highlighted the significance of biofertilizers in enhancing growth plus the yield of agricultural crops. They stimulate phytohormone production, biological nitrogen fixation, mineralizing organic phosphate, suppress pathogens.

Vejan et al. (2016) reported that PGPR is capable of increasing agricultural production through regulation of hormone, nutrition, stress resistance factors. Previously, it has been revealed that due to a reduction in the quality of soil health and impact on the environment caused by synthetic chemical fertilizers, many are now beginning to adopt the utilization of biofertilizers as alternatives to promote plant growth in sustainable farming. *Azotobacter chroococcum, Klebsiella variicola, Rhizobium larrymoorei, Klebsiella pneumonia* are known to stimulate plant development and act as biofertilizers in organic farming practice. Beg and Singh (2009) revealed that across the globe, increased productivity has been witnessed in the agricultural sector due to the adoption and utilization of biofertilizers. Raimi et al. (2017) suggested that many parts of the developing countries suffer from food shortage and low productivity due to soil pollution, consistent use of synthetic fertilizers, thus recent approach in the maintenance of sustainable agriculture incorporates the use of microbial fertilizers such as Azotobacter and many others. These microbes produce metabolites that protect the crop from pathogenic attack and boost the soil nutrients. Katiyar et al. (2015) showed that PGPR promotes systemic resistance, biofertilization, biocontrol of plant pathogens. Gupta et al. (2015) showed that PGPR increases soil fertility, suppresses phytopathogens, and enhances plant growth promotion, for the development of an eco-friendly approach in sustainable agriculture.

Paul and Nair (2008) reported that sol salinity is regulated by microbes and agricultural crops. The authors studied the mode of activity explored by PGPR in regulating salt tolerance utilizing proteome analysis. They discovered that many of the salt regulatory proteins are upregulated, thereby alleviating the high osmolarity and generating inhibitory metabolites when plant growth-promoting rhizosphere inoculant *P. fluorescens* MSP-393 is applied. Kaur et al. (2016) revealed that PGPR improves crop productivity by colonizing the plant rhizosphere or endophyte, and production of beneficial biomolecules such as organic acids, phytohormones, siderophores, antibiotics, and growth regulators utilized for plant defense system against pathogenic attack. Mishra (2018) showed that crop productivity amid harsh environmental conditions such as drought, pest attack, global warming can be enhanced through the utilization of PGPR, thereby generating exopolysaccharides as biocontrol agents. Jiménez-Gómez et al. (2017) suggested that bacterial inoculants portends the capability to produce positive results on crop yields without any adverse effects. The utilization of PGPR has increased over the years due to the huge demand placed on agricultural products across the globe. Di Benedetto et al. (2017) highlighted a few modes of action by which PGPR could induce plant growth such as nutrient uptake, production of hormones, inhibition of pathogens. Sharma et al. (2019) suggested that in the developing countries where huge monies are spent on procuring synthetic chemicals to improve agricultural productivity resulting in increased negative environmental and health impact, environment friendly plant growth-promoting rhizobia should be adopted as a substitute to synthetic fertilizers for sustainable agriculture.

Dago et al. (2018) reported that improving soybean and maize crops, rhizobacteria such as *Pseudomonas fluorescence* was utilized as biofertilizer, thus a significant difference in terms of crops development was noticed compared to those that were not treated with biofertilizer. Osman and Yin (2018) reported that PGPR could influence plant growth through a number of mechanisms like nitrogen fixation, enzyme synthesis, bioremediation of contaminants, production of volatile organic compounds. Khan et al. (2020) revealed that plant growth regulators like putrescine and salicylic acid are important components of plant development. The authors thus investigated the role of the regulators on chickpea grown in sandy soil. It was observed that chickpea plants integrated with plant growth regulators showed increased capacity for drought tolerance and they recommended that its utilization increases agricultural productivity. Zuluaga et al. (2020) highlighted the relationship between plants and their microbiome for growth-promoting agricultural productivity through alleviating stress by utilizing bacterial inoculants. Kuan et al. (2016) showed that in maize plants, plant growth-promoting rhizobacteria bacterial strains isolated from maize roots at two different locations in Malaysia provided an alternative to increasing crop yield. The strains evaluated in this study are *Klebsiella sp.* Br1, *Acinetobacter sp.* S3r2, and *Bacillus pumilus* S1r1. The authors revealed that all the strains showed positive results for phosphate solubilization, auxin production, and increased N2 fixation. Lengai and Muthomi (2018) suggested that many important biopesticides are derived from plants, microorganisms, and insects utilized in the management of pest and disease conditions. The authors disclosed that biopesticides are now being considered as potential alternatives to synthetic chemical pesticides due to the huge negative impact on the environment and human-caused by these synthetic chemicals. Thus the physiochemical characteristics of the biopesticides like biodegradability, low toxicity, less expensive, eco-friendly nature give them a greater advantage over synthetic chemicals.

Yadav and Yadav (2019) revealed that actinobacteria can be developed and used as biofertilizers for sustainable agriculture to improve plant growth and soil physiology. This class of bacteria (*Acidimicrobiia, Coriobacteriia, Actinobacteria, Nitriliruptoria, Thermoleophilia, and Rubrobacteria*) possesses huge biological characteristics with multifarious plant growth-promoting attributes. Today, actinobacteria are significantly explored for bio-inoculants for different crop growth-promoting effects. Kawalekar (2013) reported that biofertilizers are utilized for proper plant growth while minimizing the use of synthetic fertilizers and also to promote soil health in a cost-effective way. Srivastava and Singh (2017) revealed that PGPR is now a widely recognized approach in agriculture due to the green revolution as an alternative strategy to synthetic chemicals. PGPR such as *Achromobacter, Azospirillum, Azotobacter, Acetobacter, Chryseobacterium, Bacillus, Flavobacterium, Klebsiella, Enterococcus, Pseudomonas, Serratia, Paenibacillus*, and *Rhizobium* act as bioremediation, biodegradation, biocatalyst, biofertilizers, biocontrol/biopesticide agents in sustainable agricultural practice. Garg and Chandel (2010) predicted change in environmental conditions caused by anthropogenic activities, thereby affecting soil, air, and water agro-ecosystems. Thus reversing this trend will demand the incorporation of natural beneficial microbes in maintaining plant productivity and soil fertility like crop plants with arbuscular mycorrhizal. Different genes, chemical structures, and signal transduction pathways are activated to facilitate water/nutrients uptake, alleviation of abiotic soil stresses, disease protection, and increasing crop production/yield. Goswami et al. (2016) showed that the population explosion has placed a huge demand for food and other agricultural produce. Thus, today PGPR offers a promising approach in sustainable agriculture to enhance soil microbial flora and promote plant growth. Gopalakrishnan et al. (2015) revealed that many challenges are witnessed with modern agricultural practices such as climate change, soil fertility impairment, increased pests, and insect attacks. Thus the use of biofertilizers, biopesticides as plant growth promoters is gathering massive attention among different stakeholders to provide a sustainable approach for agricultural practice.

Jacoby et al. (2017) reported that a plant-rich ecosystem is made up of diverse microorganisms providing support for mineral nutrition, metabolic activities, plant shape, and defense mechanism. Mahmood et al. (2016) showed that through seed inoculation, beneficial microbes perform a significant function in the growth of the plant, soil fertility, and environmental health. Tuhuteru et al. (2016) carried out a study to obtain the most effective isolate in PGPR as biological fertilizers. They observed that the isolates were able to stimulate increase seed growth, increase the chlorophyll content with other physiochemical properties. Singh (2018) and Rifat et al. (2012) reported that PGPR like *Azospirillum brasilense, Azospirillum amazonense, Azospirillum lipoferum, Bacillus tropicalis, Acetobacter diazotrophicus, Bacillus borstelensis, Herbaspirillum seropedicae, Herbaspirillum rubrisubalbicans, Klebsiella sp., Rahnella aquatilis, Enterobacter sp., Herbaspirillum seropedicae, Paenibacillus azotofixans,* and *Bacillus circulans* enhanced crop growth through nitrogen fixation, production of hormones, production of enzymes and cytokinins, increased resistance to stress, solubilization, and mineralization of other nutrients. Zerihun et al. (2019) revealed that PGPR stimulates plant growth and protects plants stress factors. Thus the authors carried out a study to identify and characterize plant growth promoter bacteria colonizing the rhizosphere during the flowering phase for generating bioinoculant. The observed that the PGPR can be utilized as biofertilizers, biocontrol, and biopesticides to improve crop yield and productivity.

Kour et al. (2020) showed that beneficial microbes utilized as biofertilizers bring important nutrients from the soil to the plants to improve the quality and yield. Utilizing microbial bioinoculants represents an important part of sustainable agriculture. These microbes are known to colonize the plant epiphytic, rhizospheric, and endophytic system, thus regulating nutrients uptake, production of plant growth hormones and enzymes, and fixation of nitrogen. Bechtaoui et al. (2019) reported that over the years, greater attention has been placed on the role of PGPR as biofertilizer, thus the authors evaluated the application of biofertilizer bacteria plus rhizobial strains on the production of plant crops in Morocco. Their ability to solubilize complex mineral phosphorus was also investigated together with the ability to generate different biomolecules. They discovered that the combined strains displayed the most beneficial effects which significantly stimulate plant growth, hence they suggested that rhizobacterial inoculation could be utilized as potential biofertilizers.

Backer et al. (2018) stated that phytomicrobiome microbes are linked with plant tissues, thus providing a wide range of benefits to plants such as nutrients acquisition, improving soil texture, regulating extracellular molecules, activating different signals, and ultimately facilitating plant growth. The authors established that inoculating plants with PGPR could stimulate crop growth which can also improve plant tolerance for stresses by stimulation of systemic resistance. Bhat et al. (2019) highlighted the importance of plant growth-promoting rhizosphere as a capable tool for eco-friendly and ecological crop production. Amaya-Gómez et al. (2020) showed that rhizobacteria are capable of improving plant nutrients, regulate phytohormones, suppress diseases, and enhance plant survival.

Deepmala et al. (2016) demonstrated that the current soil management strategy involves the use of biofertilizers like phyla actinobacteria, firmicutes, proteobacteria, and bacteroidetes to facilitate growth and improve the biomass improvement of seedling germination, plant health, vigor, height, nutrient content of shoot tissues, shoot weight, early bloom, increase nodulation in legumes, improve chlorophyll content. Yadav et al. (2017) and Kenneth et al. (2019) suggested that rhizobacteria colonize extracellular and intracellular rhizoenvironment as biocontrol, biostimulation, and biofertilization. They revealed that to achieve self-sufficiency in agriculture, the utilization of genetically modified microbes must be deployed to enhance soil–plant–microbial interaction and develop soil flora and fauna. Orhan et al. (2006) studied the effects of two Bacillus strains on organically grown primocane fruiting raspberry. They discovered that the application of bacteria significantly enhanced the affected soil pH and nutritional contents, promoted growth, increased the yield, of the raspberry plant under organic farming conditions.

Adedeji et al. (2020) revealed the African continent is the worst hit in terms of global food insecurity due to poor economy, land degradation which is threatening the productivity in agriculture. They suggested that sustainable eco-friendly strategies like plant growth-promoting bacteria should be adopted to increase agricultural productivity, reduce environmental pollution, and improve the economy. García-Fraile et al. (2015) and Kalayu (2019) demonstrated that several rhizospheric bacterial strains like Bacillus, Rhizobium, Pseudomonas, Aspergillus, Penicillium, possess plant growth-promoting properties like phytohormones, stress resistance, and improve nutrients uptake through phosphate-solubilizing microbes. Agbodjato et al. (2015) discovered that maize rhizospheres contain a huge amount of diverse microorganisms like *B. polymyxa, B. anthracis, B. pantothenticus, B. circulans, B. thuringiensis, P. cichorii, P. syringae, P. putida,* and *Serratia marcescens* with a high rate of ammonium and hydrogen cyanide production, thus suggesting that these rhizobacteria could be utilized as biological fertilizers in promoting maize production.

Noumavo et al. (2016) highlighted the benefits of PGPR in promoting plant growth and development such as exopolysaccharides production, siderophores production, phosphate solubilization, phytostimulation, systemic resistance, production of antibiotics, enzymes, and nutrients uptake. Paul and Lade (2014) revealed that the arid and semi-arid regions are salt-stressed agricultural unproductive areas. Therefore, PGPR is one of the alternative solutions to enhance agricultural productivity through rhizobacteria counteracting the osmotic stress and enhancing plant growth. This approach will enhance resistance to diseases, nutrient uptake, stress tolerance, hydration, biocontrol of phytopathogens, chlorophyll content, increasing K^+ concentration, solubilization of mineral phosphate, osmolyte accumulation, salinity tolerance, and synthesis of antioxidative enzymes. Bhardwaj et al. (2014) and Pahari et al. (2017) recommended that due to the consistent application of synthetic chemicals over the years, the soil has been contaminated, thus the authors investigated the role of eleven bacteria isolate on some contaminated soil in promoting plant growth activities. Vibha and Madhu (2015) highlighted the role of various regulatory chemicals secreted within the vicinity of the rhizosphere such as biofertilizers, biocontrol, and biostimulants. They revealed that through genetic engineering, many of the biomolecules are incorporated into field practices in agriculture to enhance productivity. Mishra and Dash (2014) revealed that the economy of India thrives on agricultural practices, and fertilizer is a major contributing factor. Over the years, the application of synthetic fertilizers has endangered the ecosystems, plants, humans, animals, and soil, hence naturally grown biofertilizers are beginning to receive attention for sustainable agriculture economic development.

Several scientists have investigated plant growth-promoting rhizobacteria as biocontrol agents through local antagonism to soil-borne pathogens, nitrogen fixation, production of phytohormones, phosphate solubilization, nutrient mobilization, or by induction of systemic resistance against pathogens for improved cropping systems. They suggested that many bacteria inoculants displayed significant plant growth-promoting properties (Romero-Perdomo et al. 2019; Beneduzi et al. 2012; Ramprasad et al. 2014). Sinha et al. (2014) and Ahirwar et al. (2019) applied bacterial inoculants such as Azotobacter, phosphorus solubilizing bacteria, Clostridium pasteurianum, Azospirillum, vesicular arbuscular mycorrhiza to stimulate microbial activity as biofertilizers in organic farming. Many of the biomolecules are converted into powerful biofertilizers, bio-herbicides, biopesticides, bio-insecticides, viral-based bio-insecticides, and fungal based bio-insecticides utilizing microbial biotechnology.

Bashan et al. (2014) and Khatoon et al. (2020) discovered that one of the important components of soil health is PGPR with multiple ecological functions in the rhizosphere soil producing phytohormones, innate immunity, and other metabolites. Cummings (2009) showed that PGPR could improve the yield of graminaceous crops through genetically engineered strains. They revealed in their study that the physicochemical and biological features of the soils are also a major contributing factor through the direct relationship between plant–microbial organism to facilitate, phosphate solubilization, phytohormones, hydrogen cyanide production, biological nitrogen fixation, stress and biocontrol activity, antibiotic fabrication, siderophore production, synthesis of antifungal metabolites.

20.11 Specific Samples of Beneficial Microorganisms that Could Lead to Sustainable Agriculture and the Environment

20.11.1 *Rhizobium spp*

Tiwari et al. (2017) recounted the role of *Azotobacter* sp. and Rhizobium on plant growth, chlorophyll contents, nodule appearance, and carbohydrate content. The authors revealed that *Rhizobium sp.* and *Azotobacter* have friendly associations for field application most especially for sustainable agriculture. Poonia (2011) demonstrated that most Rhizobium can provide nitrogen for plant physiological needs essential for growth and development and act as a biofertilizer, thereby decreasing the use of synthetic agrochemicals. Today, agricultural production is challenged by diverse environmental and climatic factors affecting soil health and fertility. PGPR act with legumes resulting in enhanced nutrients through nitrogen fixation, systemic resistance, tolerance to stress, production of phytohormones, and solubilizing phosphates in the plant root exudates. Zeffa et al. (2019) investigated the role of *Azospirillum brasilense* as PGPR in promoting nitrogen use efficiency in maize. They suggested that rhizobia form root nodules that fix nitrogen in symbiotic legumes, thus performing the experiment in nonlegumes would be a useful way of increasing productivity in agricultural practice particularly among the resource-poor countries. From their finding, it was revealed that biomolecules like auxins, abscisic acids, cytokinins, lumichrome, lipo-chitooligosaccharides, vitamins, and riboflavin produced by rhizobia may be responsible for the plant growth property, phosphorus uptake in maize, millet, and sorghum.

Environmental factors like heat, salinity, and drought are known to alter crop growth and other soil physiological processes. In order to mitigate these effects, PGPR has been suggested to constantly minimize the negative impact of environmental stresses. Rhizobacteria have been reported to significantly improve grain yield (Bashan and de-Bashan 2010). Azospirillum–plant interaction has been shown to cause single phytohormone activity, nitrogen fixation, collections of small-sized molecules or enzymes, multiple phytohormones, increased membrane activity, the proliferation of the root system, mobilization of minerals, increased water plus mineral uptake, elimination of environmental stressors in plants, biocontrol of phytopathogens. Fukami et al. (2018) highlighted and attributed the plant growth-promoting bacteria role of genus Azospirillum towards tolerance of biotic and abiotic stresses, mediated by phytohormones through ethylene/jasmonic acid signaling pathway (Foyer et al. 2019) revealed that symbiotic nitrogen fixation is a major mechanism of legume–rhizobia relationship which may further be enhanced by arbuscular mycorrhiza.

Naveed et al. (2015) and Patel et al. (2017) revealed that rhizobia–legume and nonlegumes symbiosis for biological nitrogen fixation is changing agricultural practices. Through the development of root nodules, rhizobia fix nitrogen from the atmosphere. Bankole et al. (2019) and Borges et al. (2019) suggested that symbiotic relationship with plant roots such as rhizobia, actinomycetes, mycorrhizal fungi, diazotrophic bacteria provides an opportunity for biofertilizer, biostimulation, and biocontrol mechanism. Datta et al. (2015) and Mabrouk and Belhadj (2012) showed that rhizobium is a gram negative bacterium·linked with a symbiotic relationship with the roots of leguminous/nonleguminous plants containing granules of β-hydroxybutyrate. There are different classes of rhizobium such as Rhizobium, Mesorhizobium, Bradyrhizobium, Sinorhizobium, and Azorhizobium with efficient plant growth-promoting ability like phytohormones, phosphate solubilization, siderophores production, and hydrogen cyanide production. A symbiotic process exists in rhizobia with leguminous plants with energy in the form of nutrients, then fixing dinitrogen from the atmosphere for plant uptake and subsequent reduction of dinitrogen into ammonia making use of 16 molecules of ATP with a complex set of enzymes and releasing various chemicals by the root cells into the soil (Abhinav et al. 2015).

The application of *Rhizobium leguminosarum* and *Pseudomonas* jesseni P10 as a PGPR to support the growth and nodulation abilities of *Lens culinaris* Medik has been documented (Iqbal et al. 2012). These bacteria are also known for producing a plant growth enzyme, 1-aminocyclopropane-1-carboxylate deaminase (ACC deaminase) and these PGPR bacteria have also been isolated from plastic enriched compost (PEC) (Iqbal et al. 2012).

Yadegari and Rahmani (2010) reported that two bacteria, *Pseudomonas fluorescens* P-93 and *Azospirillum lipoferum* S-21 elicited appreciable plant promoting activities on the seeds of the bean plant (*Phaseolus vulgaris*) especially when co-cultured with two *Rhizobium* strains either individually or in combination. In an in vitro study, Flores-félix et al. (2013) sequestered *Rhizobium leguminosarum* strain PEPV 16 from the nodules of *Phaseolus vulgaris*. The authors revealed that the root nodule bacteria displayed some PGPR properties and was able to add a significant upsurge of N and K uptake in Carrot and Lettuce plants. Co-inoculation of *Rhizobium* spp. and *Enterobacter cloacae and Pseudomonas* spp PGPR strains were recorded to have reduced Cu. stress in *Vicia faba* than when compared to the non-inoculated plants (Fatnassi et al. 2015). Aamir et al. (2013) reported that the co-treatment of Mung Bean using Rhizobium and PGPRs containing ACC deaminase significantly enhanced the nodulation and growth of the plant. They observed that this was done by reducing the stress associated with salinity. Afzal and Bano (2008) also compared the outcome of inoculating together and singly, Rhizobium, and a phosphate-solubilizing bacterium with or without phosphate fertilizer, on *Triticum aestivum* plant (Wheat). The authors discovered that there was a 29% increase in growth and a significant improvement in plant morphology when a dual combination with P fertilizer was used when compared to those treated without fertilizers. This study revealed that dual treatment with a phosphate fertilizer is very important for plant growth and subsequent wheat plant yield.

20.11.2 *Azospirillum spp*

Saikia et al. (2010) and Sahu et al. (2017) reported that the growing human population across the globe demands a novel strategy to increase agricultural productivity so as to meet the growing demand. The authors pointed attention to the utilization of microbes like *Azospirillum spp*, micro-aerophillic microorganisms as biofertilizers to enhance the development. *Azospirillum spp* act as nitrogen-fixing bacteria, enhance seed germination, increase proton flux, facilitate seedling growth, phosphorus solubilization, generation of phytohormones like indole-3-acetic acid, sequestration of iron, enhance photosynthetic pigments, increase dry matter partitioning, plant growth promoters, restoration of vegetation in a harsh environment, alleviate stresses, and increase seed quality (Fukami et al. 2018).

Cassan and Diaz-Zorita (2016) reported that *Azospirillum sp.* is a PGPR that can colonize several plant species to fix nitrogen, produce metabolite, and several phytohormones like siderophore, nitric oxide, abscisic acid, ethylene biocontrol of phytopathogens, gibberellins, phosphate solubilization indole-3-acetic acid. Mehnaz et al. (2007) and Fukami et al. (2018) exhibited that genus *Azospirillum spp* confer to plants stresses tolerance, signaling molecules activation such as jasmonic acid/ ethylene pathway, osmotic adjustment, mediate antioxidants, and detoxification of oxidative stress. Pereyra et al. (2007) showed that Azospirillum spp portends the potential to improve plant development as well as secretion of antimicrobial activity and other secondary metabolites that help in stimulating phytohormones, and in production of biofertilizers (Barassi et al. 2007).

Azospirillum spp. has been recorded to resist stress conditions and has been a good PGPR in agriculture (Diaz-Zorita and Fernandez-Canigia, 2009; Bashan and de-Bashan 2010). Couillerot et al. (2010) monitored the inoculant presence and quantity of the PGPR *Azospirillum lipoferum* CRT1 in the rhizosphere of maize seedlings using real-time polymerase chain reaction (PCR) method. García-Fraile et al. (2015) compared the in vitro drought tolerance and PGPR qualities of 36 different strains of *Azospirillum* including a strain Az39 which was regularly used in Argentina for inoculating and planting of Maize. It was discovered that strain Az19 had the highest drought, salt stress resistant, and PGPR qualities out of all isolated strains. The co-inoculation of *Azospirillum* strains with other PGPR has been a plus in the field of agriculture. This synergistic combination greatly improves nutrient availability and stimulates each other's physiological and biochemical systems, leading to an improved plant growth rate. Previously, it was observed that at inoculation of 10^7 in its stationary phase, there was effective coaggregation of *Azospirillum* with other PGPR. The authors also recorded that other factors like pH and temperature also enhanced coaggregation at 5 and 35 to 40 ° C, respectively. In an experiment performed by previously, researchers had inoculated the PGPRs, *Azospirillum brasilense*, and *Bacillus sphaericus* with 33% nitrogen fertilizer to determine the growth yield and productivity of banana plantlets. It was recorded that nutrients were significantly increased and an early flowering at 3 weeks was observed. It was also discovered that the physical features of the banana fruits, when compared to the control of the experiment were considerably improved. The amount of nitrogen (N) and phosphorus (P) nutrients required by plants for their growth cannot be overemphasized. The use of NP fertilizers and their effects on the environment is of great concern. Nitrate serves as a pollutant of groundwater and the gradual loss of phosphorus from the soil through runoff finds its way to surface waters. Alternate eco-friendly methods in improving plant growth should be made available to help in the preservation of the environment. Ejaz et al. (2020) performed an experiment to determine the effectiveness of PGPRs in enhancing the development, produce, and quality of plants. They co-inoculated the pea plant (*Pisum sativum* L.) with nitrogen-fixing *Azospirillum* strain and a phosphorus solubilizing *Agrobacterium tumefaciens* strain at a different percentage of reduced nitrogen and phosphorus fertilizer (60, 65, 70, 75, 80%, and the proposed dose 100%) and compared it with a nitrogen–phosphorus (100%) fertilizer without any inoculum. It was discovered that the PGPR co-inoculation with 75% NP treated pea plant presented a 55% plant growth and development when compared to the 100% NP which was not inoculated with both rhizobacterium. They concluded that the co-inoculation of important NP rhizobacterium strains can be beneficial to the environment and also a cost-effective choice.

Scientists have identified *Azospirillum* strains as effective stimulators in the rhizosphere aiding root exudation and development. Baudoin et al. (2009) in an experiment investigated the genotypic construction of the rhizobacterial population present on maize seedlings grown in the field after its inoculation with *Azospirillum lipoferum* CRT1. They revealed that there was an alteration in the native bacteria present in the rhizosphere at days 7 and 35 and they concluded that treatment of seed with *Azospirillum lipoferum* CRT1 increased the presence of different bacteria in the field soil.

20.11.3 *Bacillus spp*

Metin et al. (2014) reported that *Bacillus megaterium* strain*, B. subtilis* strain, and *Pantoea agglomerans* strain can act as PGPR to improve seedling quality and growth in cabbages. Radhakrishnan et al. (2017) revealed that genetic and environmental factors greatly affect crop productivity and yield, thus *Bacillus* and *Pseudomonas spp* are now being utilized to facilitate plant growth by inducing physiological changes such as exopolysaccharides and siderophores secretion, pathogenic microbial control, water transport, nutrient uptake, and production of other several active metabolites like chitosanase, cellulase, protease, glucanase, hydrogen cyanide, and lipopeptides. Bacillus spp. release ammonia from nitrogenous organic matter through nifH gene, thus produce nitrogenase, fix atmospheric nitrogen, enhance plant growth plus yield by delaying senescence, iron-chelating properties generated through siderophore production which help to solubilize iron from minerals plus organic compounds in rhizospheres.

Raaijmakers and Mazzola (2012) and Malviya et al. (2012) reported that Bacillus species have been proposed to increase crop yield and quality, root colonization, chlorophyll content. Barea (2015) demonstrated that exploiting the agroecosystem of soil microbial host seems to be a promising approach. Thus the authors investigated the role of sustainable and organic agricultural production through the utilization and management of soil microorganisms such as *Bacillus spp*. The plant-linked microbiome has been revealed to ameliorate the negative impacts of stress factors, increase crop productivity. Villarreal-Delgado et al. (2017) showed that the genus Bacillus is widely distributed in the agro-ecosystems. The authors further analyzed and discovered that the microbe's mechanism of action involves secretion of toxins, phytopathogens suppression, antibiotics, siderophores, induced systemic resistance, and lytic enzymes. Jamal et al. (2018) revealed that *Bacillus amyloliquefaciens* Y1 strain was studied to identify its role on soil properties, rhizosphere bacterial flora, pepper seedling growth, plus soil enzyme activities.

From their results, it was observed that *B. amyloliquefaciens* Y1 displayed a positive role on soil fertility and recommended for biofertilizer application.

Hashem et al. (2019) reported that many microbes can stimulate plant growth and replace chemical fertilizers or pesticides. PGPR has been revealed to induction of systemic resistance, competitive omission, and antibiosis. The authors discovered that *Bacillus subtilis* exhibits these characteristics by secreting secondary metabolites, cell-wall degrading enzymes, enhances nitrogen fixation, hormones, antioxidants defense enzymes secretion, and solubilizes soil phosphorus, production of siderophores and exopolysaccharides. The authors suggested that multidisciplinary approaches such as molecular biology, physiology, biotechnology should be adopted to harness the beneficial properties of many of these plant growth-promoting rhizosphere.

Alooa et al. (2019) revealed that rhizospheric bacteria improve soil fertility and promote plant growth by producing enzymes like glucanases, chitosanases, and chitinases, siderophores, and antibiotics like pyoluteorin, zwittermicin A, and oomycin. *Bacilli rhizobacteria* are known to offer unique functions and properties such as biofertilization, bioprotection, and phytostimulation. The authors inoculated tomato seedlings with cell suspensions of B. subtilis and discovered that shoot and root growth are enhanced, increased seedling vigor was noticeable in the leaf area of the plants and higher levels of phytohormones are secreted.

It has been discovered that the genus *Bacillus* has been documented to be highly effective for phosphate-solubilizing capability. *Bacillus amyloliquefaciens* having PGPR traits and able to induce resistance to *Rhizoctonia solani* and *Fusarium solani in* in vitro has been detected in the potato rhizosphere. The restoration of the effects of salinity stress in the root system of the soybean plant (*Glycine max* L.) has been known to be carried out by the PGPR *Bacillus firmus* SW5. As a consequence, this bacterium significantly improved plant quality, yield, and antioxidant defense systems. The size and texture of tomato plants (*Lycopersicon esculentum* Mill, cv Rio Fuego) cultivated under greenhouse conditions and exposed to inoculated cultures of the PGPR *B. subtilis* BEB-13bs strain introduced at the plant root were investigated. The control system showed no effect on the plant yield but there was a considerable increase in the yield after inoculation with the PGPR *B. subtilis* strain. The authors opined that the PGPR *Bacillus subtilis* BEB-13bs strain had a positive impact on the fruit quality and yield of the cultivated tomatoes.

Lim and Kim (2013) investigated the effect of multi-functional PGPR *Bacillus licheniformis* K11 on the drought resistance attribute of the pepper plant (*Capsicum annuum* L.). The authors observed that in using a control, after a 15 day period, the pepper plants exposed to drought stress did not survive, while those inoculated with the PGPR *B. licheniformis* K11 strain survived. They reported the presence of pathogenesis-related protein 10 (*CaPR-10*), dehydrin-like protein (*Cadhn)* (cytoplasmic small heat shock protein class I) *sHSP*, and (vacuolar H+-ATPase) *VA* stress proteins genes in *C. annuum* L. inoculated with *B. licheniformis* K11. They concluded that *B. licheniformis* K11 was a good agent that could be applied as biofertilizer for the better productivity of the plant. Probanza et al. (2002) studied the effect of two different species of PGPR *Bacillus*; *B. licheniformis* CECT 5106 and *B. pumilus* CECT 5105, respectively, when used individually or in combination to treat *Pinus pinea* plant seedlings. The authors observed an improved growth and modification in microbial populations present in the rhizosphere when used individually than when used as a consortium. They concluded that the respective PGPR *Bacillus* species could not function effectively as a consortium due to competition in the rhizosphere of the affected plant.

20.11.4 *Serratia spp*

Rhizospheric borne *Serratia* strains isolated from the plant; *Nothofagus alpine* was reported to exhibit the ability to promote growth in the diameter of the plant (Martínez et al. 2018). The authors also observed the growth promoting ability of the bacteria with respect to the root collar, biomass, its height, nitrogen, and chlorophyll content of *N. alpine* plantlets. They discovered that all the isolated strains also caused increased production of plant metabolites such as 1-aminocyclopropane-1-carboxylic acid deaminase and indole acetic acid, and also increased nitrogen fixing capacity of the plant. They concluded that *Serratia* strains are PGPR which can be utilized as biofertilizers when applied in plant nurseries.

Earlier experiments have demonstrated that an appreciable decline in the oxidative stress markers with an increment in salinity stress tolerance in maize (*Zea mays* L.) by rhizospheric *Serratia liquefaciens* KM4, thereby impacting positively on the overall phenotypic and genotypic receptiveness of the plant. The genome of PGPR *Serratia marcescens* CDP-13 cultured from a plant, *Capparis decidua* was sequenced and the bacterium was known for its ability to reduce the impact of physical and biological stress on the host plant. The sequenced *Serratia marcescens* CDP-13 was discovered to have significant traits of PGPRs which include considerable growth in the saline concentration of up to 6%, aiding the improvement of wheat grown under high salt concentration. The study concluded that *Serratia marcescens* CDP-13 has the probability to enhance salt stress and act as a substitute for pesticides. Researchers have conducted a study on the usage of rhizospheric borne plant growth-promoting bacterium; *S. nematodiphila* PEJ1011 to establish the growth-promoting effect of gibberellin (GA) on pepper (*Capsicum annuum L*). It was detected that *S. nematodiphila* PEJ1011 improved the low-temperature effect on *Capsicum annuum L,* helping the plant adapt to low-temperature stress. The combined effect of the PGPR *Pseudomonas fluorescens* and *Serratia marcescens* to protect and prevent the symptoms of the cucumber mosaic cucumovirus on the cotyledon of *Cucumis sativus* and *Lycopersicum esculentum* was studied by Raupach et al. (1996). Using the enzyme-linked immunoabsorbent assay to detect the presence of the viral antigen, the authors discovered that the symptoms on plants were significantly reduced after treatment with PGPR on the seedlings and did not develop any noticeable symptoms till the end of the experiment. They concluded that the use of different genus of PGPR possesses the capability to control viral diseases of plants.

However, the mutual effect of two different concentrations of genistein (0 and 20μM) and either of *Serratia proteamaculans* or *Serratia liquefaciens* was used to determine the growth yield, fixation of nitrogen, and nodulation of soybean types (*Bradyrhizobium japonicum*). There was no significant difference recorded in using both of the PGPR with genistein but a difference was recorded when genistein was used separately.

20.11.5 *Pseudomonas spp*

Genus *Pseudomonas*is an important rhizobacterium involved in the growth and proper development of plants and known for its phosphate soulibilizing capability. A comparative study of the effectiveness of two different methods (microcapsules and liquid) utilized in the inoculation of different strains of PGPR *Pseudomonas putida*; FA-8, FA-56, and FA-60 in Tomato (*Lycopersicon esculentum*) was performed and the study was also aimed at evaluating the growth enhancing effect of the bacterium on the tomato plant. The authors observed that using the microcapsule inoculation method, *Pseudomonas putida* FA-56 produced the highest indole acetic acid (IAA) at 23.02μg mL^{-1} revealing a significant increment in all physiological characteristics and bacterial population in the plant rhizosphere. It was confirmed that the inoculation method of PGPR using microcapsules was a good substitute to chemical fertilizers, thereby promoting biofertilizers.

Bakker et al. (1986) studied the potential application of *P. putida* WCS358 for the treatment of potato seed tubers (*Solanum tuberosum* L.cv Bintje) at long and short term crop rotation. The *P. putida* WCS358 had siderophores and were also known nonproducing Tn5 transposon mutants. The authors observed that potato seed treatment with siderophore producing Tn5 transposon mutant in long crop rotations yielded no significant yield, but a significant yield of 13% after 86 days of short crop rotation was recorded when the siderophore producing Tn5 wildtype was used in the experiment. They reported that the presence of siderophores in PGPR was a necessary requirement for a good potato tuber yield.

Audenaert et al. (2002) demonstrated that salicylic acid (SA) a known siderophore metabolite elicited by the PGPR *P. aeuriginosa*7NSK2 was not a sufficient determinant of induced resistance in *Pseudomonas aeuriginosa* 7NSK2 exposed to *Botrytis cinerea*, but rather additional metaboliteswhich included Pyochelin and pyocyanin. An evaluation of the growth increment of Sorghum was done using some Arbscular mycorrhizal fungi; *Glomus fasciculatum* and *Glomus aggregatum* together with forty (40) known different fluorescent *Pseudomonas* spp., individually and together. Criteria employed for selection of the different fluorescent *Pseudomonas* spp., was their position of a single and/or numerous PGPR quality known to aid plant growth. It was documented that *Pseudomonas* spp. P10 and P13 which exhibited PGPR attributes such as exopolysaccharide, IAA, gibberellic acid, siderophore, and phosphate solubilization had a better outcome with Abscular mycorrhizae. Whereas *Pseudomonas* spp. P38 that posses the ability to produce hydrogen cyanide (HCN) and also an effective solubilizer of phosphate showed no significant effect. They concluded that the various PGPR features of different *Pseudomonas* strains may not be sufficient to present them as PGPR. It has been reported that the PGPR *Pseudomonas aurantiaca* SR1 was found to colonize the root of two (2) cereal crops: wheat (*Triticum aestivum* L) and maize (*Zea mays*). The researchers observed that when the PGPR *P. aurantiaca* SR1 was inoculated during propagation, plant growth was enhanced even when used with or without fertilizers. The biological fertilizer has in turn aided the alleviation of inorganic nitrogen pollution.

Previously, scientists have demonstrated that different strains of *Pseudomonas* sp. assisted in phosphorus fertilization and uptake of nutrients in *T. aestivum* L. under both field and greenhouse conditions. Mirza et al. (2006) reported that the growth of rice was positively aided by nitrogen-fixing *Pseudomonas* strains that have been known to be a potential PGPR inoculant. Previously, it has been observed that the presence of heavy metals in soil negatively affected the expression of PGPR traits of the respective *Pseudomonas* strains which were previously reported to have good growth-promoting traits.

20.11.6 *Stenotrophomonas spp*

Stenotrophomonas is one of the PGPR, known for its multiple traits and adaptability. Stenotrophomonas is known for its ability to be a good PGPR and is also isolated and characterized from the rhizosphere of different crops (Kumar and Audipudi 2015; Patel and Saraf 2017). This PGPR is also extensively involved in nitrogen and sulfur biogeochemical cycles. In the delivery of PGPR as bioinoculants into the rhizosphere and on seedlings for plant growth and improvement, the form in which it is added is very important. Kumar et al. (2019) isolated six strains of PGPR, known to be of the genus Stenotrophomonas from five various plants (*Solanum tuberosum, Triticum aestivum, Bacopa monnieri, Zea mays,* and *Aloe barbadensis*). They discovered that they had the ability to reduce nitrogen to ammonia when compared to *Azotobacter chroococum. S. rhizophilia* was recorded to have survived in various liquid carriers and was concluded that these PGPRs can be used in the production of liquid biofertilizer. In a recent study, researchers have investigated the reductive ability of Stenotrophomonas on Cr (IV). The authors revealed that the PGPR was extremely resistant to chromium and had a 92.5% reduction in Cr(IV) to Cr(III) within 28 h. They concluded that the rhizobacterium would serve in bioremediation of chromium polluted soils.

Alavi et al. (2013) studied the process linked with stress in the rhizosphere. The improvement of cucumber resistance to cucumber green mottle mosaic virus has been recorded by the PGPR *Stenotrophomonas maltophilia* HW2 (Li et al. 2016). It was also discovered that in 3 days *S. maltophilia* repressed the phenotypic expression of the viral protein on the leaf of the cucumber, making a good biological control agent in sustainable agriculture.

20.12 Specific Examples of Beneficial Microorganism Involved in the Maintainaece of Soil Health

Tahat et al. (2020) reported that soil health is linked with sustainable agriculture providing abundant nutrients for plant development. Recently, organic farming utilizing PGPR has been accepted as an alternative to synthetic chemicals due to its adverse effects on soil fertility and plant physiology. The authors revealed that these plant promoters could affect the level of plant composition, productivity, soil integrity, soil nutrient cycling, and its sustainability. They showed that organic farming systems are known to increase soil nutrient mineralization and microorganism large quantity and diversity as well as soil physical features. Conservation tillage, enhanced soil fungi abundance, earthworm diversity, environmental factors like physical, chemical, biological facilitate dynamic soil–rhizosphere–plant systems and stability for agricultural sustainability.

Rafiquea et al. (2017) reported that microbial application for plant growth facilitates soil health and fertility compared to chemical fertilizer. An adequate supply of phosphorus to plant is a significant indication for soil health to satisfy crop nutritional requirements. Phosphate-solubilizing microorganisms in the soil microbial communities such as *Clostridium pasteurianum, Rhodobacter, cyanobacteria, Methanogens, Bacillus mucilaginous, Bacillus circulans, Bacillus megaterium, Pseudomonas striata, Bacillus subtilis* facilitate soil management strategies for eco-friendly soil fertility enhancement, controlled soil pH.

Medina and Azcón (2010) reported that enhancing the capability of soil microorganisms for the inhibition of pathogens is an important strategy for sustainable agriculture such as regulation of plant defense activity, plant hormone signaling crosstalk, development of soil microbe–plant insect relationship. Hirsch et al. (2013) highlighted the importance of soil microorganisms in alleviating the negative impacts of osmotic stressors like salinity and drought. Many land areas across the globe are increasingly being polluted with many contaminants, thus raising the level of salinity and pH. Therefore, for the plant to cope with this osmotic stress, an appropriate mechanism must be developed such as water uptake capacity, activation of the antioxidant system, transpiration rates, maintenance of ionic homeostasis, and lowered reactive oxygen species generation.

20.12.1 Mycorrhizal Associations

Steffen et al. (2020) reported that crop production in agricultural practice is constantly being influenced by many environmental and climatic factors, thus scientists are always searching for ways to improve crop production through organic farming. The associations between arbuscular mycorrhizal fungi and plants have been evaluated for many years. The role of mycorrhizal on plant growth under stressful conditions was studied on the quality plus performance of crops. The rhizosphere characteristic feature is a major determinant of plant performance, thus arbuscular mycorrhizae are one of the common types of symbiotic relationships between rhizosphere microorganisms and plant roots. The mycorrhizal inoculants have been revealed to stimulate plant growth and development through advanced biotechnology integrated with multidisciplinary knowledge of biochemistry, microbiology, molecular biology, immunology, cell biology, enzymology, genetics, bioengineering, physiology, biophysics, chemical engineering, mathematics, to continually utilize microorganisms and their components to stimulate plant metabolism with maximum efficiency. Recent advances in the field of arbuscular mycorrhiza and their impact on plant quality, environmental protection, and biodiversity conservation need further evaluation.

Meena et al. (2017) revealed that exploring the symbiotic relationship of arbuscular mycorrhiza and crop tolerance to an unfavored environment is a sustainable approach in organic farming that needs further elucidation. The present agricultural practice is no longer sustainable due to the high cost of fertilizers, global warming due to synthetic agrochemical use, poor soil fertility as a result of constant tillage, and application of chemicals. It has been documented that Mycorrhiza fungal could exhibit a symbiotic relationship with plant roots such as *Chenopodiaceae, Caryophyllaceae, Amaranthaceae, Polygonaceae, Commelinaceae, Brassicaceae, Cyperaceae, Juncaceae* in the rhizosphere with the ability to supplied different amounts of phosphorus. Arbuscular mycorrhizas are now seen to influence plant community development, water relations, nutrient uptake and above-ground productivity, tolerant of adverse conditions, microbial biofertilizers, optimized microbiomes, biocontrol microbes, soil matching microbe-crops for different soil types.

Jakhar et al. (2017) reported that mycorrhizae display essential functions in plant growth, soil fertility, and plant protection, also their filamentous networks promote bi-directional nutrient movement. Oruru and Njeru (2016) reported that Arbuscular mycorrhizal fungi display a vital function in biological soil fertility, plant protection, and nutrition. Barea et al. (2011) and Surendirakumar et al. (2019) showed that mycorrhizal facilitate resilience of plant host against environmental stresses like nutrient deficiency, soil disturbance, and drought as biofertilizers, biostimulants, and bioprotectors.

20.13 Conclusion and Future Recommendation

Therefore, this chapter intends to provide detailed information on comprehensive information on sustainable biotechnology tools that could boost an increase in agricultural and food production. The application of genetic engineering and advanced biotechnology tool such as synthetic biology could help in the identification of a novel consortium that could help in the achievement of sustainable organics agriculture. Detailed facts on microorganisms that could influence soil health in promoting plant growth, as biological fertilizers, biological control agent, wastelands restoration, and bioremediations were also highlighted. The application of metabolomics could help to identify beneficial metabolites that could lead to the management of pests and diseases which are biotic factors that mitigate the increase in agricultural production as well as those that could help in regulating abiotic stress like salinity and drought that affect increased in agricultural production.

References

Aamir M et al (2013) Co-inoculation with rhizobium and plant growth promoting rhizobacteria (PGPR) for inducing salinity tolerance in mung bean under field condition of semi-arid climate. Asian J Agric Biol 1(1):7–12

Adedeji AA, Häggblom MM, Babalola OO (2020) Sustainable agriculture in Africa: plant growth-promoting rhizobacteria (PGPR) to the rescue. Sci Afr 9:e00492. https://doi.org/10.1016/j.sciaf.2020.e00492

Adetunji CO, Anani OA (2020) Bio-fertilizer from trichoderma: boom for agriculture production and management of soil- and root-borne plant pathogens. In: Mishra P, Mishra RR, Adetunji CO (eds) Innovations in food technology. Springer, Singapore, pp 245–256

Afzal A, Bano A (2008) Rhizobium and phosphate solubilizing bacteria improve the yield and phosphorus uptake in wheat (Triticum aestivum). Int J Agric Biol 10(1):85–88

Agbodjato NA, Noumavo PA, Baba-Moussa F, Salami HA, Sina H, Sèzan A, Bankolé H, Adjanohoun A, Baba-Moussa L (2015) Characterization of potential plant growth promoting rhizobacteria isolated from maize (Zea mays L.) in central and northern Benin (West Africa). Hindawi Publishing Corporation Appl Environ Soil Sci 2015:901656, 9 pages. https://doi.org/10.1155/2015/901656

Ahirwar NK, Singh R, Chaurasia S, Chandra R, Prajapati S, Ramana S (2019) Effective role of beneficial microbes in achieving the sustainable agriculture and eco-friendly environment development goals: a review. Front Environ Microbiol 5(6):111–123. http://www.sciencepublishinggroup.com/j/fem. https://doi.org/10.11648/j.fem.20190506.12

Ajao AT, Oluwajobi AO, Olatayo VS (2011) Bioremediation of soil microcosms from auto-mechanic workshops. J Appl Sci Environ Manag 15(3):473–477

Alavi P et al (2013) Root-microbe systems: the effect and mode of interaction of stress protecting agent (SPA) Stenotrophomonas rhizophila DSM14405T. Front Plant Sci 4:1–10. https://doi.org/10.3389/fpls.2013.00141

Alexander A, Chong KP (2014) The impact of biological control agents on soil microbial communities in oil palm plantation soils. Am-Eurasian J Sustain Agric 8:8–17

Alooa BN, Makumba BA, Mbega ER (2019) The potential of bacilli rhizobacteria for sustainable crop production and environmental sustainability. Microbiol Res 219:26–39

Amaya-Gómez CV, Porcel M, MesaGarriga L, Gómez-Álvarez MI (2020) A framework for the selection of plant growth promoting rhizobacteria based on bacterial competence mechanisms. Appl Environ Microbiol 86:e00760–e00720. https://doi.org/10.1128/AEM.00760-20

Anani OA, Olomukoro JO (2019) Assessment of metal accumulation and bioaccumulation factor of some trace and heavy metals in freshwater prawn and crab. IntechOpen. https://doi.org/10.5772/intechopen.88103

Anani OA, Olomukoro JO, Ezenwa IM (2020a) Limnological evaluation in terms of water quality of Ossiomo River, Southern Nigeria. Int J Conserv Sci 11(2):571–588

Anani OA, Olomukoro JO, Enuneku AA (2020b) Geospatial mapping, environmetrics and indexing approach for a tropical river sediment in Southern Nigeria. Pak J Sci Ind Res Ser A Phys Sci 63A(3):176–187

Audenaert K, De Meyer G, Höfte M (2002) Abscisic acid determines basal susceptibility of tomato to Botrytis cinerea and suppresses salicylic acid dependent signaling mechanisms. Plant Physiol 128:491–501

Barassi CA, Sueldo RJ, Creus CM, Carrozzi LE, Casanovas EM, Pereyra MA (2007) Dynamic soil, dynamic plant. Global science books Azospirillum spp., a dynamic soil bacterium favourable to vegetable crop production. Dyn Soil Dyn Plant 1(2):68–82

Beneduzi A, Ambrosini A, Passaglia LMP (2012) Plant growth-promoting rhizobacteria (PGPR): their potential as antagonists and biocontrol agents. Genet Mol Biol 35:41044–41051

Backer R, Rokem JS, Ilangumaran G, Lamont J, Praslickova D, Ricci E, Subramanian S, Smith DL (2018) Plant growth-promoting rhizobacteria: context, mechanisms of action, and roadmap to commercialization of biostimulants for sustainable agriculture. Front Plant Sci 9:1473. https://doi.org/10.3389/fpls.2018.01473

Bakker PAHM, Lamers JG, Bakker AW, Marugg JD, Weisbeek PJ, Schippers B (1986) The role of siderophores in potato tuber yield increase by Pseudomonas putida in a short rotation of potato. Neth J Plant Pathol 92:249–256

Bankole SO et al (2019) Potential significance of beneficial microbes for sustainable soil management and plant utilization. Acta Sci Microbiol 2(6):88–95

Barea JM (2015) Future challenges and perspectives for applying microbial biotechnology in sustainable agriculture based on a better understanding of plant-microbiome interactions. J Soil Sci Plant Nutr 15(2):261–282

Barea JM, Palenzuela J, Cornejo P, Sánchez-Castro I, Navarro-Fernández C, Lopéz-García A, Estrada B, Azcón R, Ferrol N, Azcón-Aguilar C (2011) Ecological and functional roles of mycorrhizas in semi-arid ecosystems of Southeast Spain. J Arid Environ 75:1292–1301

Bashan Y, de-Bashan LE (2010) How the plant growth-promoting bacterium azospirillum promotes plant growth—a critical assessment. Adv Agron 108(2):77–136. https://doi.org/10.1016/S0065-2113(10)08002-8

Bashan Y, de-Bashan LE, Prabhu SR, Hernandez J-P (2014) Advances in plant growth-promoting bacterial inoculant technology: formulations and practical perspectives (1998–2013). Plant Soil 378:1–33. https://doi.org/10.1007/s11104-013-1956-x

Baudoin E et al (2009) Impact of inoculation with the phytostimulatory PGPR Azospirillum lipoferum CRT1 on the genetic structure of the rhizobacterial community of field-grown maize. Soil Biol Biochem 41(2):409–413. https://doi.org/10.1016/j.soilbio.2008.10.015

Bechtaoui N, Raklami A, Tahiri A-I, Benidire L, El Alaoui A, Meddich A, Göttfert M, Oufdou K (2019) Characterization of plant growth promoting rhizobacteria and their benefits on growth and phosphate nutrition of faba bean and wheat. 2019. Published by The Company of Biologists Ltd. Biol Open 8:bio043968. https://doi.org/10.1242/bio.043968

Beg MA, Singh JK (2009) Effects of biofertilizers and fertility levels on growth, yield and nutrient removal of green gram (Vigna radiata) under Kashmir conditions. Indian J Agric Sci 79 (5):388–390

Bengtsson J, Ahnstrom J, Weibull AC (2005) The effects of organic agriculture on biodiversity and abundance: a meta-analysis. J Appl Ecol 42:261–269. https://doi.org/10.1111/j.1365-2664.2005.01005.x

Berg G, Rybakova D, Grube M, Koberl M (2016) The plant microbiome explored: implications for experimental botany. J Exp Bot 67:995–1002. https://doi.org/10.1093/jxb/erv466

Bhardwaj D, Ansari MW, Sahoo RK, Tuteja N (2014) Biofertilizers function as key player in sustainable agriculture by improving soil fertility, plant tolerance and crop productivity. Microb Cell Factories 13:66. http://www.microbialcellfactories.com/content/13/1/66

Bhat MA et al (2019) Plant growth promoting rhizobacteria (PGPR) for sustainable and eco-friendly agriculture. Acta Sci Agric 3(1):23–25

Boelling D, Groen AF, Sørensen P, Madsen P, Jensen J (2003) Genetic improvement of livestock for organic farming systems. Livest Prod Sci 80:79–88

Borges CS, de Sá EL, Muniz AW, Filho O (2019) Potential use of rhizobium for vegetable crops growth promotion. Afr J Agric Res 14(8):477–483. https://doi.org/10.5897/AJAR2017.12885

Cassan F, Diaz-Zorita M (2016) Azospirillum sp. in current agriculture: from the laboratory to the field. Soil Biol Biochem 103:117–130

Cassán F, Perrig D, Sgroy V, Masciarelli O, Penna C, Luna V (2009) Azospirillum brasilense Az39 and Bradyrhizobium japonicum E109, inoculated singly or in combination, promote seed germination and early seedling growth in corn (Zea mays L.) and soybean (Glycine max L.). Eur J Soil Biol 45:28–35. https://doi.org/10.1016/j.ejsobi.2008.08.005

Chakrabarty T, Akter A, Saifullah ASM, Sheikh MS, Bhowmick AC (2014) Use of fertilizer and pesticide for crop production in agrarian area of Tangail district, Bangladesh. Environ Ecol Res 2(6):253–261. https://doi.org/10.13189/eer.2014.020605

Costa C, Garcia-Lestion J, Costa S, Coelho P, Silva S, Pingqarilho M, Valdiqlesias V, Mattei F, Dali Armi V, Bonassi S, Laffon B, Snawder J, Teixeira JP (2014) Is organic farming safer to farmers' health? A comparison between organic and traditional farming. Toxicol Lett 230:166–176

Couillerot O et al (2010) Development of a real-time PCR method to quantify the PGPR strain Azospirillum lipoferum CRT1 on maize seedlings. Soil Biol Biochem 42(12):2298–2305. https://doi.org/10.1016/j.soilbio.2010.09.003

Cummings SP (2009) The application of plant growth promoting rhizobacteria (PGPR) in low input and organic cultivation of graminaceous crops; potential and problems. Environ Biotechnol 5 (2):43–50

Dago ND, Diarrassouba N, Saraka MDY, Moroh J-LA, Fofana IJ, Baba-Moussa L, Coulibaly A (2018) Predicting maize and soybean crops dry biomass through rhizobacteria microorganisms activity on foliar bio-fertilizer in an arid agro-climate: a multiple linear regression analysis. Afr J Microbiol Res 12(34):835–848. https://doi.org/10.5897/AJMR2018.8899

Di Benedetto A, Corbo MR, Campaniello D, Cataldi MP, Bevilacqua A, Sinigaglia M, Flagella Z (2017) The role of plant growth promoting bacteria in improving nitrogen use efficiency for sustainable crop production: a focus on wheat Nilde. AIMS Microbiol 3(3):413–434. https://doi.org/10.3934/microbiol.2017.3.413

Datta A, Singh RK, Kumar S, Kumar S (2015) An effective and beneficial plant growth promoting soil bacterium "Rhizobium": a review. Ann Plant Sci 4(1):933–942

de Souza R, Ambrosini A, Passaglia L (2015) Plant growth-promoting bacteria as inoculants in agricultural soils. Genet Mol Biol 38(4):401–419. https://doi.org/10.1590/S1415-475738420150053

dos Santos RM, Diaz PAE, Lobo LLB, Rigobelo EC (2020) Use of plant growth-promoting rhizobacteria in maize and sugarcane: characteristics and applications. Front Sustain Food Syst 4:136. https://doi.org/10.3389/fsufs.2020.00136

Ejaz S et al (2020) Effects of inoculation of root-associative azospirillum and agrobacterium strains on growth, yield and quality of pea (Pisum sativum L.) grown under different nitrogen and phosphorus regimes. Sci Hortic 270:109401. https://doi.org/10.1016/j.scienta.2020.109401

Escoba O, Hue NV (2007) Current developments in organic farming. Recent Res Dev Soilsci 2:29–62

Fatnassi IC et al (2015) Comptes Rendus Biologies impact of dual inoculation with rhizobium and PGPR on growth and antioxidant status of Vicia faba L. under copper stress. C R Biol 338:241–254

Fernando PC (2017) Pesticides, environment, and food safety. Food Energy Secur 6(2):48–60

FiBL & IFOAM (2016) The world of organic agriculture statistics & emerging trends. Research Institute of Organic Agriculture FiBL & IFOAM Organics International. https://shop.fibl.org

Flores-félix JD et al (2013) Use of rhizobium leguminosarum as a potential biofertilizer for Lactuca sativa and Daucus carota crops. J Plant Nutr Soil Sci 176(6):876–882. https://doi.org/10.1002/jpln.201300116

Foley JA et al (2011) Solutions for a cultivated planet. Nature 478:337–342

Food and Agriculture Organization of the United Nations (2013) Food wastage footprint: impacts on natural resources: summary reports. FAO, Rome

Food and Agriculture Organization of the United Nations (2015) How to feed the world in 2050. https://www.fao.org

Foyer CH, Nguyen H, Lam H-M (2019) Legumes—the art and science of environmentally sustainable agriculture. Plant Cell Environ 42:1–5

Francis CA (2013) Reference module in earth systems and environmental sciences, organic farming

Fukami J, Cerezini P, Hungria M (2018) Azospirillum: benefits that go far beyond biological nitrogen fixation. AMB Expr 8:73. https://doi.org/10.1186/s13568-018-0608-1

García-Fraile P, Menéndez E, Rivas R (2015) Role of bacterial biofertilizers in agriculture and forestry. 2(3):183–205. https://doi.org/10.3934/bioeng.2015.3.183

Garg A, Balodi R (2014) Recent trends in agriculture: vertical farming and organic farming. Adv Plants Agric Res 1(4):142–144

Garg N, Chandel S (2010) Arbuscular mycorrhizal networks: process and functions. A review. Agron Sustain Dev 30(3). https://doi.org/10.1051/agro/2009054. Springer Verlag/EDP Sciences/INRA. hal-00886481

Gopalakrishnan S, Sathya A, Vijayabharathi R, Varshney RK, Laxmipathi Gowda CL, Krishnamurthy L (2015) Plant growth promoting rhizobia: challenges and opportunities. Biotech 5:355–377. https://doi.org/10.1007/s13205-014-0241-x

Goswami D, Thakker JN, Dhandhukia PC (2016) Portraying mechanics of plant growth promoting rhizobacteria (PGPR): a review. Cogent Food Agric 2:1127500. https://doi.org/10.1080/23311932.2015.1127500. Soil and Crop Sciences, Review Article

Gupta G, Parihar SS, Ahirwar NK, Snehi SK, Singh V (2015) Plant growth promoting rhizobacteria (PGPR): current and future prospects for development of sustainable agriculture. J Microb Biochem Technol 7:2. https://doi.org/10.4172/1948-5948.1000188

Hashem A, Tabassum B, Abd_Allah EF (2019) Bacillus subtilis: a plant-growth promoting rhizobacterium that also impacts biotic stress. Saudi J Biol Sci 26:1291–1297

Hayat R, Ali S, Amara U, Khalid R, Ahmed I (2010) Soil beneficial bacteria and their role in plant growth promotion: a review. Ann Microbiol 60(4):579–598. https://doi.org/10.1007/s13213-010-0117-1

Hole DG, Perkins AJ, Wilson JD, Alexander IH, Grice PV, Evans AD (2005) Does organic farming benefit biodiversity? Biol Conserv 122:113–130. https://doi.org/10.1016/j.biocon.2004.07.018

Hirsch PR, Mauchline TH, Clark IM (2013) Culture-independent molecular approaches to microbial ecology in soil and the rhizosphere. In: de Bruijn FJ (ed) Molecular microbial ecology of the rhizosphere, vol 1. Wiley Blackwell, Hoboken, NJ, pp 45–55. https://doi.org/10.1186/s13568-018-0608-1

Iqbal MA, Khalid M, Shahzad SM, Ahmad M, Soleman N, Akhtar N (2012) Integrated use of rhizobium leguminosarum, plant growth promoting Rhizobacteria and enriched compost for improving growth, nodulation and yield of lentil (Lens culinaris Medik.). Chil J Agric Res 72: 104–110

Jacoby R, Peukert M, Succurro A, Koprivova A, Kopriva S (2017) The role of soil microorganisms in plant mineral nutrition—current knowledge and future directions. Front Plant Sci 8:1617. https://doi.org/10.3389/fpls.2017.01617

Jakhar SR, Kumar S, Jangir CK, Meena RS (2017) The role of mycorrhizal relationship in sustainable manner towards plant growth and soil fertility. Indian J Agric Allied Sci 3(4):19–24

Jalota SK et al (2018) Understanding climate change impacts on crop productivity and water balance. Elsevier, Amsterdam

Jamal Q, Lee YS, Jeon HD, Kim KY (2018) Effect of plant growth-promoting bacteria Bacillus amyloliquefaciens Y1 on soil properties, pepper seedling growth, rhizosphere bacterial flora and soil enzymes. Plant Protect Sci 54:129–137

Javaid A (2011) Effects of biofertilizers combined with different soil amendments on potted rice plants. Chil J Agric Res 71(1):157–163

Jiménez-Gómez A, Celador-Lera L, Fradejas-Bayón M, Rivas R (2017) Plant probiotic bacteria enhance the quality of fruit and horticultural crops. AIMS Microbiol 3(3):483–501. https://doi.org/10.3934/microbiol.2017.3.483

Kawalekar JS (2013) Role of biofertilizers and biopesticides for sustainable agriculture. J Biol Innov 2(3):73–78

Kalayu G (2019) Phosphate solubilizing microorganisms: promising approach as biofertilizers. Hindawi Int J Agron 2019:4917256, 7 pages. https://doi.org/10.1155/2019/4917256

Katherine K, Hendrik W (2010) Environmental impacts of agricultural technologies. Evans School Policy Analysis & Research (EPAR). EPAR Brief No. 65

Katiyar D, Hemantaranjan A, Singh B (2015) Chitosan as a promising natural compound to enhance potential physiological responses in plant: a review. Indian J Plant Physiol 20(1):1–9

Kaur H, Kaur J, Gera R (2016) Plant growth promoting Rhizobacteria: a boon to agriculture. Int J Cell Sci Biotechnol 5:17–22

Kenneth OC, Nwadibe EC, Kalu AU, Unah UV (2019) Plant growth promoting rhizobacteria (PGPR): a novel agent for sustainable food production. Am J Agric Biol Sci 14:35–54. https://doi.org/10.3844/ajabssp.2019.35.54

Khalid A, Tahir S, Arshad M, Zahir ZA (2004) Relative efficiency of rhizobacteria for auxin biosynthesis in rhizosphere and non-rhizosphere soils. Aust J Soil Res 42:921–926. https://doi.org/10.1071/SR04019

Khan N, Bano AMD, Babar A (2020) Impacts of plant growth promoters and plant growth regulators on rainfed agriculture. PLoS One 15(4):e0231426. https://doi.org/10.1371/journal.pone.0231426

Khatoon Z, Huang S, Rafique M, Fakhar A, Kamran MA, Santoyo G (2020) Unlocking the potential of plant growth-promoting rhizobacteria on soil health and the sustainability of agricultural systems. J Environ Manag 273:111118

Kour D, Rana KL, Yadav AN, Yadav N, Kumar M, Kumar V, Vyas P, Dhaliwal HS, Saxena AK (2020) Microbial biofertilizers: bioresources and eco-friendly technologies for agricultural and environmental sustainability. Biocatal Agric Biotechnol 23:101487. https://doi.org/10.1016/j.bcab.2019.101487

Krey T, Vassilev N, Baum C, Eichler-Löbermann B (2013) Effects of long-term phosphorus application and plant-growth promoting rhizobacteria on maize phosphorus nutrition under field conditions. Eur J Soil Biol 55:124–130. https://doi.org/10.1016/j.ejsobi.2012.12.007

Kuan KB, Othman R, Abdul Rahim K, Shamsuddin ZH (2016) Plant growth-promoting rhizobacteria inoculation to enhance vegetative growth, nitrogen fixation and nitrogen remobilisation of maize under greenhouse conditions. PLoS One 11(3):e0152478. https://doi.org/10.1371/journal.pone.0152478

Kumar N, Audipudi V (2015) Exploration of a novel plant growth promoting bacteria Stenotrophomonas maltophilia AVP27 isolated from the chilli rhizosphere soil. Int J Eng Res Gen Sci 3(1):265–276

Kumar A et al (2019) Stenotrophomonas: a versatile diazotrophic bacteria from the rhizospheric soils of Western Himalayas and development of its liquid biofertilizer formulation. Vegetos 32 (1):103–109. https://doi.org/10.1007/s42535-019-00013-8

Lal R (2016) Climate change and agriculture, 2nd edn

Lehman RM, Cambardella CA, Stott DE, Acosta-Martinez V, Manter DK, Buyer JS, Maul J, Smith J, Collins H, Halvorson J, Kremer R, Lundgren J, Ducey TF, Jin VL, Karlen DL (2015) Understanding and enhancing soil biological health: the solution for reversing soil degradation. Sustainability 7:988–1027. https://doi.org/10.3390/su7010988

Lengai GMW, Muthomi JW (2018) Biopesticides and their role in sustainable agricultural production. J Biosci Med 6:7–41. https://doi.org/10.4236/jbm.2018.66002

Li H et al (2016) Stenotrophomonas maltophilia HW2 enhanced cucumber resistance against cucumber green mottle mosaic virus. J Plant Biol 59:488–495. https://doi.org/10.1007/s12374-016-0246-6

Li S, Peng M, Liu Z, Shah S (2017) The role of soil microbes in promoting plant growth. Mol Microbiol Res 7(4):30–37

Lim J, Kim S (2013) Induction of drought stress resistance by multi-functional PGPR bacillus licheniformis K11 in pepper. Plant Pathol J 29(2):201–208

Mabrouk Y, Belhadj O (2012) Enhancing the biological nitrogen fixation of leguminous crops grown under stressed environments. Afr J Biotechnol 11(48):10809–10815. https://doi.org/10.5897/AJB10.2170

Mäder P et al (2002) Soil fertility and biodiversity in organic farming. Science 296:1694–1697

Mahmood A, Turgay OC, Farooq M, Hayat R (2016) Seed biopriming with plant growth promoting rhizobacteria: a review. FEMS Microbiol Ecol 92(8):fiw112

Malviya MK, Sharma A, Pandey A, Rinu K, Sati P, Palni LMS (2012) Bacillus subtilis NRRL B-30408: a potential inoculant for crops grown under rainfed conditions in the mountains. J Soil Sci Plant Nutr 12(4):811–824

Martínez OA, Encina C, Tomckowiack C, Droppelmann F, Jara R, Maldonado C, Muñoz O, García-Fraile P, Rivas R (2018) Serratia strains isolated from the rhizosphere of raulí (Nothofagus alpina) in volcanic soils harbour PGPR mechanisms and promote raulí plantlet growth. J Soil Sci Plant Nutr 18:804–819

Masciandaro G, Macci C, Peruzzi E, Ceccanti B, Doni S (2013) Organic matter-microorganism-plant in soil bioremediation: a synergic approach. Rev Environ Sci Biotechnol 12(4):399–419. https://doi.org/10.1007/s11157-013-9313-3

Medina A, Azcón R (2010) Effectiveness of the application of arbuscular mycorrhiza fungi and organic amendments to improve soil quality and plant performance under stress conditions. J Soil Sci Plant Nutr 10:354–372

Meena KK, Sorty AM, Bitla UM, Choudhary K, Gupta P, Pareek A et al (2017) Abiotic stress responses and microbe-mediated mitigation in plants: the omics strategies. Front Plant Sci 8: 172. https://doi.org/10.3389/fpls.2017.00172

Mehnaz S, Weselowski B, Lazarovits G (2007) Azospirillum canadense sp. nov., a nitrogen-fixing bacterium isolated from corn rhizosphere. Int J Syst Evol Microbiol 57:620–624

Meier MS et al (2015) Environmental impacts of organic and conventional agricultural products–are the differences captured by life cycle assessment? J Environ Manag 149:193–208

Metin T, Melek E, Ertan Y, Adem G, Kenan K, Recep K, Dursun A (2014) Plant growth-promoting rhizobacteria improved growth, nutrient, and hormone content of cabbage (Brassica oleracea) seedlings. Turk J Agric For 38:327–333. https://doi.org/10.3906/tar-1308-62. © TÜBİTAK

Mirza MS et al (2006) Molecular characterization and PCR detection of a nitrogen-fixing Pseudomonas strain promoting rice growth. Biol Fertil Soils 43(2):163–170. https://doi.org/10.1007/s00374-006-0074-9

Mishra VK (2018) Plant growth promoting rhizobacteria (PGPR) mediated temperature, drought and pesticide stress tolerance of crop plants through multidisciplinary approach. Am J Res Commun 6(3):1–9

Mishra P, Dash D (2014) Rejuvenation of biofertilizer for sustainable agriculture and economic development. Cons J Sustain Dev 11(1):41–61

Mondelaers K, Aertsens J, Van Huylenbroeck G (2009) A meta-analysis of the differences in environmental impacts between organic and conventional farming. Br Food J 111:1098–1119. https://doi.org/10.1108/00070700910992925

Muller A, Schader C, Scialabba NE, Brüggemann J, Isensee A, Erb K, Smith P, Klocke P, Leiber F, Tolze MS, Niggli U (2017) Strategies for feeding the world more sustainably with organic agriculture. Nat Commun 8:1290. https://doi.org/10.1038/s41467-017-01410-w

Naveed M, Mehboob I, Hussain MB, Zahir ZA (2015) Perspectives of rhizobial inoculation for sustainable crop production. In: Plant microbes symbiosis: applied facets. Springer, India, pp 209–239. https://doi.org/10.1007/978-81-322-2068-8_11

Nehra V, Choudhary M (2015) A review on plant growth promoting rhizobacteria acting as bioinoculants and their biological approach towards the production of sustainable agriculture. J Appl Nat Sci 7(1):540–556

Nielsen KM (2019) Organic farming. In: Encyclopedia of ecology (second edition). Elsevier, Amsterdam

Noumavo PA, Agbodjato NA, Baba-Moussa F, Adjanohoun A, Baba-Moussa L (2016) Plant growth promoting rhizobacteria: beneficial effects for healthy and sustainable agriculture. Afr J Biotechnol 15(27):1452–1463. https://doi.org/10.5897/AJB2016.15397

Nunes J, Araujola SF, Nunesal M, Carneiroaa A, Tsaib M (2012) Impact of land degradation on soil microbial biomass and activity in northeast Brazil. Pedosphere 22(1):88–95

Ojo OI, Olajire-Ajayi BL, Dada OV, Wahab OM (2015) Effects of fertilizers on soil's microbial growth and populations: a review. Am J Eng Res 4(7):52–61

Onder M, Ceyhan E, Kahraman A (2011) Effects of agricultural practices on environment. international conference on biology. In: International conference on biology, environment and chemistry, vol 24. IPCBEE, pp 28–32. http://ipcbee.com/vol24/6-ICBEC2011-C00015

Orhan E, Esitken A, Ercisli S, Turan M, Sahin F (2006) Effects of plant growth promoting rhizobacteria (PGPR) on yield, growth and nutrient contents in organically growing raspberry. Sci Hortic 111:38–43

Oruru MB, Njeru EM (2016) Upscaling arbuscular mycorrhizal symbiosis and related agroecosystems services in smallholder farming systems. Hindawi Publishing Corporation BioMed Res Int 2016:4376240, 12 pages. https://doi.org/10.1155/2016/4376240

Osman NI, Yin S (2018) Isolation and characterization of pea plant (Pisum sativum L.) growth-promoting rhizobacteria. Afr J Microbiol Res 12(34):820–828. https://doi.org/10.5897/AJMR2018.8859

Otutumi AT, Oliveira TS, Mendoça ES, Lima JBF (2004) Qualidade do solo em sistemas de cultivo agroecológicos no município de Tauá-CE. In: Oliveira TS et al (eds) Solo e água: aspectos de uso e manejo com ênfase no semi-árido nordestino. Departamento de ciências do solo, UFC, Fortaleza, CE, ix, 31 pp

Owens K, Feldman J, Kepner J (2010) Wide range of diseases linked to pesticides. Pestic You 30: 13–21

Pahari A, Pradhan A, Priyadarshini S, Nayak SK, Mishra BB (2017) Isolation and characterization of plant growth promoting rhizobacteria from coastal region and their effect on different vegetables. Int J Sci Environ Technol 6(5):3002–3010

Patel T, Saraf M (2017) Exploration of novel plant growth promoting bacteria stenotrophomonas maltophilia MTP42 isolated from the rhizospheric soil of Coleus forskohlii. Int J Curr Microbiol App Sci 6(11):944–955

Patel R, Rathor G, Bhargav BH, Singh K, Bharti D (2017) Study and effect of rhizobium bacteria and culture suspension isolated from root nodules at Nimad area. Pharma Innov J 6(9):368–371

Paul D, Lade H (2014) Plant-growth-promoting rhizobacteria to improve crop growth in saline soils: a review. Agron Sustain Dev 34(4):737–752. https://doi.org/10.1007/s13593-014-0233-6. Springer Verlag/EDP Sciences/INRA. hal-01234840

Paul D, Nair S (2008) Stress adaptations in a plant growth promoting rhizobacterium (PGPR) with increasing salinity in the coastal agricultural soils. J Basic Microbiol 48:378–384

Pereyra MA, González RL, Creus CM, Barassi CA (2007) Root colonization vs. seedling growth, in two Azospirillum-inoculated wheat species. Cereal Res Commun 35:1621–1629

Perez-Montano F, Alias-Villegas C, Bellogin RA, del Cerro P, Espuny MR, Jimenez-Guerrero I et al (2014) Plant growth promotion in cereal and leguminous agricultural important plants: from microorganism capacities to crop production. Microbiol Res 169:325–336. https://doi.org/10.1016/j.micres.2013.09.011

Poonia S (2011) Rhizobium: a natural biofertilizer. Int J Eng Manag Res 1(1):36–38. ISSN No.: 2250-0758

Probanza A et al (2002) Pinus pinea L. seedling growth and bacterial rhizosphere structure after inoculation with PGPR Bacillus (B. licheniformis CECT 5106 and B. pumilus CECT 5105). Appl Soil Ecol 20:75–84

Qessaoui R, Bouharroud R, Furze JN, El Aalaoui M, Akroud H, Amarraque A, VanVaerenbergh J, Tahzima R, Mayad EH, Chebli B (2019) Applications of new rhizobacteria pseudomonas isolates in agroecology via fundamental processes complementing plant growth. Sci Rep 9: 12832. https://doi.org/10.1038/s41598-019-49216-8

Raaijmakers JM, Mazzola M (2012) Diversity and natural functions of antibiotics produced by beneficial and plant pathogenic bacteria. Annu Rev Phytopathol 50:403–424

Radhakrishnan R, Hashem A, Abd_Allah EF (2017) Bacillus: a biological tool for crop improvement through bio-molecular changes in adverse environments. Front Physiol 8:667. https://doi.org/10.3389/fphys.2017.00667

Rafiquea M, Sultanc T, Ortasb I, Chaudharya HJ (2017) Enhancement of maize plant growth with inoculation of phosphate-solubilizing bacteria and biochar amendment in soil. Soil Sci Plant Nutr 63(5):460–469. https://doi.org/10.1080/00380768.2017.1373599

Raimi A, Adeleke R, Roopnarain A (2017) Soil fertility challenges and biofertiliser as a viable alternative for increasing smallholder farmer crop productivity in sub-Saharan Africa. Cogent Food Agric 3:1400933. https://doi.org/10.1080/23311932.2017.1400933

Ramprasad D, Debasish SD, Bhattar SB (2014) Plant growth promoting Rhizobacteria – an overview. Eur J Biotechnol Biosci 2(2):30–34

Raupach GS, Liu L, Murphy JF, Tuzun S, Kloepper JW (1996) Induced systemic resistance in cucumber and tomato against cucumber mosaic virus using plant growth-promoting rhizobacteria. Plant Dis 80:891–894

Reganold JP, Wachter JM (2016) Organic agriculture in the twenty-first century. Nat Plants 2:1–8

Riggs PJ, Chelius MK, Iniguez AL, Kaeppler SM, Triplett EW (2001) Enhanced maize productivity by inoculation with diazotrophic bacteria. Funct Plant Biol 28:829–836. https://doi.org/10.1071/PP01045

Romero-Perdomo F, Ocampo-Gallego J, Camelo-Rusinque M, Bonilla R (2019) Plant growth promoting rhizobacteria and their potential as bioinoculants on Pennisetum clandestinum (Poaceae). Rev Biol Trop 67(4):825–832

Sachidanand B, Mitra N, Kumar V, Roy R, Mishra B (2019) Soil as a huge laboratory for microorganisms. Agric Res Tech Open Access J 22(4):ARTOAJ.MS.ID.556205

Sahu PK, Gupta A, Sharma L, Bakade R (2017) Mechanisms of Azospirillum in plant growth promotion. Sch J Agric Vet Sci 4(9):338–343

Saikia SP, Dutta SP, Goswami A, Bhau BS, Kanjilal PB (2010) Chapter 16: Role of Azospirillum in the improvement of legumes. In: Microbes for legume improvement. Springer-Verlag, Wien, pp 389–408. https://doi.org/10.1007/978-3-211-99753-6_16

Sangkumchaliang P, Huang WC (2012) Consumers' perceptions and attitudes of organic food products in northern Thailand. Int Food Agribus Manag Rev 15:87–102

Sathya A, Vijayabharathi R, Gopalakrishnan S (2016) Soil microbes: the invisible managers of soil fertility. In: Singh DP et al (eds) Microbial inoculants in sustainable agricultural productivity. Springer, New Delhi, pp 1–16. https://doi.org/10.1007/978-81-322-2644-4_1

Schader C, Stolze M, Gattinger A (2012) Environmental performance of organic farming. In: Boye JI, Arcand Y (eds) Green technologies in food production and processing. Springer, New York, pp 183–210

Seneviratne G, Jayakody K, Weerasekara M, Someya T, Ryuda N (2011) Microbial biofertilizer application versus compost use in agriculture: soil health implications. In: Miransari M (ed) Soil microbes and environmental health. Nova Science Publishers, pp 81–117

Seufert V, Ramankutty N, Foley JA (2012) Comparing the yields of organic and conventional agriculture. Nature 485:229–232

Sharma N, Singhvi R (2017) Effects of chemical fertilizers and pesticides on human health and environment: a review. Int J Agric Environ Biotechnol 10(6):675–679

Sharma K, Sharma S, Rajendra Prasad S (2019) PGPR: renewable tool for sustainable agriculture. Int J Curr Microbiol App Sci 8(1):525–530

Singh I (2018) Plant growth promoting rhizobacteria (PGPR) and their various mechanisms for plant growth enhancement in stressful conditions: a review. Eur J Biol Res 8(4):191–213. https://doi.org/10.5281/zenodo.1455995

Singh A, Vaish B, Singh RP (2016) Eco-restoration of degraded lands through microbial biomass: an ecological engineer. Acta Biomedica Scientia 3(1):133–135

Sinha RK, Valani D, Chauhan K, Agarwal S (2014) Embarking on a second green revolution for sustainable agriculture by vermiculture biotechnology using earthworms: reviving the dreams of Sir Charles Darwin. Int J Agric Health Saf 1:50–64

Skoufogianni E, Solomou A, Molla A, Martinos K (2015) Organic farming as an essential tool of the multifunctional agriculture. IntechOpen, London. https://doi.org/10.5772/61630

Smith LG, Kirk GJD, Jones PJ et al (2019) The greenhouse gas impacts of converting food production in England and Wales to organic methods. Nat Commun 10:4641. https://doi.org/10.1038/s41467-019-12622-7

Smith DL, Praslickova D, Ilangumaran G (2015) Inter-organismal signaling and management of the phytomicrobiome. Front Plant Sci 6:722. https://doi.org/10.3389/fpls.2015.00722

Srivastava R, Singh A (2017) Plant growth promoting rhizobacteria (PGPR) for sustainable agriculture. Int J Agric Sci Res (IJASR) 7(4):505–510. ISSN (P): 2250-0057; ISSN (E): 2321-0087

Steffen RB, Steffen GPK, Saldanha CW (2020) Importance of mycorrhization under heat stress conditions: technological tools to stimulate mycorrhizal colonization. J Agric Environ Sci 9 (1):8–14. https://doi.org/10.15640/jaes.v9n1a2

Surendirakumar K, Pandey RR, Muthukumar T (2019) Influence of indigenous arbuscular mycorrhizal fungus and bacterial bioinoculants on growth and yield of Capsicum chinense cultivated in non-sterilized soil. J Agric Sci 157(1):31–44. https://doi.org/10.1017/S0021859619000261

Tahat MM, Alananbeh KM, Othman YA, Leskovar DI (2020) Soil health and sustainable agriculture. Sustainability 12:4859. https://doi.org/10.3390/su12124859

Tan KZ, Radziah O, Halimi MS, Khairuddin AR, Shamsuddin ZH (2015) Assessment of plant growth-promoting rhizobacteria (PGPR) and rhizobia as multi-strain biofertilizer on growth and N2 fixation of rice plant. AJCS 9(12):1257–1264

Tiwari S, Chauhan RK, Singh R, Shukla R, Gaur R (2017) Integrated effect of rhizobium and azotobacter cultures on the leguminous crop black gram (Vigna mungo). Adv Crop Sci Tech 5 (3):1–9. https://doi.org/10.4172/2329-8863.1000289

Tsvetkov I, Atanassov A, Vlahova M, Carlier L, Christov N, Lefort F, Rusanov K, Badjakov I, Dincheva I, Tchamitchian M, Rakleova G (2018) Plant organic farming research–current status and opportunities for future development. Biotechnol Biotechnol Equip 32(2):241–260

Tuhuteru S, Sulistyaningsih E, Wibowo A (2016) Effects of plant growth promoting rhizobacteria (PGPR) on growth and yield of shallot in Sandy coastal land. Ilmu Pertanian (Agric Sci) 1 (3):105–110. Available online at http://journal.ugm.ac.id/jip.doi.org/10.22146/ipas.16349

Tuomisto H, Hodge I, Riordan P, Macdonald D (2012) Does organic farming reduce environmental impacts?–a meta-analysis of European research. J Environ Manag 112:309–320

Ulm F, Avelar D, Hobson P, Penha-Lopes G, Dias T, Maguas C, Cruz C (2019) Sustainable urban agriculture using compost and an open-pollinated maize variety. J Clean Prod 212:622–629

United Nations (2015) Population pyramids of the world from 1950 to 2100. World population prospects. https://un.org

Vejan P, Abdullah R, Khadiran T, Ismail S, Boyce AN (2016) Role of plant growth promoting rhizobacteria in agricultural sustainability—a review. Molecules 21:573. https://doi.org/10.3390/molecules21050573

Vessey JK (2003) Plant growth promoting rhizobacteria as biofertilizers. Plant Soil 255:571–586. https://doi.org/10.1023/A:1026037216893

Villarreal-Delgado MF, Villa-Rodríguez ED, CiraChávez LA, Estrada-Alvarado MI, Parra-Cota FI, De los Santos-Villalobos S (2017) The genus Bacillus as a biological control agent and its implications in the agricultural biosecurity. Rev Mex Fitopatol 36(1):95–130. https://doi.org/10.18781/R.MEX.FIT.1706-5

Yadav N, Yadav AN (2019) Actinobacteria for sustainable agriculture. J Appl Biotechnol Bioeng 6 (1):38–41

Yadav AN, Verma P, Singh B, Chauahan VS, Suman A, Saxena AK (2017) Plant growth promoting bacteria: biodiversity and multifunctional attributes for sustainable agriculture. Adv Biotechnol Microbiol 5(5):AIBM.MS.ID.555671. https://doi.org/10.19080/AIBM.2017.05.5556671

Yadegari M, Rahmani HA (2010) Evaluation of bean (Phaseolus vulgaris) seeds' inoculation with rhizobium phaseoli and plant growth promoting rhizobacteria (PGPR) on yield and yield components. Afr J Agric Res 5(9):792–799

Yanakittkul P, Aungvaravong C (2019) Proposed conceptual framework for studying the organic farmer behaviors. Kasetsart J Soc Sci 40:491–498

Yanakittkul P, Aungvaravong C (2020) A model of farmers intentions towards organic farming: a case study on rice farming in Thailand. Heliyon 6(2020):e03039

Yu H, Ling N, Wang T, Zhu C, Wang Y, Wang S et al (2019) Responses of soil biological traits and bacterial communities to nitrogen fertilization mediate maize yields across three soil types. Soil Till Res 185:61–69. https://doi.org/10.1016/j.still.2018.08.017

Zeffa DM, Perini LJ, Silva MB, de Sousa NV, Scapim CA, Oliveira ALM et al (2019) Azospirillum brasilense promotes increases in growth and nitrogen use efficiency of maize genotypes. PLoS One 14(4):e0215332. https://doi.org/10.1371/journal.pone.0215332

Zerihun T, Birhanu G, Genene T, Adey F, Solomon C, Tesfaye A, Fasil A (2019) Isolation and biochemical characterization of plant growth promoting (PGP) bacteria colonizing the rhizosphere of Tef crop during the seedling stage. Biomed J Sci Tech Res 14(2):MS.ID.002534

Zuluaga MYA, Lima Milani KM, Azeredo Gonçalves LS, Martinez de Oliveira AL (2020) Diversity and plant growth-promoting functions of diazotrophic/N-scavenging bacteria isolated from the soils and rhizospheres of two species of Solanum. PLoS One 15(1):e0227422. https://doi.org/10.1371/journal.pone.0227422

Advances in Microbial Applications in Safeguarding of Plant Health: Challenges and Future Perspective

21

Pooja Sharma and Surendra Pratap Singh

Abstract

Advantageous plant-associated microbes play a key role in promoting plant health and development in both regulated and natural environments. The broad range of plant-associated microbes has the ability to improve artificial agricultural production. There is strong evidence that plants shape the microbial profiles, primarily through exudates. These microbes have evolved unique adaptations to survive in the rhizospheric niche. Different plant favorable microorganisms solubilize soil nutrients and minerals, reduce environmental stress tolerance, suppress pathogens, promote plant growth and yield, and may also be a possible solution to increasing yields. This chapter demonstrated the communication of microbes and plants that enhance nutrition, summarizing the latest expertise in several fields of study which can overlap to enhance our understanding of the molecular mechanism that explains this process.

Keywords

Plant health · Signaling molecule · Microbes · Rhizospheric · Secondary metabolites

P. Sharma
Department of Environmental Microbiology, School for Environmental Sciences, Babasaheb Bhimrao Ambedkar Central University, Lucknow, Uttar Pradesh, India

S. P. Singh (✉)
Plant Molecular Biology Laboratory, Department of Botany, Dayanand Anglo-Vedic (PG) College, Chhatrapati Shahu Ji Maharaj University, Kanpur, India

R. Soni et al. (eds.), *Microbiological Activity for Soil and Plant Health Management*, https://doi.org/10.1007/978-981-16-2922-8_21

21.1 Introduction

The diversity of microbes is an essential aspect of global biological diversity. They are found in diverse habitats such as water, air, plant surfaces, food, and the human body microorganisms are found (Prussin and Marr 2015; Rosenberg and Zilber-Rosenberg 2016). Current scientific advancements have demonstrated which is significant for microbes still are unexplored, and their functions are mostly not known in the investigation of microbial diversity. Healthy plants cohabit with different microorganisms in nature, like bacteria, fungi, archaea, and protists, which form complex microbial consortiums and impact plant growth and development. In addressing large agricultural issues such as plant health protection, crop productivity, and soil health management, and environmental issues like bioremediation of soil and water from pollutants, many microbial applications are widely known. Currently, collections of microbial culture contain more than one million distinct strains (http:/www.wdcm.niq.ac.jp) and therefore attest to the attempts made to preserve biodiversity and to the need to make available these services to the public. This is generally accepted in which microbes appear to be related to others never alone. The core plant microbiome is characterized as microorganisms that are linked to a certain plant species or genotype, regardless of soil and environmental conditions (Toju et al. 2018). Edwards et al. 2015 reported as a representative of the rice core microbiome that bacteria belong especially to Deltaproteobacteria, Alphaproteobacteria, and Actinobacteria. Among plants and soil microbes, lengthy close relationships between various biological species like symbiosis and pathogenesis are predominant.

Owing to the favorable effects of associations in natural and agricultural environments, symbiotic associations between plants and soil microorganisms such as arbuscular mycorrhizal fungi, legume nitrogen bacteria or the water fern Azolla, and the cyanobacterium Anabaena azollae have become extensively reported (Igiehon and Babalola 2017; Bhuvaneshwari and Singh 2015; Mus et al. 2016). The phyllosphere is the aerial portion of plants, mostly the layer of the herb, and is an area in bacteria that form biofilms or wider aggregates that are commonly inhabited (Baldotto and Olivares 2008). The epiphytic microbiome community is protected from a stressful environment by biofilm formations, extracellular polymeric substances (EPS), and enzyme production (Remus-Emsermann et al. 2014; Müller et al. 2016). However, bacterial diversity depends on many factors such as plant types, geographical distance, and environmental factors (Remus Emsermann et al. 2014; Copeland et al. 2015; Laforest-Lapointe et al. 2016). Plants affected soil microbial profile in the vicinity of their proliferation roots through exudation of roots (Walker et al. 2003). The structure and properties of the root exudates are determined by the plant genotype, health, developmental stage, and fitness. Root exudation contributes to changes in pH and redox gradients, along with mucilage, lost cap, cortical cells, and chemicals in soil, that help to build the microbial diversity surrounding root (Lareen et al. 2016). Applications of chemical fertilizers, chemical pesticides, chemical insecticides, and chemical herbicides have increased crop production greatly; however, these activities were never safe and generate contamination to the environment and soil. In the food chain, chemical fertilizers were incorporated. Many chemically synthesized pesticides have been removed by regulatory authorities and the higher permissible residue level has been reduced, and there is a need for environmentally sustainable alternatives (Glare et al. 2012). A soil region near the plant root is called a rhizosphere, and this area is recommended as a microbial hot spot such as fungi and bacteria. A significant variety of substances are released by plant growth-promoting rhizobacteria (PGPR) that enhance the overall plant vigor. These reports encourage the use of these substances to elevate agricultural production. During different plant growth stages, contact between plants and microbes releases many signaling molecules that play a vital role in cell growth. Bacterial and fungal phytopathogens were never limited exclusively to the infection of aerial or root tissues, so connectivity between the shoot and the root will give the plant a survival gain and possibly restrict or avoid infection. In communication between pathogen and plants, signaling played a key role last few decays. At the same time, more previous attempts are being made to discover the signals involved in plant connection with non-pathogenic microbes, particularly those that increase plant growth.

Endophytes were also microbes that could be detected at a specific time in the tissues of an otherwise healthy plant host (Schulz and Boyle 2005). The population of fungi that live in living plant tissue despite showing indications of disease in their host are defined as fungal endophytes (Rodriguez et al. 2009). These are the main members of the endophytic community that live completely inside the tissues of plants and can be associated with roots, stems, and leaves. Almost every plant seems to have or does have at least one or more endophytic fungi in the environment. Endophytic fungi can act as plant growth regulators even though they secrete a wide range of phytohormones to help host plants (Bilal et al. 2018). Besides, the literature survey showed that fungi maintain plant growth through the production of substantial enzymes such as 1-aminocyclopropane-1-carboxylic acid deaminase (ACCD), urease, catalase, etc., phosphate solubilization, siderophore, and indole-3-acetic acid (IAA) formation and phytopathogenic antagonism (Wakelin et al. 2004; Glick 2014). A secondary metabolite is a chemical compound generated by fungal species (Frisvad et al. 2008). Secondary metabolism is frequently associated with sporulation mechanisms in microorganisms, like fungi. The endophytic fungi live in the internal tissues of the living plants without any visible disease (Strobel et al. 2004). Plant endophytes enter the tissues of their hosts that make different environments during abnormal conditions and ability to change the actual living environment by producing bioactive metabolites with various features and structural features (Strobel and Daisy 2003). Endophytic fungi produce richest source of natural bioactive compounds which is applicable in agriculture, medicine, and food industry (Strobel et al. 2004). Liu (2011) reported that bioactive metabolites isolated *Pestalotiopsis fici* (CGMCC3.15140) from the stable branches of *Camellia sinensis* in Hangzhou, China (Liu 2011). Therefore, while some gene clusters appear to be passed laterally for several of these secondary metabolites, others are likely to have emerged separately. Frisvad et al. (2004) reported that Arabic acid was produced by some fungal species in the *Penicillium series, Penicillium subgenus, Verrucosa* while penicillic acid was produced by all species in the *Circumdati, Aspergillus,* respectively.

Many fungi were fast-growing and have a high regenerative capacity but are short-live, for example, *Chytridiomycota, Zygomycota* species, yeasts, and rapidly growing ascomycetous fungi, like *Neurospora spp.* And they appear low secondary metabolite producers conversely, they contain hormones and carotenoids (Galagan et al. 2003). Small amounts of gibberellic acid were observed from *Sphaceloma manihoticola, N. crassa, Phaeospharia, Gibberella fujikuroi, A. Brazilian, Rhizobium phaseoli, and Ascomycetes*, as well as from *Azospirillum lipoferum.* This chapter offers recent developments in plant growth promotion and sustainable environment by microbial utilization. Usually, three pathways are proposed to understand microbial activity in plant growth and development, (1) control plant hormonal signaling, (2) outcompete pathogenic microbes strains, and (3) increase soil-borne nutrient bioavailability (Verbon and Liberman 2016; Mendes et al. 2013; Van der Heijden et al. 2008).

21.2 Diversity of Suitable Microbes for Plants

The opportunity to produce beneficial microbial communities in agricultural soils is provided by understanding the concepts of microbe–microbe and plant–microbe interaction. Overground plant tissues, like vegetative foliar parts, leaves, and flowering parts, offer unique environments for the variety of endophytes and epiphytes, even though there are important variations in endospheres and phyllosphere bacteria ecology. In the endospheres and phyllosphere, at the level of genus and species, various microorganisms are identified. For instance, structural analysis of grapevine phyllosphere microbiota revealed that *Pseudomonas, Sphingomonas, Frigoribacterium, Curtobacterium, Bacillus, Enterobacter, Acinetobacter, Erwinia, Citrobacter, Pantoea,* and *Methylobacterium* were the prevalent genera, while *Ralstonia, Burkholderia, Pseudomonas, Staphylococcus, Mesorhizobium,* and *Propionibacterium* are its predominant genera (Zarraonaindia et al. 2015; Kecskemeti et al. 2016). Wallace et al. 2018 have recently analyzed the maize leaf microbiome through 300 different lines of maize and discovered sphingomonas and methylobacteria as the prevalent taxa. They also revealed that the microbial structure of the phyllosphere was primarily influenced by abiotic factor. Seed-associated bacteria were recently studied and found to consist mostly of Proteobacteria, Actinobacteria, Bacteroidetes, and Firmicutes (Barret et al. 2015). Overground plant microbiome usually originates predominantly from the seed, air, and soil and adapt and survive on or within the tissue where many factors form the makeup of the community, including soil, atmosphere. Plants deliberately attract their microbes from surrounding microbial sources like the soil, the phyllosphere about the external environment, the atmosphere, the spermosphere, and the carposphere, including the aerial plant ecosystem the leaf surface. Root microbiome was mainly transported horizontally, e.g., they are extracted from the soil environment, which includes remarkably diverse microorganisms dominated by Acidobacteria, Verrucomicrobia, Bacteroidetes, Proteobacteria, Planctomycetes, and Actinobacteria (Fierer 2017). Furthermore, transmission of microbes is vertically through seeds. Seeds are also an essential source of microbes that proliferate in the roots of plants that are growing (Hardoim et al. 2012). Plants through their root system include soil microbiota with special ecological niches that colonize the rhizosphere, roots, and, to some point, above-ground parts (Hartmann et al. 2009). Research briefly demonstrated root-driven improvements in the composition of the microbial communities of the wheat rhizosphere and observed a tenfold greater abundance in the rhizosphere of actinobacteria, pseudomonads, oligotrophs, and copiotrophs relative to bulk soil (Donn et al. 2015). Similarly, the *Brachypodium distachyon* rhizosphere was controlled by *Sphingobacteriales*, *Xanthomonadales*, and *Burkholderiales,* whereas the order Bacillales controlled the bulk land (Kawasaki et al. 2016). Root exudates like phenolics, organic acids, fatty acids, amino acids, plant growth regulators, nucleotides, carbohydrates, putrescine, vitamins, and sterols are reported to stimulate the microbial diversity surrounding roots rhizosphere effect (Mendes et al. 2013). The penetration into root tissues of bacterial endophytes also occurs via passive mechanisms emergence points, and by active mechanisms (Compant et al. 2005). Various groups of bacterial taxa can invade root tissues, such as Acidobacteria, Actinobacteria, Firmicutes, Verrucomicrobia, Proteobacteria, Bacteroidetes, Chloroflexi, Planctomycetes, and Gemmatimonatedetes, the frequently present in grapevine roots (Burns et al. 2015; Samad et al. 2017; Faist et al. 2016).

21.2.1 The Relation Between Plant and Microorganism

Microorganisms play an important role in plant development by affecting their physiology and growth. In 1978, the term plant growth-promoting bacteria (PGPB) were described by Scroth and Kloepper. Most PGPB are fluorescent *Pseudomonas* members (Glick 1995). A significant and persistent threat to food production and ecological stability worldwide is plant-pathogenic microorganisms. PGPB and plant growth-promoting fungi, both mutualistic and symbiotic, are in the rhizosphere, free-living is linked to many, but not all, species of plants and is found in many habitats. A better understanding of the various aspects of disease suppression by these biocontrol agents has been provided by findings into the processes of plant growth promotion by PGPB. Now, much of the attention has been on rhizobacterial free-living strains, especially *Pseudomonas* and *Bacillus*. The sustained analysis also holds for endophytic bacteria as is the case with associative nitrogen-fixing PGPB on sugarcane, the ability to produce biocontrol agents that can be self-perpetuating by colonizing hosts and being transmitted to progeny (Boddey et al. 2003) and *Burkholderia phytofirmans*, a nonsymbiotic endophyte bacterium (Sessitsch et al. 2005). A variety of many other PGPR bacteria, like organisms, are now known to be *Azotobacter, Azospirillum, Burkholderia, Acetobacter*, and *Bacilli* virtually any non-deleterious free-living bacteria that could directly or indirectly stimulate plant growth can be listed as PGPR (Glick 1995). Several other PGPRs, including *Clostridium, Flavobacterium, Hydrogenophaga, Achromobacter, Serratia, Staphylococcus Arthrobacter, Azoarcus, Enterobacter, Frankia,*

Microcoleous, *Phyllobacterium*, *Streptomyces*, and *Vibrio*, *Kluyvera*, have been recently described (Ahmad et al. 2008a). PGPR can either directly or indirectly encourage plant growth. Direct pathways would include the capacity to produce ethylene, gibberellins, indoleacetic acid, cytokinins, and asymbiotic N_2 fixation, and solubility of nutrients like phosphates. Indirect mechanisms involve degrading of enzymes of the extracellular cell wall like b-1, 3-glucanase, antagonism of chitinase phytopathogens, development of siderophores and phytopathogens, synthesis of antibiotics, and cyanide (Ahmad et al. 2008b). The function of different plant-associated microorganisms, for example, legume-rhizobium association, mycorrhiza's function in promoting plant development, etc. free-living fungi were also engaged in supporting plant growth through another pathway like phosphate solubilization. The function of free-living fungi examined from Indian soil for their multiple potential PGP behaviors revealed that several phosphate-solubilizing fungi do have other advantageous properties, like plant growth hormone production, many extracellular enzymes, tolerance to heavy metals, biosorption ability as well as a help to plant growth improvement if used as inoculants for wheat and chickpea (Imran 2009). Although several members of the microbiota of the rhizosphere were important to enhance plant development, the rhizosphere is often colonized by plant harmful microbes attempting to crack via the defensive microbial shield and overcome the inherent mechanisms of plant protection to trigger the disease. Rudrappa associates with root exudates as signaling molecules, which have shown that root-realized malic acid recruits the beneficial soil like bacteria *Bacillus subtilis* to the root, and this relationship plays an important role in plant defense against foliar pathogen *Pseudomonas syringae* (Rudrappa et al. 2008).

21.3 Application of Microbial Signaling in Plant Health

Plants are multicellular sessile organisms for growth, rely on metabolic and developmental changes. To sense, the environment and other organisms produced a network of signaling one or more specific molecular, cellular, or developmental changes pathways. Plants host different microbial groups defined by the plant microbiota on and within their tissues. Whereas bacteria and fungi have become continuously shown throughout the past 150 years to enhance plant growth and suppress plant pathogens, this expertise in agriculture biotechnology has yet to be widely utilized (Berg 2009). However, how entire microbiome communities able to carry out both growth-promoting and development behaviors interfere with plant fitness remain largely unknown. Signaling has become a key focus in phytopathology for several years during plant–pathogen interactions although more recent attempts have been taken to explore the signaling associated with plant interaction to non-pathogenic microorganisms, particularly those which improve plant growth. Microbial cells, including those that colonize internal tissues and those that adhere to external surfaces, are replete with plants. Microorganism biostimulants work by different pathways, including plant hormone production, root system elongation, and expansion mediated systemic resistance or systemic acquired resistance, 4-hydroxyphenyl acetic acid production of lytic enzymes, and antibiotics, and

1-aminocyclopropane-1-carboxylate-deaminase (ACC-deaminase) production in rhizosphere plants. The future of field-applied microbial inoculants depends mainly on both biotic and abiotic factors. The earliest model based on metabolic interactions was devised for the plant-microbe system, particularly to legume-rhizobia symbiosis (Pfau 2013). Through exchange reactions identified from the literature review, the metabolic pathway system of the nitrogen-fixing symbiont *Sinorhizobium meliloti* was then linked to the root tissue model. The unique microbial strains that contain metabolic pathways beneficial for plant food were starting to be discovered in recent genomic research (Müller et al. 2016). The fungus producing different molecules like antibiotics and elicitation of defense responses during environmental stress; all these are helpful for the growth of fungus. The favorable habits of rhizosphere microorganisms can also be associated with the production of auxins or the solubilization of nutrients that may affect the initiation of lateral roots, the growth of lateral roots, or both developmental processes, contributing to the formation of root systems with enhanced explorative ability. Cytokinins could be released by microorganisms and their production has been well documented and linked with increased plant growth by plant growth-promoting rhizobacteria (Arkhipova et al. 2005). The beneficial effect of cytokinins on plant-wide growth was reported by the characterization of genes associated with the sensing and regulation of cytokinins (Ortíz-Castro et al. 2009). It has been shown that three sensor histidine kinases, CRE1/AHK4/WOL, AHK2, AHK3, function as cytokinin receptors (Kakimoto 2003). The same group has shown that B. subtilis GB03 VOCs via the regulation of endogenous sugar, ABA signaling, improve respiration potential through growing photosynthetic efficiency and chlorophyll content in *Arabidopsis*. Besides, in reaction to high exposures of UV-B radiation, stress resistance can occur in the induction of signaling molecules like nitric oxide (NO) and calcium ions (Ca^{2+}), and abscisic acid (ABA) in plant and animal cells (Tossi et al. 2012). Also, NO, which modulates the actions of cellular and extracellular proteins in different classes of organisms, introduces important physiological functions (Medinets et al. 2015). Furthermore, nitric oxide could play a signaling role to improve microbial biofilm formation, which offers plant growth multiple biochemical and physiological benefits to plant (Qurashi and Sabri 2012; Medinets et al. 2015). Beneficial soil bacteria have resistance from a wide variety of foliar diseases via activating plant defenses, thus increasing the vulnerability of a plant to pathogen damage. Besides, complex environmental stress can account for some of the differences found in field trials, and it will also be useful to consider more thoroughly how plant–microbe interaction is affected (Smith et al. 2015).

21.4 Microbial Role in Crop Improvement

Plants have a variety of interactions with these soil-dwelling organisms, covering the full spectrum of ecological possibilities like exploitative, competitive, neutral, commensal, and mutualistic. After all, lengthy concern has been shown in classifying the favorable ecological relationships which encourage plant growth and adaptability. In the second half of the nineteenth century, for example, mycorrhizal fungi and the bacteria found in nodulated legumes were both known as root symbionts. In most developing countries, agriculture is the major economic bacteria, activity, with more than 50% of the population involved. In natural environments, fungi, protozoa, actinomycetes, and algae are colonized by all plants within soil structures and their rhizospheric zone. However, 95% of all the microbes which colonize are bacteria (Glick 2012). Consistent maintenance of soil fertility is important because of the need to expand agricultural productivity to satisfy the food needs of the ever-increasing world population. These were all primarily rhizobia in *the Allorhizobium, Azorhizobium, Bradyrhizobium, Mesorhizobium, Rhizobium,* and *Sinorhizobium* genera, that entry into symbiotic associations with their host legume host plants and fix atmospheric dinitrogen within distinct root nodules for this reason. The agricultural revolution in agriculture involving high yielding varieties and agrochemicals has been caused by low world crop production due to low soil moisture, low nutrient resources, erosion risk, low pH, high phosphorus fixation, low soil organic matter levels, aluminum toxicity pests, and diseases, weeds, and loss of soil biodiversity. Low crop productivity, low soil fertility, adverse economic returns, food poisoning, soil depletion of biodiversity, and significant environmental threats have resulted from the continuous use of fertilizers, pesticides, and herbicides. Microorganisms additives like plant growth-promoting rhizobacteria, rhizobium, and arbuscular mycorrhizal fungi could be used as biofertilizers to increase the supply and absorption of soil nitrogen, phosphorus, and potassium. All bacteria and fungi inoculants demonstrate the capacity for use in the formation and stability of aggregate particles and, therefore, improvement of soil structure. Furthermore, *Pseudomonas* spp., *Bacillus* spp., *Trichoderma* spp., *Mycorrhizas*, and *Streptomyces* spp. are the most widely used microorganisms as biofertilizers, bio-control, and bioremediation. Plant-associated microbiota always secrete different molecules that both actively and indirectly impacting plant health (Bednarek et al. 2010). First is changing the physical and mechanical characteristics of their surrounding environment to increase supply for themselves and related plants like metabolites, iron acquisition siderophores, and phosphorus acquisition enzymes, respectively. Second, pathogenic agents such as antibiotics and antimicrobial proteins are antagonistic. Third, host cells were prepared for eventual invasions, such as external pH regulation to promote pathogenesis and molecules which regulate symbiotic relationship or quorum sensing. Fourth, numerous virulence factors (or effectors) are secreted by pathogenic microbes in the cytoplasm of host cells in plant–pathogen interspaces and to attenuate host defense responses and facilitate pathogen proliferation by advanced secretion systems (Martin and Kamoun 2012). Rhizobacteria and plant growth encourage bacteria that, as in most cases the effect is produced by bacteria living on or within plant roots, were classified as bacteria that have an extremely advantageous impact on plant growth by directly or indirectly pathways. It is possible to identify certain plant growth-promoting bacteria or rhizobacteria as biocontrol agents and biofertilizers (Glick 2012). The bacteria in the soil will significantly improve several microbes' processes and improve the absorption of food in a way that is easy for plants to assimilate. Based on their existence and work, these could be categorized as N_2 fixing, phosphate solubilizing, phosphate mobilizing, or micronutrient biofertilizers. *Azotobacter chroococcum* and *Glomus fasciculatum* inoculation of lettuce has improved the vegetable's overall phenolic compounds, carotenoids, and anthocyanins content (Baslam et al. 2011). The influence of Arbuscular Mycorrhizal Fungi (AMF) in antioxidant biosynthesis has been documented (Carlsen et al. 2008; Nisha and RajeshKumar 2010; Eftekhari et al. 2012). The secondary metabolites like phenols, tannins, ortho-dihydroxy, flavonoids, and alkaloids synthesis enhanced after inoculation with microbial consortium *G. mosseae*, *Bacillus coagulans,* and *Trichoderma viride* to *Calamus thwaitesii* and *Begonia malabarica* (Lakshmipathy et al. 2002; Selvaraj et al. 2008). Many nutrients such as nitrogen (N), phosphorus (P), and sulfur (S) are found in natural environments and are thus minimally bioavailable for plants in organic molecules. Plants depending on the microbial community of soil like fungi and bacteria contain the metabolic pathway to depolymerize and mineralize organic components of S, P, and N. These macronutrients are particularly important as fertilizer in most modern agricultural practices.

21.5 Future Perspectives

Microbes associate with several plant species and have protective effects, including increased development and decreased susceptibility to diseases caused by plant-pathogenic bacteria, fungi, viruses, and nematodes. Previous research also confirmed that numerous endophytes were also intended to promote plant growth from pollution and environments like *Sebacina vermifera, Piriformospora indica,* and various species of *Colletotrichum sp. and Penicillium sp.* (Redman et al. 2011). A greater understanding of the pathways of microbial signaling is a secret to progress in the manipulation of a positive plant–microbe relationship for improved control of diseases and alleviation of crop stress through crop rotations, paving the way for sustainable development. Plant and rhizospheric protective bacteria, like endophytes and plant rhizobacteria that promote growth stimulate plant growth through a wide range of pathways. Some of the plant growth-promoting rhizobacteria and endophytes' direct and indirect benefits are (1) phosphorus solubilization absorption, (2) biocontrol, (3) hormone development, (4) nitrogen fixation, (5) increased nutrient, (6) increased immunity to biotic and abiotic (7) stress symbiotic and resistance interaction facilitation (Smith et al. 2015). Free-living microbial, including *Trichoderma* genus filamentous fungi and several plant growth-promoting rhizobacteria, is capable of suppressing soil-borne plant diseases and promoting plant growth via various mechanisms, like phytohormone production, mycoparasitism, and plant-pathogen competence, organic matter decomposition, and mineralization. In the rhizosphere, different compounds produced from the root system produce a special environment. The growth of plants is affected by microbial community structure through multiple pathways, like nitrogen fixation by different classes of proteobacteria which improved resistance of biotic and abiotic stress due to the involvement of endophytic microbes, and direct and indirect benefits of plant-promoting rhizobacteria. Both primary metabolites like carbohydrates, proteins, organic acids, etc., and secondary metabolites flavonoids, phenol, phytohormones assist in plant growth are also the broad range of signaling compounds produced by plants. On the other hand, secondary metabolites or volatile compounds called phytohormones are released by microorganisms, that might function directly or indirectly to stimulate plant immunity or control plant growth and morphogenesis. Also, plant stressful situations play a significant role in the secretion of rhizosphere signaling compounds, and a better knowledge of the relationships among environment plant stresses and signaling may assist in the production of innovations that are using plant signaling to alleviate crop stress. Further, complex environmental factors can reason for some of the differences found in field tests, and a much clearer understanding of how environmental factors affect plant–microbe interaction would have been beneficial. The following approaches for plant growth-promoting fungus metabolites are future research needs (Tables 21.1 and 21.2).

Table 21.1 Role of microbes enhance growth of plant and tolerance

Microbes	Plants	Effects	Reference
Bacillus thuringiensis	*Lavandula dentata*	Enhanced plant growth, nutrient content, and biomass	Armada et al. (2016)
Rhizophagus irregularis	*Solanum lycopersicum* L	Improve photosynthetic rate and increased conductivity of root	Calvo-Polanco et al. (2016)
Azospirillum brasilense Sp245	*Triticum*	Increased growth and easy conduction of water	Pereyra et al. (2012)
Rhizobium	*Phaseolus vulgaris*	Promoted plant weight	Yanni et al. (2016)
Arbuscular mycorrhizal fungi Bacillus thuringiensis	*Zea mays*	Improved nutrient content and water transport protein	Armada et al. (2015)
Rhizophagus intraradices Bacillus megaterium Pseudomonas putida	*Trifolium repens*	Increased plant nutrient and water contents, reduced stomatal conductance and stress enzyme activities	Ortiz et al. (2015)
Pseudomonas putida MTCC5279 (RA)	*Cicer arietinum*	Reduced the expression of stress response gene, maintained water content, osmolyte, membrane structure, and germination rate f the plant	Tiwari et al. (2016)
Azospirillum spp. (Az19)	*Zea mays*	Improve the growth and productivity	Garcia et al. (2017)
Rhizobium	*Phaseolus vulgaris*	Promoted plant weight	Yanni et al. (2016)

Table 21.2 The role of microbes as against for biocontrol agents is defined as having mode of action

S. N.	Microbes	Results	Reference
1.	*Fusarium verticillioides*	Variety in the percentage reduction of lesions in maize	Bacon and Hinton (2011)
2.	*Rhizoctonia solani*	Diverse levels of sensitivity	Faltin et al. (2004)
3.	*Ophiostoma novo-ulmi*	In vitro growth inhibition less marked	Díaz et al. (2013)
4.	*Agrobacterium tumefaciens*	Biotype 3 are resistant to the antibiotic agrocin 8	van Zyl et al. (1986)
5.	*Sclerotinia sclerotiorum*	1 strains less susceptible to infection	Huang et al. (2011)
6.	*Erwinia amylovora*	23 strains exhibit resistance to at least one phage	Schnabel and Jones (2001)
7.	*Helminthosporium solani*	Diverse levels of sensitivity	Rivera-Varas et al. (2007)
8.	*Streptomyces scabies*	Diverse levels of sensitivity	Otto-Hanson et al. (2013)
9.	*Cryphonectria parasitica*	Difference in tolerance	Peever et al. (2000)

21.6 Conclusion

For millions of years, plants and microorganisms have been coexisting. Awareness of the modes of action of microbial inoculants will play a key role in the growth and development of plants. Microorganisms inoculant technology will provide the future population with safe food protection. Plants sustain a dynamic relationship with their rhizospheric communities, which would be essential for defensive mechanisms to assimilate nutrients, evolve, and activate it. In this chapter, the role of the microbial community involved in the interactions between plants and beneficial microorganisms has been significant. Ultimately, several strategies to redirect or reshape the microbiome of the rhizosphere in support of microbes that help assess growth and health are illustrated. The important impact of plant resistance to abiotic and biotic stresses is that it will boost agricultural productivity and production to raise animals and humans. Therefore, microbes signaling in soil and plant–microbe interactions could undoubtedly contribute to sustainable agricultural practices developing increasingly cost-effective and eco-friendly farming techniques (Fig. 21.1).

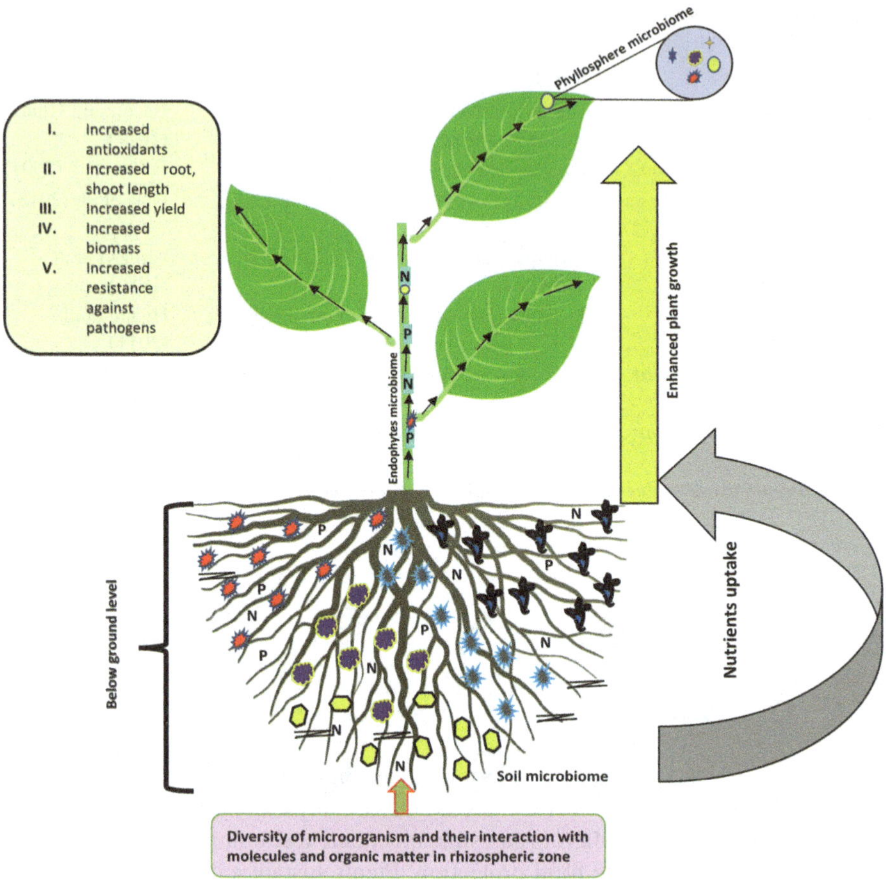

Fig. 21.1 Role of different microbial community in plant developments and nutrients uptake and distribution

Acknowledgments We are grateful to the Science and Engineering Research Board (SERB), a statutory body under the Department of Science and Technology (DST), Government of India for the Start-up Research Grant (SRG/2019/001940).

References

Ahmad F, Ahmad I, Khan MS (2008a) Screening of free-living rhizospheric bacteria for their multiple plant growth-promoting activities. Microbiol Res 163(2):173–181

Ahmad, F., Ahmad, I., Aqil, F., Khan, M. S., & Hayat, S. (2008b) Diversity and potential of nonsymbiotic diazotrophic bacteria in promoting plant growth. *Plant-Bacteria Interactions: Strategies and Techniques to Promote Plant Growth*, 81–109

Arkhipova TN, Veselov SU, Melentiev AI, Martynenko EV, Kudoyarova GR (2005) The ability of bacterium Bacillus subtilis to produce cytokinins and to influence the growth and endogenous hormone content of lettuce plants. Plant Soil 272(1–2):201–209

Armada E, Azcon R, Lopez-Castillo OM, Calvo-Polanco M, Ruiz Lozano JM (2015) Autochthonous arbuscular mycorrhizal fungi and Bacillus thuringiensis from a degraded Mediterranean area can be used to improve physiological traits and performance of a plant of agronomic interest under drought conditions. Plant Physiol Biochem 90:64–74

Armada E, Probanza A, Roldan A, Azcon R (2016) Native plant growth promoting bacteria Bacillus thuringiensis and mixed or individual mycorrhizal species improved drought tolerance and oxidative metabolism in Lavandula dentate plants. J Plant Physiol 192:1–2

Bacon CW, Hinton DM (2011) In planta reduction of maize seedling stalk lesions by the bacterial endophyte Bacillus mojavensis. Can J Microbiol 57(6):485–492

Baldotto LEB, Olivares FL (2008) Phylloepiphytic interaction between bacteria and different plant species in a tropical agricultural system. Can J Microbiol 54(11):918–931

Barret, M., Briand, M., Bonneau, S., Préveaux, A., Valière, S., Bouchez, O., .et al. (2015). Emergence shapes the structure of the seed microbiota. Appl Environ Microbiol, 81(4), 1257–1266

Baslam M, Garmendia I, Goicoechea N (2011) Arbuscular mycorrhizal fungi (AMF) improved growth and nutritional quality of greenhouse-grown lettuce. J Agric Food Chem 59:5504–5515. https://doi.org/10.1021/jf200501c

Bednarek P, Kwon C, Schulze-Lefert P (2010) Not a peripheral issue: secretion in plant-microbe interactions. Curr Opin Plant Biol 13(4):378–387

Berg G (2009) Plant-microbe interactions promoting plant growth and health: perspectives for controlled use of microorganisms in agriculture. Appl Microbiol Biotechnol 84:11–18

Bhuvaneshwari K, Singh PK (2015) Response of nitrogen-fixing water fern Azolla biofertilization to rice crop. 3 Biotech 5(4):523–529

Bilal L, Asaf S, Hamayun M, Gul H, Iqbal A, Ullah I, Lee IJ, Hussain A (2018) Plant growth-promoting endophytic fungi *Aspergillus fumigatus* TS1 and *Fusarium proliferatum* BRL1 produce gibberellins and regulate plant endogenous hormones. Symbiosis 76(2):117–127

Boddey RM, Urquiaga S, Alves BJ, Reis V (2003) Endophytic nitrogen fixation in sugarcane: present knowledge and future applications. Plant Soil 252(1):139–149

Burns KN, Kluepfel DA, Strauss SL, Bokulich NA, Cantu D, Steenwerth KL (2015) Vineyard soil bacterial diversity and composition revealed by 16S rRNA genes: differentiation by geographic features. Soil Biol Biochem 91:232–247

Calvo-Polanco M, Sanchez-Romera B, Aroca R, Asins MJ, Declerck S, Dodd IC, Martinez-Andujar C, Albacete A, Ruiz-Lozano JM (2016) Exploring the use of recombinant inbred lines in combination with beneficial microbial inoculants (AM fungus and PGPR) to improve drought stress tolerance in tomato. Environ Exp Bot 131:47–57

Carlsen S, Understrup A, Fomsgaard I, Mortensen A, Ravnskov S (2008) Flavonoids in roots of white clover: interaction of arbuscular mycorrhizal fungi and a pathogenic fungus. Plant Soil 302:33–43. https://doi.org/10.1007/s11104-007-9452-9

Compant S, Reiter B, Sessitsch A, Nowak J, Clément C, Barka EA (2005) Endophytic colonization of Vitis vinifera L. by plant growth-promoting bacterium Burkholderia sp. strain PsJN. Appl Environ Microbiol 71(4):1685–1693

Copeland JK, Yuan L, Layeghifard M, Wang PW, Guttman DS (2015) Seasonal community succession of the phyllosphere microbiome. Mol Plant-Microbe Interact 28(3):274–285

Díaz G, Córcoles AI, Asencio AD, Torres MP (2013) In vitro antagonism of Trichoderma and naturally occurring fungi from elms against O phiostoma novo-ulmi. For Pathol 43(1):51–58

Donn S, Kirkegaard JA, Perera G, Richardson AE, Watt M (2015) Evolution of bacterial communities in the wheat crop rhizosphere. Environ Microbiol 17(3):610–621

Edwards J, Johnson C, Santos-Medellín C, Lurie E, Podishetty NK, Bhatnagar S et al (2015) Structure, variation, and assembly of the root-associated microbiomes of rice. Proc Natl Acad Sci 112(8):E911–E920

Eftekhari M, Alizadeh M, Ebrahimi P (2012) Evaluation of the total phenolics and quercetin content of foliage in mycorrhizal grape (Vitis vinifera L.) varieties and effect of postharvest drying on quercetin yield. Indus. Crop Product 38:160–165. https://doi.org/10.1016/j.indcrop.2012.01.022

Faist H, Keller A, Hentschel U, Deeken R (2016) Grapevine (Vitis vinifera) crown galls host distinct microbiota. Appl Environ Microbiol 82(18):5542–5552

Faltin F, Lottmann J, Grosch R, Berg G (2004) Strategy to select and assess antagonistic bacteria for biological control of Rhizoctonia solani Kühn. Can J Microbiol 50(10):811–820

Fierer N (2017) Embracing the unknown: disentangling the complexities of the soil microbiome. Nat Rev Microbiol 15(10):579–590

Frisvad JC, Andersen B, Thrane U (2008) The use of secondary metabolite profiling in chemotaxonomy of filamentous fungi. Mycol Res 112(2):231–240

Frisvad JC, Frank JM, Houbraken JAMP, Kuijpers AF, Samson RA (2004) New ochratoxin A producing species of Aspergillus section Circumdati. Stud Mycol 50(1):23–43

Galagan JE, Calvo SE, Borkovich KA, Selker EU, Read ND, Jaffe D, FitzHugh W, Ma LJ, Smirnov S, Purcell S, Rehman B (2003) The genome sequence of the filamentous fungus *Neurospora crassa*. Nature 422(6934):859–868

Garcia JE, Maroniche G, Creus C, Suarez-Rodriguez R, Ramirez-Trujillo JA, Groppa MD (2017) In vitro PGPR properties and osmotic tolerance of different Azospirillum native strains and their effects on growth of maize under drought stress. Microbiol Res 202:21–29

Glare T, Caradus J, Gelernter W, Jackson T, Keyhani N, Köhl J, Marrone P, Morin L, Stewart A (2012) Have biopesticides come of age? Trends Biotechnol 30:250–258

Glick BR (1995) The enhancement of plant growth by free-living bacteria. Can J Microbiol 41 (2):109–117

Glick BR (2012) Plant growth-promoting bacteria: mechanisms and applications. Scientifica 2012:963401. https://doi.org/10.6064/2012/963401

Glick BR (2014) Bacteria with ACC deaminase can promote plant growth and help to feed the world. Microbiol Res 169(1):30–39

Hardoim PR, Hardoim CC, Van Overbeek LS, Van Elsas JD (2012) Dynamics of seed-borne rice endophytes on early plant growth stages. PLoS One 7(2):e30438

Hartmann A, Schmid M, Van Tuinen D, Berg G (2009) Plant-driven selection of microbes. Plant Soil 321(1–2):235–257

Huang Y, Xie X, Yang L, Zhang J, Li G, Jiang D (2011) Susceptibility of Sclerotinia sclerotiorum strains different in oxalate production to infection by the mycoparasite Coniothyrium minitans. World J Microbiol Biotechnol 27(12):2799–2805

Igiehon NO, Babalola OO (2017) Biofertilizers and sustainable agriculture: exploring arbuscular mycorrhizal fungi. Appl Microbiol Biotechnol 101(12):4871–4881

Imran M (2009) Interaction of heavy metals with indigenous isolates of free living rhizospheric fungi and their plant growth promoting potentials (Doctoral dissertation, Aligarh Muslim University)

Kakimoto T (2003) Perception and signal transduction of cytokinins. Annu Rev Plant Biol 54 (1):605–627

Kawasaki A, Donn S, Ryan PR, Mathesius U, Devilla R, Jones A, Watt M (2016) Microbiome and exudates of the root and rhizosphere of Brachypodium distachyon, a model for wheat. PLoS One 11(10):e0164533

Kecskemeti E, Berkelmann-Löhnertz B, Reineke A (2016) Are epiphytic microbial communities in the carposphere of ripening grape clusters (Vitis vinifera L.) different between conventional, organic, and biodynamic grapes? PLoS One 11(8):e0160852

Laforest-Lapointe I, Messier C, Kembel SW (2016) Tree phyllosphere bacterial communities: exploring the magnitude of intra-and inter-individual variation among host species. PeerJ 4: e2367

Lakshmipathy R, Chandrika K, Gowda B, Balakrishna AN, Bagyaraj DJ (2002) Response of Calamus thwaitesii var. canaranus wilde to inoculation with Glomus mosseae, Bacillus coagulans and Trichoderma harzianum. J Soil Biol Ecol 22:16–21

Lareen A, Burton F, Schäfer P (2016) Plant root-microbe communication in shaping root microbiomes. Plant Mol Biol 90:575. https://doi.org/10.1007/s11103-015-0417-8
Liu L (2011) Bioactive metabolites from the plant endophyte Pestalotiopsis fici. Mycology 2(1):37–45
Martin F, Kamoun S (eds) (2012) Effectors in plant–microbe interactions. Wiley-Blackwell, Oxford
Medinets S, Skiba U, Rennenberg H, Butterbach-Bahl K (2015) A review of soil NO transformation: associated processes and possible physiological significance on organisms. Soil Biol Biochem 80:92–117
Mendes R, Garbeva P, Raaijmakers JM (2013) The rhizosphere microbiome: significance of plant beneficial, plant pathogenic, and human pathogenic microorganisms. FEMS Microbiol Rev 37 (5):634–663
Müller DB, Vogel C, Bai Y, Vorholt JA (2016) The plant microbiota: systems-level insights and perspectives. Annu Rev Genet 50:211–234
Mus F, Crook MB, Garcia K, Costas AG, Geddes BA, Kouri ED et al (2016) Symbiotic nitrogen fixation and the challenges to its extension to nonlegumes. Appl Environ Microbiol 82 (13):3698–3710
Nisha MC, RajeshKumar S (2010) Influence of arbuscular mycorrhizal fungi on biochemical changes in Wedilla chinensis (Osbeck) Merril. Anc Sci Life 29:26
Ortiz N, Armada E, Duque E, Roldan A, Azcon R (2015) Contribution of arbuscular mycorrhizal fungi and/or bacteria to enhancing plant drought tolerance under natural soil conditions: effectiveness of autochthonous or allochthonous strains. J Plant Physiol 174:87–96
Ortíz-Castro R, Contreras-Cornejo HA, Macías-Rodríguez L, López-Bucio J (2009) The role of microbial signals in plant growth and development. Plant Signal Behav 4(8):701–712
Otto-Hanson LK, Grabau Z, Rosen C, Salomon CE, Kinkel LL (2013) Pathogen variation and urea influence selection and success of Streptomyces mixtures in biological control. Phytopathology 103:34–42. https://doi.org/10.1094/PHYTO-06-12-0129-R
Peever TL, Liu Y-C, Cortesi P, Milgroom MG (2000) Variation in tolerance and virulence in the chestnut blight fungus hypovirus interaction. Appl Environ Microbiol 66:4863–4869. https://doi.org/10.1128/AEM.66.11.4863-4869.2000
Pereyra MA, Garcia P, Colabelli MN, Barassi CA, Creus CM (2012) A better water status in wheat seedlings induced by Azospirillum under osmotic stress in related to morphological changes in xylem vessels of the coleoptile. Appl Soil Ecol 53:94–97
Pfau T (2013) Modelling metabolic interactions in the legume-rhizobia symbiosis Doctoral dissertation. Aberdeen University
Prussin AJ, Marr LC (2015) Sources of airborne microorganisms in the built environment. Microbiome 3(1):78
Qurashi AW, Sabri AN (2012) Bacterial exopolysaccharide and biofilm formation stimulate chickpea growth and soil aggregation under salt stress. Braz J Microbiol 43:1183–1191
Redman RS, Kim YO, Woodward CJ, Greer C, Espino L, Doty SL, Rodriguez RJ (2011) Increased fitness of rice plants to abiotic stress via habitat adapted symbiosis: a strategy for mitigating impacts of climate change. PLoS One 6(7):14823
Remus-Emsermann MN, Lücker S, Müller DB, Potthoff E, Daims H, Vorholt JA (2014) Spatial distribution analyses of natural phyllosphere-colonizing bacteria on Arabidopsis thaliana revealed by fluorescence in situ hybridization. Environ Microbiol 16(7):2329–2340
Rivera-Varas VV, Freeman TA, Gudmestad NC, Secor GA (2007) Mycoparasitism of Helminthosporium solani by Acremonium strictum. Phytopathology 97:1331–1337. https://doi.org/10.1094/PHYTO-97-10-1331
Rodriguez RJ, White JF Jr, Arnold AE, Redman ARA (2009) Fungal endophytes: diversity and functional roles. New Phytol 182(2):314–330
Rosenberg E, Zilber-Rosenberg I (2016) Microbes drive evolution of animals and plants: the hologenome concept. MBio 7(2)
Rudrappa T, Czymmek KJ, Paré PW, Bais HP (2008) Root-secreted malic acid recruits beneficial soil bacteria. Plant Physiol 148(3):1547–1556

Samad A, Trognitz F, Compant S, Antonielli L, Sessitsch A (2017) Shared and host-specific microbiome diversity and functioning of grapevine and accompanying weed plants. Environ Microbiol 19(4):1407–1424

Schnabel EL, Jones AL (2001) Isolation and characterization of Five Erwinia amylovora bacteriophages and assessment of phage resistance in strains of Erwinia amylovora. Appl Environ Microbiol 67:59–64. https://doi.org/10.1128/AEM.67.1.59-64.2001

Schulz B, Boyle C (2005) The endophytic continuum. Mycol Res 109(6):661–686

Selvaraj T, Rajeshkumar S, Nisha MC, Wondimu L, Tesso M (2008) Effect of Glomus mosseae and plant growth promoting rhizomicroorganisms (PGPR's) on growth, nutrients and content of secondary metabolites in Begonia malabarica Lam. Maejo Int J Sci Technol 2:516–525

Sessitsch A, Coenye T, Sturz AV, Vandamme P, Barka EA, Salles JF et al (2005) Burkholderia phytofirmans sp. nov., a novel plant-associated bacterium with plant-beneficial properties. Int J Syst Evol Microbiol 55(3):1187–1192

Smith DL, Subramanian S, Lamont JR, Bywater-Ekegärd M (2015) Signaling in the phytomicrobiome: breadth and potential. Front Plant Sci 6:709. https://doi.org/10.3389/fpls.2015.00709

Strobel G, Daisy B (2003) Bioprospecting for microbial endophytes and their natural products. Microbiol Mol Biol Rev 67(4):491–502

Strobel G, Daisy B, Castillo U, Harper J (2004) Natural products from endophytic microorganisms. J Nat Prod 67(2):257–268

Tiwari S, Lata C, Chauhan PS, Nautiyal CS (2016) Pseudomonas putida attunes morphophysiological, biochemical and molecular responses in Cicer arietinum L. during drought stress and recovery. Plant Physiol Biochem 99:108–117

Toju H, Peay KG, Yamamichi M, Narisawa K, Hiruma K, Naito K et al (2018) Core microbiomes for sustainable agroecosystems. Nat Plants 4(5):247–257

Tossi V, Cassia R, Bruzzone S, Zocchi E, Lamattina L (2012) ABA says NO to UV-B: a universal response? Trends Plant Sci 17:510–517

Van Der Heijden MG, Bardgett RD, Van Straalen NM (2008) The unseen majority: soil microbes as drivers of plant diversity and productivity in terrestrial ecosystems. Ecol Lett 11(3):296–310

van Zyl FG, Strijdom BW, Staphorst JL (1986) Susceptibility of Agrobacterium tumefaciens strains to two agrocin-producing agrobacterium strains. Appl Environ Microbiol 52(2):234–238

Verbon EH, Liberman LM (2016) Beneficial microbes affect endogenous mechanisms controlling root development. Trends Plant Sci 21(3):218–229

Wakelin SA, Warren RA, Harvey PR, Ryder MH (2004) Phosphate solubilization by *Penicillium spp.* closely associated with wheat roots. Biol Fertil Soils 40(1):36–43

Walker TS, Bais HP, Grotewold E, Vivanco JM (2003) Root exudation and rhizosphere biology. Plant Physiol 132:44. https://doi.org/10.1104/pp.102.019661

Wallace JG, Kremling KA, Kovar LL, Buckler ES (2018) Quantitative genetics of the maize leaf microbiome. Phytobiomes J 2(4):208–224

Yanni Y, Zidan M, Dazzo F, Rizk R, Mehesen A, Abdelfattah F, Elsadany A (2016) Enhanced symbiotic performance and productivity of drought stressed common bean after inoculation with tolerant native rhizobia in extensive fields. Agric Ecosyst Environ 232:119–128

Zarraonaindia I, Owens SM, Weisenhorn P, West K, Hampton-Marcell J, Lax S et al (2015) The soil microbiome influences grapevine-associated microbiota. MBio 6(2)

Rhizosphere Modelling and Nanotechnology: New Outlooks in Sustainable Agriculture

22

Khushboo Dasauni, Deepa Bisht, and Tapan K. Nailwal

Abstract

Plant has exhausted necessary assets to create and maintain rhizosphere, viz. plants and microbial communities in the soil. Rhizosphere modelling will provide a better knowledge of function, assembly and interaction of plants and microbiome community. Holistically rhizosphere can be engineered in small scale to promote plant health and crop improvement programmes. Nanotechnology had a significant impact on various agricultural and environmental challenges, such as energy constraints and sustainable use of natural resources. Nano-fertilizers, nanopesticides and herbicides are more beneficial due to slow release, targeted delivery and low dose of agrochemicals with high biological efficiency and negligible risk of overdose. Although there are some major drawbacks like toxicity, cost and entrance of NPs in food chain, nanomaterial based sensors are becoming popular for early detection of viral diseases and portability. Nano-remediation of different types of pollutants may detect and remove environmental contaminants. This chapter introduces different types of rhizosphere models and advanced nanotechnology-based tools and techniques to revolutionize agriculture and ecological sector.

Keywords

Rhizosphere modelling · Nano-fertilizers · Nanosensors · Nano-remediation · Microbiome

K. Dasauni · D. Bisht · T. K. Nailwal (✉)
Department of Biotechnology, Sir J. C. Bose Technical Campus, Kumaun University, Bhimtal, Uttarakhand, India

R. Soni et al. (eds.), *Microbiological Activity for Soil and Plant Health Management*,
https://doi.org/10.1007/978-981-16-2922-8_22

22.1 Introduction

The development of plant root takes place in a very mobility confined condition with different types of natural assets like air, water, microbes, nutrients, etc. (Hinsinger et al. 2009). So, nature's adaption leads to the creation of a microenvironment around this root region by plants for its own assist. The complex nature of rhizosphere which mainly refers to the soil zone around each single growing plant's root which includes interacting biotic and abiotic zones (Walker et al. 2003). The biotic rhizospheric zone which consists of microbes and different animal community with variations in many biochemical gradients. On the other hand, abiotic comprises the abiotic or non-living components of soil assets. Both zones impact the science of living organisms and its interaction with matter. It plays a crucial role as it is a prime pathway for all water, nutrients, respiration activities of plant root along with interactive association with various soil microbial communities. It is a well-known reported fact that plants invest a lot of their own resources in growth and upkeep of rhizosphere. Though, clear and precise knowledge of rhizosphere and its interaction is still not fully understood and needs to be explored (Czarnes et al. 2000; de León-González et al. 2006; York et al. 2016). Rhizosphere modelling is a tool to study rhizosphere holistically. Rhizosphere modelling works at milli-metre scales and scaling up of the information to study the zones of plants roots. There are various models of rhizosphere which works on various aspects such as nutrient and water uptake, pH, carbon flow, microbial interaction, etc. (York et al. 2016).

Sustainable agriculture is a vital necessity for food security and distribution in a developing country. Presently nanotechnology has efficient applications in every domain of science. In agriculture, nanotechnology is also playing important role in fields such as sensors, plant growth, development and protection. Its applications include nano-fertilizers, nano-carriers, nanopesticides, monitoring of plant health, genetic engineering of plants, etc. Lab scale research in this field is very promising but their applicability is a drawback and limiting factor. Moreover, a major drawback of this amalgamation is Nanoparticles (NPs) can cause toxicity in the environment and food chain. This chapter introduces different types of rhizospheric models studied and nano based agricultural practices and their applicability for future generations (Scott et al. 2016; Chen et al. 2014).

22.2 Rhizosphere Modelling

The rhizosphere modelling is very explanatory for studying rhizosphere in small scale and impactful at a larger scale. It will lead to extensive knowledge of how all elements of this system is created, disassembled and interacted with one another. Rhizosphere is spatially and temporally divided into different scales (York et al. 2016).

The dynamic nature of rhizosphere is due to the seasonal variation experienced by the plants on different times of the year. Soil acidity is an important factor to be considered and for studies always functionally active rhizosphere is examined as it

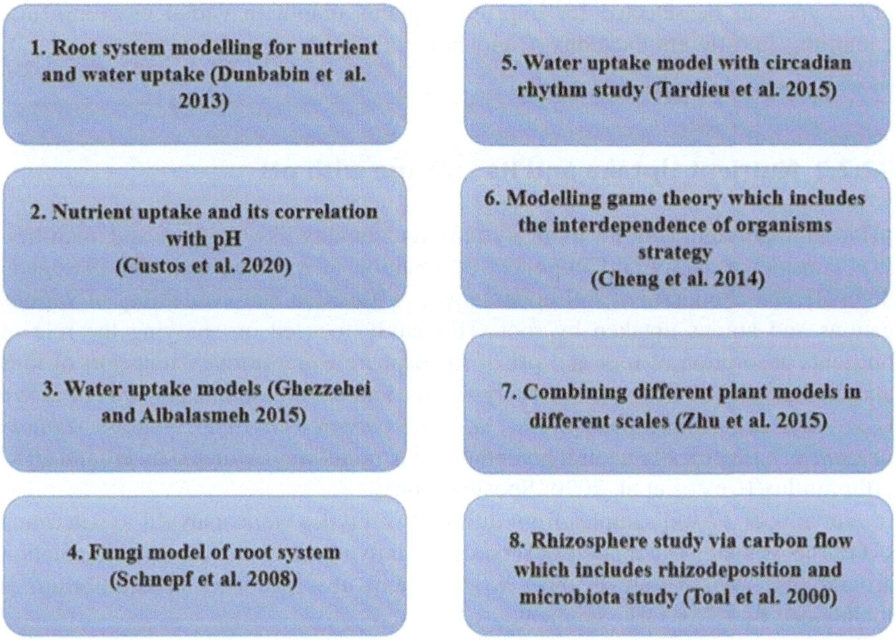

Fig. 22.1 Different examples of rhizosphere models studied by various researchers

does not include the soil zones around dead roots (Jones et al. 2004). Bisporic region of the dead and decayed root is site for the synthesis of new roots (Han et al. 2015). Transitory changes occur in this zone very frequently.

Integrating these models using computational tools is a very efficient method and leads to comprehensive study which is not possible in wet labs. Few studied examples of rhizosphere modelling as shown in Fig. 22.1.

22.2.1 Three-dimensional Root System Framework for Studying Nutrient and Water Uptake

Three-dimensional model of root system provides in-depth knowledge of the soil system along with the roots surrounding ecological niche, their elements, interaction and factors governing its physiology and nature. It provides information in all the fronts which is difficult to study and quantify (Dunbabin et al. 2013).

The dependence of root phenotype on environmental conditions surrounding it can be addressed by root soil modeling. In this study root components were studied virtually using models from their site of origin. Six models, namely RootTyp, ROOTMAP, SimRoot, R-SWMS, SPACSYS and RootBox were studied for evaluating structure, function and efficiency of rhizosphere. Furthermore, root phenotype can be selected for improving plants output in varied environmental conditions in different locations (Dunbabin et al. 2013).

22.2.2 Nutrient Uptake and Its Linkage with pH

pH is an important factor for living system functionality like enzymes and microbes. It is considered as a vital component of zonation of root and soil. The biological diversity and chemistry of soil impact the role, function and availability of various cations and anions uptaken by root. This study focused on studying the role of nutrients absorption of root and pH of rhizosphere using primary transport of ions and their gradients. Comparative analysis was done at rhizosphere zone pH of maize (*Zea mays L.*), Alpine pennycress (*Noccaea caerulescens*) and ryegrass (*Lolium perenne L.*). Electrically neutral potential of the model was regulated by H^+ and OH^- efflux-influx (Custos et al. 2020; Sposito 2008).

Variants of 37 soil samples from different soil types were analysed to determine pH of rhizosphere. The models observed alkalinity of rhizosphere caused fluctuation around pH ~3.3. Nitrate is most vital nutrient absorbed and its equilibrium is maintained by H^+ or OH^- gradient.

Rate of nitrate diffusion water potential is synergistic with the alkaline nature of rhizosphere. Moreover, root density is directly proportional to the uptake of both ion concentration and pH of rhizosphere. This study demonstrates the function of mineral nutrients and their correlation with rhizosphere pH for future studies. This is a model for studying pH alteration in the rhizosphere (Custos et al. 2020).

22.2.3 Water Potential and Uptake Model

Plants carry out rhizodeposition of a lot of organic carbon at expense of its own natural resources and energy. Rhizodeposits play a crucial role in enhancing microbial communities and nutrients movement. But nowadays it has been reported by several researchers that rhizodeposits also play crucial role in water uptake. Their (Ghezzehei and Albalasmeh 2015) study demonstrated the root system model to examine role of rhizodeposits potential for water gradient. This model has three pillars for study; firstly, rhizodeposits concentrations decrease as we move away from root zone. Secondly, rhizodeposits are mixed and attached with different size soil particles and lastly the potential of rhizodeposits to combine water molecules.

Artificial substitute of these rhizodeposits like polygalacturonic acid (PGA), sand media glass-beads, etc., were used. Dew point potentiometry was used for confirming water retention on larger scale. This model is a clear and precise representation of advantages when water uptake potential is more and rhizodeposits are confined in limited soil zone (Ghezzehei and Albalasmeh 2015; Gao et al. 2011).

22.2.4 Role of Mycorrhizal Fungi on Phosphorous Uptake in Modelling Study

Mathematical modelling for determining the external growth of mycelium of Arbuscular mycorrhiza (AM) fungi and its effect on phosphate uptake by plant roots. Researchers explain P transport in the soil and its uptake by both plant roots and fungal component in a small scale. Various regions of mycelium were examined for their active uptake and deterioration of P in spatial and temporal distributed regions of the soil. When the uptake is more concentrated in the tip region the amount of uptake is less. It is also validated with literature data and quantitative estimation was performed to enhance our knowledge of mycorrhizal symbiosis (Schnepf et al. 2008).

22.2.5 Circadian Rhythm Under Varying Abscisic Acid (ABA) Concentration

There are changes in stomatal aperture, leaf growth, transpiration, hydraulic conductance and amount of abscisic acid in sap of xylem during different times in a day. A dynamic study in model allows the amalgamation of different environmental variants affecting microbial diversity in different time zones. It also includes internal and external factors that affect the rate of hydraulic conductance and circadian rhythm of plant cycle.

Leaf water potential denotes tissues which react more slowly to environmental determinants than xylem water potential and growth. And their amalgamation with water and ABA variations which influence genetically affecting parameters.

This model can be integrated with population and whole plant model (Tardieu et al. 2015).

22.2.6 Modelling Game Theory

Plant roots and rhizodeposits surrounding it influence the mineralization process of carbon (C) and nitrogen (N). Priming of rhizosphere is a very important and distinct phenomenon. In this study sunflower and soyabean from vegetative and reproductive phase of plant cycle were used.

C mineralization using C-13 radiolabelling technique and N-15 for nitrogen were used. The positive rhizosphere priming effect was more in sunflower in contrast to soyabean. The rhizosphere priming effect was linked with root activity and quantity of rhizodeposits. Biomass of microbial community and enzyme activity was monitored. The results conclude that mutualistic relationship between plant and microbial community under different ecological conditions (Zhu et al. 2014).

22.2.7 Combining Plant Models

A complete change from isolated branches to completely integrating models can be done with the help of computer science. In silico plants is a presentation of various models which connects many domains like cell structure, its organization, gene linking and studying varied metabolic pathway. The integration of models is a necessary requirement.

Plants in silico (Psi) provides quantitative data of gene at functional gene level or developmental phase in which whole plant, crop system or ecology can be analysed. Although the integration of rhizosphere models is an important step in modelling study for larger scale implications of the data and information (Zhu et al. 2016).

22.2.8 Rhizosphere Study Via Carbon Flow

Rhizosphere study through carbon flow models inculcates rhizodeposits and microbiome population variability. Rhizosphere plays vital function in the nutrient cycling and ecosystem stability. The carbon released by the plant in the form of low molecular mass compound in soil biotic community. Various molecular biology techniques were used to differentiate bacterial metabolic pathways. Carbon being the vital macro element of the soil which is employed in basic building plant system and processes.

Mechanism based collaborative study was performed and future elemental factors were studied. Different carbon fluxes were examined to study carbon flow using sensitivity analysis. Many growth, death and maintenance models using integrated study were also discussed. There was formation of simple carbon flow model to analyse the rate of carbon flow and how it affects the physiology of soil surrounding it on a small scale. It was concluded from the study that factors that influence that rate of carbon deposition in long term are biomass production and exudate rate (Toal et al. 2000; York et al. 2016).

With implementation of all modelling studies a better understanding for holistic rhizosphere can be created for enhanced agricultural yield and productivity. These small size models are very significant part for creating a better understanding of rhizosphere holistically which includes its elements and conditions. Rhizosphere alteration using modelling studies is the future for enhancing yield in agriculture and judicious use of resources.

22.3 Nanotechnology Boosting Sustainable Agricultural Practices

Nanotechnology provides new frontier for agricultural practices, enhanced nutrient uptake, improving the ability of pest management, climate changing factors and decreasing ecological impact of agriculture. An efficient sustainable agriculture is need of the hour for coping with increasing population. The field of nano sciences can control agriculture at field level by restricting nutrients in the soil (Gruère 2012; Mukhopadhyay 2014). It can be used as water monitoring tool to assess the quality of water and pesticide for agricultural applications (Prasad et al. 2014).

The application of this field in agriculture is a key factor for sustainable agricultural practices. In food industry, which are agriculture dependent, applications of biosensors, nanotubes, fullerenes, controlled delivery, nanofiltration, delivery etc. were researched. (Ion et al. 2010; Sabir et al. 2014). It was proved to be excellent tool in management of drug delivery, resources at field level which enhances soil fertility. Use of biomass and waste generated from agricultural fields along with food packaging, processing these nanoparticles are versatile (Floros et al. 2010).

Nanosensor technology is widely used in agriculture due to their sensitive results and environmental perturbation monitoring. Contamination of land and water can be monitored using these devices (Ion et al. 2010). These nanoparticles are also used in bioremediation processes as an additive not directly (Dixit et al. 2015). In agriculture used in soil to restore its potential. Its an essential aspect to remove nanoparticles interaction among one other to improve soil quality (Ion et al. 2010; Dixit et al. 2015).

This branch also facilitates improvement in food quality and quantity, lesser investment of agricultural assets and enhanced uptake of nano-size nutrients from the soil. The goal of nanobiotechnology in agriculture is reduction in artificial fertilizers, lesser nutrient leaching and better pathogen protection with higher yield. Nano based tool and technologies are very efficient in diagnosing plant diseases and improving nutrient uptake by plants. The use of nanotechnology in particular practices like nano-fertilizers, nanopesticides, environment monitoring is very competent. The major lag is always the implementation of this nano technology at a large scale. Some newly proposed strategies can be used to combat the limitations in agricultural field (Usman et al. 2020; Prasad et al. 2017).

22.3.1 Various Types of Nanoparticles and their Significance in Agriculture

Nanomaterials (NMs) are diversely used worldwide and have some ecological limitations (Prasad et al. 2014). Parameters of NPs have effect on toxicity and chemical properties like shape, size, surface area, behaviour and agglomeration and segregation of artificial NPs (Ion et al. 2010). This is a cause for so differential variation in properties of NPs even with same chemical properties.

Box 22.1 Nano based materials and their role in agriculture

Nano based materials	Important role in agriculture
1. Carbon Nanotubes (CNTs)	• CNT based nano-sponges with iron and sulfur enhance efficacy of water contaminants uptake, fertilizers, pesticides, oil and pharma-based drugs (Camilli et al. 2014)
2. Quantum dots (QDs)	• QDs are used for live root imaging in whole plant systems to study known physiological processes (Hu et al. 2010; Das et al. 2015)
3. Nanorods	• The gold nanorods are phytotoxic at high concentration (Wan et al. 2014) and are reported to transport auxin 2,4-D (Nima et al. 2014)
4. Micro nano encapsulation	• Micro and NPs for drug delivery, protection and enhancing bioavailability of nutraceuticals (Ozdemir and Kemerli 2016). Drugs like peptides is nanoencapsulated (Puglisi et al. 1995; Hildebrand and Tack 2000)

Currently, nanoparticles are implied in various agricultural fields as shown in Box 22.1. Carbon nanotubes (CNTs) are cylindrical form of graphene sheet. It can be of two types: (SWNTs) single-walled nanotubes and multi-walled nanotubes (MWNTs). Recently, fluorescently labelled nanoparticles (NPs) and quantum dots (QDs) have been used in labelling various plant proteins (Pyrzynska 2011; Chahine et al. 2014; Prasad et al. 2017).

Overexploitation of these CNTs will open new opportunities for various nano devices (Raliya et al. 2013). Nano agrochemicals can be targeted with CNTs for effective delivery. Quantum dots (QDs) have exceptional properties as compared to other organic compounds. They are also applicable in bioimaging and biosensing phenomenon (Bakalova et al. 2004). Nano and micro encapsulation are used for preserving objects in harsh environmental conditions, limited release and specific targeting (Ezhilarasi and Swarup 2012; Ozdemir and Kemerli 2016). Nanorods are plasmonic multifunctional materials which used as sensors and size-based energy regulation process. It can lead to precise field responses (Bakalova et al. 2004).

22.3.2 Nano-fertilizers

Nano-fertilizers are now commercialized products available in the market but still some major industries are not working on them. Nano-fertilizers include nano zinc, iron, cadmium and titanium dioxide, QDs, nanorods, etc., which enhance its efficiency. Usage, target and toxicity studies of varied nanoparticles were researched in agriculture like Al_2O_3, ZnO, TiO_2, CeO_2, etc. (Nayak et al. 2015). Zn micronutrient is a very limiting factor affecting production (Sadeghzadeh 2013). Metal oxide NPs can be used after radiolabelling by bombardment of protons (Llop et al. 2014). The

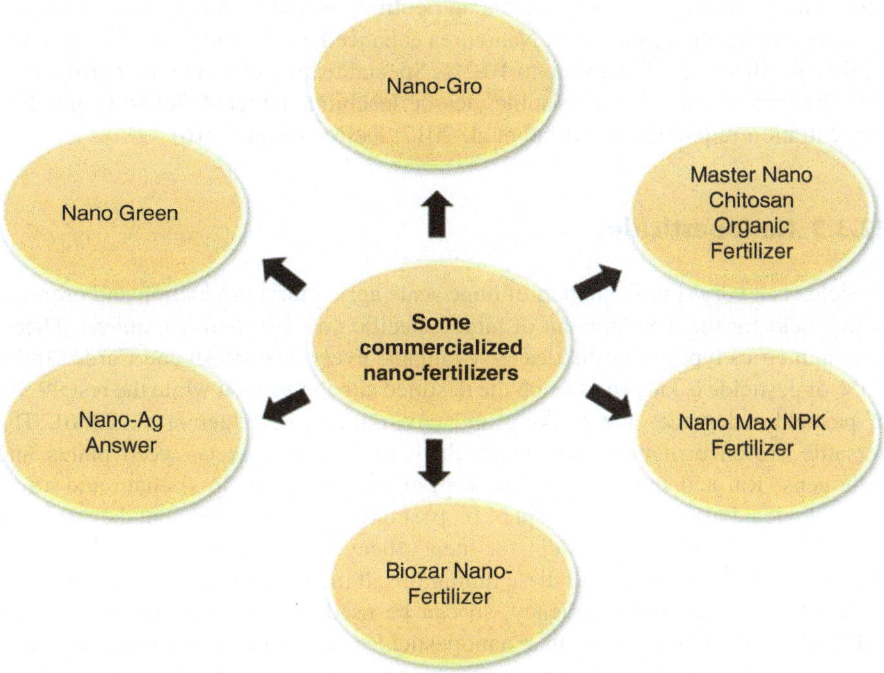

Fig. 22.2 Some nanofertilizer commercialized by small scale industries

various techniques used for uptake and target by Raman chemical imaging spectroscopy, confocal laser scanning microscopy and ion beam microscopy (Marzbani et al. 2015). Nanofertilizers are macro and micronutrients or either associated with minerals for nutrients (Kah et al. 2018). They can be formed by encapsulation of nutrients inside nanoparticles. They improve quality, quantity, are lesser costs and highly efficient.

Cell culture media and proteins were used for studying agglomeration rate, potential of metal oxide NPs variants (Llop et al. 2014; Marzbani et al. 2015; Helar and Chavan 2015). Since toxicity is a major concern but nano-fertilizers are attracting attention due to biocompatibility and unique properties. Thus, use of nano-fertilizers is more efficient over traditional fertilizers (Kandasamy and Prema 2015). Nanotechnology dependent fertilizers are a boon to the agricultural production for fulfilling needs of tremendously growing population day by day (Prasad et al. 2017).

Overapplication of fertilizers and pesticides is hazardous to environment causing pollution. So, targeted delivery and release of fertilizers is an important aspect. Conventional methods are less reliable in terms of ecological pollution (Prasad et al. 2017) (Fig. 22.2).

The Bio-fortification of some crops was done by ZnO, CuO and B_2O_3 to introduce drought tolerance in soyabean plants. Precision agriculture focuses on reduction in application of artificial pesticide, more nutrient regulation and enhanced agricultural productivity. Sustainable agriculture includes energy and resources utilization in a judicious manner. Nano urea enhanced grain yield by 44.5% although modern fertilizer can enhance up to 10.2%. Special feature of nano based fertilizer is their limited release, biocompatible, lesser leaching, targeted delivery and low concentration requirement (Prasad et al. 2017; DeRosa et al. 2010).

22.3.3 Nanopesticides

Pesticide is a very significant part of huge scale agricultural production and progress in this field for the development of target specific novel system is required. Therefore, numerous types of pesticides are evaluated every year (Resh and Cardé 2009). 0.1% of pesticide is known to reach the destined site in the plant while the rest 99.9% of pesticides degrades the ecology and environment (Carriger et al. 2006). The versatile presence of pesticides made them tolerant to insects, weed plants and pathogens (Rai and Ingle 2012). These pesticides can enter food chain and affect the ecological balance. The concept of biopesticides came into light but they are very slow and environment dependent for their efficiency which is a major drawback. Nanopesticides work above all these limitations. It includes targeted release, low rate of degradation and high efficiency, so can be used for a longer duration of time (Chhipa 2017). It concludes that nanopesticides are more environment friendly, sustainable with higher efficacy over conventional pesticides (Kah et al. 2019; Usman et al. 2020). Nano-sized materials can be released in the field in any form like partially or completely dissolved in solvent. Additionally, they require less energy and resources. They are better adapted for pest management, required in low amount, highly efficient and reduced extensive labour.

Nanopesticides have harmful impact on human health and welfare (Raliya et al. 2013) which are as follows: (a) skin can absorb pesticides due to their extremely small size, (b) with respiration can reach heart and lungs, (c) material is not always environmental friendly, etc. Popular examples of some nanopesticides are as follows: imidacloprid (IMI) is a potent insecticide affecting pests systematically. In vitro studies revealed the use of IMI and found to be 100% mortal for *Coleoptera, Tenebrionidae*. In addition, nano-IMI can be degraded in the presence of light which proves its eco-friendly nature. Nanopesticides do not affect the soil microbiome around the rhizosphere region of the soil (Guan et al. 2008). Other example is permethrin in the nano-formulated form against *Aedes aegypti* (Kumar et al. 2013a, b).

Chlorfenapyr linked with silica nanoparticles is known to cause double insecticidal performance than with soil matter particles (Song et al. 2012). Above results revealed the future prospect of using these NPs as pesticides which are effective agents for management of pests.

Nanosilica was reported for its efficacy against pest post harvesting. Antimicrobial activity of NPs is effective against viral, bacterial and fungal pathogens. Silica based silver NPs are known to degrade harmful fungi like *Colletotrichum gloeosporioides* and *Rhizoctonia solani*.

Anti-fungal activity against silver NPs was studied in *Raffaelea* sp. attacking oak trees (Park et al. 2006; Kim et al. 2009). Weedicides are important for sustainable agriculture. Nanoherbicide depends on biodegradable polymeric material which enhances the ability of herbicides. The nanoparticles encapsulated with atrazine were proved on *Brassica* spp. enhanced herbicidal property and reduces its mobile property (Pereira et al. 2014).

Various inorganic NPs were studied (Ag, Cu, Al_2O_3, ZnO, TiO_2 and ZrO_2) for toxicity analysis on earthworms and Cu, Ag and TiO_2 were found to be toxic. The reproducing ability of earthworm was reduced due to Ag (Heckmann et al. 2011). The earthworms' behaviour is also known to be affected due to Ag NPs.

The poly(epsiloncaprolactone) nanocapsules loaded herbicides in encapsulated form decrease the toxicity to *Pseudokirchneriella subcapitata* and *Prochilodus lineatus* although enhanced toxicity in *Daphnia similis* (Clemente et al. 2014; Arduini et al. 2016). Detailed and practical knowledge of nanoparticles risk assessment techniques are required at functional and cellular level so that these can be used on a higher scale (Pandey et al. 2018; Tiwari et al. 2020).

22.3.4 Nanobiosensors

Biosensors are amalgamation of receptor-transducer which are applied to monitor physical or chemical properties of medium in the presence of organic material to detect biological entities (Sun et al. 2007). Nanobiosensors are advanced biosensors which use transducer for detection of trace elements in nature. They contain nanoparticles for detection through transducer which act as asignalling agent to detect single or multi enzyme complexes. The nanoparticles popularly used in these sensors are noble metals or quantum dots. The transduced signal can be of optical, magnetic or electrical form. The elements recognized are proteins, antibodies, aptamers, etc. Solo compounds for detection can be pesticides, nutrients and soil parameters and complexes are metalloids like Pb, Hg, Cr, etc. Nanobiosensors are implied to various agricultural fields like water, soil, pesticides, fertilizers management, etc. They are highly sensitive, stable, have easy detection technology and high surface area makes them more effective over old sensors (Scognamiglio 2013). They work on the mechanism of switch which can be either in OFF/ON form (Arduini et al. 2016;Usman et al. 2020).

Early disease detection and trace pollutants detection can be useful for avoiding loss in full seasonal produce. Use of nanochips makes them more portable than others for detection kits at large scale (Ullah et al. 2018). In the class of different pesticides organo-phosphates, atrazines, carbamates are detected even in trace quantity because they are heterogeneous among soil particles. They can be detected by using nanosensors attached with piezoelectric transducers using antigen–antibody (Ag-Ab) interaction studies for blocking enzyme action (Ivask et al. 2002; Přibyl et al. 2006). The cost of these biosensors vary account to the type of material used in its construction and effective working is about 10% of active molecules from a mixture. Their efficiency improvement is a whole new domain of research (Liu et al.

2011). They are popularly used in detection of soil and water pollutants. Although they can be used for quantification of soil's composition such as total organic substances, carbon, sodium chloride, nitrate, they are limited to lab studies (Antonacci et al. 2018).

An example of nanosensor is ceramic coated material integrated with Ag/Pd electrodes or with graphene oxide (Liu et al. 2011). They exploit the properties of nanomaterials. They are nowadays applied for monitoring early plant pathogenic diseases and soil water pollutants (Hu et al. 2014). This is a whole new system of agriculture for sustainable development which requires a lot of guidelines and processing for field scale application. The major drawback is entrance of these nano elements in food chain which limits their application. So, a holistic research is required for more detailed interactive study of ecosystem.

22.3.5 Nano-bioremediation

The process of remediation of different types of pollutants in the nature can be using NPs.

It reduces their movement and convert them to lesser toxic alternatives or by carrying out their transformation. Various types of NPs are used for removal of harmful pollutants in nature. So, far it was used in three branches (1) for Soil pollution, (2) Organic contaminants and (3) Inorganic contaminants (Usman et al. 2020).

22.3.5.1 Nano-Bioremediation of Soil Pollution

Nano based materials are used in the remediation of polluted soil as NPs are of unique properties due to their shape and size. There is a huge amount of research going on in this field as they can carry out biosorption, desorption, transformation, complexation, precipitation, oxidation and reduction to alter its toxic potential and making it eco-friendly (Guerra et al. 2018). The main phenomenon which governs bioremediation in soil is sorption and desorption process of remediation (Hamid et al. 2020; Usman et al. 2020).

The methods include mobility enhancing agents for increasing the soil pollutants and reducing their movement in the soil and making them more available for purification processes. Some of these also act as carriers which bind with pollutants present in the soil which reduces their probability of entering in food web (Robinson et al. 2009). Magnetically, some contaminants can be separated; for examples, soil particles with metalloids in the presence of NPs (Boente et al. 2018a, b). These methods are both in situ and ex situ favourable but within a confined zone. So, in this field we require NPs which can adhere to the contaminants for a longer period of time for their removal. In this respect magnetic NPs work best due to their high adherence properties than other forms of nanomaterials. It is a more reliable, easy, less labour intensive and efficient method. Their easy separation at the end of the whole process makes them such a useful entity (Ajmal et al. 2018; Fu et al. 2018).

For example, NPs based on the polyacrylamide modified magnetite NPs can be used to prevent soil erosion by 90% and leaching is confined by 82% (Zheng et al. 2020). In situ methods are more reliable and efficient for pollutant removal in the soil (Hamid et al. 2020). The amalgamation of nanotechnology and biology can be helpful in the remediation of highly polluted soil.

22.3.5.2 Nano-Bioremediation of Organic Contaminants

Organic pollutants are very abundant in nature causing pollution, and nano sciences have great potential for removal of these pollutants. They can be remediated first by dechlorination and sequential dehalogenation for effective removal by NPs (Bokare et al. 2010). They researched the potential applications of NPs like Pd and Fe and their bioremediation potential of 2,4,4′-trichloro-2′-hydroxydiphenyl ether (TCS) polluted soil and water. It was reported that Pb and Fe can be reductively dechlorinated in anaerobic conditions with formation of product. Le et al. (2015) examined the ability of oxidation reduction process by NPs for effective remediation. Many variants of compounds like biphenyl were dechlorinated by NPs such as Pb and Fe. And in the next step of the process toxicity can be regulated *Burkholderia xenovorans* of the leftover biphenyls. NPs are not found to be harmful for *Burkholderia xenovorans*. Moreover, NPs help in enhancing bioavailable property of pollutants so that their degradation can be easily assessed. Manipulation in membrane properties also helps in easy degradation of organic pollutants (Gong et al. 2018). Wu et al. (2020) examined the role of Ni and Fe NPs on toxicity studies of polybrominated diphenyl ethers (PBDEs). If NMs are applied on plant parts, they reduced their phytotoxicity against PBDEs. This study is the demarcation that if NPs and bioremediation processes it will lead to a reduction of toxicity in soil contaminants (Usman et al. 2020).

22.3.5.3 Nanomaterial Assisted Remediation of Inorganic Contaminants

Nanoparticles are essentially used for enhancing the phytoremediation potential of plants against heavy metals in the contaminated soil. Phytoremediation includes processes like phytovolatilization, phytosequestration, phytostabilization, phytodegradation, etc. It is a very environment friendly and low-cost technique. Nano-titanium oxide on Cd was examined by Singh and Lee (2016) in soyabean plants. It prevents the deposition of Cd and protected the plants from oxidative damage due to Cd overload. The role of (NHAP) Nano-hydroxyapatite and nano-sized carbon black (NCB) on the lead (Pb) phytoremediation by extraction studies on ryegrass plant has been studied by some scientific groups (Liang et al. 2017).

NMs can help in deposition of heavy metals in plants upper parts by changing their cell wall properties (Hu et al. 2014) and cotransport and translocation (Hu et al. 2014). Similarly, Cd and graphene oxide can have harmful effect on Microcystis aeruginosa by modulating reactive oxygen species generation (Tang et al. 2015).

22.4 Hurdles to Overcome in Nanotechnology for Its Application in Plant Agriculture

The most important drawback which limits the use of nanotechnology in fields is majorly divided into two frontiers which are as follows: (a) Large-scale production and (b) safety principles. Many tactics have also been suggested to conquer these barriers (Hofmann et al. 2020).

22.4.1 Large-Scale Production

Important barriers are recognized which are disturbing the process of actualization of large potential of nanotechnology application in agriculture. Scale up to delivery at field scale. There is insufficient information and research about targeted delivery at field scale. Mainly soil and leaves applications are methods in old agricultural practices and chemicals around 50% reach their target like leaf, root or pest. So, the solution to this problem is coating NPs with directing biological molecules or adjusting surface size and charge, can enhance absorption percentage and follow targeting to well destined plant cell compartments and organelles, like the nucleus, mitochondria and chloroplasts (Mitter et al. 2017).

Improved protection of plant and variable distribution of nutrients while at the same time conserving the beautiful natural resources (like water, energy, raw materials, etc.) which are lost in conventional methods. On the other hand, ammonia, nitrogen dioxide and the greenhouse gases enhance nitrogen distribution to crops. Formulations which enhance leaf attachment and precision spraying can be done by foliar delivery. NPs made for agricultural soil have been researched in laboratory scale, but further research of concentrations is still needed applications. Presently, soil utilization cost which was for conventional agricultural linked chemicals, the financial and embodied resource costs of many nanoparticles are very high beyond practical usage at large scale. The leaf-based uptake of nanoparticles is appreciated in recent times.Recent research explained almost 100% absorption of foliar-applied gold nanomaterials in wheat crop was found to be very efficient. On the other hand, utilizing help of polyvinylpyrrolidone coating shows that leaf application as an efficient alternative for cost management.The major point is how to deliver these NPs at field level needs an outlook and research (Hofmann et al. 2020).

22.4.2 Safety Principles

Agriculture is staple diet source for such a large population and should be cautiously controlled for safety of food. Nano dependent technology can be very harmful for human beings and mother earth. It is not about assessment but also a major global concern. For example, in many countries, soil additives, fertilizers will suffer regulatory hurdles as compared to pesticides and genetically modified crops. An example of ribonucleicacid interference (RNAi), technique at nano-scale requires high cost and skilled labour. If nano-size particles treatment is limited to seed level, their exposure and toxicity can be reduced.

The role of NPs cannot be avoided at small and large scale and different phases of plant growth and development. Evaluative and extensive procedure of elemental traces should be performed (Hofmann et al. 2020).

22.5 Conclusion

Nanotechnology has wide range of applications in agriculture like nano-fertilizers, nanopesticides, nanobiosensors, bioremediation techniques, etc. However, the harmful effects of NPs are a major concern. Toxic effects of nanoparticles on plant growth and development have been reported. The NPs application specially targeted site delivery, fertilizer use and efficacy should be enhanced at a larger scale. More field studies should be performed for effective applications of necessary data and implementation. The limitations of field trials should be removed by effective integration of different branches of research. Standardized methods and procedure should be followed for potential applications of these NMs on a large scale. For sustainable agriculture the entrance of nano sciences in agriculture is a necessary requirement for feeding such a large growing population. Toxicity studies and their routes can be examined for making them more environment friendly and removing all the barriers of food chain entrance. If these methods will be successful, it can positively enhance our economy as India GDP is largely influenced by agriculture. Academic and industries working together by joining hands in collaboration is a way of making the path for nanotechnology in agriculture (Hofmann et al. 2020).

Similarly, rhizosphere modelling studies are a new frontier for studying dynamic and versatile nature of rhizosphere. Manipulation of rhizosphere by modelling studies can assist yield enhancement along with their use in phytoremediation. Advanced imaging, modelling with in silico tools can be used as tool for rhizosphere. Their detailed knowledge can assist in modulating biotic and abiotic factors required for yield enhancement in agriculture (York et al. 2016).

References

Ajmal Z, Muhmood A, Usman M, Kizito S, Lu J, Dong R, Wu S (2018) Phosphate removal from aqueous solution using iron oxides: adsorption, desorption and regeneration characteristics. J Colloid Interface Sci 528:145–155

Antonacci G, Reed JE, Lennox L, Barlow J (2018) The use of process mapping in healthcare quality improvement projects. Health Serv Manag Res 31(2):74–84. https://doi.org/10.1177/0951484818770411

Arduini F, Cinti S, Scognamiglio V, Moscone D (2016) Nanomaterials in electrochemical biosensors for pesticide detection: advances and challenges in food analysis. Microchim Acta 183(7):2063–2083

Bakalova R, Ohba H, Zhelev Z, Nagase T, Jose R, Ishikawa M, Baba Y (2004) Quantum dot anti-CD conjugates: are they potential photosensitizers or potentiators of classical photosensitizing agents in photodynamic therapy of cancer? Nano Lett 4(9):1567–1573

Boente C, Sierra C, Martínez-Blanco D, Menéndez-Aguado JM, Gallego JR (2018a) Nanoscale zero-valent iron-assisted soil washing for the removal of potentially toxic elements. J Hazard Mater 350:55–65

Boente C, Sierra C, Martínez-Blanco D, Menéndez-Aguado JM, Gallego JR (2018b) Nanoscale zero-valent iron-assisted soil washing for the removal of potentially toxic elements. J Hazard Mater 350:55–65

Bokare V, Murugesan K, Kim YM, Jeon JR, Kim EJ, Chang YS (2010) Degradation of triclosan by an integrated nano-bio redox process. Bioresour Technol 101(16):6354–6360

Camilli L, Pisani C, Gautron E, Scarselli M, Castrucci P, D'Orazio F, Passacantando M, Moscone D, De Crescenzi M (2014) A three-dimensional carbon nanotube network for water treatment. Nanotechnology 25(6):065701

Carriger JF, Rand GM, Gardinali PR, Perry WB, Tompkins MS, Fernandez AM (2006) Pesticides of potential ecological concern in sediment from South Florida canals: an ecological risk prioritization for aquatic arthropods. Soil Sediment Contam 15(1):21–45

Chahine NO, Collette NM, Thomas CB, Genetos DC, Loots GG (2014) Nanocomposite scaffold for chondrocyte growth and cartilage tissue engineering: effects of carbon nanotube surface functionalization. Tissue Eng A 20(17-18):2305–2315

Chen H, Seiber JN, Hotze M (2014) ACS select on nanotechnology in food and agriculture: a perspective on implications and applications. J Agric Food Chem 62(6):1209–1212

Chhipa H (2017) Nanofertilizers and nanopesticides for agriculture. Environ Chem Lett 15 (1):15–22

Clemente Z, Grillo R, Jonsson M, Santos NZ, Feitosa LO, Lima R, Fraceto LF (2014) Ecotoxicological evaluation of poly (ε-caprolactone) nanocapsules containing triazine herbicides. J Nanosci Nanotechnol 14(7):4911–4917

Custos JM, Moyne C, Sterckeman T (2020) How root nutrient uptake affects rhizosphere pH: a modelling study. Geoderma 369:114314

Czarnes S, Hallett PD, Bengough AG, Young IM (2000) Root-and microbial-derived mucilages affect soil structure and water transport. Eur J Soil Sci 51(3):435–443

Das S, Wolfson BP, Tetard L, Tharkur J, Bazata J, Santra S (2015) Effect of N-acetyl cysteine coated CdS: Mn/ZnS quantum dots on seed germination and seedling growth of snow pea (Pisum sativum L.): imaging and spectroscopic studies. Environ Sci 2(2):203–212

De León-González F, Celada-Tornel E, Hidalgo-Moreno CI, Etchevers-Barra JD, Gutiérrez-Castorena MC, Flores-Macías A (2006) Root–soil adhesion as affected by crop species in a volcanic sandy soil of Mexico. Soil Tillage Res 90(1-2):77–83

DeRosa MC, Monreal C, Schnitzer M, Walsh R, Sultan Y (2010) Nanotechnology in fertilizers. Nat Nanotechnol 5(2):91

Dixit R, Malaviya D, Pandiyan K, Singh UB, Sahu A, Shukla R, Singh BP, Rai JP, Sharma PK, Lade H, Paul D (2015) Bioremediation of heavy metals from soil and aquatic environment: an overview of principles and criteria of fundamental processes. Sustain For 7(2):2189–2212

Dunbabin VM, Postma JA, Schnepf A et al (2013) Modelling root–soil interactions using three–dimensional models of root growth, architecture and function. Plant Soil 372:93–124. https://doi.org/10.1007/s11104-013-1769-y

Ezhilarasi GA, Swarup KS (2012) Network partitioning using harmony search and equivalencing for distributed computing. J Parallel Distrib Comput 72(8):936–943

Floros JD, Newsome R, Fisher W, Barbosa-Cánovas GV, Chen H, Dunne CP, German JB, Hall RL, Heldman DR, Karwe MV, Knabel SJ (2010) Feeding the world today and tomorrow: the importance of food science and technology: an IFT scientific review. Compr Rev Food Sci Food Saf 9(5):572–599

Fu H, Yang Y, Zhu R, Liu J, Usman M, Chen Q, He H (2018) Superior adsorption of phosphate by ferrihydrite-coated and lanthanum-decorated magnetite. J Colloid Interface Sci 530:704–713

Gao Y, Yang Y, Ling W, Kong H, Zhu X (2011) Gradient distribution of root exudates and polycyclic aromatic hydrocarbons in rhizosphere soil. Soil Sci Soc Am J 75(5):1694–1703

Ghezzehei TA, Albalasmeh AA (2015) Spatial distribution of rhizodeposits provides built-in water potential gradient in the rhizosphere. Ecol Model 298:53–63

Gong X, Huang D, Liu Y, Peng Z, Zeng G, Xu P, Cheng M, Wang R, Wan J (2018) Remediation of contaminated soils by biotechnology with nanomaterials: bio-behavior, applications, and perspectives. Crit Rev Biotechnol 38(3):455–468

Gruère GP (2012) Implications of nanotechnology growth in food and agriculture in OECD countries. Food Policy 37(2):191–198

Guan YF, Pearce RC, Melechko AV, Hensley DK, Simpson ML, Rack PD (2008) Pulsed laser dewetting of nickel catalyst for carbon nanofiber growth. Nanotechnology 19(23):235604

Guerra FD, Attia MF, Whitehead DC, Alexis F (2018) Nanotechnology for environmental remediation: materials and applications. Molecules 23(7):1760

Hamid Y, Tang L, Hussain B, Usman M, Lin Q, Rashid MS, He Z, Yang X (2020) Organic soil additives for the remediation of cadmium contaminated soils and their impact on the soil-plant system: a review. Sci Total Environ 707:136121

Han E, Kautz T, Perkons U, Uteau D, Peth S, Huang N, Horn R, Köpke U (2015) Root growth dynamics inside and outside of soil biopores as affected by crop sequence determined with the profile wall method. Biol Fertil Soils 51(7):847–856

Heckmann LH, Hovgaard MB, Sutherland DS, Autrup H, Besenbacher F, Scott-Fordsmand JJ (2011) Limit-test toxicity screening of selected inorganic nanoparticles to the earthworm Eisenia fetida. Ecotoxicology 20(1):226–233

Helar G, Chavan A (2015) Synthesis, characterization and stability of gold nanoparticles using the fungus Fusarium oxysporum and its impact on seed. Int J Recent Sci Res 6:3181–3318

Hildebrand GE, Tack JW (2000) Microencapsulation of peptides and proteins. Int J Pharm 196 (2):173–176

Hinsinger P, Bengough AG, Vetterlein D, Young IM (2009) Rhizosphere: biophysics, biogeochemistry and ecological relevance. Plant Soil 321(1-2):117–152

Hofmann T, Lowry GV, Ghoshal S, Tufenkji N, Brambilla D, Dutcher JR, Gilbertson LM, Giraldo JP, Kinsella JM, Landry MP, Lovell W (2020) Technology readiness and overcoming barriers to sustainably implement nanotechnology-enabled plant agriculture. Nat Food 1(7):416–425

Hu Y, Li J, Ma L, Peng Q, Feng W, Zhang L, He S, Yang F, Huang J, Li L (2010) High efficiency transport of quantum dots into plant roots with the aid of silwet L-77. Plant Physiol Biochem 48 (8):703–709

Hu X, Kang J, Lu K, Zhou R, Mu L, Zhou Q (2014) Graphene oxide amplifies the phytotoxicity of arsenic in wheat. Sci Rep 4(1):10

Ion AC, Ion I, Culetu A, Gherase D (2010) Carbon-based nanomaterials. Environmental applications. Romania

Ivask A, Virta M, Kahru A (2002) Construction and use of specific luminescent recombinant bacterial sensors for the assessment of bioavailable fraction of cadmium, zinc, mercury and chromium in the soil. Soil Biol Biochem 34:1439–1447

Jones DL, Hodge A, Kuzyakov Y (2004) Plant and mycorrhizal regulation of rhizodeposition. New Phytol 163(3):459–480

Kah M, Kookana RS, Gogos A, Bucheli TD (2018) A critical evaluation of nanopesticides and nanofertilizers against their conventional analogues. Nat Nanotechnol 13(8):677–684

Kah M, Tufenkji N, White JC (2019) Nano-enabled strategies to enhance crop nutrition and protection. Nat Nanotechnol 14(6):532–540

Kandasamy S, Prema RS (2015) Methods of synthesis of nano particles and its applications. J Chem Pharm Res 7(3):278–285

Kim SW, Kim KS, Lamsal K, Kim YJ, Kim SB, Jung M, Sim SJ, Kim HS, Chang SJ, Kim JK, Lee YS (2009) An in vitro study of the antifungal effect of silver nanoparticles on oak wilt pathogen Raffaelea sp. J Microbiol Biotechnol 19(8):760–764

Kumar RS, Shiny PJ, Anjali CH, Jerobin J, Goshen KM, Magdassi S, Mukherjee A, Chandrasekaran N (2013a) Distinctive effects of nano-sized permethrin in the environment. Environ Sci Pollut Res 20(4):2593–2602

Kumar RS, Shiny PJ, Anjali CH, Jerobin J, Goshen KM, Magdassi S, Mukherjee A, Chandrasekaran N (2013b) Distinctive effects of nano-sized permethrin in the environment. Environ Sci Pollut Res 20(4):2593–2602

Le TT, Nguyen KH, Jeon JR, Francis AJ, Chang YS (2015) Nano/bio treatment of polychlorinated biphenyls with evaluation of comparative toxicity. J Hazard Mater 287:335–341

Liang SX, Jin Y, Liu W, Li X, Shen SG, Ding L (2017) Feasibility of Pb phytoextraction using nano-material assisted ryegrass: results of a one-year field-scale experiment. J Environ Manag 190:170–175

Liu X, Wang R, Xia Y, He Y, Zhang T (2011) LiCl-modified mesoporous silica SBA-16 thick film resistors as humidity sensor. Sens Lett 9(2):698–702

Llop J, Estrela-Lopis I, Ziolo RF, González A, Fleddermann J, Dorn M, Vallejo VG, Simon-Vazquez R, Donath E, Mao Z, Gao C (2014) Uptake, biological fate, and toxicity of metal oxide nanoparticles. Part Part Syst Charact 31(1):24–35

Marzbani P, Afrouzi YM, Omidvar A (2015) The effect of nano-zinc oxide on particleboard decay resistance. Maderas 17(1):63–68

Mitter N, Worrall EA, Robinson KE, Xu ZP, Carroll BJ (2017) Induction of virus resistance by exogenous application of double-stranded RNA. Curr Opin Virol 26:49–55

Mukhopadhyay SC (2014) Wearable sensors for human activity monitoring: a review. IEEE Sensors J 15(3):1321–1330

Nayak AP, Yuan Z, Cao B, Liu J, Wu J, Moran ST, Li T, Akinwande D, Jin C, Lin JF (2015) Pressure-modulated conductivity, carrier density, and mobility of multilayered tungsten disulfide. ACS Nano 9(9):9117–9123

Nima ZA, Lahiani MH, Watanabe F, Xu Y, Khodakovskaya MV, Biris AS (2014) Plasmonically active nanorods for delivery of bio-active agents and high-sensitivity SERS detection in planta. RSC Adv 110:64985

Ozdemir M, Kemerli T (2016) Innovative applications of micro and nanoencapsulation in food packaging. In: Encapsulation and controlled release technologies in food systems. Wiley, London, pp 333–378

Pandey S, Giri K, Kumar R, Mishra G, Rishi RR (2018) Nanopesticides: opportunities in crop protection and associated environmental risks. Proc Natl Acad Sci 88(4):1287–1308

Park G, Farrar CR, di Scalea FL, Coccia S (2006) Performance assessment and validation of piezoelectric active-sensors in structural health monitoring. Smart Mater Struct 15(6):1673

Pereira AE, Grillo R, Mello NF, Rosa AH, Fraceto LF (2014) Application of poly (epsilon-caprolactone) nanoparticles containing atrazine herbicide as an alternative technique to control weeds and reduce damage to the environment. J Hazard Mater 268:207–215

Prasad R, Kumar V, Prasad KS (2014) Nanotechnology in sustainable agriculture: present concerns and future aspects. Afr J Biotechnol 13(6):705–713

Prasad R, Bhattacharyya A, Nguyen QD (2017) Nanotechnology in sustainable agriculture: recent developments, challenges, and perspectives. Front Microbiol 20(8):1014

Přibyl J, Krejčová H, Přibylová J, Misztal I, Bohmanová J (2006) Analysis of daily gains of performance tested dual-purpose bulls using Legendre polynomials and linear splines. In: 8th WCGALP. August 13–18, Belo Horizonte, Brazil, pp 25–12

Puglisi G, Fresta M, Giammona G, Ventura CA (1995) Influence of the preparation conditions on poly (ethylcyanoacrylate) nanocapsule formation. Int J Pharm 125(2):283–287

Pyrzynska K (2011) Carbon nanotubes as sorbents in the analysis of pesticides. Chemosphere 83 (11):1407–1413

Rai M, Ingle A (2012) Role of nanotechnology in agriculture with special reference to management of insect pests. Appl Microbiol Biotechnol 94(2):287–293

Raliya R, Tarafdar JC, Gulecha K, Choudhary K, Rameshwar R, Prakash M, Saran RP (2013) Scope of nanoscience and nanotechnology in agriculture. J Appl Biol Biotechnol 1(3):41–44

Resh VH, Cardé RT (eds) (2009) Encyclopedia of insects. Academic, New York

Robinson BH, Bañuelos G, Conesa HM, Evangelou MW, Schulin R (2009) The phytomanagement of trace elements in soil. Crit Rev Plant Sci 28(4):240–266

Sabir S, Arshad M, Chaudhari SK (2014) Zinc oxide nanoparticles for revolutionizing agriculture: synthesis and applications. Sci World J 2014:925494
Sadeghzadeh B (2013) A review of zinc nutrition and plant breeding. J Soil Sci Plant Nutr 13 (4):905–927
Schnepf A, Roose T, Schweiger P (2008) Impact of growth and uptake patterns of arbuscular mycorrhizal fungi on plant phosphorus uptake—a modelling study. Plant Soil 312(1-2):85–99
Scognamiglio V (2013) Nanotechnology in glucose monitoring: advances and challenges in the last 10 years. Biosens Bioelectron 47:12–25
Scott NR, Chen H, Schoen R (2016) Sustainable global food supply. Springer, New York, pp 651–668. https://doi.org/10.1007/978-3-319-07052-0_43
Singh J, Lee BK (2016) Influence of nano-TiO2 particles on the bioaccumulation of Cd in soybean plants (Glycine max): a possible mechanism for the removal of Cd from the contaminated soil. J Environ Manag 170:88–96
Song G, Gao Y, Wu H, Hou W, Zhang C, Ma H (2012) Physiological effect of anatase TiO2 nanoparticles on Lemna minor. Environ Toxicol Chem 31(9):2147–2152
Sposito G (2008) The chemistry of soils. Oxford University Press, Oxford
Sun YP, Li XQ, Zhang WX, Wang HP (2007) A method for the preparation of stable dispersion of zero-valent iron nanoparticles. Colloids Surf A 308(1-3):60–66
Tang Y, Tian J, Li S, Xue C, Xue Z, Yin D, Yu S (2015) Combined effects of graphene oxide and Cd on the photosynthetic capacity and survival of Microcystis aeruginosa. Sci Total Environ 532:154–161
Tardieu F, Simonneau T, Parent B (2015) Modelling the coordination of the controls of stomatal aperture, transpiration, leaf growth, and abscisic acid: update and extension of the Tardieu–Davies model. J Exp Bot 66(8):2227–2237
Tiwari E, Mondal M, Singh N, Khandelwal N, Monikh FA, Darbha GK (2020) Effect of the irrigation water type and other environmental parameters on CeO2 nanopesticide–clay colloid interactions. Environ Sci: Processes Impacts 22(1):84–94
Toal ME, Yeomans C, Killham K, Meharg AA (2000) A review of rhizosphere carbon flow modelling. Plant Soil 222(1-2):263–281
Ullah N, Mansha M, Khan I, Qurashi A (2018) Nanomaterial-based optical chemical sensors for the detection of heavy metals in water: recent advances and challenges. TrAC Trends Anal Chem 100:155–166
Usman M, Farooq M, Wakeel A, Nawaz A, Cheema SA, Ashraf I, Sanaullah M (2020) Nanotechnology in agriculture: current status, challenges and future opportunities. Sci Total Environ 6:137778
Walker TS, Bais HP, Grotewold E, Vivanco JM (2003) Root exudation and rhizosphere biology. Plant Physiol 132(1):44–51
Wan Y, Li J, Ren H, Huang J, Yuan H (2014) Physiological investigation of gold nanorods toward watermelon. J Nanosci Nanotechnol 14(8):6089–6094
Wu F, You Y, Werner D, Jiao S, Hu J, Zhang X, Wan Y, Liu J, Wang B, Wang X (2020) Carbon nanomaterials affect carbon cycle-related functions of the soil microbial community and the coupling of nutrient cycles. J Hazard Mater 390:122144
York LM, Carminati A, Mooney SJ, Ritz K, Bennett MJ (2016) The holistic rhizosphere: integrating zones, processes, and semantics in the soil influenced by roots. J Exp Bot 67 (12):3629–3643
Zheng M, Huang Z, Ji H, Qiu F, Zhao D, Bredar AR, Farnum BH (2020) Simultaneous control of soil erosion and arsenic leaching at disturbed land using polyacrylamide modified magnetite nanoparticles. Sci Total Environ 702:134997
Zhu B, Gutknecht JL, Herman DJ, Keck DC, Firestone MK, Cheng W (2014) Rhizosphere priming effects on soil carbon and nitrogen mineralization. Soil Biol Biochem 76:183–192
Zhu XG, Lynch JP, LeBauer DS, Millar AJ, Stitt M, Long SP (2016) Plants in silico: why, why now and what? An integrative platform for plant systems biology research. Plant Cell Environ 39 (5):1049–1057

New Strategies for Commercialization of Microbial Technologies

23

Pritam Patil and Angana Sarkar

Abstract

Since the last few decades, many discoveries have been taken place in the field of microbiology and its applications in various fields like agriculture, pharmaceuticals, sewage treatments, healthcare, etc. But the main problem is to commercialize these new technologies on large scale for welfare of society. Many hurdles are present in the way of this commercialization, foremost being the lack of awareness, use of same old strategies which are laborious, time consuming, expensive and also lack of resources. Therefore, there is a need of new technologies to be introduced for commercialization. For example, governments of many countries have taken initiative against the problem of antimicrobial resistance (AMR), i.e. microbial resistance to antibiotics. In this approach, they conducted a national consultation to give a platform for researchers to meet with industries and share their technologies. Many governments have started giving funding for new technological projects. Vendors also have started new strategies to promote new technologies by providing new platforms, resources. In the field of agriculture, new technologies like recombinant crops for insect resistance, disease resistance are developed but are not commercialized due to lack of strategy. For this, main strategy is to make farmers aware about the new developments. In this chapter, such new strategies developed by governments, industries and other organizations are discussed.

Keywords

Microbial technologies · Commercialization · Technology transfer

P. Patil · A. Sarkar (✉)
National Institute of Technology, Rourkela, Odisha, India
e-mail: sarkara@nitrkl.ac.in

R. Soni et al. (eds.), *Microbiological Activity for Soil and Plant Health Management*, https://doi.org/10.1007/978-981-16-2922-8_23

23.1 Introduction

Ever speculated, why milk turns into curd or why dough gets bubbles after keeping it for overnight or why there is smell from ripen vegetable. In all these phenomena, one thing is common, i.e. microorganisms. And the field which studies these microorganisms is microbiology and application of this field to various other fields using proper technology is called as microbial technology. Since after the discovery of microscope by Antonie van Leeuwenhoek in 1676, field of microbiology emerged as potential field of study (Lane 2015). Then Robert Hooke made his first recorded observation of moulds in fruits. Then after that field of bacteriology emerged when Ferdinand Cohn observed bacteria B*acillus* and *Beggiatoa*. Then Louis Pasteur considered as father of modern microbiology after his contribution of the process of pasteurization and vaccine of anthrax, fowl cholera and rabies. He also gave the theory of spontaneous generation. Robert Koch is considered as father of medical microbiology for his contribution of germ theory of disease which states that every disease is caused by a pathogenic organism. He was the first one to isolate pure culture of bacteria (Madigan and Martinko 2006). After their discoveries, microbiology became a field of extensive research. And now we know that, behind production of curd, *Lactobacillus* is responsible. Bubbles formed in the batter are due to formation of carbon dioxide by microbes during their growth. Hence from our home to various industrial products like pharmaceuticals, beverages, etc., microbiology is wide spread (https://lib.guides.umbc.edu/c.php?g=836720&p=6561140#s-lg-box-wrapper-24463833).

In a paper, the time of microbiology has been divided into 3 golden eras. The first golden era, i.e. second half of the nineteenth century, discovery of pathogens which caused diseases in humans and animals takes place. Also, bacterial physiology which contains their cultivation techniques, identification, their classification was done. Also, there was some research done on medical microbiology like viruses were identified, concept of vaccination, etc. In the second golden era, i.e. first half of the twentieth century, research focussed mainly on bacterial genetics. Their genetic material, mechanism of gene expression, etc., were identified. Also, research was done on the membrane transport in microbes. In the third golden era, i.e. the second half of the twentieth century, research focusses on genetic manipulation, microbial ecology, role of microbes in chronic diseases. Also, research was done on microbial physiology, i.e. mechanism of their signal transduction, interaction between microbes were found out (Maloy 2006).

Nowadays, microbiology is applied to almost each and every field like agriculture, geology, genetic engineering, food industry, beverage, fuel industry, medical and health care, etc. Also, microbiology is spread into various environments from forests to hot springs, from soil to volcanos, from rivers to oceans. And due to this feature, there are vast environments which are still remain to discover for their microbiological research (https://www.labmanager.com/ask-the-expert/ask-the-expert-what-s-changed-in-microbiology-over-the-past-decade-7288).

As microbiology is widespread in almost every field, many products are being formed every day. But, only production of a product does not solve its purpose and is not valuable unless it is commercialized in the market using a proper procedure of commercialization. And to make this process smooth and easy, many acts have been developed. Commercialization of a product starts when a business identifies the need of market and uses scientific and engineering aspects to fulfil that need (Michael et al. 1990). For the convenience of both, consumer and producer, many rules and regulations have been developed like consumer safety act, patent acts, intellectual property rights, acts for genetically modified organism or plants, etc. To ensure the safety of a new invention, an act is made that is The Patent Act, 1970. This act encompasses all the necessary guidelines and measures that need to take to avoid larceny of a person's ideas or invention. This act also includes whole process of patenting a product or an idea in India, so that whole process will be smooth and hustle free (http://www.ipindia.nic.in/writereaddata/Portal/IPOAct/1_31_1_patent-act-1970-11march2015). Then, to ensure safety of consumer, The Consumer Protection Act, 1986, was made. By this act, consumer can complaint regarding a product to government so that proper testing and checking of that product shall be done. This act gives consumer some power to make sure their safety. The revision made in 2019, made consumer to complaint regarding a product from anywhere in India through consumer affairs portal (https://consumeraffairs.nic.in/acts-and-rules/consumer-protection). For genetically modified organism or plants, there has been separate act made so that there will be a control over use, manufacture or sale of genetically modified organisms or plants. This act is Environment Protection act, 1989. And there has been made some additions to it for genetic modifications. Then Recombinant DNA guidelines, 1990 was made in addition to first law (Ahuja 2018).

In this chapter, we will be focussing on advancements in microbial technology, basic process of commercialization of a product made from using various microbial technologies, its current scenario, problems associated with current strategies and therefore some new strategies that can be made to make this process simple and effective.

23.2 Advancements in Microbial Technologies

Microbiology, due to its potential and possible applications in almost every field, has made many advancements in recent years. Also, due to large improvements in genetic engineering, bioprocess technologies, microbial technology has gained much of the interest. After the invention of microscope by Antonie van Leeuwenhoek in 1676, microbiology has never seen any setbacks. As the advancements are being made, similarly commercialization of microbial products also taking place at high rates. And still, this field is attracting many researchers from the globe, which is helping in more and more development. Most of the development is made in the field of applied microbiology in agriculture, health care, fermentation technologies, food technology, etc. And nowadays, advancement is going on in the field of microbes in nanotechnology, sustainable energy sources, etc.

Agriculture is the most benefitted field due to advancements in microbial technology. As, many new beneficial microbes have been identified. Also there have been many advancements in their detection, identification and application technologies. Microbes are identified and applied in many ways like biopesticides, biofertilizers, nitrogen fixing bacteria, plant growth promoting rhizobacteria, etc. In the field of agriculture microbiology, major advancement has been occurred in analysing and understanding the interaction between plants and microbes (Bhattacharyya and Jha 2012). Along with that, development has been done in production technologies. Fermentation, bioprocess technology and recombinant DNA technology are the main areas, whose developments have great impact on development in agriculture microbiology. Advancement in fermentation technology has enabled production of biofertilizers and biopesticides at both small and large scales (Borkar 2015). And due to these advancements, commercialization has become easy. Both solid state and liquid fermentation technologies have developed, but liquid fermentation is the most widely used production technology due to its properties like flexibility, wide range of raw materials, easy scale up, high productivity, etc. (Hölker et al. 2004).

Another field that has been most benefitted is health care. Many new drugs, detection techniques, vaccines are being developed day by day. Nowadays many drugs are being produced using microbes as production factories, such field is called biopharma. The first application of microbiology in the field of heath care was the antibiotics when Alexander Fleming discovered Penicillin G from the mould of *Penicillium* genus (Tan and Tatsumura 2015). Since then, antibiotics are produced using various fungi, bacteria, etc. Nowadays, most of the focus is on recombinant DNA technology, which helps in maximizing the productivity, ease of the process, specificity, etc. Many human proteins are produced in microbes using rDNA technology. *Escherichia coli* is the most commonly used host for production, as it is the most studied microorganism (Maya et al. 2017). rDNA technology in the field of microbial technology flourished after the discovery of Humulin, which is human insulin produced by using E. coli (Celeste et al. 2012). In the field of clinical microbiology also many advancements have been done like developments of kits for early and less time-consuming detection of a disease. New technologies are also used in this field to have precise, correct and rapid diagnostic of a disease, like mass spectrometry, polymerase chain reaction (currently being used in detection of COVID-19 detection), cell sorters, chromatographic technologies, fluorescence technologies, etc. For example, MALDI-TOF spectrometry is used in clinical microbiology in the twenty-first century to identify microbial species. It can precisely differentiate between closely related species (Oviañoa and Rodríguez-Sánchez 2020). There is a new concept evolving these days as mobile microbiology. In this concept, molecular methods are used to detect microbes that cause infectious disease. This will help us in early detection and rapid treatments (Sanguinetti et al. 2020).

There have been many developments in the field of sustainable energy development. As non-renewable sources are being depleted day by day, there is a need of new methods to produce fuels. Algae are being used for production of biofuels due to their high fat storing capacity. Many countries have already approved algae for their use in production of biofuels. Biodiesel is the main fuel produced using algal biomass (Lin et al. 2011). As algae being photosynthetic, require very less cost and resources for their growth and also have high biomass production rates along with that fat storing capacity is more in algae than other microbes. After the processing, algal biomass can be utilized as cattle feed (GuanHua et al. 2010). Hence there have been many developments to produce biofuels using algal biomass. Firstly, developments in its production are done so to have higher biomass, e.g. use of ponds instead of large glass vessels, photobioreactors, media optimization for the large tanks, etc. Many developments are made in purification also, as microalgae is mostly used its purification is time consuming and very expensive. So, new techniques like centrifugation, electrochemical harvesting, etc., are used. Process optimization is also done to have higher fat content. Then, detection techniques are also improved like Chromatography (Miao and Wu 2006). In the context of clean energy, hydrogen is considered as an important source when it is produced by the process of green technology which is the most environmentally way to produce it. Hence, microbial fuel cells are in the phase of development and results are showing promising values (Maeda et al. 2012). Also, biogas is another option for clean energy, which is produced by microbes after anaerobic fermentation of agricultural wastes. It is a good alternative for fossil fuels which involves production of carbon dioxide and other greenhouse gases responsible for air pollution (Rasimphi et al. 2019). Bioremediation, i.e. removal of pollutants using biological entities, mostly microbes, is a new field that is developing these days. As it does not produce any harmful substances but also produces value added products (Dangi et al. 2019).

23.3 Steps of Commercialization of Microbial Product: Lab to Market

As shown in Fig. 23.1, a basic process of commercialization starts when the producer thoughts of a product. For example, a product is a drug to be used in healthcare. Then steps will be as follows:

First, isolation of specific microbes has to be done which will have capacity of producing that product. Then screening for the microbe with high production ability has to be done. Screening can be done with various methods. After selection of a specific microbe, process has to be optimized for the production. If there is need to modify the microbe genetically, then specific modifications have to be done. Then after that, again optimization of process parameters for genetically modified microbes will be done. Then, lab production will be done. And then, process will be transferred to the pilot scale. After studying on the pilot scale, product has to be analysed for various aspects of safety and then registration of the product will be done for its further production at industrial scale. Then drug will be released in the market and post market study will be done so that its efficacy and market survival will be studied. In this way, commercialization of a microbial product is done.

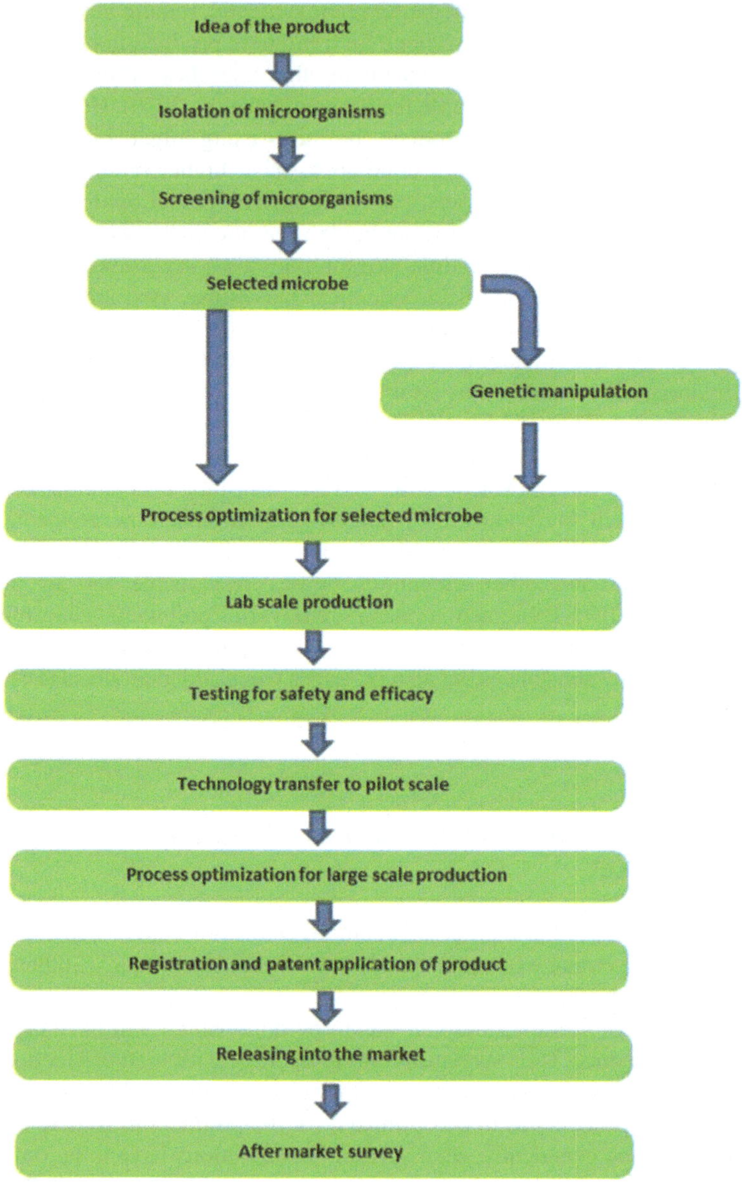

Fig. 23.1 Steps of commercialization of microbial product

23.4 Current Scenario of Commercialization

Currently many products are being commercialized world-wide and many of them are produced by using microbes. Many antibiotics, drugs, health care products, beverages, etc., are being produced by microbes. Table 23.1 gives an overview of some products synthesized using microbes and are used in various fields. Nowadays, due to increase in the awareness about microbiology, products developed from microbes have gathered much interest and are also readily accepted by consumer. In India, from the last decade, many companies have commercialized their products and technologies as shown in Table 23.2.

Many products have been developed in the field of agriculture, specially biopesticides and bio stimulants. As the microbes do not show any side effects towards the growth of plant and also do not have any harmful effects to the soil after prolonged use. Hence their use had been increased from last few years (Ravensberg 2011). Also, wide variety of microbes can be used like fungi, bacteria, algae, etc., hence products are also of wide variety. Fungi can be used for disease control, pest control, etc. Bacteria can be used for pest control, improvement of growth of plant, improving soil quality, etc. (Vega et al. 2009). Some viruses are also used for control of insects. For example, baculoviruses are used to control attack of caterpillars and codling moths. In some countries, yeasts are also used to control diseases (Ravensberg 2014). In India, during last few years there are many products that have been commercialized, for example, NEMATOX by Sri Biotech India Ltd. is a formulation of *Paecilomyces lilacinus* to control nematode. India is the 12th largest producer of chemical pesticides and these are extensively used to fulfil our ever-increasing demands. India has the Insecticide Act, 1968 section 9(3) to regulate production and commercialization of biopesticides. Under this section, 287 pesticides were registered, out of which, only 4.2% pesticides are biopesticides (Vendan 2016). Currently, more focus is on environmental safety, so considering this objective European Union and some other countries have reviewed their regulations. European countries have established a new law 91/414/CEE, which reduced the number of active substances from 900 to 400 (http://europa.eu.int/comm/food). In the United States, pesticide registration is governed by Environmental Protection Agency (EPA), which is under the Federal Insecticide, Fungicide and Rodenticide Act, 1988. This is to prevent adverse effects on consumers.

Some examples from the field of heath care are given in Table 23.2 like Compactin is a product used as an agent to lower the cholesterol level in the body (Blunt et al. 2014). Taxol is a compound which has antitumour and antifungal activities and is derived from *Sorangium cellulosum* (Newman et al. 2000). Hence, microbiology has very wide applications in many fields.

Table 23.1 Examples of products produced using microbial technology in various fields and their applications

S. no.	Product	Field	Microorganism	Application	Reference
1	Amylases	Food technology	*Aspergillus niger*	Replacement of chemical additives for treatment of wheat flour	Adejuwon et al. (2017)
2	Amylases	Food technology	*Bacillus subtilis*		Salman et al. (2016)
3	Lipase	Food technology	*Rhizomucor miehei*	Hydrolysis of milk fat. Enhances the aroma of beverages	Sharma et al. (2001)
4	Cellulases	Food technology	*Cladosporium sphaerospermum*	Degradation of plant biomass	Andersen et al. (2016)
5	Single cell protein	Food technology	*Saccharomyces, Candida and Rhodotorula spp.*	Supplementary protein	Patelski et al. (2015)
6	Citric acid	Chemical microbiology	*Aspergillus niger*	Food additive	Wang et al. (2016)
7	Acetone, butanol	Chemical microbiology	*Clostridium spp.*	Industrially important acids	Wang et al. (2001)
8	Biogas	Environmental microbiology	*Methanosarcina barkeri, M. frisius and Methanobacterium formicicum*	Sustainable fuel	Satpathy et al. (2016)
9	Peroxidases	Environmental microbiology	*P. syringae*	Removal of phenols during wastewater treatment	Kampmann et al. (2014)
10	Polyphenol oxidases	Environmental microbiology	*Arthromyces ramosus*	Removal of bi-phenols during wastewater treatment	
11	Tyrosinases	Environmental microbiology	*Agaricus bisporus*	Removal of chloro-phenols during wastewater treatment	
12	Laccases	Environmental microbiology	*P. cinnabarinus*	Degradation of benzopyrene	Rama et al. (1998)
13	Gold nanoparticles	Nanotechnology	*Pedomicrobium sp.*	Synthesis of gold nanoparticles inside the cell through bioaccumulation	Narayanan and Sakthivel (2010)

14	Octahedral Au nanoparticles	Nanotechnology	*Bacillus subtilis*	Synthesis of Au nanoparticles inside the cell through reduction	Thirumurugan et al. (2012)
15	Ag nanoparticles	Nanotechnology	*Klebsiella pneumoniae,* *E. coli,* *E. cloacae*	Extracellular synthesis of silver nanoparticles by using nitro reductase enzyme	Narayanan and Sakthivel (2010)
16	Polygalacturonase	Waste treatment	*Zygoascus hellenicus*	It is enzyme from class pectinase, which is used to degrade plant biomass	de Lima Damasio et al. (2010)
17	Pikromycin	Heath care	*S. venezuelae*	Antibiotic	Jung et al. (2006)
18	Ieodoglucomide C	Heath care	*Bacillus licheniformis*	Antifungal	Tareq et al. (2015)
19	Daunorubicin	Health care	*Streptomyces peucetius*	Anticancer	Giddings and Newman (2013)
20	Cahuitamycins	Biomedical	*Streptomyces gandocaensis*	Inhibition of *Acinetobacter baumannii* biofilms	Park et al. (2016)
21	Avermectins	Health care	*Streptomyces avermitilis*	Drug used treatment of onchocerciasis and lymphatic filariasis	Shen (2015)
22	Lovastatin	Health care	*Monascus ruber, Aspergillus terreus*	Anticancer	Lin et al. (2008)
23	Violacein	Textile	*Chromobacterium violaceum*	Blue pigments used in dyeing industry	Choi et al. (2015)
24	Cyclodextrin	Cosmetic	*Microbacterium terrae KNR 9*	Used to reduce the volatility of esters in perfumes and room freshener gels	Rajput et al. (2016)
25	L-lactate	Health care	*L. paracasei*	Inhibits chylomicron secretion from enterocytes and promotes lipid storage in enterocytes	Araújo et al. (2020)

Table 23.2 Some commercialized microbial products in India in recent times

S. no.	Product	Field	Organization	Application
1	Rivastigmine Transdermal patch	Health care	Sparsha Pharma International, Hyderabad	Used in efficient drug delivery. Reduces the amount of residual drug
2	Clot Specific Streptokinase	Health care	Symmetrix Biotech, Mumbai & CSIR, Chandigarh	Thrombolytic therapeutic protein
3	Hepatitis B and& DTP vaccine	Health care	Shantha Biotechnics, Hyderabad	Combined vaccine for hepatitis B and DTP
4	Revac-B	Health care	Bharat Biotech International, Hyderabad	Vaccine for hepatitis B
5	Enoxaparin	Heath care	Gland pharma, Hyderabad	Low molecular weight heparin used in bypass and other surgeries
6	GRANT	Agriculture	Sri Biotech Laboratories India, Hyderabad	Formulation of *Bacillus thuringiensis* to control pests
7	NEMATOX	Agriculture	Sri Biotech Laboratories India, Hyderabad	Formulation of *Paecilomyces lilacinus* to control nematodes
8	TEEKA	Agriculture	Sri Biotech Laboratories India, Hyderabad	Formulation of *Bacillus pumilus* to improve absorption of nutrients from soil
9	Symbion-N	Biofertilizers	T. Stanes & Company Ltd.	Consortium of soil borne microbes to increase intake of nutrient by plants
10	Cyclosporin A	Health care	Novartis, India	Immunosuppressant in organ transplant derived from *Tolypocladium inflatum*

23.5 Problems Associated with Commercialization of Microbial Products

As microbial products are very useful in day-to-day life but their commercialization has many problems associated with it. As products are extracted from microorganisms, so their safety and efficacy have to be tested vigorously, hence regulatory issues are associated with it. In the United States, European countries, there is no legal definition for biofertilizers. Every organism which has potential use in agricultural applications is considered as possible product. But in India, there is a legal framework on use and commercialization of biofertilizers. The Ministry of Agriculture introduced order in 2006 to include biofertilizers under essential commodities act, 1966 (Malusa and Vassilev 2014). Hence, due to this law there have been many restrictions on manufacture and marketing standards. (Sekar et al. 2016). In biopesticides based on *Bacillus thuringiensis,* there are many issues to commercialize them. As the plant has to be modified genetically or its formulation is made and applied on plants. But there are many regulations for commercialization of biopesticide which makes the process difficult and time consuming thus expensive. Hence regulations invisibly act as barriers for the development and commercialization of the biopesticides (Blok et al. 2006). In India, due to lack of knowledge about biopesticides and false information, market of biopesticides is low, thus making it difficult to commercialize a product (Sansinenea 2016). In case of healthcare products like drugs, antioxidants, anti-inflammatory products antibiotics, strict regulations are present as they are meant to be inserted in human body by some means. Hence strict regulations regarding their safety and efficacy are maintained. For example, astaxanthin is an antioxidant having higher efficacy than beta-carotene, is produced by some microbes and some microbes engineered to do so. But till date, only naturally occurring astaxanthin is approved by regulatory authorities for human consumption. As 'natural' and 'natural identical' label has to be given to the microbially produced astaxanthin (Capelli et al. 2013). And genetically modified organisms have higher production rate but regulations for genetically modified organism and its use in production of some products are very strict and registration of such product is expensive hence many producers do not involve in commercialization of such products (Zhang et al. 2020). Hence there are many problems associated with commercialization of microbial products in every field and market, whether it is native or genetically modified.

23.6 Need of New Strategies

Although microbial products are being used day by day, still there is need to improve and develop new strategies. Currently governments and people of many countries, like India, do not consider use of genetically modified plants in agriculture. They consider them to be against the moral and ethical values. Still, biotechnological technologies are not being used in considerable amount in microbiology. Many countries have stringent rules for commercialization of a microbial product than other chemically produced products. For example, in case of astaxanthin, many researchers have produced it in many microbes. But still, only naturally occurring astaxanthin is commercialized. Astaxanthin produced by genetically modified organisms is not accepted by regulatory authorities and also by the consumer, although they have high TRY (titre, rate, yield) values. Same is the case with other nutraceuticals, cosmetic products (Novoveská et al. 2019). Furthermore, many farmers consider chemical pesticides rather than microbial pesticides, knowing that chemicals have harmful effects on soil. This is due to lack of awareness regarding the microbial pesticides. Government has a big role in this process, as they should encourage people to go for microbial pesticides rather than chemical. Also, there are no stringent rules regarding the minimum cell count of microbes in the formulation (Villaverde et al. 2014). Hence, many pesticides do not work when applied to crops. Hence there should be strict rules and regulations for the same and also there should be an authority which will monitor all these processes.

In case of healthcare field, there are very stringent rules applied to any products. This is necessary because those products are directly ingested by the consumer so they have to be 99.99% pure and without any contaminations. But the time required by these products to reach to the market is very high. On a regular basis, a vaccine takes around 8–9 years before it could reach the consumer (https://bioprocessintl.com/2019/frameworks-and-strategies-for-commercialization-success-in-the-biopharmaceutical-ecosystem/). So, this time can be reduced by using new strategies. Many companies face the problem of finding representatives on whom a drug can be tested. So, there should be a process by which people can willingly participate in this process.

Hence to answer all these problems in commercialization, there is need to develop new strategies and implement them in practice. So that the whole process becomes smooth and less expensive.

23.7 Future Prospects

Although, microbial technologies have developed on a large extent so far. But still there is lack of awareness, research, discoveries in this field. This field has wide applications in many other fields but is limited due to less research and techniques. Also, many new rules and regulations should be formed to encourage this field. As this field has no harmful effects on the nature and human health, hence this should be encouraged. New strategies to commercialize products are to be made and practiced. If microbial technology gets more developed, then there will many new products that can be discovered from locations that were never researched before. Many new organisms with new properties beneficial to human and also to nature can be identified. But on the contrary part, there should be strict rules on use of such techniques. Because genetic engineering can be used for many harmful applications which can be dangerous to us and mainly to nature.

References

Adejuwon AO, Oluduro AO, Olutiola PO, Segal SJ (2017) Partial purification of α-amylase expressed by a tropical mutant strain of Aspergillus niger IFE 08. Int J Microbiol Res 3:001–004. https://doi.org/10.18685/ejmbr(3)1_ejsr-16-024

Ahuja V (2018) The regulation of genetically modified organisms in India. BMC Proc:14. https://doi.org/10.1186/s12919-018-0106-0

Andersen B, Poulsen R, Hansen GH (2016) Cellulolytic and xylanolytic activities of common indoor fungi. Int Biodeterior Biodegradation 107:111–116. https://doi.org/10.1016/j.ibiod.2015.11.012

Araújo J, Tazi A, Defranoux O, Vichier-Guerre S, Nigro G, Licandro H, Demignot S, Sansonetti P (2020) Fermentation products of commensal bacteria alter enterocyte lipid metabolism. Cell Host Microbe 27:314–316. https://doi.org/10.1016/j.chom.2020.01.028

Bhattacharyya PN, Jha DK (2012) Plant growth-promoting rhizobacteria (PGPR): emergence in agriculture. World J Microbiol Biotechnol 28:1327–1350. https://doi.org/10.1007/s11274-011-0979-9

Blok A, Jensen M, Kaltoft P (2006) Regulating pesticide risks in Denmark: expert and lay perspectives. J Environ Policy Plan 8:309–330. https://doi.org/10.1080/15239080601084687

Blunt JW, Copp BR, Keyzers RA, Munro MHG, Prinsep MR (2014) Marine natural products. Nat Prod Rep 31:160–258. https://doi.org/10.1039/c3np70117d

Borkar SG (2015) Microbes as biofertilizers and their production technology. Wood Head Publishing, New Delhi, pp 7–153. https://doi.org/10.1201/b18879

Capelli B, Jenkins U, Cysewski GR (2013) Role of astaxanthin in sports nutrition. In: Bagchi D, Nair S, Sen CK (eds) Nutrition and enhanced sports performance. Academic Press, San Diego, pp 465–471. https://doi.org/10.1016/b978-0-12-396454-0.00048-5

Celeste C, Quianzon I, Cheikh (2012) History of insulin. J Commun Hosp Intern Med Perspect:2. https://doi.org/10.3402/jchimp.v2i3.19081

Choi S, Kim S, Lyuck S, Kim S, Mitchell R (2015) High-level production of violacein by the newly isolated *Duganella violaceinigra* str. NI28 and its impact on *Staphylococcus aureus*. Sci Rep 5:15598. https://doi.org/10.1038/srep15598

Dangi AK, Sharma B, Hill RT, Shukla P (2019) Bioremediation through microbes: systems biology and metabolic engineering approach. Crit Rev Biotechnol:79–98. https://doi.org/10.1080/07388551.2018.1500997

de Lima Damasio AR, da Silva TM, Maller A et al (2010) Purification and partial characterization of an exo-polygalacturonase from Paecilomyces variotii liquid cultures. Appl Biochem Biotechnol 160:1496–1507. https://doi.org/10.1007/s12010-009-8682-0

Giddings LA, Newman DJ (2013) Microbial natural products: molecular blueprints for antitumor drugs. J Ind Microbiol Biotechnol 40:1181–1210. https://doi.org/10.1007/s10295-013-1331-1

GuanHua H, Chen F, Wei D, Zhang W, Chen G (2010) Biodiesel production by microalgal biotechnology. Appl Energy 87:38–46. https://doi.org/10.1016/j.apenergy.2009.06.016

Hölker U, Höfer M, Lenz J (2004) Biotechnological advantages of laboratoryscale solid-state fermentation with fungi. Appl Microbiol Biotechnol:175–186. https://doi.org/10.1007/s00253-003-1504-3

Jung WS, Lee SK, Hong JSJ, Park SR, Jeong SJ, Han AR et al (2006) Heterologous expression of tylosin polyketide synthase and production of a hybrid bioactive macrolide in Streptomyces venezuelae. Appl Microbiol Biotechnol 72:763–769. https://doi.org/10.1007/s00253-006-0318-5

Kampmann M, Boll S, Kossuch J, Bielecki J, Uhl S, Kleiner B et al (2014) Efficient immobilization of mushroom tyrosinase utilizing whole cells from Agaricus bisporus and its application for degradation of bisphenol A. Water Res 57:295–303. https://doi.org/10.1016/j.watres.2014.03.054

Lane N (2015) The unseen world: reflections on Leeuwenhoek (1677) 'concerning little animal'. Philos Trans R Soc Lond Ser B Biol Sci 370(1666):20140344. https://doi.org/10.1098/rstb.2014.0344

Lin YL, Wang TH, Lee MH et al (2008) Biologically active components and nutraceuticals in the Monascus-fermented rice: a review. Appl Microbiol Biotechnol 77:965–973. https://doi.org/10.1007/s00253-007-1256-6

Lin L, Cunshan Z, Vittayapadung S, Xiangqian S, Mingdong D (2011) Opportunities and challenges for biodiesel fuel. Appl Energy 88:1020–1031. https://doi.org/10.1016/j.apenergy.2010.09.029

Madigan M, Martinko J (2006) Brock biology of microorganisms. Pearson Education, London, p 1096. https://doi.org/10.1002/1521-3846(200111)21:4%3C369::aid-abio369%3E3.0.co;2-7

Maeda T, Sanchez-Torres V, Wood TK (2012) Hydrogen production by recombinant Escherichia coli strains. Microb Biotechnol 5:214–225. https://doi.org/10.1111/j.1751-7915.2011.00282.x

Maloy S (2006) The era of microbiology: a Golden Phoenix. Int Microbiol 2006:1–7. https://doi.org/10.1002/9780471729259.mcfores00

Malusa E, Vassilev N (2014) A contribution to set a legal framework for biofertilisers. Appl Microbiol Biotechnol 98:6599–6607. https://doi.org/10.1007/s00253-014-5828-y

Maya V, Idalia N, Bernardo F (2017) Escherichia coli as a model organism and its application in biotechnology. In: Escherichia coli - recent advances on physiology, pathogenesis and biotechnological applications. IntechOpen, London. https://doi.org/10.5772/67306

Miao X, Wu Q (2006) Biodiesel production from heterotrophic microalgal oil. Bioresour Technol 97:841–846. https://doi.org/10.1016/j.biortech.2005.04.008

Michael T, Summe G, Uttal B (1990) Commercializing technology: what the best companies do. Harv Bus Rev 7:154–163. https://doi.org/10.1016/0737-6782(90)90083-q

Narayanan KB, Sakthivel N (2010) Biological synthesis of metal nanoparticles by microbes. Adv Colloid Interf Sci 156:1–13. https://doi.org/10.1016/j.cis.2010.02.001

Newman DJ, Cragg GM, Snader KM (2000) The influence of natural products upon drug discovery. Nat Prod Rep 17:215–234. https://doi.org/10.1039/a902202c

Novoveská L, Ross ME, Stanley MS, Pradelles R, Wasiolek V, Sassi J-F (2019) Microalgal carotenoids: a review of production, current markets, regulations, and future direction. Mar Drugs 17(11):640. https://doi.org/10.3390/md17110640

Oviañoa M, Rodríguez-Sánchez B (2020) MALDI-TOF mass spectrometry in the 21st century clinical microbiology laboratory. Clin Microbiol Infect. https://doi.org/10.1016/j.eimc.2020.02.027

Park SR, Tripathi A, Wu J, Schultz PJ, Yim I, Mcquade TJ et al (2016) Discovery of cahuitamycins as biofilm inhibitors derived from a convergent biosynthetic pathway. Nat Commun 7:10710. https://doi.org/10.1073/pnas.1716245115

Patelski P, Berlowska J, Dziugan P, Pielech-Przybylska K, Balcerek M, Dziekonska U (2015) Utilisation of sugar beet bagasse for the biosynthesis of yeast SCP. J Food Eng 67:32–37. https://doi.org/10.1016/j.jfoodeng.2015.03.031

Rajput K, Patel K, Trivedi U (2016) A novel cyclodextrin glucanotransferase from an alkaliphile microbacterium terrae KNR 9: purification and properties. 3 Biotech 6:168. https://doi.org/10.1007/s13205-016-0495-6

Rama R, Mougin C, Boyer FD, Kollmann A, Malosse C, Sigoillot JC (1998) Biotransformation of benzo[a] pyrene in bench scale reactor using laccase of Pycnoporus cinnabarinus. Biotechnol Lett 20:1101–1104. https://doi.org/10.1023/A:1005387016390

Rasimphi TE, Tinarwo D, Gitari WM (2019) Potential of biogas technology in achieving the sustainable developmental goals: a review through case study in rural South Africa. In: Hussain CM (ed) Handbook of environmental materials management. Springer, Cham. https://doi.org/10.1007/978-3-319-73645-7

Ravensberg W (2011) A roadmap to the successful development and commercialization of microbial pest control products for control of arthropods. Springer, Dordrecht. https://doi.org/10.1007/978-94-007-0437-4_7

Ravensberg W (2014) Commercialisation of microbes: present situation and future prospects. In: Principles of plant-microbe interactions. Springer, Cham, pp 309–317. https://doi.org/10.1007/978-3-319-08575-3_32

Salman T, Kamal M, Ahmed M, Siddiqa SM, Khan RA, Hassan A (2016) Medium optimization for the production of amylase by Bacillus subtilis RM16 in shake-flask fermentation. Pak J Pharm Sci 29:439–444

Sanguinetti M, Seme K, Poljak M (2020) Mobile microbiology: an evolving concept in diagnosis of infectious diseases. Clin Microbiol Infect:409–410. https://doi.org/10.1016/j.cmi.2019.12.016

Sansinenea E (2016) Regulatory issues in commercialization of *Bacillus thuringiensis*-based biopesticides. In: Agriculturally important microorganisms. Springer, Singapore, pp 69–80. https://doi.org/10.1007/978-981-10-2576-1_4

Satpathy P, Steinigeweg S, Cypionka H, Engelen B (2016) Different substrates and starter inocula govern microbial community structures in biogas reactors. Environ Technol 37:1441–1450. https://doi.org/10.1080/09593330.2015.1118559

Sekar J, Raj R, Prabavathy V (2016) Microbial consortial products for sustainable agriculture: commercialization and regulatory issues in India. In: Agriculturally important microorganisms. Springer, Singapore, pp 107–132. https://doi.org/10.1007/978-981-10-2576-1_7

Sharma R, Chisti Y, Banerjee UC (2001) Production, purification, characterization, and applications of lipases. Biotechnol Adv 19:627–662. https://doi.org/10.1016/S0734-9750(01)00086-6

Shen B (2015) A new golden age of natural products drug discovery. Cell 163:1297–1300. https://doi.org/10.1016/j.cell.2015.11.031

Tan SY, Tatsumura Y (2015) Alexander Fleming (1881–1955): discoverer of penicillin. Singap Med J 56:366–367. https://doi.org/10.11622/smedj.2015105

Tareq FS, Lee HS, Lee YJ, Lee JS, Shin HJ (2015) Ieodoglucomide C and ieodoglycolipid, new glycolipids from a marine-derived bacterium *Bacillus licheniformis* 09IDYM23. Lipids 50:513–519. https://doi.org/10.1007/s11745-015-4014-z

Thirumurugan A, Ramachandran S, Tomy NA, Jiflin GJ, Rajagomathi G (2012) Biological synthesis of gold nanoparticles by *Bacillus subtilis* and evaluation of increased antimicrobial activity against clinical isolates. Korean J Chem Eng 29:1761–1765. https://doi.org/10.1007/s11814-012-0055-7

Vega F, Goettel M, Blackwell M (2009) Fungal entomopathogens: new insights on their ecology. Fungal Ecol 2:149–159. https://doi.org/10.1016/j.funeco.2009.05.001

Vendan S (2016) Current scenario of biopesticides and eco-friendly insect pest management in India. South Indian J Biol Sci 2:268–271. https://doi.org/10.22205/sijbs/2016/v2/i2/100315

Villaverde J, Sevilla-Morán B, Sandín-España P, López-Goti C, Alonso-Prados J (2014) Biopesticides in the framework of the European Pesticide Regulation (EC) no. 1107/2009. Pest Manag Sci 70:2–5. https://doi.org/10.1002/ps.3663

Wang Z, Zhuge J, Fang H, Prior BA (2001) Glycerol production by microbial fermentation: a review. Biotechnol Adv 19:201–223. https://doi.org/10.1016/S0734-9750(01)00060-X

Wang L, Cao Z, Hou L, Yin L, Wang D, Gao Q et al (2016) The opposite roles of agdA and glaA on citric acid production in Aspergillus niger. Appl Microbiol Biotechnol 100:5791–5803. https://doi.org/10.1007/s00253-016-7324-z

Zhang C, Chen X, Too H (2020) Microbial astaxanthin biosynthesis: recent achievements, challenges, and commercialization outlook. Appl Microbiol Biotechnol 104(13):5725–5737. https://doi.org/10.1007/s00253-020-10648-2

Techniques for Improving Microbial Inoculants as a Tool for Sustainable Development

24

Abhishek Walia, Chayanika Putatunda, Ravi Kant Bhatia, Preeti Solanki, and Shruti Pathania

Abstract

Increased use of chemical fertilizers to support crop production has resulted in global soil, water, and air pollution. It is generally agreed that the only solution is to improve the exploitation of beneficial bacteria that play a key role in increasing the supply of inaccessible minerals to plants in uninterrupted soil conditions. Main attention is paid to biotechnologies that are not completely used as tools for the production of biofertilizers, such as microbial co-immobilization and co-cultivation. Biotechnological processing and combined use of active microorganisms/organic compounds (biostimulants) such as plant extracts and exudates, not only promotes plant growth and development but also plant-bacterial interactions. This chapter focuses on the improvement in the techniques for development and formulation of microbial inoculants as well as the most significant potential and innovative strategies in this area are discussed.

A. Walia (✉)
Department of Microbiology, College of Basic Sciences, CSKHPKV, Palampur, Himachal Pradesh, India
e-mail: abhishek@hillagric.ac.in

C. Putatunda
Department of Microbiology, Om Sterling Global University, Hisar, Haryana, India

R. K. Bhatia
Department of Biotechnology, Himachal Pradesh University, Shimla, Himachal Pradesh, India

P. Solanki
Multi-Disciplinary Research Unit, Pt. B. D. Sharma Institute of Medical Sciences, Rohtak, Haryana, India

S. Pathania
Dr. YSP University of Horticulture and Forestry, Nauni, Himachal Pradesh, India

R. Soni et al. (eds.), *Microbiological Activity for Soil and Plant Health Management*, https://doi.org/10.1007/978-981-16-2922-8_24

Keywords

Bioinoculants · Biofertilizers · Carriers · Formulations

24.1 Introduction

Chemical fertilizers have been widely used to achieve more yields in recent years. This intense implementation has raised questions about agro-environmental contamination and has led to a decline in the quality of goods. Excessive use of chemicals like phosphate and nitrogen fertilizers not only accelerates water eutrophication and acidification of soils, but groundwater, rivers, and the environment can also be polluted. In addition, chemical-based agriculture has had a detrimental effect on beneficial soil microbial ecosystems, dramatically diminishing microbial biodiversity. Environmental problems such as clean water pollution, energy conservation, and soil erosion compel farmers to adopt planting methods that have a lower impact on the environment. In this context, reduced use of chemical fertilizers with increased use of organic fertilizers should be considered as compulsory mitigation measures from agricultural activities. In recent years, a variety of organic fertilizers have been launched, which also serve as natural stimulators for plant growth and development (Khan et al. 2009). A specific category of fertilizer of this type consists of products based on plant growth-promoting microorganisms (PGPM). Arbuscular mycorrhizal fungus (AMF), plant growth-promoting rhizobacteria (PGPR), and nitrogen-fixing rhizobia, commonly classified as PGPR, are considered beneficial for plant nutrient cultivation in three major categories of microorganisms (Gousterova et al. 2008). Depending on their application, different microbial inoculants may be classified into various groups, even if the precise meaning of these categories is still not clear. However, the biofertilizer group is most commonly used in products containing microorganisms that increase the availability and absorption of plant nutrients (such as rhizobia and mycorrhizal fungi) (Mehta et al. 2015; Guleria et al. 2014). Increased interest developed in the use and application of these products is growing due to the increase in the efficiency of nutrient absorption and the demands of society for more green production technologies, increasing the cost of agrochemicals. In addition, secondary positive effects of biofertilizers and phytostimulators that would improve their utility as bio-inoculants. A suitable inoculum, that is, a microbial fertilizer (biofertilizer) containing organic matter that can colonize the rhizosphere and increase plant growth, should be developed to maximize the benefits provided by PGPM. Multifunctional biofertilizers have been produced to decrease the application of chemical fertilizers by around 1/3–1/2 (Sharma et al. 2015; Walia et al. 2013). However, it should be observed that microbial fertilizers have not been usually accepted by farmers, since their beneficial effects are also difficult to duplicate. It is very much strongly clear that if microbial fertilizers are not correctly prepared, formulated, and/or administered, no advantages can be extracted from the biofertilizer. Due to better biotechnological production systems, numerous private firms are now selling commercial inoculants on the foreign market, both citing a large improvement in production and product efficiency for a wide-ranging variety of crops. Quite often, however, inoculants are of low consistency.

There have been examples of inoculant products sold in both developed and developing countries that do not have rhizobia or show high levels of contamination with other species (Herrmann and Lesueur 2013). Clearly, the failure of biofertilizers to perform their basic functions on any application scale indicates problems associated with the production and construction of the inoculum. There are several steps in the production of a viable inoculant: (1) Selection of a suitable culture and isolation of an efficient and successful indigenous multifunctional microorganism for a particular soil-plant system (or crop variety); (2) characterization of the selected microorganisms in an ideal medium with suitable conditions for cultivation and growth; (3) multiplication of microbial mass; (4) choice of carrier; (5) creation of a method of formulation to ensure microbial persistence, even under stressful conditions, in soil environments; (6) field application studies; (7) large-scale industrial-level studies and production; (8) transfer of biotechnological know-how to the industrial level in the production of biofertilizers; and (9) construction of a system for quality control and storage (Vassilev et al. 2015; Walia et al. 2017). In order for the process to lead to the quality microbial fertilizer you want, each of these steps requires equal attention. This chapter provides an overview of present-day studies and prevailing technology for improving microbial fertilizer manufacturing and soil bio-fertilization, with a specific emphasis on each individual phase of the process.

24.2 Techniques for Improving Microbial Inoculant Formulations

The agrochemicals like the conventional fertilizers and pesticides, which have been used indiscriminately, may have led to short-term gains in agricultural productivity. However, in the long run, these chemicals have not only affected the soil fertility and environment, but also taken a toll on human health (Magauzi et al. 2011). On the other hand, due to the ever increasing world population and rapidly diminishing arable land, the burden on the agriculture sector is piling up. So, an effective microbial inoculant not only needs to be environmentally suitable but also should be capable of increasing the plant productivity under the field conditions. Although many microbial strains show promise in the lab scale experiments, not many of them actually prove to be effective under field conditions (Vassilev et al. 2015). Several factors like the harsh environmental conditions, unfavorable soil characteristics as well as competition from the indigenous microflora might be responsible for this (Mehta et al. 2019; Kumar et al. 2015). So, for development of an effective microbial inoculant, not only there is requirement of an efficient microbial culture, but also it is imperative to design a proper formulation. Arora et al. (2010) have suggested that a good formulation of microbial inoculant is associated with higher number as well as increased activity of the inoculated microbes in soil. Thus improving the formulations of the biofertilizers is absolutely essential for the development of microbial inoculants which are not only safe for the environment but are also capable of meeting the requirements of the farmers (Bashan et al. 2014). Keeping this in mind, various approaches like the use of carrier materials as well as cell immobilization have been proposed. Based on the physical state, the bio-inoculant formulation may be liquid or solid. The formulations may also be various forms like concentrated dry or wet dust, granules, and briquettes (Stamenković et al. 2018). The details of these approaches are being discussed in the following sections.

24.2.1 Solid Carrier Material as Inoculant Formulation

An ideal carrier material is expected to allow the microbial culture to remain alive during the entire shelf life and deliver it in appropriate physiological state, hence the carrier is expected to possess adequate moisture holding capacity, pH-buffering capacity, appropriate level of porousness, etc. It is also expected to be eco-friendly, non-toxic, easy to sterilize, easy to handle and store, and keeping in mind the product economics, it should also be cheap and readily available (Rivera-Cruz et al. 2008; Malusá et al. 2012a, b; Mehta et al. 2019). The choice of carrier is also governed by the type of microbe and the required level of its viability as well as the mode of application (Stamenković et al. 2018). The solid carriers have been broadly grouped into four major categories, viz. various types of soils (like inorganic soil, coal, peat, etc.), plant waste materials (like spent mushroom manure, farmyard manure, wheat bran, soymeal, etc.), various inert materials (like vermiculite, perlite, rock phosphate, etc.), and lyophilized microbes and oil dried bacteria (Sahu and Brahmaprakash 2016). The carriers have been used individually as well as in conjunction with others. The dried microbial cultures may also further be mixed with other carriers or may be directly applied. In these, some cryoprotectants like mannitol, microcrystalline cellulose as well as a carbon source or cellular protectant like glucose, sucrose, maltose, trehalose, molasses, glycerol, etc. are supplemented to enhance the effectiveness and shelf life of the inoculant (Garcia-Fraile et al. 2015).

Peat has been used quite commonly as a carrier earlier and provides nutrient rich and protective environment to the microbes but there are many disadvantages associated with peat (Mahanty et al. 2016). The variations in the composition of peat, availability issues as well as difficulty in sterilization leading to higher probability of contamination are some of the shortcomings (Malusá et al. 2012a, b). Talc is another carrier which has been used by many workers. Sinha et al. (2018) reported better control of *Fusarium oxysporum* f. sp. *capsici* as well as enhancement in plant height, root length, and yield in case of 1% talc-based formulation of *Trichoderma harzianum* in chilli plants. Basheer et al. (2019) reported that a formulation of talc, carboxymethyl cellulose, and calcium carbonate containing novel endophytic *Bacillus* sp. CaB_5 had a positive effect on the growth of cow pea and lady's finger plants. Also, stable microbial count was observed in the formulation even up to 45 days. Another carrier which has been gaining importance is the biochar. Biochar, which is the solid residue left after pyrolysis of biomass, is rich in nutrients, has good water holding capacity and is already used as soil amendment (Egamberdieva et al. 2018).

Pastor-Bueis et al. (2019) reported that a formulation of autochthonous *Rhizobium leguminosarum* bv. *phaseoli* strain based on biochar and perlite proved to be very promising in case of common bean.

Apart from these various plant based carriers or wastes of agricultural and industrial sectors have also been found to been assessed by different researchers. This approach also provides a better way of waste management. Wang et al. (2015) reported that mixture of wheat husk and 20% perlite resulted in enhancement of available phosphorus content in case of a biofertilizer containing *Aspergillus niger*. Wheat bran based formulation of fungal endophytes resulted in better plant growth as compared to talc-based formulation (Mastan et al. 2019). Arif et al. (2017) have reported that application of nitrogen enriched compost supplemented with plant growth-promoting rhizobacteria led to enhancement in seed quality and nutrient use efficiency in sun flower. Adeoye et al. (2019) found a formulation containing de-chromed tannery waste and egg shells to be suitable for *Bradyrhizobium japonicum*. Paliya et al. (2019) reported sludge ash to be very promising carrier material for *Rhizobium* inoculant formulation. The shelf life of the formulation was found to be quite long (150 days) and it also resulted in higher nodulation and pronounced enhancement in the growth of lentil plants.

Another new approach which has been suggested by some workers for improving the shelf life as well as decreasing the contamination of microbial inoculants is the fluid bed drying (Brahmaprakash and Sahu 2012). The technique is already commonly used in food industry. Sahu et al. (2013) have reported that in case of a microbial consortium subjected to fluid bed drying, survival of up to 180 days without any contamination was there.

24.2.2 Liquid Carrier Material as Inoculant Formulation

Liquid microbial inoculants form another important category. As per Mahanty et al. (2016), these may be "based on aqueous (broth cultures), mineral or organic oils, oil in water, or polymer-based suspensions." However, apart from the nutrients for microbes, these formulations generally have some other additives which perform various roles like act as cell protectants, help in stability of the formulation, better adhesive properties, protection after application etc. (Sahu and Brahmaprakash 2016). The liquid carriers allow easier handling, easier supplementation of appropriate additives, as well as easier application (Herrmann and Lesueur 2013; Bashan et al. 2014; Kaur and Kaur 2018). Also, the higher cell concentrations in case of liquid formulations usually lower quantity application leads to good efficiency (Mahanty et al. 2016). Some researchers have also reported longer shelf lives and lesser probability of contamination (Gopal and Baby 2016). Liquid formulations which are based on broth cannot be expected to offer protection to the microbes especially when exposed to the harsh field conditions, that is why the role of protectants becomes even more significant. Polysaccharides like resins like gum arabic, Carboxymethylcellulose, polyvinyl alcohol derivatives, glycerol, sucrose, etc. are some of the commonly used additives (Shaikh and Sayyed 2015, John et al. 2011: Mehta et al. 2013). Another advantage of the liquid formulations containing appropriate additives lies in the fact that they not only allow development of higher microbial concentrations, but also promote the formation of cysts and spores, which in turn promotes better survival of the microbes under stressful conditions. Vendan and Thangaraju (2007) reported one hundred per cent conversion of *Azospirillum* vegetative cells to cysts within 96 hours in a cyst inducing minimal salt medium. They also reported survival of cysts up to fourteen months. Moreover, the formulation also allowed survival of the microbes under high temperature as well as desiccation.

Santosh (2015) assessed the survival of the bacteria *Rhizobium*, *Azotobacter*, *Azospirillum*, and *Bacillus megaterium* in Yeast extract mannitol broth, Waksman medium No.77 broth, Dobereiner's malic acid broth with NH_4Cl (1g/l), and Pikovskaya broths, respectively, supplemented with various cell protectants (alone and in combinations) glycerol (0.5%), polyvinylpyrrolidone (PVP, 0.5%), polyethylene glycol (0.5%), gum arabic (0.5%), and sodium alginate (0.1%). It was concluded that in case of all the bacteria, the formulation with 0.5% PVP and 0.5% glycerol, maximum increase in the shelf life was observed. Neneng (2020) reported a liquid formulation comprising of 15% sugar and 85% coconut water containing two microbial consortia (designated as KHY and IBT, comprising of cellulolytic microbes, nitrogen-fixing bacteria, and phosphate solubilizing bacteria) proved to be the best.

24.2.3 Application of Cell Immobilization for Bio-inoculant Formation

Another approach for bio-inoculant formulation which is gaining popularity is the cell immobilization or the bio-immobilization technique. In this technique, the cells are mixed with appropriate polymer allowed to undergo solidification leading to the immobilization of the cells within the matrix. The cells may be further allowed to grow and then dried (Sahu and Brahmaprakash 2016). The media may further be supplemented with nutrients to enhance the shelf life of the product The greatest advantage of this approach lies in the fact that the matrix provides temporary protection to the microbes from both harsh field conditions as well as competition with indigenous microflora, and allows controlled release of microbial cells (Sahu and Brahmaprakash 2016; Stamenković et al. 2018). Some of the polymers which have been used for immobilization include agar, methoxy pectin, gellan, xanthan, alginate, and carrageenan (Vassilev et al. 2014, 2020).

Although there are several advantages of immobilization approach, there are some constraints also. One of the major problem is the higher production cost owing to the additional cost of polymeric carrier as well as labor and handling charges (Sahu and Brahmaprakash 2016; Bashan et al. 2016). Moreover, the lower stability and mechanical strength of the polymeric carriers may also impede the large-scale application of this technique. So, some researchers have suggested that this approach can be applied for co-immobilization of more than one type of microbes together (Hickert et al. 2014), which will bring down the cost of production. For improving the characteristics of the polymer, Kekez et al. (2016) have proposed the use of different polymers together (like levan and polystyrene) which combine the best features of both types of materials lead to improvement in the characteristics of the final product. Other possible solution could be looking for cheaper alternative polymers (Vassilev et al. 2020). Nano-additives might also offer some novel solutions by improving the shelf life, survival or delivery of the final product (Jampílek and Králová 2017; Prasad et al. 2017). However, all these aspects call for more in-depth investigations. In the conventional immobilization approach, a variety of additives have been used for various purposes, some of which are being discussed in the following sections.

24.2.3.1 Additives Effect on the Efficacy of Immobilized Inoculants

Whole cell immobilization techniques, applicable to both viable and non-viable cell-systems, can be broadly divided into:

1. Adsorption: physical process, done using solid/water insoluble additives.
2. Aggregation/ cross-linking/covalent bonding: there is covalent bond formed between the cells and the carrier, no matrix or capsule is used.
3. Confinement or encapsulation: cells are enclosed in a semipermeable membrane or capsule.
4. Entrapment: cells do not form any bond with the matrix, are merely trapped in it. Matrix is formed by polymerization.

The effects of immobilization of cells are observed on its physiology, mobility and its interaction with its microenvironment, including the carrier/additives acting as the support material. The chemistry of these interactions is still being studied to define the mechanisms that will predict the immobilized cell's behavior in terms of its physical and metabolic characteristics. The selection of immobilization technique will depend upon these bio-thermo-dynamic principles (Karel et al. 1985).

The criteria for development and selection of additives during the preparation of cell-based immobilized microbial inoculants are (Fig. 24.1):

a. The additives need to be economically sustainable in terms of the overall cost of the process development. The additives that act as bulking agents decrease the operational and processing cost.
b. The product must have high potential for microbial colonization in the soil and/or plant, as well as survival of microbes during storage.
c. The additive must be compatible and stable for use in both agricultural soil and plants.
d. There should be no deleterious impact of the product/additive on the nutritional status and microflora of the soil, and vice versa, preferably having benefitting potential for both soil and plant.

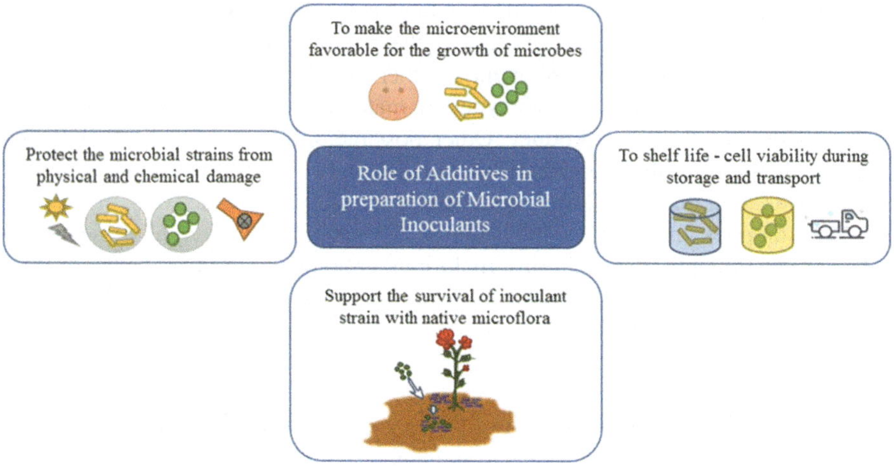

Fig. 24.1 Role of additives in preparation of immobilized microbial inoculants

For making the laboratory prepared and greenhouse tested bio-inoculants work under field conditions, their commercial sustainability and cell viability needs to be enhanced, through the role of additives. Hence, the microbial inoculant once introduced into the soil, should be able to survive and multiply in the presence of the existing microbiome of the soil and plant, and sustain a viability (10^6–10^7 cells/ plant) at which the beneficial effects may be obtained (Bashan et al. 2016; Vassilev et al. 2020).

The carriers/ additives define the type of microbial formulation prepared for immobilized cell inoculants—liquid, cell free, solid, or gel based, which are summarized in Table 24.1.

Common methods being used more often for cell immobilization in preparation of microbial inoculants are given below:

a. *Gel-cell immobilized inoculants*: This technique of bio-immobilization utilizes water soluble polymers like methoxy pectin, agar, gums (gellan, xanthan, locust bean), carrageenan and alginate as carriers, additives or encapsulation media for microbial inoculants to be applied as biofertilizers in soil-plant systems (Bashan 1998; Vassilev et al. 2014).
b. *Microencapsulation*: In this technique core material (cell/spore) is coated or entrapped within a polymer forming microspheres (size: 1–1000 μm). It increases capacity for cell loading, cell survival, and production rate. Matrix is semipermeable and cells are physically confined/ protected from external environment, keeping the internal microenvironment suitable for microbial survival (Rathore et al. 2013).
c. *Interfacial polymerization*: It is a method used for microencapsulation that utilizes the polymerization of two different reactive monomers that form the

Table 24.1 Types of microbial inoculant formulations based on carriers/additives

1. Liquid inoculants (Bashan et al. 2016; Malusà et al. 2016): a. The microbial cultures are prepared in aqueous, oil based or polymeric carriers/ additives b. Advantage: ↑dispersion capacity, stability, and viscosity of cell suspension c. Disadvantage: ↓ viability after inoculation in soil, as liquid additives get dispersed in soil	3. Solid formulations (Adholeya and Das 2012; Malusá et al. 2012a, b; Sahu and Brahmaprakash 2016): a. Carriers/ additives may be organic, inorganic, granular, powdery or solid forms b. Basis of classification is particle size or mode of application c. Example: peat, vermiculite, compost, perlite, agro-industrial wastes, calcium sulfate, rock phosphate, polysaccharides d. Advantage: ↑ cell viability and ↓ cost Disadvantage: Require more research for commercial applications.
2. Cell-free formulations (post-biotic) (Bashan et al. 2016; Mendes et al. 2017; Vassilev et al. 2017): a. It contains the beneficial extracts from microbial culture, including their metabolic by-products b. The products may include antibiotics, lytic enzymes, siderophores, toxins as insecticide/pesticide, or solubilised phosphates. Example: filtrates from fermentation broths c. Advantage: ↑ plant growth and protection d. Disadvantage: effects last for limited period only, and require repeated inoculation	4. Solid state fermentations (Vassilev and Mendes 2018): a. Uses agro-industrial wastes (solid substrates + liquid wastes) b. Advantages: two microbial strains may be co-cultivated enrichment, ↑ soluble P, ↑ bioactivity control c. Disadvantages: requires standardization 5. Gel-cell immobilized inoculants: Most recent technology, having advantages of both liquid and solid formulations

polymeric film at the point of contact / interface of their respective immiscible phases (Perignon et al. 2014).

d. *Nano-encapsulation*: It is the latest technique of cell entrapment or encapsulation, having strong potential for the future of immobilized microbial inoculants. The capsule is biologically or chemically derived and is ultrathin (<100 nm) yet sturdy. It can be used even for the non-spore forming cells. It has the advantage of cyto-compatibility and mimics the naturally occurring cell-in-shell structures. These "nanoshells" may be phenolics-based or silica-based coatings, resulting in environmentally resistant hybrids. They have the potential of cost effectiveness, increased shelf life, and protection of the microbial strain from environmental stressors. As mentioned previously, these newer age technologies need further evaluation for their effects on cells' characteristics as well as on the environment (Jampílek and Králová 2017; Park et al. 2016; Prasad et al. 2017).

The preparation of the immobilized microbial cell inoculations requires various kinds of additives or carriers or supports that are summarized in Table 24.2.

Complex formulations can be designed containing more than one type of additive to enhance the stability of the cell-shell system and increase the inoculum efficacy, like skim milk and clay materials, chitosan-starch formulation, alginate-chitosan beads, bentonite and kaolin in alginate-glycerol encapsulations, alginate-bentonite-starch microencapsulations, perlite-alginate microbeads in $CaCl_2$—paraffin emulsions, skim milk—alginate cell beads, bentonite-skim milk—alginate

Table 24.2 Types of additives and their characteristics used in preparation of immobilised microbial inoculants

S. No.	Additives	Characteristics	Examples	Microbial strains
1.	Polysaccharide or clay mineral combinations (Bashan et al. 2002; He et al. 2017; Liffourrena and Lucchesi 2018; Zohar-Perez et al. 2003)	• ↑ protection to the inoculum • Ensure controlled release of cells into environment • Act as bulking agents • Protection from UV damage • Make thicker shell for the cell • ↑ porosity • ↑ Solid content/ biomass, protein content, etc. • ↑ Cell gel mechanical stability. • ↑ Microbial colonization	Pyrophyllite, bentonite and kaolin	*Pantoea agglomerans*, *Trichoderma harzianum*, *Raoultella planticola*, *Pseudomonas putida*, *Arabidopsis thaliana*
2.	Skim milk (Bashan et al. 2002; Power et al. 2011; Vassileva et al. 1999)	• ↑ Cell viability and cell number during storage and after inoculation • ↑ Rate of cell release into environment • ↑ Plant mycorrhizae • ↑ Microbial colonization • ↑ P-solubilising activity	3–10% skim milk	*Azospirillum brasilense*, *Pseudomonas fluorescens*, *Enterobacter* sp.
3.	Starch (Bashan et al. 2002; Dunkle and Shasha 1989; Kim et al. 2005; Stinson et al. 2003; Vassilev et al. 2020)	• ↓ Physical stress to inoculum • ↑ Cell viability • Protection from UV • ↑ Plant health and compatibility with microflora • Used for both bacterial and fungal strains • Porosity of beads ↑ with cell activity in time	Dried starch beads or capsules with liquid core	*Bacillus thuringiensis*, *Metarhizium anisopliae*, *Beauveria bassiana*, *Fusarium oxysporum*, *Muscodor albus and Muscodor roseus*
4.	Chitin and chitosan (oligosaccharide)	• ↑ Cell multiplication	Chitin in dry olive wastes,	*Bacillus subtilis*, *Penicillium*

(continued)

Table 24.2 (continued)

S. No.	Additives	Characteristics	Examples	Microbial strains
	(Berger et al. 2014; Muxika et al. 2017; Perez and Francois 2016; Vassilev et al. 2020)	• Antifungal, Antibacterial, non-toxic, easily modifiable, chelating agent, and biodegradable • ↑ Plant health • ↑ P-solubilising activity • Bio-control activity • Better storage and farm application • ↑ Stress tolerance and antioxidant activity • ↑ Osmoregulation in plants	chitosan-$CaCl_2$ solution	*janthinellum, A. brasilense and P. fluorescens*
5.	Sugars (Mishra et al. 2013; Morgan et al. 2006)	• Protection of cells from osmotic stress, physical damage, drying • ↑ Cell viability and cell multiplication rate • ↑ Encapsulation and storage efficiency	Glucose, sucrose, trehalose	*Raoultella terrigena, B. bassiana*

formulation, carrageenan—skim milk and bentonite-montmorillonite formulation, and so on (Vassilev et al. 2020).

24.2.3.1.1 Other Additives Besides Starch and Sugars

Humic Acid

Humic acid (HA) is a complex mixture of stable compounds that is soluble in alkaline solution formed by microbial decomposition of organic life (plant and animals) in the soil (Ekin 1883). They contain a larger reservoir of organic Nitrogen and Carbon. Humic acid is generally solubilized at higher pH. Humic substances (HS) has both direct and indirect effect on the plant growth, in which they directly interact with the plant root and activates the physiological processes in the plant, whereas indirect effects involves its pH buffering action, mobilizing nutrients and increase in water retention rate (Pukalchik et al. 2018). Direct effect and indirect effect of humic substances on plants were already studied on various agricultural crops (Ekin 2019; Baldotto et al. 2010; Schmidt et al. 2007; Pishchik et al. 2016). However, an integrated approach has been adopted by the researchers both in

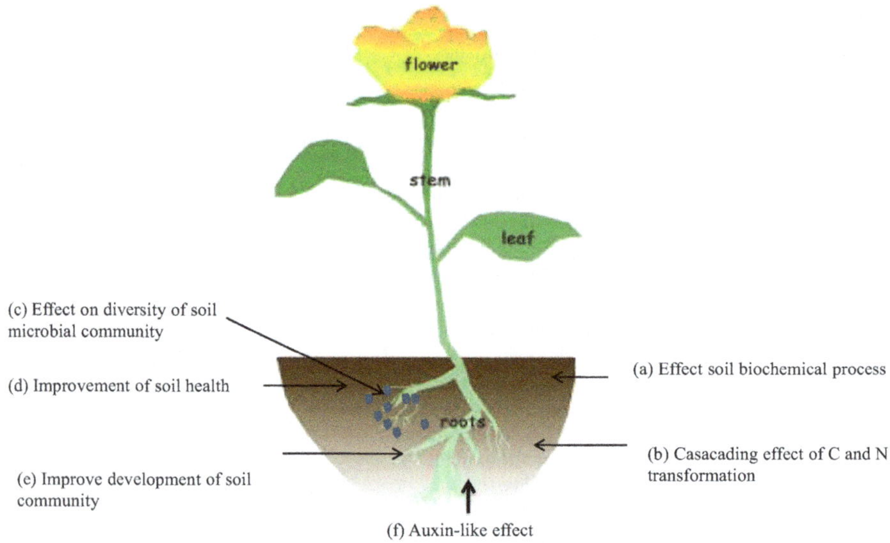

Fig. 24.2 Application of humic substances (**a**) By controlling soil enzymes that effects the supply of C and N to soil and process of organic matter decomposition. (**b**) Redox functional group of HA act as electron donor for bacterial respiration. (**c**) Effect bacterial community more than soil fungi and actinomycetes. (**d**) Increasing the density of earthworm in soil, by mitigated the damage inflicted by nickel and deltamethrin on DNA, protein and lipid membrane. (**e**) It activates molecular signaling for the transition from growth form to dormant form in the growth media. (**f**) Induction of ATPase activity in the plasma membrane, help in root elongation

developed and developing countries that include combined application of humic substances and plant growth-promoting rhizobacteria (PGPR) in promoting plant growth and yield under glass house and field condition (Nardi et al. 2017; Dobbss et al. 2010). An enhanced interaction of rhizospheric bacteria with the plant enhances nutrition supply to the plant, tolerance to abiotic stress, and improvement of crop quality traits. Humic acid and PGPRs are known to enhance the plant biomass (root and shoot) formation and stimulating soil microflora (Fig. 24.2).

A study conducted by Schoebitz et al. (2016) on combined application of PGPRs and humic acid in promoting plant growth of blueberry, demonstrated that synergestic effect was observed on agronomical parameters of plant. H^+-ATPase activity in the plant induces in the presences of humic subst coances, that further energies secondary ion transporters and promote nitrate uptake, which is facilitated by ion channels (Canellas and Olivares 2017). Another morphological benefit, induced within the plant by humic substance addition is lateral roots emergence. The endophytic bacteria enter the plant through infection sites (root tips, root cracks, and lateral root) increasing the root formation (lateral roots), root hair density and length. Thereby, providing more opportunity for the PGPRs to enter the plant and promote its growth and development (Nardi et al. 2017). Similarly, an integrated approach adapted for safflower production under semiarid conditions using two plant growth-promoting rhizobacteria (*Bacillus megaterium* and *Bacillus subtilis*) and humic acid by Ekin (1883). A synergism between improved nutritional status and plant stimulation factor leads to increase in plant height, stem diameter, number of branches, capitula per plant, whereas *B. subtilis* helps in nitrogen fixation and *B. megaterium* enhances phosphorous solubilization. Availability of nutrients to the plant enhances by architectural and biochemical changes in the root system induced by humic substances that enhances the root surface area (Canellas and Olivares 2017). Humic acid directly effects the generation of reactive oxygen species within the plant that helps inducement of root growth and the spread of lateral root (Cordeiro et al. 2011). Plant constitutes the border cell that acts as a first living boundary for the plant that helps in environmental sensing and regulates the interaction of rhizospheric microbes in the soil ecosystem. The mucilage compound secreted by the humic acid within the soil helps in protecting this border cells and enhances their viability. Similarly, Ahmed et al. (2010) studied the combined effect of plant growth-promoting bacteria and humic acid to improve the canola nourishment and yield. PGPR application resulted in a greater tolerance level of canola plants for *B. brassicae* and some other pathogenic microorganism by accelerating plant growth (Pineda et al. 2010), whereas humic acid helps in improving the phenotypic characteristic improvement. Some research has shown that humic acid application could increase the growth indices such as the dry weight, fresh weight, and shoot length of maize (Eyheraguibel et al. 2008) and pepper (Gulser et al. 2010).

Protein Hydrolysate (PH)

Reducing the agriculture impact on human and soil health and simultaneously enhancing the yield of crop are major challenges that we are facing today. Biostimulant (animal or plant origin) is a promising and sustainable solution adopted by the researchers to deal with the solution. PHs are "mixtures of polypeptides, oligopeptides, and amino acids or individual amino acid (glutamate and proline) that are manufactured from protein sources using partial hydrolysis (Schaafsma 2009). It can be by-product of animal (i.e., leather, viscera, feathers, blood) or plant origin (i.e., vegetable by-products) and in biomass of dedicated legume crops (i.e., seeds, hay) that has undergone hydrolysis either by using chemicals (with strong acids or alkalis) and/or enzymatic hydrolysis (Maini 2006; Schiavon et al. 2008; Du Jardin 2015; Halpern et al. 2015). It is a sustainable and environmentally friendly approach that utilizes the waste from agro-industries. Currently, plant derived protein produced from enzymatic hydrolysis, whereas mostly market available protein hydrolysate produced using chemical hydrolysis (Colla et al. 2014) i.e. collagen from leather by-products in Europe, India and China; fish by-products in United States. PHs play key roles through the modulation of plant molecular and physiological processes that trigger growth, increase the yield of plant, and help the plant to cope with the alleviated abiotic stress of crops (i.e. salinity, heavy metal, thermal, nutrient stress, and water stress) (Calvo et al. 2014; Yakhin et al. 2017; Botta 2013; Cerdán et al. 2009; Colla et al. 2013, 2014; Ertani et al. 2013). Protein hydrolysate enhances the yield and phenotypic characteristic of plant, i.e. increased shoot, root biomass, and productivity. PHs improve the rhizospheric microbial community and its enzymatic activities by improving micronutrient mobility (i.e. Fe, Zn, Mn, and Cu) and

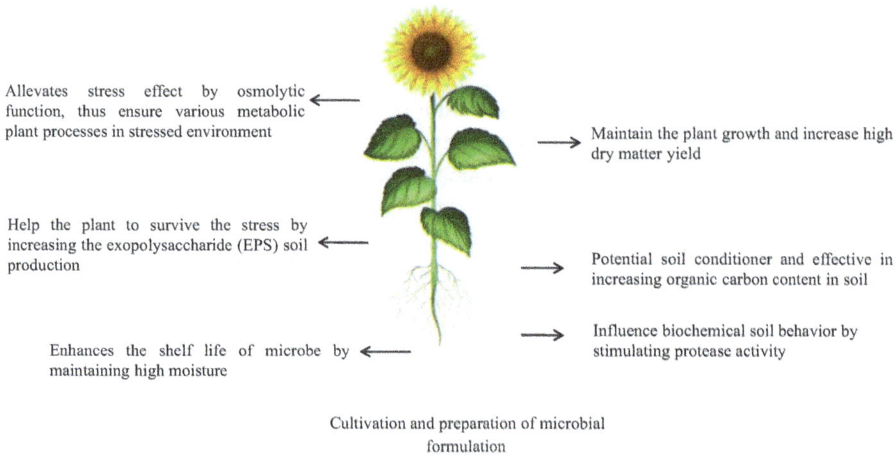

Fig. 24.3 Benefits of glycerol with PGPR in enhancing plant growth

also modifications in the root architecture of plants, in particular root length, density and number of lateral roots (Fig. 24.2). Direct effects of PHs on plants include stimulation of carbon and nitrogen metabolism, as well as regulation of N uptake mediated by key enzymes involved in the N assimilation process and regulation of the activity of three enzymes involved in the tricarboxylic acid cycle (citrate synthase, isocitrate dehydrogenase, and malate dehydrogenase) (Colla et al. 2014).

Cerdán et al. (2009) demonstrated the effectiveness of two commercial protein hydrolysate using root and foliar application to improve the alkalinity tolerance in tomato plant. The effectiveness and mode of action of biostimulant (protein hydrolysate) vary depending upon the origin and mode of action. Its effectiveness on plant vary upon the species, cultivars, phonological stages, growing conditions, concentration, and permeability to leaf. Similarly, Botta (2013) uses animal based protein hydrolysate (obtained by enzymatic hydrolysis) under controlled environmental condition on lettuce to deal with the cold stress conditions. An enhanced shoot growth, root growth and enhanced stomatal conductance were observed in protein hydrolysate treated lettuce as compared to untreated plants. An improved photosynthetic efficiency, enhancement of chlorophyll content, increase in higher photosynthetic efficiency, and improvement of carotenoid were some physiological changes observed in protein hydrolysate treated ryegrass in comparison with untreated plants (control) grown under controlled environmental condition of 36 °C. An increase in secondary metabolite production (e.g., terpenes, glucosinolates) on lettuce plants were observed by Lucini et al. (2015) after foliar application of plant derived PH that was grown under saline conditions. Its because of secondary metabolites production that activates the defense pathway within the plant under stress condition that improves the salinity tolerance of lettuce plant (Fig. 24.3).

Similarly, foliar application of two types of natural plant biostimulants i.e., legume-derived protein hydrolysate or tropical plant extract on (*Diplotaxi stenuifolia* (L.) DC.) was studied under glasshouse condition. It was observed that plant biostimulant improves the physiological mechanism, enhances the growth of perennial wall, vegetative growth in the plant. An enhanced yield, color and photosynthetic status, improvement in quality, increase in Ca, P, phenols and ascorbic acid content and antioxidant activity are some of the characteristics observed in the treated plant.

Glycerol

Glycerol is an economical polyols, formed as side product during trans-esterification of triglycerides with the monovalent alcohol to fatty acid alkyl esters (Vassilev et al. 2016). A larger amount of glycerol as a waste was generated from biodiesel industries, whose disposal is an environmental concern. An alternative approach has to be adopted by the biologist to efficiently convert this waste into high added value products. Because of ease of availability and cheap in price, it play a multifunctional role in the development of microbial inoculants. Glycerol is widely used as biocontrol agent, preparation and cultivation of PGPR. It is applied as a supplement of carbon source for the growth media and high water holding capacity thereby protecting the microbial cell from desiccation (Fig. 24.4). Carbon sources have a differential influence on the wide spectrum of antibiotic produced by the varied biocontrol strains (Duffy and Défago 1999). Using the same approach Siddiqui and Shaukat (2003) reported that nematicidal activity of *P. aeruginosa* strain IE-6S+ is largely influenced by physiological factors such as carbon sources and mineral nutrition (Zinc and glycerol). Zinc is among the most essential element, as it act as a catalyst for varied enzymes and proteins secreted by the microbes and also influences their cell membrane integrity.

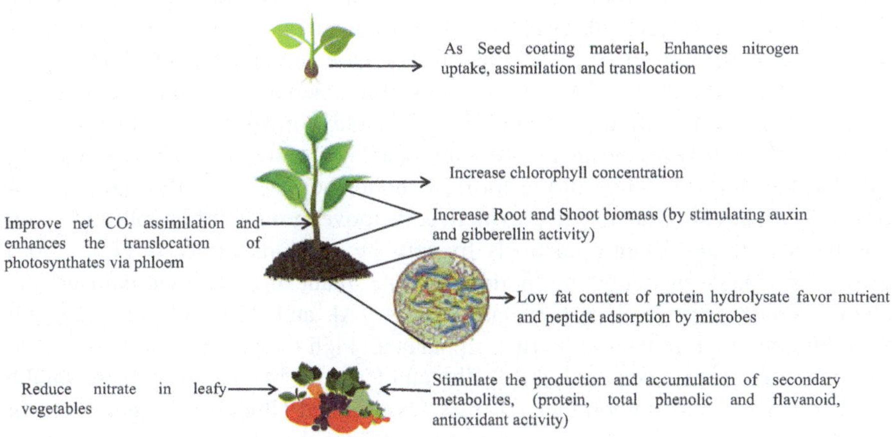

Fig. 24.4 Benefits of protein hydrolysate as biostimulant for plant

The success of bio-inoculant technology focuses on increasing the shelf life and preparation of user friendly formulations. The carrier based bio-inoculants suffer from high contamination, low field performance, and short shelf life (Hegde 2002).

Therefore, for the success of bio-inoculant technology the production of quality inoculants with increased shelf life and user friendly formulation is the approach need to be adopted by the researchers. The efficiency of liquid formulation prepared using PGPR *P. fluorescens* and glycerol against Fusarium wilt under glasshouse and field conditions was studied by Manikandan et al. (2010). Glycerol in the liquid formulation was able to preserve the cell count of microbes even after 150 days, as it holds a high amount of water and protect the cell from the desiccation effect by slowing the drying rate, cell membrane and stabilize enzymes (Fillinger et al. 2001). A significant reduction in mycelial growth of *A. solani* and *F. oxysporum* f. sp. Lycopersici was observed under both glasshouse and field conditions. The adhesive nature of liquid formulation (Kundu and Gaur 1981) also attributed to greater number of antagonist cell adhesion to the seed surface and inhibition of plant pathogens. A microbial formulation is required that increases the shelf life of the microbial for its better establishment and controlled delivery within the plant. An inorganic based formulation was prepared by Sarma et al. (2011) using fluorescent pseudomonad strains and *Piriformospora indica* (Pi) under field and greenhouse conditions. Formulation was prepared by adding glycerol and as a source of carbon for keeping the cell viable, carboxymethylcellulose (CMC), that act as an adhesive and inorganic carrier (Vermiculite). The addition of prepared formulation enhances the growth of plant growth, i.e. shoot and root formation and simultaneously control the incidence of wilt disease caused by *Fusarium oxysporum* F.sp. lycopersici in tomato plants under glass house and field conditions.

Silicon

Silicon (Si) is among the most abundant element in the earth's crust following oxygen on the basis of mass or number of atoms (Ma 2005). Silicon constitute a category of biostimulants that are non-essential for the plant nutrition but have the potential to modify physiological processes of plant, along with the benefits of growth and development. Silicon called as a quasi-essential element, as their presence stimulates the plant growth, whereas there deficiency leads to physical abnormality in the plant (Rafi and Epstein 1997; Ma and Yamaji 2008). Silica provides rigidity to the cell as it deposited in the epidermal layer of the cell wall. It is generally translocated through xylem in the form of monosilicic acid $Si(OH)_4$ (Barber and Shone 1996) having tendency to polymerize. Its move through the apoplast of xylem into the leaves, and taken up actively by cells i.e. trichomes (Savvas et al. 2002). Silicon is capable of mitigating biotic stress, i.e. plant diseases, pest damage, and abiotic stress such as heavy metal toxicity (Al and Mn), salinity, drought, waterlogging, high radiation, nutrient imbalance, high temperature, wounding, and freezing (Van Bockhaven et al. 2013; Zhu and Gong 2014). Level of water soluble silica in the soil varies upon the soil type. Oxisols and Ultisols are among the soil type that contains low level of silica, whereas Histosols contains higher level of Si that improves pest and disease tolerance of agricultural crops (Haynes 2014).

Guleria et al. (2014) found that root application of Si to wheat plant enhances its resistance to green-aphids (*Schizaphis graminum* (Rond.) and reduces powdery mildew severity (Guleria et al. 2014) as compared to its foliar applications. It also helps in reduction in grazing of wheat plants by rabbits (Cotterill et al. 2007) and slug feeding on wheat seedlings (Griffin et al. 2015). Si also provide mechanical benefits to the plant cell wall by forming phytoliths that strengthens the stems and reduce lodging (Ma 2004; Liang et al. 1996). Si is deposited in the cell wall upon polymerization of SiO_2 (Savvas and Ntatsi 2015), acting as a physical barrier for the pest penetration and stimulates natural defense mechanisms (Pilon-Smits et al. 2009). Liang et al. (2005) demonstrated that Si adsorption by root can be either active or passive flow that directly depends upon the concentration of Si in the soil solution and accumulation efficiency of plant species. Rains et al. (2006) demonstrated that the active uptake of Si by rice, wheat or other species involves a similar mechanism, however, the difference in concentration is only in the degree of their activity. Raven (2001) studied the accumulation of Si in rice and its effect upon its metabolic process. It was estimated that the permeability of plasma membrane and passive flow against a concentration gradient, responsible for the Si accumulation within the plant species, that directly effects its growth and development. Some microbes can also solubilize rock K by chelating silicon ions (Parmar and Sindhu 2013), therefore PGPR in combination with silica can promote plant growth and also reduces the addition of N and P fertilizer into the soil, thereby maintaining soil health and quality (Adesemoye et al. 2009).

24.2.3.2 Metabolites

Metabolites are intermediates and metabolism products, developed by microbes, intended to manage and influence different internal and surrounding activities. These were classified into primary and secondary metabolites. Primary metabolites are considered to be necessary for proper growth in microorganisms, whereas secondary metabolites do not play a role in formation, growth, and reproduction and are usually developed during the steady-state growth process. They also inhibit unhealthy species, as well as their own growth and production (Arora et al. 2010). The regulation of metabolites, along with microbes promoting plant growth, can be used even more effectively in agricultural applications compared to traditional cell-based formulations. Inoculant formulations containing microbes or their metabolites can be used to supply nutrients or bioregulators that directly or indirectly impact plant growth through a range of mechanisms (Lugtenberg and Kamilova 2009). Microbes of the genera *Azospirillum*, *Azotobacter*, *Bacillus*, *Bradyrhizobium*, *Mesorhizobium*, *Pseudomonas*, *Rhizobium*, *Trichoderma*, and mycorrhizal fungi are commonly used in inoculum formulations. Various metabolite formulations can be used to overcome the problems associated with cell-based formulations (Arora 2015). Without funding from the government and non-government sectors, the desired results to increase the fertility of agricultural fields could not be achieved by overcoming various environmental issues and the need for sustainability. Use of metabolites and additives, in combination with cells, making the encapsulated formulation more reliable and effective. Rhizobial inoculants mixed with flavonoids provide better nodulation, nitrogen fixation, and ability to tolerate abiotic stress (Oldroyd 2013). Lipo-chitooligosaccharides (LCOs) play a significant role for the symbiotic relationship of rhizobia with the roots of the legume and have beneficial impact on crop yields (Adholeya and Das 2012). Similarly, mycorrhizae secreted factors that activate typical symbiosis (SYM) pathways in the host plant and induce mycorrhizae interaction with plants (Maillet et al. 2011).Various PGPRs, secrets exopolysaccharides (EPS) that help in nodulation, root colonisation under stress conditions, neutralise toxins and serve as a carbon source (Tewari and Arora 2014). EPS can be easily produced at large scale and can be used in EPS modified bioformulation to protect the cells. EPS may also play an important role in defending plant roots under abiotic stress (Sandhya and Ali 2015). EPS as a carrier help to protect the microbe from radiation, extreme pH, osmotic shock, desiccation etc. and even from the predators (Seneviratne et al. 2011). EPS can therefore be used effectively as a bioformulation for slow release of bioprotective carrier for better adaptation PGP microbes in soil. It has been reported that the addition of tryptophanic precursors stimulate the PGP to increase the production of IAA and enhancement of root hair formation, plant biomass, grain yield, and even pathogen control (Zahir et al. 2010). The use of phytohormones such as gibberlic acid, cytokinins along with chitin as additives in formulations along with the inoculant microbe has improved plant growth and biocontrol by supplying carbon and enhancing chitinase function. A number of agriculturally important microbes are known to develop secondary metabolites that act as antibiotics and antifungal compounds (Haas and Defago 2005). Likewise, the *Bacillus* sp. antibiotics such as zwittermycin and kanosamine are also documented (Raaijmakers et al. 2002). Research also indicates the usage metabolites based formulations, especially in the case of some human pathogens such as *Pseudomonas aeruginosa*, *Bacillus* sp. Fungi such as *Purpureocillium lilacinum* (Khare and Arora 2010; Cavello et al. 2015) and most of the secondary metabolites are considered to have anti-phytopathogenic activities.

However, owing to the cost of processing, most of the metabolites are not recommended and have not been developed in pure form in field trials. But in the near future, and advancement in technological tools it is very likely to make these metabolites use commercially in various applications of agriculture (Zacky and Ting 2015). These problems need to be resolved before the manufacture and implementation of cell-free formulations on a wide scale. Some of the metabolites are already in use in bioformulation, which are cheaply available; other such additives have high value for their inputs, which are needed at very low concentrations (Cavello et al. 2015). Such metabolite-based formulations can be useful even in varied geographic and climatic areas, making them holistic and more appropriate to both the agro-market and end-users. The updated formulations containing bacterial cells and metabolites would be a sustainable and durable technology that enhances the efficiency of agricultural process and their productivity manifold. These future bioformulations can be described as amalgamation of microbe(s)-promoting plant growth, carrier and additives that support the microbe in one way or another (Zahir et al. 2010).

24.2.3.3 Bioencapsulation

Bioencapsulation entails covering and preserving microorganisms that have been poured into the soil for gradual and sustained release of PGPR in order to maximize their effects on plant health and reduce their costs (Kim et al. 2012). Microbial encapsulated cells under suitable condition can be stored at room temperature for a long time by reducing the likelihood of diminished survival and this can be done by adding important nutrients in the encapsulated matrix required for bacterial growth. Bioencapsulation of PGPR have number of beneficial effects such as regulated release of bacteria into the soil, protection of microorganisms from environmental stress, and reduction of contamination that may happen during storage and transport (John et al. 2011). Bioencapsulation of the microbial cells can be done by introducing the active ingredient into the matrix and stabilizing it with chemical or physico-chemical process. Different encapsulation methods have their own pros and cons thus, the selection of effective bioencapsulation strategy depend upon the type of strain, expense, processing conditions, etc. (Yeo et al. 2001; Rekha et al. 2007). PGPR plays an important role in improving sustainable agriculture to increase the yield of crops (Bashan et al. 2004; Rivera-Cruz et al. 2008) by development of various phytohormones that enhance root growth, water absorption and nutrients (Dobbelaere et al. 2001; Spaepen et al. 2007), non-symbiotic nitrogen fixation (Pedraza 2008) and even solubilization of phosphate (Rodriguez et al. 2006; Zaidi et al. 2009). Direct inoculation of PGPR without encapsulation could not be so beneficial because the microorganisms are susceptible to environmental variations and other stresses etc. but can be overcome through bioencapsulation (Wu et al. 2012; Schoebitz et al. 2013). Encapsulated beads used in the farm must be the size of seed to make it easy for farmer to use in the fields just like sowing and can be carried out at the same time and with the same seed drill near to the soil, favouring the potency of microbial inoculants (Yeo et al. 2001). Most of the bioencapsulation studies carried out only in the laboratories and under controlled conditions at a very small scale. However, to generate a significant volume of inoculant, the use of advanced bioencapsulation devices in large-scale field trials are needed (Tewari and Arora 2014). In the future, the implementation of various new formulation techniques will be developed as per the need of the agriculture sector to convert the laboratory-scale experiments into a reality for fields.

24.3 Impact of the Gel-Forming Polysaccharides on Plant Health

Plant beneficial microorganisms released into the soil slowly and steadily if those are encapsulated in suitable polymers matrixes such as alginate, carrageenan etc. (Bashan et al. 2004). The beneficial impact of many polysaccharides on plants used in encapsulation technologies are focused on the synergistic behaviour of growth regulators, osmolytes and polysaccharides used (Bhattacharyya and Jha 2012). There is clear evidence that polysaccharide gels play a significant role in abiotic stress protection mechanisms for PGPR (Vassilev et al. 2012). The use of

Table 24.3 Showing various micro-organisms and their encapsulating agents along with benefits and their health effect on plants

Gel-forming agent	Microorganism	Beneficial effects	Health effects on plant	Reference
Pyrophyliite	*Pseudomonas cepacia*	Bulking agent	• Protection from biotic and abiotic stress by providing a better micro-environment in rhizosphere • Slow release of microbial cells and nutrients • Controlled release triggered by environmental conditions • Healthy growth of cell in a micro-fermenter • Isolation of active ingredients during application • Attract and kill pathogens • Two or more bio-control agents can be encapsulated together • Maintain extended metabolic activities • Targeted delivery • Worker protection	Rekha et al. (2007)
Chitin	*Trichoderma harzianum*	Reduce UV transmission		Zohar-perez et al. (2003)
Bentonite	*Raoultella planticola*	Continuous cell release		He et al. (2015)
Perlite	*Pseudomonas putida*	Cell-gell stability		Liffourrena and Lucchesi (2018)
Skim milk	*Azospirilum brasilense*	Increase cell number in bead		Bashan et al. (2002)
Montmorilonite	*Enterobacter* sp.	Higher cell servival rate		Oldroyd (2013)
Clay minerals	*Pseudomonas aeruginosa*	Higher plant growth promotion		Tewari and Arora (2014)
Starch	*Fusarium oxysporum*	Extended self-life		Ivanova et al. (2005)
Olive waste	*Pencillium janthinellum*	Biocontrol activity		Vassilev et al. (2012)
Humic acid	*Bacillus subtilis*	Higher survival rate		Young et al. (2006)
Trehlose	*Raoultella terrigena*	Desiccation protection		Schoebitz et al. (2013)
Protein hydrolysate	*Bacillus salmalaya*	High encapsulation index		Vejan et al. (2018)

exopolysaccharide in encapsulation helps bacteria to grow under harsh conditions (Sa et al. 2019). These provide hydrated microenvironment that supports all living conditions for the microbial cells for their efficient growth and development (Chang et al. 2004) beside acting as signal molecules controlling plant production and defense (Larskaya and Gorshkova 2015). Various researchers studied the effect of g-irradiation of polysaccharides such as chitosan, k-carrageenan alginate, etc. and showed positive impact on plants growth and development (Hien et al. 2012) and even some of them like oligochitosan, on gamma-irradiation, acts as growth stimulator and antimicrobial agent under drought conditions (Muley et al. 2019). Alginate gel beads improve the cell viability and stability of *Lactobacillus fermentum* when exposed to particular environmental conditions (Liao et al. 2019) (Table 24.3).

Inclusion of oligosaccharides in formulations can be expected in the near future for better understanding of the relationship among carriers-additives-microorganisms-soil-plant systems that is necessary for identifying strategies for their use (Bhattacharyya and Jha 2012). Study on creation of suitable microenvironment inside the capsules also is considered to promote encapsulation that supports the overall use of these encapsulated carriers in the agriculture. Research scientists are consistently using new approaches and modifying the existing strategies that have already been proven track record in the biotechnological fields (Cavello et al. 2015; John et al. 2011). Further advancement in immobilization methodologies should be focused on multidisciplinary scientists having expertise in microbiology/biotechnology, plant physiology/pathology, and agricultural engineers to provide effective, safe, economically reasonable, and easy-to-use complex biotechnological products for plant growth and health that can revolutionize the agriculture sector.

24.4 Conclusion

The production, formulation, and application of microbial biofertilizers are the major steps in biofertilizer technology. This brief analysis of certain biotechnological methods used in the development and formulation of microbial inoculants shows the great potential of additional unexploited features. Co-cultivation and co-immobilization/encapsulation of compatible plant-beneficial microorganisms is a highly promising strategy, which for several reasons, is expected to be studied in the near future. Therefore, the co-cultivation of two or more different microorganisms in free and/or immobilized form provides an optimal solution that leads to a significantly enhanced production of both constitutively present and cryptic compounds that are not usually contained in the producing strain's axenic cultures. The approach of developing synthetic (artificial) microbial systems will lead to the development of new methods for applying (formulating) a consortium of microorganisms with different functional properties to maintain, sustain, or improve soil fertility in order to better study different interactions in a soil–plant environment. The aforementioned biotechnological tools may therefore be viewed as the first step in this direction. Using the products of different microorganisms and plant hormones, growth regulators, derivatives of humic acid, oligosaccharides, and algal extracts, biocontrol substances and vitamins, further advantages could be achieved. Solid-state fermentation processes provide an excellent tool for the co-cultivation and research of compatible microorganisms, their associations, improvements in the metabolite profile and secretion of novel products, as solid medium cultures are typically carried out on the Petri dish scale and certain cultivation conditions make certain review impossible. Both SSF and immobilized-cell systems can be effectively used in order to explore the range of stress tolerance of plant-beneficial microorganisms and their metabolic function. The results of such studies will result in biofertilizer products adapted to drought stress, low/high temperatures, and/or resistant to high salt concentrations and low/high pH. Therefore, when evaluating the existing and unrevealed potential of the biotechnological methods used in the development and formulation of traditional and novel inoculant products, we should always take into account the functional relationship between the large number of players in the exciting game called biofertilizer technology. However, recent advances in technology-related microbial research, plant–pathogen interactions, genomics, and genetic modification will also help to refine the protocols needed for biofertilizer use. Therefore, the efficacy of biofertilizers depends on the development of novel strategies which, by means of advanced and improved techniques, are linked to the functions of various beneficial bacteria and their proper application to fields.

References

Adeoye DO, Mukhtar B, Ameh AO, Yusuf AA (2019) Production of *Rhizobium* bio-fertilizer carrier from tannery solid wastes. Nigerian J Eng 26:1–7

Adesemoye AO, Torbert HA, Kloepper JW (2009) Plant Growth-promoting rhizobacteria allow reduced application rates of chemical fertilizers. Microb Ecol 58:921–929

Adholeya A, Das M (2012) Biofertilizers: potential for crop improvement under stressed conditions. In: Tuteja N, Gill SS, Tuteja R (eds) Improving crop productivity in sustainable agriculture. Wiley, New York, pp 183–200

Ahmed H, Nesiem AH, Hewedy HR, Sallam H (2010) Effect of simulation compounds on growth, yield and chemical composition of snap bean plants grown under calcareous soil conditions. J Am Sci 6:552–569

Arif MS, Shahzad SM, Riaz M, Yasmeen T, Shahzad T, Akhtar MJ et al (2017) Nitrogen-enriched compost application combined with plant growth-promoting rhizobacteria (PGPR) improves seed quality and nutrient use efficiency of sunflower. J Plant Nutr Soil Sci 180:464–473

Arora NK (2015) Microbes symbiosis: applied facets. Springer, New Delhi, p 383

Arora NK, Khare E, Maheshwari DK (2010) Plant growth promoting rhizobacteria: constraints in bioformulation, commercialization and future strategies. In: Maheshwari DK (ed) Bacteria and plant health. Springer, Cham, pp 97–116

Baldotto LEB, Baldotto MA, Canellas LP, Bressan-Smith R, Olivares FL (2010) Growth promotion of pineapple 'vitória' by humic acids and *Burkholderia* spp. during acclimatization. Rev Bras Ciênc Solo 34:1593–1600

Barber DA, Shone MGT (1996) The absorption of silica from aqueous solutions by plants. J Exp Bot 17(3):569–578

Bashan Y (1998) Inoculants of plant growth-promoting bacteria for use in agriculture. Biotechnol Adv 16:729–770

Bashan Y, Hernandez JP, Leyva LA, Bacilio M (2002) Alginate micro beads as inoculant carriers for plant growth-promoting bacteria. Biol Fertil Soils 35:359–368

Bashan Y, Holguin G, de Bashan LE (2004) Azospirillum-plant relationships: physiological, molecular, agricultural, and environmental advances (1997-2003). Can J Microbiol 50:521–577

Bashan Y, de Bashan LE, Prabhu SR, Hernandez JP (2014) Advances in plant growth-promoting bacterial inoculant technology: formulations and practical perspectives (1998–2013). Plant Soil 378:1–33

Bashan Y, de Bashan LE, Prabhu SR (2016) Superior polymeric formulations and emerging innovative products of bacterial inoculants for sustainable agriculture and the environment. In: Singh HB, Sarma BK, Keswani C (eds) Agriculturally important microorganisms - commercialization and regulatory requirements in Asia. Springer, Berlin, pp 15–46

Basheer J, Ravi A, Mathew J, Krishnakutty RE (2019) Assessment of plant-probiotic performance of novel endophytic *Bacillus* sp. in talc-based formulation. Probiot Antimicro Prot 11:256–263

Berger LR, Stamford TC, Stamford-Arnaud TM, de Alcântara SR, da Silva AC, da Silva AM, do Nascimento AE, de Campos-Takaki GM (2014) Green conversion of agroindustrial wastes into chitin and chitosan by *Rhizopus arrhizus* and *Cunninghamella elegans* strains. Int J Mol Sci 15 (5):9082–9102

Bhattacharyya PN, Jha DK (2012) Plant growth-promoting rhizobacteria (PGPR): emergence in agriculture. World J Microbiol Biotechnol 28:1327–1350

Botta A (2013) Enhancing plant tolerance to temperature stress with amino acids: an approach to their mode of action. Acta Hortic 1009:29–35

Brahmaprakash GP, Sahu PK (2012) Biofertilizers for sustainability. J Indian Inst Sci 92(1):37–62

Calvo P, Nelson L, Kloepper JW (2014) Agricultural uses of plant biostimulants. Plant Soil 383:3–41

Canellas LP, Olivares FL (2017) Production of border cells and colonization of maize root tips by *Herbaspirillum seropedicae* are modulated by humic acid. Plant Soil 417:403–413

Cavello IA, Crespo JM, Garcia SS, Zapiola JM, Luna MF, Cavalitto SF (2015) Plant growth promotion activity of keratinolytic fungi growing on a recalcitrant waste known as hair waste. Biotechnol Res Int 2015:952921

Cerdán M, Sánchez-Sánchez A, Oliver M, Juárez M, Sánchez-Andreu JJ (2009) Effect of foliar and root applications of amino acids on iron uptake by tomato plants. Acta Hortic 830:481–488

Chang WS, Van de Mortel M, Nielsen L, Nino de Guzman G, Li X, Deaker R, Roughley RJ, Kennedy IR (2004) Legume seed inoculation technology-a review. Soil Biol Biochem 36:1275–1288

Colla G, Svecova E, Rouphael Y, Cardarelli M, Reynaud H, Canaguier R et al (2013) Effectiveness of a plant–sderived protein hydrolysate to improve crop performances under different growing conditions. Acta Hortic 1009:175–179

Colla G, Rouphael Y, Canaguier R, Svecova E, Cardarelli M (2014) Biostimulant action of a plant-derived protein hydrolysate produced through enzymatic hydrolysis. Front Plant Sci 5:448

Cordeiro FC, Catarina CS, Silveira V, De Souza SR (2011) Humic acid effect on catalase activity and the generation of reactive oxygen species in corn (Zea mays). Biosci Biotechnol Biochem 75(1):70–74

Cotterill JV, Watkins RW, Brennon CB, Cowan DP (2007) Boosting silica levels in wheat leaves reduces grazing by rabbits. Pest Manag Sci 63:247–253

Dobbelaere S, Croonenborghs A, Thys A, Ptacek D, Vanderleyden J, Dutto P et al (2001) Responses of agronomically important crops to inoculation with *Azospirillum*. Aust J Plant Physiol 28:871–879

Dobbss LB, Canellas LP, Olivares FL (2010) Bioactivity of chemically transformed humic matter from vermicompost on plant root growth. J Agric Food Chem 58:3681–3688

Du Jardin P (2015) Plant biostimulants: definition, concept, main categories and regulation. Sci Hortic 196:3–14

Duffy BK, Défago G (1999) Environmental factors modulating antibiotic and siderophore biosynthesis by *Pseudomonas fluorescens* biocontrol strains. Appl Environ Microbiol 65:2429–2438

Dunkle RL, Shasha BS (1989) Response of starch-encapsulated *Bacillus thuringiensis* containing ultraviolet screens to sunlight. Environ Entomol 18:1035–1041

Egamberdieva D, Hua M, Reckling M, Wirth S, Bellingrath-Kimura SD (2018) Potential effects of biochar-based microbial inoculants in agriculture. Environ Sustain 1:19–24

Ekin D (1883) Co-application of humic acid and bacillus strains enhances seed and oil yields by mediating nutrient acquisition of safflower (*Carthamus tinctorius* L.) plants in a semi-arid region. Appl Ecol Environ Res 18(1):1883–1900

Ekin Z (2019) Integrated Use of humic acid and plant growth promoting rhizobacteria to ensure higher potato productivity in sustainable agriculture. Sustain For 11:3417

Ertani A, Schiavon M, Muscolo A, Nardi S (2013) Alfalfa plant-derived biostimulant stimulate short-term growth of salt stressed *Zea mays* L. plants. Plant Soil 364:145–158

Eyheraguibel B, Silvestre J, Morard P (2008) Effects of humic substances derived from organic waste enhancement on the growth and mineral nutrition of maize. Bioresour Technol 99 (10):4206–4212

Fillinger S, Chaveroche MK, van Dijck P, de Vries R, Ruijter G, Thevelein J et al (2001) Trehalose is required for the acquisition of tolerance to a variety of stresses in the filamentous fungus *Aspergillus nidulans*. Microbiologia 147:1851–1862

Garcia-Fraile P, Menendez E, Rivas R (2015) Role of bacterial biofertilizers in agriculture and forestry. AIMS Bioeng 2(3):183–205

Gopal KS, Baby A (2016) Enhanced shelf-life of *Azospirillum* and PSB through addition of chemical additives in liquid formulations. Int J Sci Environ Technol 5:2023–2029

Gousterova A, Nustorova M, Christov P, Nedkov P, Neshev G, Vasileva-Tonkova E (2008) Development of a biotechnological procedure for treatment of animal wastes to obtain inexpensive biofertilizer. World J Microbiol Biotechnol 24(11):2647–2652

Griffin M, Hogan B, Schmidt O (2015) Silicon reduces slug feeding on wheat seedlings. J Pest Sci 88(1):17–24

Guleria S, Sharma K, Walia A, Chauhan A, Shirkot CK (2014) Population and functional diversity of phosphate solubilizing bacteria from Apricot (*Prunus Armeniaca*) of mid and high regions of Himachal Pradesh. Bioscan 9(2):1435–1443

Gulser F, Sonmez F, Boysan S (2010) Effects of calcium nitrate and humic acid on pepper seedling growth under saline condition. J Environ Biol 31(5):873–876

Haas D, Defago G (2005) Biological control of soil-borne pathogens by fluorescent pseudomonads. Nat Rev Microbiol 3:307–319

Halpern M, Bar-Tal A, Ofek M, Minz D, Muller T, Yermiyahu U (2015) The use of biostimulants for enhancing nutrient uptake. Adv Agron 130:141–174

Haynes RJ (2014) A contemporary overview of silicon availability in agricultural soils. J Plant Nutr Soil Sci 177:831–844

He Y, Wu Z, Tu L, Han Y, Znahg G, Li C (2015) Encapsulation and characterization of slow-release microbial fertilizer from the composites of bentonite and alginate. Appl Clay Sci 109:68–75

He Y, Wu Z, Tu L, Shan C (2017) Effect of encapsulated *Pseudomonas putida* Rs-198 strain on alleviating salt stress of cotton. J Plant Nutr 40:1180–1189

Hegde SV (2002) Liquid biofertilizers in Indian Agriculture. Biofert News Lett 12:17–22

Herrmann L, Lesueur D (2013) Challenges of formulation and quality of biofertilizers for successful inoculation. Appl Microbiol Biotechnol 97:8859–8873

Hickert LR, Cruz MM, Dillon AJP, Fontana RC, Rosa CA, Záchia Ayub MA (2014) Fermentation kinetics of acid–enzymatic soybean hull hydrolysate in immobilized-cell bioreactors of *Saccharomyces cerevisiae*, *Candida shehatae*, *Spathaspora arborariae*, and their co-cultivations. Biochem Eng J 88:61–67

Hien NQ, Nagasawa N, Tham LX, Yoshii F, Dang VH, Mitomo H et al (2012) Growth-promotion of plants with depolymerized alginates by irradiation. Radiat Phys Chem 59:97–101

Ivanova E, Teunou E, Poncelet D (2005) Alginate based macrocapules as inoculants carriers for production of nitrogen biofertilizers. In: Proceedings of the Balkan scientific conference of Biology, Plovdiv, pp 90–108

Jampílek J, Králová K (2017) Nanomaterials for delivery of nutrients and growth-promoting compounds to plants. In: Prasad R, Kumar M, Kumar V (eds) Nanotechnology: an agricultural paradigm. Springer, Singapore, pp 177–226

John RP, Tyagi RD, Brar SK, Surampalli RY, Prévost D (2011) Bioencapsulation of microbial cells for targeted agricultural delivery. Crit Rev Biotechnol 31:211–226

Karel SF, Libicki SB, Robertson CR (1985) The immobilization of whole cells: engineering principles. Chem Eng Sci 40(8):1321–1354

Kaur R, Kaur S (2018) Biological alternates to synthetic fertilizers: efficiency and future scopes. Indian J Agric Res 52(6):587–595

Kekez B, Gojgić-Cvijović G, Jakovljević D, Pavlović V, Beškoski V, Popović A, Vrvić MM, Nikolić V (2016) Synthesis and characterization of a new type of levangraft-polystyrene copolymer. Carbohydr Polym 154:20–29

Khan W, Rayirath UP, Subramanian S et al (2009) Seaweed extracts as biostimulants of plant growth and development. J Plant Growth Regul 28(4):386–399

Khare E, Arora NK (2010) Effect of indole-3-acetic acid (IAA) produced by Pseudomonas aeruginosa in suppression of charcoal rot disease of chickpea. Curr Microbiol 61:64–68

Kim YJ, Park HG, Yang YL, Yoon Y, Kim S, Oh E (2005) Multifunctional drug delivery system using starch-alginate beads for controlled release. Biol Pharm Bull 28(2):394–397

Kim IY, Pusey PL, Zhao Y, Korban SS, Choi H, Kim KK (2012) Controlled release of *Pantoea agglomerans* E325 for biocontrol of fire blight disease of apple. J Control Release 161:09–15

Kumar A, Guleria S, Mehta P, Walia A, Chauhan A, Shirkot CK (2015) Plant growth promoting traits of Phosphate solubilizing rhizobacteria isolated from seabuckthorn growing in cold desert region of trans-Himalayas and evaluating their potential on growth of tomato seedlings. Acta Physiol Plant 37(3):1–12

Kundu BS, Gaur AC (1981) Carrier studies on phospho-microorganisms as single and mixed inoculants. Indian J Agric Sci 54:252–255

Larskaya IA, Gorshkova TA (2015) Plant oligosaccharides-Outsiders among elicitors? Biochemistry 80:881–900

Liang Y, Shen Q, Shen Z, Ma T (1996) Effects of silicon on salinity tolerance of two barley cultivars. Plant Nutr 19:173–183

Liang YC, Jin SI, Römheld V (2005) Silicon uptake transport is an active process in *Cucumis Sativus*. New Phytol 167(3):797–804

Liao N, Luo B, Gao J, Li X, Zhao Z, Zhang Y et al (2019) Oligosaccharides as co-encapsulating agents: effect on oral *Lactobacillus fermentum* survival in a simulated gastrointestinal tract. Biotechol Lett 41:263–272

Liffourrena AS, Lucchesi GI (2018) Alginate-perlite encapsulated *Pseudomonas putida* A (ATCC 12633) cells: preparation, characterization and potential use as plant inoculants. J Biotechnol 278:28–33

Lucini L, Rouphael Y, Cardarelli M, Canguier R, Kumar P, Colla G (2015) The effect of a plant-derived biostimulant on metabolic profiling and crop performance of lettuce grown under saline conditions. Sci Hortic 182:124–133

Lugtenberg B, Kamilova F (2009) Plant-growth-promoting rhizobacteria. Annu Rev Microbiol 63:541–556

Ma JF (2004) Role of silicon in enhancing the resistance of plants to biotic and abiotic stresses. Soil Sci Plant Nutr 50:11–18

Ma JF (2005) Plant root responses to three abundant soil minerals: silicon, aluminum and iron. Crit Rev Plant Sci 24:267–281

Ma JF, Yamaji N (2008) Functions and transport of silicon in plants. Cell Mol Life Sci 65:3049–3057

Magauzi R, Mabaera B, Rusakaniko S, Chimusoro A, Ndlovu N, Tshimanga M et al (2011) Health effects of agrochemicals among farm workers in commercial farms of Kwekwe district, Zimbabwe. Pan Afr Med J 9:26

Mahanty T, Bhattacharjee S, Goswami M, Bhattacharyya P, Das B, Ghosh A, Tribedi P (2016) Biofertilizers: a potential approach for sustainable agriculture development. Environ Sci Pollut Res 24(4):3315–3335

Maillet F, Poinsot V, Andre O, Puech-Pages V, Haouy A, Gueunier M et al (2011) Fungal lipochitooligosaccharide symbiotic signals in arbuscular mycorrhiza. Nature 469:58–63

Maini P (2006) The experience of the first biostimulant, based on amino acids and peptides: a short retrospective review on the laboratory researches and the practical results. Fertil Agrorum 1:29–43

Malusá E, Paszt LS, Ciesielska J (2012a) Technologies for beneficial microorganisms inocula used as biofertilizers. Sci World J 2012:1–12

Maluså E, Sas-Paszt L, Ciesielska J (2012b) Technologies for beneficial microorganisms inocula used as biofertilizers. Sci World J 2012:491206

Malusà E, Pinzari F, Canfora L (2016) Efficacy of biofertilizers: challenges to improve crop production. In: Singh D, Singh H, Prabha R (eds) Microbial inoculants in sustainable agricultural productivity. Springer, New Delhi, pp 17–40

Manikandan RD, Saravanakumar L, Rajendran T, Samiyappan RR (2010) Standardization of liquid formulation of *Pseudomonas fluorescens* Pf1 for its efficacy against Fusarium wilt of tomato. Biol Control 54:83–89

Mastan A, Rane D, Dastager SG, Vivek Babu CS (2019) Development of low-cost plant probiotic formulations of functional endophytes for sustainable cultivation of *Coleus forskohlii*. Microbiol Res 227:126310

Mehta P, Walia A, Chauhan A, Kulshrestha S, Shirkot CK (2013) Phosphate solubilization and plant growth promoting potential by stress tolerant *Bacillus* sp. isolated from rhizosphere of apple orchards in trans Himalayan region of Himachal Pradesh. Ann Appl Biol 163:430–443

Mehta P, Walia A, Shirkot CK (2015) Functional diversity of phosphate solubilizing plant growth promoting rhizobacteria isolated from apple trees in the Trans Himalayan region of Himachal Pradesh, India. Biol Agric Hortic 31(4):265–288

Mehta P, Sharma R, Putatunda C, Walia A (2019) Endophytic fungi: role in phosphate solubilization. In: Singh B (ed) Advances in endophytic fungal research. Fungal biology. Springer, Cham, pp 183–209

Mendes G, Galvez A, Vassileva M, Vassilev N (2017) Fermentation liquid containing microbially solubilized P significantly improved plant growth and P uptake in both soil and soilless experiments. Appl Soil Ecol 117:208–211

Mishra S, Kumar P, Malik A (2013) Preparation, characterization, and insecticidal activity evaluation of three different formulations of *Beauveria bassiana* against *Musca domestica*. Parasitol Res 112(10):3485–3495

Morgan CA, Herman N, White PA, Vesey G (2006) Preservation of micro-organisms by drying; a review. J Microbiol Methods 66(2):183–193

Muley AB, Shingote PR, Patil AP, Dalvi SG, Suprasanna P (2019) Gamma radiation degradation of chitosan for application in growth promotion and induction of stress tolerance in potato (*Solanum tuberosum* L.). Carbohydr Polym 210:289–301

Muxika A, Etxabide A, Uranga J, Guerrero P, de la Caba K (2017) Chitosan as a bioactive polymer: processing, properties and applications. Int J Biol Macromol 105:1358–1368

Nardi S, Ertani A, Ornella F (2017) Soil root cross talking: the role of humic substances. J Plant Nutr Soil Sci 180:5–13

Neneng L (2020) Formulation of liquid biofertilizer for enhance of soil nutrients in peat land. Budapest Int Res Exact Sci J 2:314–322

Oldroyd GE (2013) Speak, friend, and enter: signalling systems that promote beneficial symbiotic associations in plants. Nat Rev Microbiol 11:252–263

Paliya S, Mandpe A, Kumar S, Kumar MS (2019) Enhanced nodulation and higher germination using sludge ash as a carrier for biofertilizer production. J Environ Manag 250:109523

Park JH, Hong D, Lee J, Choi IS (2016) Cell-in-shell hybrids: chemical nanoencapsulation of individual cells. Acc Chem Res 49(5):792–800

Parmar P, Sindhu SS (2013) Potassium solubilization by rhizosphere bacteria: influence of nutritional and environmental conditions. J Microbiol Res 3:25–31

Pastor-Bueis R, Sánchez-Cañizares C, James EK, González-Andrés F (2019) Formulation of a highly effective inoculant for common bean based on an autochthonous elite strain of *Rhizobium leguminosarum* bv. *phaseoli*, and genomic-based insights into its agronomic performance. Front Microbiol 10:2724

Pedraza RO (2008) Recent advances in nitrogen-fixing acetic acid bacteria. Int J Food Microbiol 125:25–35

Perez JJ, Francois NJ (2016) Chitosan-starch beads prepared by ionotropic gelation as potential matrices for controlled release of fertilizers. Carbohydr Polym 148:134–142

Perignon C, Ongmayeb G, Neufeld R, Frere Y, Poncelet D (2014) Microencapsulation by interfacial polymerisation: membrane formation and structure. J Microencapsul 2014:1–15
Pilon-Smits EAH, Quinn CF, Tapken W, Malagoli M, Schiavon M (2009) Physiological functions of beneficial elements. Curr Opin Plant Biol 12:267–274
Pineda A, Zheng SJ, Van Loon JA, Pieterse MJ, Dicke M (2010) Helping plants to deal with insects: the role of beneficial soil-borne microbes. Trends Plant Sci 15:507–514
Pishchik VN, Vorobyov NI, Walsh OS, Surin VG, Khomyakov YV (2016) Estimation of synergistic effect of humic fertilizer and *Bacillus subtilis* on lettuce plants by reflectance measurements. J Plant Nutr 39:1074–1086
Power B, Liu X, Germaine KJ, Ryan D, Brazil D, Dowling DN (2011) Alginate beads as a storage delivery and containment system for genetically modified PCB degrader and PCB biosensor derivates of *Pseudomonas fluorescens* F113. J Appl Microbiol 110:1351–1358
Prasad R, Bhattacharyya A, Nguyen QD (2017) Nanotechnology in sustainable agriculture: recent developments, challenges, and perspectives. Front Microbiol 8:1014
Pukalchik M, Mercl F, Terekhova VA, Tlustoš P (2018) Biochar, wood ash and humic substances mitigating trace elements stress in contaminated sandy loam soil: evidence from an integrative approach. Chemosphere 203:228–238
Raaijmakers JM, Vlami M, de Souza JT (2002) Antibiotic production by bacterial biocontrol agents. Antonie Van Leeuwenhoek 81:537–547
Rafi MM, Epstein E (1997) Silicon deprivation causes physical abnormalities in wheat (*Triticum aestivum* L.). J Plant Physiol 151:497–501
Rains DW, Epstein E, Zasoski RJ, Aslam M (2006) Active silicon uptake by wheat. Plant Soil 280 (1):223–228
Rathore S, Desai PM, Liew CV, Chan LW, Heng PWS (2013) Microencapsulation of microbial cells. J Food Eng 116(2):369–381
Raven JA (2001) Silicon transport at the cell and tissue level. Plant Sci 8:41–55
Rekha PD, Lai WA, Arun AB, Young CC (2007) Effect of free and encapsulated *Pseudomonas putida* CC-FR2-4 and *Bacillus subtilis* CC-pg104 on plant growth under gnotobiotic conditions. Bioresour Technol 98:447–451
Rivera-Cruz MC, Narcía AT, Ballona GC, Kohler J, Caravaca F, Roldán A (2008) Poultry manure and banana waste are effective biofertilizer carriers for promoting plant growth and soil sustainability in banana crops. Soil Biol Biochem 40:3092–3095
Rodriguez H, Fraga R, Gonzalez T, Bashan Y (2006) Genetics of phosphate solubilization and its potential applications for improving plant growth-promoting bacteria. Plant Soil 287:15–21
Sa C, Cardosos P, Figueira E (2019) Alginate as a feature of osmotolerance differentiation among soil bacteria isolated from wild legumes growing in Portugal. Sci Total Environ 681:312–319
Sahu PK, Brahmaprakash GP (2016) Formulations of biofertilizers – approaches and advances. In: Singh D, Singh H, Prabha R (eds) Microbial inoculants in sustainable agricultural productivity. Springer, New Delhi, pp 179–198
Sahu PK, Lavanya G, Brahmaprakash GP (2013) Fluid bed dried microbial inoculants formulation with improved survival and reduced contamination level. J Soil Biol Ecol 33:81–94
Sandhya V, Ali SZ (2015) The production of exopolysaccharide by *Pseudomonas putida* GAP-P45 under various abiotic stress conditions and its role in soil aggregation. Microbiology 84:512–519
Santosh GP (2015) Formulation and shelf life of liquid biofertilizer inoculants using cell protectants. Int J Res Biol Agric Technol 7:243–247
Sarma MVRKV, Kumar K, Saharan R, Srivastava AK, Sharma A, Prakash V, Bisaria SV (2011) Application of inorganic carrier-based formulations of fluorescent pseudomonads and *Piriformospora indica* on tomato plants and evaluation of their efficacy. J Appl Microbiol 111:456–466
Savvas D, Ntatsi G (2015) Biostimulant activity of silicon in horticulture. Sci Hortic 196:66–81

Savvas D, MAnos G, Kotsiras A, Souvaliotis S (2002) Effects of silicon and nutrient-induced salinity on yield, flower quality, and nutrient uptake of gerbera grown in a closed hydroponic system. J Appl Bot 76:153–158

Schaafsma G (2009) Safety of protein hydrolysates, fractions thereof and bioactive peptides in human nutrition. Eur J Clin Nutr 63:1161–1168

Schiavon M, Ertani A, Nardi S (2008) Effects of an alfalfa protein hydrolysate on the gene expression and activity of enzymes of TCA cycle and N metabolism in *Zea mays* L. J Agric Food Chem 56:11800–11808

Schmidt W, Santi S, Pinton R, Varanini Z (2007) Water-extractable humic substances alter root development and epidermal cell pattern in Arabidopsis. Plant Soil 300:259–267

Schoebitz M, Lopez MD, Roldan A (2013) Bioencapsulation of microbial inoculants for better soil–plant fertilization-a review. Agron Sustain Dev 33:751–765

Schoebitz M, López MD, Serri H, Martínez O, Zagal E (2016) Combined application of microbial consortium and humic substances to improve the growth performance of blueberry seedlings. J Soil Sci Plant Nutr 16:1010–1023

Seneviratne G, Jayasekara APDA, De Silva MSDL, Abeysekera UP (2011) Developed microbial biofilms can restore deteriorated conventional agricultural soils. Soil Biol Biochem 43:1059–1062

Shaikh SS, Sayyed RZ (2015) Role of plant growth-promoting rhizobacteria and their formulation in biocontrol of plant diseases. In: Arora NK (ed) Plant microbes symbiosis: applied facets. Springer, New Delhi, pp 337–351

Sharma R, Walia A, Chauhan A, Shirkot CK (2015) Multi-trait plant growth promoting rhizobacteria from tomato rhizosphere and evaluation of their potential as bioinoculants. Appl Biol Res 17(2):1–12

Siddiqui IA, Shaukat SS (2003) Role of salicylic acid in *Pseudomonas aeruginosa* Strain IE-6S+-mediated induction of systemic resistance against *Meloidogyne javanica* in tomato. Phytopathol Mediterr 42(3):268–274

Sinha A, Harshita SR, Verma A (2018) Bioefficacy of *Trichoderma harzianum* and *Trichoderma viride* against *Fusarium oxysporum* f. sp. *capsici* causing wilt disease in chilli. J Pharmacog Phytochem 7:965–966

Spaepen S, Vanderleyden J, Remans R (2007) Indole-3-acetic acid in microbial and microorganism-plant signaling. FEMS Microbiol Rev 31:425–448

Stamenković S, Beškoski V, Karabegović I, Lazić M, Nikolić N (2018) Microbial fertilizers: a comprehensive review of current findings and future perspectives. J Agric Res 16:e09R01

Stinson AM, Zidack NK, Strobel GA, Jacobsen BJ (2003) Mycofumigation with *Muscodor albus* and *Muscodor roseus* for control of seedling diseases of sugar beet and verticillium wilt of eggplant. Plant Dis 87(11):1349–1354

Tewari S, Arora NK (2014) Multifunctional exopolysacccharides from *Pseudomonas aeruginosa* PF23 involved in plant growth stimulation, biocontrol and stress amelioration in sunflower under stress conditions. Curr Microbiol 69:484–494

Van Bockhaven J, De Vleesschauwer D, Höfte M (2013) Towards establishing broad-spectrum disease resistance in plants: silicon leads the way. J Exp Bot 64:1281–1293

Vassilev N, Mendes G (2018) Solid-state fermentation and plant beneficial microorganisms. In: Pandey A, Larroche CH, Soccol C (eds) Current developments in biotechnology and bioengineering, current advances in solid-state fermentation. Elsevier, Amsterdam, pp 402–416

Vassilev N, Eichler-Lobermann B, Vassileva M (2012) Stress tolerant P-solubilizing microorganisms. Appl Microbiol Biotechnol 95:851–859

Vassilev N, Mendes G, Costa M, Vassileva M (2014) Biotechnological tools for enhancing microbial solubilization of insoluble inorganic phosphates. Geomicrobiol J 31:751–763

Vassilev N, Vassileva M, Lopez A, Martos V, Reyes A, Maksimovic I, Eichler-Löbermann B, Malusà E (2015) Unexploited potential of some biotechnological techniques for biofertilizer production and formulation. Appl Microbiol Biotechnol 99(12):4983–4996

Vassilev N, Malusa E, Requena AR, Martos V, López A, Maksimovic I, Vassileva M (2016) Potential application of glycerol in the production of plant beneficial microorganisms. J Ind Microbiol Biotechnol 9:44–45

Vassilev N, Eichler-Löbermann B, Flor-Peregrin E, Martos V, Reyes A, Vassileva M (2017) Production of a potential liquid plant bio-stimulant by immobilized *Piriformospora indica* in repeated-batch fermentation process. AMB Express 7(1):106

Vassilev N, Vassileva M, Martos V, Garcia del Moral LF, Kowalska J, Tylkowski B, Malusá E (2020) Formulation of microbial inoculants by encapsulation in natural polysaccharides: focus on beneficial properties of carrier additives and derivatives. Front Plant Sci 11:270

Vassileva M, Azcon R, Barea JM, Vassilev N (1999) Effect of encapsulated cells of Enterobacter sp. on plant growth and phosphate uptake. Bioresour Technol 67:229–232

Vejan P, Abdullah R, Khadiran T, Ismail S (2018) Encapsulation of *Bacillus salmalaya* 139SI using double coating bio-polymer technique. Lett Appl Microbiol 68:56–63

Vendan RT, Thangaraju M (2007) Development and standardization of liquid formulation for *Azospirillum* bioinoculant. Acta Microbiol Immunol Hung 54:167–177

Walia A, Mehta P, Chauhan A, Shirkot CK (2013) Effect of *Bacillus* sp. strain CKT1 as inoculum on growth of tomato seedlings under net house conditions. Proc Natl Acad Sci 84(1):144–155

Walia A, Guleria S, Chauhan A, Mehta P (2017) Endophytic bacteria: role in phosphate solubilization. In: Maheshwari DK, Annapurna K (eds) Endophytes: crop productivity and protection, sustainable development and biodiversity. Springer, Cham, pp 1–33

Wang H, Liu S, Zhai L, Zhang J, Ren T, Fan B, Liu H (2015) Preparation and utilization of phosphate biofertilizers using agricultural waste. J Int Agric Adv 14:158–167

Wu Z, Guo L, Qin S, Li C (2012) Encapsulation of R. planticola Rs-2 from alginate-starch-bentonite and its controlled release and swelling behavior under simulated soil conditions. J Ind Microbiol Biotechnol 39:317–327

Yakhin OI, Lubyanov AA, Yakhin IA, Brown PH (2017) Biostimulants in plant science: a global perspective. Front Plant Sci 7:2049

Yeo Y, Back N, Park K (2001) Microencapsulation methods for delivery of protein drugs. Biotechnol Bioprocess 6:213–230

Young CC, Rekha PD, Lai WA, Arun AB (2006) Bioencapsulation of plant growth promoting bacteria in alginate beads enriched with humic acid. Biotechnol Bioeng 1:76–83

Zacky FA, Ting ASY (2015) Biocontrol of *Fusarium oxysporum* f. sp. cubense tropical race 4 by formulated cells and cell free extracts of *Streptomyces griseus* in sterile soil environment. Biocon Sci Technol 25:685–696

Zahir ZA, Shah MK, Naveed M, Akhter MJ (2010) Substrate-dependent auxin production by *Rhizobium phaseoli* improves the growth and yield of *Vigna radiate* L. under salt stress conditions. J Microbiol Biotechnol 20:1288–1294

Zaidi A, Khan MS, Ahemad M, Oves M (2009) Plant growth promotion by phosphate solubilizing bacteria. Acta Microbiol Immunol Hung 56:263–284

Zhu Y, Gong H (2014) Beneficial effects of silicon on salt and drought tolerance in plants. Agron Sustain Dev 34:455–472

Zohar-Perez C, Chernin L, Chet I, Nussinovitch A (2003) Structure of dried cellular alginate matrix containing fillers provides extra protection for microorganisms against UVC radiation. Radiat Res 160:198–204

Bioinoculants for Agricultural Sustainability 25

Manali Singh, Shruti Bhasin, Neha Madan, Deep Chandra Suyal, Ravindra Soni, and Dipti Singh

Abstract

Bioinoculants in agriculture are employed to promote crop growth and development mainly by mobilizing soil nutrients and tolerance to abiotic stresses. As the population is increasing, the demand for agricultural products also increases which can be achieved by manipulating or introducing enhance agricultural methods. In recent years, the application of chemical growth regulators, fertilizers, and pesticides has increased that has imposed serious threats to humankind. Therefore, bioinoculants are being targeted for achieving an environment-friendly and economic alternative for sustainable farming. The plant–microbiome associations play a critical part in plant's growth and development by changing over inaccessible nutritional components into an accessible form. In conclusion, bioinoculants based sustainable agricultural frameworks must be essential to address food security, biosafety, and soil management.

M. Singh (✉) · N. Madan
Department of Biotechnology, Invertis Institute of Engineering and Technology (IIET), Invertis University, Bareilly, India

S. Bhasin
Department of Biotechnology, Banasthali Vidyapith, Jaipur, Rajasthan, India

D. C. Suyal
Department of Microbiology, Akal College of Basic Sciences, Eternal University, Baru Sahib, Sirmaur, Himachal Pradesh, India

R. Soni
Department of Agricultural Microbiology, College of Agriculture, Indira Gandhi Krishi Vishwa Vidyalaya, Raipur, Chhattisgarh, India

D. Singh
Department of Genetics, IARI, New Delhi, India

R. Soni et al. (eds.), *Microbiological Activity for Soil and Plant Health Management*,
https://doi.org/10.1007/978-981-16-2922-8_25

Keywords

Bioinoculant · Biofertilizers · Sustainable farming · Beneficial microorganisms

25.1 Introduction

Bioinoculant or known as microbial inoculants are formed by beneficial microorganisms that indirectly or directly help the host plant by promoting plant growth and increasing nutrient availability. Microbial species show symbiosis or free-living relationships within the plant root system primarily due to the deposition of photosynthetic carbon in them (Joshi et al. 2019) and enhance plant growth by changing over inaccessible nutritional components into accessible form (Bhardwaj et al. 2014; Imam et al. 2017; Rawat et al. 2019). Microorganisms contain latent or live cells for zinc, phosphate, potassium solubilization, and nitrogen-fixing that is effective within the host plant's rhizospheric root (Patil and Solanki 2016). When it is utilized as a soil application or as seedling root dip or as seed treatment it rapidly multiplies the microbes present which indirectly helps to increase the population in the rhizosphere. Bioinoculants help in bioremediation, seed treatment by framing a uniform covering of inoculants over seeds (Dangi et al. 2019). Humankind has consistently been worried about food production to meet the expanding population demand and, for quite a while, the arrangement was to extend agriculture to new regions. In any case, this situation has changed in late many years, first because of restrictions of unexplored cultivable land, yet additionally fortified by the improvement of innovations that permit better returns, notwithstanding expanding environmental concerns, promoting agricultural works on targeting accomplishing sustainable production (Singh et al. 2020). In this manner, despite the fact that the worldwide interest for food keeps on expanding, the ideas of agribusiness manageability, recuperation of corrupted territories, and alleviation of natural effects are increasing. Bioinoculant is used as biofertilizers when added to soil it promotes the growth of a plant by increasing drought resistance, mobilization for soil nutrients, and acts as biocontrol. It is needed for sustainable agriculture development for the improvement of crop productivity. As the population is increasing the demand for agricultural products also increases which can be achieved by manipulating or introducing enhanced agricultural methods. Plant uses microbiota for the improvement and growth in different manners like plant growth-promoting fungi, plant growth-promoting rhizobacteria, phytostimulation (phytohormones communicated by an organism like *Azospirillum*) that legitimately advances the development of plants, by phytoremediation cycle (*Pseudomonas, Trichoderma, Bacillus cereus*) that secures plant against harmful organism and substantial metals (Tang et al. 2020). Bioinoculants which are used for enhancing the plant yield and growth are arbuscular mycorrhizal fungi, rhizobacteria (naturally occurring), *Piriformospora indica* fungus, etc. (Meena et al. 2017). It plays a role to solubilize the phosphorus present in the soil, produce phytohormones on the stimulation of plant growth, in mitigation of abiotic stress, in ameliorating biotic stress (Figs. 25.1, 25.2 and 25.3).

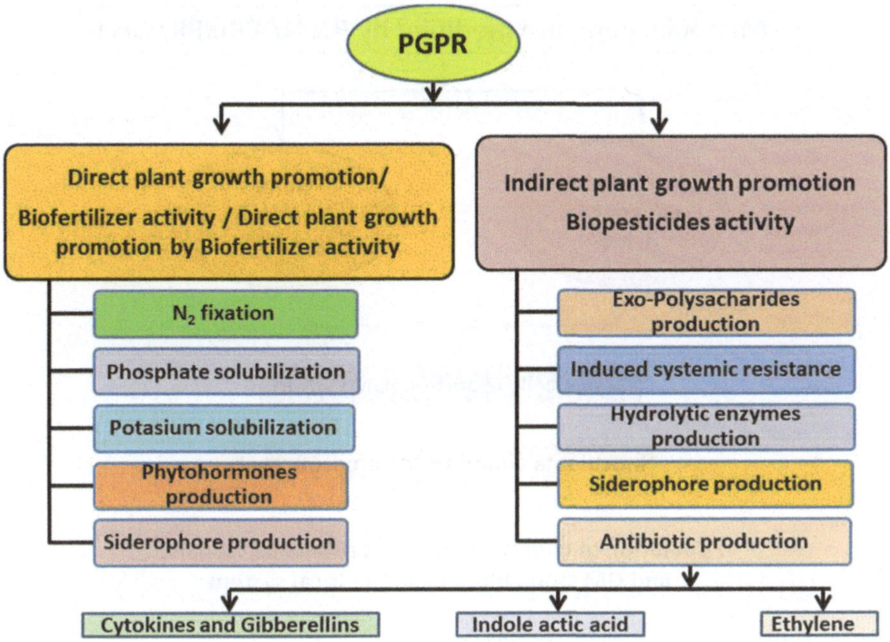

Fig. 25.1 Role of PGPR in crop production

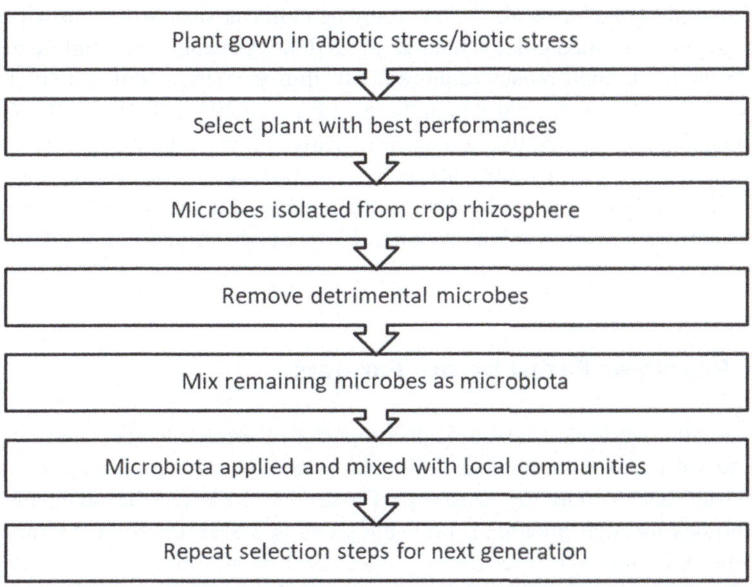

Fig. 25.2 Isolation and selection of best indigenous microbes

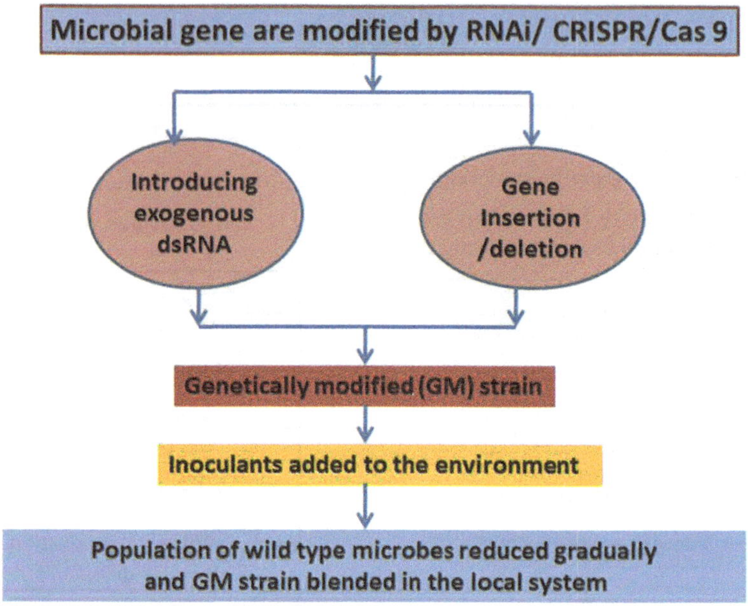

Fig. 25.3 Utilisation of genetically modified inoculants

Nowadays, natural raw materials are preferred for producing secondary phytochemicals (Bhasin et al. 2019). Ashwagandha is one of the most powerful herbs in Ayurvedic medicine. (Datta et al. 2017). Ashwagandha and Neem have revered herb due to their bioactive ingredients, thus we need to cultivate it at a large scale to fulfill the commercial demands (Singh et al. 2017; Singh et al. 2018a, b; Gupta et al. 2019). Their cultivation can be greatly enhanced by using bioinoculants. Production of ENA (extra neutral alcohol) from molasses of sugarcane employs the addition of microbes in the Aero tiller machine operated daily to mix the effluent and to aerate the press mud to stimulate microbial activity (Mishra et al. 2019; Singh et al. 2019).

25.2 Problems Faced by the Farmers

Agro-synthetic overflow is a significant supporter of surface-water pollution. Abundance and wrong use of chemical fertilizers bring about contamination of streams, ocean, groundwater, and soil washing. Those chemicals can be taken inside the human bodies through breathing, oral ingestion, and skin invasion. Despite widespread microbial inoculants technology utilization in agriculture, it also faces some challenges. Microbial inoculants have been applied (for the most part in research) in the types of fluids (as root dips, soaks, sprays) or as dry formulations with the immense success recorded, yet the greater part of these strategies is not practicable for an enormous scope. This is because a huge sum is needed for the ideal use of the inoculants (Callaghan 2016). PGPR is profoundly specific and focused on dissimilar to chemical inputs that are the expansive range of products. It just affects a specific or focused on a living being. Thus this brings about the irregularity of value and adequacy under field conditions including different living beings act at the same time (Timmusk et al. 2017). It is a genuine challenge to keep up the reasonability of microorganisms under microbial formulations (Callaghan 2016). It has been observed that the microbial viability fluctuated significantly in the treated seeds along with the treatment strategy and capacity temperature. Expanded endurance of microbial inoculants at surrounding stockpiling conditions is prescribed for microbial inoculants to turn out to be essential for standard agribusiness. The storage of microbial inoculants is also expensive and sometimes may become dangerous due to poisonous biocontrol agents (Callaghan 2016).

25.3 Government Initiatives to Control Losses in the Crop Cultivation

Recent studies have reported that biofertilizers can improve the food quality by modifying the antioxidants, chlorophyll, and phenolics content (Khalid et al. 2017). *Azotobacter chroococcum* and *Glomus fasciculatum* are known to enhance the carotenoids, absolute phenolics, and anthocyanins in the lettuce. Numerous microorganisms can be used as biopesticides as they exhibit antibacterial and antifungal properties (Rani et al. 2018). Such bioinoculants are the potential source to be used as biocontrol agents. They produce extracellular hydrolytic enzymes, compete with other pathogenic microbes, and induce systemic acquired resistance (Rani et al. 2018). Moreover, they are also known to regulate few plant hormones, viz. auxin and ethylene. By doing so, they protect the plants from the parasitic attack. *Colletotrichum coccodes* is well known mycoherbicide of velvetleaf. Similarly, *Trichoderma* offers good biocontrol potential against several pathogens and thus commercially available in the market (Harman et al. 2004).

25.4 Impact of Climate Variability on Agriculture

There is an adverse effect of environmental stress on plant development, production, and growth (Koskey et al. 2017). Microorganisms are reported to alleviate the climatic stress, viz. drought stress, cold stress, and salt stress efficiently when inoculated in the crops (Benidire et al. 2017; Youseif et al. 2017). Drought is abiotic stress which affects plant productivity due to water imbalance. This dry period often occurs and the disband of rainfall affects the area that relies on rainfall for water. With some modification, plants can overcome drought stress at various levels such as cellular, metabolic, and morphological levels. Bioinoculants reduce the drought stress in the supplement to their natural mechanism. Arbuscular mycorrhizal fungi can confer resistance to drought stress to its host plants by symbiosis process. One of the major threats to the earth's surface is the increase in salt content in the soil which can affect plant growth and yield. It is caused by natural processes such as mineral weathering. In saline soil, there is high electric conductivity with low pH. Higher the concentration of Cl^- and Na^+ ions in soil leads to lower yield and growth of the plant. Various physiological processes of plants are affected by low osmotic potential due to which there is a decrease in plant water uptake. To alleviate salt stress various microorganisms can be used such as endophytic fungi, arbuscular mycorrhizal fungi.

25.5 Biotic Factors and their Effects on Crop Production

The crops are greatly affected by biotic factors. These factors may be categorized as negative (due to plant pathogens) or positive (primarily mediated by plant growth-promoting microorganisms).

25.5.1 Negative Effects

Human activities have greatly affected the environment that has ultimately influenced the natural balance and ecosystem productivity (Asawa et al. 2019a, b). Among the negative effects, MLN (maize lethal necrosis), a disease of the maize observed in African countries could be an important example. The causative agent is a synergistic interaction between maize chlorotic mottle virus (MCMV) and sugarcane mosaic virus (SCMV). It is known to reduce the crop yield at very high rate (Karanja et al. 2018). Russian wheat aphid (RWA) is known to affect barley, wheat, and other cereal grains worldwide, especially in Asia, Africa, Europe, the Middle East, and America. The visual symptoms include stunting, necrosis, chlorosis, wilting, rolled leaves, trapped awns, yellowish, whitish, and/or purple leaf markings.

25.5.2 Positive Effects

Apart from negative effects, several microorganisms positively affect plant growth and enhance crop yield. Such microorganisms are known as plant growth-promoting rhizobacteria (PGPR).

25.5.2.1 PGPR

PGPRs (plant growth-promoting rhizobacteria) are the microorganisms associated with the plant's rhizosphere. They are the important contributors towards enhanced soil fertility and productivity. They minimize the adverse effects originated from biotic as well abiotic factors. Moreover, they help in the removal of the agrochemicals and the xenobiotics from the agroecosystems. Several studies have been done to assess the beneficial effects of PGPR on the different crops including cereals, vegetables, pulses, horticultural crops, etc. Moreover, efforts have also been done to evaluate their potential in agroforestry and micro-phytoremediation. The potential PGPR candidates are *Azospirillum, Azotobacter, Bacillus, Pseudomonas, and Rhizobium* (Suyal et al. 2014; Tomer et al. 2017; Kumar et al. 2018; Rajwar et al. 2018). Application of PGPR has shown a significant reduction in the need of chemical fertilizers and other chemical amendments. Following major mechanisms are employed by the microorganisms to promote plant growth and development.

25.5.2.1.1 Production of Growth Regulators

PGPR produces many different types of antibiotics that combat against phytopathogens and mainly include zwittermycin A, streptomycin, oligomycin A, butyrolactones, oomycin A, kanosamine, phenazine-1-carboxylic acid, pyrrolnitrin, pyoluteorin, xanthobaccin, viscosinamide, and 2,4-diacetyl phloroglucinol (2,4-DAPG).

25.5.2.1.2 Increased Uptake of Minerals and Soil Fertility

The PGPRs are well known for their potential P solubilizing ability (mainly rock phosphate) by producing phosphatase and organic acids and mineralize the unavailable fixed phosphate and make it available to plant roots. Soils contain a large reservoir of total phosphorus (P) but a tiny proportion of this total is supplied to the plants. The plants only absorb phosphorus in two forms, that is, the monobasic ($H_2PO_4^{-1}$) and the diabasic ($H_2PO_4^{-2}$).

25.5.2.1.3 Induced Systemic Resistance (ISR)

It represents the plant's defense ability to exclude or overcome the pathogen attacks. The ISR mechanism steps are (1) linked to growth promotion, that is, developmental escape, (2) reduced symptom expression, that is, physiological tolerance, (3) microbial antagonisms, and (4) induction of biochemical-resistance. It can be represented by reinforcement of the cell wall and production of pathogen-related proteins. PGPR promotes plant growth by taming growth-restricting conditions in an indirect mode. Moreover, they produce pathogen-antagonistic substances to protect the host plants with the help of ISR mechanisms.

25.5.2.1.4 Siderophore Production

Some living-organisms (fungi, bacteria, grasses, etc.) are involved in the secretion of high-affinity iron-chelating compounds which are referred as siderophores and known as the strongest soluble Fe3+ binding agents. The enzymes used in the key physiological processes of the plants, viz. photosynthesis, respiration, and metabolism are greatly dependent on the iron as a co-factor.

However, its bio-availability is limited due to low solubility under neutral pH conditions. In such Fe-limiting environments, PGPRs are known to be excellent producers of siderophores which are selective ferric ion chelators. Application of pyoverdine produced by Pseudomonas in the soils is advantageous to fusarium wilts.

25.5.2.2 Adaptation Strategies

The intricate and solid connection between soil degradation, environmental change, and food weakness is a worldwide test. Sustainable agricultural frameworks must be essential to any plan to address environmental change and changeability, improve inexhaustible freshwater flexibility and quality, re-establish degraded soils and biological systems, and advance food security. These difficulties are being exacerbated by expanding population and diminishing per capita of land that is capable to produce a crop and sustainable freshwater gracefully, the expanding recurrence of outrageous occasions, the diminishing strength of agroecosystems, and expanding income and well-off way of life with developing an inclination towards meat-based eating regimen and diminishing soil quality and use effectiveness of sources of info. Turning around these descending spirals infers the usage of demonstrated innovations, for example, protection horticulture, coordinated supplement of the executives, accuracy agribusiness, agroforestry frameworks, and so forth. The rebuilding of debased soil and desertified biological systems and the formation of positive soil and environment C financial plans are significant. Urban agriculture and green rooftops can decrease the vitality impression of creation chains for metropolitan and non-urban zones and upgrade the reusing of side-effects. Researchable needs to incorporate supportable land use and soil/water the executive's alternatives, reasonable soil administration, and usual way of doing things towards installments to land directors for the provisioning of environment administrations.

25.6 Microbial Inoculants

Inoculants have their creation living microorganisms that fit for profiting the improvement of various plant species. These are diazotrophs, rhizobia, and others that colonize the plant's surfaces and make the positive associations. The beneficial interaction of legumes rhizobia enhances the biological nitrogen fixation, which regularly can completely fulfill the plant's requirement on nitrogen. Also, other diazotrophic microscopic organisms, for example, *Azospirillum*, set up non-obligate associations with the plants and provide nitrogen to them (Santos et al. 2019). Recent innovations and advancements are based on the applications of microbial consortia that can enhance plant growth through various mechanisms (Chibeba et al. 2015). Such microbial consortia which have proved their potential are strongly recommended to the farmers (Bulegon et al. 2017; Ferri et al. 2017). The first commercially available microbial inoculant "Nitragin," was developed by USA in 1896. It utilized gelatin as a carrier. Later, it was supplemented by peat to reduce the microbial death rate. In 1956 first inoculants were introduced in the industry, in Argentina, the inoculants industry was established in 1980 where reinoculation of soybean was increased. In 2000 Brazil has made the first liquid inoculants of the microorganisms. In India, more than 100 bioinoculants industries were established in 2011. In 2014 first Brazilian inoculants were registered for co-inoculation. In 2016 China reaches 842 deposited patents of inoculants and Ghana establishes its first inoculants industry. In 2019 Brazil sold 70 million doses of inoculants per year (Santos et al. 2019). In any case, these days, expanded interest for food, enthusiasm for practical horticulture, and expanding awareness about pest and pathogen that posesprotection from agrochemicals is exponentially raising the attention of the world (Santos et al. 2017; Gundi et al. 2018).

25.7 Bioinoculated Crops

25.7.1 Soybean

Glycine max (Soybean) is an economically important, annual, herbaceous, dicotyledonous legume crop contributing to major portion of the edible oil worldwide. Eastern Asia Argentina, Brazil, and the USA are the main soybean producers worldwide. The success of the soybean crop lies in the biological nitrogen fixation. The excessive application of agrochemicals, viz. xenobiotics, pesticides, fungicides may kill the natural symbionts of the crop and thus greatly reduce the biological nitrogen fixation.

25.7.2 Maize

Zea mays (maize) is the world's third most cultivated cereals and native to Central America. It is used for human consumption, animal feed, and several industrial applications and thus is in great demand. The USA, China, and Brazil are among the top producer and consumers of this crop. In recent years, it is being targeted for producing biofuel. Maize is known to harbor several PGPB, especially the genus *Azospirillum*. This genus is being explored for maize production worldwide.

25.7.3 Rice

Oryza sativa (rice) is native to Asia and cultivated since 120 million years. Presently, this cereal crop is grown all over the world and covers approximately 11% of the agricultural area worldwide (Singh et al. 2018a, b). It is a most common and staple food for majority of the population in the world. The production of this crop is greatly affected by abiotic as well as biotic factors. More importantly, climate change and vulnerability towards pathogens have seriously affected its production worldwide.

25.7.4 Faba Beans

Faba beans (Vicia faba L.) are used in China for 5000 years, however, their exact origin is still unknown. Presently, this crop is grown all over the world due to its adaptability towards a range of climatic and edaphic conditions. Brazil, China, Egypt, Italy, Mexico, Spain, Morocco, the UK, and Russia are the main producers and consumers. In recent years, climate change has severely affected its production.

Several plant growth-promoting microorganisms are employed to improve its production.

25.7.5 Vegetables

Tomato, potato, onion, and several other vegetable crops are consumed by millions of people in their diet, as ingredients of hot dishes, consumed as salads, for the manufacture of many products they also used as a raw material with great application in the industry, mainly sauces. These crops are known to benefitted by several PGPR, especially *Azospirillum, Bacillus, and Pseudomonas* (Heuvelink 2018; Lima et al. 2018).

25.7.6 Legumes

Legumes are cultivated and consumed by a majority of the population worldwide. They are known to harbor diversified rhizobial strains. Among these, Cicer arietinum (Chickpea) is highly nutritive and most studied legume crop that is known to be inhabited by *Mesorhizobium* strains. In recent years, isolation, characterization, and development of novel rhizobial strains associated with the pulses are being targeted.

25.8 Bioinoculant Formulations

It is a mixture of selected microbial strain and a suitable carrier. The carrier substance offers safe and steady application of the inoculants. It should have sufficient shelf life, biodegradable, and should not be contaminated easily (Sahu et al. 2018; Bhattacharyya et al. 2020). Several bio-formulation techniques are used now days; among which some are being discussed below:

25.8.1 Cell Immobilization

Earlier macro-alginate beads of size 2-5 mm were used for the bacterial strain and mycorrhiza fungi (*Streptomycetes* sp.) immobilization using kaolin-alginate formulation. For 15 weeks starch and talcum powder were added and formulated to increase *Streptomycetes* sp. survival rate. It was accounted for that co-immobilization of the two microorganisms could give preferred outcomes over a single microorganism.

25.8.2 Encapsulation

In this process, microorganisms are entangled in polymeric matrix to improve effectivity and stability. Initially, it involves the retention of usable ingredients into a matrix (liquid or solid), the subsequent step comprises the solution splashing on solid particles, and at the end adjustment by physico-chemical processes and a chemical cycle of polymerization. This process ensures organisms under abiotic and biotic stress conditions. *Pseudomonas* fluorescence upgrades the endurance of microorganisms by colonizing *Beta vulgaris* and producing 2,4-diacetyl phloroglucinol, an antifungal metabolite. Similarly, miniature alginate beads having *Beauveria bassiana* are used to control *Solenopsis invicta.*

25.9 Conclusion

In recent years, scientific and commercial attention is being paid to the isolation and development of novel bioinoculants due to their additional benefits over agrochemicals. Although long-term storage of the seeds inoculated with biopesticides has raised some issues. But the recent developments and innovations emerge a ray of hope. Further, microorganisms have specific requirements which must be taken into consideration before developing them as bioinoculants. Conclusively, production of microbial inoculants seems additional benefits to the farmers for environmental safety besides plant growth and development. Therefore, their exploration should be encouraged to achieve enhanced crop production and agricultural sustainability.

References

Asawa P, Varshney N, Gupta S, Pandey V, Singh M (2019a) Advances in ZIKA virus host-cell interaction: current knowledge and future perspectives. Euro J Biotechnol Biosci 7(3):45–49

Asawa P, Pandey VN, Singh V, Kashyap S, Gupta I, Saumya SM (2019b) Efforts to reduce greenhouse gases, air pollution and climate change: global and regional perspectives. Int J Agri Environ Biores 4(3):1–10

Benidire L, Lahrouni M, El Khalloufi F, Göttfert M, Oufdou K (2017) Effects of *Rhizobium leguminosarum* inoculation on growth, nitrogen uptake and mineral assimilation in *Vicia faba* plants under salinity stress. JAST 19(4):889–901

Bhardwaj D, Ansari MW, Sahoo RK, Tuteja N (2014) Biofertilizers function as key player in sustainable agriculture by improving soil fertility, plant tolerance and crop productivity. Microb Cell Factories 13:66

Bhasin S, Singh M, Singh D (2019) Review on bioactive metabolites of Withania somnifera. (L.) Dunal and its pharmacological significance. J Pharma Phytochem 8(3):3906–3909

Bhattacharyya C, Roy R, Tribedi P, Ghosh A, Ghosh A (2020) Biofertilizers as substitute to commercial agrochemicals. In: Majeti N, Vara P (eds) Agrochemicals detection, treatment and remediation Butterworth-Heinemann, United Kingdom, pp 263–290

Bulegon LG, Guimarães VF, Klein J, Batisttus AG, Inagaki AM, Offmann LC, Souza AKP (2017) Enzymatic activity, gas exchange and production of soybean co-inoculated with *Bradyrhizobium japonicum* and *Azospirillum brasilense*. Aust J Crop Sci 11:888–896

Callaghan MO (2016) Microbial inoculation of seed for improved crop performance: issues and opportunities. Appl Microbiol Biotechnol 100:5729–5746

Chibeba AM, Guimarães MDF, Brito OR, Nogueira MA, Araujo RS, Hungria M (2015) Co-inoculation of soybean with *Bradyrhizobium* and *Azospirillum* promotes early nodulation. Am J Plant Sci 6:1641–1649

Dangi AK, Sharma B, Hill RT (2019) Bioremediation through microbes: systems biology and metabolic engineering approach. Crit Rev Biotechnol 39(1):79–98

Datta A, Jain G, Avashthi H, Singh M, Agrawal S (2017) Molecular docking of Withanolides from Withania Somnifera against vimentin protein. Indi Res J Genet Biotech 9(4):609–612

Ferri GC, Braccini AL, Anghinoni FBG, Pereira LC (2017) Effects of associated co-inoculation of *Bradyrhizobium japonicum* with *Azospirillum brasilense* on soybean yield and growth. Afr J Agric Res 12:6–11

Gundi JS, Santos MS, Oliveira ALM, Nogueira MA, Hungria M (2018) Development of liquid inoculants for strains of *Rhizobium tropici* group using response surface methodology. Afr J Biotechnol 17:411–421

Gupta A, Ansari S, Gupta S, Narwani M, Gupta M, Singh M (2019) Therapeutics role of neem and its bioactive constituents in disease prevention and treatment. J Pharma Phytochem 8 (3):680–691

Harman GE, Lorito M, Lynch JM (2004) Uses of s spp. to alleviate or remediate soil and water pollution. In: Laskin AI, Bennett JW, Gadd San Diego GM (eds) Advances in applied microbiology. Elsevier Academic Press, London, pp 313–330

Heuvelink E (2018) Tomatoes: crop production science in horticulture. CABI Publishing, Boston

Imam J, Shukla P, Prasad Mandal N, Variar M (2017) Microbial interactions in plants: perspectives and applications of proteomics. Curr Protein Pept Sci 18(9):956–965

Joshi D, Chandra R, Suyal DC, Kumar S, Goel R (2019) Impact of bioinoculants *Pseudomonas jesenii* MP1 and *Rhodococcus qingshengii* S10107 on *Cicer arietinum* yield and soil nitrogen status. Pedosphere 29(3):388–399

Karanja J, Derera J, Gubba A, Mugo S, Wangai A (2018) Response of selected maize inbred germplasm to maize lethal necrosis disease and its causative viruses (sugarcane mosaic virus and maize chlorotic mottle virus) in Kenya. Open Agric J 12:2015–2226

Khalid M, Hassani D, Bilal M, Asad F, Huang D (2017) Influence of bio-fertilizer containing beneficial fungi and rhizospheric bacteria on health promoting compounds and antioxidant activity of Spinacia oleracea L. Bot Stud 58:35

Koskey G, Mburu SW, Njeru EM, Kimiti JM, Ombori O, Maingi JM (2017) Potential of native rhizobia in enhancing nitrogen fixation and yields of climbing beans (*Phaseolus vulgaris* L.) in contrasting environments of eastern Kenya. Front Plant Sci 8:1–12

Kumar S, Suyal DC, Bhoriyal M, Goel R (2018) Plant growth promoting potential of psychrotolerant Dyadobacter sp. for pulses and finger millet and impact of inoculation on soil chemical properties and diazotrophic abundance. J. Plant Nutr 41(8):1035–1046

Lima NSA, Vogel GF, Fey R (2018) Rates of application of *Azospirillum brasilense* in tomato crop. Rev Agric Neotrop 5:81–87

Meena VS, Meena SK, Verma JP (2017) Plant beneficial rhizospheric microorganism (PBRM) strategies to improve nutrients use efficiency: a review. Ecol Eng 107:8–32

Mishra AK, Mishra P, Singh AVV, Singh M (2019) Production of ENA (Extra Neutral Alcohol) from molasses of sugarcane. Int J Adv Sci Res 4(3):56–60

Patil HJ, Solanki MK (2016) Microbial inoculant: modern era of fertilizers and pesticides. In: Pratap Singh D (ed) Microbial inoculants in sustainable agricultural productivity. Springer, New Delhi, pp 319–343

Rajwar J, Chandra R, Suyal DC, Tomer S, Kumar S, Goel R (2018) Comparative phosphate solubilizing efficiency of psychrotolerant *Pseudomonas jesenii* MP1 and *Acinetobacter* sp. ST02 against chickpea for sustainable hill agriculture. Biologia 73(8):793–802

Rani A, Singh R, Kumar P, Shukla G (2018) Pros and cons of fungicides: an overview. Int J Eng Sci Res Technol 6:112–117

Rawat N, Sharma M, Suyal DC, Singh DK, Joshi D, Singh P, Goel R (2019) Psyhcrotolerant bio-inoculants and their co-inoculation to improve *Cicer arietinum* growth and soil nutrient status for sustainable mountain agriculture. J Soil Sci Plant Nutr 19(3):639–647

Sahu PK, Gupta A, Singh M (2018) Bioformulation and fluid bed drying. A new approach towards an improved biofertilizer formulation. In: Sengar RS, Singh A (eds) Eco-friendly agro-biological techniques for enhancing crop productivity. Springer, Singapore, pp 47–62

Santos MS, Hungria M, Nogueira MA (2017) Production of polyhydroxybutyrate (PHB) and biofilm by *Azospirillum brasilense* aiming at the development of liquid inoculants with high performance. Afr J Biotechnol 16:1855–1862

Santos MS, Nogueira MA, Hungria M (2019) Microbial inoculants: reviewing the past, discussing the present and previewing an outstanding future for the use of beneficial bacteria in agriculture. AMB Expr 9:205

Singh M, Shah P, Punetha H, Gaur AK, Kumar A, Agrawal S (2017) Isolation and quantification of a potent anti cancerous compound, Withaferin a from the aerial parts of *Withania Somnifera* (Ashwagandha). Ad Plant Sci 30(II):231–235

Singh M, Shah P, Punetha H, Agrawal S (2018a) Varietal comparison of withanolide contents in different tissues of *withania somnifera* (l.) Dunal (Ashwagandha). Int J Life-Sci Sci Res 4 (3):1752–1758

Singh PK, Venkatesan K, Swarnam TP (2018b) Rice genetic resources in tropical islands. In: Sivaperuman CA, Singh AA, Jaisankar I (eds) Biodiversity and climate change adaptation in tropical Islands. Academic Press, London, pp 355–386

Singh N, Iqbal Z, Ansari TA, Khan MA, Ali N, Khan A, Singh M (2019) The portent plant with a purpose: Aloe vera. J Pharma Phytochem 8(3):4124–4130

Singh M, Poddar NK, Singh D, Agrawal S (2020) Foliar application of elicitors enhanced the yield of Withanolide contents in *Withania somnifera* (L.) Dunal (variety, Poshita). 3Biotech 10:157

Suyal DC, Shukla A, Goel R (2014) Growth promotory potential of the psychrophilic diazotroph *Pseudomonas migulae* S10724 against native *Vigna radiata* (L.) Wilczek. 3Biotech 4 (6):665–668

Tang Y, Kang H, Qin Z, Zhang K, Zhong Y, Li H, Mo L (2020) Significance of manganese resistant *Bacillus cereus* strain WSE01 as a bioinoculant for promotion of plant growth and manganese accumulation in *Myriophyllum verticillatum*. Sci Total Environ 707:135867

Timmusk S, Behers L, Muthoni J, Muraya A, Aronsson AC (2017) Perspectives and challenges of microbial application for crop improvement. Front Plant Sci 8:49

Tomer S, Suyal DC, Rajwar J, Yadav A, Shouche Y, Goel R (2017) Isolation and characterization of phosphate solubilizing bacteria from Western Indian Himalayan soils. 3Biotech 7(2):1–8

Youseif SH, El-Megeed FHA, Saleh SA (2017) Improvement of faba bean yield using *Rhizobium/Agrobacterium* inoculant in low-fertility sandy soil. Agronomy 7:1–2

The manufacturer's authorised representative in the EU is Springer Nature Customer Service Centre GmbH, Europaplatz 3, 69115 Heidelberg, Germany. If you have any concerns regarding our products, please contact ProductSafety@springernature.com

Printed and bound by CPI Group (UK) Ltd, Croydon, CR0 4YY
30/09/2026
02225289-0019